Preventing Chemical Weapons

Arms Control and Disarmament as the Sciences Converge

Preventing Chemical Weapons

Arms Control and Disarmament as the Sciences Converge

Edited by

Michael Crowley
University of Braford, UK
Email: m.j.a.crowley1@bradford.ac.uk

Malcolm Dando
University of Braford, UK
Email: mrdando@bradford.ac.uk

and

Lijun Shang
University of Braford, UK
Email: l.shang1@bradford.ac.uk

ROYAL SOCIETY OF CHEMISTRY

Print ISBN: 978-1-78262-649-7
PDF ISBN: 978-1-78801-009-2
EPUB ISBN: 978-1-78801-472-4

A catalogue record for this book is available from the British Library

The Royal Society of Chemistry is a charity, registered in England and Wales, Number 207890, and a company incorporated in England by Royal Charter (Registered No. RC000524), registered office: Burlington House, Piccadilly, London W1J 0BA, UK, Telephone: +44 (0) 207 4378 6556.

Visit our website at www.rsc.org/books

Printed in the United Kingdom by CPI Group (UK) Ltd, Croydon, CR0 4YY, UK

Foreword

The life and chemical sciences are in the midst of a period of rapid and revolutionary transformation. The advances in science and technology, including those that result from convergence at the intersection of chemistry and biology, and progress in parallel disciplines and technologies, notably nanoscience and nanotechnology, are expected to bring about a wide range of societal benefits. In medicine, this could yield better diagnostic tools, prophylaxes and treatments, with, for example, an increasing ability to tailor drugs to the needs of individual patients; in agriculture, more environmentally hardy and disease resilient crops are being developed and farmed; energy production employing new bio-fuels and processes is being explored; and there are also advances towards more efficient, safer, environmentally friendly and sustainable production of consumer and industrial materials specifically designed to meet particular needs.

Inevitably, such transformational scientific and technological developments potentially have malign applications, and the dangers of such misuse have potentially profound implications for arms control and disarmament. Scientific advance could lead to the discovery or development of novel chemical agents, a greater ability to deliver these chemical agents to specific targets within the human body, and new or enhanced means of chemical agent dispersal over increasingly wider areas effecting greater numbers of people; eventually, such developments may even lead to the capability, or perceived capability, to conduct new forms of chemical warfare.

Such concerns are exacerbated by the unstable nature of the international security environment at the present time, with continuing warfare in the Middle East, threatened conflict in the Korean peninsula, the large-scale displacement and migration of people across the globe, the growth of extremist groups, and continuing terrorist attacks in many countries. This instability is likely to continue for many years and could fuel the further use of diverse

Preventing Chemical Weapons: Arms Control and Disarmament as the Sciences Converge
Edited by Michael Crowley, Malcolm Dando and Lijun Shang

Published by the Royal Society of Chemistry, www.rsc.org

toxic chemicals as improvised weapons, a desire by certain States to retain and employ existing stockpiles of chemical weapons, as well as increased interest in developing more advanced systems of chemical weapons.

Utilising a multi-disciplinary approach, and drawing upon an international group of experts, this book analyses current and near-term predictions of developments in relevant chemical and life sciences, and assesses the risks of their potential misapplication in the development of chemical weapons either through State programmes or by non-State actors such as armed opposition groups, terrorist organisations or criminal networks. The book analyses the current capabilities, limitations and failures of the existing relevant international arms control and disarmament instruments—notably the Chemical Weapons Convention—in preventing the development and use of chemical weapons. Through the employment of an innovative Holistic Arms Control methodology, the authors also look beyond the bounds of such treaties, to explore the full range of international law, international agreements and regulatory mechanisms potentially applicable to weapons employing toxic chemical agents, in order to develop recommendations for more effective routes to combat their proliferation and misuse.

This book is a call to action for the international scientific and governmental communities, underlining the vital importance of their active protecting and nurturing of the prohibition against poison and chemical weapons; and of building effective and responsive measures to ensure that the rapid advances in chemistry and the life sciences are safeguarded from malign use and are instead employed for the benefit of us all.

Diana Anderson
University of Bradford

Contents

Section I: Introduction

Preventing Chemical Weapons: Arms Control and Disarmament as the Sciences Converge
Edited by Michael Crowley, Malcolm Dando and Lijun Shang

Published by the Royal Society of Chemistry, www.rsc.org

Section II: The Chemical and Biological Weapons Prevention and Disarmament Regime Today

Section VI: Conclusion

Section I

Introduction

CHAPTER 1

Introduction: Preventing the Re-emergence of Chemical Weapons

MICHAEL CROWLEY*, MALCOLM DANDO AND LIJUN SHANG

University of Bradford, Bradford, West Yorkshire, BD7 1DP, UK
*E-mail: m.j.a.crowley1@bradford.ac.uk

1.1 The Context

Since its entry into force in 1997, the Chemical Weapons Convention (CWC) and its associated implementing body, the Organisation for the Prohibition of Chemical Weapons (OPCW),[1] have become the most important mechanism for global action in combating the use of chemical weapons, whether employed against armed forces or civilian populations, under any circumstances.

On 10th December 2013, the OPCW received the Nobel Peace Prize for its ongoing activities overseeing the destruction of declared chemical weapons stockpiles in Syria, and for its continuing efforts supporting CWC implementation throughout the world. In his Nobel Lecture, the OPCW Director General, Ambassador Ahmet Üzümcü, described the incredible achievements and continuing ambition of the Chemical Weapons Convention and the attendant control regime: "The OPCW has been overseeing the elimination of an entire category of weapons of mass destruction. Our task is to consign chemical weapons to history, forever. A task we have been carrying

Preventing Chemical Weapons: Arms Control and Disarmament as the Sciences Converge
Edited by Michael Crowley, Malcolm Dando and Lijun Shang

Published by the Royal Society of Chemistry, www.rsc.org

out with quiet determination, and no small measure of success ... we are hastening the vision of a world free of chemical weapons to reality."[2]

To date, 192 States, covering 98% of the world's population, have joined the CWC, and only four countries have yet to take affirmative action on the treaty (Israel which has signed but not ratified the Convention; and Egypt, North Korea and South Sudan which are all non-signatories). Since coming into force, over 6700 inspections have taken place at over 3100 chemical weapons related sites and at over 3600 industrial sites throughout the world.[3]

The primary focus of the OPCW, to date, has been the identification and destruction of all existing chemical weapons arsenals and production facilities around the world. By January 2018, following the complete verified destruction of all declared Libyan and Russian chemical weapons stocks,[4] approximately 96% of the world's declared stockpiles of over 72000 metric tonnes of chemical agents have been eliminated, and almost all former chemical weapons production facilities have been verified as destroyed or converted for peaceful purposes. These hard-won accomplishments—which clearly demonstrate what can be achieved by multi-lateral disarmament—have proven critically important for safeguarding international peace and security. Consequently, all outstanding destruction activities must be completed with the full support and encouragement of the international community, in word and by deed. In particular, full attention must now be given to ensuring that the delayed and subsequently re-scheduled verified US destruction of its remaining chemical weapons stockpiles is completed safely as planned by 2023. In addition, the OPCW must investigate and address all concerns regarding potential undeclared chemical weapons related activities, such as the recent allegations of development, stockpiling and use of Novichok chemical weapons.

The most serious and challenging situation currently facing the Organisation concerns the ongoing chemical attacks in Syria. Although the OPCW–UN (Organisation for the Prohibition of Chemical Weapons–United Nations) Joint Mission successfully removed and destroyed all declared chemical weapons stockpiles, there are continuing concerns that the Syrian Government did not provide the UN and OPCW with full and accurate details of its chemical weapons programme and stockpiles. In addition, reports of the use in Syria of sarin and improvised chemical weapons employing chlorine and other toxic chemicals have continued. Where evidence of the development, possession and use of such chemical weapons is discovered, the OPCW must act swiftly to halt all such activities. Full cooperation in these processes must be given by the Syrian Government and all parties to the conflict in Syria.

Along with its ongoing activities in supporting the full current implementation of the CWC, the OPCW is now engaged in a difficult and potentially turbulent process of review, planning, and structuring of the Organisation to prevent any future re-emergence of chemical weapons. Although the future priorities of the OPCW have yet to be collectively agreed by its Member States, an initial "vision paper" circulated by the OPCW Director-General in 2015 stated that: "The operating environment of the ... OPCW will change significantly over the coming decade. To stay relevant, the focus of the Organisation's activities will progressively have to be shifted from disarmament of chemical weapons to preventing their re-emergence."[5]

An essential element in the OPCW's effort to prevent the re-emergence of chemical weapons will be its responses to complex and evolving challenges concerning the appropriate regulation of toxic chemicals. One critical set of challenges arises from the continuing rapid developments in, and convergence of, chemistry, biology, and other relevant scientific disciplines and associated technologies; and the potential danger that certain developments may be employed for the production of existing and new types of chemical weapons.[6] The potential effects of such scientific and technological developments are amplified and influenced by the continuing growth and patterns of globalisation of commercial chemical and biotechnology industries, the increased availability of materials, and the 'democratisation' of expertise arising from trade and communications.[7]

Another set of challenges relates to the changing nature of the external environment, within which the OPCW operates, and in particular, the evolving security environment where "conflict is no longer framed in the context of opposing military alliances in a bipolar world", but instead the "borderlines between war, civil war, large-scale violations of human rights, revolutions and uprisings, insurgencies and terrorism as well as organized crime are blurred."[8] In this unstable environment, the number and range of actors who may be interested in acquiring and using chemical weapons is hard to determine, though a particular concern has been the employment of chemical weapons by State and Non-State actors, including terrorist organisations, against civilian populations, most recently in Iraq and Syria. Another concern is the apparent desire from certain military, security and police forces for delivery systems employing toxic chemicals for use in a variety of law enforcement operations. This has included a growing range of systems capable of delivering significant amounts of riot control agents over wide areas and/or extended ranges. In addition, certain States have explored the utility of so-called incapacitating chemical agent weapons. Unease about such developments is exacerbated by concerns that in certain situations, the lines separating law enforcement, counterinsurgency, and low-level conflict may become blurred.[9]

It is currently unclear in what ways the OPCW will seek to transform itself into an organisation able to respond effectively to these multi-faceted challenges, and how this transformation will be achieved. However, it is clear that the 4th Review Conference of the CWC to be held in November 2018 will be a very important forum where these issues will be discussed, if not finally resolved. Consequently, the 4th Review Conference, the long term processes initiated there, and subsequent follow up activities, are key foci of this book.

1.2 Concept for the Book

The 4th Review Conference of the CWC will have to address many issues and determine how best to respond to, and balance, the competing demands upon the OPCW's limited resources. Of critical importance will be how the Organisation and its Member States address the very rapid advances in relevant science and technology (S&T), and the potential threats of their employment in new and existing forms of weapons employing chemical agents. Indeed, the Review Conference is specifically tasked under the CWC to examine such long-term issues of concern and to "take into account any relevant scientific and technological developments."[10]

Given the wide range and diverse nature of S&T developments, spanning the chemical and life sciences, and increasingly encompassing many other disciplines notably nanotechnology, engineering and informatics, the Review Conference is likely to require a broad input of objective technical information in order for it to properly assess the scope of the advances, their potential impact for the regime, and to determine how the Organisation should respond to them. Thus, in addition to the official technical papers that will be produced by the Technical Secretariat, the Scientific Advisory Board, and individual State Parties, for consideration by the OPCW in preparation for and during the Review Conference, it seemed useful to bring together a group of scientists—chemical, biological and social—to produce a wide-ranging set of contributions focussing on different aspects of relevant S&T advances and exploring how the OPCW and broader international community should attempt to respond to them. In addition to informing and stimulating State and civil society thinking on these issues, it is hoped that this publication will inform the relevant OPCW structures and mechanisms tasked with addressing these issues in the longer term.

1.3 Holistic Arms Control

For many years, the governmental and non-governmental arms-control communities have sought to construct strategies to combat the development, proliferation and subsequent use of chemical and biological weapons by State and non-State actors. Recognising that reliance upon a single disarmament or arms-control agreement alone would not guarantee success, scholars have explored a number of concepts seeking to broaden the range of possible regulatory mechanisms. Utilising and building upon such work, particularly the concepts of "preventative arms control"[11] and "webs of prevention (or protection)",[12] a "holistic arms control" (HAC) framework for regulation, has been developed[13] and in this book is adapted and applied to frame an exploration of effective strategies to address the threats arising from the misapplication of the rapid advances in the chemical and life sciences and, more broadly, to prevent the re-emergence of chemical weapons.

Whilst the proposed HAC analytical framework concentrates upon existing arms-control and disarmament measures (in this case primarily the Chemical Weapons Convention, but also the Biological and Toxin Weapons Convention, the BTWC), it attempts to widen the range of applicable mechanisms for regulation, and also the nature of the actors involved in such regulatory measures.

As originally envisaged, HAC consists of the following three stage process:

- *Stage one:* constitutes an examination of the nature of the weapons and technology that are to be controlled and explores the current and potential future scenarios of application, together with the attendant national and human security concerns of inappropriate use. During this stage the potential relevance of advances in science and technology is assessed.
- *Stage two:* constitutes an analysis of the full range of potentially applicable international law, (arms control, disarmament and other) instruments and attendant control regimes; highlighting strengths, weaknesses and ambiguities in these mechanisms. The potential roles of relevant civil society actors are also explored.
- *Stage three:* following an analysis of information derived from stages one and two, a comprehensive strategy is developed to strengthen existing mechanisms and/or introduce new mechanisms to facilitate effective regulation or prohibition of the weapon or weapons-related technology of concern.

1.4 The Structure of the Book

The conception, development and organisation of this book have been informed by the HAC analytical framework. The book consists of 21 chapters divided into six sections. In Section I, following this introductory first chapter is Chapter 2 by Paul Rogers which provides an overview of the very different international security system that has evolved since the end of the East–West Cold War and how asymmetric warfare within this novel system may well encourage the use of forms of chemical warfare by parties involved.

Section II then gives an overview of the nature and health of the contemporary chemical weapons prevention and disarmament regime today, with contributions by Ralf Trapp on the Chemical Weapons Convention (Chapter 3), Jeremy Littlewood on the Biological and Toxin Weapons Convention (Chapter 4) and Michael Crowley on United Nations mechanisms designed to combat the development, acquisition and use of chemical weapons (Chapter 5), and wider international legal constraints on the weaponisation of toxic chemicals (Chapter 6).

Section III forms the major part of the book with a series of chapters on the advances in relevant science and technology. Chapter 7 by Trapp considers the continuing convergence of the sciences of chemistry, biology and nanotechnology and the implications of these advances. In Chapter 8, Malcolm Dando describes some of the developments in our understanding of the operation of the central nervous system and how this may open up the possibility of the design and weaponisation of novel chemical agents. Kathryn Nixdorff, in Chapter 9 then discusses the advances being made in our capabilities in targeting chemical agents to designated sites in living systems.

Advances in anti-plant and anti-animal biological and chemical agents are dealt with respectively by Simon Whitby in Chapter 10 and Piers Millet in Chapter 11; contemporary trends in toxic chemical dispersal and delivery mechanisms by Crowley in Chapter 12. The section ends with a technical review of advances in our capabilities for prevention and treatment of traditional chemical agents by Xinbo Zhou, Wu Zhong, Pengbin Bu and Lijun Shang in Chapter 13, and of the challenges to the current export control system particularly produced by the changing nature of the marketing of relevant goods through the internet, authored by Philippe Mauger and Ray Zilinskas in Chapter 14.

The implications of these advances for the chemical weapons prevention and disarmament regime and options for how they should be addressed are then the subject of the chapters in Section IV. Chapter 15 by Trapp deals with the implications for the CWC, and Chapter 16 by Littlewood with the implications for the BTWC.

Given the importance of a wide range of expertise, experiences and viewpoints being available to the 4th CWC Review Conference and the OPCW more broadly, Section V is devoted to an investigation of the ways in which civil society has attempted to assist in combating the development and use of chemical weapons, and of how it might continue such efforts in the future, particularly in the context of rapid scientific and technological developments. In Chapter 17 Katie Bowman, Alistair Hay and Jo Husbands detail the initiatives of the non-governmental chemical community, Tatyana Novossiolova examines the non-governmental life science community's efforts in Chapter 18, and Crowley and Vivienne Nathanson explore the importance of medical ethics and the role of the non-governmental medical community in Chapter 19. In Chapter 20 Crowley, Catriona Mcleish and James Revill then consider the work of informed activist civil society as a whole.

The final Section VI of the book comprises the concluding Chapter 21 by the editors and Trapp. This seeks to employ the HAC framework as a means of analysing the contributions of the preceding chapters and synthesising them into a proposed integrated approach which would prioritise the most appropriate and effective mechanisms to address potential threats arising from the rapid advances in the chemical and life sciences, and more generally, the dangers of the (re-)emergence of existing or new forms of weapons employing toxic chemicals. Given the central importance of the CWC, the chapter incorporates recommendations for both States and civil society to strengthen OPCW action in these areas, at the 4th Review Conference and beyond.

On behalf of all of the authors we would like to thank the external reviewers – Stefan Mogl, Julian Perry Robinson and John Walker – who kindly read through the final drafts of the chapters and offered us such helpful comments. The authors and editors are, of course, responsible for any remaining errors.

References

1. For an Online Version of the Text of the Chemical Weapons Convention (CWC), See Organisation for the Prohibition of Chemical Weapons (OPCW), *Convention on the Prohibition of the Development, Production, Stockpiling and Use of Chemical Weapons and on Their Destruction*, https://www.opcw.org/fileadmin/OPCW/CWC/CWC_en.pdf, accessed 24 July 2017.
2. A. Üzümcü, *Working Together for a World Free of Chemical Weapons, and Beyond*, Nobel Peace Prize Lecture, 2013.
3. OPCW website, *Statistics*, http://www.opcw.org, accessed 9 April 2018.
4. OPCW, *OPCW Director-General Commends Major Milestone as Russia Completes Destruction of Chemical Weapons Stockpile under OPCW Verification*, 27 September 2017; OPCW, *OPCW Director-General Praises Complete Destruction of Libya's Chemical Weapon Stockpile*, 11 January 2018.
5. OPCW Technical Secretariat, *The OPCW in 2025: Ensuring a World Free of Chemical Weapons, S/1252/2015*, 6 March 2015.
6. OPCW Technical Secretariat, *Report of the Advisory Panel on Future Priorities of the Organisation for the Prohibition of Chemical Weapons, S/951/2011*, 25 July 2011, paragraphs 21–22.
7. OPCW Technical Secretariat, *Report of the Advisory Panel on Future Priorities of the Organisation for the Prohibition of Chemical Weapons, S/951/2011*, 25 July 2011, paragraphs 16–19.
8. OPCW Technical Secretariat, *Report of the Advisory Panel on Future Priorities of the Organisation for the Prohibition of Chemical Weapons, S/951/2011*, 25 July 2011, paragraph 11.
9. OPCW Technical Secretariat, *Report of the Advisory Panel on Future Priorities of the Organisation for the Prohibition of Chemical Weapons, S/951/2011*, 25 July 2011, paragraph 13.
10. OPCW, *Convention on the Prohibition of the Development, Production, Stockpiling and Use of Chemical Weapons and on Their Destruction*, n.d., Organisation for the Prohibition of Chemical Weapons (OPCW), Article VIII (22), https://www.opcw.org/fileadmin/OPCW/CWC/CWC_en.pdf, accessed 24 July 2017.
11. See for example: J. Altmann, Preventive arms control: concept and design, in *Military Nanotechnology: Potential Applications and Preventive Arms Control*, ed. J. Altmann, Routledge, London, 2006.
12. See for example: *A Web of Prevention: Biological Weapons, Life Sciences, and the Governance of Research*, ed. C. McLeish and B. Rappert, Earthscan, London, 2007.
13. M. Crowley, *Chemical Control: Regulation of Incapacitating Chemical Agent Weapons, Riot Control Agents and Their Means of Delivery*, Palgrave, Basingstoke, 2016.

CHAPTER 2

The Changing Nature of the Chemical and Biological Weapons Threat

PAUL ROGERS

Department of Peace Studies, Bradford University, Bradford, West Yorkshire, BD7 1DP, UK
*E-mail: p.f.rogers@bradford.ac.uk

2.1 Introduction

In February 1993, an extreme paramilitary group placed a very powerful truck bomb in an underground car park up against one of the main support pillars of the North Tower of the New York World Trade Centre (WTC). The bomb was detonated by means of a timing device after those responsible had left the Centre. They were later apprehended and sentenced to long terms of imprisonment. When the bomb detonated, it did considerable damage and came close to bringing down the 21 storey Vista Hotel which connected the North and South Towers of the Centre. Following emergency engineering work, the Vista Hotel survived, as did the North Tower. Six people died, and many hundreds were affected by smoke inhalation.

The purpose of the attack was to cause the greatest loss of life of any act of war since the atomic bombing of Nagasaki on 9 August 1945 – to bring down the North Tower across the Vista Hotel and into the South Tower, devastating

Preventing Chemical Weapons: Arms Control and Disarmament as the Sciences Converge
Edited by Michael Crowley, Malcolm Dando and Lijun Shang

Published by the Royal Society of Chemistry, www.rsc.org

the entire complex. Given that this was in the middle of the working day, and that the collapse would have been almost instant, the death toll would have been in the region of 30 000 people.

Eight years later, both of the towers of the Trade Centre were destroyed at the start of the working day in the notorious 9/11 attacks, as part of which the headquarters of the US Department of Defense, the Pentagon in Washington DC, was also hit. In the 9/11 attacks the nineteen perpetrators gave their own lives along with almost 3000 people, marking this action out from most other examples of paramilitary attacks in the preceding decade which had not seen large scale mass casualty attacks. There were some exceptions during that decade and earlier, but most paramilitary and political attacks across much of the world had had very specific purposes such as political assassination, the destruction of powerful symbols or economic damage. There were, though, some other acts designed deliberately to cause serious casualties. Some were devastating in their impact but almost all, as 9/11 and the 1993 WTC attack showed, relied on conventional explosives rather than the use of weapons of mass destruction.

Since 9/11, international conflicts have been dominated by the evolution of paramilitary violence and the attempts to control it, largely by military means. What has been commonly termed as the "War on Terror" is, at the time of writing, just into its 16th year with no end in sight. It represents a very different security outlook to the previous fifty years, most of which were dominated by the Cold War confrontation and it is this changed outlook which will be the main emphasis in this chapter, which examines the security context that relates to the possible use of chemical and biological weapons in the coming years. It is particularly relevant given that the most recent clear examples of the use of chemical weapons have been during the bitter civil war in Syria which started in 2011.

The development, occasional use and extensive efforts on the control of chemical and biological weapons (CBW) at the level of State involvement over the past half century have been the subject of many books and articles, providing much information available to interested readers. In the current context though, this book is being written at a time of necessarily changed attitudes to possible future CBW use because of the 9/11 experience and its consequences, what has happened since and what appears plausible in the coming years. The experience of the Syrian civil war and the conflict involving ISIS is particularly relevant here as there have been a number of examples of chemical agents being used. This may have involved the use of pre-existing CW stocks held by the Assad regime or the use of industrial chemicals such as chlorine as CW agents, and the individual instances may not necessarily have amounted to mass casualty attacks, but the issue here is that such CW use has not been seen at such an intensity since the actions of the Saddam Hussein regime in Iraq in early 1988.

The central problem that is relevant here is the rise of sub-State groups that may be able to develop CBW systems, however crude, they may have the intention of using them in mass casualty attacks and may be sufficiently motivated to put their own lives at risk. While such a circumstance is not new, it has been rare up until now, but must be accepted as possible. As such it indicates that we will have to change our approach to the control of such systems, quite probably reassessing how we still see the control of paramilitary groups as a primarily military process. Much of this relates specifically to Al-Qaeda, ISIS and related movements and the chapter will be structured as follows.

Firstly, the examples of mass casualty attacks already referred to in this introduction will be supplemented by four further examples of paramilitary violence that shed light on the capabilities of pre-9/11 movements. This provides a necessary context for understanding both the capabilities and motivations of more recent paramilitary groups. Secondly, the actual experience of the War on Terror will be reviewed, with particular reference to the aims and motivations of the main movements – Al-Qaeda and ISIS. Thirdly, the extent to which such groups may be models for future paramilitary movements will be discussed. Fourthly, the circumstances in which existing and successor groups might acquire and use CBW will be tentatively explored and, finally, a conclusion will point to how these changing international circumstances require an evolution in our approach to CBW control. Before continuing, we have to recognise that modern-day extreme movements that might wish to embark on mass casualty attacks will not distinguish fundamentally between chemical, biological, radiological or nuclear devices. The concern of this volume is with CBW systems but the arguments in this chapter do apply to radiological and nuclear devices, though the latter would commonly involve significantly greater technical prowess.

2.2 Mass Casualty and Mass Effect Attacks

Prior to 9/11, four examples of paramilitary attacks stand out, as follows.

2.2.1 LTTE and the Bombing of the Colombo War Trade Centre

By 1995, the bitter war between the Sri Lankan Government and the Tamil Tiger separatists (LTTE) had been under way for twelve years, but towards the end of that year, the Sri Lankan Army had made considerable gains against the LTTE in the north of the island, especially against the LTTE's stronghold of Jaffna. The cost to both sides was huge, but especially to the LTTE and to the Tamil civilians. The movement responded with a series of attacks against economically significant targets such as the main airport and oil refineries, with this culminating in a devastating mass casualty attack in the capital city of Colombo. On 31 January 1996, a powerful truck bomb was driven up to the entrance of the Sri Lankan Central Bank and detonated with appalling results. At least a hundred people were killed and 1400 injured, and several key buildings in Colombo's central business district were seriously damaged.

This incident was characterised by three elements – the damage to an economically and politically significant target, the willingness to cause very large numbers of civilian casualties, and the use of a suicide bomber. Significantly, the use of suicide or "martyr" bombers by the LTTE was not primarily a matter of religious conviction, but instead stemmed from the ability of the movement to persuade some members that the cause was sufficiently noble to demand this ultimate sacrifice.

2.2.2 The 1994 Attempted Attack on Paris

A year before the Colombo bombing another kind of attack was prevented which would, if it had been successful, have caused huge civilian casualties, this time in Paris. Like Colombo, this also involved perpetrators prepared to give their lives, in this case members of an extreme Islamist group with origins in Algeria. On 24 December 1994, an Air France Airbus A300 bound for Paris was hijacked and landed at Marseilles where refuelling was demanded. It was later learnt that the aim of the hijackers was to crash the fuel-laden plane on the centre of Paris, killing themselves, all 239 passengers and crew, and hundreds of people on the ground. A French counter-terror commando unit intervened at Marseilles, killing the hijackers and freeing all of the passengers and crew, except for three who were killed by the hijackers.

2.2.3 The Provisional IRA and the London Bombings, 1992–97

By early 1992, the bitter conflict in Northern Ireland was at a stalemate. The Provisional IRA (PIRA) could not force Northern Ireland into a united Ireland, but the combined forces of the Northern Ireland security forces and the British Army were not able to quell the rebellion. PIRA then commenced a campaign of systematic attacks on economic targets in Great Britain, with the primary focus on London, then in competition with Frankfurt to be the leading financial centre in Europe. The General Election of 9 April 1992 was won narrowly by the Conservative Party under John Major, with the Northern Ireland conflict not featuring strongly in the election campaign. The following day PIRA detonated two large van bombs in London, one damaging the major road interchange between the M1 motorway and London's North Circular Road and the other in the centre of London's financial district. Neither was designed to cause mass civilian casualties, although three people were killed and scores injured, but the City of London bomb was detonated in the middle of a high-rise business district and did about £1 billion in damage.

This was the start of a five year campaign, interrupted by a short ceasefire, in which numerous attacks were staged on many economically significant targets, including a second large bomb in the centre of London, another in the secondary business district of Canary Wharf, a large bomb in the retail centre of Manchester, an attempt to disrupt the whole of the electricity supply to Greater London, and numerous disruptions of road and rail communications. Although not acknowledged by successor governments, the campaign led to a willingness to engage in back-channel discussions with PIRA leaders and, after the 1997 election, a determined and ultimately successful attempt to negotiate a settlement to the conflict. In the current context, the significance of the PIRA campaign is the capability to use crude large explosive devices to have substantial political effects.

2.2.4 Aum Shinrikyo

In the context of this book, the most relevant example of an extremist group was the Aum Shinrikyo sect that staged a chemical weapons attack on the Tokyo subway system on 30 March 1995. The attack used sarin nerve agent and the intention was to cause death and chaos on a huge scale as part of the movement's apocalyptic vision for the future. Three subway lines and 15 stations over a distance of 12 km were affected with 12 people killed and 5500 made ill. Even so, the attack failed to achieve its much greater intended impact, the explanation being that the agent was impure or that the perpetrators had misjudged the dispersal potential of the subway's ventilation system. The Aum Shinrikyo incident has three implications in the current context – it was a chemical weapons (CW) attack, it was an example of how CW could be used in a restricted environment, and it was manufactured by a wealthy and determined group able to operate in a clandestine manner within a modern industrial State, but without controlling substantial territory of its own.

These examples together give some indication of the abilities of oppositional sub-State groups when acting against modern urban-industrial States. They illustrate a trend that was apparent before 9/11, and all that followed, and demonstrate a change in the relative capabilities of State and sub-State groups. This trend was appositely noted by a former US Navy submarine commander Roger Barnett back in 1992,[1] when he noted the impact of changes in technology on the increased ability of the "weak to take up arms against the strong". Only one of these examples involved CBW and the concern of this chapter is whether that element of capability will increase in significance. In order to discuss this, it is helpful to summarise the recent experience of the growth of major Islamist movements and then analyse the wider context which strongly indicates that there are wider global trends that will make extreme movements more common.

2.3 Relevant Elements of the Context of the War of Terror

The Al-Qaeda movement has its origins in a group of foreign fighters who went to Afghanistan to aid the mujahidin in their war against the Soviet occupying forces in the mid-1980s, motivated by an Islamist outlook and an intense desire to see foreign invaders repulsed. The motivations of these fighters were rooted much more in this religious identity and conviction than the ethnic and nationalist motivations of most Afghan mujahidin. Many, including Osama bin Laden, came from Saudi Arabia and had been much influenced by the Wahhabi tradition of austere Islam and the teachings of the Egyptian Sayyid Qutb, who had been executed by the Nasser regime in 1966. Their puritanical version of Islam evolved into a worldview demanding the fundamental reform of Islam, the replacement of State governance by Islamist rule, and the belief that they were the bringers of the final and purist form of the belief. It was very much a minority view within the Muslim world, but by the end of the war with the Soviets and their withdrawal from Afghanistan, it had acquired a world vision.

By the early 1990s, the aim was to attack and undermine unacceptable regimes across the Arab world and through into South Asia, with these including the Kingdom of Saudi Arabia, which was seen as entirely unsuited to being the Guardian of the Two Holy Places. Beyond this "near enemy" of regional regimes was the "far enemy" of the United States, and its western allies, with their support for the regimes, and also for Israel. During the 1990s, the Al-Qaeda movement developed, attacking western military units in Saudi Arabia and then consolidating its activities in Afghanistan where it came to the aid of the Taliban movement as it sought to unify the country by defeating the North Alliance of warlords. At the end of the 1990s, the Al-Qaeda movement had supporters in many countries, and in 2000–2001, one group, mainly comprising young Saudi professionals based in Frankfurt, conceived the idea of a symbolic attack on the far enemy, an idea that was translated into the 9/11 attacks.

Whilst a popular view of Al-Qaeda at the time of 9/11 was that they were an insane, illogical and highly dangerous fringe group, only the term "dangerous" is accurate. In practice, and it is a crucial point in the context of this book, Al-Qaeda should be seen as a transnational revolutionary movement, but one that differs from most in its religious foundations. Revolutionary movements seek radical change in societies, often, but by no means always, involving violence, and they tend to be commonly rooted in political ideologies, although nationalist and ethnic elements may be prominent. Most such movements believe that there is the potential to achieve success within the lifetimes of the leaders and the most dedicated activists. They do not look to successor generations, and their timescales are typically measured in years or, at most, a few decades.

A movement rooted in a religious identity may take a very different view and may see itself as part of a much greater continuum that will certainly be measured in decades, and quite possibly, centuries. Thus, the Al-Qaeda movement had as its short-term aim the replacement of unacceptable regimes in the region with "true" Islamist rule as part of the creation of a regional caliphate – a politico-religious entity in full control of a society. This would certainly take decades, but beyond this was the view that such a caliphate was the starting point for a global transformation towards what was seen as the only basis for human living, a process that might take a century or more. To repeat a point made earlier, this outlook may have attracted some sympathy from followers of the Wahhabi tradition of Sunni Islam but was anathema to the great majority of Muslims of all perspectives.

At the core of such a movement in the current context are three elements. One is an eschatological dimension, looking beyond this life and seeing a clear continuum with what comes after. Because of this, the leadership and any core strategic thinkers do not expect to see success in their lifetimes. The second follows on from this and means that individual acts that may require self-martyrdom—suicidal actions—fit in naturally with this outlook, as in the case of the 9/11 attackers, all nineteen of whom died in the attacks.

Contrary to general belief, suicide attacks and suicide actions are not limited to extreme Islamist movements. Some, such as the LTTE Tamil Tiger movements in Sri Lanka, have had adherents sufficiently dedicated to the mainly secular cause to determinedly give their own lives, and there is a history of other religious orientations taking extreme actions. One of the most notorious examples was the Jonestown "revolutionary suicide" in North Western Guyana in 1978 when 918 people died, mostly by taking cyanide, although some were murdered. They were almost all Americans and followers of the neo-Christian preacher Jim Jones and his Peoples Temple organisation that had settled in Guyana but had come under such pressure that its members destroyed themselves on his orders.

The final element at the core of movements rooted in an extreme interpretation of a religious belief is an outlook that sees non-believers as threats to the success of the revolution. As such, they may in some circumstances be expendable. For example, the killing of nearly 3000 people in the 9/11 attacks, or the willingness to kill ten times that number if the 1993 WTC attack had succeeded, were both examples of this. Moreover, such an outlook extends to co-religionists who do not share the vision, and it may consider such apostates even more dangerous and therefore especially worthy of the harshest treatment.

In relation to the potential use of chemical and biological weapons, all three of these elements are relevant. Planning and acting at almost indefinite timescales means that seeking CBW systems need not be a short-term matter but can be planned and prepared over many years. Being willing to cause the deaths of thousands of people means that, if available, CBW systems might readily be used, and if adherents are willing to give their lives in pursuit of the cause, this makes preparation and delivery easier. Furthermore, such an outlook might not even be deterred by the prospect of an overwhelming response. Thus, if a successful mass casualty attack was carried out against a nuclear-armed State, leading to many thousands of people killed, then that State might retaliate with a nuclear strike on the centre of the movement. For the leaders of the movement themselves, that might be a part of the calculation on the basis that it would so anger and inspire potential supporters that the end result might be assumed to be a stronger and more vigorous movement. We should not assume that the normal logic of deterrence will apply to an eschatologically orientated movement.

2.4 Wars in Afghanistan and Iraq

After 9/11, the United States moved rapidly to respond to Al-Qaeda by successfully terminating the Taliban regime using a combination of aerial bombardment, Special Forces and the employment of Northern Alliance militias as ground troops. By late January 2002, just four months after 9/11, President Bush could report success in his State of Union Address to Congress and could stretch the war on terror against Al-Qaeda to a much more global enterprise against an axis of evil of States such as Iraq, Iran and North Korea that were deemed to be supporting terrorist movements, while also working to develop weapons of mass destruction.

The termination of the Saddam Hussein regime in Iraq began in March 2003 and the regime collapsed within three weeks. President Bush was able to deliver his "mission accomplished" speech three weeks later on 1 May 2003 when the United States had seemed to have recovered from the shock of 9/11. It had successfully terminated regimes in Afghanistan and Iraq, opened military bases in both countries, expanded its military connections in Central Asia and the Balkans, and was in a position to oversee the transition of governance in Iraq towards a pure free market pro-western approach.

Within two years the situation had changed radically, with a full-scale urban insurgency in place in Iraq and a deteriorating security environment in Afghanistan as the Taliban emerged as a renewed threat to a largely dysfunctional government in Kabul. By 2011, the US and its coalition partners had poured 140 000 troops into the country to little effect, and most had been withdrawn by late 2016, even though the Taliban and other armed opposition groups controlled close to a third of the rural areas. In Iraq, the shadow war fought by Special Forces against the hardest of the opposition groups, Al-Qaeda in Iraq (AQI) appeared initially to work, and President Obama was able to withdraw all significant combat units by the end of 2011.

In practice, the survivors of AQI went on to form Islamic State in Iraq and Syria (ISIS – one of several names for the movement) which rapidly took control of much of Northern Iraq and Syria, declaring a new caliphate at the main mosque in Iraq's second city of Mosul in July 2014 and basing its leadership headquarters at Raqqa in northern Syria. The significance of this in the current context is that this was an extreme movement which took control of substantial territory with upwards of six million people under its control. Stable territorial control is a huge aid to planning and executing a major war and allows, given time, the possibility of devoting resources to developing CBW. In the event, the United States rapidly organised a coalition of western and regional States to commence an intense air war against ISIS, also serving to aid Iraqi Army and militia units in their aim to retake territory from ISIS. That was extended to Syria amidst OPCW–UN (Organisation for the Prohibition of Chemical Weapons–United Nations) reports that both ISIS and the Assad regime were using chemical weapons, either legacy systems from Syrian arsenals, or agents such as chlorine diverted from industrial use.

Overall, by late 2016, the wars in Afghanistan, Iraq and Libya had killed some 300 000 people, injured hundreds of thousands more, displaced millions and cost around $3000 billion. Controlling movements such as Al-Qaeda and ISIS with tens of thousands of boots on the ground had been abandoned, to be replaced by what is commonly termed "remote control warfare" employing sustained air power including armed drones, Special Forces, private military and security companies, and other low-profile forces only partially open to independent political scrutiny.[2]

2.5 Implications for Future Security

In its own right, the experience of the last fifteen years demonstrates that we have entered an era of largescale paramilitary movements with considerable transnational potential. If any such group can in the future evolve a reasonably competent economic and technical infrastructure, then the capability to develop CBW systems will be there. Indeed, there are already two pointers to this. The first is that, as we have seen, the Aum Shinrikyo group in Japan in the early 1990s had sufficient resources and secure locations to be able to produce a basic nerve agent, and the same group also worked with anthrax for use as a biological agent. The second relates to a State but is also indicative. During the 1980s, the Saddam Hussein regime developed a broadly based CBW programme involving the largescale production of nerve agents and also a number of toxins and biological agents. Iraq was a mid-level technical power, but a sustained programme over several years was enough to provide several fully weaponised and deployable systems including spray bombs and ballistic missile warheads. The weaponisation was part of a much wider arms production capability, but the agent development and production processes were more basic and might not be beyond the capabilities of a determined paramilitary group, provided that it had stable control of territory, access to technical knowledge and capabilities, and a motivation to develop and use such systems.

Perhaps the most important element in this, is that the Saddam Hussein regime appears to have focused its efforts on producing CBW systems for two different functions. One would be specifically for use against Iran, and the other as a deterrent force against States such as Israel and the United States, CBW being actually deployed for possible use against the US-led coalition in early 1991 after the invasion of Kuwait. It follows that what is relevant to consider is whether the kind of paramilitary movement that has been in evidence in the past fifteen years, and which might consider CBW systems relevant to its operations, is likely to be a more common feature of the next two or three decades. There are two levels at which to consider this – the extent to which paramilitary movements are more likely in the future and whether it is likely that some may acquire CBW aspirations.

2.6 Revolts from the Margins

ISIS, Al-Qaeda and other loosely related movements should be seen as part of a much wider issue, the development of a trend towards what might loosely be called "revolts from the margins". An extreme Islamist approach is by no means the fundamental issue for the coming decades, the risk being that we are moving into an "age of insurgencies" rather than a "clash of civilisations" between the West and the Islamic world – towards a global environment of fragility, instability, increasing violence and irregular war. Thus, while ISIS may be a major security problem, the real drivers of current and global insecurity are quite different – deepening socio-economic divisions and the relative marginalisation of most peoples across the world, while the entire world faces the prospect of deep and lasting environmental constraints, especially climate disruption.

The issue of relative marginalisation stems from the evolving failure of the neoliberal economic system since it became dominant between 1975 and 1990. This system has produced variable economic growth, but the fruits of that growth have been remarkably concentrated in the hands of about one fifth of the global population – a transnational elite community rather than a clear rich country/poor country dynamic. In various measures, whether income or of wealth, that fifth of the global population has about 85% to 90% of the total, and there is no sign of this decades long trend being reversed.

It is a circumstance made more immediate because a whole raft of welcome improvements in education and literacy are having the effect of making far more people aware of their own marginalisation and they are unwilling to accept it. This was one of the underlying drivers of the Arab Awakening that developed during 2011 and has also been a driver for neo-Maoist movements in South and South East Asia, most notably the persistent Naxalite rebellion in India. The combination of major wealth–poverty differentials with the improvements in education produce this combination of anger and resentment that can readily fuel radical and extreme movements, but it is made worse because the wealth–poverty divide is actually widening year by year. It is now being further exacerbated by the impact of climate disruption, with clear indications that the combination will produce formidable insecurity, at least on present trends.

Although there is still a climate change denial community, this is diminishing in strength in the face of massive evidence that the climate is being altered by anthropogenic actions, principally the release of carbon dioxide through the rapid combustion of fossil carbon in the form of gas, oil, and especially coal. It is doubly significant in terms of the effect on human organisation and wellbeing because of two factors – the rate of warming is accelerating and its effects are globally asymmetric. On the latter issue, there is mounting evidence that tropical and sub-tropical regions are warming faster than temperate regions and that rainfall is tending to decrease over these landmasses while increasing over northern latitudes and the oceans.

This "drying-out" of the tropics, which has been recognised as a trend for over two decades, has huge socioeconomic implications in that the ecological carrying capacity of some of the world's most important croplands is decreasing.

Over the past decade, we have seen a marked increase in the attempts by economically marginalised peoples to migrate to more favourable environments, a trend exacerbated after 2011 by large-scale refugee flows from areas of intense post 9/11 conflict. In response, there has been marked social and political resistance to such pressures, most notably in Europe, the United States and Australia, and the current political trends suggest that the intensity of this response will grow as climate disruption increases in impact. Yet, one cannot close the castle gates in a globalised world, the likely result being even more anger and resentment from what should correctly be seen as the majority margins.

In an analysis from nearly twenty years ago it was suggested that:[3]

> "What should be expected is that new social movements will develop that are essentially anti-elite in nature and draw their support from people on the margins. In different contexts and circumstances they may have the roots in political ideologies, religious beliefs, ethnic, nationalist or cultural identities, or a complex combination of several of these. They may be focussed on individuals or groups but the most common feature is an opposition to existing centres of power. They may be sub-State groups directed at the elites in their own State or foreign interests, or they may hold power in States in the South, and will no doubt be labelled as rogue States as they direct their responses towards the North. What can be said is that, on present trends, anti-elite action will be a core feature of the next 30 years – not so much a clash of civilisations, more an age of insurgencies."

Even so, there is a presumption that security can best be controlled, when other methods fail, by resorting to military responses. This is greatly aided by the power and influence of what Eisenhower called the military–industrial complex but is better described as the military–industrial–academic–bureaucratic complex. At its crudest level, what is sometimes termed the "control paradigm" might better be termed "liddism" – keeping the lid on problems rather than understanding their causes and manifestations.

In this context, ISIS, Al-Qaeda, Boko Haram, the Naxalites and others, should be seen as markers of what could be to come. Thus, what may be likely on present trends is what the geographer and politician Edwin Brooks[4] warned of over forty years ago, that what we had to avoid is a dystopic future of a "crowded glowering planet of massive inequalities in wealth, buttressed by stark force yet endlessly threatened by desperate people in the global ghettoes." Our concern here, though, is whether that danger will be prevented and, if not, to consider the risk of extreme movements developing and using CBW systems.

2.7 Responses to the New Security Challenges

If the widening wealth–poverty gap and the resulting mass marginalisation in an environmentally constrained world is to be avoided, a societal transformation will be required over the next two decades. This will require the superseding of the current neo-liberal economic system with forms of economic organisation that place a premium on cooperation, emancipation and greater equity, all within the context of a rapid transition to ultra-low carbon economies. There are already many signs of change including a much greater recognition of the fundamental inadequacies of the neo-liberal system. Close to a billion people are members of cooperatives and mutuals, the work of the New Economics Foundation[5] and other think tanks are paving the way for many other changes and a number of significant political movements are predicated on critiquing the current system.

There have also been impressive and rapid developments in a wide range of renewable energy technologies, not least those applied to tropical and sub-tropical environments, and especially through major developments in solar, wind, wave and sub-sea systems. There is also impressive progress under way in developing reliable energy storage technologies and there is immense scope for conserving energy throughout the current systems of economic and social organisation, not least relating to transport and space heating.

Even so, the degree of political and social inertia in responding to the urgency of the many necessary changes does not inspire great confidence, especially as most political leaderships, both at governmental and intergovernmental levels, are simply not prepared to accept the rapid rate of societal transformation that is needed. Too often there are impressive statements that fail to be backed up with actual change. It may well be that major catastrophes may catalyse more rapid action, in which case the many developments summarised above could have a rapid and very welcome effect. For the purposes of the present volume, though, it is right to assume that the prospect of Brooks's "crowded glowering planet" should be considered seriously likely, making it necessary to face the risk of new actors exploiting CBW capabilities.

2.8 The Risk of CBW Development and Use in the New Security Context

This final section will work on the basis that current security trends indicate that extreme paramilitary movements will continue to develop and, given the factors just discussed, will increase both in frequency and intensity. The most worrying model will be the kind of movement that ISIS has represented – a determined movement that looks beyond this life and has a defended and stable territorial base, giving it scope for CBW development. If, in the future, such an extreme paramilitary organisation was to acquire CBW systems, it might do so for two different reasons. One would be to act as a deterrent against much more powerful forces intent on attacking it and the territory it controls. This might not have too great an effect on modern armed forces with CBW protection but could be highly effective against civilian populations in the State or coalition attacking it, thus deterring an attack.

It could be argued that such an action would be self-defeating since it might well bring about the large-scale destruction of the movement through the use of nuclear retaliation, but if a movement has an outlook that is eschatological, then that argument does not stand up. As we have seen, a movement that regards itself as part of a grand history that transcends mere temporal existence might welcome such a fate in the belief that it will result in a much greater uprising from a wider and energised support base against its enemies.

A second purpose would be the direct projection of its own power and authority, most likely against unprotected civilians. Given that it has in recent years proved possible for a paramilitary movement to kill many thousands of people with non-CBW systems, the relative value of such a posture would be more in terms of the level of fear inculcated with any CBW use. Such is the concern over such systems in the public mind in many countries, particularly in the West, that the effect could be profound. Thus, if a CW or biological weapons (BW) attack on a shopping centre or indoor arena caused many hundreds of deaths, the fear impact might actually be much greater than an action involving more conventional systems such as the coordinated use of high explosives. The simplest definition of terrorism is the threat or use of actions that cause fear among a much larger population than is directly affected, and the public fear of both chemical and biological weapons would add substantially to this impact.

Here again, it might seem illogical for an extreme movement to carry out such an attack given the likely devastating consequences, but what has to be appreciated is that if there are movements that do not follow the usual patterns of inter-State behaviour, especially if they see themselves as part of a plan that transcends ordinary human existence, then normal behaviour does not apply. The Aum Shinrikyo movement, in planning and undertaking its attack on the Tokyo subway, did not appear concerned about the consequences for it if the attack had succeeded and caused many hundreds of deaths, since its view of human futures looked to the action fomenting a wider conflict with an apocalyptic dimension.

2.9 Conclusions

In practice, the kind of circumstance being considered here is uncomfortably plausible. Thus, given the underlying drivers of insecurity, it is highly unlikely that movements such as Al-Qaeda and ISIS are close to the point of suppression and extinction. Other movements will form and we should therefore assume that there is a new context relating to quite novel forms of chemical and biological weapons proliferation, a context that has scarcely been recognised so far.

Responses are required at four levels.

- First, there should be an overall recognition that the threat of use of chemical and biological weapons may have receded with the ending of the Cold War, but it is now reappearing in a new guise.
- Second, because of this there should be renewed efforts to strengthen existing arms control regimes for both categories of weapons. This should include the recognition that methods to research and develop chemical and biological agents have been made easier and have come to require less expensive research and development (R&D) capabilities.
- Thirdly, much more attention should be focused on those factors that enable movements such as ISIS and the Naxalites to gain and maintain strong recruiting bases, since the lesson of over fifteen years of the war on terror is that conventional military approaches have not worked.
- Finally, the evolving security environment of revolts from the margins in a divided and constrained world must be countered by healing those divisions and overcoming those constraints. The risk of CBW proliferation involving sub-State movements should be a powerful additional motive for action.

References

1. R. W. Barnett, Regional conflict: requires naval forces, *Proceedings of the US Naval Institute*, 1992.
2. http://www.remotecontrolproject.org/.
3. P. Rogers, *Losing Control: Global Security in the 21st Century*, Pluto Press, London, 2000.
4. E. Brooks, The Implications of Ecological Limits to Development in Terms of Expectations and Aspirations in Developed and Less Developed Countries, Chapter in: Anthony Vann and Paul Rogers, Human Ecology and World Development, Plenum Press, London and New York, 1974.
5. http://www.neweconomics.org/.

Section II

The Chemical and Biological Weapons Prevention and Disarmament Regime Today

CHAPTER 3

The Chemical Weapons Convention – Past Success, Current Challenges

R. TRAPP

Independent Disarmament Consultant, 116 Route de la Contamine, Chessenaz-74207, France
*E-mail: ralf.trapp@gmail.com

3.1 Introduction

The Chemical Weapons Convention (the Convention or CWC) is a cornerstone agreement in international disarmament. Negotiated in a particular historical context—the end of the Cold War—many of its provisions on the elimination of chemical weapons stockpiles and production facilities, and of the verification of these measures, reflect the experience with the chemical weapons programmes of that period and previous decades.

At the same time, its comprehensive and unqualified prohibition of chemical weapons, and of any military preparation for their use, are a continuation and expansion of the principles enshrined in international humanitarian law, in particular the 1925 Geneva Protocol. The Convention's norm against the re-acquisition of chemical weapons capabilities—whether in the form of "traditional" chemical weapons known from past State programmes or in the form of new types of agents or dissemination systems—and the related national implementation requirements and international verification measures were

Preventing Chemical Weapons: Arms Control and Disarmament as the Sciences Converge
Edited by Michael Crowley, Malcolm Dando and Lijun Shang

Published by the Royal Society of Chemistry, www.rsc.org

intended to make the Convention future-proof, consistent with its unlimited duration and its proclaimed objective: for the sake of all mankind to completely exclude the possibility of the use of chemical weapons.

The aspirations of the States Parties of the CWC go well beyond merely "dealing with the past". The norm they have subscribed to is meant to be comprehensive and lasting, and therefore must be at the same time principled in regulatory approach, and adaptable to change in practical application. Much progress towards a chemical weapons free world has been made since the entry into force of the Convention in 1997, and the efforts of the Organisation for the Prohibition of Chemical Weapons (OPCW) and its Member States aimed at ridding the world of the menace of chemical warfare have been acknowledged by the award of the Nobel Peace Prize in 2013.

As the Fourth Review Conference of the Convention is drawing nearer, the OPCW is looking back at some impressive achievements towards global chemical weapons disarmament. Most of the world's offensive chemical weapons programmes have been declared to the OPCW, closed down, and they have been, or are being, eliminated. At the time of writing, 96% of the world's declared stockpile of 72 304 metric tonnes of chemical agent had been destroyed, and almost all former chemical weapons (CW) production facilities have been verified as destroyed or converted for peaceful purposes. 192 States have joined the Convention; only 4 countries (Israel as a signatory State; the DPRK (Democratic People's Republic of Korea), Egypt and South Sudan as non-signatories) have yet to take affirmative action on the treaty.[1] A robust verification system has been put in place that has demonstrated its dependability for providing confidence in chemical weapons disarmament, including under extreme conditions such as the Syrian armed conflict—implemented in the context of the OPCW–UN (Organisation for the Prohibition of Chemical Weapons–United Nations) Joint Mission established under UN Security Council resolution 2118(2013) and subsequent OPCW missions—or the situations in Libya and Iraq. Investigative measures have successfully been executed under challenging conditions, such as the OPCW's contribution to the UN Secretary-General's investigation of the use of chemical weapons in Syria in 2013, the subsequent OPCW Fact-finding Mission in Syria, and the OPCW's contribution to the OPCW–UN Joint Investigative Mechanism. The Convention's system to verify the non-production of chemical weapons in the chemical industry and the consistency of the industry's activities with the obligations undertaken under the Convention, and to inspect other relevant civilian and military facilities, is being applied successfully and creates transparency and confidence in the legitimacy of chemical activities of the States Parties and of chemical companies operating in their territories. The level of confidence that this routine verification system provides depends on whether all States Parties with declarable facilities submit their declarations in a timely and complete manner. To this end, and in order to implement the other requirements of the Convention, States Parties have put in place legislative, regulatory and administrative measures (although this process is by no means complete).

However, the approach of the Fourth Review Conference also signals a need to review the strategic direction taken by the OPCW. In 2011, the Director-General's Advisory Panel on the future priorities of the OPCW already pointed out that "the OPCW needs to prepare for a transition from mandates and efforts primarily characterised by the elimination of chemical weapons stockpiles and production facilities to an agency that will have as its main task to ensure that the menace of chemical warfare and the use of toxic chemicals for hostile purposes will never reappear, and that international cooperation and assistance in the field of peaceful uses of chemistry can flourish."[2] The panel also highlighted that the external environment within which the OPCW was operating was undergoing significant changes – a new and evolving security environment, rapid advances in science and technology (with a new biological revolution getting under way), changes in the industrial landscape and trade patterns are all creating new opportunities, but also new risks.

These advances in science and technology and in their industrial application are today often subsumed under the concept of "convergence". In the context of implementing the CWC, convergence has been described as "integrative and collaborative trends in the life sciences that bring together theoretical concepts, experimental techniques and knowledge of different science and engineering disciplines at the intersection of chemistry and biology. Such interdisciplinary approaches often revolutionise scientific discovery and open up new areas of application of science and technology in society. The benefits of convergence can be huge, but it can also create new risks to safety and security, including the existing arms control regimes."[3] Convergence, then, can be a truly transitional force that can affect the Convection with regards to the strength of its prohibitions and its scope, as well as the way in which it is being implemented.

Member States of the OPCW have been addressing these transitional pressures for several years, including in ambassadorial retreats and in the Executive Council. The Technical Secretariat, too, has been looking at these strategic challenges, including in its Medium-term Plan for the period of 2017–2021.[4] This plan highlights, *inter alia*, the possibility of other States joining the Convention as possessor States, the need for a rapidly deployable surge capacity to deal with non-routine matters, such as challenge inspections or requests for investigations of and responses to the use of chemical weapons, chemical weapons risks associated with non-State actors (especially terrorist groups, but also organised crime and individuals), and advances driven by an increasing convergence of scientific disciplines, scientific collaboration at both national and international levels, and innovative funding strategies. Enabling technologies are expected to permeate into daily life and industrial application. In its Medium-term Plan, the Technical Secretariat identified three strategic enablers for a holistic approach towards preventing the re-emergence of chemical weapons: enhanced resilience and adaptive capacity, improved results-based management, and strengthened knowledge-management structures and processes.

The careful management of the response to these transitional pressures, and to those emanating from the changing external environment within which the Convention is operating, will be critical for the OPCW to adapt to change by design rather than default. The relevance of the OPCW's actions for the security of its Member States as well as for the wider disarmament and cooperation agenda of the world community are at stake, and so are the efficiency, effectiveness and impact of the OPCW's work. This chapter will provide an overview of the key provisions of the Convention and how they relate to today's world, look at success stories related to the convention's implementation, and discuss the challenges that are emerging, with particular emphasis being paid to advances in science and technology.

3.2 Key Concepts and Provisions of the Chemical Weapons Convention

3.2.1 Elimination of Chemical Weapons Programmes

The permanent and complete elimination of the different elements that make up a chemical weapons programme is at the core of the disarmament objectives of the Convention and defines its object and purpose. CW disarmament, of course, is a process that only *starts* with the elimination of past stockpiles, doctrines, production capacities, development establishments and other aspects of a chemical weapons programme – once complete, the state of chemical weapons disarmament must be maintained and any re-acquisition of chemical weapons prevented.

These basic undertakings are set out in Article I. The article creates a comprehensive fabric of provisions, covering not merely the prohibition of the main elements of a State chemical weapons programme (development, production, stockpiling, military preparation for use and actual use, use of riot control agents as a means of warfare) and the elimination of CW stockpiles and previous CW production facilities, but also of activities that could lead to the proliferation of chemical weapons capabilities (transfer of chemical weapons; assistance, encouragement or inducement of anyone to engage in acts that are prohibited by the Convention).

The specific provisions concerning the declaration, inspection, and verified destruction of chemical weapons and CW production facilities are set out in Articles III, IV and V and the corresponding Parts of the Verification Annex (Parts I through V). These provisions, further elaborated by the understanding adopted over time by the OPCW in the form of decisions of the Conference of the States Parties, spell out the principles and targets for the declaration and destruction of chemical weapons and CW production facilities, and of the verification of their declaration and destruction. They establish the Order of Destruction (destruction sequence and intermediate and final destruction targets) of chemical weapons and CW production facilities, and they reaffirm the importance of using destruction technologies that would do no harm to people and the environment, including the banning of certain methods of chemical weapons destruction.[5] They also establish principles and conditions for the conversion for peaceful purposes of former CW production facilities.[6]

The Verification Annex further details the principles and methods of the systematic verification of the declarations of chemical weapons and CW production facilities, which are essential for confirming the baseline for the destruction of chemical weapons and production facilities and their specialized equipment. These data provided the reference point for the subsequent reporting of destruction progress, and for the OPCW to verify and eventually confirm the completeness of the destruction/elimination of all declared CW stockpiles and production equipment and buildings, and to confirm that no diversion has taken place. The concept of the verification of chemical weapons and their destruction was built around the principle of full accountancy for each declared item (weapons, containers, specialised equipment).

3.2.2 Preventing the Re-emergence of Chemical Weapons

The Convention establishes a permanent and comprehensive norm against the use of toxic chemicals as chemical weapons and against any military preparations to this end. Whilst its provisions on the elimination of existing CW programmes have been designed against the characteristics of existing CW arsenals, its provision to prevent the re-emergence of chemical weapons is open-ended and covers all conceivable types of chemical weapons - past, present or future. Critical to this approach is the use of a "General Purpose Criterion", both in the definition of a chemical weapon (Article II) and the obligations of States Parties to take measures to prevent toxic chemicals and their precursors from being used as chemical weapons (Article VI). Whilst the verification system of the Convention is constructed largely around the chemicals listed in the Schedules, its prohibitions and national implementation requirements are based on the intended purpose of a chemical – *all* toxic chemicals and their precursors are considered to be chemical weapons *unless* they are intended for purposes not prohibited by the Convention (the Convention does not use the concept of "permitted purposes" as there are many other laws that permit or restrict the use of chemicals for legitimate purposes which apply alongside the Convention – hence the reference to purposes not prohibited by the Convention). Those purposes not prohibited are specifically defined in the Convention, and there is an additional requirement that the types and quantities of the toxic and precursor chemicals must be consistent with those purposes not prohibited which they are intended for.

The prohibitions of the Convention, as well as its provisions, that aim at preventing the re-emergence of chemical weapons bind the States that are parties of the Convention. However, States are not the only actors to consider in the context of chemical warfare and this is why the Convention obliges States Parties to adopt laws that ensure that the prohibitions which they have accepted as States also apply to anyone who comes under their jurisdiction—natural and legal persons alike—and to extend that jurisdiction to its nationals abroad (Article VII). The Convention also requires States Parties to adopt other regulatory and administrative measures they need to fully implement the Convention, for example, with regards to the submission of industry declarations or the provision of access for OPCW inspectors to private premises as required for the conduct of on-site inspections.

At the international level, the Convention establishes a verification system composed of data monitoring and routine inspections in certain types of chemical facilities (mostly, but not exclusively, these are located in the chemical industry), and of special inspections that States Parties can invoke in the case of a suspected use of chemical weapons (investigations of alleged use under Articles IX or X), or to resolve non-compliance concerns (challenge inspection under Article IX). Whilst the Technical Secretariat is tasked to implement these verification measures and to inform the policy-making organs of the OPCW about these activities and results, the Executive Council supervises the verification activities and promotes the implementation of the Convention and compliance with its provisions (Article VIII).

An important feature of the Convention is its system of adapting administrative or technical treaty provisions when this becomes necessary to ensure the viability of the Convention. Procedures for treaty amendments are a common feature of international treaties, but rarely are they used in practice when agreements deal with security and disarmament. But, the Convention needs tools to adapt certain provisions to changes in the implementation environment, in order to keep up with advances in science, technology, industrial chemicals manufacturing, and more broadly speaking, the utilisation of chemicals in society. This is why, in addition to the normal treaty amendment procedure, the Convention contains a procedure for adapting administrative and technical provisions to new requirements (paragraphs 4 and 5 of Article XV). This simplified amendment procedure ("change") also applies to the Annex on Chemicals, including the Schedules. Furthermore, with regards to industry inspections, the Convention makes provision for a comprehensive review of the entire industry verification system at the first Review Conference (Verification Annex, Part IX, paragraph 26), and requires the first and any subsequent Review Conferences to review the operations of the Convention, including to take account of scientific and technological developments (Article VIII, paragraph 22).

The "change procedure" has been used twice by the OPCW, first in 1999 to adapt the transfer notification provision for Schedule 1 chemicals to facilitate urgent small-scale transfers of saxitoxin for medical/diagnostic purposes (this was necessary to ensure that the toxin would be available as a reference standard for certain food safety tests), and again in 2000 to allow the conversion of former CW production facilities for purposes not prohibited in the case of States that became part of the Convention after the six year period following its entry into force.[7]

This interlocking system of definitions, national implementation, international verification, science and technology review, and adaptation of technical and administrative provisions provides a legal and procedural framework for the States Parties to respond to any future challenges to the Convention and to prevent the re-emergence of chemical weapons in whichever form. What matters then, is the political will of the States Parties to use this framework effectively.

3.2.3 Resilience to the Hostile Uses of Toxic Chemicals

The Convention also makes provision for strengthening the resilience of States Parties against the use of chemical weapons. Article X reaffirms their right to continue activities to protect themselves against chemical weapons, and it encourages cooperation between the parties as well as support by the OPCW for their efforts to enhance their protective capacity. The Convention, furthermore, provides for a mechanism for investigating allegations of the use or threat of use of chemical weapons, or of the use of riot control agents as a means of warfare, and it establishes a system of measures that can be used to provide assistance to the victims should chemical weapons be used.

These provisions recognise that despite the nearly universal adherence to the Convention, there remain risks of chemical warfare against which States Parties need to protect themselves (posed by States that are not party that have chemical weapons programmes, or by actors other than States that may acquire toxic chemicals and use them as weapons, such as terrorists, criminals, or individuals, and also as a deterrent against non-compliance).

Beginning with the sarin attack by Aum Shinrikyo in Tokyo in 1995, reinforced by the attacks of September 11 (2001) and the subsequent anthrax letter attacks against the United States, and again underlined by the uses of chemical weapons during recent insurgencies and armed conflicts in the Middle East (Iraq, Libya and Syria), the threats associated with non-State actors have received growing attention. The response of the international community included such measures as the adoption of UN Security Council Resolution 1540 (2001), the establishment of several Security Council Committees (including its Counter-terrorism Committee and its 1540 Committee), the adoption of a counter-terrorism strategy in 2006, and the development of coordination mechanisms such as the Counter-terrorism Implementation Task Force (CTITF) to coordinate the activities and capacities of the different UN and associated organisations that have mandates in the area of fighting terrorism. All these initiatives and strategies also address the threats related to the use of weapons and materials that may be suitable for mass destruction (or mass intimidation) purposes, and hence there was a need for the OPCW to establish its relationship with these organisations and initiatives and clarify its own contribution to this struggle. Following the attacks on September 11, the OPCW adopted a decision on its contribution to the fight against terrorism and established a working group on terrorism issues. It co-chairs the CTITF's working group on preventing and responding to weapons of mass destruction (WMD) terrorist attacks, which in 2011 issued its first report on interagency coordination in the event of a terrorist attack using chemical or biological weapons or materials.[8] In 2014, the CTITF initiated a new project on ensuring effective inter-agency interoperability and coordinated communication in case of chemical and/or biological attacks. As part of this new initiative, the OPCW hosted a table-top exercise in January 2017, and there are plans to eventually also conduct field exercises.[9]

3.2.4 Chemicals for Peace

As a counterbalance to the prohibitions and regulatory requirements that the Convention establishes, including with regards to the adoption of national implementation measures and the acceptance of international verification, the Convention offers States Parties the prospect of deeper and more intensive international cooperation in the field of peaceful uses of chemistry. Article XI provides a general framework for States Parties to cooperate in the peaceful application of chemistry, to enhance the exchange of chemical information, equipment, goods and technologies, and it establishes principles that they should apply in their economic and trade relations in the chemical field. This includes a review of their export control measures in the chemical field to render them consistent with the Convention – measures that have been adopted since the 1980s in response to the threats of proliferation of chemical weapons programmes and their actual use in war. The Convention does not prescribe the outcome of these reviews, and the balance it establishes is between enhanced cooperation in the field of peaceful uses of chemistry on the one hand, and actual implementation of the national and international measures aimed at ensuring chemical weapons disarmament and non-proliferation on the other.

Expanding beyond this general framework provided by the Convention itself, the OPCW has developed an international cooperation programme that aims at facilitating these exchanges between States Parties, and that implements a range of targeted capacity building programmes. These include today the Associate Programme (industry-related national implementation and promotion of good practices), the Analytical Skills Development Course (disseminating practical experience in analytical chemistry relevant to national implementation), the Conference Support Programme (facilitating scientific exchanges by rendering financial support to scientific conferences), the Equipment Exchange Programme (facilitating the exchange of second-hand but fully functional scientific equipment to publicly funded laboratories and academic institutions of States Parties), Industry Outreach activities (sensitising States Parties to modern approaches in chemical safety and security), a Fellowship Programme (allowing scientists and engineers from developing countries to conduct specialised research in the laboratories of advanced countries), a Laboratory Assistance Programme (improving the technical competence of analytical laboratories), a Research Projects Support Programme (financially assisting small-scale research projects in targeted countries in a range of scientific fields relevant to the Convention and its implementation), the Article XI Workshop (which in 2010 provided a platform for developing new ideas to fully implement the requirements of Article XI and has been followed up by a number of regional workshops), and a dedicated Africa programme that implements targeted projects that respond to the specific needs of African States Parties.[10]

This programme portfolio recognises the link between enhancing the technical and regulatory capacity of States Parties to implement the requirements of the Convention effectively and in full, and the effect that this ought to have on facilitating scientific, technological and trade relationships between States Parties in the chemical field. It needs to be seen in conjunction with the OPCW's support to States Parties with respect to meeting their national implementation requirements, such as with regards to support for the development and enactment of legislation and regulations, the support for customs authorities (Forum for Customs Authorities), and the training of staff from National Authorities.[11]

3.2.5 Managing Advances in Science and Technology

Chemical warfare can be conceptualised as a military application of chemical and related sciences, requiring a science and technology, as well as industrial base, for all elements of a chemical weapons programme, from research and development to production and use, and today, also disarmament. The acquisition of chemical weapons requires the testing of the chemical, physical and toxicological properties of chemical compounds and mixtures being screened as candidate CW agents, the development of a range of countermeasures (medical treatments and prophylaxis, physical barriers for the respiratory tract and the skin, means of decontamination and so on) as protections against chemical agents, the development of processes and technologies to manufacture the agents on an industrial scale, the design and manufacture of weapons systems that effectively disseminate them for a range of tactical uses (such as for air or surface contamination), and the development of additives of tactical mixtures such as stabilisers to enable long-term storage of the weapons. The effective use of chemical weapons (as well as the protection against them) requires an understanding of atmospheric chemistry and physics and of the behaviour of agent clouds in different types of terrain and under different weather conditions. Medical countermeasures require a thorough understanding of the pharmacology and pharmacokinetics of the agents and treatments, with respect to both their acute and long-term effects.

Science and technology therefore underpin many key provisions of the Convention, from the definitions provided in Article II to understandings built into the verification procedures of the Verification Annex and the selection criteria of the Annex on Chemicals and its Schedules of Chemicals. As a consequence, advances in science and technology can significantly alter the implementation environment of the Convention, and they may even affect the scope of its prohibitions, which is why the Convention makes provision for the establishment of a Scientific Advisory Board (SAB) to provide advice to the Director-General, and through him to the States Parties and the OPCW policy-making organs, about advances in science and technology and how they may affect the operation of the Convention (Article VIII, paragraph 21(h) and 45).

The terms of reference of the SAB were adopted by the Conference of the States Parties in late 1997, and at the same time the Conference requested the Director-General to establish the SAB, and it tasked the board to address a first set of technical issues (ricin production and saxitoxin transfers).[12] The Technical Secretariat has supported the work of the SAB since its inception, and since 2013 a dedicated position of science policy advisor was included in the Technical Secretariat's staffing table to render dedicated support to the SAB.

3.3 CWC Implementation – Successes Stories and Challenges

3.3.1 Overview

When the Convention entered into force in 1997, a few uncertainties remained about how exactly its implementation would work in practice. Although the Preparatory Commission had laid the ground for most of the practical implementation tasks ahead, it had also left behind more than a hundred unresolved issues, some of which were important for the smooth implementation of the verification and other provisions of the Convention. Despite these initial gaps in the procedures and guidelines (many of which were subsequently addressed in the work of facilitators of the Committee of the Whole and subsequently the Executive Council), the Technical Secretariat began implementing its verification and other tasks from day one, completed the campaign of initial inspections at CW and Schedule 1 facilities as foreseen by the Convention and, beginning in the autumn of 1997, started phasing in industry inspections at Schedule 2, and later, Schedule 3 facilities. When the First Review Conference met in 2003,[13] it concluded with satisfaction that the OPCW had established a verification system that met the requirements of the Convention. This included a declarations system, that despite some difficulties with regard to completeness and inconsistencies, had been improving over time, and a functioning system of on-site inspections that included well-trained inspectors, approved equipment that met the requirements of OPCW inspection conduct, inspection procedures (compiled in the Secretariat's Inspection Manual) that worked well in the field, a network of designated laboratories that regularly demonstrated their professional capacity and competence to undertake off-site analysis of environmental samples collected by inspection teams should that be required, a still-growing Central OPCW Analytical Database (OCAD) that could be used in on-site analyses by OPCW inspection teams, as well as specialised software that allowed inspectors to operate the analytical instruments in open as well as "blind" mode. It also noted the progress that had been made in the other programme areas (international cooperation, assistance and protection, legislative and administrative support for national authorities, including training) as well as in relation to the policy and managerial systems that supported the work of the States Parties and the policy-making organs. Subsequent review conferences confirmed these conclusions and noted further progress in key areas, such as the optimisation of verification procedures and the use of on-site monitoring systems leading to smaller inspection team sizes and other economy savings in inspection conduct, the establishment and operation of a Verification Information System, the development of a dedicated electronic declarations management system for National Authorities and of tools for electronic declarations submission, and the step-by-step establishment of a capability to analyse biomedical samples (eventually leading to the conduct of proficiency tests in this field and, beginning in 2016, the establishment of a second network of designated laboratories, this time for biomedical samples).

3.3.2 Elimination of Chemical Weapons Stockpiles and Programmes

Much progress has been made by the possessor States Parties on the elimination of their chemical weapons stockpiles and production facilities, even though essentially all possessor States Parties have experienced delays at one point or another, and many of them have needed to request extensions of intermediate, and in some cases final, destruction deadlines. In 2013, the Technical Secretariat reported to the Third Review Conference that "[I]n total, 69 430.336 metric tonnes (MTs) of Category 1 chemical weapons have been declared by the possessor States Parties. Of this amount, as at 31 December 2012, 54 620 MTs (78.67%) of Category 1 chemical weapons have been destroyed under strict verification by the Secretariat. In addition to Albania, which completed the destruction of all its declared chemical weapons in July 2007, two other States Parties (A State Party and India) completed the destruction of all their chemical weapons in July 2008 and March 2009, respectively. Libya has destroyed 51.15%, the Russian Federation has destroyed 70.10%, and the United States of America 89.75% of their respective declared quantities."[14] At the time of writing, Russia and Libya had completed the elimination of their declared stockpiles, and so had Iraq who encapsulated two bunkers containing the remnants of its former CW arsenal. The percentage of chemical weapons destroyed stands at 96%.[15]

Three issues deserve particular attention in the context of the elimination of chemical weapons stockpiles: the delays in the destruction programmes of Russia and the United States which took both States Parties past the ultimate extended destruction deadline foreseen under the Convention; the investigations of the use of chemical weapons in Syria and the elimination of the Syrian chemical weapons programme at the time of a raging armed conflict in that country; and the removal of the remainder of Libya's chemical weapons from the territory of Libya for destruction abroad. The delays in the Russian and US destruction programmes beyond the Convention's 2012 deadline posed significant legal and political challenges – States Parties had to decide about the approach to take (for example, amendment of the Convention or invocation of the Convention's procedures to deal with non-compliance and to redress situations of non-compliance), in a manner that would serve the object and purpose of the Convention, ensure the earliest possible completion of these destruction programmes, and avoid political strain that could tear the OPCW apart. The issue was resolved by the Conference with a decision in 2011[16] that avoided terminology akin to a non-compliance procedure and left the setting of the final destruction targets to the two possessor States Parties concerned, but otherwise closely followed the principles of the Convention for redressing a non-compliance situation.

The operations of the OPCW in Syria (in close collaboration with the United Nations) have taken the organisation into unchartered territory in several ways. The 2013 investigation of the alleged use of chemical weapons under the mandate of the UN Secretary-General's Mechanism to Investigate Allegations of the Use of Chemical and Bacteriological Weapons is a procedure envisaged under the Convention (Verification Annex Part XI). However, previous investigations of this kind were never conducted under the conditions of a raging internal armed conflict. The SGM mission took almost half a year before it was able to enter the country, and within days it was confronted with investigating an incident of a large-scale use of sarin nerve agent against Ghouta near Damascus, killing hundreds of civilians, including women and children.[17] This led to the agreement of the Russian–US framework to eliminate the Syrian chemical weapons stockpile, the accession of Syria to the Convention, and subsequent inter-locking decisions by the OPCW Executive Council and the UN Security Council that resulted in the removal of Syria's chemical weapons stockpile from Syria for destruction abroad (including at high sea in the Mediterranean Sea aboard US vessel Cape Ray that had been fitted out as a floating CW destruction facility), and the functional disablement and subsequent destruction of Syria's stationary and mobile CW production facilities (including mixing and filling facilities).[18,19] These disarmament measures were implemented by a Joint Mission of the UN and the OPCW from October 2013 to September 2014, by which time the declared chemical weapons materials had all been removed and destroyed. The remaining CW production facilities were subsequently destroyed by the OPCW under its own mandate, with support by the United Nations.[20] But whilst this disarmament programme was making progress despite the complicated security conditions in Syria, reports began to appear about the alleged use of toxic chemicals (chlorine as well as mustard agent; government and opposition were accusing one another of being the instigator); furthermore, there were suggestions that the Syrian declaration contained inconsistencies and gaps. The OPCW addressed these issues with two separate mechanisms that the Director-General implemented under his general authority, with the agreement of the Syrian government, which were subsequently endorsed by the Executive Council: the Fact-finding Mission (FFM) to investigate the allegations of the use of improvised chemical weapons in Syria (the OPCW also investigated an alleged use of mustard agent in Iraq under a separate arrangement with the Iraqi government), and the Declaration Assessment Team (DAT) to resolve the uncertainties with regard to the Syrian CW declaration (which led to a number of additional facilities and activities being declared by Syria). Finally, the question of attribution of responsibility for the uses of chemical weapons in Syria was taken up by the UN Security Council, which established a special mechanism (the OPCW–UN Joint Investigative Mechanism (JIM)) under resolution 2235(2015). The activities of the JIM were extended by the Security Council under resolution 2319(2016) for one year, but a further extension could not be agreed and its mandate ended in November 2017.

These operations in Syria have demonstrated that the principles of the Convention can be applied even under circumstances that are highly challenging, and that were not explicitly envisaged when the treaty was drafted. Some of the rules needed adaptation, which required leadership by the Director-General and the Executive Council, and political will as well as a sense of common purpose on the part of the policy-making organs (the failure to extend the JIM mandate, on the other hand, showed the consequences of a lack thereof). The Syria operations also resulted in a much closer working relationship between the United Nations and the OPCW, at the policy and legal levels (between the political organs as well as the two Secretariats) but equally important, at the operational and planning level. The OPCW is not a UN entity and had not been designed to implement the kinds of undertakings that it was requested to put into effect in Syria. These activities exceeded its resources and required certain competencies that the OPCW did not have, and where it needed to rely on the capacity and expertise of the United Nations, including with respect to security management and in-country knowledge and logistics. The Joint Mission did require close cooperation on a range of issues and joint planning of field activity between the UN and the OPCW. Many of the details for this unprecedented operation had to be worked out "on the fly" as the operation unfolded. However, the Syria operation required not merely the close cooperation between the UN and the OPCW; it was only possible because of the strong financial and in-kind assistance by many States Parties that took on responsibilities and tasks that would normally fall onto the possessor State Party.

The removal in 2016 of the remainder of Libya's CW stockpile for destruction abroad was another example of the willingness of the OPCW to find innovative solutions to ensure that even under complicated circumstances, the objectives of the Convention can be met and its requirements implemented. This stockpile had been discovered and declared after the fall of the Gadhafi regime in 2011 and was located in Ruwagha, a remote location in the south of the country. In the summer of 2015, Libya informed the OPCW of security challenges as well as environmental risks associated with its remaining CW stocks. In July 2016, after some private diplomacy and quiet practical preparations, Libya informed the OPCW that it had moved all of its remaining Category 2 chemical weapons to a temporary storage site in the north of the country, and requested assistance and support to destroy these remaining chemical weapons at a specialised waste treatment facility outside Libya. The Council granted this request[21] and the Director-General subsequently prepared a modified plan for the destruction of these chemical weapons at an appropriate destruction facility abroad, together with detailed verification arrangements.[22–24] In his report to the UN Security Council on the implementation of these measures, the Director-General of the OPCW provided some details on how this operation was implemented with the help of a number of States Parties (Germany providing a destruction facility (the Gesellschaft zur Entsorgung von chemischen Kampfstoffen und Rüstungsaltlasten (GEKA mbH)) which otherwise is involved in the destruction of old chemical weapons in Germany; Italy assisting with the transport of samples required to be analysed before the materials could be imported into Germany; the UK undertaking said analysis at its designated laboratory; and Denmark providing a maritime transport vessel (the Ark Futura) to move the materials from the Libyan port of Misrata to the port of Bremen in Germany),[25] subject to systematic verification by the Secretariat. Several countries made financial support available for a special trust fund established by the OPCW for this operation (Canada, Finland, New Zealand and the USA), or supported other aspects of the removal operations (UK), and the EU was considering an offer to assist with the clean-up at the Ruwagha site.[26]

In addition to the elimination of chemical weapons stockpiles, an important aspect of the Convention's undertaking is the destruction of the remnants of past CW programmes in the form of old and abandoned chemical weapons (OCW, ACW). In its most recent annual report, the OPCW stated that since the entry into force of the Convention, four States Parties have reported ACW on their territory and one (Japan) had declared that it had abandoned chemical weapons in the territory of China. 16 States Parties have declared OCW since the entry into force.[27] Issues related to the destruction of old and abandoned chemical weapons (as well as with regard to the management of the risks associated with sea-dumped chemical weapons) are going to remain relevant for States Parties for many more years, and the OPCW as the global depository of knowledge with regard to the destruction of chemical weapons materials will be expected to contribute to dealing with these legacy issues for many years to come.

3.3.3 Preventing the Re-emergence of Chemical Weapons

"Preventing the re-emergence of chemical weapons" is a concept that has become increasingly prominent in recent years – a reflection of the beginning of a transition of the OPCW from its original focus on the verified elimination of CW programmes to its long-term mandate of ensuring the continuing state of chemical weapons disarmament. It is an awkward set of words as it suggests a degree of automatism with regard to how chemical weapons programmes are initiated, when in fact these are distinct political and strategic decisions by governments (or non-State actors). Such decisions, amongst other factors, could emanate from weak or unenforced norms or a disregard for such norms, perceived advantages from acquiring chemical weapons in an evolving security environment, temptations caused by new scientific discoveries that might create opportunities, or developments in the margins of what the treaty allows that may inadvertently undermine its norms (such as the acquisition of new types of riot control agents that become integrated into operational concepts at the intersection of law enforcement and military combat operations). Measures to prevent the future acquisition of chemical weapons must therefore take into account a range of potential actors, scenarios and drivers, and devise robust long-term strategies to counter such attempts in a holistic manner.

Consequently, the Convention approaches the prevention of the re-emergence of chemical weapons through a combination of several provisions that back each other up: strong prohibitions and their enforcement at the national level by all States Parties; an effective international verification system that can adapt to advances in science, technology, and industrial manufacturing; effective protections against chemical weapons to make re-acquisition of chemical warfare capabilities less attractive; and a governance system that effectively addresses compliance concerns and that manages compliance in the interest of achieving the objectives of the Convention.

National Implementation. The domestication of the legal norms of the Convention by the States Parties has made some progress since the entry into force of the Convention, but weaknesses in national implementation by a large number of States Parties have remained an issue of concern ever since the First Review Conference recognised them; it decided to call for the adoption of an OPCW Action Plan on Article VII Implementation.[28] This action plan was designed for a 2 year period and led to subsequent follow-up measures.[29] Early indicators for weaknesses in national implementation systems were the delays in initial declarations due under Articles III and VI and the apparent incompleteness of Article VI declarations of many States Parties. A failure to declare, of course, undercuts the routine verification regime of the Convention and removes facilities or even entire States Parties from the application of verification measures. There are also persistent inconsistencies in transfer declarations (with regards to the notifications and annual declaration of Schedule 1 transfers, and with regards to the export–import data contained in aggregate national data for Schedule 2 and 3 chemicals) which indicate differences and gaps in the national systems for the collection of exports and imports data. At the time of writing, the latest public data on the status of implementation of these requirements, dated September 2015,[30] showed that whilst 99% of the States Parties had established or designated a National Authority, only 61% had covered all legislative requirements under the Convention in their domestic laws and regulations; another 30% had covered some of them. 71% of the States Parties had actually submitted the text of their laws to the OPCW, as required. The gaps in legislation are most prominent in the areas of declarations and reporting of scheduled chemicals, including with regards to transfers of scheduled chemicals. With regards to the submission of declarations, at the beginning of 2017 one State

Party remained that had yet to submit its initial declaration under Article VI. The overall quality and timeliness of the submission of Article VI annual declarations had significantly improved over time by 2017,[31] yet it is not clear from these data whether the assessment of the Technical Secretariat is that the declarations are also complete.

An analysis of the underlying reasons for these delays that the Technical Secretariat included in its reporting highlighted 5 factors: (a) difficulties in enacting legislation; (b) difficulties in identifying declarable activities; (c) difficulties in collecting data; (d) difficulties in preparing declarations; and (e) logistical difficulties in regard to transmitting the declarations.[31] These factors would suggest that there also is an assumption that the declarations are incomplete, and/or that some States Parties with declarable activities and facilities have not submitted any declarations.

Whilst the efforts of the OPCW to bring all the remaining States not party into the realm of the Convention have resulted in significant progress in universal adherence to the Convention, with only 4 countries today remaining outside, the efforts to increase the share of States Parties that have a full set of legislative and administrative measures in place to fully implement the Convention have yet to yield similar results. Universalization efforts have been out-pacing efforts to ensure full and effective implementation at the national level, despite the calls from the OPCW policy making organs to do more, and despite the numerous efforts by States Parties and the Secretariat to provide technical assistance to States Parties that lack experience or capacity to meet their national implementation requirements. It is important that the Technical Secretariat continues with its support programmes, such as (*inter alia*):

- Training courses for National Authorities;
- Bilateral technical assistance visits;
- Support in legislative drafting, including through its web-based Legislative Assistance Support Tool (LAST);[32]
- Thematic and technical workshops;
- Its forum for customs authorities which includes e-Learning modules and a series of presentations on practical implementation issues;[33]
- Its efforts to cooperate with the World Customs Organisation to identify the globally most traded scheduled chemicals through the Harmonised System (HS);
- Support and advice with regards to identifying declarable facilities and activities and with the submission of declarations;
- Support for outreach activities to other stakeholders;
- The development of electronic tools for declaration preparation and submission (the Electronic Declaration Tool for National Authorities (EDNA)[34] and the Secure Information Exchange System (SIX));[35] and
- Regional and subregional as well as annual meetings of National Authorities to share experiences and promote good practices in national implementation of the Convention.

But, of equal importance will be to use the forthcoming Fourth Review Conference as an opportunity to increase the political momentum in support of further improvements to national implementation of the Convention.

One of the pitfalls in this respect remains the noticeable fragmentation in the international system that deals with chemicals management, with regards to both safety and security. Requirements for national legislative, enforcement, administrative management and other controls emanate from a range of instruments and mechanisms in the fields of environmental protection, occupational safety and protection of people, security/counter-terrorism/fighting organised crime, arms control and non-proliferation, including with regard to non-State actors, transportation safety and security, the protection of critical infrastructure, occupational as well as populations safety, and so forth. There has been a hope that with the move towards a more integrated approach to chemicals management (under the Strategic Approach to International Chemicals Management—SAICM—a policy framework to foster the sound management of chemicals), national capacities in this complex field could be integrated and enhanced more quickly and effectively, and that the political visibility and support for these efforts would be strengthened.[36] Better integration of chemicals safety and security, at least with regards to arms control and disarmament, has yet to be accomplished on a broader scale.

Verification. A second set of the Convention's provisions that aim at the prevention of the re-emergence of chemical weapons relates to its verification system, both with regards to verification in the chemical industry and challenging inspection to clarify compliance concerns. The Third Review Conference in 2013 reaffirmed that the full, effective and non-discriminatory implementation of Article VI was essential for the realisation of the object and purpose of the Convention, and in this context, characterised the Convention's verification system as an important element of the Convention. The verification system "should continue to be improved in a manner consistent with the Convention in response to advances in science and technology…".[37]

The very fact that the OPCW has been routinely implementing inspections in private chemical industry on a worldwide basis is a significant achievement. Whilst the chemical industry has been well acquainted over the years with national and international inspections in the context of product and manufacturing safety and compliance with quality standards, inspections that are directed at providing transparency and confidence that chemical companies are not involved in the manufacturing of chemical weapons or sell their products to others to those ends, had no precedent before the Convention entered into force. Inspections of Schedule 2 facilities began as early as the autumn of 1997, Schedule 3 inspections shortly thereafter, and OCPF inspections in April 2000, as anticipated by the Convention. The number of inspections conducted by the OPCW over the years in chemical plant sites is impressive: in August 2015, the OPCW reached a milestone by conducting its 3000th inspection of a facility declared under Article VI. In 2003, the OPCW conducted 132 Article VI inspections per year; that number had risen to 241 by 2015.[38] The distribution of inspections reflected the number of declared facilities liable to be inspected, the relative distribution of these chemical facilities among States Parties, and the risk perceptions associated with the different types of chemicals involved (Schedule 1, 2 or 3 chemicals, respectively, or discrete organic chemicals).

Progress has also been made in a range of technical and procedural areas that affect the economy as well as quality of industry verification. Examples include the increasing number of sequential inspections (a methodology originally implemented in CW facilities to ensure that the front-loading of initial inspections at such facilities could be managed; but today this methodology has become more broadly accepted in chemical industry inspections despite some States Parties still stating confidentiality concerns; when in 2003 only 8 sequential inspections had been implemented, this number stood at 59 in 2015, six of which were carried out sequentially in two countries),[39] the development and implementation of electronic tools for the selection of plant sites for inspections under the different Article VI sub-regimes, as well as the Verification Information System, and the use of sampling and analysis in industry inspections.

There are nevertheless limits to the degree of confidence in compliance that industry inspections can deliver. Given the restrictions set out in the Convention on annual inspection numbers per plant site and per State Party with regards to the combined number of inspections of Schedule 3 and OCSP facilities, and taking into account the unequal distribution of chemical plant sites among States Parties and the limited overall number of chemical industry inspections per year that States Parties are willing to authorise through the OPCW programme and budget each year (241 in 2015), the primary effect of conducting industry inspections will always be one of gradually enhancing confidence in treaty compliance, rather than providing full accountability as is accomplished in systematic verification concepts that are used in the verification of chemical weapons destruction. Given the numbers involved (the number of States Parties and of industry facilities liable to inspection, and of inspections that can be conducted per year), the degree of assurance that the system of industry inspections can deliver is nevertheless comparable to other risk-based routine control systems. In 2015, the inspection frequencies in relation to the total number of facilities liable to inspection were reported as follows: 41% for Schedule 1 facilities, 22% for Schedule 2 facilities, 5% for Schedule 3 facilities, and 4% for OCPFs.[38] Whilst for Schedule 1 facilities, that frequency comes close to what would be aimed at in systematic verification, the inspection frequencies for Schedule 3 and OCSP facilities are in the general range of what would be expected a reasonable standard of assurances provided by, for example, product safety, product quality, or customs compliance inspection schemes.

However, industry verification is more than just inspections of industry facilities. It also includes data monitoring of declared information not verified on-site, and the analysis of data available to the Technical Secretariat. The discussion of national implementation issues above has already highlighted that despite improvements, significant discrepancies remain in data sets that should in fact corroborate each other over time – differences in export and import data between States Parties involved in such transfers, or the absence of declarations of industry facilities when transfer data and other information suggest that there should be some declarable facilities. Significant efforts have been made by the OPCW to improve the timeliness of annual declaration submission, and by the Third Review Conference this had indeed improved the mechanics of the declaration system considerably;[40] but what about completeness and accuracy of declarations? There have been consistent efforts by States Parties as well as the Technical Secretariat to identify the reasons for discrepancies between declaration data sets and to find mechanisms to resolve them (including through consultative mechanisms and attempts to further standardise the rules used by States Parties to prepare their declarations and notifications), or to improve the completeness of industry declarations. At the same time, there has also been strong resistance by some States Parties to accept that the Secretariat use, on a more regular basis, authoritative and validated open-source data to corroborate information submitted in declarations or identify gaps and uncertainties in declaration data. The Secretariat has been assisting States Parties, upon request, with the identification of declarable industry facilities, but a more systematic use of open-source data to generate a more comprehensive set of data that could be used to demonstrate compliance has been resisted. At the Third Review Conference, the Secretariat reported that despite these efforts, "[U]ndeclared activities covered by the Convention are an issue of major concern and undermine the effectiveness of the verification regime. The reporting of such information should be seen as key to improving effectiveness."[41] In 2015, the Secretariat noted that the overall quality of annual declarations under Article VI and the rate of their timely submission were significantly improved,[42] but it also reported that considerable discrepancies in the transfer declarations remained, which is one of the indicators for gaps in implementation by some States Parties, or the use of different criteria for identifying declarable activities and facilities.

Another verification aspect worth highlighting is the conclusiveness of inspection findings, and in this context, the use of sampling and analysis. During the negotiations, sampling and analysis had been seen as a key aspect of unambiguous verification of compliance. Based on the work of the Preparatory Commission, the States Parties and the Technical Secretariat have built up a comprehensive system of procedures and technical capacities for sampling and on-site, as well as off-site, analysis. This includes procedures and equipment for collecting, processing and transporting samples under a rigid chain-of-custody regime, mobile laboratory equipment and databases for on-site analysis in both open and "blinded" mode (the latter can be invoked by the inspected State Party as a means of protecting confidential information; the inspector will be able to retrieve analytical information if a comparison of the test run with the data contained in the OPCW's OCAD database yields a match, but no analytical details will be revealed if there is no match), and a network of Designated Laboratories that regularly participate in official proficiency tests (today, there are two such networks in place: one for environmental samples, and a second network for biomedical samples – the latter is important in the context of investigations of alleged use). The use of sampling and analysis in industry inspections did not start until 2006, and initially it was limited to the inspection of Schedule 2 facilities.[43] There were a number of reasons for these delays: the initial focus in Schedule 2 inspections on initial inspections and the preparations of facility agreements, the time constraints in Schedule 3 and OCPF inspections that pose significant challenges to on-site analysis given set-up and analysis times, the need to build up a sufficiently large analytical database with fully validated spectra to minimise the chance of false-positive detections, the need to gain experience with the procedures for on-site sampling and analysis under field conditions (and concerns about using off-site analysis in industry inspections as a matter of routine), and cost factors, also played a role. However, there was also some reluctance on the part of a number of States Parties to see sampling and analysis be given too much prominence in industry inspections. It took some practical experience to build up confidence that sampling and analysis in industry was necessary and feasible and could be done without compromising commercial confidentiality or invoking unreasonable costs. Between 2006 and 2015, a total of 83 inspections in industry involved sampling and analysis and beginning in 2016, sampling and analysis was used in all types of industry inspections as a matter of routine.[42]

Protection against Chemical Weapons. Although the provisions of Article X on assistance and protection emphasise the response mechanism to the use or threat of use of chemical weapons, the practical measures implemented by the OPCW with regards to providing assistance and advice to States Parties in the field of strengthening their protective capacity have taken on significantly more prominence. To some extent, this was the result of some States Parties interpreting the requirement to make offers for assistance under Article X in a fairly broad manner, including in their offers, not merely a promise of what they would deliver should the assistance mechanism under Article X be triggered in the future, but making training and expertise available as a form of building capacity in OPCW Member States as a prevention and preparedness measure. Similarly, the Technical Secretariat used the provision under paragraph 4 of Article X to render advice and assistance to States Parties that wished to strengthen their protective capacity against chemical weapons as a basis for developing a programme of regular technical support in this area. After September 11 (2001), and supported by subsequent Review Conferences, the Technical Secretariat with the support of certain States Parties expanded this programme further. A recent report on the status of implementation of Article X[44] gives an overview on the measures taken in preparation of emergencies (contributions to the OPCW Voluntary Fund for Assistance, today standing at approximately 1.5 million Euros; assistance procurement agreements between the OPCW and three individual States

Parties; declarations of the kind of assistance that States Parties would make available if called upon in response to a chemical weapons use; the establishment of an Assistance Coordination and Assessment Team to support the needs assessment and the coordination of international support in response to a chemical weapons use) as well as in the area of capacity building for strengthening capacity in States Parties for the protection against chemical weapons. These activities include basic and advanced training courses in the area of protection, regional and subregional training courses and dedicated courses for instructors as well as first responders, as well as emergency response exercises, training for field laboratories, field sampling and analysis, and other types of laboratory training. The training programme is aimed at developing standing capabilities at regional levels in the form of reference teams of first responders, and there is a dedicated programme to strengthen cooperation with Africa which amongst others includes subregional action plans leading to detailed chemical threat assessments and national protection programmes of the participating countries. The OPCW also supports activities under an international Support Network for Victims of Chemical Weapons and a related Trust Fund, which resulted amongst others in a Guidebook for Practitioners on Medical Management of Chemical Warfare Casualties. The OPCW has also enhanced its coordination with other international agencies that have mandates in the prevention of and response to incidents involving toxic chemicals (the European Union (EU), the Organisation for Economic Co-operation and Development (OECD), the World Health Organisation (WHO), the United Nations Interregional Crime and Justice Research Institute (UNICRI), the United Nations Office for the Coordination of Humanitarian Affairs (UNOCHA), the International Committee of the Red Cross (ICRC), the International Criminal Police Organization (INTERPOL) and others), has participated in international exercises, and it co-chairs the CTITF's working group on response to terrorist incidents involving weapons of mass destruction and related materials.[8,45] The OPCW is also making additional efforts to improve its monitoring and evaluating systems for these programmes.

All in all, the protection and assistance programmes implemented by the OPCW have come a long way from the experimental approach that the OPCW adopted after its entry into force, to a professional capacity building approach based on clear needs assessments, the implementation of high-quality training which uses principles and approaches designed to ensure sustainable outcomes, increasingly more realistic exercises to validate results, and systems to monitor and evaluate results. The programme is today in a stage where it demonstrably helps States Parties to enhance their protective capacity, thus contributing to the deterrence effect of good preparedness and protection against chemical weapons threats.

Compliance Management. In its 20 years of existence, the OPCW has had several occasions when compliance issues needed to be addressed and ways and means of ensuring or re-establishing full compliance had to be found. Whilst the Technical Secretariat has played a key role in providing and verifying information needed in such compliance actions, and successive

Director-Generals have made efforts to assist States Parties and the PMOs to find solutions that could command broad support in the OPCW whenever compliance issues emerged, the primary responsibility for addressing and managing compliance rests with the Executive Council.

The Executive Council has been reluctant to label situations that called for measures to redress deficiencies as "non-compliance". Its preferred choice of words was to "ensure or improve implementation" or something similar – perhaps in an attempt to avoid polarization of the Council and to maintain unity among its members. This is also reflected in the decision-making culture of the Executive Council: different from the Conference of the States Parties who should whenever possible take its decisions by consensus, the Convention expects the Council to take swift action on compliance issues and when necessary take its decisions by vote; but in practice the Council has shown a strong preference for consensus decisions even when that meant delaying decisions or avoiding them altogether. It has been argued that this departure from the decision-making rules set forth by the Convention undermines the ability of the Council to effectively manage compliance, as foreseen by the Convention.[46]

Nevertheless, there are a number of examples that show that the OPCW has effective ways of addressing compliance concerns and re-establishing compliance, sometimes using informal procedures and processes to build up consensus before issues were addressed formally by the Council or the Conference, sometimes relying on initiatives carried forward by the Technical Secretariat under the authority given by the Convention to the Director-General. Examples include the initiatives to encourage full implementation of the requirement by States Parties to submit initial declarations, enact penal and implementing legislation and adopt other measures to implement the Convention including establishing or designating a National Authority (support measures implemented by the Technical Secretariat as well as the Article VII Action Plan and subsequent follow-up measures), the measures taken by the Technical Secretariat to help States Parties identify declarable facilities, the decision on the extended destruction deadline for chemical weapons beyond the Convention's final deadline of 2012, the Council's decision on the elimination of the Syrian chemical weapons programme, the Director-General's initiative to establish a Fact-Finding Mission and a Declaration Assistance Team to clarify allegations of the use of improvised chemical weapons in Syria as well as of concerns about the accuracy and completeness of Syria's CW declarations, or the Council decisions on the removal of the remaining CW stockpile from Libya for destruction abroad. Another example for how diverging views about compliance have been accommodated by relying on informal processes as well as programme activities implemented by the Technical Secretariat was the issue of the full implementation of Article XI.

What characterised all these situations was that the Council and the Director-General, respectively, exercised leadership, took the object and purpose of the Convention as the baseline for arriving at difficult decisions, and took the mandate and responsibility which the Convention assigned to them in

Article VIII seriously. In all of these cases, they clearly recognised the ultimate purpose of these informal and formal processes: to find effective ways of re-establishing full compliance as early as possible.

Fact-finding and challenge inspections. An important tool for compliance management and the resolution of compliance concerns through cooperation and fact-finding—and one that has never been used in practice—is challenge inspection. Successive Review Conferences have on the one hand encouraged the Technical Secretariat to continue its efforts to maintain a high state of readiness to implement challenge inspections should the mechanism be invoked, but at the same time have either expressed satisfaction that States Parties have hitherto not resorted to using this tool[47] or stressed the value of bilateral consultations amongst the parties concerned.[48] A recent commentary on the Convention has noted that this preference of States Parties for relying on less intrusive means of resolving doubts about compliance might suggest that challenge inspection "is or should only be subsidiary in nature", but that at the same time the OPCW practice of maintaining a high degree of readiness to implement challenge inspection points to these statements being a matter of policy rather than affecting the importance (in law) of challenge inspection.[49] This commentary argues that the practicability and operability of Article IX must be ensured in order to provide the necessary confidence in treaty compliance, and that this is even more important against the background that States are increasingly reluctant to refer to established mechanisms of fact-finding in situations of conflict (not necessarily armed conflict),[50] and instead tend to rely on *ad hoc* mechanisms.

The FFM and the DAT established during the OPCW's Syria missions—already alluded to above—are examples in case. These are two *ad hoc* fact-finding mechanisms to resolve non-compliance concerns in circumstances of conflict and political tension that avoid the use of regular mechanisms – either those of the CWC itself, or the special inspection mechanism set up by the Executive Council, and sanctioned by the UN Security Council, to ensure the completeness of Syria's declaration, in paragraph 2(c)(d) of decision EC-M-33/Dec.1. On the one hand, these *ad hoc* mechanisms can be seen as a sign of flexibility and adaptability in the manner in which the OPCW manages compliance issues. They have subsequently been endorsed by the OPCW policy-making organs and (implicitly) by the Security Council. However, there remains a concern that a proliferation of *ad hoc* procedures in preference over the regular mechanisms of the Convention may eventually undermine the legal norm. By their very nature, *ad hoc* mechanisms are devised against a particular context rather than as a set of rules that would apply to each and every State Party. There may be reasons for relying on *ad hoc* mechanisms in particular circumstances when the object and purpose of the Convention cannot be furthered otherwise and the regular mechanisms cannot be implemented, but decisions to this end should not be driven by political convenience, and it must be clear that when extraordinary circumstances revert to normal, the mechanisms and rules of the Convention must govern again.

This, of course, raises the question of whether the OPCW is ready to implement a "normal" challenge inspection. At the level of policy, the answer would have to be found in the broader context of how the policy-making organs deal with compliance concerns – and the experience in this regard is mixed. However, at the practical level, there is the question of how a Technical Secretariat and its verification and inspection officers can maintain and demonstrate their ability to effectively implement an inspection mechanism that in so many ways differs from their daily routine of inspecting declared facilities under pre-arranged conditions that are stable and predictable, non-confrontational in nature, and part of an ongoing routine, rather than a one-off. This is a question about maintaining technical skills that are unique to certain challenge inspection scenarios, but also about the depth and strength of the competencies within the Technical Secretariat that are needed in these special inspections, the mind-set needed in such types of inspections, the mobilisation capacity within the Technical Secretariat to mount a politically and technically complex operation on short notice, the availability of equipment and logistical and administrative support, and the degree to which these non-routine procedures and tasks are regularly trained at all operational levels concerned.

In a certain sense, the operations in Syria were test cases for whether and how the Technical Secretariat was able to switch from a routine to a non-routine operations mode. A lessons-learned exercise of the OPCW[20] highlighted the challenges that these missions needed to address: (a) to implement a complex operation in a situation of high risk; (b) to work against very tight deadlines; (c) to master the novelty and complexity of these missions whilst having to switch from sixteen years of routine business that had created an institutional culture not suited for contingency operations—a host of tasks that had no precedence in previous OPCW verification work—and taking on tasks that normally fall to the inspected State Party, such as ensuring security and safety, (d) to work in synchronisation with the United Nations and other partners, and (e) to collaborate with a large number of stakeholders including Member States. The report also recognised that the OPCW had never before operated in such a politically charged, ambitious and complex context (and of course this sort of undertaking is implemented with constant exposure to the media and public as well as political scrutiny).

One of the key observation of this review was the recognition of the importance of the human factor (strong leadership, effective teamwork, political sensitivity, situational awareness, technical competence) and the need to further improve the OPCW's mechanisms with regards to human resource management, knowledge transfer, and maintaining critical competencies either in house or through effective ramping-up mechanisms. The Syria operation exposed some of the weaknesses that had built up over time, including the lack of depth in competence and operational capacity within the Secretariat, and the risks associated with situations where mission success critically depended on single individuals and their performance/availability, or where the Secretariat had to operate out of *ad hoc* structures (in this case a

Core Group composed of officers with the particular experience led by the Director-General) rather than its normal institutional command and control structures.

In addition, many practical lessons were learned in the areas of legal and administrative support, operational concepts, information management, duty of care, safety and security management, verification conduct in complex and fluid situations, and with regards to specific skill sets, investigation methods and equipment. Before the Syria missions, the Technical Secretariat had already concluded that "[C]hemical weapons-related knowledge and expertise needs to be retained by the Organisation, combined with a rapidly deployable surge capacity in case of requests for [challenge inspections] and [investigations of alleged use], or to provide support for United Nations-led operations."[51]

Many of the lessons from the Syria Missions have now been built back into the Secretariat's systems, plans and procedures. The Technical Secretariat reports regularly about the state of preparedness for the conduct of challenge inspections and investigations of alleged use. In the most recent report available at the time of writing, the Technical Secretariat reported that it had established, in 2016, a Rapid Response and Assistance Mission (RRAM) to enhance its capacity to deploy teams in response to alleged uses of chemical weapons, including to investigate such cases and deliver assistance to victims.[52] The activities conducted and coordinated by this new unit included mandatory training required under the OPCW's Quality Management System but also specialised training for experienced inspectors and inspection team leaders for conducting non-routine inspections, including a field exercise and a newly developed module on contingency operations training. More attention than in the past is being paid to skill sets that are important for the conduct of investigative missions, including (a) forensics awareness, (b) interaction with other international actors in response and assistance situations, (c) the transfer of experience from the missions in Syria with regard to command, control, coordination and information management, (d) the management of large deployed teams, and (e) executive risk management. The Secretariat is also adapting standard operating procedures, phasing in new items of equipment and improving its coordination with relevant UN departments and offices (Office for Disarmament Affairs (ODA), Office for the Coordination of Humanitarian Affairs (OCHA), Department of Safety and Security (DSS), and others) that are important partners at the operational level.

The ability of the OPCW to implement the challenge inspection tool also depends on whether and how the States Parties themselves implement the administrative arrangements, which the Convention requires them to take. In this regard, the First Review Conference had already noted significant deficiencies,[53] and that situation has only gradually improved. Significant gaps remain in the preparations by States Parties for hosting challenge inspections, in a number of areas: visa issuance and/or the acceptance of UN Laissez-Passer (only 68% of the States Parties are implementing the requirements of the CWC); the designation of points of entry (only 136 States Parties have done so); the issues of standing diplomatic clearance numbers for inspection aircraft (only 32 States Parties have permanently assigned such numbers, another 40 have set up expedited procedures for their issuance, and 37 assign them on an annual basis); and the assignment of radiofrequencies for inspection team communications (98 States Parties have provided them and another 6 have put in place expedited procedures for obtaining these radiofrequencies when needed).[54] These gaps would not make the challenge inspection tool practically unusable, but they would add unnecessary strain and create possible tension and delays in situations when at the technical level, the deterrence as well as compliance-assurance effect of challenge inspections rest critically on the assumption that the procedures of the scheme can actually be applied as set out in the Convention.

Decision Making and Political Issues. However, shortcomings also remain at the political level, which may perhaps be even more significant than the gaps in providing the necessary standing arrangements. Despite the positive examples discussed above, the OPCW's record of compliance management also shows examples of avoidance of critical decision-making and resistance on debating important matters relevant to compliance. Examples include the reluctance to address the issue of the use of incapacitating chemical agents (ICA)—more recently described as Central Nervous System (CNS) acting chemicals—for law enforcement purposes; the cases of commercial availability of certain types of wide-area dispersion systems for riot control agents (RCA) that appear to be inconsistent with the prohibition of the Convention on the use of RCA as a method of warfare; or the inability of the Council to agree on guidelines for how to deal with situations such as the discovery and *in situ* destruction of remnants of chemical weapons in conflict zones on the territory of States Parties.

The ICA issue was initially brought to the attention of the OPCW by the International Committee of the Red Cross (ICRC) in the margins of the First Review Conference. Although a number of States Parties raised concerns about the matter, the OPCW policy-making organs avoided addressing the legal and practical implications of the possibility that ICAs might be developed and stockpiled for law enforcement purposes, and the debate about this issue took place mostly outside the OPCW.[55–57] Only much later did the issue re-emerge in the OPCW, but rather than developing agreed guidance on the matter, the policy making organs left the matter to individual statements of interpretation and policy tabled by States Parties. Whilst such an approach is certainly useful to confirm State practice, it does not substitute for clear guidance by the Executive Council or a decision by the Conference. Similarly, when the Executive Council's attention was drawn to the fact that companies from a number of States Parties were commercially advertising large area RCA delivery systems,[58] States Parties and the Executive Council showed little appetite to discuss whether or not some of these delivery systems were compatible with the prohibitions under Article I of the Convention not to use RCA as a method of warfare (and hence might indicate that national penal laws enacted to enforce the prohibitions of the CWC might not be working as expected). And, when diverging views emerged in the Executive Council about exactly what the US and the UK should have done when their forces searched for, discovered and destroyed *in situ* chemical weapons left behind after the occupation of Iraq, a facilitation process was initiated in the Executive Council to develop guidelines for any future such situations, but it never led to such guidelines being adopted.[59] The reasons for that lack of follow-up are not apparent from public record, but the argument that these were situations that the CWC had not foreseen and where, hence, the rules of the Convention would not apply is less than convincing. The purpose of a convention such as the CWC is not to foresee every conceivable circumstance in the future and provide a rule for it, but to establish principles and objectives that can be used to set such rules where there are none.

The difficulty of managing diverging perceptions about compliance came to the fore when the Security Council established the Joint Investigation Mechanism (JIM) of the OPCW and the United Nations to follow up on findings of the OPCW FFM about the use of toxic chemicals (mustard agent, chlorine) in Syria, and to identify "those individuals, entities, groups, or governments responsible for any use of chemicals as weapons, including chlorine or any other toxic chemical"[60] so that they be held accountable. When the JIM published its third and fourth reports,[61,62] it became clear that some States Parties found it difficult to accept some of its findings – notwithstanding the fact that they had previously accepted the investigation methodology that the JIM described in detail in its first two reports.[63,64] For example, whilst the Russian Federation accepted the JIM's conclusion that ISIS had used mustard agent on one occasion, it challenged the JIM's conclusions regarding the use by Syrian armed forces of barrel bombs that released chlorine, at three other locations, in Idlib province.[65] The investigative approach and the nature of the evidence used by the JIM to substantiate its findings in all four confirmed cases was the same, so the criticism levelled against the JIM's methodology should also have disqualified its conclusion about ISIS. The same occurred when the JIM published its final report in 2017, concluding *inter alia* that Syria had used Sarin in Khan Sheikhun on 4 April 2017. The manner in which the debate in the Executive Council led into polarization along political lines rather than an attempt to find common ground on the basis of the object and purpose of the Convention showed a lack of collective responsibility by the Council to uphold the values and principles of the Convention. At the time of writing, a similar controversy had erupted when the United Kingdom informed the Council about the Skripal affair in March 2018, attributing the source of the agent used to Russia. This sliding back from multilateral problem resolution into power politics is not unique to the OPCW, and the evolving global security and diplomatic environment may even render such trends more significant in the years to come. However, the experience of the CWC as well as that of other treaties and mechanisms underlines the dangers involved in a loss of common purpose by the political organs of an international organisation, and the consequential inability to manage compliance issues in the interest of the States community as a whole and of the stability of the instrument in case. The upcoming Review Conference must be used as an opportunity to re-establish that common sense of purpose and direction within the OPCW.

Finally, the way in which the OPCW manages the advances in science, technology, industry and trade of chemical products, equipment and technologies will be important for the prevention of the re-emergence of chemical weapons. This issue is discussed further in the next section of this chapter.

3.4 The Operation of the CWC and Science and Technology

Science and technology permeate all aspects of the CWC – its scope, the way in which its provisions are being implemented, the manner in which the Technical Secretariat verifies compliance (creating new challenges to the verification system and at the same time providing new verification tools), the role that chemical warfare as well as chemical defence play (or could play) in armed conflict, and the opportunities that there are for peaceful cooperation in the use of chemistry for the benefit of society. All previous review conferences have recognised the importance of monitoring and assessing advances in science and technology, and the profound impact that changes in the chemical industry have on the implementation of the Convention.

This underscores the importance of science and technology advice for the smooth operation of the Convention. Different international mechanisms and treaties used a variety of approaches to render such advice – governmental experts as part of national delegations; groups of experts brought together to study and report on particular issues; formal and informal arrangements with bodies representing science, engineering and industry; the invitation of academic and industry experts to formal negotiations as "guests", "observers" or invited speakers, and in some cases even as negotiation partners; the organisation of side events on science and technology alongside diplomatic conferences including panel discussions, presentations and poster sessions; and formal advisory boards. Also, Secretariats of international organisations often have an inherent scientific and technical competence and entertain relations with relevant science and technology bodies, and they support Member States with technical advice and information.

The Chemical Weapons Convention provides a formal institutional approach to science and technology advice, in the form of the Scientific Advisory Board. The SAB advises the Director-General and in doing so enables him to render specialised advice to the States Parties and the OPCW policy-making organs. The SAB already began work during the first year of CWC implementation, being tasked at that time to render advice on one issue that had been left unresolved by the Preparatory Commission (the application of the Convention's provisions to ricin), and to take up a second issue that had emerged during the practical implementation of the CWC transfer rules for Schedule 1 chemicals and that would eventually require a treaty amendment using the change procedures (notification of transfers of saxitoxin for diagnostic purposes; this was related to the need for reference standards for the screening of bivalve molluscs for food safety and export purposes). In subsequent years, several other issues that the negotiations among States Parties had been unable to resolve ("production by synthesis" of discrete organic chemicals and biological manufacturing processes; low concentration guidelines for Schedule 2 and 3 chemicals; captive use of Schedule 1 chemicals, the significance of CAS registry numbers from a regulatory perspective, and others) were handed to the SAB for advice. This "piecemeal approach" carried a certain risk that the SAB could come under pressure to provide advice that would suit certain political ends (risking the independence and objectivity of its work), or that the recommendations it provided would not meet the expectations of all key actors and would then be "filtered" through procedural mechanisms that could be used to qualify the SAB advice provided to the Director-General. By and large, the OPCW managed to avoid these pitfalls, but at the expense of many recommendations re-appearing in successive SAB reports over many years without States Parties being able to agree on consequential decisions or implementation guidelines.

The work programme of the SAB during its initial years was thus composed of specific issues given to it by the Director-General or the Conference of the States Parties, for which the SAB would establish temporary working groups to bring in specialised technical and scientific expertise or, in some cases, such as the saxitoxin issue, organise SAB work in the margins of relevant scientific conferences, thus increasing its reach-back capability into the wider scientific community. The work in these early years was task driven and in this narrow sense quite effective (although some of the recommendations did not lead to follow-up action by the policy making organs), but it was also somewhat fragmented, lacked breadth in scope and it was often short-term in nature. This was an indication that SAB advice was sought, in those early years, mainly for operational rather than strategic reasons.

This approach changed when the SAB began discussing its contribution to the First CWC Review Conference. In 2000, the SAB "considered that it would be useful to identify areas of science and technology that should be analysed in some detail, with respect to how current developments will impact on the scientific foundations of the Convention. Areas that may deserve detailed study could include chemical analysis, equipment and instruments, biosynthesis and other trends in chemical manufacturing, biotechnology, remote sensing, nano-technology (both with respect to analysis and synthesis), and bioassays". The Board considered that, as a next step, the areas to be studied in detail needed to be further clarified and that this should be done in consultation with both the OPCW and "*scientific institutions and associations outside the OPCW*" (emphasis added).[66] In the following year, the SAB established contact with the International Union of Pure and Applied Chemistry (IUPAC), following an exchange of letters with the Secretary-General and the President of IUPAC in which both officials expressed IUPAC's wish to contribute to the review process. The SAB also discussed the possible involvement of

National Academies of Sciences, and it highlighted the issue areas it considered particularly relevant for this review, and included therein:[67]

- With regards to the CWC definitions and new threats: toxicology, pharmacology, molecular genetics, genetic engineering, and receptor research;
- With regards to chemical production methods: biosynthesis and process technology including some emerging methods of chemical synthesis (*e.g.* solid phase chemistry);
- Whether and how scientific developments affect the interface between the Convention and the Biological and Toxin Weapons Convention, for example with respect to toxins and related molecules;
- With regards to effective CWC implementation: destruction technologies, analytical chemistry, bioassays, remote sensing, and the miniaturisation of instrumental analysis, including the use of biosensors and nano-technology.

Without using the term "convergence", this was the beginning of the SAB engagement with convergence issues at the intersection of chemistry and biology. It was also the start of a partnership between the SAB and IUPAC that has been beneficial for both organisations: it gave the OPCW access to the breadth and depth of IUPAC's expertise in its Divisions and more than 50 affiliated national academies and chemical societies, and for IUPAC it helped to increase and demonstrate the relevance of its work for society, in line with its strategic plan, adopted in 1988 and further refined in subsequent years.[68]

Convergence became an overarching theme in the SAB's contributions to all three Review Conferences. In this work, the SAB could build and expand on the input from international conferences and studies organised by IUPAC as well as some States Parties, including the IUPAC/OPCW workshop in Zagreb in 2007,[69] the workshop co-hosted by the US National Academy of Science and Chinese Academy of Science held in Beijing in 2010,[70] and the IUPAC workshop in Spiez in 2012.[71] In 2011, the Director-General established the SAB's Temporary Working Group on Convergence, which held a series of meetings and in 2014 submitted its report to the Director-General and the OPCW.[72] The findings and recommendations of the SAB were subsequently carried forward in a new series of international convergence workshops, organised by Spiez Laboratory in Switzerland. The first Spiez Convergence workshop took place in October 2014,[3] and a second Spiez Convergence workshop was held in September 2016.[73]

This series of international studies and workshops has highlighted the pace with which science and technology have been making progress at the intersection of chemistry and biology. The international landscape of science collaborations is changing profoundly, with the emergence of virtual laboratories and remote collaborations of research and development centres that come together based on their specific expertise and capacity rather than geographical location. This is facilitated by the internet which has become a place to manage projects and organise collaborations, a repository of data, and a marketplace for specialised materials and equipment including custom-made biomolecules such as DNA and proteins. New techniques and equipment have been disseminating rapidly on a global scale, and in some cases have revolutionised certain research activities, such as the spread of CRISPR/Cas as a new and transformational tool for gene editing. Automation of laboratory tasks has made research much more efficient and reproducible. However, these workshops have also increased the understanding of the limitations of these new technologies, and have helped to render impact assessments more balanced, making projections less-alarmist and more grounded in reality.

As these advances in research and development begin to manifest themselves in industrial technology and manufacturing, they will have an impact on the operation of the Convention and the way in which some of its provisions are being applied in practice. The changing industrial landscape will affect, in one way or another, the way in which national implementation systems function, and international verification measures will also have to take these changes into account.

When the Technical Secretariat published its latest Medium-term Plan for the OPCW for 2017–2020, it included among its planning assumption that "[A]dvances in science and technology will continue to evolve at an accelerated pace, driven by an increasing convergence of scientific disciplines, scientific collaboration at both national and international levels, and innovative funding strategies. Enabling technologies will be adopted into daily life and become operational across a broad range of industries and sectors. Especially relevant for the Convention is their use in chemical production and disposal applications. ... The chemical industry will increasingly integrate new technologies into routine use, requiring the Technical Secretariat ... to have familiarity with a broader ranges of production equipment and processes."[74] This led it to strengthen, as one of its medium-term goals, the OPCW's capability to monitor scientific and technological developments of relevance to the Convention, with emphasis on the effective implementation of the Convention's verification system as well as the widening spectrum of relevant toxic chemicals that need to be monitored. It is worth noting that the Medium-term Plan not only makes reference to the advice expected from the SAB, but also to augmented in-house scientific and technological resources, specialised staff, information tools for handling large data sets, and contacts with relevant international organisations, academia, and chemical industry. This in-house science monitoring and evaluation capacity should be combined with augmented state-of-the-art methods and technologies for sampling and analysis, training, and capabilities in chemical forensics, incorporating advice from the SAB and in consultation with the OPCW's network of designated laboratories, including for the analysis of toxins as well as biomedical and other samples. Furthermore, the Secretariat pointed out that the OPCW needed to have in place the necessary tools, procedures, and skilled and experienced staff to achieve the four medium-term verification goals (re-balancing the verification regime from disarmament to preventing the re-emergence of chemical weapons; augmented routine verification based on risk management; enhanced capability for contingency operations; and stronger science and technology monitoring), in particular for data monitoring, analysis and interpretation, and independently verifying the accuracy and veracity of information received.[4]

It needs to be understood that this Medium-term Plan and its goals reflect the thinking of the Technical Secretariat and the Director-General. This document is not subject to formal approval by the policy-making organs and this gives the Technical Secretariat and the Director-General a degree of flexibility to frame issues and possible problem solutions in terms that the policy-making organs may not (yet) be in a position to endorse, but it also means that these goals and measures do not bind the States Parties. Whether and how States Parties will eventually decide to adapt the rules of implementing the Convention to changes in the external environment—be it with regard to which specific activities and facilities should be declared and verified or how the verification system should evolve, for example with regards to any changes in the Schedules or the further development of the verification system for OCPFs—will not merely depend on the merits of the proposals developed by the Technical Secretariat, but also on the political, economic and security considerations of the States Parties and their ability to take timely and meaningful decisions to these ends.

3.5 The Future – Staying Relevant and Implementing Change

3.5.1 The Changing CWC/OPCW Environment

When the Director-General's Advisory Panel on Future OPCW Priorities, in 2011, looked at the changing environment within which the OPCW and the CWC are operating, it observed that the conditions that prevailed during the negotiations no longer characterise the Convention's external environment. Conflict was no longer framed in the context of opposing military alliances in a bipolar world. The number of inter-State conflicts had declined, yet the level of violence had not. The borderlines between war, civil war, large-scale violations of human rights, revolutions and uprisings, insurgencies and terrorism as well as organised crime were blurred. In addition to traditional military forces, non-State actors had appeared on the battlefield, including paramilitary groups, warlords and their militias and volunteers, mercenaries and private military companies, terrorists and criminal groups.[2] The ongoing armed conflicts in Northern Africa and the Middle East serve as reminders that these trends have not been reversed.

This evolving security environment and the broader involvement of non-State actors as combatants create a certain potential for the weakening of the role of humanitarian law, and in general can undermine the prohibitions of illegal weapons such as chemical weapons if these types of weapons are perceived by some actors to have military or psychological utility. It may also create demands for new types of weapons in the hand of States, including non-lethal weapons that may be seen as weapons of choice for certain types of military operations ("peace support operations" in current NATO terminology) in densely populated areas or in scenarios where combatants and civilians blend together. The Director-General's Advisory Panel observed that as "distinctions between law enforcement, counter-terrorism, counter-insurgency and low-intensity warfare may get blurred, ... certain types of chemical weapons such as incapacitants may appear to offer tactical solutions to operational scenarios where civilians and combatants cannot easily be separated or distinguished."[75]

The evolving security environment may also entail that the OPCW may have to more regularly operate in non-permissive environments such as civil wars or insurgencies, either as an investigative body after allegations of the use of (improvised or military) chemical weapons, or as part of an assistance operation to victims of chemical warfare, or in a context similar to its operations in Syria and Libya when existing chemical weapons needed to be removed from a conflict scenario to avoid them being used by any of the conflict parties.

At the same time, the pace of progress in science and technology is likely to increase rather than slow down, driven by expectations that these advances may bring about solutions to many societal problems, ranging from managing global warming to securing access to food and clean water, improving public health and life expectancy as well as quality of life, and providing a range of goods and services that involve solutions and materials emerging from chemistry, biology, and the life sciences. Strong drivers within the sciences are expected to lead to an even stronger convergence in scientific concepts and methods, and growing global collaborations supported by enabling technologies in such areas as communications and information management, information repositories, modelling and simulations are likely to promote the global diffusion of these new scientific methods and technological capability. The chemical and biotechnology industries will be important drivers of this progress and integrate the emerging technologies, where economically interesting, into their routine operations.

These trends in science, technology, industrial manufacturing and trade are likely to leave their marks on the way international collaborations will be initiated and managed, with fewer roles for governments and their ministries and agencies involved in technical and industrial collaborations, and more emphasis on initiatives coming from the private sector and non-governmental actors.

Within this changing external environment, the OPCW's mandate, too, will evolve, as already described above. This evolving mandate will call for a more holistic approach, as described by the Technical Secretariat in the OPCW's 2017–2010 Medium-term Plan. The activity areas of the OPCW will range from "activities in traditional results areas of the Organisation, and will contain elements of verification, capacity development, improved national implementation, effective national protective programmes, engagement with relevant stakeholders through education and outreach activities, promotion of responsible use of science, the continued pursuit of universality, as well as the monitoring of relevant new developments in science and technology. ... achieving and measuring results will be a more complex task than was hitherto the case, cutting across the traditional programmatic areas of the Organisation. This will require enhanced resilience and adaptive capacity, including through the adoption of additional new, project-based working methods and tools; continued focus on improving results-based management (RBM); and strengthened knowledge-management structures and processes."[76]

3.5.2 Political and Cultural Challenges

These developments will pose significant challenges to the OPCW, both in terms of political culture and governance. Common aspirations of the States Parties, based on the goals set out by the Convention, will need to be transformed into agreed objectives and strategies that the policy-making organs use in their decision making, and the Technical Secretariat in its programming and results monitoring and evaluation.

At the political level, this will require the development and agreement of clear strategic objectives, and decision-making that does overcome fragmentation caused by narrow national interest and group solidarity, and that takes the object and purpose of the Convention as its primary measure. It is interesting to note in this regard that the OPCW has established an Open–Ended Working Group on the Future Priorities of the OPCW (OEWG-FP).[77] This informal mechanism is to receive, discuss, prioritise, elaborate, and integrate ideas and proposals from States Parties and the Secretariat on the future priorities of the OPCW on any aspect of the Convention. The OEWG-FP has been tasked to present a holistic, coherent, forward-looking, and action-oriented document consisting of recommendations for consideration by the Fourth Review Conference, and it should operate in a transparent, inclusive, and consensus-driven basis. The leadership of this new mechanism was asked to make full use of other existing informal mechanisms (which could, for example, be other open–ended working groups such as the one on terrorism and its sub-working group on non-State actors, of any of the facilitation processes under the different Clusters established by the Executive Council, or any of the advisory boards including the SAB or the newly-established Advisory Board on Education and Outreach) including with regards to referring topics that fall within another body's mandate to that body for discussion, feedback, and formal input.

The decision that established this new mechanism made explicit reference to a strategy paper published by the Director-General in 2015 – a vision paper entitled "The OPCW in 2025 – Ensuring a World Free of Chemical Weapons". That paper had concluded that the changing operating environment of the Convention and the progressive shift from chemical weapons disarmament to preventing their re-emergence would require new investments in a widening range of activities. The Secretariat's paper[51] specifically mentioned in this context:

- Verification to ensure continued confidence in compliance that will remain at the heart of the Organisation's work but its methods and practices will need to be adapted to changing realities (viable industry verification system, preparedness for non-routine inspections, greater emphasis on analytical capabilities);
- Capacity development to prevent the misuse of chemistry and to foster its beneficial uses (national implementation support, assistance and protection, measures to enhance chemical security);
- Engaging a wider audience and broadening existing networks to extend the OPCW's reach (more effective engagement with other international organisations, the chemical industry, think tanks, academia, the scientific and technological community and others; empowering broader participation in the OPCW's mission and leveraging the capabilities of others to achieve common goals);
- Adapt in order to remain fit for purpose (develop further essential management and governance tools in a number of areas including administration, human resources, and information and computing technology (ICT); increase the OPCW's ability to serve as a platform for States Parties and relevant stakeholders to promote the objective and purpose of the Convention).

This approach, whilst firmly based on the objectives set out in the Convention, positions the OPCW in a wider context of international organisations, mechanisms and programmes. How necessary this is in policy as well as operational terms has already become recognised during the Syria missions. It will require both informal and formal networking, as well as formalising relationships where necessary in memoranda of understanding and similar agreements and arrangements. There will also be a need to further expand the OPCW's networking with civil society organisations and non-governmental organisations (NGOs) that are active in the areas where the OPCW wants to engage. For example, work in the area of assistance and protection, in contingencies such as the removal of toxic materials from conflict zones, or in investigations of alleged use will inevitably bring the OPCW into contact with humanitarian organisations operating in the conflict zones concerned. This interaction and the necessary collaboration in the field will not be free from misunderstandings and tension unless the OPCW is making efforts to network with these potential partners, even if they are not at first glance close associates in the implementation of the OPCW's mandates.

It is hoped that the OEWG-FP will be able in the run-up to the Fourth Review Conference to strengthen the common sense of direction, priorities and purpose among the States Parties so that these policy and programmatic decisions can form a coherent and effective institutional strategy for the OPCW.

3.5.3 Operational Capacity, Competence and Knowledge

The implementation of such a strategy requires adequate operational capacity, technical and political competence, and knowledge within the Technical Secretariat. The Technical Secretariat is a relatively small staff organisation with limited depth, and a number of operational areas where its success critically depends on a small number of staff who have the required competence and technical skills. This lack of depth will be particularly important in non-routine operations and in non-permissible environments, where managing mission criticality is important.

To meet this challenge, the Secretariat will need to combine several approaches under a coherent and long-term staff policy: timely hiring of suitable staff, an ability to ramp up human capacity in critical areas with short-term hire when needed (or to back-fill human capacity in case non-routine operations create gaps because of re-assignment of staff to other functions), effective knowledge transfer and information management, training and practical exercises, strengthening the analytical capabilities of the Secretariat, collaborations with other relevant international organisations to ensure well-functioning interfaces and inter-operability, and perhaps also mechanisms for exchanging staff with specialised expertise and outsourcing mechanisms where appropriate.

This leads to the question of whether it is again time to review the way in which the OPCW's tenure policy is being implemented. Staff rotation is important to bring fresh ideas, up-to-date expertise and a wide range of skills into the Organisation; to avoid sterility and getting stuck in the past; and if implemented carefully, it can be a motivator for professional excellence and high performance. But the way the tenure policy is being implemented needs to balance these benefits against the need to maintain institutional knowledge, the ability of the Organisation to function as a depository of knowledge in the field of chemical weapons disarmament, and the need to maintain at all times a critical mass with regards to essential competencies and skills.

This OPCW role as a global depository of knowledge in the area of chemical weapons disarmament, from verification and destruction to investigations and incident response, also requires that the OPCW maintain and further expand the knowledge base it has in-house and through its advisory boards, partners and networks. The upcoming Review Conference is an opportunity for the OPCW to renew these partnerships and pledge continuity in its broad-based governance approach that involves all relevant stakeholders and develops long-term strategies based on engagement, education and public diplomacy.

3.6 Convergence and Disarmament – Conclusions

The theme of this book is convergence of chemistry and biology, and the implication of convergence for arms control and disarmament. This chapter has provided a general overview of the key features of the Chemical Weapons

Convention, drawn attention to some of the lessons learned in its implementation, and looked at key challenges ahead. As other chapters of this book will discuss in more depth, advances in science and technology have a profound impact on the operations of the Convention, from affecting its scope to changing its implementation environment to creating new opportunities in such diverse areas as verification of treaty compliance and investigation of hostile uses of toxic chemicals, protection against the effects of chemical weapons, increasing resilience in society against the adverse effects of toxic chemicals to humans and the environment, or the peaceful application of chemistry for the benefit of society. It is important that the OPCW monitors the changes in science, technology and chemical industry and trade, and understands how these advances affect the different implementation processes of the CWC.

Today's advances in science and technology are often subsumed under the concept of convergence. To some observers, this may appear to be an academic construct that may be relevant to people at the laboratory bench, but less so to practitioners in security or arms control. Convergence, however, is no longer just an intellectual curiosity – it is beginning to manifest itself in the world of industrial production, trade and commerce. It is only natural that the arms control and disarmament as well as security communities are beginning to think about how these trends may affect their mandates and objectives.

Whilst there are examples that States have brought together certain aspects of national implementation of the disarmament and security regimes pertaining to chemical and biological weapons and related materials and technologies (in terms or laws and regulations, but also in the form of national bodies that have mandates in both areas), it is unlikely that convergence in science and technology will lead to a (re)-fusion of the international arms control and disarmament regimes with regards to chemical and biological weapons, at least at the international level.[78] The Fourth Review Conference, however, is an opportunity to explore how the implementation of the Chemical Weapons Convention can be better harmonised and integrated with the work of other international organisations and mechanisms that have mandates that in one way or another overlap with the work of the OPCW – such as the Implementation Support Unit of the Biological and Toxin Weapons Convention with regards to the implementation of disarmament and arms control requirements at the intersection between chemistry and biology (including with regards to toxins); the 1540 Committee of the UN Security Council with regards to the control of chemical and biological materials, equipment and technologies of proliferation relevance; the UN Office of Disarmament Affairs, the WHO and INTERPOL with regards to the conduct of investigations of allegations of the use of chemical weapons; or the range of UN and other international organisations working in the area of disaster response that may have mandates in situations when chemical weapons have been used. The notion of convergence can be a helpful conceptual tool for thinking about and gradually adopting common and increasingly overlapping approaches to treaty implementation and norm maintenance at the international level in the field of biological and chemical safety and security.

References

1. https://www.opcw.org, last accessed February 2017.
2. OPCW, *Note by the Director General Report of the Advisory Panel on Future Priorities of the Organisation for the Prohibition of Chemical Weapons*, document S/951/2011, 2011, p. 3.
3. Spiez Laboratory and ETH Zürich, *Spiez Convergence – Report on the First Workshop*, 6–9 October 2014, p. 11.
4. OPCW, *Medium-term Plan of the Organisation for the Prohibition of Chemical Weapons 2017–2020*, documents EC-83/S/1 C-21/S/1, 2016.
5. *Convention on the Prohibition of the Development, Production, Stockpiling and Use of Chemical Weapons and on Their Destruction (Chemical Weapons Convention), Verification Annex Part IV (A)*, paragraph 13, entry into force 29 April 1997.
6. *Chemical Weapons Convention, Verification Annex Part V*, paragraphs 70–72bis, entry into force 29 April 1997.
7. *Chemical Weapons Conversion, new paragraphs 5bis to Section B of Part VI of the Verification Annex as well as 72bis to Section D of Part V of the Verification Annex, entry into force 29 April 1997.*
8. http://www.un.org/en/terrorism/ctitf/pdfs/ctitf_wmd_working_group_report_interagency_2011.pdf, last accessed February 2017.
9. https://www.un.org/counterterrorism/ctitf/en/preventing-and-responding-wmd-terrorist-attacks, last accessed February 2017.
10. https://www.opcw.org/our-work/international-cooperation/capacity-building-programmes/, last accessed February 2017.
11. https://www.opcw.org/our-work/national-implementation/, last accessed February 2017.
12. OPCW, *Report of the Second Session of the Conference of the States Parties*, document C-II/8, 1997.
13. OPCW, *Report of the First Special Session to Review the Operation of the Chemical Weapons Convention*, document RC-1/5, 2003.
14. OPCW, *Note by the Technical Secretariat Review of the Operation of the Chemical Weapons Convention Since the Second Review Conference*, document RC-3/S/1, 2013, p. 3.
15. https://www.opcw.org, last accessed February 2017.
16. OPCW, *Final Extended Deadline of 29 April 2012*, document C-16/Dec/11, 2011.
17. http://www.un.org/zh/focus/northafrica/cwinvestigation.pdf, last accessed February 2017.
18. OPCW, *Decision – Destruction of Syrian Chemical Weapons*, document EC-M-33/Dec.1, 2013.
19. United Nations, *Security Council Resolution 2118*, 2013.

20. https://www.opcw.org/fileadmin/OPCW/PDF/Lessons_learned_from_the_OPCW_Mission_in_Syria.pdf, accessed February 2017.
21. OPCW, *Decision – Destruction of Libya's Remaining Chemical Weapons*, document EC-M/52/Dec.1, 2016.
22. OPCW, *Decision – Detailed Requirements for the Destruction of Libya's Remaining Category 2 Chemical Weapons*, document EC-M-52/Dec.2, 2016.
23. OPCW, *Decision – Plan for the Destruction of Libya's Remaining Category 2 Chemical Weapons outside the Territory of Syria*, document EC-M-53/Dec.1, 2016.
24. OPCW, *Arrangements between the OPCW and the Government of Germany governing on-site inspections at the Gesellschaft zur Entsorgung von Chemcischen Kampfstoffen und Rüstungsaltlasten mbH (GEKA mbH), Munster, and at the port of disembarcation in Germany*, document EC-M-53/Dec 2, 2016.
25. United Nations, *Letter Dated 8 November 2016 from the Secretary-General Addressed to the President of the Security Council*, document S/2016/943, 2016.
26. United Nations, *Letter Dated 1 December 2016 from the Secretary-General Addressed to the President of the Security Council*, document S/2016/1014, 2016.
27. OPCW, *Report of the OPCW on the Implementation of the Convention on the Prohibition of the Development, Production, Stockpiling and Use of Chemical Weapons and on Their Destruction in 2015*, document C-21/4, 2016.
28. OPCW, *Decision – Plan of Action Regarding the Implementation of Article VII Obligations*, document C-8/Dec.16, 2003.
29. OPCW, *Decision – Follow-up T the Plan F Action Regarding Article VII Obligations*, document C-10/DEC.16, dated 11 November 2005.
30. OPCW, *Report by the Director-General – Status F Implementation of Article VII of the Chemical Weapons Convention as at 31 July 2015: Article VII Initial Measures*, document EC-80/DG.15 C-20/DG.12, 2015.
31. OPCW, *Note by the Director-General – Status Report on the Timely Submission by States Parties of Declarations under Article VI of the Chemical Weapons Convention for the Period from 1 January to 31 December 2016*, document EC-84/DG.8, 2017.
32. https://www.opcw.org/legislative-assistance-support-tool/, last accessed February 2017.
33. https://www.opcw.org/our-work/national-implementation/customs-forum/, last accessed February 2017.
34. https://www.opcw.org/our-work/non-proliferation/declarations-adviser/edna/, last accessed February 2017.
35. OPCW, *Note by the Technical Secretariat - Update on the Secure Information Exchange System*, document S/1327/2015, 2015.
36. http://www.saicm.org, last accessed Fenruary 2017.
37. OPCW, *Report of the Third Special Session to Review the Operation of the Chemical Weapons Convention*, document RC-3/3, 2013, p. 15.
38. OPCW, *Report of the OPCW on the Implementation of the Convention on the Prohibition of the Development, Production, Stockpiling and Use of Chemical Weapons and on Their Destruction in 2015*, document C-21/4, 2016, p. 8.

39. OPCW, *Report of the OPCW on the Implementation of the Convention on the Prohibition of the Development, Production, Stockpiling and Use of Chemical Weapons and on Their Destruction in 2015*, document C-21/4, 2016, p. 9.
40. OPCW, *Note by the Technical Secretariat – Review of the Operation of the Chemical Weapons Convention since the Second Review Conference*, document RC-3/S/1, 2013, p. 4.
41. OPCW, *Note by the Technical Secretariat – Review of the Operation of the Chemical Weapons Convention since the Second Review Conference*, document RC-3/S/1, 2013, p. 47.
42. OPCW, *Report of the OPCW on the Implementation of the Convention on the Prohibition of the Development, Production, Stockpiling and Use of Chemical Weapons and on Their Destruction in 2015*, document C-21/4, 2016, p. 10.
43. OPCW, *Note by the Technical Secretariat – Review of the Operation of the Chemical Weapons Convention since the Frist Review Conference*, document RC-2/S/1*, 2008.
44. OPCW, *Note by the Director-General - Status of Implementation of Article X of the Chemical Weapons Convention as at 30 June 2016*, document EC-83/DG.4/Rev.1, 2016.
45. OPCW, *Note by the Director-General - Status of the OPCW's Contribution to Global Anti-terrorism Efforts*, document EC-81/DG.7, 2016.
46. W. Krutzsch and T. Dunworth, in *The Chemical Weapons Convention – A Commentary*, ed. W. Krutzsch, E. Myjer and R. Trapp, Oxford University Press, 2014, pp. 275–276.
47. OPCW, *Report of the Second Special Session to Review the Operation of the Chemical Weapons Convention*, document RC-2/4, 2008, p. 20.
48. OPCW, *Report of the Third Special Session to Review the Operation of the Chemical Weapons Convention*, document RC-3/3, 2013, p. 21.
49. T. Marauhn, in *The Chemical Weapons Convention – A Commentary*, ed. W. Krutzsch, E. Myjer and R. Trapp, Oxford University Press, 2014, p. 327.
50. T. Marauhn, in *The Chemical Weapons Convention – A Commentary*, ed. W. Krutzsch, E. Myjer and R. Trapp, Oxford University Press, 2014, p. 328.
51. OPCW, *Note by the Technical Secretariat- the OPCW in 2025: Ensuring a World Free of Chemical Weapons*, Annex of document S/1252/2015, paragraph 18(d), 2015.
52. OPCW, *Note by the Director-General – Status of the OPCW's contribution to global anti-terrorism efforts*, document EC-87/DG.17, 2018.
53. OPCW, *Report of the First Special Session to Review the Operation of the Chemical Weapons Convention*, document RC-1/5, 2003, p. 10.
54. OPCW, *Report of the First Special Session to Review the Operation of the Chemical Weapons Convention*, document RC-1/5, 2003, pp. 5–6.
55. *International Committee of the Red Cross Incapacitating Chemical Agents – Law Enforcement, Human Rights Law and Policy Perspectives. Montreux Switzerland 24–26 April 2012*, ICRC Geneva, 2013.
56. Spiez Laboratory, *Technical Workshop on Incapacitating Agents. Spiez 8-9 September 2011*, Spiez Laboratory, 2012.

57. *Incapacitating Biochemical Weapons – Promise or Peril?* ed. A. M. Pearson, M. I. Chevrier and M. Wheelis, Lexington Books Lanham etc., 2007.
58. M. Crowley, *Drawing the Line: Regulation of "wide Area" Riot Control Agent Delivery Mechanisms under the Chemical Weapons Convention,* Bradford Non-lethal Weapons Project and Omega Research Foundation, April 2013.
59. R. Trapp and P. Walker, in *The Chemical Weapons Convention – A Commentary*, ed. W. Krutzsch, E. Myjer and R. Trapp, Oxford University Press, 2014, pp. 122–126.
60. United Nations, *Security Council Resolution S/Res/2235*, 2015, paragraph 4.
61. United Nations, *Letter Dated 24 August 2016 from the Secretary-General Addressed to the President of the Security Council (Transmitting the 3rd Report of the OPCW UN Joint Investigation Mission)*, document S/2016/738, 2016.
62. United Nations, *Letter Dated 21 October 2016 from the Secretary-General Addressed to the President of the Security Council (Transmitting the 4th Report of the OPCW UN Joint Investigation Mission)*, document S/2016/888, 2016.
63. United Nations, *Letter Dated 12 February 2016 from the Secretary-General Addressed to the President of the Security Council (Transmitting the 1st Report of the OPCW UN Joint Investigation Mission)*, document S/2016/142, 2016.
64. United Nations, *Letter Dated 10 June from the Secretary-General Addressed to the President of the Security Council (Transmitting the 4th Report of the OPCW UN Joint Investigation Mission)*, document S/2016/530, 2016.
65. Russian Federation, *Statement by H.E. Ambassador A.V. Shulgin, Permanent Representative of the Russian Federation to the OPCW at the Eighty-third Session of the Executive Council (On the Results of the Vote on the Draft Decision of the Executive Council on Syria)*, document EC-83/NAT.20, 2016.
66. OPCW, *Report of the Third Session of the Scientific Advisory Board*, document SAB-III/1, 2000, p. 6.
67. OPCW, *Report of the Fourth Session of the Scientific Advisory Board*, document SAB-IV/1, 2001.
68. https://iupac.org/who-we-are/strategic-plan/, accessed February 2017.
69. M. Balali-Mood, P. S. Steyn, L. K. Sydnes and R. Trapp, *Pure Appl. Chem.*, 2008, **80**(1), 175–200.
70. Trends in Science & Technology Relevant to the BWC, *Report of an International Workshop, Held 31 October-3*, November 2010, Beijing Nat. Academies Press, Washington DC, available from: http://www.nap.edu/catalog/13113.html.
71. K. Smallwood, R. Trapp, R. Mathews, B. Schmidt and L. K. Sydnes, *Pure Appl. Chem.*, 2013, **85**(4), 851–881.
72. OPCW, *Convergence of Chemistry and Biology – Report of the Scientific Advisory Board's Temporary Working Group*, document SAB/REP/1/14, 2014.

73. Spiez Laboratory and ETH Zürich, *Spiez Convergence Report on the Second Workshop 5–8 September 2016*, Spiez Laboratory, 2016.
74. OPCW, *Note by the Director General Report of the Advisory Panel on Future Priorities of the Organisation for the Prohibition of Chemical Weapons*, document S/951/2011, 2011, p. 2.
75. OPCW, *Note by the Director-General Report of the Advisory Panel on Future Priorities of the Organisation for the Prohibition of Chemical Weapons*, document S/951/2011, 2011, p. 5.
76. OPCW, *Medium-term Plan of the Organisation for the Prohibition of Chemical Weapons 2017–2020*, documents EC-83/S/1 C-21/S/1, 2016, pp. 2–3.
77. OPCW, *Decision - Establishment of an Open–Ended Working Group on the Future Priorities of the OPCW*, document EC-82/Dec.2, 2016.
78. R. Trapp, *Convergence at the Intersection of Chemistry and Biology – Implications for the Regime Prohibiting Chemical and Biological Weapons*, Biochemical Security 2030 Project, University of Bath, Policy Paper 6, 2014.

CHAPTER 4

The Biological and Toxin Weapons Convention

JEZ LITTLEWOOD

Norman Paterson School of International Affairs (NPSIA), Carleton University, Ottawa, Ontario, K1S 5B6, Canada
*E-mail: Jeremy.Littlewood@carleton.ca

4.1 Introduction

There is a paradox at the heart of concerns about biological and toxin weapons. On the one hand, the theoretical potential of these weapons has been demonstrated in field tests and laboratory studies.[1,2] Moreover, biological and toxin weapons have a wide range of possible uses, as even the limited empirical evidence of actual use demonstrates.[3] In addition, developing an offensive biological warfare capability is easier (but not easy) than it was in previous decades: scientific and technological developments and recent history indicate that such weapons are within the reach of determined actors. As Wheelis, Rózsa and Dando noted in 2006, the "problem is serious" and "preventing biological warfare, terrorism, or criminality is more acute now" than in the past.[4] A decade hence, the problem has not become less acute.

On the other hand, actual use of biological and toxin weapons has been limited and has failed to have a major impact on the outcome of any conflict.[5] In addition, the number of known State programmes remains low. Even accounting for the use of such weapons in assassination and covert operations by State actors[6] and by violent non-State actors with political

Preventing Chemical Weapons: Arms Control and Disarmament as the Sciences Converge
Edited by Michael Crowley, Malcolm Dando and Lijun Shang

Published by the Royal Society of Chemistry, www.rsc.org

objectives, *i.e.* terrorists, the number of entities with both an interest and capability in this realm remains low.[7–9]

Why, therefore, are both States and civil society actors concerned about such weapons? At the 2017 Munich Security Conference, the issue was brought to the fore by Bill Gates who claimed that the next epidemic could result from a terrorist actor with a computer and exploiting the possibilities inherent in synthetic biology.[10] Gates' claim provides a partial answer to the question: the nexus of science, technology and globalization that permit a convergence of knowledge, capabilities, and materials to combine in ways that will have an effect on biological and toxin weapons development, use, control, and prohibitions.[11] Beyond scientific developments and the impact of globalization, other trends and developments are also affecting biological weapons. These include, the changing nature of war and conflict, the contingent aspect of normative constraints, an international legal regime that is weaker than comparable controls and law on nuclear and chemical weapons, and the politics of biological disarmament. Taken together, the intersection of norms, politics, science, law and war provide an explanation for the increase in concerns about biological weapons over the last two decades.

In this chapter, these intersections are explored through the evolution of biological disarmament over the last 100 years. The chapter proceeds in five sections dealing with the normative constraints on biological weapons, the international legal prohibitions first on their use, and later on their development, production, and stockpiling of such weapons. Past programmes within States, the use of biological weapons by all actors, and the concerns related to terrorist interest in these weapons are considered in the next section. The chapter then proceeds to outline and assess the evolution of the Convention in context before focusing on how scientific and technological developments have been considered by BTWC States Parties. In the final, concluding section, the overall strengths and weaknesses of the anti-biological weapons regime are considered in the contemporary environment with a view to the identification of emerging concerns that affect the level of confidence that states and other actors have in the BTWC.

4.1.1 Overview

The Biological and Toxin Weapons Convention (BTWC) is now in its fifth decade. Upon agreement in April 1972, and its later entry into force in March 1975, it was heralded as the first real disarmament treaty for a weapon of mass destruction. Although the Convention, and more importantly its objective of ensuring that disease causing organisms should be prohibited from warfare, has supporters, adherents and, not insignificantly, a small epistemic community of 'Friends of the Convention',[12,13] during the Cold War it languished in the shadows of concerns about nuclear weapons, chemical weapons, and other forms of disarmament and arms control. Others referred to the agreement as one of the low-hanging fruits of arms control and the BTWC as one of a series of "insignificant agreements… [that] have been turned into insignificant treaties".[14] Few would call the BTWC insignificant nowadays, but the 1972 Convention solidified a norm against the use of biological weapons in legal form by prohibiting the development, production, and stockpiling of such weapons. Combined with the normative constraints on the use of disease as a weapon, the BTWC is a significant legal obstacle against the temptation to exploit the life sciences for hostile purposes despite being simultaneously under-resourced, and on the margins of the priorities of States Parties.

The life of the Convention has never been easy or simple. Five years after entry into force, it weathered a diplomatic spat between the United States (US) and the then Union of Soviet Socialist Republics (USSR) over the allegations related to the USSR's offensive programme. Suspicions about the USSR's activities in this realm were confirmed in the early 1990s, but remain subject to contestation in the contemporary period.[15] In addition, concerns about the assurance of compliance in general resulted in innovative procedures that contributed to a very gradual strengthening of the procedural aspects of its implementation, albeit far from complete.[16] However, 15 years after entry into force, the end of the Cold War (1989–1991) and the emergence of the Chemical Weapons Convention inspired States Parties and others to reimagine what a more robust BTWC would entail. This led to a decade of efforts to 'strengthen the Convention' between 1991 and 2001, which came to nothing in the summer of 2001 with the US rejection of a draft Protocol to the BTWC.[17] A slow recovery from its 'doldrums' ensued,[18] but a period of revitalization after 2006 proved short-lived, and in 2011, the ambitions of some actors to reinvigorate the operation of the Convention were stymied by differences related to the balance between compliance and development agendas. This produced a "modest outcome" at the Seventh Review Conference[19] in 2011, but also signaled the reemergence of a group of recalcitrant States Parties opposed to treaty evolution that involved efforts outside a multilaterally negotiated agreement. By mid-2016, on the eve of its Eighth Review Conference, there was a perception of change, but even the limited ambitions of providing the Convention with a scientific advisory panel and a slightly larger implementation support unit of five individuals were not realized in 2016: an outcome categorized as "one of disappointment and frustration".[20]

As a result, the Convention is experiencing a crisis. It is not in danger of collapse and no State Party is indicating that it will withdraw from the BTWC. Moreover, states do not openly admit to offensive biological weapons programmes, and information in the public domain reveals that very few States are suspected of activities that would be in violation of the BTWC. Nevertheless, the outcome of the Eighth Review Conference perpetuates the very modest development of the BTWC since its entry into force. A resolution to the current crisis at the December 2017 meeting is a partial response. It may slow erosion of the norm, weakening of law, and demise of the international community's collective will and ability to prevent the re-emergence of biological and toxin weapons in the arsenals of States.

4.2 Normative and Legal Evolution Pre-BTWC

Prior to the efforts of the United Kingdom (UK) in 1968 to separate chemical and biological weapons into distinct categories, it was common to see statements linking chemical and biological warfare and the generic use of poison as a weapon. This conflation, chemical *and* biological weapons was neither absolute nor overly detrimental to efforts to prohibit the use of such weapons.[21] It has origins in the evolution of the definition of chemical and biological (CB) weapons from the 1920s onwards.[22] As Goldblat and others have noted, ancient custom(s) condemned "the use of poison or poisoned weapons in war or the use of weapons causing unnecessary suffering".[23] Such terminology can be seen in the 1874 Brussels Declaration, the 1899 Hague Peace Conference and 1907 Hague Conference. Greater specificity emerged in the 1919 Versailles Treaty in Article 171 with 'asphyxiating, poisonous or other gases and all analogous liquids, materials or devices' and the same language was used in the 1922 Washington Treaty and the Preamble to the 1925 Geneva Protocol. After 1925, chemical and bacteriological weapons became the accepted term in the literature, since the actual text of the Geneva Protocol would require, "use in war of asphyxiating, poisonous or other gases, and of all analogous liquids, materials or devices and bacteriological methods of warfare".[24]

The inclusion of "bacteriological" within the Geneva Protocol is now taken for granted, but exactly how, and why, "bacteriological" was added in the 1920s is something of a mystery.[25] It also tells us something about how norms and law evolve over time. As Walker details with regards to the history of the Geneva Protocol, Poland inserted a reference to 'bacteriological' in May 1925 within a jointly drafted US, UK, Italian and Polish formulation focused initially on prohibiting chemical weapons use. However, the bacteriological reference was absent from an early June 1925 version of the emerging text, only to be reinserted following an intervention from Poland on June 8, 1925. The final text was adopted on 10 June 1925 and opened for signature on 17 June, and while 'bacteriological' is not due solely to Poland, "the Poles clearly had a leading role".[26]

Having determined who was responsible for its inclusion, Walker also explores the question of *why* it was included. As the archival research reveals, "in the early and mid-1920s... [the]... potentialities of bacteriological warfare"[27] were evident and of concern even if these weapons had not manifested themselves in conflict and war in a manner that was in any way decisive to that point.[28] Even so, biological weapons were on the agenda for exploration among violent actors: the Irish Republican Army (IRA) considered their use against the British in 1920[29] and UK intelligence reports from 1924 suggest a Soviet Union experimental exploration of biological weapons.[30] Carus also notes a number of possible incidents in other States that mixed criminal objectives, unethical experiments, and potentially purposeful use of pathogens, which are unproven, but suggest awareness of and some interest in the deliberate use of disease as a weapon.[31] Such concerns are as relevant in 2017 as they were in 1925.

The prohibition of the use of chemical and bacteriological (biological) weapons in war between "high contracting parties" inherent to the Geneva Protocol is now considered customary international law. As such, the use by any State of chemical or biological weapons is a violation of established law whether or not that State is a party to the Geneva Protocol. However, its prohibition on use left certain loopholes related to the development, production and stockpiling of such weapons and the reservations attached to the Geneva Protocol and interpretation of its text were such that it evolved into a no first use agreement. Retaliation with chemical and biological weapons, if attacked, was deemed to be permitted. Neither chemical nor biological weapons were widely employed during World War II, although they certainly were used by some protagonists, notably Japan, and offensive programmes were evident in a number of States.[32,33] Post World War II, the focus of concern shifted to nuclear weapons and the emergence of 'weapons of mass destruction' which led to the Commission on Conventional Armaments to determine that only "lethal chemical and biological weapons" were incorporated into the WMD term.[34] That 1948 distinction between lethal and non-lethal or incapacitant biological and chemical weapons did not survive in either the BTWC or the Chemical Weapons Convention (CWC) in due course, but remains potentially important in future years.

The separation, or split, of chemical and biological weapons emerged in July 1968 from the United Kingdom (UK) as a means to initiate negotiations on the BTWC.[35] Although ultimately successful, the separation was never absolute in practice given the question of toxins[36] and, more recently, the convergence of chemistry and biology.[37] Thus, although it is not overly harmful to have distinct legal and organizational regimes for biological and chemical weapons, it is equally important to recognize that the regimes overlap in scientific, political, legal, and normative dimensions. The fate of each Convention is inextricably bound to the other.

A broader perspective of the evolution of the normative and legal regime, is also provided by the deepening and widening of arms control over time, particularly over the last half of the twentieth century.[38] In this context 'deepening' represents the elaboration of mechanisms that provide for more extensive verification provisions. The 200 plus pages of the CWC from 1993 epitomize this process of deepening when compared to the four pages of the BTWC from 1972. Although intended by States, because deepening resulted from political willingness to embrace arms control, political intentions are not the only factor that affects the outcomes of arms control negotiations or the progress of such agreements once they exist.[39] Price offers a persuasive argument that the constraints against poison weapons have contingent elements which are as much to do with "the marriage of chance occurrences, fortuitous connections, and reinterpretations" that converge in ways that are not predictable, but can be exploited to develop new understandings about appropriate and legal behavior.[40] To an extent, Walker illustrates this with the Polish insertion of "bacteriological" in the Geneva Protocol. Spelling offers a further example of a collision between norms, law, and political realities in late 1960s Britain. At this time, the UK faced difficult questions nationally and internationally about its own use of CS gas in Northern Ireland, while simultaneously promoting the separation of chemical and biological weapons in the proposed BTWC, and avoiding the ire of its closest ally, the US, which was using CS munitions and herbicides in warfare in Vietnam.[41] Internal UK deliberations on the question of tear gas in 1968–69 did not have significant international implications, but Spelling correctly observes that "a minor technical and legalistic issue on a relatively unremarkable substance actually had far greater ramifications for long-term international efforts to re-enforce and extend prohibition of all forms of chemical and biological weapons" over time.[42]

Such contingencies and intersections are also evident when the US itself undertook a unilateral commitment to abandon biological weapons in 1969, because the origins of that decision also demonstrate that chance, personalities, and political realities can be as important as scientific fact, or expertise. Tucker's historical assessment of the US decision underlines that accidents involving chemical weapons, national political and public concerns over chemical weapons disposal, and the rising critiques of the US use of chemical weapons in Vietnam combined to bring about the chemical and biological review that culminated in the US unilateral decision under President Nixon.[43]

This historical review indicates that the record of the last century is relevant in the contemporary period because it demonstrates that the normative constraints against poisons that become international law arise in part from purposeful and concerted efforts of states. However, the purposeful activity of states is inextricably bound with chance occurrences, individual personalities, and contemporary political realities that involve contested normative and legal reasoning within States, as well as between them, and contested scientific expertise. Norms and laws evolve and neither are absolute. While evolution over the last one hundred years has been predominantly positive, it could be reversed over time. Thus, while the norms against chemical and biological weapons remain relatively robust today, there is no guarantee that they will be so a decade or more hence from now.

4.3 Use of Biological and Toxin Weapons in History

The definitive text covering biological weapons use in history is Carus' 2015 *The history of biological weapons use: what we know and what we don't*. As he notes in the opening paragraphs, "As is evident to anyone with more than a passing interest in the subject, there have been very few BW (biological weapons) incidents and what has occurred has had, with perhaps one significant exception, remarkably little impact."[44] Yet, as his own study concludes, "We can say that the incidence of illicit biological agent use is greater than many people realize. Equally important, we also know that the effect of these attacks has been limited. In some cases, we know that the ambitions of those seeking to perpetrate the attacks far exceeded their actual accomplishments.

We can also document that fears of intentional disease are not new."[45] Carus' study is not out of line with other works that have assessed the use and alleged use of biological weapons. Furmanski and Wheelis note that the historical evidence of alleged use post-1945 includes four cases: Chinese and North Korean claims of use of such weapons by the US during the Korean War (1950–53), Cuban allegations of use against it by the US between 1962 and 1997, the 'yellow rain' allegations against the Soviet Union between 1978 and the early 1980s, and the Sverdlovsk anthrax outbreak in 1979 in the Soviet Union.[46] As they conclude, "all the accusers believed that their accusations were correct"[47] but investigation mechanisms and transparency were lacking in all cases, even though only the Sverdlovsk outbreak has been demonstrated to be conclusively a result of an accident involving the Soviet offensive biological weapons programme. Indeed, it was only in 2016 that a robust rebuttal of the allegation that the United States had used biological weapons in the 1950–53 Korean War appeared, even though it had been debunked some time ago.[48] With regard to the terrorist use of biological and toxin weapons, Wheelis and Sugishima note that only one group "is known to have caused large numbers of casualties",[49] whereas other incidents are insufficiently documented or relate to a terrorist group that has failed to develop and use biological weapons.

Pre-World War Two, Germany and Japan initiated and used biological weapons. Germany in a sabotage programme during the First World War that has been referred to as "an ambitious programme on three continents of veterinary sabotage against neutral suppliers of its enemies and against actual and potential adversaries... [that]... began with nearly simultaneous operations in Romania, the USA, and probably Spain in the summer and autumn of 1915."[50] It subsequently included operations in Argentina and Norway, and while it may not have had explicit approval from the highest authorities of German political decision-making, it certainly had "involvement of the General Staff in direct oversight."[51] Japan's activities in Manchuria (China) from the late 1930s are better documented,[52] and Harris deems the programme, its human experimentation, use of biological weapons, and violation of international law as "astounding in its scale and cruelty."[53] As the best studies indicate, post World War II use of BW has been less well documented and demonstrated; indeed, Carus observes that although biological warfare has been subject to numerous articles and historical surveys, no study is complete, many contain inaccuracies and "all should be used cautiously".[54] Yet, even the partial history of actual use indicates interest in such weapons by a wide variety of actors: States, terrorists, and criminal groups. However, no incident has caused deaths comparable to State level employment of chemical weapons, only one post-1945 incident has caused significant casualties (the Rajneeshees in 1984), no repeated widespread use has been proven, and neither States nor terrorists have turned to such weapons on a significant scale. Indeed, a 2004 study by the World Health Organization included only three cases of verified use since 1918: by Japan in World War II, and by two non-state actors in the United States, one in 1984, the other in October 2001.[55]

4.4 Offensive Programmes

If use of biological and toxin weapons has been limited over the last 100 years, preparation for and exploration of their possible use has been more extensive. It is evident that more States have pursued an interest in biological weapons historically than have used them. The two dominant histories of offensive programmes[56,57] reveal the following States to be of principal interest: Germany; France; Japan; Poland, the Soviet Union; the United Kingdom; Canada; the United States; Iraq; and South Africa. While these are not the only States to have pursued or dabbled in biological weapons, the extent of documentary evidence permits assessment beyond speculation and rumour. None of these programmes survived past the mid-1990s – although a lack of transparency about the Soviet programme and Russia's obfuscation of certain activities and historical revisionism led some commentators to express doubts about Russian compliance with the BTWC.[58]

Post-1945, it was the US,[59] UK,[60] USSR,[61,62] and France[63] that continued biological weapons activities, and Canada remained intertwined with the UK and US programmes.[64] Iraq's programme is believed to have been initiated in the mid-1970s,[65,66] after the US, UK, France, and Canada had given up their programmes, and South Africa's programme spanned a period from the early 1980s to the early 1990s.[67]

Taken together, the two historical studies analyze offensive state programmes and the use of biological weapons throughout the last 100 years. As with Carus' assessment of use, the history of such programmes remains "obscured and complicated by false allegations, unfounded suspicions and the repeated difficulties of separating what is true from what is false"[68] in both the pre-1945 period and the post-1945 period, which still has gaps in knowledge to be filled.

What emerges from known State programmes is a convergence of factors that influenced each other and gave rise or impetus to biological and toxin weapons programmes. In addition, there is a quite remarkable, and disturbing, similarity of these factors across time even though each State-level programme was influenced by discrete, country-specific political, security, scientific cultures and capabilities.

First, States do recognize the moral and normative barriers to the use of biological weapons and the legal strictures from international agreements. Both influence deliberations about such weapons. Recognition is not the same thing as adherence or compliance with such law or undertakings, or acceptance of the normative constraints or taboo against poisons as weapons, but both normative constraints and law did have an effect on programmes. Primary, but not exclusively, among the influence was secrecy and a lack of transparency within militaries and governments. This, in turn, led to inaccurate and incomplete assessments of the threat posed by other States, which in turn fueled such programmes. Avery illustrates this with respect to the Canadian involvement in the US and UK programmes up to the late 1960s in his conclusion that defending against biological weapons "was related to effective deterrents, or, more specifically, a retaliatory BW capability... [so]... it is not surprising that Suffield scientists [in Canada] believed that in supporting the US offensive BW program they were also safeguarding Canada's national security."[69] The possibility of biological weapons being developed by an adversary, coupled with limited intelligence on the capabilities and intentions of adversaries, awareness of the scientific and technological possibilities of such weapons, and organizational political (bureaucratic) differences over an appropriate response to such weapons also influenced the UK program.[70] Thus, up to 1945 "three factors fueled the BTW programmes of the major belligerents: first, a perceived threat; second, recognition of the national vulnerability to biological attack and a belief that biological and toxin warfare is feasible...; and third, an internal momentum once the political decision had been taken which encouraged and supported national biological and toxin warfare programmes."[71] Later, studies revealed that the same factors are identifiable in the post-1945 period, at least up to 1969 for Western States. Gould and Hay identify similar attributes in South Africa, although the programme was focused not on States, but on the development, production and weaponization of "novel and questionable crowd-control and assassination weapons".[72] Large, centrally organized bureaucratic programmes appeared more successful than their haphazard, individually driven programmes in both periods,[73,74] but all were driven by assessments of feasibility, incomplete intelligence, internal bureaucratic factors, and a blurred offensive-defensive nexus, underpinned above all by fear of the consequences of large-scale biological weapons use. In this regard, Dando *et al.* capture the prevailing issues:

> "The major states that ended on the winning side after World War II had developed BW programs because such weapons were seen to be potentially important militarily for retaliation in kind, and they continued or restarted them for the same reasons. The two states definitely known to have begun offensive programs later in the century also had military reasons for their programs. Programs were terminated because of regime change (South Africa and possibly Russia), because of imposed disarmament following defeat in war (Iraq), because BW were overshadowed by nuclear weapons (the UK, France, the US), or because major partners terminated their program (Canada). Finally, when states terminated their offensive programs, because they had been kept highly secret there was generally little attention given to providing information to convince other states that the offensive programs had been terminated."[75]

Understanding the history of offensive programmes offers pointers as to how to prevent the reemergence of biological weapons in the next quarter of a century and beyond. No arms control agreement, or verification mechanism, can entirely guarantee against the covert development of biological or toxin weapons, but if the historical assessment of past programmes is correct it is evident that various factors influenced each other. The decision to develop biological weapons was multi-causal. No single variable such as security, or science, or bureaucracy provides an adequate explanation for why and how states pursued offensive biological weapons. The convergence of factors evident in pre-1945 programmes included, "anxiety regarding biological and toxin warfare activities... producing a cycle in which mirror images predominated" exacerbated by "false information, deliberately or accidently disseminated" and further complicated by perceptions that such weapons might confer an advantage.[76] Cumulatively there was an "assumption... that any method which appears to offer advantages to that nation will be vigorously pursued by that nation." Post-1945, Dando *et al.* conclude that the changing perception of biological weapons, their utility and lack of utility in conflict, alongside intelligence gaps and limitations, combined with secrecy and transparency, suspicion and confidence, and the influence of treaties, law and international collaboration, were recurrent themes on national biological weapons activities.[77] As such, interlocking and overlapping measures in combination will be required to prevent the re-emergence of biological and toxin weapons in the 21st century.

4.5 Evolution of the BTWC

Different authors and analysts have developed various approaches to outlining the evolution of the Convention. Most approaches are chronological, using developments since 1975 and the regular review conferences as way markers, for assessing the health of the Convention and the robustness of the prohibition on biological and toxin weapons. Given that the Convention is now entering its fifth decade, following its fortieth anniversary since entry into force in 2015, the chronological approach permits analysis of the evolution of the Convention in context. Disputes and differences within the BTWC are often proxies for other issues and trends within international affairs, though they do provide an indication of the challenges that the Convention faces in the real world. Even prior to the limited outcome of the 2016 Eighth Review Conference, one of the most astute analysts of the BTWC remarked in 2015, upon its fortieth anniversary, that a bright future for the Convention could not be guaranteed.[78]

Its first fifteen years, 1975 to 1990 spanned the Cold War when it was a largely forgotten Convention and far down the list of priorities of States and civil society actors. Even the occasional crisis and suspected, or alleged, violation of the BTWC was secondary to the political climate and the priorities of States.

Its second period of life was the decade 1991 to 2001 spanning the end of the Cold war and the pre-September 11, 2001 period. Within the confines of the Convention the effort to 'strengthen the Convention' with an additional Protocol that would resemble—but not mimic—the Chemical Weapons Convention (1993) dominated the politics of biological disarmament. However, two other issues maintained a strong influence on approaches to biological disarmament and a strengthened BTWC. The first related to

State programmes, including the uncovering of the Iraqi biological weapons programme,[79] the admission of South Africa that it had maintained some form of offensive biological weapons activity,[80] and the acknowledgement of non-compliance with the BTWC by the USSR in the Russian confidence building measure Form F in 1992. The second related to the emergence of new concerns about the proliferation of biological weapons to both State and non-State actors and the terminology of 'rogue states' and 'bioterrorism' that took root from the early 1990s and entered more mainstream usage from 1995.[81–83]

A third period in the life of the Convention began post-2001 following both the collapse of the BTWC Protocol negotiations in July of that year and the September 11, 2001 (9/11) attacks on the US in September. This third period remains in existence, thus spanning 15 years to date.

4.5.1 From Entry into Force to the End of the Cold War

Few arms control agreements enter into force without certain weaknesses being known, but it is important to recall that all agreements reflect what was technically and politically feasible at the time of negotiation. The first few years of the Convention's existence were unremarkable. Indeed, at the First Review Conference in 1980 a significant portion of States Parties were more vocal with respect to Article IX of the BTWC—the commitment to continue seeking an agreement on chemical weapons disarmament—than on the BTWC *per se*. As the delegation of Nigeria remarked, the Convention was working well.[84] Other States Parties were not as sanguine. In retrospect, 1980 marked the opening of what has been a continual competition of ideas related to reforming and strengthening the BTWC to fix known weaknesses. This competition was not solely related to compliance with the Convention *vis-à-vis* its weapons and disarmament architecture. From 1980 onwards, giving effect to the belief that disarmament should lead to development and operationalization of peaceful cooperation in the biological sciences has bedeviled meetings of the States.

Nevertheless, an assessment of the first period of the Convention's life indicates the following key themes. First, concerns about scientific and technological developments relevant to the BTWC. Second, attempts to enhance the architecture of compliance mechanisms within the BTWC through accumulation of politically binding agreements and additional understandings about the Convention and how it could be used in the event of both routine and extraordinary differences related to compliance. Third, the lack of any supporting infrastructure or institutions, meaning that the Convention was wholly dependent on its States Parties and support from the United Nations for its quinquennial meetings. Fourth, disputes related to the balance between compliance and peaceful cooperation that split parties on developed-developing States lines. And, fifth, the appropriation and use of quite innovative measures and understandings to ensure that the Convention remained relevant and, broadly, adhered to among States Parties. Working backwards, these can be illustrated with examples. Notably, each of these is extant in the contemporary period and should be considered as enduring challenges to the BTWC that manifest themselves throughout the life of the Convention.

4.5.1.1 Relevance and Innovation

In terms of relevance and innovation, the Second Review Conference (1986) agreed on annual information exchange—more often referred to as Confidence-Building Measures (CBMs)—in four areas.[85] The idea of CBMs can be found in a non-arms control agreement from the Cold War, namely the Helsinki Accords, but were put to use in the BTWC context as a means to address the lack of mechanisms related to implementation and the passage of information between States Parties on compliance with the BTWC. They addressed, initially, four areas related to (a) maximum containment laboratories, (b) unusual outbreaks of disease, (c) publications relevant to the BTWC, and (d) contacts between scientists. Agreed in outline in 1986, a special, technical, conference of States Parties was held in 1987 to develop the detailed mechanisms for reporting, including the format of the CBMs.[86]

The CBMs have never fulfilled their potential, but even the limited information exchange has delivered benefits to States Parties over time.[87] In hindsight, two things are of particular note here. First, the CBMs were a tangential means to respond to concerns about compliance following the outbreak of anthrax in Sverdlovsk. The response was muted insomuch that a US–USSR bilateral 'dialogue of the deaf'[88] in the early 1980s gave way to a multilateral BTWC approach that was less adversarial and focused on encouraging States Parties to provide basic information and data on activities most relevant to the BTWC. The second aspect was the determination of States Parties that they could choose to meet whenever they decided if all agreed by consensus. This latter observation may seem minor to the point of insignificance, but a narrow reading of the BTWC allowed only for a review conference five years after the Convention entered into force or a meeting to discuss an Amendment to the BTWC. Such a narrow interpretation was rejected by necessity given the decision to hold a Second Review Conference in 1986—a decision taken at the First Review Conference in 1980—and also implied by agreement in 1980 related to consultative mechanisms. The 1987 technical meeting of experts that finalized the CBMs forms was, therefore, a precursor to other meetings of experts in the BTWC context and an indicator of the preferences of States Parties in handling difficulties within the Convention. That preference is low-level, technically focused meetings intended to address specific issues within a holistic context rather than undertake wholesale changes to the operation of the BTWC. As indicated below, the single attempt to address issues within an ambitious holistic framework failed in 2001. Jumping forward to the present day, the BTWC States Parties have actually convened in some form of formal meeting every year since 1991, while in the period 1975 to 1991 only four meetings were held: the First (1980), Second (1986) and Third (1991) Review Conferences and the technical meeting on CBMs in 1987. Routine meetings of States Parties have therefore become normal practice.

4.5.1.2 Article X and Peaceful Cooperation

Article X of the Convention deals with peaceful cooperation issues. At the time of the negotiations it remains questionable how important or contentious this issue was. The negotiation record suggests the inclusion of Article X—which mirrored and made only minor changes to a similar Article IV in the Nuclear Nonproliferation Treaty (NPT)—was not contentious. However, following entry into force of the BTWC, the emergence within the United Nations Committee on Trade and Development (UNCTAD) and other fora of the United Nations of a development community pushed Article X to the fore of disputes among States Parties.[89,90] The issue is one of differing priorities for States Parties. Certain States view the BTWC as a disarmament agreement where any peaceful cooperation is secondary to the core objective of a prohibition on the development, production and stockpiling of biological weapons. Others, drawing in part from the view that disarmament should free up resources to spend on development, consider peaceful cooperation in the life sciences as both an inalienable right under the BTWC and a necessary *quid pro quo* and incentive for developing States to adhere to the Convention. As history indicates, all States must cope with natural outbreaks of disease, whereas deliberate disease has been extremely rare. An additional complicating factor emerged in the late 1980s when Article X also became a proxy dispute for issues relevant in other arms control agreements, namely export controls under the Nuclear Suppliers Group (NSG) and the emergence of the Australia Group on chemical (and from late 1991, biological) dual use materials and goods. Export controls have long been a means for States to thwart adversaries from developing their own weapons, but by the 1980s, the narrative of a cartel of technologically sophisticated (Western) States purposefully stymieing the development of others took hold. From 1986 it poisoned the relations between States in important areas.[91,92] Although the export control debate has eased in more recent years, it remains a point of contention between certain States Parties and fact and fiction about export controls are still detectable. The perceived lack of development assistance and peaceful cooperation has become an important factor in the deterioration of relations within the BTWC overall, although its origins can be traced back to 1980.

4.5.1.3 Infrastructure of Support

Unlike the NPT and its relationship with the International Atomic Energy Agency (IAEA) and the Chemical Weapons Convention and the Organisation for the Prohibition of Chemical Weapons (OPCW) relationship, the BTWC lacks an infrastructure of support. Up until 1997 it relied upon the United

Nations to support any formal meetings with a secretariat that could assist any Chair or President of such meetings. The lack of any organizational infrastructure of support for the Convention has been documented by Sims, who notes that "serious consideration" was given to the establishment of a secretariat to support the CBM process and meetings of States Parties at the Third Review Conference, as a means to address the "chronic institutional deficit".[93] As with so many BTWC proposals, it fell by the wayside and although *ad hoc* support for the BTWC emerged from 1997, the first fifteen years of the Convention were marked by a lack of infrastructure. Rather than the 2500 staff of the IAEA in Vienna and the 500 staff in the OPCW, the Convention acquired a supporting infrastructure composed of three individuals based in Geneva in 2006. This Implementation Support Unit (ISU) was agreed, formally, at the Sixth Review Conference. Like many other developments in the BTWC, the formal establishment of the ISU in 2006 confirmed and codified practice within the Convention that existed beforehand. States Parties had been paying for additional support staff for the Convention from 1997 rather than relying on the United Nations alone to provide secretarial and other support.[94]

4.5.1.4 Additional Understandings

The steady accumulation of additional understandings between parties began in 1980 and has continued to date. While the exploitation of 'latent' possibilities within the text of each article was initiated in 1980 at the First Review Conference, over time States Parties have tended to reaffirm the obligations under the BTWC at review conferences in the form of final declarations that capture their consensus view on an obligation and how it relates to contemporary events. Some issues are simply administrative, such as the recognition of Arabic as the sixth official language of the BTWC, whereas other are more substantive, such as those on Article I and scientific and technological developments that potentially affect the scope of the BTWC. As reported prior to the Eighth Review Conference, the additional understandings now cover all articles and the ISU recorded 142 such understandings developed between 1980 and 2011.[95]

4.5.1.5 Science and Technology

Concerns about scientific and technological developments have been at the forefront of at least some States Parties throughout the life of the BTWC. That such a focus exists reflects the prohibition itself, but also the explicit requirement under Article XII of the Convention that a review conference should take into account relevant scientific and technological developments. As indicated below, this occurred at each review conference and substantial external (non-governmental) commentary, assessment and review of science and technology have also been provided to States Parties over time.[96–99]

4.5.2 The Post-cold War Decade

The Third Review Conference of the BTWC convened in 1991. The USSR still existed, but the Cold War was over. The Chemical Weapons Convention was in its final stages of negotiations. Iraq had been defeated in the Gulf War (1991) and the United Nations had established the United Nations Special Commission (UNSCOM) to rid Iraq of its nuclear, chemical and biological weapons. A flurry of arms control agreements between the US and the Soviet Union in preceding years also provided impetus for a significant effort to strengthen the Convention. This began at the Third Review Conference with the establishment of the Verification Experts Group (VEREX), continued in 1994 with the decision of the Special Conference of States Parties to begin negotiations to strengthen the Convention, and took form in the negotiations of the Ad Hoc Group (AHG) between 1995 and the summer of 2001.[100]

The progress of the AHG has been well documented and its demise in 2001 remains a central point of contention among States Parties to the BTWC. As such, the issue of verification stymied agreement in 2016 at the Eighth Review Conference.[101] Why the AHG failed is complex, although the dominant explanation is that the United States denigration of the draft Protocol and decision to reject any text and outcome that the AHG might develop was the reason for the collapse of the negotiations. That assessment is factually correct, but contextually deficient. Factually correct in that the US did, on July 25, 2001, effectively end the negotiations and through its decision to remain involved in the BTWC prevented other States Parties from agreeing a Protocol without the US. Contextually deficient in that it assumes that the Protocol would have been agreed and subsequently have entered into force. Although both are reasonable assumptions, neither is certain. The fate of the Comprehensive Test Ban Treaty (CTBT) and an understanding of the fundamental differences that remained between States Parties in the summer of 2001 provide an alternative indication of a quite different future: a Protocol on paper, but not in force, or a verification agreement stripped of its most intrusive measures.[102]

The fate of the AHG is, therefore, counterfactual: but, the idea of the multilaterally negotiated, legally binding verification agreement still "enthralls some States parties, appalls others, and is used adeptly by still others for political gains" even today.[103] The dominance of the AHG in the BTWC accounts for the truncated, two week, Fourth Review Conference of 1996 which did not fail, but left many issues to the assumed outcome of a legally binding verification agreement.[104] That the effort came to nothing in the summer of 2001 means that the post-Cold War period has become a lost decade for biological disarmament. As Pearson recounted at the time, the predominant view was that "[T]here is simply no prospect of any strengthening of the biological weapons multilateral prohibition regime by any means other than the Protocol in the foreseeable future."[105] As it turned out, other means did emerge slowly over the next few years, but Pearson's perspective remains valid in part given that what has emerged over the last fifteen years has been of limited value within the BTWC itself, although not necessarily to biological disarmament as a whole.

4.5.3 Recovery, Revitalization, and Recalcitrance After 2001

Events in 2001 still effect the contemporary operation of the BTWC in a very profound way. Chronologically, three issues are of importance. First, in the seven-month period up to the end of July that saw the AHG enter its final stage of negotiations, the US decision to withdraw its support for the AHG and the process of negotiations, and the collapse of the draft Protocol. Second, the attacks on the US homeland on September 11, which killed nearly 3000 people from over 90 countries and turned a previously peripheral issue (terrorism) into the focus of US national security policy. The US, alongside allies and others who joined its various coalitions of the willing, embarked upon two wars in the three years afterwards: the first in Afghanistan, the second in Iraq. Both conflicts are still on going. The third issue is the failure of the scheduled Fifth Review Conference in December 2001. Until the last day, that review conference appeared likely to paper over disputes from the summer of 2001 and move the Convention forward with continued discussions on a variety of issues. Yet, the final afternoon involved a stand-off between the US and practically all other States Parties—including its closest allies such the UK—when the US insisted that any new programme of work had to formally replace and terminate the mandate of the AHG. Pugnacious diplomacy scuttled an emerging deal. The sole agreement was to suspend the Conference and reconvene in 2002.[106]

In the interim, a tightly crafted and spare text was agreed between key States and the reconvened Fifth Review Conference in 2002 approved what has become known as the intersessional process (ISP): meetings of experts followed by a meeting of States Parties. The central mandate of the ISPs was to "discuss and promote common understandings and effective action" on various topics as approved at subsequent review conferences.[107]

Within the BTWC, the three ISPs to date represent discrete indicators of the state of play. The first, in 2002 to 2005, was a rescue mission, intended to keep a process of dialogue open and find a way to strengthen the Convention in whatever way was possible, no matter how limited. It succeeded in setting States Parties on a path to recovery. The second, in 2007 to 2010, was a revitalization of the BTWC and in retrospect the high point of the ISP: not in terms of achievements *per se*, but in terms of operation, collegiality, and a sense of making progress to further repair the wounds of 2001 and the verification debacle that ended the AHG.[108] The third, in 2012 to 2015, was overloaded with issues, burdened by minimal time, and hamstrung by limited resources and proved to be a disappointment as revitalization met with the recalcitrant attitudes of a few States Parties.[109] Its agreement in 2011 originated from a disappointing seventh review conference; a portent of the divisions among States Parties locking them into a process that prevented going back (negotiations and verification) or moving forward to something else. After rescuing the Convention and its diplomacy in 2002, the intersessional process became a rut that States Parties have been unable to navigate their way out of.

As indicated in Table 4.1, over the course of eleven years between 2003 and 2015, when the ISP held meetings of experts and meetings of States Parties, 12 issues have been addressed. In addition, during the third ISP between 2012 and 2015, each year included three standing agenda items on national legislation, scientific and technological developments, and peaceful cooperation under Article X. The separate topics covered encompassed national legislation, security and oversight of pathogens, codes of conduct for scientists, biosafety and biosecurity regulations, education and awareness raising (Article IV), confidence building measures (Article V), investigations of alleged use (Article VI), assistance and protection in the event of actual or alleged use of biological weapons (Article VII), and peaceful cooperation among States Parties, scientific and technical exchange, disease surveillance, detection and diagnosis (Article X), as well as a general topics such as regional cooperation.

The focus of efforts has, therefore, been Articles I, IV, V, VI, VII, and X of the BTWC. Article III, on non-proliferation, was rolled into discussions on Article IV and national implementation efforts. The ISPs have been the primary area of work within the BTWC, but three review conferences have occurred in this period from 2002 to the end of 2016; the sixth (2006), seventh (2011), and eighth (2016). The Sixth Review Conference turned out to be a modest success. While it did not reshape debates in the Convention or lead to an increase in action, it contained two decisions that have had far-reaching consequences. The first was the establishment of the ISU. The second was the acceptance by States Parties of a role for other actors, such as non-governmental organizations, within implementation of the BTWC. The Seventh Review Conference built upon the second intersessional process and going into the conference in 2011, many States Parties and civil society groups were ambitious for its outcome.

The optimistic tone of the opening days was replaced by dismay in the final week following a series of proposals from Iran, India, China, Russia and Pakistan that nearly caused the conference to fail. Averted by compromise, it delivered a distinctly lesser outcome than many envisaged: "Whilst the Sixth Review Conference in 2006 had successfully demonstrated the will of the States Parties to work together to nurture and strengthen the Convention... the outcome of the Seventh Review Conference has been met with considerable disappointment because its substantive content fell so far below realistic expectations."[110] In hindsight, the difficulties experienced in 2011 should have acted as a warning to those with more ambitious agendas in 2016 and

Table 4.1 List of Topics for the BTWC Intersessional Work Programmes 2003–2015.

<table>
<tr><th>Year</th><th>Topic</th><th>Standing agenda items</th></tr>
<tr><td>2003</td><td>1. National implementation measures for the prohibitions in the convention, including penal legislation
2. National mechanisms to establish and maintain the security and oversight of pathogenic microorganisms and toxins</td><td>Not applicable</td></tr>
<tr><td>2004</td><td>1. Enhancing international capabilities for responding to, investigating and mitigating alleged use of biological or toxin weapons or suspicious outbreaks of disease
2. Strengthening and broadening national and international institutional efforts and existing mechanisms for the surveillance, detection, diagnosis and combating of infectious diseases affecting humans, animals, and plants</td><td>Not applicable</td></tr>
<tr><td>2005</td><td>1. Content, promulgation, and adoption of codes of conduct for scientists</td><td>Not applicable</td></tr>
<tr><td>2007</td><td>1. Enhance national implementation, including enforcement of national legislation, and coordination among national law enforcement
2. Regional and sub-regional cooperation on implementation of the convention</td><td>Not applicable</td></tr>
<tr><td>2008</td><td>1. Improve biosafety and biosecurity, including laboratory safety and security of pathogens and toxins
2. Oversight, education, awareness raising, and adoption and/or development of codes of conduct</td><td>Not applicable</td></tr>
<tr><td>2009</td><td>1. Enhancing international cooperation, assistance and exchange in biological sciences and technology for peaceful purposes, promoting capacity building in the fields of disease surveillance, detection, diagnosis, and containment of infectious diseases</td><td>Not applicable</td></tr>
<tr><td>2010</td><td>1. Assistance and coordination in the case of alleged use of biological or toxin weapons</td><td>Not applicable</td></tr>
<tr><td>2012</td><td>1. How to enable fuller participation in the CBMs</td><td rowspan="4">1. Strengthening cooperation and assistance under article X
2. Developments in the field of science and technology
3. Strengthening national implementation</td></tr>
<tr><td>2013</td><td>1. How to enable fuller participation in the CBMs</td></tr>
<tr><td>2014</td><td>1. How to strengthen implementation of article VII, procedures and mechanisms for the provision of assistance</td></tr>
<tr><td>2015</td><td>1. How to strengthen implementation of article VII, procedures and mechanisms for the provision of assistance</td></tr>
</table>

beyond. The Eighth Review Conference in 2016 became mired in disputes that pushed the verification issue back to the front of the agenda. While Iran has been identified as the source of the breakdown, the outcome has left many observers despondent about the prospects for the Convention in the coming years.[111]

4.5.4 Evolution in Retrospect

Over forty years, the Convention has evolved *via* small steps, and in an incremental manner.[112] It has experienced, to use Sims' pertinent observation in 1988, the vicissitudes of a treaty in force many times over its 40 plus years of life to date. Never in danger of collapse, it has nevertheless experienced many crises of confidence, diplomatic disputes, claims of non-compliance, and evolved alongside major geopolitical, scientific, and national political shifts. It has done more than simply survive these disputes and trends, but it has never thrived or been a treaty that has served as a model agreement for successful disarmament. Sims' own assessment of the BWC in 2015 noted that its main weaknesses do not lie in legal, institutional, scientific, technical, or definitional realms. Rather, because those issues have been addressed in piecemeal, the main weakness remains in deriving "a common understanding as to how to reassure one another and demonstrate that their shared commitment to biological disarmament governs what they are doing and what they allow to be done."[113] Implicit in this observation is a process of constantly evolving activity and actions that should reassure and demonstrate a commitment to biological disarmament. Historically, particularly worrisome periods of time were 1981–1986 and 2001–2005 which Sims refers to in his fortieth anniversary remarks as "bleak periods", albeit not fatal setbacks. Whether or not 2017–2021 will become another bleak period has yet to be determined.

4.6 How Scientific and Technological Developments Have Affected the BTWC

In the forty year life of the BTWC, how scientific and technological developments have affected the Convention can be illustrated with assessments 35 years apart. In the lead up to the First Review Conference in 1980, a paper on scientific and technological developments was prepared jointly by Depositary Governments (UK, USA, USSR) which noted the developments in recombinant DNA, toxin synthesis, new diseases, and industrial fermentation techniques and their development. The view of these governments was that the text of Article I was unaffected by all of these, including the ability to modify organisms and the potential for development or fabrication of substances using recombinant DNA techniques. In addition, "[A]lthough recombinant DNA techniques could facilitate genetic manipulation of micro-organisms for biological and toxin warfare purposes, the resulting agents are unlikely to have any advantages over known natural agents sufficient to provide compelling new motives for illegal production or military use in the foreseeable future. Nonetheless, developments in the ability to manipulate genetic material intentionally should be followed closely and periodically re-evaluated."[114] In a similar manner, the ability to synthesize toxins was considered "unlikely to facilitate the production of toxins in militarily significant quantities."[115] Finally, in the concluding section of the report, the Depositaries noted that "Scientific and technological developments have not created ambiguities or fundamentally new possibilities that could be exploited to violate covertly or bypass the Convention."[116]

Jumping forward to 2016, the most authoritative review of scientific and technological developments was by a non-governmental entity, the Inter Academies Panel. Although this did not preclude, or supersede, individual or joint papers by States Parties to the BTWC, the conclusions of the report provided the basis for many assessments. Notwithstanding the detailed technical assessment report, the executive summary noted that "since the Seventh Review Conference, there had been no further novel developments that could enable activities inconsistent with the aims and objectives of the BWC... [And]... Likewise, there were no developments that would not be covered by the treaty or additional supplementary understandings"[117] reached by States Parties in review conferences to date.

The abovementioned documents represent views, albeit informed, of individual States or non-governmental entities. They suggest, quite strongly, that the Convention has withstood the test of time and scientific developments. States Parties have collectively reached a consensus agreement on various issues, including scientific and technological developments, at review conferences in the Final Declarations. Such consensus views are called "additional understandings" or "agreements" which either "(a) interpret, define or elaborate the meaning or scope of a provision of the Convention; or (b) provide instructions, guidelines or recommendations on how a provision should be implemented."[118] Those related to Article I of the Convention are instructive of how States Parties have considered scientific and technological developments and how States Parties have responded to such developments. Since the ISU prepared document for the Eighth Review Conference on additional understandings includes all aspects up to 2016 it serves as a repository of decisions, understandings and agreements.

At the Second Review Conference it was determined "that the scope of Article I covers scientific and technological developments relevant to the Convention" and this generic language was refined at the Sixth, Seventh, and most recently the Eighth Review Conference when States Parties determined "that Article I applies to all scientific and technological developments in the life sciences and in other fields of science relevant to the Convention."[119] Nevertheless, specific emerging developments were also addressed, for example, between 1986 and 1996 at the Second, Third, and Fourth Review Conferences, when States Parties noted:

> "Conscious of apprehensions arising from relevant scientific and technological developments, *inter alia*, in the fields of microbiology, genetic engineering and biotechnology, and the possibilities of their use for purposes inconsistent with the objectives and the provisions of the Convention, reaffirmed that the undertaking given by States Parties in Article I applies to all such developments. The Fourth Review Conference supplemented the list of scientific and technological developments with molecular biology... and any applications resulting from genome studies."[120]

Moreover, building on consensus decisions at the Second, Third, and Fourth Review Conferences, the Sixth, Seventh, and Eighth Review Conferences also determined "that the Convention is comprehensive in its scope and that all naturally or artificially created or altered microbial and other biological agents and toxins, as well as their components, regardless of their origin and method of production and whether they affect humans, animals or plants, of types and in quantities that have no justification for prophylactic, protective or other peaceful purposes, are unequivocally covered by Article I."[121] In addition, the use of agents and toxins "in any way and under any circumstances... that is not consistent with prophylactic, protective or other peaceful purposes, is effectively a violation of Article I" and States Parties over the last decade have also reaffirmed their "determination... to condemn any use of biological agents or toxins for other than peaceful purposes, by anyone at any time."[122]

The central issue here is that while scientific and technological developments affect the BTWC, they do not circumvent or bypass the Convention. The General Purpose Criterion of Article I has remained sufficiently robust to capture any such developments. States Parties have been consistent in their assessment of the issue. These seemingly simple words are backed-up by understandings and agreements reached by consensus and repeated over time. Indeed, no scientific and technological development that would permit the development, production, stockpiling or use of any biological agent or toxin against humans, animals, or plants that is not justified for prophylactic, protective or other peaceful purposes, has been deemed to be or understood to be outside the General Purpose Criterion of the BTWC. Advances in science and technology have not made the Convention obsolete or irrelevant in terms of its scope.

This is reassuring, but alone is an insufficient bulwark against biological weapons, as the history of offensive programmes illustrates. It is also evident from the 2012–2015 ISP that some States Parties reject the view that scientific and technological developments require a more timely and urgent review within the Convention and necessitate action within and outwith the BTWC.[123]

Such views are in stark contrast to the prevailing professional and informed scientific view, as detailed in the IAP report:"Scientific advances reported at the conference could also facilitate almost every step of a biological weapons programme and technological barriers to acquiring and using a biological weapon have been conspicuously eroded since the Seventh Review Conference."[124] The central observation from such reports is that it is not Article I of the Convention or the General Purpose Criterion inherent to that it is the problem; rather, it is the challenges posed by scientific and technological developments to implementation of the BTWC at the national and international level that poses concerns. Hence, the proposal for some form of a scientific advisory mechanism for the BTWC that featured prominently in non-governmental and state party proposals for the Eighth Review Conference.[125,126]

At the national level, the additional understandings reached at review conferences related to Article IV of the Convention have called for the adoption and implementation of legislative, regulatory and administrative measures to ensure fulfillment of the obligations under the Convention, enhance domestic compliance with it, ensure both biosafety and biosecurity, enhance understanding of the prohibitions *via* education and awareness raising activities, and provide information on implementation of the BTWC to other parties. In addition, regional and other collective initiatives "towards strengthening the implementation of" the BTWC, as well as providing "assistance, upon request, to other states parties" that enhance and strengthen implementation of the Convention have been consistent within Final Declarations.[127] Although broadly successful, this approach has left gaps in coverage at the national, regional and international level because compliance with obligations is a complex, expensive, and time consuming issue. Non-compliance, for example, a lack of national implementation legislation, may not be due to purposeful activity by a State, but a result of a lack of resources, technical capacity, awareness, and differing priorities within States.[128,129]

However, offensive biological weapons activity and use of such weapons is a second order priority for most States in the world given that it is natural outbreaks of disease that pose the most pressing safety and security problem to them. The relationship between global health and international security is not a new issue. Indeed, the US noted in 2013 five sources of health security risks: the acquisition and use of biological weapons by states or non-state actors; risks posed by advances in scientific capabilities; the emergence and spread of drug resistant pathogens; globalization and the vulnerabilities posed to human health and food supply; and, newly emerging diseases.[130] These five different challenges have disease as their common connection, which should be a priority and in which fora epitomizes the dispute over the balance between disarmament and peaceful cooperation in the BTWC. And, as indicated, this dispute has become a consistent block on further progress on strengthening the BTWC.

4.6.1 Responding to Scientific and Technological Developments

In practice, scientific and technological developments have affected the BTWC in a number of ways. First, they have been subject to periodic review as required by Article XII of the Convention at each review conference. In terms of awareness and action to respond to concerns, States Parties have been proactive and have a good record of confirming the General Purpose Criterion and the scope of the BTWC. Second, has been the response beyond words *via* new means of national implementation, regional or like-minded coordination, such as common control lists for dual-use goods and materials, and attempts to promote the adoption of implementation measures. This activity occurs within the BTWC but is equally undertaken in other fora such as the Australia Group, under the United Nations Security Council

Resolution 1540 Committee, and in the World Health Organization and Global Health Security Initiative among others. Here, the record of achievement is less impressive for States Parties as a whole. Third, scientific and technological developments have brought into the realm of BTWC diplomacy and operation, many more actors beyond States. The Royal Society of the UK, the National Academy of Sciences of the United States, and the IAP have all played a role in assessing scientific and technological developments and bringing issues to the attention of states parties. Outside the Convention, the synthetic biology community has developed its own guidelines for safeguarding dual-use knowledge and information[131] in the absence of State leadership. Such activities and agreements, albeit not legally binding, have emerged despite, rather than because of, States Parties to the BTWC, given that the latter's record for action is extremely modest.

This mixed assessment provides both reassurance and concern about the future. Reassurance in that for any State Party, or States Parties collectively, to break with over 40 years of consensus understanding and confirmation that the General Purpose Criterion of Article I remains comprehensive and valid would be a significant political development. Put simply, to argue that a scientific and technological development would not be covered by the BTWC would require a State to undertake a purposeful and concerted campaign while simultaneously turning attention to the intentions of the particular State. On paper, the normative constraints and legal undertaking appear both robust and supported. However, concerns are evident. Exhortations at review conferences impose few concrete commitments for most States Parties to the BTWC. Final Declarations are not legally binding and turning consensus language from a review conference into a collective response by States Parties has never been easy and is becoming ever more difficult.

As a result, concerns about biological and toxin weapons are neither unfounded nor overblown, even if anticipated and feared advanced and newer weapons have yet to emerge or be used for hostile purposes.[132] The trends, however, are clear, as elucidated by the IAP report: the barriers to bioweapons are falling. Although scientific development is but one aspect of any decision to turn to biological weapons and that technologically deterministic approaches to the militarization of biology are simple and incomplete,[133,134] these developments have significant implications for Articles I, III and IV of the BTWC. At this stage, these cutting-edge capabilities are known to be "expensive and complicated to acquire and deploy successfully... [and]... Making use of them for prohibited purposes would probably require the resources of a State"[135] but that aspect is less reassuring in the context of the history of biological weapons programmes. First, because biological weapons programmes have remained hidden and cloaked in secrecy throughout history. Second, because as Dando noted in 2001, it would be a mistake to view the problem of scientific developments through the lens of "*current military* applications of biotechnology" without awareness of emergent capabilities to manipulate a wider range of cognitive and other functions within the human body.[136] One of the central lessons from past offensive programmes is that perceptions of feasibility effect decision making in States.

As far as is known, the various factors that have historically led States to pursue biological weapons are not converging or influencing one another to the extent that a State Party to the BTWC is considering a biological weapons programme today. However, the implications of recent (and anticipated) developments cast doubt upon the view that existing and emerging technologies do not provide a 'compelling advantage' to any actor seeking to produce them illegally, use them in a future conflict, or use them for purposes other than war in counterterrorism, peacekeeping, or other intra-state conflict, law enforcement, or non-international armed conflict environments.[137–139]

4.7 Conclusion: the BTWC in Context

In the BTWC meeting of December 2017 States Parties did agree to make progress on issues of substance and procedure for the period before the next review conference in 2021. Another ISP has been set in motion. There is no doubt that the Eighth Review Conference was a disappointment, but it was not a fatal setback to biological disarmament. Of equal note, an awareness of past offensive biological and toxin weapons programmes, when combined with the current security environment raises legitimate concerns about the future attraction of these weapons. Assessing the Convention in a longer timeframe suggests two significant recurring issues. First, and positively, the Convention has remained resilient in the face of a rapidly changing world. The normative constraints on biological weapons have not weakened significantly, as indicated by the lack of use of such weapons and the reaffirmation of the scope of the BTWC and its applicability to scientific and technological developments. Second, less positively but nonetheless reassuring, the BTWC has survived and withstood numerous crises over time and its history points to it being a treaty under constant pressure from a variety of angles throughout its life.

The failure by States Parties to push forward implementation practices collectively has also not prevented an on-going strengthening of biological disarmament by groups of States Parties acting in concert with each in other fora. This creates a patchwork of controls and mechanisms that inevitably leave some States, regions, and areas with vulnerabilities related to the potential exploitation of the life sciences for hostile purposes. However, variation in the breadth and depth of implementation of the BTWC is preferable to inaction or only lowest common denominator agreements as a result of consensus decision-making within the Convention. A slow and incremental evolution of the BTWC is the usual practice because a decision by one or very few States to prevent, block, or stymie progress is a result of short-term political views taking precedence over medium term assessments of the balance of interests of all States Parties. The coming years may simply repeat the pattern of diplomatic crisis, political compromise, and further minimal progress within the Convention as a means to withstand another 'bleak period' within the life of the BTWC. Whether or not that is sufficient to cope with the advances of science and technology and the changing nature of conflict is open to debate, but historical experience indicates minimal attention to normative constraints and an under-developed treaty will be insufficient to withstand a convergence of different elements that are affecting the norms, politics, science, and law surrounding biological disarmament. Therefore, substantive concerns about the relevance and resilience of the BTWC in future years are legitimate. Unless States Parties to the BTWC undertake concerted purposeful action in the coming decade both within, and under the Convention, as well as continue activities outside the umbrella of the Convention in terms of export controls, national implementation, assistance and protection, as well as safeguarding scientific and technical developments, the fifth decade of the BTWC's existence may witness its gradual demise into irrelevancy.

References

1. *Deadly Cultures Biological Weapons Since 1945*, ed. M. Wheelis, L. Rózsa and M. Dando, Harvard University Press, Cambridge, 2006, pp. 5–7.
2. R. M. Atlas and M. Dando, *Biosecur. Bioterror.*, 2006, **4**, 276.
3. W. S. Carus, *Health Secur.*, 2015, **13**(4), 1.
4. *Deadly Cultures Biological Weapons Since 1945*, ed. M. Wheelis, L. Rózsa and M. Dando, Harvard University Press, Cambridge, 2006.
5. W. S. Carus, *Health Secur.*, 2015, **13**(4), 1.
6. G. D. Koblentz, *Contemp. Secur. Pol.*, 2013, **34**(3), 501–525.
7. J. Tucker, *Toxic Terror*, The MIT Press, Cambridge, 2001.
8. C. Loeb, Jihadists and weapons of mass destruction, in *Jihadists and Biological and Toxin Weapons*, ed. G. Ackerman and J. Tansett, CRC Press, Boca Raton, 2009, pp. 153–172.
9. V. H. Asal, G. A. Ackerman and R. K. Rethemeyer, *Stud. Conflict Terrorism*, 2012, **35**(3), 229.
10. E. MacAskill, *The Guardian*, 2017.
11. K. C. Bansak and J. B. Tucker, Governance of emerging dual-use technologies, in *Innovation, Dual Use, and Security*, ed. J. B. Tucker, The MIT Press, Cambridge, 2012, pp. 306–329.
12. N. A. Sims, *The Diplomacy of Biological Disarmament: Vicissitudes of a Treaty in Force*, St. Martin's Press, New York, 1988.
13. D. Feakes, Global civil society and biological and chemical weapons, in *Global Civil Society Yearbook 2003*, ed. M. Kaldor, H. Anheier and M. Glasius, Oxford University Press, Oxford, 2003, pp. 87–118.
14. E. Young, *A Farewell to Arms Control?*, Penguin Books, Harmandsworth, 1972, p. 135.
15. M. Leitenberg and R. A. Zilinskas, *The Soviet Biological Weapons Program*, Harvard University Press, Cambridge, 2012.
16. N. A. Sims, *The Evolution of Biological Disarmament*, Oxford University Press, Oxford, 2001, p. 70.

17. J. Littlewood, *The Biological Weapons Convention*, Ashgate, Aldershot, 2005.
18. N. A. Sims, *Disarmament Diplomacy*, (70), 2003, http://www.acronym.org.uk/old/archive/dd/dd70/70op2.htm.
19. G. S. Pearson and N. A. Sims, *Report from Geneva, Review No. 35*, 2012.
20. M. David, E. Konovalova and C. Bertherat, *Trust and Verify*, 2017, vol. 155, p. 1.
21. J. P. Zanders, *J. Conflict Secur. Law*, 2003, **8**(2), 391.
22. J. Goldblat, *The Problem of Chemical and Biological Warfare, Volume IV CB Disarmament Negotiations, 1920–1970*, Almqvist & Wiksell, Stockholm, 1971, p. 121.
23. J. Goldblat, *The Problem of Chemical and Biological Warfare Volume IV CB Disarmament Negotiations, 1920–1970*, Almqvist & Wiksell, Stockholm, 1971, p. 17.
24. Geneva Protocol, *Protocol for the Prohibition of the Use in War of Asphyxiating, Poisonous or Other Gases, and of Bacteriological Methods of Warfare*, 1925, https://www.state.gov/t/isn/4784.htm.
25. J. Walker, *The 1925 Geneva Protocol: Export Controls, Britain, Poland and Why the Protocol Came to Include 'Bacteriological' Warfare. Harvard Sussex Program Occasional Paper Issue 5*, 2016.
26. J. Walker, *The 1925 Geneva Protocol: Export Controls, Britain, Poland and Why the Protocol Came to Include 'Bacteriological' Warfare, Harvard Sussex Program Occasional Paper Issue 5*, 2016, p. 12.
27. J. Walker, *The 1925 Geneva Protocol: Export Controls, Britain, Poland and Why the Protocol Came to Include 'Bacteriological' Warfare, Harvard Sussex Program Occasional Paper Issue 5*, 2016, p. 18.
28. W. S. Carus, *Health Secur.*, 2015, **13**(4), 1, p. 16.
29. W. S. Carus, *Health Secur.*, 2015, **13**(4), 1, p. 16–17.
30. J. Walker, *The 1925 Geneva Protocol: Export Controls, Britain, Poland and Why the Protocol Came to Include 'Bacteriological' Warfare, Harvard Sussex Program Occasional Paper Issue 5*, 2016, p. 14.
31. W. S. Carus, *Health Secur.*, 2015, **13**(4), 1.
32. R. Price, *Int. Organ.*, 1995, **49**(1), 73.
33. *Deadly Cultures Biological Weapons since 1945*, ed. M. Wheelis, L. Rózsa and M. Dando, Harvard University Press, Cambridge, 2006.
34. W. S. Carus, *Defining "Weapons of Mass Destruction," Center for the Study of Weapons of Mass Destruction, Occasional Paper No 8*, National Defense University, Washington, D.C., 2012, p. 5.
35. J. Walker, *Britain and Disarmament*, Ashgate, Aldershot, 2012, pp. 49–71.
36. J. B. Tucker, *Int. Secur.*, 2002, **27**(1), 107.
37. C. A. McLeish and R. Trapp, *Nonproliferation Rev.*, 2011, **18**(3), 527.
38. S. Croft, *Strategies of Arms Control: A History and Typology*, Manchester University Press, Manchester, 1996, pp. 20–66.
39. C. C. Flowerree, *Wash. Q.*, 1993, **13**(1), 199.
40. R. Price, *Int. Organ.*, 1995, **49**(1), 73.
41. A. Spelling, *Diplomacy & Statecraft*, 2016, vol. 27(4), p. 701.

42. A. Spelling, *Diplomacy & Statecraft*, 2016, vol. 27(4), p. 702.
43. J. B. Tucker, *Int. Secur.*, 2002, **27**(1), 107.
44. W. S. Carus, *Health Secur.*, 2015, **13**(4), 1, p. 16.
45. W. S. Carus, *Health Secur.*, 2015, **13**(4), 1, p. 16–17.
46. M. Furmanski and M. Wheelis, in *Deadly Cultures Biological Weapons since 1945*, ed. M. Wheelis, L. Rózsa and M. Dando, Harvard University Press, Cambridge, 2006, Allegations of biological weapons use, pp. 252–283.
47. M. Furmanski and M. Wheelis, in *Deadly Cultures Biological Weapons since 1945*, ed. M. Wheelis, L. Rózsa and M. Dando, Harvard University Press, Cambridge, 2006, Allegations of Biological Weapons Use, pp. 252–283.
48. M. Leitenberg, *China's False Allegations of the Use of Biological Weapons by the United States during the Korean War, CWIHP Working Paper 78*, 2016.
49. M. Wheelis and M. Sugishima, in *Deadly Cultures Biological Weapons since 1945*, ed. M. Wheelis, L. Rózsa and M. Dando, Harvard University Press, Cambridge, 2006, Terrorist Use of Biological Weapons, pp. 284–303.
50. M. Wheelis, in *Biological and Toxin Weapons: Research, Development and Use from the Middle Ages to 1945*, ed. E. Geissler and J. E. van Courtland Moon, Oxford University Press, Oxford, 1999, Biological sabotage in world war I, pp. 35–62.
51. M. Wheelis, in *Biological and Toxin Weapons: Research, Development and Use from the Middle Ages to 1945*, ed. E. Geissler and J. E. van Courtland Moon, Oxford University Press, Oxford, 1999, Biological sabotage in world war I, pp. 35–62.
52. S. Harris, in *Biological and Toxin Weapons: Research, Development and Use from the Middle Ages to 1945*, ed. E. Geissler and J. E. van Courtland Moon, Oxford University Press, Oxford, 1999, The Japanese biological warfare programme: an overview, pp. 127–167.
53. S. Harris, in *Biological and Toxin Weapons: Research, Development and Use from the Middle Ages to 1945*, ed. E. Geissler and J. E. van Courtland Moon, Oxford University Press, Oxford, 1999, The Japanese biological warfare programme: an overview, pp. 127–167.
54. W. S. Carus, *Health Secur.*, 2015, **13**(4), 1.
55. World Health Organization, *Public Health Response to Chemical and Biological Weapons*, World Health Organization, Geneva, 2nd edn, 2004, p. 35.
56. *Biological and Toxin Weapons Research, Development and Use from the Middle Ages to 1945*, ed. E. Geissler and J. E. van Courtland Moon, Oxford University Press, 1999.
57. *Deadly Cultures Biological Weapons since 1945*, ed. M. Wheelis, L. Rózsa and M. Dando, Harvard University Press, Cambridge, 2006.
58. M. Leitenberg and R. A. Zilinskas, *The Soviet Biological Weapons Program*, Harvard University Press, Cambridge, 2012, pp. 673–678.
59. J. M. van Courtland Moon, in *Deadly Cultures Biological Weapons since 1945*, ed. M. Wheelis, L. Rózsa and M. Dando, Harvard University Press, Cambridge, 2006, The US Biological Weapons Program, pp. 9–46.

60. B. Balmer, in *Deadly Cultures Biological Weapons since 1945*, ed. M. Wheelis, L. Rózsa and M. Dando, Harvard University Press, 2006, The UK Biological Weapons Program, pp. 47–83.
61. J. Hart, in *Deadly Cultures Biological Weapons since 1945*, ed. M. Wheelis, L. Rózsa and M. Dando, Harvard University Press, Cambridge, 2006, The Soviet Biological Weapons Program, pp. 132–156.
62. M. Leitenberg and R. A. Zilinskas, *The Soviet Biological Weapons Program*, Harvard University Press, Cambridge, 2012.
63. O. Lepick, in *Deadly Cultures Biological Weapons since 1945*, ed. M. Wheelis, L. Rózsa and M. Dando, Harvard University Press, Cambridge, 2006, The French Biological Weapons Program, pp. 108–131.
64. D. A. Avery, in *Deadly Cultures Biological Weapons since 1945*, ed. M. Wheelis, L. Rózsa and M. Dando, Harvard University Press, Cambridge, 2006, The Canadian Biological Weapons Program and the Tripartite Alliance, pp. 84–107.
65. G. S. Pearson, in *Deadly Cultures Biological Weapons since 1945*, ed. M. Wheelis, L. Rózsa and M. Dando, Harvard University Press, Cambridge, 2006, The Iraqi Biological Weapons Program, pp. 169–190.
66. A. E. Smithson, *Germ Gambits*, Stanford University Press, Stanford, 2011.
67. C. Gould and A. Hay, in *Deadly cultures biological weapons since 1945*, ed. M. Wheelis, L. Rózsa and M. Dando, Harvard University Press, Cambridge, 2006, The South African Biological Weapons Program, pp. 191–212.
68. Lessons from the history of biological and toxin warfare, in *Biological and Toxin Weapons Research, Development and Use from the Middle Ages to 1945*, ed. E. Geissler, J. E. van Courtland Moon, G. S. Pearson, E. Geissler and J. E. van Courtland Moon, Oxford University Press, Oxford, 1999, pp. 255–276.
69. D. A. Avery, Deadly cultures biological weapons since 1945, in *The Canadian Biological Weapons Program and the Tripartite Alliance*, ed. M. Wheelis, L. Rózsa and M. Dando, Harvard University Press, Cambridge, 2006, pp. 84–107.
70. B. Balmer, Deadly cultures biological weapons since 1945, in *The UK Biological Weapons Program*, ed. M. Wheelis, L. Rózsa and M. Dando, Harvard University Press, Cambridge, 2006, pp. 47–83.
71. E. Geissler, J. E. van Courtland Moon and G. S. Pearson, Lessons from the history of biological and toxin warfare, in *Biological and Toxin Weapons Research, Development and Use from the Middle Ages to 1945*, ed. E. Geissler and J. E. van Courtland Moon, Oxford University Press, Oxford, 1999, pp. 255–276.
72. C. Gould and A. Hay, Deadly cultures biological weapons since 1945, in *The South African Biological Weapons Program*, ed. M. Wheelis, L. Rózsa and M. Dando, Harvard University Press, Cambridge, 2006, pp. 191–212.
73. E. Geissler, J. E. van Courtland Moon and G. S. Pearson, Lessons from the history of biological and toxin warfare, in *Biological and Toxin Weapons Research, Development and Use from the Middle Ages to 1945*, ed. E. Geissler and J. E. van Courtland Moon, Oxford University Press, Oxford, 1999, pp. 255–276.

74. *Deadly Cultures Biological Weapons since 1945*, ed. M. Wheelis, L. Rózsa and M. Dando, Harvard University Press, Cambridge, 2006.
75. M. Dando, G. Pearson, L. Rózsa, J. Perry Robinson and M. Wheelis, Analysis and implications, in *Deadly Cultures Biological Weapons since 1945*, ed. M. Wheelis, L. Rózsa and M. Dando, Harvard University Press, Cambridge, 2006, pp. 355–373.
76. E. Geissler, J. E. van Courtland Moon and G. S. Pearson, Lessons from the history of biological and toxin warfare, in *Biological and Toxin Weapons Research, Development and Use from the Middle Ages to 1945*, ed. E. Geissler and J. E. van Courtland Moon, Oxford University Press, Oxford, 1999, pp. 255–276.
77. M. Dando, G. Pearson, L. Rózsa, J. Perry Robinson and M. Wheelis, Analysis and implications, *Deadly Cultures Biological Weapons since 1945*, ed. M. Wheelis, L. Rózsa and M. Dando, Harvard University Press, Cambridge, 2006, pp. 355–373.
78. N. A. Sims, *What Future for Biological Disarmament? BWC Fortieth Anniversary Event*, Geneva, http://www.unog.ch/80256EDD006B8954/(httpAssets)/5321B0C99F660EC8C1257E190050888A/$file/Statement+by+Nicholas+A+Sims.pdf.
79. A. E. Smithson, *Germ Gambits*, Stanford University Press, Stanford, 2011.
80. C. Gould and P. I. Folb, *Nonproliferation Rev.*, 2000, 7(3), 10–23.
81. J. Tucker, *Toxic Terror*, The MIT Press, Cambridge, 2001.
82. M. Ranstorp and M. Normark, *Unconventional Weapons and International Terrorism*, Routledge, London, 2009.
83. B. Hoffman, *The National Interest*, 2014, (130), pp. 61–71.
84. J. Littlewood, *The Biological Weapons Convention*, Ashgate, Aldershot, 2005, p. 17.
85. M. Chevrier and I. Hunger, *Nonproliferation Rev.*, 2000, 7(3), 24–42.
86. E. Geissler, Agreed measures and proposals to strengthen the convention, in *Strengthening the Biological Weapons Convention by Confidence-building Measures*, ed. E. Geissler, Oxford University Press, Oxford, 1990, pp. 43–70.
87. F. Lentzos, *Bull. At. Sci.*, 2011, **67**(3), 26–33.
88. N. A. Sims, *Disarmament Diplomacy*, 2001, vol. 58, pp. 15–21, http://www.acronym.org.uk/old/archive/dd/dd58/58sims.htm.
89. N. A. Sims, *The Evolution of Biological Disarmament*, Oxford University Press, Oxford, 2001, pp. 119–131.
90. J. Littlewood, *The Biological Weapons Convention*, Ashgate, Aldershot, 2005, pp. 164–169.
91. M. Moodie, *Wash. Q.*, 1995, **18**(2), 181–202.
92. R. J. Mathews, *CBW Conv. Bull.*, 2004, **66**(2004), 1–4.
93. N. A. Sims, *CBW Conv. Bull.*, 2001, **53**, 3–5.
94. J. Littlewood, The biological weapons convention, in *Unfinished Business*, ed. G. O. Faure, University of Georgia Press, Athens, 2012, pp. 107–129.

95. United Nations, *Additional Understandings and Agreements Reached by Previous Review Conferences Relating to Each Article of the Convention, Background Information Document Submitted by the Implementation Support Unit BWC/CONF.VIII/C/4*, 2016.
96. M. R. Dando, *Biological Warfare in the 21st Century: Biotechnology and the Proliferation of Biological Weapons Brassey's*, London, 1994.
97. M. Meselson, *CBW Conv. Bull.*, 2000, **48**, 16–19.
98. M. Dando, *The New Biological Weapons*, Lynne Rienner, Boulder, 2001.
99. M. R. Dando, *Neuroscience and the Future of Chemical-biological Weapons*, Palgrave Macmillan, Basingstoke, 2015.
100. J. Littlewood, *The Biological Weapons Convention*, Ashgate, Aldershot, 2005.
101. R. Guthrie, *The Eighth BWC Review Conference: A Minimal Outcome, Bio-Weapons Prevention Project, RevCon Report 16*, 2016, http://www.cbw-events.org.uk/RC16-16.pdf.
102. M. Chevrier, The biological weapons convention: the protocol that almost was, in *Verification Yearbook 2001*, ed. T. Findlay and O. Meier, VERTIC, London, 2001, pp. 79–97.
103. J. Littlewood, *Arms Control Today*, 2007, **37**(2), 103.
104. M. Weston, *NATO Rev.*, 1997, **45**(3), 33–35.
105. G. S. Pearson, *CBW Conv. Bull.*, 2001, **53**, 6–9.
106. G. S. Pearson, *CBW Conv. Bull.*, 2001, **54**, 13–26.
107. United Nations, *Final Document: Fifth Review Conference of the States Parties to the Convention on the Prohibition of the Development, Production and Stockpiling of Bacteriological (Biological) and Toxin Weapons and on Their Destruction, BWC/CONF.V/17*, United Nations, Geneva, 2002, p. 3.
108. *Improving Implementation of the Biological Weapons Convention: The 2007–2010 Intersessional Process*, ed. P. Millett, United Nations Publications, Geneva and New York, 2011, pp. 1–9.
109. Time for structural changes to make the BTWC more effective, in *Key Points for the Eighth Review Conference*, ed. G. S. Pearson, N. A. Sims and M. R. Dando, University of Bradford, Bradford, 2016, pp. 3–17.
110. G. S. Pearson and N. A. Sims, *The Biological Weapons Convention Seventh Review Conference: A Modest Outcome, Report from Geneva, Review No. 35*, 2012, p. 36, http://www.sussex.ac.uk/Units/spru/hsp/Reports%20from%20Geneva/HSP%20Reports%20from%20Geneva%20Special%20March%202012.pdf.
111. R. Guthrie, *The Eighth BWC Review Conference: A Minimal Outcome, Bio-Weapons Prevention Project, RevCon Report 16, December 1*, 2016, http://www.cbw-events.org.uk/RC16-16.pdf.
112. N. A. Sims and J. Littlewood, *Nonproliferation Rev.*, 2011, **18**, 499–511.
113. N. A. Sims, *What Future for Biological Disarmament?, BWC Fortieth Anniversary Event, March 30*, Geneva, http://www.unog.ch/80256EDD006B8954/(httpAssets)/5321B0C99F660EC8C1257E190050888A/$file/Statement+by+Nicholas+A+Sims.pdf.

114. United Nations, *Background Paper Relating to the Convention on the Prohibition of the Development, Production and Stockpiling of Bacteriological (Biological) and Toxin Weapons and on Their Destruction, BWC/CONF.I/4 Note by Secretariat 20 February*, United Nations, Geneva, 1980, p. 8.
115. United Nations, *Background Paper Relating to the Convention on the Prohibition of the Development, Production and Stockpiling of Bacteriological (Biological) and Toxin Weapons and on Their Destruction, BWC/CONF.I/4 Note by Secretariat 20 February*, United Nations, Geneva, 1980, p. 12.
116. United Nations, *Background Paper Relating to the Convention on the Prohibition of the Development, Production and Stockpiling of Bacteriological (Biological) and Toxin Weapons and on Their Destruction, BWC/CONF.I/4 Note by Secretariat 20 February*, United Nations, Geneva, 1980, p. 18.
117. Interacademy Partnership, *The Biological and Toxin Weapons Convention: Implications of Advances in Science and Technology, Conference Report, DES4023_2 December*, 2015, p. 7.
118. United Nations, *Additional Understandings and Agreements Reached by Previous Review Conferences Relating to Each Article of the Convention. Background Information Document Submitted by the Implementation Support Unit, BWC/CONF.VIII/C/4 31 May*, United Nations, Geneva, 2016, p. 1.
119. United Nations, *Additional Understandings and Agreements Reached by Previous Review Conferences Relating to Each Article of the Convention. Background Information Document Submitted by the Implementation Support Unit, BWC/CONF.VIII/C/4 31 May*, United Nations, Geneva, 2016.
120. United Nations, *Additional Understandings and Agreements Reached by Previous Review Conferences Relating to Each Article of the Convention. Background Information Document Submitted by the Implementation Support Unit, BWC/CONF.VIII/C/4 31 May*, United Nations, Geneva, 2016, pp. 4–5.
121. United Nations, *Additional Understandings and Agreements Reached by Previous Review Conferences Relating to Each Article of the Convention. Background Information Document Submitted by the Implementation Support Unit, BWC/CONF.VIII/C/4 31 May*, United Nations, Geneva, 2016, p. 4.
122. United Nations, *Additional Understandings and Agreements Reached by Previous Review Conferences Relating to Each Article of the Convention. Background Information Document Submitted by the Implementation Support Unit, BWC/CONF.VIII/C/4 31 May*, United Nations, Geneva, 2016, p. 4.
123. M. Dando, *To what Extent Was the Review of Science and Technology Made More Effective and Efficient at the 2013 Meeting of States Parties?, Policy Paper 5 BioChemical Security 2030 Project (May)*, University of Bath, Bath, 2014, p. 1.
124. Interacademy Partnership, *The Biological and Toxin Weapons Convention: Implications of Advances in Science and Technology, Conference Report, DES4023_2 (December)*, 2015, p. 7.

125. Interacademy Partnership, *The Biological and Toxin Weapons Convention: Considerations for a Science Advisory Mechanism, Meeting Summary, DES4395 (July)*, 2016.
126. United Nations, *Frameworks for Effective Oversight of Scientific Research Facilities and Awareness of Dual-use Risks Submitted by Canada, BWC/CONF.VIII/PC/WP.25 9 August*, United Nations, Geneva, 2016.
127. United Nations, *Additional Understandings and Agreements Reached by Previous Review Conferences Relating to Each Article of the Convention. Background Information Document Submitted by the Implementation Support Unit, BWC/CONF.VIII/C/4 31 May*, United Nations, Geneva, 2016, pp. 7–10.
128. A. Chayes and A. H. Chayes, *Int. Organ.*, 1993, **47**(2), 175–205.
129. J. Enia and J. Fields, *Nonproliferation Rev.*, 2014, **21**(1), 43–64.
130. United States, *Statement by Christopher Buck, Head of Delegation, United States of America Meeting of the States Parties to the Biological Weapons Convention*, United Nations, Geneva, 2013.
131. L. P. Knowles, Current dual-use governance measures, in *Innovation, Dual Use and Security*, ed. J. B. Tucker, MIT Press, Cambridge, 2012, pp. 45–66.
132. J. B. Petro, T. R. Plasse and J. A. McNulty, *Biosecur. Bioterror.*, 2003, **1**(3), 161–168.
133. S. B. Ouagrham-Gormley, *Int. Secur.*, 2012, **36**(4), 80–114.
134. K. M. Vogel, *Bull. At. Sci*, 2013, **69**(1), 43–52.
135. Interacademy Partnership, *The Biological and Toxin Weapons Convention: Implications of Advances in Science and Technology, Conference Report, DES4023_2 (December), 2015*, 2015.
136. M. Dando, *The New Biological Weapons*, Lynne Rienner, Boulder, 2001, p. 157.
137. A. Pearson, *Nonproliferation Rev.*, 2006, **13**(2), 151–188.
138. M. R. Dando, *Neuroscience and the Future of Chemical-biological Weapons*, Palgrave Macmillan, Basingstoke, 2015.
139. M. Crowley, *Chemical Control*, Palgrave Macmillan, Basingstoke, 2016.

CHAPTER 5

United Nations Mechanisms to Combat the Development, Acquisition and Use of Chemical Weapons

MICHAEL CROWLEY

University of Bradford, Bradford, West Yorkshire, BD7 1DP, UK
*E-mail: m.j.a.crowley1@bradford.ac.uk

5.1 Introduction: Role of the United Nations in Maintaining International Peace and Security

Article 1 of the founding Charter of the United Nations (UN), adopted on 26 June 1945, determines that a central purpose of the organisation will be to "maintain international peace and security, and to that end: to take effective collective measures for the prevention and removal of threats to the peace, and for the suppression of acts of aggression or other breaches of the peace."[1]

The UN body with primary responsibility in these matters, on a day-to-day basis, is the UN Security Council (UNSC), which under Article 39 of the UN Charter, shall "determine the existence of any threat to the peace, breach of the peace, or act of aggression and shall make recommendations, or decide what measures shall be taken ... to maintain or restore international peace and security."[2] In adopting the UN Charter, all States have agreed (under Article 24) that on issues of international peace and security, the Security

Preventing Chemical Weapons: Arms Control and Disarmament as the Sciences Converge
Edited by Michael Crowley, Malcolm Dando and Lijun Shang

Published by the Royal Society of Chemistry, www.rsc.org

Council acts on their behalf, and they have also agreed to be bound by its resolutions (Articles 2, 25, and 49).[3]

Under Article 41, the Security Council may decide what measures not involving the use of armed force are to be employed to give effect to its decisions, and it may call upon UN Members to apply such measures.[4] Article 41 provides a non-exhaustive list of potential sanctions which "may include complete or partial interruption of economic relations… and the severance of diplomatic relations."[4]

In certain extreme circumstances, *i.e.* where the Security Council considers that measures provided for in Article 41 would be inadequate or have proved to be inadequate, it may under Article 42 of the UN Charter, "take such action … as may be necessary to maintain or restore international peace and security." Such action "may include demonstrations, blockade, and other operations by air, sea, or land forces of [UN] Members."[5]

Although the UNSC has primary responsibility for addressing such matters, the UN General Assembly (UNGA) is also empowered under the UN Charter to: "consider and make recommendations on the general principles of cooperation for maintaining international peace and security, including disarmament"; and in addition may "discuss any question relating to international peace and security and, except where a dispute or situation is currently being discussed by the Security Council, make recommendations on it."[6] Furthermore, the UN Charter also enables both the UNGA and the UN Secretary-General (UNSG) to bring to the attention of the Security Council any matter which may threaten the maintenance of international peace and security.[6,7] Under its Charter, the UN, and the UNSC in particular, have been given great potential flexibility regarding what they can consider to be a threat or potential threat to "international peace and security", when they should intervene, and what mechanisms they can employ to address such threats.[8]

In addition to responding to immediate threats to "international peace and security" the UNSC was enjoined under Article 26 of the UN Charter to establish a wide-ranging "system for the regulation of armaments."[9] Part of this was what, by 1979, had become the Conference on Disarmament (CD) – intended to serve as the international community's single multilateral arms control and disarmament negotiating forum.[10] One of the CD's most significant successes was the negotiation of the Chemical Weapons Convention (CWC) and the consequent establishment of the Organisation for the Prohibition of Chemical Weapons (OPCW).

Although the CWC is the principle international legally binding agreement for combatting the development, acquisition, stockpiling or use of chemical weapons, it is clear from examination of relevant international law and State practice that concerns regarding such activities conducted by either State or non-State actors would constitute potential or actual threats to "international peace and security", and fall under the auspices of the UN.

Indeed, the importance of the UN in this regard is explicitly recognised in the CWC, particularly in relation to the resolution of doubts or concerns about compliance or in cases of non-compliance with the Convention.

The CWC has mechanisms enabling, and in fact obliging, States Parties and relevant treaty structures (*i.e.* the Executive Council and the Conference of States Parties) in cases of "particular gravity and urgency" to bring the issue to the attention of the UNGA and the UNSC.[11] If any CWC State Party decided to withdraw from the Convention it is required to notify the UNSC, as well as the UNSG as the depositary, in addition to the CWC States Parties and Executive Council, 90 days in advance of such action.[12] Furthermore, the UN would have primary responsibility for addressing any cases of alleged chemical weapons use involving States not Party to the CWC, though the actual investigation on the ground would be conducted by the OPCW.

This chapter will firstly explore mechanisms that currently exist under the auspices of the UN Secretary-General and those under the UNSC that can be employed to investigate the suspected use of chemical weapons by either State or non-State actors. The chapter will then examine UN mechanisms that can be employed in response to evidence of the development, acquisition or use of chemical weapons by State or non-State actors. The chapter will present an overview of the relevant instrument and associated regime, and then examine potential benefits, difficulties and limitations in their employment.

5.2 Investigatory Mechanisms

Two types of mechanism have been established, under UN auspices, for exploring reported allegations of chemical (and biological) weapons use: (a) the UN Secretary-General's Mechanism (UNSGM): a standing mechanism which is generally applicable for the investigation of all incidents of chemical (and biological) weapons use in all countries at all times; (b) *Ad hoc* country specific investigations – established by a discrete UN Resolution, and limited in time and locality.

5.2.1 UN Secretary-General's Mechanism

On 12 December 1980, the UNGA granted the UN Secretary-General a mandate under Resolution 35/144C to investigate allegations of chemical weapons use.[13] On 13 December 1982, Resolution 37/98D broadened this mandate to include "activities that might constitute a violation of the Geneva Protocol", and thus extended the scope of this authority to cover allegations of the use of biological and toxin weapons as well as chemical weapons.[14] Resolution 42/37C, 1987, expanded the mandate further still to encompass "the possible use of chemical and bacteriological (biological) or toxin weapons that may constitute a violation of the 1925 Geneva Protocol or other relevant rules of customary international law."[15] This Resolution also empowered the UN Secretary-General to launch, on his or her own authority, an immediate field investigation of any credible complaint of alleged chemical or biological weapons use.[15,16] Consequently, as Tucker has argued, the UN

Secretary-General became "the sole arbiter of which allegations to investigate and the level of effort devoted to each investigation."[17] Prior to this, a specific UNGA or UNSC Resolution formally requesting the Secretary-General to launch an investigation had been considered necessary by some. The mandate given to the Secretary-General, as enunciated in the 1987–88 Resolutions, was subsequently reaffirmed in UNGA Resolutions each year up to 1991.[18] It did not include an expiration clause and therefore continues to be in force to this day. Since 1993, however, the Mechanism had lain dormant, until it was once again employed in 2013 to investigate chemical weapons use in Syria (see Section 5.2.1.2).

5.2.1.1 Strengthening the UNSGM

In his 2006 report, Littlewood described how the UNSGM had "been allowed to atrophy".[19] At that time, the Mechanism lacked dedicated investigation equipment or resources and relied on that provided by States, international organizations or other UN bodies; lists of experts were irregularly updated; there was no collective training of potential investigatory team members; investigation guidelines and procedures had not been updated; developments in detection capabilities, inspection protocols and operating methodologies of investigation teams in other environments were not formally reviewed; laboratories were not tested and calibrated with one another to ensure accurate identification of chemical, biological and toxin agents and there was no direct line of funding to cover the costs of any investigation.[19]

However, from the mid-2000s, support for revivifying the UNSGM subsequently grew. In 2007, a stock-taking and overhaul of the UNSGM was undertaken by a group of experts from interested UN Member States and representatives of international organizations, including the World Organisation for Animal Health (OIE), the World Health Organisation (WHO) and the International Criminal Police Organisation (INTERPOL). Whilst the UN, had in 2001, already agreed a memorandum of understanding (MOU) with the OPCW;[20] similar MOUs were signed in January 2011 with the WHO for investigating the alleged use of chemical, biological and toxin weapons; and with the OIE in 2012.[21]

Efforts were made to develop an international network of diagnostic and analytical laboratories that could be utilised by the UNSGM to identify the nature of potential biological or chemical agents, and ensure they were fit for purpose. By August 2015, this network consisted of 42 laboratories nominated by Member States, including one mobile laboratory for field deployment.[21] Similarly, the roster of chemical and biological experts that could be called upon to conduct investigations had been updated, and by August 2015, included more than 240 experts in total.[21] A series of training courses, table top and full field exercises for selected experts have been conducted, in close cooperation with specialized international organizations, including the OPCW, WHO, OIE and INTERPOL.[22] In addition, the first integrated training of selected experts from the UNSGM roster and staff of the OPCW took place in September 2015.[22] The importance of continuing these activities so as to ensure that the UNSGM is able to respond quickly and effectively when requested, has been made starkly apparent following the wide spread reports of the use of chemical weapons against civilians and combatants in Syria.

5.2.1.2 Case Study: UNSGM Investigation of Chemical Weapons Attacks in Syria

From April 2012 there were reports—which increased throughout 2013—that the Syrian Government armed forces and/or the armed opposition groups had employed a range of toxic chemicals as weapons during the ongoing civil war. These reports came from a variety of sources including non-governmental organisations (NGOs), press interviews with activists and defectors, media reports, online postings of "on the ground" videos and photographs, as well as public statements and briefings by officials and representatives from a number of States.[23]

On 19 March 2013, the Syrian Government reported the alleged use of chemical weapons against its forces in the Khan Al-Asal area of the Aleppo Governorate.[24] The following day, Syria formally requested that the UN Secretary-General launch an urgent investigation under the auspices of his Mechanism.[25] The UN Secretary-General agreed to do so and contacted the OPCW and the WHO requesting their cooperation in mounting an investigation.[26]

Other governments—specifically France and the UK—came forward alleging the same and other incidents of chemical weapons use and the Secretary-General announced that all credible allegations would be investigated.[27,28] A special team for this purpose was put together comprising qualified experts from the OPCW and WHO.[29] The team was dispatched to investigate three of the reported incidents, including Khan al-Asal, arriving in Damascus on 18 August 2013. In the early morning of 21 August 2013, reports emerged that two districts of the Ghouta area of Damascus had been attacked with chemical weapons.[30] Although there were widely varying initial estimates, it became quickly clear that the attacks had resulted in large numbers of casualties and fatalities. A preliminary US Government assessment released on 30 August 2013 "determined that 1429 people were killed in the chemical weapons attack, including at least 426 children."[31]

Subsequently, "dozens" of UN Member States wrote to the Secretary-General requesting an urgent investigation into this incident making use of the Mission already in Syria.[32] On 22 August, the Secretary-General determined that the investigation team should, as a priority, investigate this incident. On 25 August, permission was obtained from the Syrian Government and the investigation mission commenced its on-site work the following day.[33]

Following its investigations, the UN Mission produced two reports; the first (interim) report was restricted to "ascertain[ing] the facts" related to "the alleged use of chemical weapons in the Ghouta area of Damascus on 21st August 2013".[34] It concluded that "on 21 August 2013, chemical weapons have been used in the ongoing conflict between parties in the Syrian Arab Republic, also against civilians, including children, on a relatively large scale."[35] The report further concluded that "the environmental, chemical and medical samples we have collected provide clear and convincing evidence that surface-to-surface rockets containing the nerve agent sarin were used in Ein Tarma, Moadamiyah and Zamalka in the Ghouta area of Damascus."[36] The second (final) report,[24] in addition to documenting the Ghouta attacks, also stated that the UN Mission collected "credible" information corroborating allegations of the use of a chemical weapon—apparently an organophosphorous compound—against soldiers and civilians in Khan Al-Asal on 19 March 2013. The report also documented alleged chemical weapons attacks in Jobar, Saraqueb, Ashrafiah Sahnaya, Bahhaariyeh and Sheik Maqsood.

The UNSGM did not identify the perpetrators of the chemical weapons attacks in Syria. However, by documenting a wealth of witness testimony and by collecting and subsequently conducting verifiable laboratory analysis of environmental and physiological samples, it was able to provide evidence that chemical weapons had been employed in Syria on a "relatively large scale". Such evidence, gathered by an objective investigation mission acting under the UN Secretary-General's auspices, could not easily be dismissed by any side to the conflict. Its findings galvanised the international community to act and played an important role in facilitating Syria's rapid accession to the CWC and the subsequent dismantling and destruction of Syria's declared chemical weapons programme and stockpiles.

5.2.1.3 *Effectiveness of the UNSGM*

Under the initial 1980 mandate, reaffirmed in 1981, the Secretary-General had authority to investigate allegations brought to him from any source of chemical weapons use. However, in 1982, the Secretary-General's authority was limited to investigating allegations raised by UN Member States, and this remains the position today. Consequently, non-State actors such as armed opposition groups or indeed civilian communities that have been the targets of an attack utilising toxic chemicals cannot directly call upon the UN Secretary-General to initiate an investigation.

One indirect route by which affected communities or other civil society actors could, in theory, attempt to trigger the UNSGM is by "request[ing] assistance from the United Nations, where a plea for aid could be presented to the President of the UN Security Council or to the UN General Assembly (UNGA)."[37] To date, however, no successful attempt has been made to trigger the Mechanism in this manner. A potentially more effective route would be for civil society to attempt to alert a receptive UN Member State and seek its assistance in bringing their allegations before the attention of the UN Secretary-General. In the case of Syria, the photographic and video footage recorded by affected communities and first responders which was widely and quickly distributed through social media, appears to have been an important factor in alerting and shaming the international community into action.

Although the UNSGM can be initiated by the UN Secretary-General at the request of a single Member State and does not require any further body—such as the Security Council or General Assembly—to sanction such action; in reality, whether this mechanism would be triggered would probably be influenced by a variety of political factors. Such factors could include the identity of the State or States requesting use of the Mechanism, the State allegedly employing the chemical weapon (and its allies), the target of such action, the circumstances surrounding the alleged use, as well as the perceived strength, independence and freedom of action of the UN Secretary-General at the time. A similar mix of political considerations would, no doubt, influence the effectiveness of the resulting investigations, and also, whether and how the UNSC or the UNGA acted upon the findings of such investigations.

Such political considerations have exacerbated the range of bureaucratic, logistical and methodological challenges faced by previous UNSGM investigations. For example, excessive delays in triggering the Mechanism, and establishing and dispatching investigatory teams, have led to concerns regarding the potential deterioration or contamination of evidence before UN investigation teams could reach sites of alleged chemical weapons attack. On other occasions, access to such sites has been limited or denied entirely.

5.2.1.4 Scope of Coverage

Since the UN Resolutions establishing and enhancing the UNSGM did not define chemical or biological weapons, or seek to list the potential agents covered, the Mechanism would clearly have a very wide scope of coverage in terms of the types of suspected agents it could investigate. Indeed, this would reflect the reality in many cases, where prior to an investigation, the nature of the agent employed may well be unknown and the circumstances surrounding its use may very likely be contested.

The UNSGM is designed to investigate the suspected use of chemical or biological agents in "violation of the 1925 Geneva Protocol or other relevant rules of customary international law."[38] Consequently, it can be argued that in addition to the suspected use of chemical or biological agents in armed conflict, the UNSGM can potentially be employed to investigate the possible use of such agents that may constitute a violation of "other relevant rules of customary international law" in other circumstances. However, the range of "relevant" rules of customary international law has not been determined by the UNGA or the UNSC. A strong case can, however, be made for its use in the suspected employment of chemical or biological agents, to commit or facilitate serious human rights abuses, whether they be in the context of international armed conflict, internal armed conflict, or in the absence of conflict.

Whilst the UNSGM could be employed to investigate the *use* of chemical and biological weapons, the Mechanism could not be utilised to study the development, stockpiling, deployment or transfer of such weapons, where no use has taken place. There may however be some room for the UNSGM to investigate such activities where they bear directly upon questions of use.

Although the UNSGM has previously been employed to investigate the use of chemical weapons, since the coming into force of the CWC in 1997, the OPCW would now take the lead in cases of the alleged use of chemical and toxin weapons against a CWC State Party. In such circumstances, and following a request from the CWC State under attack or a second CWC State Party, the OPCW could initiate the CWC investigation and assistance procedures.[39] However, the UNSGM provides a possible fall-back option which could potentially be used in certain situations where the CWC investigation and assistance mechanisms were not initiated.

One situation where the UNSGM could be triggered is when toxic chemicals have reportedly been used by and/or against CWC non-State Parties. In such cases—as recognised in the CWC Verification Annex—the UN Secretary-General would take the lead and "the Organization [OPCW] shall closely cooperate... If so requested, the Organization shall put its resources at the disposal of the [UN] Secretary-General."[40] In 2001, the OPCW and the UN further formalised and strengthened cooperation in such cases, through agreement of a memorandum of understanding.[41] Although such scenarios may initially appear unlikely given the large majority of States now party to the CWC (192), there continue to be some particular hold-out States of concern, namely Egypt, Israel and North Korea. Furthermore, in the future the importance of the UNSGM may increase if any of the existing CWC States Parties decide to withdraw from the Convention.

5.2.2 *Ad hoc* UN Investigatory Mechanisms

In addition to the UNSGM, there is potential for the UNSC or UNGA to establish *ad hoc* country-and-incident-specific investigations into the alleged use of chemical weapons in order to identify the perpetrators of such activities. From 1981 till 1989, *ad hoc* UN investigations were authorised into alleged chemical/toxin weapons use by: Vietnam in Cambodia/Kampuchia, Laos and Thailand (1981–82), the Soviet Union in Afghanistan (1982), Iraq and Iran, as well as the use by Iraq against the Kurdish population within its borders (1984–88). Following the formal establishment of the UNSGM in 1989, there has been only one subsequent UNSC authorised *ad hoc* investigation – that of the OPCW–UN Joint Investigative Mechanism (JIM) in Syria.

5.2.2.1 OPCW–UN Joint Investigative Mechanism (JIM) in Syria

Despite Syria's accession to the CWC and subsequent destruction of all declared Syrian chemical weapons, widespread reports of the use of weaponised toxic chemicals continued. In September 2014, a fact-finding mission (FFM) established by the OPCW Director General found evidence constituting "compelling confirmation" that a toxic chemical was used "systematically and repeatedly" as a weapon in villages in northern Syria in 2014.[42] The FFM report stated that "the descriptions, physical properties, behaviour of the gas, and signs and symptoms resulting from exposure, as well as the response of patients to the treatment, [led] the FFM to conclude with a high degree of confidence that chlorine, either pure or in mixture, is the toxic chemical in question."[42]

The 2013 UNSGM investigation and subsequent OPCW FFM investigations both provided some indication of the identity of those responsible for the respective chemical weapons attacks. The UNSGM report stated that samples collected in the Ghouta area provided "clear and convincing evidence that surface-to-surface rockets containing the nerve agent sarin were used";[34] the rocket dimensions suggested that they were 330 mm surface-to-surface artillery rockets.[34] Subsequently, the OPCW FFM reports referred to eyewitness testimony of helicopters dropping barrel bombs containing chlorine. In both cases, the means of delivery were known to be possessed by the Syrian military, but not the armed opposition forces. However, neither the UNSGM nor the OPCW FFM was mandated to investigate and identify the perpetrators of these crimes, nor those who facilitated them. Consequently, accusations of culpability were rife – with a number of States, notably the US, accusing the current Syrian government of orchestrating these attacks, whilst the Russian Federation blamed the insurgents. Meanwhile, reports of further attacks employing toxic chemicals continued.

On 7 August 2015, the UNSC unanimously adopted Resolution 2235 (2015),[43] which established a Joint Investigative Mechanism (JIM) of the UN and OPCW, initially for one year, to identify "to the greatest extent feasible" those responsible for the "use of chemicals as weapons" in Syria. Under UNSCR 2235, the Council also reaffirmed it would impose measures under Chapter VII of the UN Charter in response to violations of Resolution 2118 (2013) which had determined that chemical weapons use anywhere constituted a threat to international peace and security.[44]

The JIM was managed by a Leadership Panel and consisted of a political office and planning/operations office based in New York; an investigative office based in The Hague focusing on chemical and medical analysis, forensics, military ordnance analysis and other relevant information analysis; and a liaison office in Damascus which acted as the main focal point for the Syrian Government. At its full capacity, the Mechanism's core staff consisted of 22 professional staff from 16 countries. It also had seven consultants, including translators.[†]

[†]For details of the JIM and its activities see: United Nations, Security Council, First report of the Organization for the Prohibition of Chemical Weapons–United Nations Joint Investigative Mechanism, (OPCW–JIM), S/2016/142, 12 February 2016; UNSC, 2nd JIM Report, S/2016/530, 10 June 2016; UNSC, 3rd JIM Report, S/2016/738, 24 August 2016; UNSC, 4th JIM Report, S/2016/888, 21 October 2016; UNSC, 5th JIM Report, S/2017/131, 13 February 2017; UNSC, 6th JIM Report, S/2017/552, 28 June 2017; UNSC 7th JIM Report, S/2017/904, 26 October 2017.

The JIM delivered seven reports to the UNSC,[†] the first two concentrating upon its methodological approaches and its investigative activities. According to its first report, the JIM considered "the starting point of the investigation to stem from the [OPCW] fact-finding mission information". Accordingly, priority was given initially to reviewing information and evidence obtained by the FFM.[45] In addition, the JIM also gathered information from: the Syrian Government and all parties to the Syrian conflict; other UN Member States; international organizations, international and national non-governmental organizations, other entities and individuals; and open sources.[46]

A total of 116 alleged incidents of the use of chemicals as weapons were recorded in the FFM reports and annexes. The OPCW FFM investigated 29 of these incidents and in 23 "confirmed with outmost confidence exposure to a chemical substance".[47] Taking into account severity (deaths, casualties, *etc.*), the delivery method and munition, and the quality and quantity of data and information about the incidents,[48] the JIM consequently selected nine of these incidents for further investigation.[49]

In its Third Report, the JIM concluded that in three incidents there was "sufficient information to reach a conclusion on the actors involved." According to the JIM, the Talmenes attack on 21 April 2014, "was caused by a Syrian Arab Armed Forces helicopter dropping a device … and was followed by the release of [an unidentified] toxic substance that affected … large numbers of people";[50] the attack on 16 March 2015 "was caused by a Syrian Arab Armed Forces helicopter dropping a device that hit the house and was followed by the release of a toxic substance, matching the characteristics of chlorine, that was fatal to all six occupants. The remnants of the device [were] consistent with the construction of a barrel bomb."[51] In addition, the JIM concluded that Islamic State in Iraq and the Levant (ISIL) had conducted an attack on Marea on August 2015 using several artillery shells filled with sulfur mustard.[52] Subsequently, as a result of further investigation, in its Fourth Report, the JIM was able to conclude that during an attack on Qmenas on 16 March 2015, a "Syrian Arab Armed Forces helicopter dropp[ed] a device from high altitude, which hit the ground and released the toxic substance that affected the population."[53]

In its Fifth and Sixth Reports, the OPCW–UN JIM largely updated the UNSC on operational issues and its ongoing inquiries. In contrast, the Seventh Report presented the OPCW–UN JIM's findings following its analysis of evidence compiled by the OPCW FFM and its own "comprehensive investigation" into two chemical attacks. The OPCW–UN JIM declared that it was "confident that the Syrian Arab Republic [was] responsible for the release of sarin at Khan Shaykhun on 4 April 2017."[54] It also concluded that "ISIL (Islamic State of Iraq and the Levant) [had been] responsible for the use of sulfur mustard at Umm Hawsh on 15 and 16 September 2016."[55]

Regrettably, on 24 October 2017 and again on 16 November 2017, Russia vetoed UN Security Council Resolutions that would have extended the lifespan of the JIM for another year. Consequently the Mechanism's mandate expired on 17 November 2017 and it began to archive its documents and liquidate its assets.[56] Unfortunately, as of April 2018 all attempts to revive the JIM have failed due to lack of consensus among UNSC members.

5.2.2.2 Scope of the Mechanism

In comparison with previous OPCW–UN and OPCW investigations in Syria, the JIM was granted a relatively broad scope of coverage. Firstly, under UNSCR 2235, the JIM was mandated to investigate "the use of chemicals as weapons, including chlorine or any other toxic chemical". Consequently, it was not restricted to attacks employing chemical agents previously declared by Syria as part of its chemical weapons programme; instead this essentially open-ended scope allowed the JIM to investigate attacks employing *inter alia* toxic industrial chemicals, riot control agents, malodorants and incapacitating chemical agents, in addition to those chemicals listed in the Schedules of the CWC.

Similarly, the range of activities that could be investigated by the JIM and the range of actors that could potentially be found culpable was surprisingly broad. UNSCR 2235 enjoined the JIM to "identify to the greatest extent feasible individuals, entities, groups, or governments who were perpetrators, organizers, sponsors or otherwise involved in the use of chemicals as weapons…"[57] In its first report, the JIM underlined the importance of investigating the full range of activities under its mandate:

> "The [JIM] Leadership Panel strongly believes that all individuals, groups, entities or Governments that have any role in enabling the use of chemicals as weapons, for whatever reason and under any circumstances, must understand that they will be identified and made accountable for these abhorrent acts. As such, the Panel hopes that its investigation will act as a deterrent and contribute to the elimination of the use of toxic chemicals as weapons."[58]

Consequently, in its Fourth report, the JIM detailed its attempts to determine the identities of those in command and control of the helicopters—apparently from the 63rd Helicopter Brigade—involved in the three confirmed attacks carried out by the Syrian armed forces.[60] And stated that: "those with effective control in the military units … must be held accountable. Accountability must also be applied to others responsible for the use of chemicals as weapons in the Syrian Arab Republic."[62]

However, UNSCR 2235 did impose a severe temporal restriction upon the scope of the JIM investigations, namely to cases solely where the OPCW FFM "determines or has determined that a specific incident in the Syrian Arab Republic involved or likely involved the use of chemicals as weapons".[57] Consequently, all of the chemical weapons attacks investigated and confirmed by the UN SGM, notably including the large scale use of sarin in the 2013 Ghouta attack which resulted in an estimated 1400 deaths, were excluded from the JIM remit. This was a serious exclusion which undermined the credibility of the whole process.

5.2.2.3 *Effectiveness of the Joint Investigative Mechanism*

In all its reports, the JIM noted the highly unusual situation it faced, in terms of its working practices. Unlike the UN SGM, which had guidelines and procedures which it could (and indeed must) follow, "there were no precedents for the [JIM] investigation into the identification of the perpetrators, organizers, sponsors or those otherwise involved in the use of chemicals as weapons."[58,59] Consequently, "in the absence of a framework to guide its efforts in fulfilling its unprecedented mandate", the JIM developed its own "operating guidelines and procedures ... in order to ensure that its work adhered to the principles of impartiality, objectivity and independence."[60,61] As these "may be useful for future bodies tasked with similar mandates" the JIM will submit them to the UNSC at the end of its mandate.[62,63] The relative freedom of the JIM to develop its own procedures led to some important innovations.

A notable positive development was the degree to which the JIM explicitly utilised open source information and actively engaged with civil society:

> "A great deal of the information has been collected by the Mechanism by establishing a network of contacts with relevant information on the cases under investigation... As such, non-governmental organizations, the private sector, research organizations, institutes and other civil society organizations have been contacted when the investigators and/or the Leadership Panel thought that they could provide pertinent information to the investigation and thus help to fill gaps or corroborate existing information."[64]

The JIM in its reports to the UNSC outlined a range of factors that had constrained its activities and consequently limited the effectiveness of the mechanism. First were the mission specific challenges and constraints the JIM faced operating in Syria at this particular time, notably the lack of access to the locations under investigation owing to the dire security situation on the ground. This consequently limited the JIM's ability to confirm specific sites of interest, collect comparative environmental samples, identify new witnesses and physically evaluate the material of interest to it (*e.g.*, remnants).[65] Other challenges and constraints included: the time that had elapsed since the incidents were investigated (which in some cases was over two years); the lack of chain of custody for some of the material received; the secondary or tertiary nature of the source of information and material; the misleading nature of some of the information/material, including those depicting the size and nature of the incident; difficulty in finding independent sources of information that could provide access to individuals and information/material; and impact locations which were not preserved and were compromised by the time they were recorded.[66]

Secondly, the JIM reports highlighted the consequences of the Mechanism's restricted mandate as framed under UNSCR 2235 and subsequent Terms of Reference. Most notably, as discussed previously, the JIM was limited to investigating "only those cases" in which the FFM had "determined that a specific incident in [Syria] had involved or had likely involved the use of chemicals as weapons";[67] thereby excluding all the cases identified by the UNSGM. The effectiveness of the JIM was also curtailed by UNSCR 2235 limiting the JIM's duration, initially, to one year (though this was subsequently extended for a further year). In addition, the JIM was "not mandated to act and function as a judicial or quasi-judicial body…"[68] Consequently, it could not compel the submission of information or documents to it.[69] Furthermore, and critically, the JIM did not have "any authority or jurisdiction, either directly or indirectly, to make a formal or binding judicial determination of criminal liability."[70] Instead it was required to submit its reports to the UNSC for their consideration. Unfortunately, the UNSC proved itself incapable of acting on the JIM's evidence because of vetoes wielded by Russia and China.[71]

Despite these limitations, the establishment and successful employment of the JIM in Syria was a major step forward in UN mandated investigatory mechanisms. It has become an important progressive exemplar through the innovative tools and methodologies it employed and in the tenacity with which it sought to identify specific entities responsible for utilising chemical weapons. In addition, although it had no juridical authority and could not refer its findings to such bodies, the evidence that it collected, preserved, and made public can potentially be used by entities such as the prosecutor to the International Criminal Court to bring to justice those responsible for chemical weapons use in Syria.

5.3 UN Mechanisms to Address Confirmed Chemical Weapons Development, Acquisition or Use

On rare occasions, the scale and nature of the threat to international peace and security arising from the development, stockpiling and potential employment of chemical weapons by a specific State has been judged to be so grave and immediate that there has been an international consensus achieved for radical action through the UN. This has enabled the UNSC to establish, under its Chapter VII mandate, *ad hoc* free standing UN monitoring, verification and compliance enforcement mechanisms tasked with entering the target State to facilitate/oversee destruction of all chemical weapons stockpiles and elimination of chemical weapons production facilities. The associated UN Resolutions have compelled the target State to agree to such activities or face the threat of UN sanctions potentially including use of force.

5.3.1 Iraq: UN Special Commission (UNSCOM) and UN Monitoring, Verification and Inspection Commission (UNMOVIC)

Following Iraq's invasion of Kuwait, the subsequent 1990–1 Gulf War and its ending, UNSCR 687 of 3rd April 1991 [72] established the terms and conditions for the formal cease-fire between Iraq and the coalition of Member States co-operating with Kuwait. Section C of the UNSC Resolution called for the elimination, under international supervision, of Iraq's chemical, biological and nuclear weapons and ballistic missiles, together with related items and production facilities. It also called for measures to ensure that the acquisition and production of prohibited items were not resumed.

Under the Resolution, Iraq was specifically required to: "unconditionally accept the destruction, removal, or rendering harmless, under international supervision, of ... all chemical and biological weapons and all stocks of agents and all related subsystems and components and all research, development, support and manufacturing facilities."[73] Iraq was also required to submit to the Secretary-General within 15 days "a declaration of the locations, amounts and types of [all these items] and agree to urgent, on-site inspection as specified."[74]

The Resolution established the United Nations Special Commission (UNSCOM) to implement the non-nuclear provisions of the Resolution. Under its mandate, UNSCOM's activities included carrying out immediate on-site inspections of Iraq's biological, chemical and missile capabilities; taking possession for the destruction, removal or rendering harmless of all chemical and biological weapons and stocks of agents, all related sub-systems and components, and all research, development, support and manufacturing facilities; supervising the destruction by Iraq of all of its ballistic missiles with a range greater than 150 km and related major parts, and repair and production facilities; and monitoring and verifying Iraq's compliance with its undertaking not to use, develop, construct or acquire any of the items specified above.[75]

In accordance with UNSCR 687 (1991), large quantities of chemical weapons and related items and materials declared by Iraq were destroyed under UN supervision, from 1991 to 1994, mainly at the Muthanna State Establishment, Iraq's former chemical weapons complex. This destruction included 22 000 filled and over 16 000 unfilled chemical munitions, 690 tons of chemical warfare agents (both bulk and weaponized), more than 3000 tons of precursor chemicals and over 100 major items of chemical weapons production equipment.[76]

A dedicated UNSCOM team, the Chemical Destruction Group (CDG), was deployed to supervise and monitor destruction operations carried out by Iraq. The CDG also conducted regular environmental monitoring and provided decontamination and medical support to Iraqi personnel involved in the destruction operations. Over 100 experts from 25 countries served with the CDG during the period of its work in Iraq.[77]

Just one month after it was established, UNSCOM conducted its first in-country inspection in May 1991. By the end of December 1998, the Commission had fielded more than 250 inspection missions.[78] After it became apparent that Iraq's declarations were incomplete, UNSCOM sought to identify the gaps and discover the location of hidden or undeclared facilities and items. All through this process UNSCOM faced delays, evasion and obstruction from Iraq. In November 1998, Iraq completely refused to co-operate with UNSCOM, alleging that it had been used as a cover for Western intelligence activities. UNSCOM inspectors conducted their last inspections in Iraq between 8–15 December 1998, left Iraq on 16 December 1998 and never returned again.[79] UNSCOM inspectors continued to monitor relevant Iraqi activities from outside the country until UNSCOM was ended by the UN Security Council in December 1999.

UN Security Council Resolution 1284 of 17 December 1999[80] formally terminated UNSCOM and replaced it with the United Nations Monitoring, Verification and Inspection Commission (UNMOVIC). As well as inheriting UNSCOM's responsibilities, UNMOVIC was also tasked with establishing and operating "a reinforced system of ongoing monitoring and verification" and with addressing "unresolved disarmament issues."[81]

In November 2002, UNMOVIC's mandate was significantly strengthened through the unanimous adoption by the UN Security Council of Resolution 1441[82] which required that Iraq:

> "provide UNMOVIC and the International Atomic Energy Agency (IAEA) immediate, unimpeded, unconditional, and unrestricted access to any and all, including underground, areas, facilities, buildings, equipment, records, and means of transport which they wish to inspect, as well as immediate, unimpeded, unrestricted, and private access to all officials and other persons whom UNMOVIC or the IAEA wish to interview."[83]

Under UNSCR 1441, UNMOVIC (and the IAEA) had the right to declare exclusion zones prohibiting all Iraqi ground and aerial movement from areas surrounding prospective inspection sites, could "remove, destroy, or render harmless all prohibited weapons, subsystems, components, records, materials, and other related items", and could "impound or close any facilities or equipment for the production thereof."[84] As well as providing UNMOVIC with far greater powers than UNSCOM, UNSCR 1441 explicitly stated that Iraq's failure to comply "shall constitute a further material breach of Iraq's obligations", which would be reported to the UNSC for immediate assessment, with the possibility of "serious consequences".[85]

UNMOVIC inspections in Iraq commenced on 27 November 2002.[86] During just over three months of activities in Iraq, UNMOVIC conducted 731 inspections at 411 sites – of which 88 had not been previously inspected.[87] Whilst

Iraq allowed UNMOVIC access to inspection sites, it continued its tactics of delay and evasion. On 18 March 2003, following US recommendations that all UN inspectors should leave Iraq immediately for their own safety, UNMOVIC and IAEA teams were withdrawn. Bombing of Iraq by the US and UK commenced on 20 March 2003 and a ground invasion was launched shortly afterwards. UNMOVIC continued to operate with respect to those parts of its mandate it could implement outside of Iraq. It maintained a roster of over 300 experts, continued to conduct training activities and retained capacity and preparedness if called on to resume work in Iraq.[88] On 29 June 2007, the Security Council adopted Resolution 1762(2007) immediately terminating UNMOVIC's mandate.[89]

5.3.1.1 *Effectiveness of UNSCOM and UNMOVIC*

Despite varying degrees of cooperation from the Iraqi government over the time period, the UN inspectors successfully oversaw the destruction, removal or rendering harmless of Iraq's declared chemical weapons, biological weapons and missile stockpiles, and the dismantling of its associated development and production programmes. Indeed the supervision of Iraq's destruction of its chemical weapons was one of the main achievements of UN verification activities in that country, and was the first chemical weapons destruction operation supervised by an international organization. It demonstrated that, by utilization of the best international expertise available and the development of optimal operating procedures, UNSCOM was able to ensure that this dangerous operation was conducted safely and relatively rapidly. UNSCOM and UNMOVIC also implemented a monitoring system which essentially ensured that proscribed chemical weapons, biological weapons and missile programmes were not revived whilst UN inspectors remained in the country.[90]

One important component of their success was the unprecedented rights of access to Iraq's relevant facilities, personnel, documents and other materials provided to both UNSCOM and UNMOVIC inspectors. In order to effectively implement their mandates, UNSCOM and UNMOVIC consequently developed comprehensive inspection procedures and practices which were far more extensive and invasive than those previously employed under the UN Secretary-General's investigation mechanism. These included the use of satellite, high, medium and low altitude aerial imagery as well as ground photography, state of the art equipment for analysing chemical and biological samples, equipment designed for the drilling and sampling of munitions filled with live chemical or biological agents, computer forensics, ground penetrating radar, remote monitoring and many other inspection tools used for the first time in an inspection and verification process. In addition, an effective import/export monitoring system was introduced for dual-use items and materials which provided a high degree of confidence that they were not diverted to any proscribed activity.[90]

5.3.2 Syria: OPCW–UN Joint Mission

On 14 September 2013, following intense international diplomacy, led by the Russian Federation and the US,[91] Syria formally deposited its instruments of accession to the Chemical Weapons Convention with the UN, setting in train international efforts to dismantle Syria's chemical weapons programme and safely destroy its chemical weapons stockpile.

On 27 September 2013, the OPCW Executive Council adopted decision EC-M-33/DEC.1 setting out an accelerated and highly ambitious mechanism for eliminating Syrian chemical weapons. It called upon Syria, as required under the CWC, to submit a full declaration of its chemical weapons programme, including all weapons, agents, chemicals, laboratories, production plants, and storage facilities.[92] The decision also required that Syria destroy all "chemical weapons production and mixing/filling equipment" by 1 November 2013 and "complete the elimination of all chemical weapons material and equipment in the first half of 2014".[93] It also required that inspections in Syria commence from 1 October 2013.[94]

On the same day, the UNSC unanimously adopted Resolution 2118 (2013) which required that Syria "shall comply with all aspects of the decision of the OPCW Executive Council of 27 September 2013".[95] It further required that Syria "cooperate fully with the OPCW and the UN, including by complying with their relevant recommendations".[96] It specifically required that Syria accept personnel designated by the OPCW or the United Nations, provide for and ensure the security of activities undertaken by these personnel, provide these personnel with immediate and unfettered access to and the right to inspect any and all sites, and allow immediate and unfettered access to individuals that the OPCW had grounds to believe to be of importance for the purpose of its mandate.[96] The UNSC further decided that "in the event of non-compliance with this resolution, including unauthorized transfer of chemical weapons, or any use of chemical weapons by anyone in [Syria]", it would "impose measures under Chapter VII of the United Nations Charter".[97]

Syria formally acceded to the CWC on 14 October 2016 becoming the 190th State Party.[98] On 23 October 2013 as required under the Convention, Syria submitted its Initial Declaration to the OPCW detailing its possession of: chemical weapons (CW), chemical weapons production facilities (CWPF), abandoned chemical weapons (ACW), other CW-related facilities and riot control agents, and the import of chemicals for the CW program and for industrial purposes.[99] The initial declaration was subsequently repeatedly revised by Syria in response to OPCW requests for more information and clarification. The declaration included 1040 tonnes of seven Category 1 chemicals comprising chemical warfare agents, notably sulfur mustard, and chemicals used in binary chemical weapon systems; together with approximately 260 tonnes of 13 different Category 2 chemicals.[100]

On 16 October 2013, following close consultations between the UN Secretary-General and the OPCW Director-General, the OPCW–UN Joint Mission in Syria was formally established.[101] As mandated by the UN Security Council, the Joint Mission's main tasks were to oversee the timely elimination of the Syrian chemical weapons programme in the safest and most secure manner possible. To support the Joint Mission's operations, separate but complementary Trust Funds were established by the two organizations. On 1 October 2013, an OPCW–UN advance team arrived in Damascus to establish an operational base and start the preparatory Mission's activities.[102] The subsequent Joint Mission had a "light footprint" in Syria and included between 30–35 OPCW inspectors who were organised as multidisciplinary inspection teams together with support staff in Damascus.[103]

In its Declaration, Syria detailed the possession of 27 Chemical Weapons Production Facilities (CWPFs), which were rendered non-operational when it acceded to the treaty. In October and November 2013, OPCW inspectors verified that all production equipment had been destroyed. As of February 2018, the OPCW verified that 25 CWPFs had been destroyed.[‡104] Syria had also declared possession of 12 Chemical Weapons Storage Facilities. The Joint Mission inventoried all of the declared chemicals in the CWSFs and then verified that all of the declared chemicals were removed from the CWSFs, and these sites were closed.[105]

In addition to the extremely ambitious destruction schedule set by the UNSC and the OPCW EC, which was significantly quicker than any previously declared CW stockpile elimination process, the Joint Mission had to overcome further severe operational challenges. It was the first mission to be undertaken in the midst of an ongoing civil war, where inspectors had come under live fire, necessitating that all inspectors be fitted with bulletproof vests for the first time in the OPCW's history.[106] Following analysis of the security environment and informed by previous contingency planning on destruction options, it was decided, for reasons of safety and security, that the chemicals themselves should be removed from Syria and destroyed in a foreign country. The UN and OPCW initially explored whether any Mediterranean countries might be willing to receive the chemicals and have them destroyed on their territory. No country responded positively to these requests.[107] Subsequently, the destruction of precursor chemicals and the larger volume of toxic effluent from first-stage neutralization was organised in the US, UK and Germany through in-kind contribution, and in commercial chemical disposal facilities in Finland and the US.[108]

Under the agreed destruction plan, Syria's declared chemical weapons stockpile, which was held in more than 20 sites throughout the country, was transported by Syrian forces to the port of Latakia in North Western Syria.

[‡]In November 2017, the OPCW Technical Secretariat conducted an initial inspection of the last two stationary above-ground facilities; subsequently the Secretariat, together with the United Nations Office for Project Services (UNOPS), began preparations to assist Syria in the destruction of the facilities located at these two sites. See reference 104.

This process commenced on 7 January 2014 and was undertaken with advice and support from the UN Environment Programme and the WHO.[109]

The chemicals were subsequently transferred onto two freighters, the Danish *Ark Futura* and the Norwegian *Taiko*. The *Ark Futura* and the *Taiko* transported the more dangerous "Priority 1" chemicals—including sulfur mustard and the sarin precursor, methylphosphonyl difluoride—to the US freighter, the *MV Cape Ray*, for on-board neutralization in the Mediterranean Sea. This vessel had been outfitted by the US Department of Defense with two field deployable hydrolysis systems which had been designed on the basis of technology used in the US chemical weapons destruction programme to hydrolyse (and thereby neutralise) large quantities of chemical warfare agents.[107,110] The *Ark Futura* and *Taiko* also transported the less dangerous "Priority 2" precursor chemicals and the effluent from the *MV Cape Ray*'s hydrolysis process to land-based incinerators in Finland, the UK, and the US. Warships from Denmark, Norway, and the UK provided continuous security to the cargo ships until the chemicals were off-loaded. Naval vessels from China and the Russian Federation provided security to the cargo ships while in Syrian territorial waters.[107,109]

Although the OPCW–UN Joint Mission formally closed on 30 September 2014, the OPCW has continued to undertake the necessary residual activities required to fully implement UNSCR 2118. On 4 January 2016, the OPCW confirmed the complete destruction of all chemical weapons declared by Syria.[111]

5.3.2.1 *Effectiveness of the OPCW–UN Mission*

The OPCW–UN Mission and wider Syrian chemical weapons destruction process has been a remarkable example of successful multilateral disarmament operations conducted in the middle of a dangerous civil war. It has significantly reduced (but not entirely eliminated) the threat of widespread mass casualty attacks with deadly nerve agents against soldiers and civilians in Syria itself, and also significantly degraded the capability of Syria to launch chemical weapons attacks against neighbouring countries. There were reportedly "some issues of miscommunication and competing leadership early in the operation", but in the end, the OPCW–UN Joint Mission was successful, and showed that these "large, multilateral bureaucracies" can work well together.[107]

However, there were concerns regarding the limited public transparency surrounding the Mission and associated chemical demilitarization operations. This led to unnecessary suspicions about what was happening to secure and destroy Syria's chemical weapons stockpile and what risks might arise, especially for Mediterranean countries. The UN and the OPCW were constrained in the level and nature of the information they could make public as Syria reportedly requested confidentiality regarding its declaration and stockpile figures; in addition, other countries were very sensitive about their role in the operation.[107] Summary figures of the Syrian declared stockpile and monthly updated destruction figures were eventually published on the OPCW website and these helped alleviate some public anxiety and build confidence in the process. Although the US State Department and the OPCW organised a number of meetings with NGOs and other civil society stakeholders, these were by invitation-only and the participants limited. Consequently, the public outreach effort was never sufficient to satisfy concerned organizations and citizens, especially those in the Mediterranean region most likely to be affected by any accidents.[107]

The CWC requires that each State Party "meet the costs of destruction of chemical weapons it is obliged to destroy",[112] however Syria made it known that it could not afford to pay for this process. Whilst it covered some of the costs for aspects such as land-based transportation, security, and facility destruction; Syria refused to pay for out-of-country chemical destruction or for the verification work by the OPCW.[107] In October 2014, the OPCW Director General informed the Executive Council that 24 States Parties and the European Union had donated $65 million to a trust fund established for that purpose.[113]

Although the OPCW–UN Mission successfully removed and destroyed all of the chemical weapons stockpiles declared by Syria, there are continuing concerns that Syria may not have provided the UN and OPCW with full details of its chemical weapons holdings. The OPCW's Declarations Assessment Team (DAT) is currently in discussions with the Syrian Government over the apparent discrepancies in the country's declaration of its chemical weapons stockpile to the OPCW, and the process of completing the destruction of the final declared production and storage facilities. Furthermore, as detailed in Section 5.2.2.1, the OPCW FFM has documented a number of chemical attacks employing chlorine and one employing sarin, upon Syrian civilians and/or combatants during the ongoing civil war; the OPCW–UN JIM subsequently identified the Syrian Armed Forces as the perpetrators of the sarin attack and of three of the chlorine attacks. Although the OPCW-UN JIM has ceased its investigations, the OPCW FFM continues to gather evidence of alleged chemical attacks in Syria.

5.4 UN Embargoes to Halt the Development, Acquisition or Use of Chemical Weapons

UN arms embargoes are imposed by UNSC Resolutions adopted under Article 41, Chapter VII of the UN Charter. Such resolutions require at least nine of the 15 UNSC members to vote in favour, and these must include all five permanent members (P5) – China, France, Russia, the UK and the US.[114] UNSC arms embargo can be either voluntary or mandatory. Voluntary UN arms embargoes are invoked when a UNSC resolution "calls upon all States" to end the supply of arms, ammunition, military materiel and related services. A 2007 study by SIPRI argued that they are regarded as "primarily symbolic" because States are "only requested" to cease supplying to target states or groups. Furthermore, they send "mixed signals to both targets and UN member states" and consequently SIPRI argued that such embargoes had a "negligible impact on curtailing arms flows and improving target behaviour".[115]

In contrast, mandatory UN arms embargoes are invoked when the UNSC in its Resolution "decides that all States shall" prohibit the sale or supply of arms, ammunition, military equipment and related services, implement a general and complete embargo on all deliveries of weapons and military equipment, or words to similar effect. Article Two of the UN Charter specifies that "All Members shall give the United Nations every assistance in any action it takes in accordance with the present Charter, and shall refrain from giving assistance to any state against which the United Nations is taking preventive or enforcement action."[116] Consequently, all UN Member States are legally obliged to implement such embargoes.

Since the end of the Cold War the use of mandatory UN arms embargoes and other targeted sanctions has increased, and the range of intended purposes widened. The UNSC has used them as a tool to repel aggression, restore or strengthen specific governments, restore democracy, protect human rights, apply pressure to regimes supporting terrorist activities and others charged with international crimes, and combat or constrain attempts by States to develop or acquire nuclear, chemical or biological weapons and related means of delivery (NCB weapons).[8,115,117] Whereas sanctions have traditionally been used against States, the UNSC has also increasingly imposed sanctions against a range of non-State actors. Of the 26 mandatory UNSC embargoes established from 1st January 1990 till 31st December 2016 (detailed in Appendix), 10 included provisions solely against non-State actors.

5.4.1 Range of Items Controlled

Analysis of the 26 UN arms embargoes effective from 1st January 1990 to 31st December 2016 has shown that only one included chemical weapons generically within its scope. In October 2006, the UNSC adopted Resolution 1718 instituting a wide ranging arms embargo, currently in force, against North Korea that prohibited the transfer of "items, materials, equipment, goods and technology, determined by the Security Council or the [Sanctions] Committee, which could contribute to [North Korea's] nuclear-related, ballistic missile-related or other weapons of mass destruction related programmes."[118] In order to facilitate the effective implementation of this resolution, in November 2006, the UNSC and the Sanctions Committee designated a range of "chemical and biological items, materials, equipment, goods and technologies related to other weapons of mass destruction programmes" whose export to or import from North Korea is prohibited.[119] This list is regularly reviewed and revised, the latest ammendments released in September 2017.[120]

In contrast, the vast majority (25)[§] of UNSC Resolutions establishing arms embargoes in this period described the goods covered (essentially conventional weapons and certain security equipment) as: "Arms and related material of all types including weapons and ammunition, military vehicles

[§]See Appendix.

and equipment, paramilitary equipment, and spare parts for the aforementioned", or variations to that effect – which appears to capture riot control agents (RCAs) and related means of delivery within their scope. As none of these UNSC Resolutions contained an annexed list detailing the specific range of arms and related equipment included within its scope, there is considerable ambiguity over precisely which RCAs and related means of delivery would be covered by each of the embargoes. This ambiguity has been compounded by the fact that there have been no agreed definitions of key terms such as "paramilitary equipment" or "arms and related material", nor have there been any indicative lists of material covered by these terms. Furthermore, although a number of UN Member States and regional or pluri-lateral organisations (such as the EU and the Wassenaar Arrangement) have agreed and published lists of arms, and military equipment (which include RCAs) whose export is regulated, there are still no international military or paramilitary lists accepted by all UN Member States and applicable to UN embargoes generally. Instead, the scope of coverage for each embargo has and will be determined by national interpretation of the relevant descriptive terms in the UNSCR (and may also be informed by recommendations of relevant UN bodies such as UNSC Sanctions Committees). This of course presents the danger that different UN Member States will interpret the scope of arms and equipment to be embargoed differently, leading to inconsistent implementation and the potential weakening of the embargo regime.

5.4.2 Monitoring and Facilitating Implementation of Embargoes

All of the mandatory UN arms embargoes currently in force contain mechanisms and procedures for monitoring and facilitating their implementation, which as a minimum normally include a UNSC Sanctions Committee. While they do not have formal authority to make binding decisions, in practice Sanctions Committees—which are subsidiary organs of the Security Council—have been delegated substantial tasks including: seeking information from all UN Member States on embargo implementation measures; considering information concerning suspected violations; periodic reporting on violations and violators to the Security Council; considering requests for humanitarian exceptions; managing designation lists; providing guidelines for implementation; and making information publicly available.[8,115,121] The UNSC Sanctions Committees often face resource constraints and normally work within highly politicised environments. They have a varied record of activity and effectiveness – which is often related to the personnel and personalities driving each one.[122]

The work of UNSC Sanctions Committees has been increasingly aided by expert panels tasked with monitoring sanctions, assessing and investigating serious suspected violations or allegations of non-compliance, and regularly reporting their findings to the relevant Sanctions Committees. Whilst certain expert panels have been notably successful, a number have reported that they lack the authority and powers to effectively explore embargo breaches.

Furthermore, as most expert panels report to the relevant Sanctions Committee, and the release of their reports requires consensus among Committee members, this *de facto* veto power has been used to delay or suppress reports – thus potentially compromising the integrity of the process.[123]

5.4.3 Effectiveness of UN Arms Embargoes

UN bodies, States and civil society researchers, have regularly highlighted the systemic failings and limitations of UN arms embargoes. The "main finding" of the SIPRI 2007 study was that the effectiveness of UN arms embargoes depended primarily on the capacity and will of UN Member States, particularly the P5 States, (as well as arms-supplying, transit and transhipment States, and States neighbouring embargoed targets), but that such embargoes were undermined by a lack of unity and commitment by the P5 regarding implementation and end goals.[124]

Arms control and human rights organisations have regularly highlighted delays in the imposition of certain UNSC embargoes which have allowed protagonists to enter a conflict or those engaged in large scale human rights abuses to procure sufficient arms to continue their activities after the embargo is finally imposed. Furthermore, there have been a number of cases where a conflict, humanitarian or human rights crisis has developed and the UNSC has been unable or unwilling to impose an arms embargo on the offending State or non-State actor at all.

Similarly, in certain cases the UNSC has failed to impose an embargo upon a State or non-State actor where there was evidence that the entity had attempted or succeeded in acquiring, developing or employing chemical weapons. For example, the UNSC has to date been unable to impose an arms embargo to prevent combatants to the Syrian conflict from acquiring or developing chemical weapons. This has been despite long standing evidence of Syrian Government CW stockpiles, (declared holdings of which were consequently destroyed by the OPCW–UN), and the subsequent findings of the OPCW–UN JIM that Syrian armed forces had carried out chemical attacks with chlorine barrel bombs and sarin, whilst ISIL had employed sulfur mustard. In the light of the OPCW–UN JIM findings, France, the UK and the US developed a draft UNSC resolution that would have imposed measures to prevent the direct or indirect supply, sale or transfer to named individuals and entities of all arms and related materials used to deliver chemical weapons. Those materials included certain chemical agents listed on the CWC Schedules and also a wide range of additional toxic chemicals, including chlorine.[125] Unfortunately, the UNSC resolution was vetoed by both China and Russia on 28 February 2017.[126]

Other factors believed to contribute to the limited effectiveness of UN arms embargoes include: a lack of clarity in their definitions of coverage or scope; ineffective UN monitoring and enforcement mechanisms; a chronic lack of human, logistical and other resources in many developing States which are often in those regions where embargoes are imposed; ineffective border controls between the embargoed State and its neighbours; the failure to incorporate arms embargoes into national legislation and a failure by States to identify, arrest and punish arms traffickers and other embargo busters. Indeed, in certain cases, States neighbouring embargoed targets ignored or undermined embargos if they conflicted with their regional interests and instead provided financial, military or political support to the embargoed targets.[127]

5.5 UNSCR 1540: Combating the Development, Acquisition or Use of Chemical Weapons by Non-State Actors

UN Security Council Resolution (UNSCR) 1540 is intended to prevent the development, manufacture, acquisition, possession, transport, transfer or use of nuclear, chemical or biological weapons and their means of delivery (NCB weapons) by non-state actors, in particular for terrorist purposes.[128] The adoption of UNSCR 1540 was prompted by growing concerns among the international community about the limitations and weaknesses of the existing NCB non-proliferation regimes (principally the BTWC, CWC and the Non-Proliferation Treaty (NPT)) and their specific inability to combat the potential acquisition of NCB weapons by terrorist groups.

5.5.1 Obligations Under UNSCR 1540

As UNSCR 1540 was adopted under Chapter VII of the UN Charter it is legally binding upon all UN States. Under its key operative provisions, all States are required to:

- Refrain from providing any form of support to non-State actors that attempt to develop, acquire, manufacture, possess, transport, transfer or use NCB weapons;[129]
- Adopt and enforce appropriate effective laws which prohibit the manufacture, acquisition, possession, development, transport, transfer or use of NCB weapons by any non-State actor; or act as an accomplice, assist or finance them;[130]
- Establish domestic controls to prevent the proliferation of NCB weapons, including by establishing and enforcing controls over and preventing illicit trafficking of related materials through measures in material accountancy and security; physical protection; border controls; export and trans-shipment controls and related financial regulation.[131]

The Resolution also called upon (but did not require) States to:

- Promote the universal adoption and full implementation, and where necessary, strengthening of multilateral treaties to which they are parties, whose aim is to prevent NCB weapons proliferation;[132]
- Adopt national rules and regulations to ensure compliance with their commitments under the key multilateral non-proliferation treaties.[133]

5.5.2 Scope of Actors, Agents and Delivery Mechanisms Covered

UNSCR 1540 defines a non-State actor as an "individual or entity, not acting under the lawful authority of any State in conducting activities which come within the scope of this resolution."[134] This is a very open definition potentially capturing anyone acting unlawfully in the light of the Resolution's objectives, including for example criminal entities, traffickers, 'lone wolf' or terrorist organisations, unregulated private military or security companies, or armed opposition groups. It thus implies a wide scope for the legislative provisions that States could put into place, reflecting the different sectors that have the potential for NCB weapons proliferation.[135]

Although the term "chemical weapons" is regularly employed throughout UNSCR 1540, it is not defined in the Resolution, and the range of chemical agents covered under UNSCR 1540 is not listed or otherwise explicitly specified. However, the Resolution does define "related materials" as "materials, equipment and technology covered by relevant multilateral treaties and arrangements or included on national control lists, which could be used for the design, development, production or use of [NCB weapons]."[134] Such "materials" would certainly include those chemical agents listed in the Schedules of the CWC and also, arguably, those contained in the Control Lists of the Australia Group and Wassenaar Arrangement. In addition, the definition of "related materials" would also encompass the essentially open-ended General Purpose Criterion of the CWC. Consequently, UNSCR 1540 appears to cover all weapons employing toxic chemicals including industrial toxic chemicals, central nervous system CNS-acting chemicals and riot control agents, as well as the "classic" chemical warfare agents.

In contrast, UNSCR 1540 appears to cover a fairly narrow range of "means of delivery" which are defined in the Resolution as "missiles, rockets and other unmanned systems capable of delivering NCB weapons that are specially designed for such use".[134] Whilst this would include certain important chemical weapons delivery mechanisms specifically developed for long range and/or large scale agent dispersal, it would not, *per se*, cover a wide range of essentially dual use conventional weaponry and associated means of delivery that are capable of carrying a variety of payloads including, but not exclusively, chemical agents.

5.5.3 Significance of UNSCR 1540 Obligations

The obligations established under UNSCR 1540 are intended to address several critical lacunae in the existing international system for combatting the proliferation and use of NCB weapons. Firstly, it focusses on preventing proliferation of such weapons to non-State actors. In contrast, with regard to chemical weapons, the CWC establishes prohibitions on chemical weapons development, production, acquisition, stockpiling, transfer and use applicable, first and foremost, to State Parties. The references to non-State actors are secondary and derive from national implementation obligations upon CWC States. Because international law does not apply directly to individuals, non-State actors are subject only to prohibitions laid down in an often-ambiguous patchwork of domestic law.[135] Whereas certain States have sophisticated and effective legal frameworks of laws, regulations, and controls, others have little in place that could deter terrorists and traffickers from acquiring and transporting chemical weapons, related materials, and delivery mechanisms or that could prosecute them for doing so. In an effort to oblige all States to tighten their domestic legal framework, Resolution 1540 is a re-presentation of States Parties' obligations under the CWC (and BTWC and NPT).[136]

Secondly, although UNSCR1540 recognizes that most States have undertaken legally binding obligations under the CWC (and the BTWC and NPT) to combat NCB weapons proliferation, the Regulation requires them to "take additional effective measures to prevent the proliferation of [such weapons]."[137] These new measures, detailed in operative paragraphs 2 and 3, go beyond existing treaty requirements and are intended to improve the financial, security, and physical protection of sensitive materials and also border and export controls. They are in large part directed at combatting illicit trafficking, identified in the Resolution as a new dimension of proliferation that needs to be addressed.[138]

In addition, there are certain activities that CWC States Parties are already obliged to regulate under the Convention, for example, the monitoring and reporting of production of Scheduled chemicals, which are given a different (non-State actor) emphasis under UNSCR 1540, which radically alters the nature and adds to the burden of effective implementation. This arises from the fact that the CWC was negotiated primarily to combat State level militarily significant chemical weapons programmes and stockpiles, whilst UNSCR 1540 is directed against non-State actors where the quantities of toxic chemicals necessary to mount terrorist attacks against unprotected civilian populations are orders of magnitude less than those regarded as militarily significant.[139]

Thirdly, UNSCR 1540 requires that all States, even those not party to the CWC (and BTWC and NPT), shall criminalize and enforce measures against NBC weapons proliferation to and by non-State actors. Even though the CWC currently has 192 States Parties with only Egypt, North Korea, and South Sudan having neither signed nor acceded to the Convention, and Israel the sole State that has signed but not ratified the treaty, UNSCR 1540's universal application is significant. Despite the small number of non-State Parties, there is open source information indicating that at least three—Egypt, Israel and North Korea—have held stockpiles of chemical weapons,[140] with the consequent danger of the use of such weapons or proliferation to State or non-State actors. In addition, UNSCR 1540 could act as an additional last barrier to chemical weapons development and proliferation for any State that subsequently choose to withdraw from the CWC.

Fourthly, UNSCR 1540 requires that all States implement and enforce the Resolution provisions at the national level. As of July 2016, 150 CWC States Parties had adopted national implementing legislation of which 118 States Parties had legislation covering all the initial measures and 32 States Parties had legislation covering some of the initial measures. Forty two States Parties had yet to adopt legislation.[141] Many CWC States Parties appear to have taken the view that since they neither possess chemical weapons nor undertake any other related activities, that there is no need for them to divert national resources from other important issues to drafting and implementing the required legislation. However, the existence of comprehensive national implementing legislation is not only necessary for a State Party to meet fully its obligations under the CWC, but also is a vital step in preventing its territory becoming a safe haven for non-State actors seeking to obtain or produce chemical weapons.[142]

Lastly, because UNSCR 1540 has been adopted under Chapter VII of the UN Charter in response to "threats to the peace, breaches of the peace, and acts of aggression" its obligations are mandatory for all States which means that non-compliance of such obligations opens up the possibility that the UN Security Council may consider imposition of a range of penalties or sanctions.[143]

5.5.4 Mechanisms to Facilitate and Monitor Implementation

UNSCR 1540 established a "Committee of the Security Council"—comprising 15 UNSC Members—to monitor and "report to the Security Council ... on the implementation of the Resolution."[144] From 2005, its work has been supported by a Group of Experts. In 2009, the Committee established four working groups to:[145]

- Monitor State implementation;
- Act as a clearing house for implementation assistance;
- Facilitate cooperation with international, regional, and sub-regional organisations, and relevant 'anti-terrorist' UN bodies and entities;
- Promote, facilitate and participate in outreach and capacity building activities.

Under UNSCR 1540, States were called upon "to present a first report no later than six months from the adoption of this resolution ... on steps they have taken or intend to take to implement this resolution."[144] By the deadline for submissions on 28 October 2004, only 59 States (and the EU) had submitted an initial implementation report.[135] Subsequently, a critical activity of the Committee and Group of Experts has been to encourage and facilitate the submission of reports from remaining States and also from States where a significant change in the nature of implementation has occurred. These reports are subsequently analysed by the Group of Experts whose findings form the basis of the Committee's report to the Security Council on the levels and nature of implementation. As of April 2016, 17 States had yet to submit their first national report on implementation.[146] With the notable exception of North Korea, the vast majority of these States have no history of aiding terrorism or attempting to develop or acquire NBC weapons. Since they are all developing countries, their continuing non-submission is likely to be a consequence of limited available human and financial resources.

5.5.5 Current Implementation and Effectiveness

In 2011, as part of UNSCR 1977, the 1540 Committee was mandated to conduct two "comprehensive" reviews of the status of the implementation of UNSCR 1540 (2004),[147] the first of which was submitted to the Security Council in December 2016.[148] The Comprehensive Review determined that there was progress in the overall implementation of UNSCR 1540 (2004) over the past five years, with "most States" increasing such measures "especially in taking legal actions to prohibit activities of non-State actors related to [NBC] weapons and their means of delivery".[149] However, although some progress was also made in relation to accounting, security and export control measures, "it is clear, that for many States there remain significant efforts to be undertaken to address gaps in these areas."[149] Consequently, the Committee concluded that "the rate of progress confirms that accomplishing the objective of full implementation of the resolution is a long-term task that will require continuous efforts at national, regional and international levels."[150]

With regard specifically to chemical weapons, the Committee noted that there had been an "absolute increase of 15% in recorded measures covering the obligations of operative paragraph 2 in relation to [such] weapons ... and these prohibitions represent a 74% implementation rate overall" by States (Table 5.1). Of particular note was the "sizable increase" in the number of States which have a legal framework in place to prohibit the use of chemical weapons by non-State actors; this rose from 150 States in 2011 to 175 in 2016.

Table 5.1 Number of States with legal and enforcement measures in place covering paragraph 2 of UNSCR 1540 relating to chemical weapons[167].

Obligation	Legal measures	Enforcement measures
Manufacture/produce	161	142
Acquire	158	133
Possess	151	141
Stockpile/store	150	121
Develop	142	113
Transport	106	106
Transfer	159	143
Use	175	159
Means of delivery	116	93
Accomplice	159	154
Assist	166	156
Financing	166	161

Table 5.2 Number of States with legal and enforcement measures in place covering paragraph 3 of UNSCR 1540 relating to chemical weapons[168].

Obligation	Legal measures	Enforcement measures
Accounting		
Production	104	101
Use	104	99
Storage	106	102
Transport	85	78
Securing		
Production	69	64
Use	70	63
Storage	76	71
Transport	75	71
Physical protection	55	52

Similarly, 161 States now have in place a legal framework to prohibit the manufacture of chemical weapons by non-State actors compared to a 135 in 2011.[151] In addition, there was also a measurable increase in the number of States that have in place penalties and enforcement measures for these prohibitions, and the Committee identified 161 States that had in place at least one enforcement measure compared to 147 States in 2011, and 96 States in 2008.[152] The Comprehensive Review indicated that even greater advances had apparently been made with regard to measures prohibiting chemical weapons means of delivery. Whilst only 54 States had a legal framework in place prohibiting chemical weapons means of delivery to non-State actors in 2011, this had risen to 116 States in 2016. This advance was mirrored with regard to enforcement measures which rose from 52 States in 2011 to 93 States in 2016.[153]

In contrast to these advances, the developments in State implementation of obligations established under operative paragraph 3 of UNSCR 1540 relating to chemical weapons were decidedly mixed (Table 5.2). With regard to accounting for materials related to chemical weapons, there had "only been a modest increase" with 104 States having adopted legal framework measures covering the accounting of the production and use in 2016, compared to 96 States in 2011 and 64 States in 2008. Of even greater concern to the Committee was its finding that "chemical security measures have virtually remained static, improved only slightly, or in some cases, due to the interpretation of existing legislation, decreased since 2011."[154]

5.5.6 Effectiveness and Challenges Faced

Although UNSCR 1540 was adopted by the Security Council under Chapter VII of the UN Charter and is therefore binding on all UN Member States, it is not, as Crail noted, "immune to the challenge of a lack of political will."[155] Indeed the danger of NBC weapons proliferation to, and use by, non-State actors is not a priority for many States. Furthermore, there has been some political opposition to the role of the Security Council in responding to this threat. When the Resolution was initially discussed at an open Security Council meeting on 22 April 2004, 51 countries made statements concerning the draft text and the wider implications of UNSC action in this area.[156] Some expressed concern that the proposed Resolution would serve as a basis for imposing economic and potentially military sanctions against States considered to be in "non-compliance". Others questioned whether it was appropriate for the UNSC to dictate domestic law for all UN members, as well as the adoption of international non-proliferation obligations outside the traditional negotiation process, particularly when they had not been involved in drafting the Resolution.[157] Despite these important reservations, over the following years, the UNSCR 1540 regime has gradually become accepted and is now embedded into the UN and broader international community's NCB weapons non-proliferation architecture. However, as Crail has noted, whilst States cannot simply ignore a Chapter VII UNSC Resolution because they do not agree with it and there is little likelihood of outright resistance to its implementation, they may be less inclined to commit themselves to the efforts needed to adopt all of the necessary domestic legislation and create and/or train all of the enforcement bodies called for by the Resolution.[155]

In addition to questions about the Resolution's legitimacy, there were concerns, in the years immediately following its adoption, regarding the limited resources and temporary nature of mechanisms intended to monitor and facilitate its implementation – most notably the 1540 Committee which originally only had a lifespan of " no longer than two years."[158] Although the Committee's existence was subsequently extended, this was initially done in a spasmodic and limited fashion by two subsequent UN Resolutions for consecutive periods of two years.[159] In recent years there has been a substantial and continuing reinforcement of the mechanisms in place to facilitate and build capacity to implement UNSCR 1540; of particular importance being the UN Security Council's adoption on 20 April 2011 of UNSCR 1977.[160] Under this Resolution, the Security Council agreed to extend the Committee's mandate for a full ten years until 25 April 2021.[161] The Resolution also expanded the group of experts assisting the Committee – which had initially comprised just 4 members in 2005, later expanded to 8 and then 9 in 2012 – to 12 members.[162]

One of the key challenges preventing the universal implementation of UNSCR 1540 is the inability of many States to introduce the 300 plus domestic legislation and enforcement measures required under the Resolution – because of limitations in technical and legal expertise, as well as financial and human resource constraints within the country. Such constraints are intensified for States facing internal or external armed conflict, epidemic disease, climate change, poverty or famine who will devote their limited resources to their immediate existing (and often overwhelming) problems rather than to potential threats that they believe they may never face.

These concerns are exacerbated because many developing countries continue to see UNSCR 1540 as an 'unfunded mandate', since an obligation was created without concrete provisions to directly assist States in implementing the Resolution's requirements. In order to address this capacity gap a number of the major supporters of UNSCR 1540—including the US, the EU, and Japan—have actively offered assistance, whilst the 1540 Committee has developed mechanisms to facilitate increased cooperation and provision of assistance between States. However, despite these initiatives many States remain concerned about the level of resources that will be needed in order to fully implement the resolution in the long term.[163] Unless donor States and relevant international, multilateral and (sub) regional organisations—facilitated by the 1540 Committee—make concerted efforts to build capacity, adherence to 1540 will be severely limited.[164]

Furthermore, as Crail has highlighted, the Resolution does not specify that any provisions are more important than others, and neither the UNSC nor the UNSCR 1540 Committee have established a public hierarchy of priorities, either overall, or for a given State. Consequently, all States are faced by the challenges of determining which UNSCR 1540 measures need to be implemented, at a minimum, and which of these should take priority. These considerations also limit the ability of the 1540 Committee to undertake meaningful analysis of State implementation, as seen in the latest Comprehensive Review which was largely limited to tallying the number of measures that States have fulfilled without considering the question of how important each fulfilment is for the intent of the Resolution, let alone how effective these measures are or how they are being implemented in practice in each individual State. Similar challenges would also be faced by the UNSC in any attempt to determine whether a specific State is in compliance or non-compliance with these obligations. In addition, whilst UNSCR 1540 is a Chapter VII Resolution, the enforcement capabilities at the UNSC's disposal have little meaning if there are no consistent criteria for judging compliance, or if so many States have failed to comply that a substantial percentage of the UN membership would in theory face punitive action.[165]

Finally, the UNSCR 1540 regime, like all the mechanisms discussed in this and other chapters must face and respond to sometimes fast-changing and dramatic real-world developments. In its recent Comprehensive Review, the 1540 Committee took note of the increasing risks of NBC weapons proliferation in relation to non-State actors arising from "developments in terrorism" and also in the potential for misuse arising from "the rapid advances in science, technology and international commerce". Consequently, the Committee underlined the need for States to pay constant attention to these developments to assure effective implementation of the resolution.[166]

5.6 Conclusions

Although the CWC is the principle international treaty for combatting the development, acquisition, stockpiling, transfer or use of chemical weapons; the UN, and in particular the UNSC, with their responsibility to address potential or actual threats to "international peace and security" are important mechanisms by which the international community can respond to threats arising from chemical weapons. Consequently, the UNSC has, over time, created—and employed—an impressive "tool box" of measures, precedents and procedures to combat attempts by a range of States and non-State actors to develop, acquire or employ such weapons.

5.6.1 UN Mandated Investigatory Procedures

The UN Secretary-General's Mechanism (UNSGM) has become a relatively rapid and authoritative investigative procedure to establish the facts surrounding allegations of chemical (and biological) weapons use, and a recognised channel to bring such information to the Council for appropriate action. Although investigatory procedures established under the CWC would normally be applied for suspected CW attacks upon CWC States Parties, the UNSGM provides a potential avenue of last resort when such procedures are deemed inappropriate. The utility of the UNSGM was exemplified when it was successfully employed in 2013 to investigate widespread allegations of CW attacks upon civilians and combatants in the Syrian conflict. A significant advance in UN mandated investigatory procedures was the establishment in 2015 of the OPCW–UN Joint Investigative Mechanism (JIM) which, unlike previous UN investigations, was tasked with identifying specific entities it deemed to be responsible for utilising chemical weapons in Syria. Although it had no juridical authority and could not refer its findings to such bodies (but reported to the UNSC), the evidence the JIM collected and made public can potentially be used by juridical entities to bring to justice those responsible for and facilitating chemical weapons use in Syria. It is very regrettable that, following Russia's vetoes of UNSC Resolutions extending the Mechanism's lifespan, the JIM mandate expired on 17 November 2017. Given the continuing chemical attacks in Syria, it is imperative that the JIM mandate be renewed and its investigations recommence as soon as possible.

5.6.2 UN Sanctions

Amongst the sanctions that the UNSC can impose in response to threats to "international peace and security" are mandatory arms embargoes. Although such embargo regimes covering chemical weapons, their agents, precursors and related means of delivery are extremely rare, they do provide a potentially powerful mechanism for halting or at least constraining CW acquisition or development programmes, and combatting the subsequent proliferation and use of such weapons by targeted entities. Similarly, the far more commonly utilised UN embargoes covering conventional arms and related security equipment (normally including RCAs and delivery mechanisms) can limit the supply of such goods to specific State or non-State actors that have seriously misused or are likely to misuse them for grave human rights abuses, breaches of international humanitarian law or for armed aggression. However, in practice, as a result of a lack of unity and commitment by States (particularly the P5) certain UN embargoes have been undermined, delayed or have not been introduced, even when, for example, there was clear evidence of the possession and use of chemical weapons.

5.6.3 UN Enforced CW Destruction and Demilitarisation

On rare occasions (*i.e.* Iraq (1992) and Syria (2013)), the scale and nature of the threat to international peace and security arising from the development, stockpiling and potential employment of chemical weapons by a specific State has been judged to be so grave and immediate that there has been an international consensus achieved for radical action through the UN. This has enabled the UNSC to establish, under its Chapter VII mandate, *ad hoc* "free standing" UN monitoring, verification and compliance enforcement mechanisms tasked with entering the target State to facilitate/oversee the destruction of all chemical weapons stockpiles and the elimination of chemical weapons production facilities. The associated UN Resolutions have compelled the target State to agree to such activities or face the threat of UN sanctions potentially including use of force. Whilst UNSCOM/UNMOVIC in Iraq and the OPCW–UN Joint Mission in Syria had to operate in highly challenging on-the-ground and geo-political environments, in both cases these UN sanctioned processes were highly successful in verifying and subsequently destroying significant declared chemical weapons stockpiles and in dismantling declared chemical weapons production facilities.

5.6.4 UN Measures to Combat CW Proliferation and Use by Non-State Actors

Adopted by the UNSC in 2004, and despite initial reservations by certain States, UNSCR 1540 has now become established as a long-term, and potentially permanent, international mechanism to combat the proliferation and misuse of chemical (as well as biological and nuclear) weapons and their means of delivery. Its potential utility is reinforced by its focus on non-State actors which were previously inadequately addressed by then-existing NCB weapon non-proliferation regimes; its mandatory application, under Chapter VII, to all UN Member States regardless of whether they are States Parties to the CWC (BTWC or NPT); and the extensive additional measures it requires States to implement including financial regulations, export controls, border controls and material security measures. However, very significant challenges remain, arising in large part from the extremely ambitious nature of the Resolution's coverage in terms of NBC weapons and related materials, actors and activities to be regulated; and its lack of guidance to States as to which measures should be prioritised. Furthermore, although the international community in recent years has provided increased resources to improve UN monitoring and facilitate State implementation of the Resolution, many (particularly developing) States have insufficient financial, technical and human resources, legislative time or political will to introduce and implement the national measures required for effective implementation.

Appendix UN Arms Embargoes Introduced During The Period 1 January 1990–31 December 2016

Target	Entry into force/expiry	Details of resolution	Goods covered	Verification
Afghanistan (Taliban)	22/10/1999, 19/12/2000–16/01/2002	UNSCR 1076 (1996); 1333 (2000)	Arms and related material	Sanctions Committee
Taliban, Al-Qaeda, Osama Bin Laden	16/01/02	UNSCR 1267 (1999); 1333 (2000); 1388 (2002); 1390 (2002); 1452 (2002); 1526 (2002)	Arms and related material	Sanctions Committee and Monitoring Group
Angola (UNITA)	15/09/1993–09/12/2002	UNSCR 864 (1993); 1448 (2002)	Arms and related material	Sanctions Committee
Central African Republic [non-State actors]	5/12/2013	UNSCR 2127	Arms and related material	Sanctions Committee
Cote d'Ivoire	15/11/2004–28/4/2016	UNSCR 1572 (2004)	Arms and related material	UNOCI; Group of Experts and Committee
Democratic Republic of the Congo (DRC) [non-State actors]	28/07/03	UNSCR 1493 (2003) (rebels); 1596 (2005) (extends provisions to other DRC recipients)	Arms and related material	MONUC; Panel of Experts
Eritrea	12/02/99–17/05/2000, 17/05/2000–15/05/2001	UNSCR 1227 (1999); UNSCR 1298 (2000)	Arms and related material	Sanctions Committee
Eritrea	23/12/2009	UNSCR 1907 (2009)	Arms and related material	Sanctions Committee and Monitoring Group
Ethiopia	12/02/99–17/05/2000, 17/05/2000–15/05/2001	UNSCR 1227 (1999); UNSCR 1298 (2000)	Arms and related material	Sanctions Committee
Haiti	16/06/1993–29/09/1994	UNSCR 841; UNSCR 944	Arms and related material	Sanctions Committee
Iran	23/12/2006	UNSCR 1737 (2006); 1747 (2007); 1920 (2010)	Technology, contributing to nuclear weapons programme. Major conventional weapons. Prohibit export of arms	Sanctions Committee and Panel of Experts
Iraq	06/08/1990–08/06/2004	UNSCR 661 (1990); 687 (1991); 707 (1991); 1483 (2003); 1511 (2003); 1546 (2004)	Arms and related material	Multi-national forces and Government of Iraq

ISIL, Al-Qaeda, associated entities and individuals	16/1/2002	UNSCR 1390	Arms and related material	Sanctions Committee and Monitoring Team
Lebanon (non-State actors)	11/08/2006	UNSCR 1701 (2006)	Arms and related material	Government of Lebanon or (UNFIL)
Liberia	19/11/1992–26/5/2016	UNSCR 788 (1992); 1343 (2001); 1521 (2003); 1532 (2004); 1579 (2004)	Arms and related material	Committee, Panel of Experts, UNMIL
Libya	31/03/1992–12/09/2003	UNSCR 731 (1992); 748 (1992); 1192 (1998); 1506 (2003)	Arms and related material	Sanctions Committee
Libya	26/02/2011	UNSCR 1970 (2011); 1973 (2011); 2009 (2011)	Arms and related material	Sanctions Committee and Panel of Experts
North Korea	14/10/2006	UNSCR 1718 (2006); 1847 (2009); 2270 (2016)	Materials, technology, *etc.* which could contribute to WMD programmes. Arms and related material	Sanctions Committee and Panel of Experts
Rwanda (rebel groups)	16/08/1994–10/06/2008	UNSCR 918 (1994); 997 (1995); 1005 (1995); 1011 (1995)	Arms and related material	Sanctions Committee
Sierra Leone (Government), (rebels)	08/10/1997–05/06/1998, 05/06/1998–09/2010	UNSCR 1132 (1997); UNSCR 1171 (1998); 1306 (2000); 1940 (2010)	Arms and related material	UNAMSIL and Sanctions Committee
Somalia	23/01/1992	UNSCR 733 (1992); 1356 (2001); 1425 (2002); 1725 (2006); 1774 (2007); 1844 (2008)	Arms and related material	Sanctions Committee and Panel of Experts/Monitoring Group
South Africa	4/11/1977–25/5/1994	UNSCR 418	Arms and related material	Sanctions Committee
Sudan (Darfur) (non-State actors), all forces	30/07/2004	UNSCR 1556 (2004); UNSCR 1591 (2005)	Arms and related material	UN Mission in Sudan (UNMIS); Sanctions Committee and Panel of Experts
Yemen	01/06/1994	UNSCR 924 (1994); 931 (1994)	Arms and related material	None
Yemen (non-State actors)	14/4/2015	UNSCR 2216	Arms and related material	Sanctions Committee and Panel of Experts
Yugoslavia (FRY)	31/03/1998–10/09/2001	UNSCR 1160 (1998); 1367 (2001)	Arms and related material	Sanctions Committee

References

1. United Nations, *Charter of the United Nations*, San Francisco, US, 26 June 1945, Article 1.
2. United Nations, *Charter of the United Nations*, San Francisco, US, 26 June 1945, Article 39.
3. C. Ward, Building capacity to combat international terrorism: the role of the United Nations security council, *J. Conflict Secur. Law*, 2003, **8**(2), 294.
4. United Nations, *Charter of the United Nations*, San Francisco, US, 26 June 1945, Article 41.
5. United Nations, *Charter of the United Nations*, San Francisco, US, 26 June 1945, Article 42.
6. United Nations, *Charter of the United Nations*, San Francisco, US, 26 June 1945, Article 11.
7. United Nations, *Charter of the United Nations*, San Francisco, US, 26 June 1945, Article 99.
8. Security Council Report, *UN Sanctions*, *Special Research Report No. 3*, November 2013, pp. 2–3.
9. United Nations, *Charter of the United Nations*, San Francisco, US, 26 June 1945, Article 26.
10. United Nations, *Organisation for Disarmament Affairs*, https://www.un.org/disarmament/geneva/cd/an-introduction-to-the-conference/, accessed 30 January 2017.
11. OPCW, *Chemical Weapons Convention*, 1993, Article IX, paragraph 1, Article VIII, paragraph 36 and Article XII, paragraph 4.
12. OPCW, *Chemical Weapons Convention*, 1993, Article XVI, paragraph 2.
13. United Nations General Assembly, *Resolution 35/144C*, 12 December 1980.
14. United Nations General Assembly, *Resolution 37/98D*, 13 December 1982.
15. United Nations General Assembly, *Resolution 42/37C*, 30 November 1987, Paragraph 4.
16. United Nations General Assembly, *Resolution 42/37C*, 30 November 1987, Paragraph 7.
17. J. Tucker, *Putting Teeth in the Biological Weapons Convention, Issues in Science and Technology*, National Academy of Sciences, 2002.
18. United Nations General Assembly, *Resolution 46/35B*, 9 December 1991.
19. J. Littlewood, *Investigating Allegations of CBW Use: Reviving the UN Secretary-General's Mechanism, Compliance Chronicles, Number 3*, Carleton University, December 2006.
20. United Nations General Assembly, *Resolution A/Res/55/283*, 7 September 2001.
21. N. Smidovich, *The Secretary-General's Mechanism for Investigation of Alleged Use of Chemical, Biological or Toxin Weapons, UNODA Presentation, from the Geneva Protocol to the UNSG Mechanism: Lessons Learned and Opportunities*, UNIDIR, Geneva, Switzerland, 12 August 2015.
22. United Nations, *Office for Disarmament Affairs, The Secretary-General's Mechanism for the Investigation of Alleged Use of Chemical and Biological Weapons*, Factsheet, June 2013.

23. J. P. Robinson, *Alleged Use of Chemical Weapons in Syria, Harvard Sussex Program, Occasional Paper 4*, 26 June 2013.
24. United Nations, *United Nations Mission to Investigate Allegations of the Use of Chemical Weapons in the Syrian Arab Republic, Final Report*, 12 December 2013, Paragraph 5.
25. United Nations, *United Nations Mission to Investigate Allegations of the Use of Chemical Weapons in the Syrian Arab Republic, Final Report*, 12 December 2013, Paragraph 6.
26. United Nations, *United Nations Mission to Investigate Allegations of the Use of Chemical Weapons in the Syrian Arab Republic, Final Report*, 12 December 2013, Paragraph 2.
27. United Nations, *United Nations Mission to Investigate Allegations of the Use of Chemical Weapons in the Syrian Arab Republic, Final Report*, 12 December 2013, Paragraphs 7, 8, 9 & 12.
28. N. MacFarquhar, *UN to Investigate Chemical Weapons Accusations in Syria*, New York Times, 22 March 2013.
29. United Nations, *United Nations Mission to Investigate Allegations of the Use of Chemical Weapons in the Syrian Arab Republic, Final Report*, 12 December 2013, Paragraphs 28 & 29.
30. United Nations, *Frequently Asked Questions about the United Nations Mission to Investigate the Allegations of the Use of Chemical Weapons in the Syrian Arab Republic*, 20 September 2013.
31. United States, *The White House, Office of the Press Secretary, Government Assessment of the Syrian Government's Use of Chemical Weapons on August 21, 2013*, 30 August 2013.
32. United Nations, *United Nations Mission to Investigate Allegations of the Use of Chemical Weapons in the Syrian Arab Republic, Final Report*, 12 December 2013, Paragraph 35.
33. United Nations, *United Nations Mission to Investigate Allegations of the Use of Chemical Weapons in the Syrian Arab Republic, Final Report*, 12 December 2013, Paragraph 36.
34. United Nations, *United Nations Mission to Investigate Allegations of the Use of Chemical Weapons in the Syrian Arab Republic Report on the Alleged Use of Chemical Weapons in the Ghouta Area of Damascus on 21 August 2013, Note by the Secretary-General*, 16 September 2013, p. 1.
35. United Nations, *United Nations Mission to Investigate Allegations of the Use of Chemical Weapons in the Syrian Arab Republic Report on the Alleged Use of Chemical Weapons in the Ghouta Area of Damascus on 21 August 2013, Note by the Secretary-General*, 16 September 2013, paragraph 27.
36. United Nations, *United Nations Mission to Investigate Allegations of the Use of Chemical Weapons in the Syrian Arab Republic Report on the Alleged Use of Chemical Weapons in the Ghouta Area of Damascus on 21 August 2013, Note by the Secretary-General*, 16 September 2013, paragraph 28.
37. Editorial note, Kurds accuse Turkish Military of using chemical weapons, *WMD Insights*, April 2006.
38. United Nations General Assembly Resolution, *42/37C,* 30 November 1987.

39. OPCW, *Chemical Weapons Convention*, 1993, Article IX, Consultations, Cooperations and Fact Finding & Article X, Assistance and Protection against Chemical Weapons.
40. OPCW, *Chemical Weapons Convention*, 1993, Verification Annex, Part XI.
41. United Nations General Assembly, *Agreement Concerning the Relationship between the United Nations and the Organisation for the Prohibition of Chemical Weapons, Resolution, A/Res/55/283*, 7 September 2001.
42. OPCW, *OPCW Fact Finding Mission: "Compelling Confirmation" That Chlorine Gas Used as Weapon in Syria*, 10 September 2014.
43. United Nations, *Security Council Resolution 2235, S/RES/2235 (2015)*, 7 August 2015.
44. United Nations, *Security Council Resolution 2235, S/RES/2235 (2015)*, 7 August 2015, Paragraph 15.
45. United Nations, *Security Council, First Report of the Organization for the Prohibition of Chemical Weapons-United Nations Joint Investigative Mechanism, (OPCW–JIM), S/2016/142*, 12 February 2016, Paragraph 26.
46. United Nations, *Security Council, First Report of the Organization for the Prohibition of Chemical Weapons-United Nations Joint Investigative Mechanism, (OPCW–JIM), S/2016/142*, 12 February 2016, Paragraph 25.
47. United Nations, *Security Council, First Report of the Organization for the Prohibition of Chemical Weapons-United Nations Joint Investigative Mechanism, (OPCW–JIM), S/2016/142*, 12 February 2016, Paragraph 42.
48. United Nations, *Security Council, Second Report of the Organization for the Prohibition of Chemical Weapons-United Nations Joint Investigative Mechanism, (OPCW–JIM), S/2016/530*, 10 June 2016, Paragraph 4.
49. United Nations, *Security Council, Second Report of the Organization for the Prohibition of Chemical Weapons-United Nations Joint Investigative Mechanism, (OPCW–JIM), S/2016/530*, 10 June 2016, Paragraph 3.
50. United Nations, *Security Council, Third Report of the Organization for the Prohibition of Chemical Weapons-United Nations Joint Investigative Mechanism, (OPCW–JIM), S/2016/738*, 24 August 2016, Paragraph 54.
51. United Nations, *Security Council, Third Report of the Organization for the Prohibition of Chemical Weapons-United Nations Joint Investigative Mechanism, (OPCW–JIM), S/2016/738*, 24 August 2016, Paragraph 56.
52. United Nations, *Security Council, Third Report of the Organization for the Prohibition of Chemical Weapons-United Nations Joint Investigative Mechanism, (OPCW–JIM), S/2016/738*, 24 August 2016, Paragraph 58.
53. United Nations, *Security Council, Fourth Report of the Organization for the Prohibition of Chemical Weapons-United Nations Joint Investigative Mechanism, (OPCW–JIM), S/2016/888*, 21 October 2016, Paragraph 19.
54. United Nations, *Security Council, Seventh Report of the Organization for the Prohibition of Chemical Weapons-United Nations Joint Investigative Mechanism, (OPCW–JIM), S/2017/904,* 26 October 2017, Paragraph 46.
55. United Nations, *Security Council, Seventh Report of the Organization for the Prohibition of Chemical Weapons-United Nations Joint Investigative Mechanism, (OPCW–JIM),S/2017/904,* 26 October 2017, Paragraph 25.

56. United Nations, *Security Council Fails to Renew Mandate of Joint Investigative Mechanism on Chemical Weapons Use in Syria, as Permanent Member Casts Veto, Meetings coverage, Security Council, 8073rd Meeting, SC/13040*, 24 October 2017; UN, Security Council, Meetings coverage, Security Council Fails to Adopt 2 Draft Resolutions on Extending Mandate of Joint Mechanism Investigating Chemical Weapons Attacks in Syria, 8105th Meeting, SC/13072, 16 November 2017.
57. United Nations, *Security Council Resolution 2235, S/RES/2235 (2015)*, 7 August 2015, Paragraph 5.
58. United Nations, *Security Council, First Report of the Organization for the Prohibition of Chemical Weapons-United Nations Joint Investigative Mechanism, (OPCW–JIM), S/2016/142*, 12 February 2016, Paragraph 51.
59. United Nations, *Security Council, Third report of the Organization for the Prohibition of Chemical Weapons-United Nations Joint Investigative Mechanism, (OPCW–JIM), S/2016/738*, 24 August 2016, paragraph 18.
60. United Nations, *Security Council, Fourth Report of the Organization for the Prohibition of Chemical Weapons-United Nations Joint Investigative Mechanism, (OPCW–JIM), S/2016/888*, 21 October 2016, Paragraphs 27–33.
61. United Nations, *Security Council, Third report of the Organization for the Prohibition of Chemical Weapons-United Nations Joint Investigative Mechanism, (OPCW-JIM), S/2016/738*, 24 August 2016, Annex, paragraph 1.
62. United Nations, *Security Council, Fourth Report of the Organization for the Prohibition of Chemical Weapons-United Nations Joint Investigative Mechanism, (OPCW–JIM), S/2016/888*, 21 October 2016, Paragraph 52.
63. United Nations, *Security Council, Second report of the Organization for the Prohibition of Chemical Weapons-United Nations Joint Investigative Mechanism, (OPCW-JIM), S/2016/530*, 10 June 2016, paragraph 31.
64. United Nations, *Security Council, Second Report of the Organization for the Prohibition of Chemical Weapons-United Nations Joint Investigative Mechanism, (OPCW–JIM), S/2016/530*, 10 June 2016, Paragraph 18.
65. United Nations, *Security Council, Third Report of the Organization for the Prohibition of Chemical Weapons-United Nations Joint Investigative Mechanism, (OPCW–JIM), S/2016/738*, 24 August 2016, Paragraph 26, Annex 1.
66. United Nations, *Security Council, Third Report of the Organization for the Prohibition of Chemical Weapons-United Nations Joint Investigative Mechanism, (OPCW–JIM), S/2016/738*, 24 August 2016, Paragraph 27, Annex 1.
67. United Nations, *Security Council, Third Report of the Organization for the Prohibition of Chemical Weapons-United Nations Joint Investigative Mechanism, (OPCW–JIM), S/2016/738*, 24 August 2016, Paragraph 26.
68. United Nations, *Security Council, First Report of the Organization for the Prohibition of Chemical Weapons-United Nations Joint Investigative Mechanism, (OPCW–JIM), S/2016/142*, 12 February 2016, Paragraph 15.
69. United Nations, *Security Council, Third Report of the Organization for the Prohibition of Chemical Weapons-United Nations Joint Investigative Mechanism, (OPCW–JIM), S/2016/738*, 24 August 2016, Paragraph 28.

70. United Nations, *Security Council, First Report of the Organization for the Prohibition of Chemical Weapons-United Nations Joint Investigative Mechanism, (OPCW–JIM), S/2016/142*, 12 February 2016, paragraph 15.
71. R. Campos, *Russia vetoes Extension of Mission Probing Chemical Weapons Use in Syria*, Reuters, 24 October 2017.
72. United Nations, *Security Council Resolution 687, S/RES/687 (1991)*, 3 April 1991.
73. United Nations, *Security Council Resolution 687, S/RES/687 (1991)*, 3 April 1991, Paragraph 8(a).
74. United Nations, *Security Council Resolution 687, S/RES/687 (1991)*, 3 April 1991, Paragraph 9.
75. United Nations, *United Nations Special Commission, Mandate, Basic Facts*, undated.
76. United Nations Security Council, *UNMOVIC, Summary of the Compendium of Iraq's Proscribed Weapons Programmes in the Chemical, Biological and Missile Areas, S/2006/420*, 21 June 2006, Paragraph 147.
77. United Nations Security Council, *UNMOVIC, Summary of the Compendium of Iraq's Proscribed Weapons Programmes in the Chemical, Biological and Missile Areas, S/2006/420*, 21 June 2006, Paragraph 151.
78. United Nations, *United Nations Special Commission, Operational Activities, Basic Facts*, undated.
79. P. Lewis, *From UNSCOM to UNMOVIC: The United Nations and Iraq, Disarmament Forum*, 2001, vol. 2, p. 2.
80. United Nations, *Security Council Resolution 1284, S/RES/1284 (1999)*, 17 December 1999.
81. United Nations, *Security Council Resolution 1284, S/RES/1284 (1999)*, 17 December 1999, Paragraph 2.
82. United Nations, *Security Council Resolution 1441 (2002), S/RES/1441 (2002)*, 8 November 2002.
83. United Nations, *Security Council Resolution 1441 (2002), S/RES/1441 (2002)*, 8 November 2002, Paragraph 5.
84. United Nations, *Security Council Resolution 1441 (2002), S/RES/1441 (2002)*, 8 November 2002, Paragraph 7.
85. United Nations, *Security Council Resolution 1441 (2002), S/RES/1441 (2002)*, 8 November 2002, Paragraphs 4 & 13.
86. United Nations, *UNMOVIC, Chronology of Main Events*, http://www.unmovic.org/.
87. United Nations, *13th Quarterly Report of the Executive Chairman of UNMOVIC to the Security Council, S/2003/580*, 30 May 2003.
88. United Nations, *UNMOVIC, Basic Facts*, http://www.unmovic.org/, accessed 22 January 2017.
89. United Nations, *Security Council Resolution 1762 (2007), S/RES/1762(2007)*, 29 June 2007.
90. United Nations, Security Council, UNMOVIC, *Summary of the Compendium of Iraq's Proscribed Weapons Programmes in the Chemical, Biological and Missile Areas, S/2006/420*, 21 June 2006, p. 2.

91. OPCW, *Joint National Paper by the Russian Federation and the United States of America: Framework for Elimination of Syrian Chemical Weapons, EC-M-33/NAT.1*, 17 September 2013.
92. OPCW, *Decision: Destruction of Syrian Chemical Weapons, EC-M-33/DEC.1*, 27 September 2013, operative paragraph 1(a).
93. OPCW, *Decision: Destruction of Syrian Chemical Weapons, EC-M-33/DEC.1*, 27 September 2013, operative paragraph 1(c–d).
94. OPCW, *Decision: Destruction of Syrian Chemical Weapons, EC-M-33/DEC.1*, 27 September 2013, operative paragraph 2(b).
95. United Nations, *S/RES/2118 (2013) Security Council Resolution 2118 (2013)*, 27 September 2013, operative paragraph 6.
96. United Nations, *S/RES/2118 (2013) Security Council Resolution 2118 (2013)*, 27 September 2013, operative paragraph 7.
97. United Nations, *S/RES/2118 (2013) Security Council Resolution 2118 (2013)*, 27 September 2013, operative paragraph 21.
98. OPCW, *Press Release: Syria's Accession to the Chemical Weapons Convention Enters into Force*, 14 October 2013.
99. Removal and destruction of Syrian chemical weapons, *Presentation by A. Tucker Chemical Demilitarisation Branch*, OPCW, undated, slide 4.
100. Removal and destruction of Syrian chemical weapons, *Presentation by A. Tucker Chemical Demilitarisation Branch*, OPCW, undated, slide 6; Under the CWC Verification Annex, for destruction purposes Category 1 – comprises chemical weapons employing Schedule 1 chemicals, whilst Category 2 comprises chemical weapons employing all other toxic chemicals.
101. United Nations, *Joint Statement from the Organisation for the Prohibition of Chemical Weapons (OPCW) and the United Nations (UN)*, 16 October 2013.
102. OPCW–UN Joint Mission, Background http://opcw.unmissions.org/AboutOPCWUNJointMission/Background.aspx, accessed 22 January 2017.
103. Removal and destruction of Syrian chemical weapons, *Presentation by A. Tucker Chemical Demilitarisation Branch*, OPCW, undated, slide 3.
104. OPCW Executive Council, Eighty-seventh session, EC-87/DG.16, 13–16 March 2018, *Note by the Director General, Progress in the Elimination of the Syrian Chemical Weapons Programme*, 23 February 2018.
105. Removal and destruction of Syrian chemical weapons, *Presentation by A. Tucker Chemical Demilitarisation Branch*, OPCW, undated, slide 8.
106. OPCW, *OPCW Director-general Condemns Attack on UN Inspection Team*, 26 August 2013.
107. P. Walker, *Syrian Chemical Weapons Destruction: Taking Stock and Looking Ahead, Arms Control Today*, December 2014.
108. Removal and destruction of Syrian chemical weapons, *Presentation by A. Tucker Chemical Demilitarisation Branch*, OPCW, undated, slide 15.
109. OPCW–UN Joint Mission, *Frequently Asked Questions*, OPCW, undated.
110. OPCW–UN Joint Mission, *Communications Centre, Factsheet No.2: Environment and Health Protection,* undated.

111. OPCW, *Destruction of Syrian Chemical Weapons Completed*, 4 January 2016, press release.
112. OPCW, *Chemical Weapons Convention*, 1993, Article IV, paragraph 16.
113. OPCW, *Note by the Director-general: Progress in the Elimination of the Syrian Chemical Weapons Program, EC-77/DG.22*, 24 September 2014, Paragraph 14.
114. United Nations, *Charter of the United Nations*, San Francisco, US, 26 June 1945, ch. V, Article 27.3.
115. D. Fruchart, P. Holtom and S. Wezeman, *United Nations Arms Embargoes Their Impact on Arms Flows and Target Behaviour*, Stockholm International Peace Research Institute (SIPRI) & Uppsala University, 2007, pp. 2 and 50.
116. United Nations, *Charter of the United Nations*, San Francisco, US, 26 June 1945, ch. 1, Article 2, paragraph 5.
117. J. Stremlau, *Sharpening Economic Sanctions: Toward a Stronger Role for the United Nations*, Carnegie Corporation, New York, November 1996.
118. United Nations, *UNSCR 1718*, 14 October 2006, Article 8.a.(ii).e.
119. United Nations Security Council, *Letter Dated 1 November 2006 from the Chairman of the Security Council Committee Established Pursuant to Resolution 1718*, 2006, concerning the Democratic People's Republic of Korea addressed to the President of the Security Council, S/2006/853*, 7 November 2006.
120. United Nations Security Council, *Letter dated 29 September 2017 from the Chair of the Security Council Committee established pursuant to resolution 1718 (2006) addressed to the President of the Security Council*, S/2017/822, 29 September 2017; see also United Nations Security Council, *Letter Dated 4 April 2016 from the Chair of the Security Council Committee Established Pursuant to Resolution 1718*, 2006, addressed to the President of the Security Council, S/2016/308, 4 April 2016.
121. Bonn International Center for Conversion (BICC)/Bonn-Berlin Process, *Report of Expert Working Group IV, Monitoring and Enforcing UN Arms Embargoes*, January 2001, p. 5.
122. BICC/Bonn-Berlin Process, Monitoring and Enforcement of Arms Embargoes, in *Design and Implementation of Arms Embargoes and Travel and Aviation Related Sanctions*, ed. M. Brzoska, BICC, Bonn, 2001, p. 114.
123. Security Council Report, UN Sanctions, *Special Research Report No. 3*, November 2013, pp. 7–8.
124. D. Fruchart, P. Holtom and S. Wezeman, *United Nations Arms Embargoes Their Impact on Arms Flows and Target Behaviour*, Stockholm International Peace Research Institute (SIPRI) & Uppsala University, 2007, pp. 5 and 51.
125. UN Security Council, *Letter from the Permanent Representatives of France, the UK and the US to the President of the Security Council, S/2017/170*, 24 February 2017.
126. United Nations, Press Office, *Double Veto Prevents Security Council from Adopting Draft Resolution Intended to Impose Sanctions for Use of Chemical Weapons in Syria*, 28 February 2017.

127. D. Fruchart, P. Holtom and S. Wezeman, *United Nations Arms Embargoes Their Impact on Arms Flows and Target Behaviour*, Stockholm International Peace Research Institute (SIPRI) & Uppsala University, 2007, pp. 5 and 51.
128. United Nations, *Security Council, Resolution 1540* (2004), S/RES/1540 (2004), 28 April 2004.
129. United Nations, *Security Council, Resolution 1540* (2004), S/RES/1540 (2004), 28 April 2004, Operative provision 1.
130. United Nations, *Security Council, Resolution 1540* (2004), S/RES/1540 (2004), 28 April 2004, Operative provision 2.
131. United Nations, *Security Council, Resolution 1540* (2004), S/RES/1540 (2004), 28 April 2004, Operative provision 3.
132. United Nations, *Security Council, Resolution 1540* (2004), S/RES/1540 (2004), 28 April 2004, Operative provision 8(a).
133. United Nations, *Security Council, Resolution 1540* (2004), S/RES/1540 (2004), 28 April 2004, Operative provision 8(b).
134. United Nations, *Security Council, Resolution 1540* (2004), S/RES/1540 (2004), 28 April 2004, footnote.
135. O. Bosch and P. van Ham, Global non-proliferation and counter-terrorism: the role of resolution 1540 and its implications, in *Global Non-proliferation and Counter terrorism: The Impact of UNSCR 1540*, ed. O. Bosch and P. van Ham, Chatham House, London, 2007, p. 12.
136. O. Bosch and P. van Ham, Global non-proliferation and counter-terrorism: the role of resolution 1540 and its implications, in *Global Non-proliferation and Counter terrorism: The Impact of UNSCR 1540*, ed. O. Bosch and P. van Ham, Chatham House, London, 2007, pp. 10–11.
137. United Nations, *Security Council, Resolution 1540* (2004), S/RES/1540 (2004), 28 April 2004, preambular paragraph.
138. O. Bosch and P. van Ham, Global non-proliferation and counter-terrorism: the role of resolution 1540 and its implications, in *Global Non-proliferation and Counter terrorism: The Impact of UNSCR 1540*, ed. O. Bosch and P. van Ham, Chatham House, London, 2007, pp. 10, 14–18.
139. G. Manley, Restricting non-state actors' access to chemical weapons and related materials: implications for UNSCR 1540, in *Global Non-proliferation and Counter Terrorism: The Impact of UNSCR 1540*, ed. O. Bosch and P. van Ham, Chatham House, London, 2007, p. 79.
140. Egypt, Israel and North Korea Chemical Weapons Profiles on the Nuclear Threat Initiative website, http://www.nti.org/learn/countries/, accessed 25 January 2017.
141. OPCW, *Report of the OPCW on the implementation of the Convention in 2016, 22nd Session of the Conference of States Parties, C-22/4, 29 November 2017*.
142. G. Manley, Restricting non-state actors' access to chemical weapons and related materials: implications for UNSCR 1540, in *Global Non-proliferation and Counter Terrorism: The Impact of UNSCR 1540*, ed. O. Bosch and P. van Ham, Chatham House, London, 2007, pp. 75–76.

143. United Nations, *Charter of the United Nations*, San Francisco, US, 26 June 1945, ch. VII, Article 41 & 42, in particular.
144. United Nations, *Security Council, Resolution 1540* (2004), S/RES/1540 (2004), 28 April 2004, Article 4.
145. H. Veel, *1540 and the 2016 Comprehensive Review*, International Law and Policy Institute, June 2016, p. 5.
146. Central African Republic, Chad, Comoros, Equatorial Guinea, Gambia, Guinea, Guinea-Bissau, Haiti, Mali, Mauritania, Mozambique, North Korea, Solomon Islands, Somalia, Swaziland, Timor-Leste and Zimbabwe had not submitted a report by 24 April 2016, See: United Nations, *Security Council, Report of the Security Council Committee Established Pursuant to Resolution 1540* (2004), S/2016/1038, 9 December 2016, Annex VI.
147. UNSC, *Resolution 1977* (2011), SC/10228, 20 April 2011, Paragraph 3.
148. United Nations, *Security Council, Report of the Security Council Committee Established Pursuant to Resolution 1540* (2004), S/2016/1038, 9 December 2016.
149. United Nations, *Security Council, Report of the Security Council Committee Established Pursuant to Resolution 1540* (2004), S/2016/1038, 9 December 2016, Paragraph 26.
150. United Nations, *Security Council, Report of the Security Council Committee Established Pursuant to Resolution 1540* (2004), S/2016/1038, 9 December 2016, paragraph 25.
151. United Nations, *Security Council, Report of the Security Council Committee Established Pursuant to Resolution 1540* (2004), S/2016/1038, 9 December 2016, paragraph 63.
152. United Nations, *Security Council, Report of the Security Council Committee Established Pursuant to Resolution 1540* (2004), S/2016/1038, 9 December 2016, paragraph 64.
153. United Nations, *Security Council, Report of the Security Council Committee Established Pursuant to Resolution 1540* (2004), S/2016/1038, 9 December 2016, paragraph 73.
154. United Nations, *Security Council, Report of the Security Council Committee Established Pursuant to Resolution 1540* (2004), S/2016/1038, 9 December 2016, paragraph 89.
155. P. Crail, Implementing UN security council resolution 1540, a risk-based approach, *Nonproliferation Rev.*, 2006, **13**(2), 359.
156. H. Veel, *1540 and the 2016 Comprehensive Review*, International Law and Policy Institute, June 2016, p. 3.
157. O. Bosch and P. van Ham, Global non-proliferation and counter-terrorism: the role of resolution 1540 and its implications, in *Global Non-proliferation and Counter Terrorism: The Impact of UNSCR 1540*, ed. O. Bosch and P. van Ham, Chatham House, London, 2007, p. 7.
158. United Nations, *Security Council, Resolution 1540* (2004), S/RES/1540 (2004), 28 April 2004, paragraph 4.
159. UNSC, *Resolution 1673*, 27 April 2006; UNSC, *Resolution 1810*, 25 April 2008.

160. UNSC, *Resolution 1977*, 20 April 2011.
161. UNSC, *Resolution 1977*, 20 April 2011, Paragraph 2.
162. UNSC, *Resolution 1977*, 20 April 2011, Paragraph 5a.
163. *NTI, Challenges Faced by States in Implementing 1540*, http://www.nti.org/analysis/reports/1540-reporting-overview/, accessed 4 January 2017.
164. P. Crail, Implementing UN security council resolution 1540, a risk-based approach, *Nonproliferation Rev.*, 2006, **13**(2), 358.
165. P. Crail, Implementing UN security council resolution 1540, a risk-based approach, *Nonproliferation Rev.*, 2006, **13**(2), 358–359.
166. United Nations, *Report of the Committee Established Pursuant to Security Council Resolution 1540* (2004), 9 December 2016, S/AC.44/2016/OC.79, p. 2.
167. United Nations, *Report of the Committee Established Pursuant to Security Council Resolution 1540* (2004), 9 December 2016, S/AC.44/2016/OC.79, table 1.
168. United Nations, *Report of the Committee Established Pursuant to Security Council Resolution 1540* (2004), 9 December 2016, S/AC.44/2016/OC.79, table 2.

CHAPTER 6

International Legal Constraints upon the Weaponization of Toxic Chemicals

MICHAEL CROWLEY

University of Bradford, Bradford, West Yorkshire, BD7 1DP, UK
*E-mail: m.j.a.crowley1@bradford.ac.uk

6.1 Introduction

This chapter seeks to explore the full range of potentially applicable international law constraining the "weaponization of toxic chemicals". Whilst clearly covering the use of chemical warfare agents as weapons against combatants or civilian populations, this term is chosen so as to emphasise the comprehensive nature of the potential chemicals that can be employed and also the range of circumstances in which they can be employed, some of which may not be captured by the more restrictive term "chemical weapons".[†]

This chapter will review four areas of relevant international law. Firstly, it will look at the prohibitions and restrictions on the weaponization of toxic chemicals imposed by arms control and disarmament treaties (primarily the Chemical Weapons Convention, but also the Geneva Protocol and

[†]For example, States may not agree that a specific toxic chemical (*e.g.* one promoted as an incapacitating chemical agent) used as a weapon in certain circumstances (*e.g.* for law enforcement) should be considered as a "chemical weapon". However, in such cases, clearly the weaponised use of such toxic chemicals should and is constrained by relevant international law.

Preventing Chemical Weapons: Arms Control and Disarmament as the Sciences Converge
Edited by Michael Crowley, Malcolm Dando and Lijun Shang

Published by the Royal Society of Chemistry, www.rsc.org

the Biological and Toxin Weapons Convention). The chapter will then explore the application of international humanitarian law and the further constraints it imposes on the use of toxic chemicals by combatants engaged in armed conflict. This will be followed by an analysis of international human rights law and the wide-ranging obligations upon States to regulate the use of force (including weapons employing toxic chemicals) by law enforcement officials (potentially covering military and security personnel as well as police and prison officers). Whilst preceding sections concentrate upon international law regulating the activities of States, the final section will address individual responsibility and culpability for crimes involving weaponised toxic chemicals, through the employment of international criminal law.

In addition to examining relevant treaty law (which is binding only on those States that are party to the specific agreements, and only after the State has ratified the relevant instrument), the obligations based upon customary international law will be explored. Customary international law which can be considered as "a general practice accepted as law",[1] is binding on all States whether or not they are parties to the relevant treaties. Throughout this analysis, the author will highlight ambiguities in the relevant instruments or contested interpretations of law, exploring the implications for effective regulation and highlighting areas where further action is needed by the international community.

6.2 Geneva Protocol

Under the 1925 Geneva Protocol, the High Contracting Parties have acknowledged that "the use in war of asphyxiating, poisonous or other gases, and of all analogous liquids, materials or devices, has been justly condemned by the general opinion of the civilised world" and have further declared that "this prohibition shall be universally accepted as a part of International Law, binding alike the conscience and the practice of nations."[2]

The scope of the chemical (and biological) agents covered by the Protocol is considered by the majority of States to be very broad, including *inter alia* "(a) any chemical agents of warfare—chemical substances, whether gaseous, liquid or solid—which might be employed, because of their direct toxic effects on man, animals or plants."[3] However, the prohibition relates solely to their *use*; the Protocol does not address the development, production, transfer or stockpiling of such agents. Furthermore, the Protocol's prohibition on use is limited to situations of war, although it is now interpreted through customary international law to apply to both international and non-international armed conflict.[4]

Although the Geneva Protocol has widespread support, its prohibitions are somewhat weakened by reservations submitted by a number of countries when becoming Parties to the Protocol, declaring that they only regarded the non-use obligations as applying to other States Parties and that these obligations would cease to apply if the prohibited weapons were used against them. Whilst some States Parties subsequently removed such qualifications, a number still maintain their reservations on this issue.[‡]

6.3 Chemical Weapons Convention

6.3.1 General Obligations

The Chemical Weapons Convention (CWC) under Article I, prohibits the development, production, stockpiling, transfer and use of chemical weapons "under any circumstances".[5] States Parties are also prohibited from engaging in "any military preparations to use chemical weapons"[6] or to "assist, encourage or induce, in any way, anyone to engage in any activity prohibited to a State Party under this Convention."[7] In addition, Article I also requires that all existing stocks of chemical weapons[8] and chemical weapons production facilities[9] be destroyed. Although the Convention prohibits chemical weapons, it allows for the controlled peaceful use of toxic chemicals.[10] Under Article VII of the CWC, all States Parties are required to "adopt the necessary measures to implement [their] obligations under this Convention" and "not permit in any place under [their] control any activity prohibited to a State Party under this Convention."[11]

6.3.2 Definitions

6.3.2.1 Toxic Chemicals and Precursors

The comprehensive nature of the CWC in terms of substances covered is established under Article II.2 which defines toxic chemicals as: "any chemical which through its chemical action on life processes, can cause death, temporary incapacitation or permanent harm to humans or animals.[§] This includes all such chemicals, regardless of their origin or of their method of production, and regardless of whether they are produced in facilities, in munitions or elsewhere."[12]

In their 2014 Commentary on the Convention, Krutzsch and Trapp note that "the level of toxicity and the specific nature of the chemical action on life processes are not relevant to the characterization of the substance as a toxic chemical and that there is no dosage criterion."[13]

Furthermore, since the definition of toxic chemicals included "any chemical, regardless of its origin or method of production" the Convention is not restricted to chemicals manufactured by humans but would also cover those produced naturally, including certain poisons produced

[‡]For details of States Parties with reservations and text of such reservations see ref. 141.

[§]The CWC's definition of "toxic chemical" does not include toxicity against plants. Instead the CWC makes reference to herbicides only in the Preamble, where it recognises the prohibition of their use as a method of warfare.

by living organisms, *i.e.* toxins. The scope of substances captured by the CWC is further widened by the accompanying definition of precursor chemicals in Article II.3 as "[A]ny chemical reactant which takes part at any stage in the production by whatever method of a toxic chemical. This includes any key component of a binary or multicomponent chemical system."[14]

The importance of the Convention's comprehensive coverage has been widely recognised, not least by the World Health Organisation (WHO) in its 2004 report on the public health response to biological and chemical weapons.[15] The WHO report highlighted the dangers of the weaponization of a spectrum of chemical and toxin agents: "now that large-scale production processes for biologically active peptides and similar substances are undergoing rapid commercial development, bioregulators and other toxins constitute a field rich in potential weapons as well as pharmaceuticals, and in particular weapons of intense disabling or incapacitating power..."[15] It subsequently noted that "It is fortunate, therefore, that this advance in biotechnology should have coincided with the adoption of the Chemical Weapons Convention, since it places its States Parties under the express obligation to ensure that bioregulators and other toxins, like all other toxic chemicals, are used only for the purposes that the Convention does not prohibit."[15]

6.3.2.2 Riot Control Agents

Article II.7 of the Convention defines riot control agents (RCAs) as "[A]ny chemical not listed in a Schedule, which can produce rapidly in humans sensory irritation or disabling physical effects which disappear within a short time following termination of exposure."[16]

Consequently, RCAs clearly fall within the ambit of "toxic chemicals" as defined by Article II, paragraph 2, of the Convention. In June 2013, the Organisation for the Prohibition of Chemical Weapons (the OPCW) Director-General recognised that "the definition of RCAs in the Convention leaves some room for interpretation as to which chemicals can be considered as meeting the requirement specified in Article II(7)."[17] Consequently, in June 2014 the OPCW Technical Secretariat and Scientific Advisory developed a non-exhaustive and indicative list of 17 toxic chemicals that corresponded to the Article II.7 definition and could therefore be considered as RCAs and subject to declaration under the CWC.[18]

6.3.2.3 Chemical Weapons

Whilst the Convention includes, in one of its annexes, three Schedules of toxic chemicals and precursors, these are specifically "identified for the application of verification measures"[12] and are not intended to delineate the range of chemicals that may be considered to be "chemical weapons" under the Convention.[19] Instead the Convention defines "chemical weapons" under Article II.1 as:

(a) "toxic chemicals and their precursors, *except where intended for purposes not prohibited under this Convention, as long as the types and quantities are consistent with such purposes;*"[emphasis added].
(b) munitions and devices specifically designed to cause death or other harm through the toxic properties of those toxic chemicals specified in subparagraph (a), which would be released as a result of the employment of such munitions and devices;
(c) any equipment specifically designed for use directly in connection with the employment of the munitions and devices referred to in (b)."[20]

This highlighted text has become known as the General Purpose Criterion (GPC).

Krutszch and Trapp emphasised the implications of the open-ended nature of this provision:

> "Under this concept, *all* toxic or precursor chemicals are regarded as chemical weapons *unless* they have been developed, produced, stockpiled, or used for purposes *not* prohibited. The definition thus covers any toxic or precursor chemical if intended for chemical weapons purposes, irrespective of whether it has been listed in one of the Schedules." [Highlighted in the original].[21]

Indeed, the use of the toxic properties of *any* chemical as a weapon against either humans or animals—even those chemicals not originally developed or intended for such purposes, such as industrial toxic chemicals (*e.g.* ammonia or chlorine), or herbicides and defoliants (*e.g.* Agent Orange)—is captured by the GPC. Furthermore, because the GPC establishes a prohibition based on intent rather than on a limited list of toxic chemical agents, it allows the Convention to accommodate and reflect developments in science and technology; consequently, as Meselson and Perry Robinson have highlighted, "even toxic chemicals whose existence is not yet known are covered" by its provisions.[22]

To determine whether the use of a toxic chemical would be in conformity with the CWC, the intention or purpose for its use needs to be established. The range of "purposes not prohibited" under the CWC has been defined under Article II.9 as:

(a) "Industrial, agricultural, research, medical, pharmaceutical or other peaceful purposes;
(b) Protective purposes, namely those purposes directly related to protection against toxic chemicals and to protection against chemical weapons;

(c) Military purposes not connected with the use of chemical weapons and not dependent on the use of the toxic properties of chemicals as a method of warfare;
(d) Law enforcement including domestic riot control purposes."[23]

Consequently, any toxic chemicals held or utilised for purposes not provided for in Article II.9. are then chemical weapons and prohibited under the CWC. Furthermore, the use of toxic chemicals for "purposes not prohibited" would be acceptable only "as long as the types and quantities [of toxic chemicals] are consistent with such purposes."[24]

6.3.3 Use of Weapons Employing Toxic Chemicals in Armed Conflict

The use in armed conflict of any toxic chemical as weapons is absolutely prohibited under the CWC, as is their development, production, acquisition, stockpiling, retention or transfer when intended for such purposes, under Articles I and II. In addition, Article 1.5 specifically prohibits the use of "riot control agents as a method of warfare."[25]

While the vast majority of CWC States Parties hold a comprehensive interpretation of Article I.5 prohibiting all military use of RCAs in armed conflict, the US maintains a long-held position, under Executive Order 11850, that RCAs can be legitimately used *inter alia*: (a) in riot control situations including rioting prisoners of war; (b) where civilians are used to mask or screen attacks; (c) in rescue missions in remote areas, of downed aircrews and passengers, and escaping prisoners; and (d) to protect convoys from civil disturbances, terrorists and paramilitary organizations.[26] Depending on the circumstances, activities described in (b), (c), and (d) could potentially breach the Convention, and consequently the US position has been of long-standing concern to many States Parties.

Such concerns regarding divergent interpretation (and the danger of divergent implementation by States Parties) are compounded because of difficulties in determining whether a particular use of an RCA or other toxic chemical by military forces is occurring in the context or armed conflict or law enforcement. The question of exactly when a specific situation moves from civil unrest, such as rioting, requiring a law enforcement response to armed conflict is not always easy to answer, particularly in contexts approaching the lower threshold of armed conflict. Furthermore, in certain cases a situation on the ground may move back and forth between low-intensity armed conflict and law enforcement.[27] Such considerations have come to the fore, for example, with regard to the use of RCAs by military forces in occupied territories when the occupied population takes part in acts of violence and the occupying power has an obligation under International Humanitarian Law to maintain law and order. In addition, there has been growing concern over the perceived blurring of the lines between law enforcement and armed conflict through, for example, the development in military doctrine of concepts such as military operations in urban terrain (MOUT), counter-insurgency (COIN) operations and urban warfare.[27]

6.3.4 Use of Weapons Employing Toxic Chemicals in Law Enforcement

Under Article II.9.d certain toxic chemicals are permitted for "(d) law enforcement including domestic riot control purposes".[23] Whilst the CWC contains no definition of this expression, it is clear from the wording that "law enforcement" has a broader meaning than just "domestic riot control". Boothby argues that it is "best interpreted" as meaning the "enforcement of the domestic law of the territory in which the operations are taking place."[28] It is claimed—though contested—that certain States, namely the US, during the negotiations argued for a treaty wording that would allow them to carry out capital punishment utilising toxic chemicals[142¶] and this is subsequently reflected in, for example, US national legislation implementing the Convention. In addition, legal scholars have argued that the use of certain toxic chemicals (*i.e.* RCAs) would be permissible in certain law enforcement operations abroad where for example consent has been given by the host State or under authorisation of the UN Security Council.

Because neither "law enforcement" nor the types of activities covered by it have been defined under the CWC or in any clarificatory decision agreed by the Executive Council or Conference of States Parties, it is the obligation of each individual State Party to implement the relevant provisions of the Convention in good faith, taking into account "any relevant rules of international law applicable in the relations between the parties."[29] Of particular relevance are those rules and restrictions on the use of force by law enforcement officials—which can include police, security or military personnel—that arise from international human rights law and related standards (as explored in Section 6.6).

6.3.4.1 Riot Control Agents

Clearly under the CWC, RCAs can be employed for "law enforcement including domestic riot control purposes", as long as such use is in conformity with international human rights law, and provided it is consistent with the 'types and quantities' restriction of Article II.1.

However, in a number of cases of reported human rights violations, RCAs have been employed in quantities or in a manner where the agent dosages have become injurious or even fatal. In addition to contravening international

¶In contrast Robinson states that "such an association between capital punishment and law enforcement is nowhere referred to in the formal papers of the CWC negotiation". See ref. 143.

human rights law and related standards, such use also appears to be a direct breach of the *"types and quantities"* provisions of the CWC. Although UN and regional human rights bodies have documented the widespread and serious misuse of RCAs by law enforcement officials, to date, no cases of such misuse have been formally raised publicly by CWC States Parties as matters of concern in the context of the Convention.

6.3.4.1.1 RCA Means of Delivery. Since the use of "riot control agents as a method of warfare" is expressly prohibited under the Convention, States are prohibited from developing, transferring, acquiring, stockpiling or employing RCA means of delivery intended for such use in either international or non-international armed conflict. This prohibition arguably extends to RCA means of delivery intended for use in certain military operations in urban terrain (MOUT) not of a law enforcement nature; as well as certain counter-insurgency operations (COIN) or urban warfare. Consequently, any State or commercial entity manufacturing and promoting RCA means of delivery for such purposes would appear to be in contravention of the Convention.

CWC States Parties are permitted to manufacture, promote, acquire, stockpile and employ delivery systems to disseminate appropriate "types and quantities" of RCAs for "law enforcement" purposes. Consequently, there are a range of means of delivery, widely employed by police and security forces, which have a narrow dispersal area, short range, are discriminatory and emit a limited quantity of RCA, and consequently do not appear to conflict with the "types and quantities" restriction. Their use should not raise concern, provided it is consistent with the "law enforcement" purpose under the Convention, and is in accordance with the relevant human rights standards. In contrast, a range of delivery mechanisms have been developed that deliver far larger amounts of RCAs over wider areas and/or over greater distances. Certain such "wide area" RCA means of delivery—notably RCA cluster munitions, artillery shells and mortar bombs—inherently breach the CWC "types and quantities" provision and/or the prohibition on use of RCAs as a "method of warfare".

6.3.4.2 Incapacitating Chemical Agent (ICA) Weapons

Whilst there is consensus that RCAs can be employed for law enforcement, a highly contested issue currently being debated within the OPCW and amongst legal scholars is whether the CWC would allow the use of a further limited range of toxic chemicals for such purposes.

One line of treaty interpretation supported by Chayes and Meselson,[30] Krutzsch and Trapp,[31] the International Committee of the Red Cross (the ICRC),[32] and a growing number of States Parties including Australia, Canada, Germany the UK and the US, holds that only RCAs can be used for law enforcement activities under the CWC. The only possible exceptions to this restriction recognised are those toxic chemicals used for judicially sanctioned execution, and only provided such chemicals are not on the Convention's Schedule 1 list. Consequently, the use of all other toxic chemicals, including so called incapacitating chemical agents (ICA),[||] as weapons for law enforcement would be prohibited. Krutzsch and Trapp, for example, argue that "[A]ccepting, in principle, the use of certain toxic chemicals such as ICAs for law enforcement in exceptional situations, for instance during an anti-terrorist operation, would undermine the prohibition on Article I of the Convention and wrongly imply that there was a 'choice' between applying binding rules of treaty interpretation and replacing them by explanations constructed as required by military interest."[33]

A second line of opposing interpretation, as advanced by Fidler,[34] and supported by Neill[35] and the International Law and Policy Institute[27] contends that the CWC does not limit the range of toxic chemicals that can be used for law enforcement purposes to RCAs. Such interpretation may allow the use of ICA weapons for law enforcement in certain circumstances. However, Fidler notes that the "use of a toxic chemical for law enforcement purposes is still subject to the CWC requirements that the types and quantities of chemicals ... must be consistent with such permitted purposes (Article II.1 [a.])."[36] According to Fidler, this restriction therefore "requires scrutiny of the relationship between the chemical or biochemical agent and the law enforcement objective in question. The more difficult it is to control the effects of the use of a chemical or biochemical in a law enforcement operation, the more suspect such use becomes in terms of the agent being of a type or quantity consistent with a law enforcement purpose."[37] Consequently, Fidler believes that: "For domestic law enforcement, use of incapacitating agents in contexts in which the government could control neither dosage nor the exposure environment would only be legitimate in extreme law enforcement situations ... For extraterritorial law enforcement activities undertaken by military forces and sanctioned by international law, States can at present only legitimately use riot control agents, not incapacitating agents."[38]

The potential threats arising from the application of advances in science and technology to the development of new weapons purportedly for law enforcement purposes, but which employ toxic chemicals (such as ICAs) beyond the accepted RCAs, or which incorporate wide area dispersal mechanisms, have been raised by a growing number of scientific, medical and human rights organisations. To date, however, no OPCW policy making organ (PMO) has made statements regarding the application of the CWC in these areas. It is therefore left to individual States Parties to interpret the scope and nature of their obligations in these areas.

[||]Although there is no universally agreed definition of ICA weapons, for the sake of this discussion they can be considered as weapons employing toxic chemicals whose purported intended use is to cause prolonged but non-permanent disability; they include centrally acting agents producing loss of consciousness, sedation, hallucination, incoherence, paralysis, disorientation or other such effects.

6.4 Biological and Toxin Weapons Convention (BTWC)

Article I of the Biological and Toxin Weapons Convention (BTWC) establishes the principal obligations and prohibitions of the Convention, declaring that:

"Each State Party to the Convention undertakes never in any circumstances to develop, produce, stockpile or otherwise acquire or retain:

- Microbial or other biological agents, or toxins, whatever their origin or method of production, of types and in quantities that have no justification for prophylactic, protective or other peaceful purposes.
- Weapons, equipment or means of delivery designed to use such agents or toxins for hostile purposes or in armed conflict."[39]

Article 1 together with the extended understandings agreed at successive BTWC Review Conferences, make it clear that the Convention is intended to be fully inclusive with regards to the range of substances which it covers and also the potential targets (which include plants as well as animals and humans). For example, the Final Declaration of the Eighth Review Conference of 2016 stated that:

> "*...all naturally or artificially created or altered microbial and other biological agents and toxins, as well as their components, regardless of their origin and method of production and whether they affect humans, animals or plants*, of types and in quantities that have no justification for prophylactic, protective or other peaceful purposes, are unequivocally covered by Article I. ... [and that] ... *Article I applies to all scientific and technological developments in the life sciences and in other fields of science relevant to the Convention.*"[40][Emphasis added].

Since a range of toxic chemicals are either components or products of biological organisms, they fall within the scope of Article I of the Convention (as arguably do certain synthetic biologically active analogues of such substances). As well as being captured under the term "other biological agents", some toxic chemicals may also be covered by the Article I language that relates to toxins. Although toxins were not defined under the Convention nor has a definition been collectively agreed subsequently by BTWC States Parties, they are commonly taken to denote chemical compounds produced by living organisms that are toxic or harmful to another living organism.[144] In this regard, a number of BTWC States Parties including the Netherlands, Russian Federation, Sweden, UK and the US have specifically recognised the relevance of peptides, bioregulators and their analogues to the BTWC, and highlighted the potential dangers that misuse of such chemicals could pose.[145]

Although the BTWC covers a range of relevant toxic chemicals, there are ambiguities regarding the nature and scope of such coverage. For example, since the terms "hostile purposes" and "peaceful purposes" have not been defined under the Convention, it is unclear how the use of such toxic chemicals for counter-terrorist, counter-insurgency or military operations short of armed conflict would be regulated by the BTWC. To date, there have been no determinations of these issues by the BTWC States Parties at either the BTWC Review Conferences or at any Special Conference convened for the purpose. As explored in Chapter 4 of this publication, further important limitations on the value of the BTWC (and its control regime) as a tool to regulate such substances arise from its current lack of effective verification and compliance mechanisms, and also the absence of an international organization comparable to the OPCW which could coordinate such activities and facilitate implementation by States Parties.

6.5 International Humanitarian Law

6.5.1 Introduction

International Humanitarian Law (IHL) is the body of law that applies during situations of armed conflict with the aim of protecting civilians and others who are no longer participating in hostilities, and regulating the conduct of such hostilities. Amongst its provisions are those regulating the means of conflict (including the weapons employed) and also the methods of warfare (how such weapons are employed). IHL is applicable to international armed conflicts *i.e.* conflicts that arise between two or more States, even if a state of war is not recognized by one of them. It also covers all cases of partial or total occupation of the territory of a State, even if the said occupation meets with no armed resistance. In addition, IHL is applicable to armed conflicts of a non-international character, which can be considered as conflicts between the armed forces of a State and organized armed groups (or between two or more such groups) which are under responsible command, control territory and carry out sustained and concerted military operations. An armed conflict is not considered to include a situation of internal disturbances and tensions, such as riots, isolated and sporadic acts of violence and other acts of a similar nature. As noted previously, the determination of whether a specific situation should be considered as an internal armed conflict or as internal disturbance requiring law enforcement can be highly contentious. As well as regulating the activities of States in armed conflicts, IHL is applicable to armed non-State actors that meet the requisite criteria.[41]

6.5.2 Over-arching IHL Obligations Constraining the Weaponization of Toxic Chemicals

In addition to the weapons-specific agreements prohibiting the development, possession or use of chemical weapons (notably the CWC and the Geneva Protocol), there are a number of generally applicable IHL treaties that further constrain the weaponization of toxic chemicals. Of particular relevance are the four Geneva Conventions, which principally safeguard a range of "protected persons" in the power of the opposing party, including wounded combatants, prisoners of war, medical personnel and civilians;[42] and two Additional Protocols[43] which address certain aspects of the conduct of armed conflict. Key obligations deriving from these treaties or from related customary international law are discussed below.

6.5.2.1 *The Prohibition of Deliberate Attacks on Civilians, the Prohibition of Indiscriminate Weapons and of Attacks that do not Discriminate Between Civilians and Military Objectives*

The above prohibitions are considered fundamental to IHL and are covered by both treaty law (*e.g.* Additional Protocol I[44]) and customary IHL.[45] These prohibitions are applicable to both international and non-international armed conflict.[45]

The application of such prohibitions to the use of industrial toxic chemicals as weapons in Syria has been highlighted in 2017 by the Independent International Commission of Inquiry on the Syrian Arab Republic, which stated that:

> "The use of weapons in densely-populated areas which are by nature indiscriminate and whose effects cannot be limited as required by international humanitarian law is prohibited. As the dispersal pattern of gas found in chlorine bombs cannot be controlled, their use throughout residential areas in eastern Aleppo city amounts to the war crime of indiscriminate attacks in a civilian populated area."[46]

Although the International Court of Justice has stated that the principle of distinction was one of the "cardinal principles" of IHL and one of the "intransgressible principles of international customary law",[47] proponents of "less lethal" weapons employing toxic chemicals have argued that the rules of IHL, particularly regarding distinction, should be reviewed and applied in a different manner for such weapons than for conventional arms.[146] Highlighting the dangers of such an approach, Herby, Head of the Arms Unit in the Legal Division of the ICRC, noted that in discussions of potential scenarios for the use of ICA weapons, "situations in which civilians are interspersed with combatants are consistently mentioned … It is likely that the use of incapacitants will lower the threshold for attacks that affect civilians and combatants without distinction, with an inherent risk that this rule [prohibiting indiscriminate attacks] will be undermined."[48] The employment of RCAs against mixed populations of civilians and combatants, as permitted under existing US policy and regulations,[26] raises related concerns.

6.5.2.2 *The Prohibition of the Employment of Means and Methods of Warfare of a Nature to Cause Superfluous Injury or Unnecessary Suffering (SIRUS)*

The SIRUS prohibition has been enunciated in a number of IHL treaties,[147] and is considered to be part of customary IHL.[49] A review of military manuals by the ICRC has shown that a number of States such as Australia, France and Germany consider that the use of chemical weapons can cause unnecessary suffering.[49] However, there is no international consensus regarding an objective means of determining what constitutes "superfluous injury or unnecessary suffering", nor on the criteria which can be used to judge whether specific weapons potentially breach the SIRUS prohibition.

In 1997, the ICRC's SIRUS Project proposed that what constituted superfluous injury and unnecessary suffering should:

"[B]e determined by design-dependent, foreseeable effects of weapons when they are used against human beings and cause[d]:

- specific disease, specific abnormal physiological state, specific abnormal psychological state, specific and permanent disability or specific disfigurement; or
- field mortality of more than 25% or hospital mortality of more than 5%; or
- Grade 3 wounds as measured by the Red Cross wound classification; or,
- effects for which there is no well-recognized and proven treatment."[50]

Criteria 1, 2 and 4, in particular, appear to be of potential applicability for States considering the legality of the weaponization of certain toxic chemicals. The ICRC subsequently raised the potential dangers that ICA weapons, even though promoted as "less lethal" weapons, could violate the SIRUS prohibition if they resulted in effects such as "lifelong epileptic convulsions, permanent damage to internal organs, long-term and severe vomiting, or an extended coma".[51]

6.5.2.3 *The Protection of Persons Considered* Hors de Combat

Common Article 3 of the Geneva Conventions stated that "Persons taking no active part in the hostilities, including members of armed forces who have laid down their arms and those placed *hors de combat* by sickness, wounds, detention, or any other cause, shall in all circumstances be treated humanely."[52] In addition, certain actions against such persons are prohibited, including "(a) violence to life and person, in particular murder of all kinds, mutilation, cruel treatment and torture; (b) taking of hostages; (c) outrages upon personal dignity, in particular humiliating and degrading treatment ...".[52] There is also a positive obligation to ensure that "the wounded and sick shall be collected and cared for".[53] These obligations are also deemed to be part of customary international humanitarian law.[148]

The ICRC has highlighted its concerns that the use of certain ICA weapons in armed conflict would "make it difficult or impossible to determine when a combatant is 'out of action' and thereby to be afforded protection and assistance. An incapacitated combatant would probably not appear to be injured and may be unable to show a sign of surrender."[54] Consequently, the ICRC has argued that: "It would be difficult to train soldiers to distinguish whether an enemy were incapacitated or remained a threat. The resulting combination of incapacitants and lethal force could significantly increase the lethality of armed conflicts."[54]

Further disquiet has been voiced regarding the potential use of certain toxic chemicals in the coercive treatment of enemy combatants rendered *hors de combat*. The Royal Society has highlighted the importance of IHL in constraining such activities, specifically noting that the prohibition against "outrages upon personal dignity, in particular humiliating and degrading treatment [as enunciated under Additional Protocol I] could extend ... to the use of neuropharmacological agents to control or alter behaviour".[55] In addition, the Royal Society argued that "the use of neuropharmacological agents or stimulation technologies to coerce or interrogate prisoners of war" would also be prohibited under the Third Geneva Convention.[56]

Such concerns have also been addressed by certain States. For example, the *US Army Field Manual* whilst describing what it considers legitimate techniques for interrogation of prisoners of war, stresses that:

> "The psychological techniques and principles in this manual should neither be confused with, nor construed to be synonymous with, unauthorized techniques *such as brainwashing, physical or mental torture, including drugs that may induce lasting or permanent mental alteration or damage*. Physical or mental torture and coercion revolve around eliminating the source's free will, and are expressly prohibited by [the Geneva Conventions]."[57](Emphasis added).

6.5.2.4 *Requirement to Respect and Ensure Respect of International Humanitarian Law*

The application of a range of neuropharmacological agents could potentially impair or degrade the cognitive ability of enemy combatants (or indeed your own soldiers), with the consequent dangers of their committing serious breaches of IHL. Under Common Article 1 to the four Geneva Conventions, and Article 1 of Additional Protocol I, "The High Contracting Parties undertake to respect and to ensure respect" for the relevant treaties "in all circumstances".[58] The Royal Society has argued that "degrading the cognitive abilities of an adversary such that they are unable to distinguish between military targets and civilians, which often require a high degree of concentration, will undermine this requirement".[59] This is because such cognitive impairment could easily result in an unintended attack on one's own civilians or other persons or places specifically protected by law. Furthermore, the Royal Society has contended that "Such attacks could not be prosecuted because the perpetrators will have been rendered mentally incapable of being responsible for the offences."[149]

6.5.2.5 *Prohibition of Methods or Means of Warfare Intended to Cause Widespread, Long-term and Severe Damage to the Natural Environment; Prohibition on the Deliberate Destruction of the Natural Environment as a Form of Weapon*

Additional Protocol I prohibits the use of "methods or means of warfare which are intended, or may be expected to cause, widespread, long-term and severe damage to the natural environment".[60] According to the ICRC, State practice has established this rule as a norm of customary international law applicable in international, and arguably also in non-international armed conflicts.[61] However certain States, notably the US, are "persistent objectors" and others would dispute the customary status of the rule.

In addition, there is extensive State practice prohibiting the deliberate modification of the natural environment as a form of weapon. Under the Convention on the Prohibition of Military or Any Other Hostile Use of Environmental Modification Techniques (ENMOD), States Parties will not "engage in military or any other hostile use of environmental modification techniques having widespread, long-lasting or severe effects as the means of destruction, damage or injury to any other State Party". Nor will they assist, encourage or induce others to engage in such activities.[62] The difference between this provision and the one in Additional Protocol I is that the latter prohibits attacks on the environment as such, regardless of the means used, whereas the ENMOD Convention refers to the deliberate manipulation of natural processes that could produce phenomena such as hurricanes, tidal waves or changes in climate. According to the ICRC, whilst it is unclear whether the provisions in the ENMOD Convention are now customary, there is sufficiently widespread, representative and uniform practice to conclude that the destruction of the natural environment may not be used as a weapon.[63]

Clearly, both these rules are potentially applicable to the development and use of weapons employing toxic chemicals, including those targeting plants as well as animals, if they were intended to be employed on a sufficient scale. Indeed, the Final Declaration of the Second Review Conference of the Parties to the ENMOD Convention reaffirmed that the military and any other hostile use of herbicides as an environmental modification technique is a prohibited method of warfare "if such a use of herbicides upsets the ecological balance of a region, thus causing widespread, long-lasting or severe effects as the means of destruction, damage or injury to another State Party".[64]

6.5.3 Obligations to Review "New" Weapons Under International Humanitarian Law

Under Article 36 of Additional Protocol I to the Geneva Conventions, all High Contracting Parties that are engaged "in the study, development, acquisition or adoption of a new weapon, means or method of warfare" are "under an obligation to determine whether its employment would, in some or all circumstances, be prohibited by this Protocol or by any other rule of international law applicable to the High Contracting Party".[65] As of April 2018, there were 174 States party to Additional Protocol I, which are explicitly bound by this obligation. In addition, the *ICRC Guide* has stated that this requirement is "arguably" one that applies to all States, regardless of whether or not they are party to Additional Protocol I.[66] The obligation upon all States to review new weapons has been voiced by a number of pluri-lateral and multilateral bodies, such as the Conference of the Convention on Certain Conventional Weapons.[67] The importance of such legal reviews has also been regularly highlighted at successive International Conferences of the Red Cross and Red Crescent.[150] Lawand has argued that: "[E]stablishing national mechanisms to review the legality of new weapons is *especially relevant and urgent in view of emerging new weapons technologies such as* directed energy, *[chemical] incapacitants, behaviour change agents*, acoustics and nanotechnology, to name but a few" (emphasis added).[68] However, despite the widespread recognition of States' obligations to conduct legal reviews of new weapons, in 2016 Giacca, Legal Advisor at the ICRC, stated that only "between 15 and 20 States are known to conduct their own legal reviews"; in contrast, "many others rely on manufacturer information and on the reviews conducted by others."[69]

6.6 International Human Rights Law

6.6.1 Introduction

Although international human rights law (IHRL) does not specifically address the use of discrete arms or security equipment, it is certainly of great relevance to the employment of all such weapons, including those that may employ toxic chemicals, as it regulates the use of force by law enforcement officials and other agents of the State. An important strength of IHRL is its applicability in a broad range of circumstances where such weapons might be considered including domestic policing operations, non-international conflicts (whether or not the State recognized it as such), and to those aspects of an international conflict occurring in territory under a State's jurisdiction. The importance of this breadth of coverage has been highlighted by Hampson, who has noted that "[S]tates frequently refuse to characterize an internal armed conflict as such, preferring to call it criminal or terrorist activity. In such a situation, they can hardly challenge the applicability of human rights law."[70] While several human rights norms may be applicable to the regulation of weapons employing toxic chemicals, the rights to life and to freedom from torture and other ill-treatment together with associated obligations on the restraint of force, are the most relevant.**

6.6.2 Protection of the Right to Life and Restrictions on the Use of Force

The 'inherent' right to life is enshrined in many international,[152] and regional, human rights instruments.[153] The UN Human Rights Committee, the body that monitors the implementation of the International Covenant on Civil and Political Rights (ICCPR), has stated that the right to life is "the supreme right from which no derogation is permitted even in time of public emergency which threatens the life of the nation".[71] The Committee has further stated that States Parties to the ICCPR should take measures not only to prevent and punish deprivation of life by criminal acts, "but also to prevent arbitrary killing by their own security forces".[72] The Committee has argued that the right to life, in the case of the ICCPR at least, is a right "which should not be interpreted narrowly".[71]

Guidance to States on their attendant obligations to restrain and govern the use of force in law enforcement is provided by the 1990 UN Basic Principles on the Use of Force and Firearms by Law Enforcement Officials (UNBP),[73] and the 1979 UN Code of Conduct for Law Enforcement Officials (UNCoC).[74] These two instruments specify that the use of force must be proportionate, lawful, accountable and necessary. Under Principle 5 of the UNBP, law enforcement officials are required to "exercise restraint in such use and act in proportion to the seriousness of the offence and the legitimate objective to be achieved; minimize damage and injury, and respect and preserve human life; ensure that assistance and medical aid are rendered to any injured or affected persons at the earliest possible moment."[75]

Furthermore, under Principle 2 of the UNBP "[G]overnments and law enforcement agencies should develop a range of means as broad as possible and equip law enforcement officials with various types of weapons and ammunition that would allow for a differentiated use of force and firearms. These should include the development of non-lethal incapacitating weapons for use in appropriate situations, with a view to increasingly restraining the application of means capable of causing death or injury to persons."[76]

It is clear that the use of weapons employing a range of toxic chemicals, such as 'classic' chemical warfare agents (*e.g.* mustard gas, sarin, VX), industrial toxic chemicals (*e.g.* ammonia, chlorine) or toxins (*e.g.* aflatoxin, botulinum toxin, ricin, SEB), that have caused widespread injury and death would be a serious violation of international human rights law, notably contravening State obligations to protect life and restrict use of force. In contrast, certain so called 'non-lethal'

**Other potentially applicable rights include those to liberty and security; to freedom of opinion, expression, association and assembly; to human dignity; and to health. For further discussion see ref. 151.

or 'less lethal' weapons[††] employing a restricted range of toxic chemicals (in limited quantities) and purportedly intended to incapacitate but not cause permanent harm, have been employed by law enforcement personnel.

6.6.2.1 Application to Riot Control Agents

RCAs are widely employed by law enforcement officials throughout the world for activities such as the dispersal of assemblies posing an imminent threat of serious injury. When used in accordance with manufacturers' instructions and in a lawful, proportionate and discriminate manner, in line with international human rights standards, RCAs can provide an important alternative to other applications of force more likely to result in injury or death, notably firearms. Indeed, in certain circumstances States have been held accountable for killings by security forces that had not been provided with adequate 'less lethal' alternatives to firearms, including RCAs.[154]

However, regional and UN human rights monitoring bodies and international non-governmental human rights organizations have regularly expressed concern regarding reports of the employment of RCAs as part of the indiscriminate, excessive or lethal use of force by law enforcement officials, particularly in crowd control situations. A specific recurring concern has been the employment of RCAs in excessive quantities or in confined spaces where the targeted persons cannot disperse and where the toxic properties of the agents can lead to serious injury or death, particularly to vulnerable individuals.[155]

In 2003, the UN Special Rapporteur on Torture, van Boven, stated that "[C]hemical agents provided for 'crowd-control' purposes are prone to abuse if used against demonstrators in an indiscriminate manner. Precise practical guidelines regarding the circumstances in which such chemical agents may be used, as well as information regarding their effects on specific categories of persons such as children, pregnant women and persons with respiratory problems, are said often to be lacking."[77]

In 2012, the UN Special Rapporteur on the Rights to Freedom of Peaceful Assembly and of Association, Kiai, stated: "[W]ith regard to the use of tear gas, the Special Rapporteur recalls that gas does not discriminate between demonstrators and non-demonstrators, healthy people and people with health conditions." He also "warn[ed] against any modification of the chemical composition of the gas for the sole purpose of inflicting severe pain on protestors and, indirectly, bystanders."[78]

RCAs are also employed as a means to subdue prisoners and maintain order in correctional centres, prisons, police stations and other places of detention. Human rights bodies have raised concerns about the appropriateness of such application. For example, the European Committee for the Prevention of Torture and Inhuman or Degrading Treatment or Punishment

[††]The nature, scope and application of the terms 'non-lethal' or 'less lethal' weapons is contested. Although elaborated by certain national, pluri-lateral and civil society bodies, there is currently no internationally agreed definition of such terms.

(CPT) has stated that tear gas is a "potentially dangerous substance" and that "only exceptional circumstances can justify [its] use ... inside a place of detention—but never in a confined space such as a cell—for control purposes, and such exceptional use should be surrounded by appropriate safeguards".[79]

6.6.2.2 *Considerations Regarding Means of Delivery and Dispersal of RCAs*

A range of projectiles and delivery systems which have a narrow dispersal area and emit a limited quantity of agent (*e.g.* hand-thrown or weapons-launched RCA canisters and grenades or hand-held RCA spray disseminators) are widely employed by law enforcement officials, for example in public order situations. If such devices have been properly tested and trialled, their use should not raise undue concerns as long as it is in strict accordance with the relevant human rights standards, specifically, UNBP, UNCoC, and the UN Standard Minimum Rules for the Treatment of Prisoners,[80] and in conformity with, national deployment guidelines.

However, the kinetic impact safety of certain weapons-launched pyrotechnic RCA grenades and canisters may be of concern given limitations in their accuracy and their potential to cause trauma; furthermore, direct firing of such devices at individuals, particularly at short ranges, has led to a number of serious injuries and deaths. Consequently, this issue has been regularly highlighted by the UN and regional human rights bodies,[‡‡] as well as by international non-governmental human rights organizations.

In contrast to the foregoing, a range of delivery mechanisms have been developed for crowd control and dispersal that deliver far larger amounts of RCAs over wider areas than could be delivered by hand-held sprays and the like. Some forms of such "wide area" RCA means of delivery (*e.g.* certain multi-barrel RCA launchers) may have utility in large-scale law enforcement situations provided that they are employed in strict conformity with human rights standards. However, a number of these devices raise questions concerning proportionality and/or about the feasibility of their discriminate use, with the consequent danger of affecting bystanders, in potential contravention of human rights law and standards. In addition, a further range of 'wide area' RCA means of delivery have been developed and promoted which are inherently unacceptable for employment in law enforcement activities, contravening relevant use of force standards (as well as being in apparent breach of the CWC). These include munitions containing RCAs which have military utility, such as cluster munitions, aerial bombs, mortar rounds and artillery shells.

6.6.2.3 *Application to Incapacitating Chemical Agent Weapons*

Proponents of ICA weapons have long advocated their development and use in certain law enforcement scenarios, notably largescale hostage situations, where there is a need to rapidly and completely incapacitate individuals or

---

[‡‡]See ref. 173 and 174 for examples of two judgements by the European Court of Human Rights (ECtHR).

a group without causing death or permanent disability. However, the legitimacy of such weapons in any form of law enforcement is highly contested. The ICRC has argued against the development and use of ICA weapons, highlighting the grave dangers of their employment in practice and the extremely limited circumstances under which the use of such "potentially lethal force" should even be considered, given States obligations under IHRL:

"In light of the certainty that bystanders will also come to harm, the question to be asked is whether such a means is absolutely necessary to save the lives of those who are threatened, that is whether there are any other means available that would achieve the same aim while posing less of a danger to life; and whether this is an unavoidable measure of last resort, the State having exhausted all feasible less harmful means before it resorts to this means."[81]

Furthermore, legal scholars have warned that even if such conditions are met, the obligations upon States under IHRL to protect the right to life still apply, with contingent severe constraints upon employment of the ICA weapon as well as the requirement to take appropriate remedial measures. Fidler has argued that "[T]he inability to control dosage or exposure environment in extreme law enforcement emergencies heightens governmental responsibility to ensure all precautions are taken to minimize harm to innocent people and to provide immediate and adequate medical attention to those exposed and perhaps adversely affected."[82]

6.6.2.3.1 European Court of Human Rights: Finogenov and Others *vs.* Russia. Case law in this area is limited. To date, there has been only one well-documented instance of an ICA weapon agent employed in such extreme law enforcement situations. In October 2002, Russian security forces employed an ICA weapon to free 900 hostages held by heavily armed Chechen separatists in the Dubrovka theatre in Moscow. The Russian security forces pumped the aerosolised ICA[156] into the theatre, putting hostages and hostage-takers into a 'deep sleep' and then stormed the theatre killing all of the Chechen fighters. Although the hostages were freed, 125 hostages died due to the effects of the ICA weapon and an unknown number of former hostages suffered long-term injury.

In August 2003, a group of 64 former hostages and relatives filed a complaint before the European Court of Human Rights (ECtHR) claiming that their right to life (protected under Article 2 of the European Convention on Human Rights [ECHR]) had been violated by the actions of the Russian authorities. In its ruling, given on 20 December 2011,[83] the Court rejected the Russian Federation authorities' assertions that the ICA weapon had been harmless, declaring that: "it is safe to conclude that the gas remained a primary cause of the death of a large number of the victims".[84]

Nonetheless, in this particular case, the Court believed there had been a real, serious and immediate risk of mass human losses and the authorities had every reason to believe that a forced intervention had been "the lesser evil". The Court therefore concluded that, in the circumstances, the authorities' decision to end the negotiations and resolve the hostage crisis by force by using an ICA weapon and storming the theatre had not been disproportionate and had not, as such, breached Article 2 of the ECHR.[85] However, the Court concluded that the rescue operation had been inadequately prepared and carried out, for example due to delays in hostage evacuation, lack of appropriate on-site medical treatment and equipment, and failure to notify medical personnel at the receiving hospitals of the nature of the ICA weapon employed. Consequently, the Court found that, as a whole, the Russian authorities had not taken all feasible precautions to minimize the loss of civilian life, in violation of Article 2.[86]

The ICRC subsequently raised concerns about the Court's judgement,[81] and there are certainly aspects of this ruling that would benefit from further legal analysis, including the inexplicable failure of the Court to assess the legality of the use of the ICA weapon under the CWC;[157] the Court's willingness to make a judgement without knowledge of the identity, nature and consequently (predicted) effects of the ICA weapon employed; and the Court's claims that it was impossible for it "to establish whether or not the gas was a 'conventional weapon' or to identify the rules for its use".[87]

However, such concerns notwithstanding, the Court's ruling may well set important precedents, guide future judgements, and influence State practice regarding the development and use of such weapons. It is notable that the Court considered the use of the ICA weapon as definitely falling under the ambit of the ECHR and that such use could potentially be a violation of Article 2 of the Convention. It appears that the use of such ICA weapons would only be considered consistent with the Convention in very limited, extreme situations, and only when such use occurred in conjunction with proper planning and the provision of adequate medical care and remedial support (both immediately and in the longer term). Although the Court, in this first instance, gave the Russian Federation some "margin of appreciation"[§§] regarding its decision to use an ICA weapon with then unknown effects, the consequences of such action have now been shown to include the deaths of a large proportion of the hostages and long-term injury for many of the survivors. Furthermore, subsequent technical studies and reports by medical and scientific bodies have highlighted the extreme difficulties of ensuring that such outcomes will not be repeated given narrow safety margins of potential ICAs, the inability to control "dose" in tactical situations and provide immediate medical care.[158] Consequently, it is debatable whether this or another court would give such discretion to a State for the use of an ICA weapon in similar situations in the future.

Finally, and with more general application, this case has highlighted the potential utility of regional human rights mechanisms as a means that victims of human rights violations (including those inappropriately targeted with ICA weapons, RCAs or other toxic chemicals) can employ to hold the relevant authorities to account. Although the relevant courts[¶¶] have a number

[§§]This term refers to the space for manoeuvre that the European Court of Human Rights is willing to grant national authorities in fulfilling their obligations under the ECHR.

[¶¶] The African Court on Human and Peoples' Rights, the European Court of Human Rights, and the Inter-American Courts of Human Rights.

of limitations and only have jurisdiction over those States that have ratified the relevant Conventions and Protocols, they deliver legally binding judgements. Furthermore, as Hampson has noted: "[An] important feature of these enforcement mechanisms is that, almost uniquely in international law, they can be directly or indirectly triggered by individuals ... who are far more likely to bring such complaints than are foreign States, particularly where the claim concerns something that the State has done on its own territory."[88]

6.6.2.4 *Employment of Toxic Chemicals in Judicial Executions*

Although the inherent right to life is enshrined and protected under IHRL and the arbitrary deprivation of life is prohibited, State-sanctioned judicial executions are not currently forbidden, though they are subject to certain restrictions. For example, the ICCPR prohibits imposition of the death penalty "for crimes committed by persons below eighteen years of age" and specifies that it "shall not be carried out on pregnant women".[89] In China, Guatemala, the Maldives, Papua New Guinea, Taiwan, Thailand, Vietnam, and the US, the intravenous administration of a lethal dose of certain pharmaceutical toxic chemicals ("lethal injection") is provided for as a method of execution.

In addition to the moral and human rights arguments against the death penalty *per se*, international and regional human rights bodies, nongovernmental human rights organizations and medical associations have raised specific concerns about execution by lethal injection. These include the attempts of death penalty proponents to promote lethal injections as pain-free and humane, thereby masking the long-term suffering inflicted on prisoners through the entire death penalty process; the potential for this method to cause intense, sometimes prolonged physical suffering, potentially breaching a State's obligations not to commit torture with the danger that such suffering may be hidden due to the action of a paralysing agent in the lethal mixture;[90] and the involvement (in contravention of medical ethics) of health personnel, directly or indirectly, in the execution process.[91]

6.6.3 Prohibition Against Torture and Other Cruel, Inhuman or Degrading Treatment or Punishment

The prohibition against torture and other cruel, inhuman or degrading treatment or punishment (CIDTP) is recognized in a wide range of international,[92] and regional,[159] human rights agreements, and is a customary norm, applicable at all times and in all circumstances, including in armed conflict. It is one of the few human rights for which no derogation has been permitted.[160]

Torture is defined under Article 1 of the UN Convention against Torture as:

> "Any act by which severe pain or suffering, whether physical or mental, is intentionally inflicted on a person for such purposes as obtaining ...

information or a confession, punishing him..., or intimidating or coercing him or a third person, or for any reason based on discrimination of any kind, when such pain or suffering is inflicted by or at the instigation of or with the consent or acquiescence of a public official." It should be noted that this definition is narrower than most relevant international and regional treaties, which normally require only that a State agent intentionally inflicts severe pain or suffering, without the need to show instigation, consent or acquiescence.[161]

Elements of CIDTP have been defined in other relevant legal texts. For example, "degrading treatment" has been defined by the European Commission of Human Rights as treatment or punishment that "grossly humiliates the victim before others or drives the detainee to act against his/her will or conscience".[93]

6.6.3.1 *Application to Psychoactive Incapacitating Chemical Agents*

In its report exploring the potential use and misuse of neuroscience, the Royal Society highlighted the European Commission of Human Rights' definition of degrading treatment and considered it to be of "particular importance in considering the potential applications of neuroscience that could, for example, manipulate behaviour or thought processes".[93,94] Consequently, the Royal Society stated that "the use of potential militarised agents including noradrenaline antagonists such as propranolol to cause selective memory loss, cholecystokinin B agonists to cause panic attacks, and substance P agonists to induce depression could all be considered violations of the prohibition against degrading treatment". The weaponization of such toxic chemicals would also clearly violate the CWC.[162]

Fidler has stated that "non-consensual, non-therapeutic use of any chemical or biochemical against detained individuals would constitute degrading treatment and could, constitute cruel or inhumane treatment and perhaps even torture".[95] He did, however, believe that there may be situations where use of such chemicals might be compatible with IHRL: where the detained person posed an immediate, violent threat to himself (*e.g.* attempting suicide) or for safety and order in the detention facility (*e.g.* attacking guards or participating in riots).[95]

In addition to the prohibitions against torture and CIDTP, further important potential constraints upon the non-consensual application of psychotropic drugs to detainees relate to obligations to ensure respect for the detainee's right to freedom of opinion. For example, Article 19 of the ICCPR declares that "everyone shall have the right to hold opinions without interference".[96] In his legal commentary to the Convention, Nowak stated that this provision consequently "obligates the States Parties to refrain from any interference with freedom of opinion (by indoctrination, 'brainwashing', influencing the conscious or unconscious mind with psychoactive drugs or other means of manipulation) and to prevent private parties from doing so".[97]

6.6.3.1.1 Use of Truth Serums in Interrogation. Certain States have employed psychoactive incapacitating chemical agents (such as sodium thiopental, sodium amytal or scopolamine) as so-called 'truth drugs' or 'truth serums' against detainees without their consent, for the purposes of interrogation by law enforcement officials.[‖‖] Such practices are contrary to IHRL, prohibiting CIDTP. In addition, the UN Body of Principles for the Protection of All Persons under Any Form of Detention or Imprisonment (UN Principles for the Protection of Detainees), addressing the issue of interrogation, stated that "[N]o detained person while being interrogated shall be subject to violence, threats or methods of interrogation which impair his capacity of decision or his judgement."[98]

Further constraints or outright prohibitions upon the use of psychoactive chemicals in interrogation have been established at the national level in certain States. In May 2010, the Indian Supreme Court ruled that the administration of such chemicals to detainees—a practice known in India as "narco-analysis"—without their consent during interrogation violated the Indian Constitution and was illegal as it constituted cruel, inhuman or degrading treatment.[99] Concerns have also been raised by jurists in the United States, and the use of 'truth serums' or 'truth drugs' is not recognized as an authorized method of interrogation by US courts. Under US law, confessions made under the influence of 'truth serums' are not considered as 'voluntary' and are consequently inadmissible as evidence.[100] The employment of such chemicals to facilitate interrogations also appears to be contrary to the Inter-American Convention to Prevent and Punish Torture. Article 2 of this Convention defines torture as including "the use of methods upon a person intended to obliterate the personality of the victim or to diminish his physical or mental capacities, even if they do not cause physical pain or mental anguish."[101]

6.6.3.2 Application to Riot Control Agents

Despite the absolute prohibition on torture and CIDTP, the misuse of RCAs for such purposes by law enforcement officials has been reported by regional and UN human rights bodies and international human rights nongovernmental organisations. Presenting the results of a 2003 study on the trade in security equipment that could be used for torture and CIDTP,[102] the UN Special Rapporteur on Torture, stated that:

> "The allegations of torture that he has received from all regions of the world have involved instruments such as *[inter alia]* ... chemical control substances

[‖‖]For further discussion and case studies see ref. 163

(*e.g.* tear gas and pepper spray). While some of the cases have involved the use of equipment which is inherently cruel, inhuman or degrading, and would *per se* breach the prohibition of torture, the vast majority [including RCAs] have involved the misuse of those instruments, legitimate in appropriate circumstances, to inflict torture or other forms of ill-treatment."[103]

In certain cases, RCAs have reportedly been employed as a means of inflicting 'collective punishment' upon groups of individuals or crowds. Other cases of concern involved the use of hand-held irritant sprays against individual prisoners and detainees in a targeted fashion.[163]

A potentially significant development in relevant human rights case law was the July 2012 ECtHR final judgement regarding the employment of tear gas upon Ali Güneş, a high-school teacher who took part in a demonstration on 28 June 2004 against a NATO summit meeting being held in Istanbul. According to Mr Güneş, although he was unarmed and participating peacefully with colleagues in the demonstration at a site sanctioned by the authorities, police grabbed him by the arms, sprayed him with tear gas and beat him up, following which they took him to a police station in which they kept him for 11 hours. A prosecutor saw him after that and ordered his release.[104] In its final judgement, the Court considered "that the unwarranted spraying of the applicant's face in the circumstances described ... must have subjected him to intense physical and mental suffering and was such as to arouse in him feelings of fear, anguish and inferiority capable of humiliating and debasing him".[105] The Court, therefore, concluded "that by spraying the applicant in such circumstances the police officers subjected him to inhuman and degrading treatment within the meaning of Article 3 of the Convention."[105] The Court's ruling is important, given the bearing it may have on future judgements on the use of RCAs in such circumstances.

6.6.4 Obligations to Review and Monitor the Use of 'Less Lethal' Weapons

To fulfil their obligations under international human rights law and to ensure the responsible use of force by law enforcement officials, States are encouraged to implement review mechanisms to ensure that so-called 'non-lethal' or 'less lethal' weapons—including RCAs and related means of delivery—developed or otherwise acquired, are consistent with such obligations. UNBP Principle 3 states that "development and deployment of 'non-lethal' incapacitating weapons should be carefully evaluated in order to minimize the risk of endangering uninvolved persons, and the use of such weapons should be carefully controlled."[106]

A number of UN human rights bodies and mechanisms have highlighted the importance of the effective regulation of the introduction and employment of such weapons. The UN Human Rights Council in its Resolution 25/38 of April 2014 "underline[d] the importance of thorough, independent and scientific testing of non-lethal weapons prior to deployment to establish their lethality and the extent of likely injury, and of monitoring appropriate training and use of such weapons".[107]

The UN Special Rapporteur on extrajudicial, summary or arbitrary executions, Heyns, examined the application of relevant international human rights law and standards to the development and employment of 'less lethal' weapons and unmanned systems. In his August 2014 report to the UN General Assembly, he stated that "[A] process involving States and the international community, in addition to civil society, is needed to set out how the standards set by the [UNBP] and the relevant jurisprudence should be applied to the scenarios created by the new technology." The Special Rapporteur argued that "[M]inimum standards need to be set for the development of weapons and their use, and good practices need to be identified."[108] Among the areas requiring greater specificity, the Special Rapporteur highlighted: "accuracy required of a projectile" and "the safe levels of a chemical irritant to be delivered by an aerosol spray". The Special Rapporteur also argued that greater specificity was needed with regard to "where such devices could be used (*e.g.* tear gas grenades should not be used in closed spaces…)" and argued that "The new technologies may require that monitoring of force is mandatory in many cases."[109]

Despite the recommendations of human rights bodies and the continuing widespread well-documented misuse of certain 'less lethal' weapons (including RCAs and associated means of delivery) in human rights violations (sometimes resulting in serious injury or death), there are currently no internationally accepted procedures for evaluating new 'less lethal' weapons, for ensuring adequate training in their use or effectively controlling and monitoring their subsequent employment.

6.7 International Criminal Law

6.7.1 Introduction

The preceding analysis has concentrated upon international law intended to constrain and influence the activities of States, and indirectly through national implementing legislation apply to individuals. An important alternative and complementary approach is to directly address individual responsibility and culpability through the employment of international criminal law. According to Cassese "[I]nternational crimes are breaches of international rules entailing the personal criminal liability of the individuals concerned."[110] Although there is no definitive list of such crimes under international law, it has been widely accepted that war crimes, crimes against humanity and genocide would be included.

The reach of international criminal law and attendant judicial mechanisms is potentially broad, allowing those directly or indirectly responsible for such crimes—whether they control or are members of State security or armed forces, armed opposition groups, terrorist or criminal organisations or other non-State entities—to be brought to justice, even if their own States were unwilling or unable to do so. In practice, however, there are considerable legal, logistical and political barriers to be overcome to make this potential recourse a reality. Notwithstanding such hurdles, in certain situations international criminal law could be applied by national courts or international courts to those involved in the use of, and possibly the transfer of, certain weapons employing toxic chemicals.

6.7.2 Rome Statute of the International Criminal Court

The International Criminal Court (ICC), governed by the Rome Statute,[111] is the first permanent, treaty based, independent international criminal court established to help end impunity for the perpetrators of the most serious crimes of concern to the international community.[164] The ICC came into being on the 1st July 2002 when the Rome Statute entered into force, following its ratification by 60 States Parties. As of April 2018, there were 123 States Parties to the ICC, with a further 15 having signed but not ratified the Rome Statute.[165]

Pursuant to the Rome Statute, the ICC Prosecutor can initiate an investigation on the basis of a referral from any State Party or from the UN Security Council. In addition, Prosecutors can initiate investigations *proprio motu* (*i.e.* on their own initiative) on the basis of information on crimes within the jurisdiction of the Court received from individuals or organizations.[112]

The scope of the Statute's coverage in terms of criminal responsibility is broad. A person, regardless of their official capacity, can be held criminally responsible if they committed or attempted to commit the crime; but also if they ordered, solicited or induced the crime; aided, abetted or otherwise assisted in the crime or "in any other way intentionally and knowingly contributed to the commission or attempted commission of the crime".[113] Furthermore, crimes within the jurisdiction of the ICC are not subject to any statute of limitations.[114]

The Rome Statute has asserted jurisdiction over genocide (defined as "acts committed with intent to destroy, in whole or in part, a national, ethnical, racial or religious group");[115] crimes against humanity (defined as "acts when committed as part of a widespread or systematic attack directed against any civilian population, with knowledge of the attack");[116] and war crimes (incorporating grave breaches of the Geneva Conventions and other serious violations of IHL).[117] The Statute's definition of "war crimes" specifically includes:

> (xvii)"Employing poison or poisoned weapons;
> (xviii)Employing asphyxiating, poisonous or other gases, and all analogous liquids, materials or devices."[118]

Although attempts were made during the Rome Statute negotiations to incorporate a specific reference under war crimes to "chemical weapons" as defined and prohibited under the CWC, these failed due to linkage with parallel attempts to include a prohibition on nuclear weapons. Instead the first definition (xvii) enunciates a norm codified in the Second Hague Convention, whilst the second (xviii) is derived from the Geneva Protocol. Consequently, there has been disagreement within the legal community as to the range of weapons employing toxic chemicals whose use in armed conflict would be considered, *per se*, a war crime under the Rome Statute.

Some guidance in this matter can be obtained from considering the "Elements of Crimes", adopted by the Assembly of States Parties to "assist the Court in the interpretation and application of articles ... consistent with the Statute."[119] The Elements of Articles 8(2) (b) (xviii) and (xvii) both include the specification, *inter alia*, that "the substance was such that it causes death or serious damage to health in the ordinary course of events, through its toxic properties."[120] Consequently, the weaponization of a wide range of toxic chemicals including those listed under the three Schedules of the CWC together with a range of non-Scheduled chemicals such as certain industrial toxic chemicals, CNS acting chemicals and toxins would be covered. The use of such toxic chemicals as weapons in an international armed conflict would fall within the scope of Article 8(2) (b) (xvii) and/or 8(2) (b) (xviii), and would inherently be considered a war crime, with the possibility that those responsible for such acts might be tried before the Court. The possible applicability of the Statute with regards to the use of RCAs as a method of warfare, *per se*, however, is contested.

In addition, even if a toxic chemical failed to fall within the scope of Articles 8(2) (b) (xvii) and (xviii), its use in *specific* circumstances could potentially be covered, if it were employed for actions prohibited as "war crimes" by the Rome Statute, including "torture or inhuman treatment",[121] or to "wilfully caus[e] great suffering, or serious injury to body or health".[122] Furthermore, the largescale use of such toxic chemicals may also be covered if employed to commit genocide[115] or crimes against humanity.[123]

Whilst the use of a wide range of toxic chemicals as weapons in armed conflict would be considered a crime under the Rome Statute, initially the scope of ICC jurisdiction in this area was restricted solely to the use of such weapons in international armed conflicts. In order to address this limitation, the First Review Conference in 2010 adopted a resolution to extend coverage to include the use of such weapons in non-international armed conflict.[124] This extension is applicable for each State Party that ratifies the change;[125] as of April 2018, only 36 States have done so.[166] However, crucially this extension would arguably also apply in cases referred to the ICC by the UN Security Council regardless of whether the activities concerned took place in a State that ratified the Resolution or indeed was a Party to the Rome Statute.

Despite this welcome development, the scope of ICC applicability remains constrained. The ICC would only have jurisdiction over cases involving the *use* of weapons employing toxic chemicals in armed conflict, but not cases solely involving the development, production, acquisition, stockpiling or transfer of such weapons *per se*; engaging in military preparations to use such weapons; or assisting, encouraging, or inducing others to engage in activities prohibited to State Parties under the CWC, short of weapons use.[126] Furthermore, cases involving the use of such weapons in law enforcement operations (or terrorist acts) would not normally be covered unless they were deemed to be acts of genocide or crimes against humanity.

The Court can only prosecute crimes committed on or after the date of its establishment,[127] and it will not admit cases that are being, or have already been, investigated or prosecuted by a State which has jurisdiction over the case, unless the State is genuinely unwilling or unable to carry out the investigation or prosecution.[128] In addition, the Court's jurisdiction is normally limited to nationals of States Parties that have ratified the Statute, or where a crime has been committed on the territory of a State Party, the exception being where a case has been referred to it by the UN Security Council. To date, the UNSC has made use of its referral power on just two occasions (both unrelated to the use of toxic chemicals), with regard to the Darfur region of Sudan in 2005, and for Libya in 2011. In contrast, the UNSC has been unwilling or unable to agree upon such a referral on numerous other instances when there was convincing evidence of widespread and serious international crimes and little likelihood of the perpetrators being brought to justice by national courts.

6.7.2.1 Case Study: Attempts to Ensure Accountability for International Crimes in Syria

UN and international human rights organisations have extensively documented the widespread and serious violations of international humanitarian law and international human rights law amounting to crimes against humanity and war crimes committed by both the Syrian Government forces and their allies and, also, by armed opposition groups.

Syria is not a signatory to the Rome Statute, consequently a referral by the UN Security Council is the only way that the ICC would be able to exercise jurisdiction over these serious international crimes and bring the perpetrators to justice. On 12 May 2014, France circulated the draft of a proposed UNSC Resolution on Syria which incorporated such a referral, by the Security Council, of Syria to the prosecutor of the ICC.[129] This UNSC Resolution would have given the ICC jurisdiction over alleged crimes against humanity and war crimes, including those committed using toxic chemicals, regardless of the identity of the perpetrators.

Backing for such a referral was widespread, with 58 countries issuing a statement of support on 19 May 2014[130] and over 100 non-governmental organizations from around the world urging the Security Council to approve the French-proposed Resolution.[131] Such a referral had previously been recommended by both the UN High Commissioner for Human Rights[132] and the UN Commission of Inquiry on Syria.[133] However, on 22 May 2014 the Resolution was defeated. Although 13 members of the Security Council including France, the UK and the US voted in favour; China and the Russian Federation exercised their veto power.[134]

In response to continuing widespread and serious breaches of international human rights law and international humanitarian law committed in Syria, (which included further use of weapons employing toxic chemicals) and the continuing failure of the UNSC to establish effective accountability mechanisms, the UN General Assembly decided to act. On 21 December 2016, the UNGA adopted a Resolution[135] to establish an "[I]nternational, Impartial and Independent Mechanism to Assist in the Investigation and Prosecution of Those Responsible for the Most Serious Crimes under International Law Committed in the Syrian Arab Republic since March 2011". The Independent Mechanism will operate under the auspices of the UN in close cooperation with the Independent International Commission of Inquiry on Syria. It will "collect, consolidate, preserve and analyse evidence of violations of international humanitarian law and human rights violations and abuses", and States, civil society and the UN system are requested to fully cooperate with it.[135]

The Independent Mechanism is also tasked with "prepar[ing] files in order to facilitate and expedite fair and independent criminal proceedings, in accordance with international law standards, in national, regional or international courts or tribunals that have or may in the future have jurisdiction over these crimes, in accordance with international law."[135] Although the UNGA did not have the authority to directly hold the perpetrators of international crimes in Syria to account through referral to the ICC, the Independent Mechanism established by this Resolution can initiate objective investigations into and preserve evidence of international crimes, thereby increasing the likelihood that the perpetrators of such crimes will be held accountable in the future. An important subsequent development has been the launch on 23 January 2018 of the "International Partnership Against Impunity for the Use of Chemical Weapons". Chaired by France, and currently comprising around 30 countries, this intergovernmental initiative is a forum for cooperation intended to: gather, compile and retain information on those who use chemical weapons; facilitate sharing of this information with relevant bodies so that perpetrators will be held accountable; support multilateral actions to sanction them; publish names of all individuals and entities subject to sanctions; help States to implement information collection and introduce appropriate national legislation to prosecute perpetrators of chemical attacks. Although the Partnership intends to combat chemical weapons use globally, its current focus is upon Syria with participating States specifically reaffirming their "common determination" to support the UNGA mandated Independent Mechanism investigating serious crimes in Syria.[175]

6.7.3 International and Internationalised Courts and Tribunals

From the end of the Cold War, a number of international or internationalised legal entities have been established to deal with international crimes – notably war crimes, crimes against humanity and/or genocide.[167] Firstly, in the early 1990s, two *ad hoc* international criminal tribunals—the International Court for The Former Yugoslavia (ICTY), and the International Criminal Tribunal for Rwanda (ICTR)—were set up by the UN Security Council, in response to ethnic cleansing and other atrocities, which occurred during the conflicts in those countries. Both tribunals were established as measures adopted under Chapter VII of the UN Charter and both were set up outside of the territory in conflict.[168]

In the aftermath of subsequent conflicts involving serious violations of international criminal law, several States decided not to try persons allegedly responsible for atrocities before their domestic court nor in international courts, but instead preferred to resort to a criminal court combining national and international structures and laws. Such bodies have been variously called hybrid, mixed or internationalised courts or tribunals. There exist a variety of such courts, each of them having their own unique characteristics. The differences concern the political background, the legal bases for their establishment, and their structure and jurisdiction. The common characteristic of all such bodies is the existence of an international element, which varies in scope and intensity.[167]

These courts and tribunals are constrained by the terms of their constituent instruments with regard to the crimes they can try, and their jurisdiction may be limited geographically and to certain time periods.[169,170] It is possible that the alleged use of weapons employing toxic chemicals to conduct war crimes, crimes against humanity and even genocide could fall within the jurisdiction of such (existing or future) courts and tribunals, but this would be dependent on the specific nature of the alleged crimes, as well as that of the court or tribunal itself. To date, one case—the Anfal trial—involving the use of chemical weapons has been prosecuted in such a court.

6.7.3.1 *Case Study: The Anfal Trial of Iraqi Officials*

The Iraq High Tribunal was established as an internationalized national court in 2003, following the overthrow of Saddam Hussein and his Ba'athist regime. It is limited in jurisdiction to trials of Iraqi nationals for crimes including genocide, war crimes and crimes against humanity, committed during the regime era (1968 to 2003). On 21 August 2006, the Anfal trial opened before the Tribunal with seven former Iraqi high officials charged with crimes against humanity and war crimes, and two of them (former Iraqi president Saddam Hussein and his cousin and former regional military commander, Ali Hassan Al-Majid) with the additional charge of genocide.[136,137] The charges related to the defendants' involvement in the Anfal military operation against the Iraqi Kurdish population initiated in 1988. During this operation, an estimated 182 000 people lost their lives, entire villages were razed, mass executions were carried out with the dead buried in mass graves, and survivors were forcibly relocated to detention camps. Chemical warfare agents were used repeatedly as part of the Anfal campaign – notably in an attack against the Kurdish village of Halabja in March 1988 where approximately 5000 villagers were killed by a mixture of mustard gas and nerve agents. During the trial, the testimony of numerous witnesses and victims of the chemical attacks were heard, and documentary evidence and film footage linking the defendants to the chemical weapons attacks was presented.

Saddam Hussein was executed for other crimes before his trial on these charges was concluded. However, Ali Hassan Al-Majid was convicted of war crimes and genocide for ordering the chemical attacks on the Kurdish population. He was sentenced to death and executed in January 2010. Former defence minister, Sultan Hashem, was also found guilty for ordering the Halabja chemical attack and sentenced to 15 years' imprisonment. The trial was important in documenting the use of weaponised toxic chemicals in activities that the Tribunal considered as amounting to war crimes, crimes against humanity and genocide. However, there were serious concerns expressed about the fairness of the proceedings and the lack of an adequate appellate process.[137]

6.7.4 National Courts

In the case of international crimes, any State is free to try a suspected perpetrator providing that they have established the requisite jurisdiction in national law and that the suspect cannot successfully claim sovereign immunity. Certain categories of grave crimes that violate international law—notably crimes against humanity, genocide, torture and war crimes—are subject under international law to 'universal jurisdiction' allowing the court of any State to try persons for crimes committed outside its territory even when not linked to the State by the nationality of the suspect or the victims or by harm to the State's own national interests.

Whilst it has previously been possible in the domestic context to initiate criminal prosecutions for certain international crimes, the coming into force of the Rome Statute and the establishment of the ICC provided an important impetus for a large number of countries to introduce changes to their laws to ensure that they were in compliance with international obligations and the tenets of the Rome Statute.[138] In those States where national law dealing with war crimes has incorporated, or made reference to, the relevant definition set out in the Rome Statute; the use of at least a significant proportion of weapons employing toxic chemicals in armed conflict would *per se* be considered a war crime.

To date, there has been one case—the Van Anraat trial—where a national court has employed international criminal law to try an individual accused of complicity in international crimes involving the use of weaponised toxic chemicals.

6.7.4.1 Case Study: Van Anraat Trial – Complicity in War Crimes

On 23 December 2005, the District Court of The Hague, Criminal Law Section, found Dutch businessman Frans van Anraat guilty, under the 1952 Wartime Offences Act, of complicity in war crimes committed by the former Iraqi regime. He was acquitted of a second charge: complicity in genocide.[169,171,172] The Court sentenced van Anraat to the maximum penalty of 15 years' imprisonment (subsequently increased to 17 years), while concluding that the sentence was insufficient due to the severe repercussions of the chemical attacks and the nature of the crime.[169,171,172] During the 1980s, Mr Van Anraat was Saddam Hussein's most important supplier of chemicals used for the production of mustard gas. According to the Court, van Anraat's involvement in supplying chemicals to Iraq was an essential contribution to the chemical weapons programme of Saddam Hussein's regime.[169,171,172]

The Court determined that Mr van Anraat knowingly and intentionally supplied chemicals to the former Iraqi regime, which were used to produce chemical weapons employed by Iraq in Iraqi Kurdistan and in Iran.[139] Mr van Anraat was convicted of complicity in war crimes since his deliveries facilitated the attacks on the Kurdish population and made the carrying out of the regime's ambitions considerably easier.[169,171,172] As Tabassi and Dhavle note, this case is important in that criminal proceedings were initiated against a businessman for complicity in war crimes (and genocide); generally these economic actors have largely been shielded from such criminal charges, even though their role regarding gross violations of international human rights and humanitarian law has been raised repeatedly.[140]

6.7.5 Conclusions

Whilst the Chemical Weapons Convention (CWC) and associated customary law is the *lex specialis* prohibiting States from developing, stockpiling and using chemical weapons, there are a number of additional international agreements and relevant international laws further constraining the weaponization of toxic chemicals. Consequently, States should give full and careful consideration to the application of such agreements and international law; firstly because of the direct obligations that arise from such agreements and law which may either prohibit or severely restrict the development, stockpiling, transfer or use of weapons employing toxic chemicals, and secondly because relevant international law should inform the implementation of those areas of the CWC where treaty language is ambiguous and interpretation is currently contested.

In terms of armed conflict, the CWC prohibitions upon the development and use of chemical weapons are further underpinned by associated obligations derived from international humanitarian law (IHL) treaties and customary international law. Indeed, the use of weaponised toxic chemicals in armed conflict would potentially breach relevant IHL prohibitions (such as those forbidding SIRUS, indiscriminate weapons and deliberate attacks on civilians). In addition, IHL also provides protection to prisoners of war or others considered *hors de combat*, for example, prohibiting the use of toxic chemicals against them in forced interrogations, torture or human experimentation. Furthermore, IHL provides added protection for the civil population, for example, prohibiting the largescale use of toxic chemicals (such as herbicides) which have widespread, long lasting and severe effects on the environment. However, the potential utility of IHL to the regulation of toxic chemicals is curtailed due to limitations in investigation and enforcement procedures, and also to extremely low levels of State implementation of

Article 36 legal reviews of new weapons. Furthermore, IHL is only applicable to situations of armed conflict. Whilst much relevant IHL would extend to non-international armed conflicts, there may well be disagreements as to whether a particular situation is indeed a non-international armed conflict, with the relevant State instead claiming to be involved in law enforcement activities against criminals or terrorist organizations. However, even in such disputed situations, State action would be constrained by a second body of law, international human rights law (IHRL).

IHRL covers the full 'use of force' spectrum from law enforcement activities through to armed conflict, including 'grey areas' such as counter-terrorist, counter-insurgency and military operations outside of armed conflict where use of weapons employing toxic chemicals have been proposed. While several human rights norms may be applicable to the regulation of such weapons, the rights to life; and to freedom from torture and other cruel, inhuman or degrading treatment or punishment; together with attendant obligations on the restraint of force, are the most relevant. An important aspect of human rights law is that there are a number of international and regional mechanisms to monitor adherence to relevant treaties. In addition, certain treaties provide for the possibility of individual petition which can result in legally binding judgements for those States party to the relevant instrument (as was seen in the ECtHR rulings on the use of an ICA weapon by the Russian Federation, and of tear gas and related means of delivery by Turkey). However, it must be recognized that these monitoring and enforcement mechanisms have limited preventative value (though they may have some deterrent effect), as they would only be initiated after a potential misuse of such weapons has occurred. Furthermore, there are currently no internationally accepted standardised procedures, under IHRL, for evaluating new weapons employing toxic chemicals, or for monitoring their subsequent use at a national level.

Whilst the preceding analysis concentrated upon international law constraining the activities of States, an important alternative and complementary approach is to directly address individual responsibility and culpability through the employment of international criminal law (ICL). The application of ICL and attendant judicial mechanisms to crimes involving weaponised toxic chemicals can play an important role in strengthening and enforcing the prohibition against such activities, allowing those responsible for such crimes, whether conducted by State forces, armed opposition groups, terrorists or other non-State actors, to be prosecuted. The bringing to justice of those directly or indirectly responsible for such crimes, even if their own States were unwilling to do so, would potentially have important deterrent effects. Of the existing judicial bodies, the International Criminal Court—with its attendant review and amendment mechanisms—affords an important route for concerted international action and normative development in the long term. However, its potential utility is currently constrained, for example due to problems with establishing jurisdiction, the unwillingness of certain States to prosecute suspects on their territory, and the failure of the UNSC to refer cases to the Court *in extremis*. Consequently, no case involving the weaponization of toxic chemicals has been referred to the Court, despite widespread international pressure to do so in response to the chemical attacks in Syria. However, two cases involving direct or indirect complicity in the use of weaponised toxic chemicals have been successfully tried, one through a national court and one through an internationalised court, and such mechanisms may provide alternative routes for action in the short to medium term.

Acknowledgements

The author would like to acknowledge and sincerely thank Professor Francoise Hampson for her expert advice, and for reviewing and commenting upon draft versions of this chapter.

References

1. International Court of Justice, *Statute of the International Court of Justice*, 1945, Article 38(1)(b).
2. *Protocol for the Prohibition of the Use in War of Asphyxiating, Poisonous or Other Gases, and of Bacteriological Methods of Warfare (Geneva Protocol)*, 17th June 1925, available from ICRC, Treaties, States Parties and Commentaries, https://ihl-databases.icrc.org/ihl/INTRO/280?OpenDocument, accessed 8 April 2018.
3. United Nations, *UNGA Resolution 2603 a (XXIV) of 16th December 1969*.
4. J. Henckaerts and L. Doswald-Beck, *Customary International Humanitarian Law Volume 1: Rules*, Cambridge University Press, 2005, pp. 263–265.
5. OPCW, *Chemical Weapons Convention*, 1993, Article I.
6. OPCW, *Chemical Weapons Convention*, 1993, Article I.1.c.
7. OPCW, *Chemical Weapons Convention*, 1993, Article I.1.d.
8. OPCW, *Chemical Weapons Convention*, 1993, Article I.3.
9. OPCW, *Chemical Weapons Convention*, 1993, Article I.4.
10. OPCW, *Chemical Weapons Convention*, 1993, Articles II(9) (a), V(14), and XI(2) (c).
11. OPCW, *Chemical Weapons Convention*, 1993, Article VII.
12. OPCW, *Chemical Weapons Convention*, 1993, Article II.2.
13. W. Krutzsch and R. Trapp, Article II: Definitions and Criteria, in *The Chemical Weapons Convention: A Commentary*, ed. W. Krutzsch, E. Myjer and R. Trapp, Oxford University Press, Oxford, 2014, p. 82.
14. OPCW, *Chemical Weapons Convention*, 1993, Article II.3.
15. World Health Organisation, *Public Health Response to Biological and Chemical Weapons: WHO Guidance*, 2004, http://www.who.int/csr/delibepidemics/biochemguide/en/index.html, accessed 1 June 2017.
16. OPCW, *Chemical Weapons Convention*, 1993, Article II.7.
17. OPCW, *Report of the Twentieth Session of the Scientific Advisory Board SAB-20/1*, 14 June 2013.
18. OPCW, *Technical Secretariat, Office of Strategy and Policy, S/1177/2014*, 1 May 2014.

19. J. Walker, The origins of the CWC Schedules and Industry Verification regime: the concept of risk, in *Security and the Non-proliferation of Chemical Weapons, Seminar Proceedings*, ed. H. Mashadi, K. Parej, P. Runn and R. Trapp, 11–12 April 2011.
20. OPCW, *Chemical Weapons Convention*, 1993, Article II.1.
21. W. Krutzsch and R. Trapp, Article II: Definitions and Criteria, in *The Chemical Weapons Convention: A Commentary*, ed. W. Krutzsch, E. Myjer and R. Trapp, Oxford University Press, Oxford, 2014, p. 77.
22. M. Meselson and J. P. Robinson, *New Technologies and the Loophole, Editorial, Chemical Weapons Convention Bulletin 23*, March 1994, Harvard Sussex Program, p. 1.
23. OPCW, *Chemical Weapons Convention*, 1993, Article II.9.
24. OPCW, *Chemical Weapons Convention*, 1993, Article II.1.a.
25. OPCW, *Chemical Weapons Convention*, 1993, Article I.5.
26. G. Ford, *Executive Order 11850 – Renunciation of Certain Uses in War of Chemical Herbicides and Riot Control Agents*, 8 April 1975, 40 FR 16187 CGR, 1971–1975 Comp., 980.
27. International Law and Policy Institute, Chemical weapons and law enforcement under international law, *ILPI weapons of mass destruction project*, December 2016, http://nwp.ilpi.org/?p=5667, accessed 1 June 2017.
28. W. Boothby, *Weapons and the Law of Armed Conflict*, Oxford University Press, 2nd edn, 2016, p. 119.
29. UN, *Vienna Convention on the Law of Treaties 23 May 1969, 1155 UN Treaty Series (UNTS) 331*, Article 31 (3) (a)–(c).
30. A. Chayes and M. Meselson, *Proposed Guidelines on the Status of Riot Control Agents and Other Toxic Chemicals under the Chemical Weapons Convention, Chemical Weapons Convention Bulletin 35*, Harvard Sussex Program, March 1997, pp. 13–18.
31. W. Krutzsch and R. Trapp, Article II: Definitions and Criteria, in *The Chemical Weapons Convention: A Commentary*, ed. W. Krutzsch, E. Myjer and R. Trapp, Oxford University Press, Oxford, 2014, pp. 73–104.
32. ICRC, *Toxic Chemicals as Weapons for Law Enforcement: A Threat to Life and International Law?*, September 2012.
33. W. Krutzsch and R. Trapp, Article II: Definitions and Criteria, in *The Chemical Weapons Convention: A Commentary*, ed. W. Krutzsch, E. Myjer and R. Trapp, Oxford University Press, Oxford, 2014, p. 102.
34. D. Fidler, Incapacitating Chemical and Biochemical Weapons and Law Enforcement under the Chemical Weapons Convention, in *Incapacitating Biochemical Weapons*, ed. A. Pearson, M. Chevrier and M. Wheelis, Lexington books, 2007, p. 174.
35. D. Neill, Riot control and incapacitating chemical agents under the chemical weapons convention, *Defence R&D Canada Centre for Operational Research and Analysis*, June 2007.
36. D. Fidler, Incapacitating Chemical and Biochemical Weapons and Law Enforcement under the Chemical Weapons Convention, in *Incapacitating Biochemical Weapons*, ed. A. Pearson, M. Chevrier and M. Wheelis, Lexington books, 2007, p. 174.

37. D. Fidler, Incapacitating Chemical and Biochemical Weapons and Law Enforcement under the Chemical Weapons Convention, in *Incapacitating Biochemical Weapons*, ed. A. Pearson, M. Chevrier and M. Wheelis, Lexington books, 2007, p.1 75.
38. D. Fidler, Incapacitating Chemical and Biochemical Weapons and Law Enforcement under the Chemical Weapons Convention, in *Incapacitating Biochemical Weapons*, ed. A. Pearson, M. Chevrier and M. Wheelis, Lexington books, 2007, p. 185.
39. *Convention on the Prohibition of the Development, Production and Stockpiling of Bacteriological (Biological) and Toxin Weapons and on Their Destruction (BTWC)*, 1972, available from ICRC, Treaties, States Parties and Commentaries, https://ihl-databases.icrc.org/applic/ihl/ihl.nsf/INTRO/450?OpenDocument, accessed 1 June 2017.
40. United Nations, *BTWC 8th Review Conference Final Document*, 2017, Article I, paragraphs 1 & 2.
41. A. Clapham, *Human Rights Obligations of Non-State Actors*, Oxford University Press, Oxford, 2006, p. 271.
42. Geneva Conventions (I)–(IV), *Adopted by Diplomatic Conference for the Establishment of International Conventions for the Protection of Victims of War*, Geneva, 12 August 1949.
43. (a) *Protocol Additional to the Geneva Conventions of 12th August 1949, and Relating to the Protection of Victims of International Armed Conflicts [Protocol I]*, 8 June 1977; (b) *Protocol Additional to the Geneva Conventions of 12th August 1949, and relating to the Protection of Victims of Non-International Armed Conflicts [Protocol II]*, 8 June 1977.
44. *Protocol Additional to the Geneva Conventions of 12th August 1949, and Relating to the Protection of Victims of International Armed Conflicts [Protocol I]*, 8 June 1977, Articles 48 and 51.
45. J. Henckaerts and L. Doswald-Beck, *Customary International Humanitarian Law Volume 1: Rules*, Cambridge University Press, 2005, Rules 1 [pp. 3–8], 7 [pp. 25–29], 9, 10 [pp. 25–34], 11 and 12 [pp. 37–43].
46. United Nations General Assembly, Human Rights Council, *Report of the Independent International Commission of Inquiry on the Syrian Arab Republic, A/HRC/34/64*, 2 February 2017, p. 31.
47. International Court of Justice, *Nuclear Weapons Case, Advisory Opinion, 1996, as Cited by OPCW, Chemical Weapons Convention*, 1993, Article I.1.c., Rule 1, p. 5.
48. P. Herby, Protecting and reinforcing humanitarian norms: the way forward, in *Incapacitating Biochemical Weapons*, ed. A. Pearson, A. M. Chevrier and M. Wheelis, Lexington books, 2007.
49. J. Henckaerts and L. Doswald-Beck, *Customary International Humanitarian Law Volume 1: Rules*, Cambridge University Press, 2005, Rule 70, pp. 237–244.
50. *The SIRUS Project, Towards a Determination of Which Weapons Cause "Superfluous Injury or Unnecessary Suffering"*, ed. R. Coupland, ICRC, Geneva, 1997, p. 23.

51. P. Herby, Protecting and reinforcing humanitarian norms: the way forward, in *Incapacitating Biochemical Weapons*, ed. A. Pearson, A. M. Chevrier and M. Wheelis, Lexington books, 2007, p. 286.
52. Geneva Conventions (I)–(IV), *Adopted by Diplomatic Conference for the Establishment of International Conventions for the Protection of Victims of War*, Geneva, 12 August 1949, Common Article 3.1.
53. Geneva Conventions (I)–(IV), *Adopted by Diplomatic Conference for the Establishment of International Conventions for the Protection of Victims of War*, Geneva, 12 August 1949, Common Article 3.2.
54. Statement of the International Committee of the Red Cross, *First CWC Review Conference*, 28th April–9th May 2003.
55. Royal Society, Science Policy Centre, *Brain Waves Module 3, Neuroscience, Conflict and Security*, RS publications, London, February 2012, p. 19.
56. (a) Royal Society, Science Policy Centre, *Brain Waves Module 3, Neuroscience, Conflict and Security*, RS publications, London, February 2012, p. 19; (b) *Specifically Under: Third Geneva Convention*, 1949, Article 17.
57. US Army, *Field Manual FM 2–22.3 (FM 34–52): Human Intelligence Collector Operations*, Headquarters, Department of the Army, September 2006, p. 102.
58. Geneva Conventions (I)–(IV), *Adopted by Diplomatic Conference for the Establishment of International Conventions for the Protection of Victims of War*, Geneva, 12 August 1949, Common Article 1; *Protocol Additional to the Geneva Conventions of 12th August 1949, and Relating to the Protection of Victims of International Armed Conflicts [Protocol I]*, 8 June 1977, Article 1.
59. Royal Society, Science Policy Centre, *Brain Waves Module 3, Neuroscience, Conflict and Security*, RS publications, London, February 2012, p. 20.
60. *Protocol Additional to the Geneva Conventions of 12th August 1949, and Relating to the Protection of Victims of International Armed Conflicts [Protocol I]*, 8 June 1977, Article 35(3).
61. J. Henckaerts and L. Doswald-Beck, *Customary International Humanitarian Law Volume 1: Rules*, Cambridge University Press, 2005, Rules 45, p. 151.
62. *Convention on the Prohibition of Military or Any Other Hostile Use of Environmental Modification Techniques*, 18 May 1977, Articles 1 & 2.
63. J. Henckaerts and L. Doswald-Beck, *Customary International Humanitarian Law Volume 1: Rules*, Cambridge University Press, 2005, Rules 45, p. 156.
64. *Second Review Conference of the Parties to the ENMOD Convention, Final Declaration*, 17 September 1992, paragraph 633.
65. *Protocol Additional to the Geneva Conventions of 12th August 1949, and Relating to the Protection of Victims of International Armed Conflicts [Protocol I]*, 8 June 1977, Article 36.
66. ICRC, *A Guide to the Legal Review of New Weapons, Means and Methods of Warfare, Measures to Implement Article 36 of Additional Protocol I of 1977*, January 2006, p. 4.

67. *Final Declaration of the Second Review Conference of the States Parties to the Convention on Certain Conventional Weapons*, Geneva, 21 December 2001, CCW/CONF.II/2, p. 11.
68. K. Law and, *Reviewing the Legality of New Weapons, Means and Methods of Warfare*, International Review of the Red Cross, December 2006, vol. 88, (864), p. 926.
69. ICRC, *Legal Review of New Weapons: Scope of the Obligation and Best Practices*, Humanitarian Law and Policy, http://blogs.icrc.org/law-and-policy/2016/10/06/legal-review-new-weapons/, accessed 1 June 2017.
70. F. Hampson, International Law and the Regulation of Weapons, in *Incapacitating Biochemical Weapons*, ed. A. Pearson, M. Chevrier and M. Wheelis, Lexington books, 2007, pp. 244–245.
71. Human Rights Committee, *General Comment No. 6: The Right to Life*, 30 April 1982, paragraph 1.
72. Human Rights Committee, *General Comment No. 6: The Right to Life*, 30 April 1982, paragraph 3.
73. United Nations, *Basic Principles on the Use of Force and Firearms by Law Enforcement Officials, Adopted by the Eighth United Nations Congress on the Prevention of Crime and the Treatment of Offenders*, Havana, Cuba, 7 September 1990.
74. United Nations, *Code of Conduct for Law Enforcement Officials, Adopted by United Nations General Assembly Resolution 34/169*, 17 December 1979.
75. United Nations, *Basic Principles on the Use of Force and Firearms by Law Enforcement Officials, Adopted by the Eighth United Nations Congress on the Prevention of Crime and the Treatment of Offenders*, Havana, Cuba, 7 September 1990, Principle 5 (a)–(c).
76. United Nations, *Basic Principles on the Use of Force and Firearms by Law Enforcement Officials, Adopted by the Eighth United Nations Congress on the Prevention of Crime and the Treatment of Offenders*, Havana, Cuba, 7 September 1990, Principle 2.
77. UN, Human Rights Commission, *Report of the Special Rapporteur on Torture, Theo Von Boven, E/CN.4/2003/69*, 13 January 2003, paragraph 12.
78. UN, *Report of the Special Rapporteur on the Rights to Freedom of Peaceful Assembly and of Association, Maina Kiai, Human Rights Council, Twentieth Session*, 21 May 2012, A/HRC/20/27, paragraph 35.
79. *European Committee for the Prevention of Torture and Inhumane or Degrading Treatment or Punishment, CPT/Inf (2009) 13*, 19 March 2009, paragraph 92.
80. UN, *Standard Minimum Rules for the Treatment of Prisoners, Approved by the Economic and Social Council by its Resolutions 663 C (XXIV) of 31 July 1957 and 2076 (LXII) of 13 May 1977*.
81. ICRC, *Toxic Chemicals as Weapons for Law Enforcement: A Threat to Life and International Law?*, September 2012, p. 3.
82. D. Fidler, Incapacitating Chemical and Biochemical Weapons and Law Enforcement under the Chemical Weapons Convention, in *Incapacitating Biochemical Weapons*, ed. A. Pearson, M. Chevrier and M. Wheelis, Lexington books, 2007, pp. 171–194.

83. European Court of Human Rights (ECtHR), *Finogenov and Others v. Russia, Application Numbers 18299/03 and 27311/03, Judgement*, 20 December 2011.
84. European Court of Human Rights (ECtHR), *Finogenov and Others v. Russia, Application Numbers 18299/03 and 27311/03, Judgement*, 20 December 2011, paragraph 203.
85. European Court of Human Rights (ECtHR), *Finogenov and Others v. Russia, Application Numbers 18299/03 and 27311/03, Judgement*, 20 December 2011, paragraph 236.
86. European Court of Human Rights (ECtHR), *Finogenov and Others v. Russia, Application Numbers 18299/03 and 27311/03, Judgement*, 20 December 2011, paragraph 266; for further discussion see paragraphs 243–262.
87. European Court of Human Rights (ECtHR), *Finogenov and Others v. Russia, Application Numbers 18299/03 and 27311/03, Judgement*, 20 December 2011, paragraph 229.
88. F. Hampson, International Law and the Regulation of Weapons, in *Incapacitating Biochemical Weapons*, ed. A. Pearson, M. Chevrier and M. Wheelis, Lexington books, 2007, p.243.
89. *International Covenant on Civil and Political Rights*, adopted 16 December 1966, Article 6.(5).
90. Amnesty International, *Execution by Lethal Injection, ACT 50/007/2007, October 2007; Report of the UN Special Rapporteur on Torture, Juan Mendez*, 9 August 2012, A/67/279.
91. World Medical Association, *Resolution on Physician Participation in Capital Punishment, Adopted by the 34th World Medical Assembly Lisbon, Portugal, 28 September–2 October 1981; International Council of Nurses, Torture, Death Penalty and Participation by Nurses in Executions, Adopted 1998, Revised 2006.*
92. (a) UN, *Universal Declaration of Human Rights, General Assembly Resolution 217 A*, 10 December 1948, Article 5; UN, *International Covenant on Civil and Political Rights*, adopted 16 December 1966, Article 7 (b) UN, *Convention against Torture and Other Cruel, Inhuman or Degrading Treatment or Punishment, Adopted by UNGA Resolution 39/46*, 10 December 1984.
93. European Commission of Human Rights, *Greek case (Cited in Vol. II, Ch. 32, §1339), as Cited in J. Henckaerts and L. Doswald-Beck, Customary International Humanitarian Law Volume 1: Rules, Cambridge University Press, 2005*, Rule 90.
94. Royal Society, Science Policy Centre, *Brain Waves Module 3, Neuroscience, Conflict and Security*, RS publications, London, February 2012, p. 24.
95. D. Fidler, Incapacitating Chemical and Biochemical Weapons and Law Enforcement under the Chemical Weapons Convention, in *Incapacitating Biochemical Weapons*, ed. A. Pearson, M. Chevrier and M. Wheelis, Lexington books, 2007, p. 176.
96. UN, *International Covenant on Civil and Political Rights*, adopted 16 December 1966, Article 19.

97. M. Nowak, *U.N. Covenant on Civil and Political Rights: CCPR Commentary*, Kehl, N. P. Engel, 2nd edn, 2005, p. 340.
98. UN, *Body of Principles for the Protection of All Persons under Any Form of Detention or Imprisonment*, UN General Assembly, 43/173, 9 December 1988, Principle 21.
99. Supreme Court of India, *Criminal Appellate Jurisdiction, Selvi v. State of Karnataka & Anr., Criminal Appeal Number 1267 of 2004*, 5 May 2010.
100. See US Supreme Court, *Townsend v. Sain, 372 U.S. 293*, 1963, pp. 307–309.
101. Organisation of American States, *Inter-American Convention to Prevent and Punish Torture*, 9 December 1985, Article 2.
102. UN Commission on Human Rights, *Study on the Situation of Trade in and Production of Equipment Which is Specifically Designed to Inflict Torture or Other Cruel, Inhuman or Degrading Treatment, its Origin, Destination and Forms, Submitted by Theo Van Boven, Special Rapporteur on Torture, Pursuant to Resolution 2002/38 of the Commission on Human Rights, UN Doc. E/CN.4/2003/69*, 13th January 2003.
103. UN, *Report of the Special Rapporteur on the Question of Torture*, Theo Van Boven, E/CN.4/2005/62, 15th December 2004, paragraph 13.
104. ECtHR, *Ali Güneş v. Turkey, application number 9829/07, Judgement*, 10 April 2012, This judgement became final on 10 July 2012, paragraphs 5–10.
105. ECtHR, *Ali Güneş v. Turkey, application number 9829/07, Judgement*, 10 April 2012. This judgement became final on 10 July 2012, paragraph 43.
106. United Nations, *Basic Principles on the Use of Force and Firearms by Law Enforcement Officials, Adopted by the Eighth United Nations Congress on the Prevention of Crime and the Treatment of Offenders*, Havana, Cuba, 7 September 1990, Principle 3.
107. UN, Human Rights Council, *Resolution 25/38*, 11th April 2014, paragraph 15.
108. UN General Assembly, *Report of the Special Rapporteur on Extrajudicial, Summary or Arbitrary Executions, Christof Heyns, A/69/265, 6th August 2014*, paragraph 75.
109. UN General Assembly, *Report of the Special Rapporteur on Extrajudicial, Summary or Arbitrary Executions, Christof Heyns, A/69/265, 6th August 2014*, paragraph 75.
110. (a) A. Cassese, *International Criminal Law*, Oxford University Press, Oxford, 2003, p. 23, as cited by; (b) S. Oñate, B. Exterkate, L. Tabassi and E. van der Borght, *Lessons Learned: Chemicals Trader Convicted of War Crimes, Hague Justice Journal*, 2007, vol. 2, (1), p. 37.
111. International Criminal Court, *Rome Statute of the International Criminal Court*, 17th July 1998, A/CONF.183/9.
112. International Criminal Court, *Rome Statute of the International Criminal Court*, 17th July 1998, A/CONF.183/9, Article 13.
113. International Criminal Court, *Rome Statute of the International Criminal Court*, 17th July 1998, A/CONF.183/9, Article 25.
114. International Criminal Court, *Rome Statute of the International Criminal Court*, 17th July 1998, A/CONF.183/9, Article 29.

115. International Criminal Court, *Rome Statute of the International Criminal Court*, 17th July 1998, A/CONF.183/9, Article 6.
116. International Criminal Court, *Rome Statute of the International Criminal Court*, 17th July 1998, A/CONF.183/9, Article 7.
117. International Criminal Court, *Rome Statute of the International Criminal Court*, 17th July 1998, A/CONF.183/9, Article 8.
118. International Criminal Court, *Rome Statute of the International Criminal Court*, 17th July 1998, A/CONF.183/9, Articles 8 (2) (b) (xvii) and (xviii).
119. International Criminal Court, *Elements of Crimes, Official Records of the Assembly of States Parties to the Rome Statute of the International Criminal Court (ASP)*, First session, New York, 3rd–10th September 2002, ICC-ASP/1/3, part II.B.
120. International Criminal Court, *Elements of Crimes, Official Records of the Assembly of States Parties to the Rome Statute of the International Criminal Court (ASP)*, First session, New York, 3rd–10th September 2002, ICC-ASP/1/3, Articles 8(2) (b) (xvii) and (xviii), Elements.
121. International Criminal Court, *Rome Statute of the International Criminal Court*, 17th July 1998, A/CONF.183/9, Article 8(2) (a) (ii).
122. International Criminal Court, *Rome Statute of the International Criminal Court*, 17th July 1998, A/CONF.183/9, Article 8(2) (a) (iii).
123. International Criminal Court, *Rome Statute of the International Criminal Court*, 17th July 1998, A/CONF.183/9, Article 7.
124. Resolution RC/Res.5, *Amendments to Article 8 of the Rome Statute*, http://www.icc-cpi.int/iccdocs/asp_docs/Resolutions/RC-Res.5-ENG.pdf, accessed 1 June 2017.
125. International Criminal Court, *Rome Statute of the International Criminal Court*, 17th July 1998, A/CONF.183/9, Article 121, paragraph 5.
126. L. Tabassi, *Impact of the CWC: progressive development of customary international law and evolution of the customary norm against chemical weapons*, CBW Conventions Bulletin, March 2004, ((63),), p. 2.
127. International Criminal Court, *Rome Statute of the International Criminal Court*, 17th July 1998, A/CONF.183/9, Articles 11 and 12.
128. International Criminal Court, *Rome Statute of the International Criminal Court*, 17th July 1998, A/CONF.183/9, Article 17.
129. *Full Draft: UN Resolution Referring Syria to the ICC*, available at: https://justiceinconflict.org/2014/05/12/full-draft-un-resolution-referring-syria-to-the-icc/.
130. *Public Statement: 58 States Call on UN Security Council to Adopt French Resolution Referring Syria to the ICC*, 19 May 2014, available at: https://www.hrw.org/news/2014/05/19/public-statement-58-states-call-un-security-council-adopt-french-resolution.
131. *Statement by Civil Society Organizations on Need for Justice*, 15 May 2014, available at: https://www.hrw.org/news/2014/05/15/syria-groups-call-icc-referral.
132. *UN Human Rights Chief Renews Call on Security Council to Refer Syria to ICC*, UN News Centre, 2 July 2012.

133. UN, Human Rights Council, *Report of the Independent International Commission of Inquiry on the Syrian Arab Republic, A/HRC/22/59*, 5 February 2013.
134. UN, *Russia, China block Security Council referral of Syria to International Criminal Court*, UN News, 22 May 2014.
135. *United Nations General Assembly, Resolution A/RES/71/248, 66th plenary meeting, 21 December 2016.*
136 *International Centre for Transitional Justice, the Anfal Trial and the Iraqi High Tribunal Updates 1–3*, 2006, available at: http://ictj.org/publications, accessed 1 March 2017.
137. J. Trahan, A Critical Guide to the Iraqi High Tribunal's Anfal Judgement: Genocide Against the Kurds, *Mich. J. Int. Law*, 2009, **30**, 305–407.
138. J. Rikhof, Fewer Places to Hide? The Impact of DomesticWar Crimes Prosecutions on International Impunity, *Crim. Law Forum*, 2009, **20**(1), 8.
139. S. Oñate, B. Exterkate, L. Tabassi and E. van der Borght, *Lessons Learned: Chemicals Trader Convicted of War Crimes, Hague Justice Journal*, 2007, vol. 2(1), p. 23.
140. L. Tabassi and A. Dhavle, Article VII: national implementation measures, in *The Chemical Weapons Convention—a Commentary*, ed. W. Krutzsch, E. Myjer and R. Trapp, Oxford University Press, Oxford, 2014, pp. 225–227.
141. ICRC Treaties website, ICRC, Treaties, States Parties and Commentaries, https://ihl-databases.icrc.org/ihl/INTRO/280?OpenDocument.
142. See for example: International Law and Policy Institute, *Chemical weapons and law enforcement under international law, ILPI weapons of mass destruction project*, December 2016, http://nwp.ilpi.org/?p=5667, accessed 1 June 2017.
143. J. P. Robinson, Non Lethal Warfare and the Chemical Weapons Convention, *Further HSP submission to the OPCW Open-Ended Working Group on Preparations for the Second CWC Review Conference*, 24 October 2007.
144. World Health Organisation, *Public Health Response to Biological and Chemical Weapons: WHO Guidance*, 2004, http://www.who.int/csr/delibepidemics/biochemguide/en/index.html, Annex 2: Toxins.
145. See relevant national contributions to the: *Background Information Document on Scientific and Technological Developments Relevant to the Biological Weapons Convention, Prepared for the Sixth Review Conference*, available from: http://www.opbw.org.
146. See for example: G. Fenton, Current and Prospective Military and Law Enforcement Use of Chemical Agents for Incapacitation, in *Incapacitating Biochemical Weapons*, ed. A. Pearson, M. Chevrier and M. Wheelis, Lexington books, 2007, pp. 103–120.
147. See for example: *Protocol Additional to the Geneva Conventions of 12th August 1949, and Relating to the Protection of Victims of International Armed Conflicts [Protocol I]*, 8 June 1977, Article 35.
148. J. Henckaerts and L. Doswald-Beck, *Customary International Humanitarian Law Volume 1: Rules*, Cambridge University Press, 2005, Rule 87, pp. 306–308; Rule 110, pp. 400–403; Rule 111, pp. 403–405.

149. For an example of this principle see: *Statute of the International Criminal Court (1998), Article 31, Paragraphs 1(a) and (B), as Cited in Royal Society, Science Policy Centre, Brain Waves Module 3, Neuroscience, Conflict and Security*, RS publications, London, February 2012, p. 20.
150. See for example: *30th International Conference of the Red Cross and Red Crescent*, Geneva, 26th–30th November 2007, Paragraph 19, Resolution 3.
151. M. Crowley, *Chemical Control: Regulation of Incapacitating Chemical Agent Weapons, Riot Control Agents and Their Means of Delivery*, Palgrave Macmillan UK, Basingstoke, November 2015, ch. 8.
152. See for example: UN, *International Covenant on Civil and Political Rights*, adopted 16 December 1966, Article 6.
153. See for example: *European Convention on the Protection of Human Rights and Fundamental Freedoms*, signed 4 November 1950, Article 2.
154. See for example: European Court of Human Rights, *Gulec v. Turkey, 54/1997/838/1044*, 27 July 1998.
155. M. Crowley, *Chemical Control: Regulation of Incapacitating Chemical Agent Weapons, Riot Control Agents and Their Means of Delivery*, Palgrave Macmillan UK, Basingstoke, November 2015, ch. 3, 4, 8 and 12.
156. For identification of the ICAs employed, see: J. Riches, R. Read, R. Black, N. Cooper and C. Timperley, Analysis of Clothing and Urine from Moscow Theatre Siege Casualties Reveals Carfentanil and Remifentanil Use, *J. Anal. Toxicol.*, 2012, **36**, 647–656.
157. European Court of Human Rights (ECtHR), *Finogenov and Others v. Russia, Application Numbers 18299/03 and 27311/03, Judgement*, 20 December 2011, paragraphs 162–164 and 228–229.
158. (a) British Medical Association, *The use of drugs as weapons*, May 2007; (b) Royal Society, Science Policy Centre, *Brain Waves Module 3, Neuroscience, Conflict and Security*, RS publications, London, February 2012.
159. See for example: *European Convention for the Prevention of Torture and Inhuman or Degrading Treatment or Punishment*, Strasbourg, 26 November 1987.
160. (a) UN, *International Covenant on Civil and Political Rights*, adopted 16 December 1966, Articles 4 and 7; (b) UN, Human Rights Committee, *General Comment Number 29: States of Emergency (Article 4), UN Doc. CCPR/C/21/Rev.1/Add.11*, 21 August 2001, paragraph 7.
161. UN, *Convention against Torture and Other Cruel, Inhuman or Degrading Treatment or Punishment*, adopted by UNGA Resolution 39/46, 10 December 1984, Article 1(1). It should be noted that this definition is narrower than most relevant international and regional treaties, which normally require only that a State agent intentionally inflicts severe pain or suffering, without the need to show instigation, consent or acquiescence.
162. Royal Society, Science Policy Centre, *Brain Waves Module 3, Neuroscience, Conflict and Security*, RS publications, London, February 2012, p. 24, note 78. The weaponisation of such toxic chemicals would also clearly violate the CWC.

163. M. Crowley, *Chemical Control: Regulation of Incapacitating Chemical Agent Weapons, Riot Control Agents and Their Means of Delivery*, Palgrave Macmillan UK, Basingstoke, November 2015, ch. 8.
164. International Criminal Court, About the Court, ICC website, https://www.icc-cpi.int/about, accessed April 2018.
165. UN Treaty Collection, available at: https://treaties.un.org/Pages/ViewDetails.aspx?src=IND&mtdsg_no=XVIII-10&chapter=18&clang=_en, accessed 8 April 2018.
166. UN Treaty Collection website for ratifying States, https://treaties.un.org/Pages/ViewDetails.aspx?src=IND&mtdsg_no=XVIII-10-a&chapter=18&clang=_en, accessed 8 April 2018.
167. M. Zgonec-Rožej, Historical Development and the Establishment of the International Courts and Tribunals, in *International Criminal Law Manual, International Bar Association*, May 2010.
168. (a) United Nations, *International Criminal Tribunal for the Former Yugoslavia*, available at: http://www.icty.org/sections/AbouttheICTY; (b) United Nations, *International Criminal Tribunal for Rwanda*, available at: http://www.unictr.org/en.
169. J. Rikhof, Fewer Places to Hide? The Impact of Domestic War Crimes Prosecutions on International Impunity, *Criminal Law Forum*, 2009, **20**(1), 4–8.
170. M. Zgonec-Rožej, Historical Development and the Establishment of the International Courts and Tribunals, in *International Criminal Law Manual, International Bar Association*, May 2010, pp. 55–61 and 65–76.
171. The Hague Justice Portal, Frans van Anraat, available at: http://www.haguejusticeportal.net/index.php?id=6411, accessed 1 June 2017; L. Tabassi and A. Dhavle, Article VII: national implementation measures, in *The Chemical Weapons Convention—a Commentary*, ed. W. Krutzsch, E. Myjer and R. Trapp, Oxford University Press, Oxford, 2014, pp. 225–227.
172. (a) M. Zwanenburg and G. den Dekker, Prosecutor v. Frans van Anraat, *Am. J. Int. Law*, 2010, **104**(1), 86–94; (b) H. G. van derWilt, Genocide, Complicity in Genocide and International v. Domestic Jurisdiction: Reflections on the van Anraat Case, *J. Int. Crim. Justice*, 2006, **4**, 239–257.
173. ECtHR, *Case of Abdullah Yaşa and Others v. Turkey, application number 44827/08, judgement*, 16th July 2013.
174. ECtHR, *Death during a Demonstration: Turkey Must Regulate the Use of Tear-gas Grenades, Press Release Issued by the Registrar of the Court*, ECHR 227 (2014), 22nd July 2014.
175. International Partnership Against Impunity for the Use of Chemical Weapons, *Declaration of Principles*, 23 January 2018, https://www.noimpunitychemicalweapons.org/-en-.html.

Section III

Advances in Chemistry and Biology

CHAPTER 7

Convergence of Chemistry and Biology, and Nanotechnology

R. TRAPP

Independent Disarmament Consultant, 116 Route de la Contamine, Chessenaz-74207, France
*E-mail: ralf.trapp@gmail.com

7.1 Introduction

The use of poison as weapon is as old as our knowledge of poison. The nature of chemical warfare, however, changed fundamentally at the turn of the 19th century with the emergence of the chemical industry. New technologies and equipment enabled the cheap, industrial-scale production of toxic chemicals and their efficient dissemination on the battlefield, creating a scientific and technological potential for chemical warfare on a scale hitherto unknown.

Although used only rarely, chemical weapons remained a menace throughout the 20th century, not merely to soldiers on the battlefield but even more so to civilian populations. New agents were developed, and huge arsenals of chemical weapons were amassed by the Soviet Union and the USA during the Cold War. Several other countries also acquired chemical weapons – eight States declared current chemical weapons stockpiles when they joined the Chemical Weapons Convention (the CWC): Russia, the United States, India, a State not disclosed by the Organisation for the Prohibition of Chemical Weapons (the OPCW), Albania, Iraq, Libya, and Syria. Of the few States that have yet to join the CWC, at least one is believed to maintain a chemical

Preventing Chemical Weapons: Arms Control and Disarmament as the Sciences Converge
Edited by Michael Crowley, Malcolm Dando and Lijun Shang

Published by the Royal Society of Chemistry, www.rsc.org

weapons (CW) stockpile today (North Korea). In recent decades, terrorist organisations also have shown an interest in acquiring a chemical warfare capability; after the 1995 use of sarin gas by the Aum Shinrikyo in Japan, terrorist organisations involved in the conflict in the Middle East also used toxic chemicals in combat and in attacks on civilian targets. A joint OPCW–UN investigation in Syria recently confirmed that, in addition to attacks it attributed to Syrian government forces, ISIL (Daesh) was the only belligerent party that could have been responsible for a mustard gas attack in Marea (Aleppo province, Syria) in August 2015.[1]

International efforts to ban chemical weapons resulted in the adoption of the Chemical Weapons Convention (CWC), which entered into force in 1997. The CWC is designed to ensure the lasting and comprehensive prohibition of all chemical weapons – never under any circumstances shall States Parties develop, acquire or use chemical weapons and all their stockpiles were to be eliminated within prescribed time frames. Its national implementation and international verification mechanisms were to ensure that all States Parties honour their treaty commitments. At the same time, these provisions largely mirror past chemical weapons State programmes. Paradigm shifts in science, technology and industry, and the emergence of new actors such as terrorist or criminal groups, may challenge the way in which the CWC functions or even test some of its design principles. This is why the CWC contains provisions for a regular review of its operations, including assessing the impact of advances in science and technology, and mechanisms to adapt CWC provisions if required by changes in its implementation environment.

Such a paradigm shift may be about to occur in the life sciences, prompted by a combination of science convergence and advances in key enabling technologies.

7.2 Convergence in the Life Sciences

7.2.1 Advances at the Intersection of Chemistry and Biology

Science can advance in different directions: deepening the knowledge and understanding of natural processes and phenomena by narrowing the focus of investigation using ever more specialised tools and concepts (specialisation), or bringing together diverse knowledge, concepts and techniques from multiple fields of science to broaden the understanding of natural phenomena, develop new theoretical concepts, and devise new investigative approaches (convergence). Both approaches complement each other, yet convergence is more likely to lead to non-linear progress, sparking revolutionary change.

At the turn of the 21st century, convergence has become a prominent feature of the life sciences. A recent working paper submitted to the Meeting of the States Parties to the Biological Weapons Convention describes convergence as "an integrative and collaborative approach in the life sciences that brings together theoretical concepts, experimental techniques as well as knowledge of different (science and engineering) disciplines at the crossroads of chemistry and biology."[2]

An example of convergence is the "omics" – studies of all constituents collectively of a set of data/biomolecules such as genes (genomics), lipids (lipidomics), RNA (transcriptomics), proteins (proteomics), metabolites (metabolomics), the brain's structural and functional connections (connectomics), the cellular systems of an organism (cytomics), and more. Advances in the "omics" have made it possible to edit genetic instructions, providing the tools for correcting genetic defects or directing organisms to produce molecules that are alien to them. These tools can be used to create molecules that are similar to those created by nature but different in terms of the building blocks used (*e.g.*, amino acids). At the same time, advances in bioinformatics, simulation and modelling help life scientists understand how biological systems work and how their processes can be augmented, interrupted or otherwise interfered with. Gradually, the life sciences are moving from a descriptive to a predictive approach that attempts to work from first principles. Today, scientists can model simple biological systems or components of more complex ones. For the time being, rational design and comprehensive predictability of complex biological systems still remain beyond reach. There is a lack of reliable biological data and standardised protocols, and biological systems are dynamic and complex. It has been observed that "[T]o understand dynamic systems, the data ideally should be time resolved at a sufficient scale, be deep enough to cover all the components of the system, broad and complete enough to cover the extent of the cell model, and cheap enough to be feasible."[3] Huge efforts have been made towards a more holistic, integrated approach but past emphasis on expanding the existing data sets on biological functions and components, rather than on ensuring their quality and reproducibility, has led to uncertainties about the validity of many of the data, resulting in researchers spending much time validating and integrating existing data sets. Also, there remain gaps in understanding: the transcriptome is not fully understood, and research has shown that proteome organisation cannot be explained by the organisation of the genome.[4]

Furthermore, when science is advancing at a high pace and new technologies spread rapidly into a variety of areas of potential application, it is often difficult to predict accurately what their impact will be as they mature. New technologies have to compete with existing technologies and solutions, and their limitations will only become apparent as they are used in practice. Two recent examples for new technologies, the impact of which on arms control was over-estimated, were micro processing and additive manufacturing (3D printing). Whilst the former remains an interesting tool for new products and process development, and has to a degree changed features in parts of the chemical industry, its actual impact on CWC verification appears (at least at this stage) to be modest. As for additive manufacturing, it has become clear that whilst it is an excellent tool for fast prototyping and repair, it is not for large-scale industrial manufacturing of critical parts where performance to a high standard is essential.[4]

Nevertheless, the life sciences are making rapid progress and the advances at the interface of chemistry and biology will bring about a range of changes in society, affecting such diverse fields as energy production, food production, medicine, information technology, the production of consumer goods, new materials and much more.

Inevitably, such changes will also affect the CWC and the way in which it operates. They will create new opportunities for treaty implementation but also may challenge some of its basic design criteria. Four technological trends will be discussed below in more detail: the emergence of biological production technologies, the production of complex biological compounds, the possible emergence of new types of chemical agents, and advances in the protection against toxic chemicals.

7.2.2 Biological Production Technologies for Chemical Products

Biological processes have become more widely used in the manufacturing of chemical products, from high-volume organic products such as alcohols and urea to high-value fine chemicals and pharmaceuticals. The OPCW's Scientific Advisory Board (SAB) noted that "[B]ulk and fine chemicals are being produced increasingly using biologically mediated processes, *e.g.* by microbial fermentation or using enzymes as catalysts. It is estimated that approximately 10% of chemical production volume will use such processes by 2020."[5]

With regard to their relevance for the implementation of the CWC, the SAB identified the following three processes in particular:

- Fermentation technology and metabolic engineering;
- Enzymes/biocatalysts;
- Synthetic biology.

Although these projections were perhaps optimistic in scale, in light of changes in the price of crude oil, biologically mediated chemicals production is no longer a niche business. The range of products today manufactured by biologically mediated processes reaches from small hydrocarbons, alcohols, amino and other organic acids, to polymers and complex molecules such as peptides.[3] The manufacturing capacity for biomaterials (including biopolymers) in Europe, for example, is expected to reach 1.7 million tonnes in 2020 (it stood at 700 000 tonnes per year in 2011). Sales of bio-based products in the EU were at 35.8 million USD in 2013; whilst their overall share of chemical products in 2013 had reached 6.26% in the EU, in some European countries it had climbed as high as 38.9% (Denmark) or 21.72% (Sweden).[6]

One of the drivers is the continuing pressure to move towards using renewable resources. Biomass is the only renewable carbon source other than carbon dioxide, and therefore remains an attractive alternative to crude oil or natural gas. There are a number of routes for converting biomass into chemical products, starting from simple sugars to starchy polymeric biomass and lignocelluloses. The fermentation of sugars is a well-established industrial process. Bioconversion of starchy biomass and celluloses is more challenging. Starches first need to be depolymerised through enzymatic hydrolysis. Starches as well as sugars, of course, are also food sources and are therefore not attractive for manufacturing platform chemicals.

Lignocellulose, on the other hand, is an abundant raw material, but it is recalcitrant and not easy to convert; its bioconversion requires pre-treatment (steam explosion, sulphuric acid treatment, pre-treatment using ammonia or lime), and the resulting product contains a variety of sugars at low concentrations and significant lignin ballast. Research is under way to develop consolidated bioprocessing technologies that can overcome these shortcomings, thereby reducing cost and simplifying the process. Such consolidated lignocellulose bioconversion processes have yet to be scaled up to industrial production levels, however.[4]

How these trends towards biologically mediated processes will evolve will of course also depend on the evolution of the prices of raw materials, services and equipment. The current low crude oil price puts strong pressures on biotechnological chemicals manufacturers, forcing them to look for increased efficiencies, new types of enzymes, and high-value products and solutions rather than finding new applications for established products.[7]

In a Working Paper to the 2015 meeting of the States Parties of the Biological Weapons Convention (BWC), Switzerland observed that in the future, chemicals would be produced "by traditional chemical methods, with the help of biological catalysts such as enzymes, or through the specially designed metabolic process of a self-replicating organism or an organism-like system. Drivers for pursuing a particular method of production will be based on economic, environmental and other factors."[2]

In the meantime, efforts to increase the efficiency of industrial biological processes continue. Directed evolution is being used to optimise the functionality of both natural and artificial enzymes (bio-catalysts) in an attempt to create catalysts with predicted reactivity and selectivity.[3] Modelling remains a challenge (electrostatics, dynamics, tertiary structures) and directed evolution adds significant time to the development process. At the same time, efforts are being made to use experience from past research and development to create "toolboxes" for speeding up process development and up-scaling, and to catalogue and automate repetitive development steps such as the synthesis of DNA sequences. In addition to saving costs and development time, these efforts also contribute to better quality control.

Another field of study that is finding its way into industrial manufacturing is synthetic biology. Genomics, transcriptomics, proteomics and metabolomics together provide a set of tools that can be used to design organisms with desired properties.[3] These tools have been applied already in several industrial projects, and the capability to redesign or manipulate organisms and to engineer improved enzymes is growing.[8] It is critical for a synthetic biology approach to have comprehensive and accurate information on the cellular machinery to be used in the process, and this remains a challenge. Only with such accurate data can one rationally approach the modification of an organism to meet the requirements of an industrial process, from ensuring the desired quality and yield of the product to ensuring the survival of the microorganism in industrial production equipment.

Another interesting approach is the plant-based production of chemicals. An example is the production of engineered proteins for pharmaceutical purposes, which combines flexible and rapid engineering methods with the benefits of high volume expression for protein isolation or seed-based long-term storage.[9]

In addition to the manufacturing of small molecules, significant progress has been made in the manufacturing of complex biomolecules such as toxins and bioregulators. Toxins are poisonous chemicals that are produced by living organisms. They often contain proteins (neurotoxins, cytotoxins) but there are also certain toxins that are small molecules, such as the saxitoxins. Natural metabolic pathways have been published for saxitoxin, ricin and other toxins, and their *in vitro* biosyntheses have been described in the literature. At present, obtaining toxins from their natural sources is still simpler than using engineered metabolic pathways, but with advances in enabling technologies that may change,[6] partly as the result of a growing interest in toxins for applications in research, medicine and agriculture.[3]

Bioregulators, also, are naturally occurring chemicals. They have not evolved as weapons for defence or to kill prey, but to ensure the proper functioning of vital physiological systems in living organisms. They are modulators used by the nervous, endocrine and immune systems, and include both relatively simple molecules such as hormones and neurotransmitters, as well as complex proteins, polypeptides and nucleic acids. At concentrations significantly different from their normal physiological levels, bioregulators can exert a range of effects, both lethal and incapacitating, including mood alteration, physiological imbalances, other physiological reactions (pain, rise in temperature, sensual impairment), and death.

Bioregulators have attracted much attention in physiological research and as potential medicines. Alongside this interest, the methods of industrial-scale production of bioregulators such as peptides have also evolved. Today, chemical synthesis remains the most widely used technique, but other methods have also been proposed, including microwave synthesis or the use of recombinant microorganisms or transgenic organisms.[10]

To understand how the emergence of a biosynthetic industry may affect the functioning of the CWC, it is important to assess whether the products and processes associated with them are relevant to what a chemical weapons programme of the future might require. Traditional chemical weapons programmes depended on an industrial capability to manufacture cheaply, and at an industrial scale, precursors and toxic chemicals that were suitable for chemical weapons purposes. Many of the agents acquired in State programmes were closely related to products manufactured by the chemical industry (*e.g.*, pesticides), and so were the technologies used for their production. This association between industrial chemical processes and materials, and the materials and manufacturing technologies required in a chemical weapons programme, formed the conceptual basis for the design of the CWC's verification and national implementation systems.

With regard to the production of scheduled chemicals, the CWC States Parties have agreed that the provisions of the CWC on national implementation and international verification cover biologically mediated and biological production processes.[11] This is consistent with the provision in Article II, paragraph 2, of the Convention which clarifies that the term "toxic chemicals" includes all such chemicals *regardless of their origin or method of production* (emphasis added). The emergence of biological and biologically mediated technologies for the manufacturing of scheduled chemicals would therefore not affect the legal norm, nor the application of the verification provisions of the CWC – what matters for national implementation, declaration and verification is that a scheduled chemical is actually produced above a certain threshold, not how it was acquired.

But, the application of the CWC is not limited to the production of scheduled chemicals. Certain toxic chemicals, even though not listed in the Schedules, may nevertheless have utility as chemical warfare agents (examples include the fentanyl derivatives used to break up the Moscow theatre siege, or the Novichok agent used in the attack on Sergei and Yulia Skripal).[12,13] Also, certain production equipment, even though not specifically designed for the production of a scheduled chemical, may nevertheless be used to that end (an example being multipurpose chemical plants/units). Part IX of the Verification Annex, dealing with discrete organic chemicals (DOC), was included in the CWC as a safety net to address such latent capabilities. But the scope of that Part is qualified by reference to "production by synthesis", and States Parties have been unable to agree whether biological and biologically mediated processes are covered by this provision. State practice, as a consequence, remains inconsistent: some States Parties declare and open up for international inspection plants that manufacture DOC by biological processes whilst others are have decided not to apply the CWC's provisions to these types of plants. This causes uneven treaty implementation and runs contrary to the concept of maintaining a level playing field for the chemical industry. It also poses the question as to whether significant chemical production capabilities relevant to chemical weapons arms control remain outside the reach of the CWC verification system (or, alternatively, whether the verification system is being swamped with biochemical plants that have little or no CWC relevance).

How significant this issue is with regard to preventing the re-acquisition of chemical weapons will also depend on which types of chemicals are involved – the traditional agents and precursors as they are known from past State-level CW programmes, or novel types of agents not as yet reflected in the CWC's Schedules of Chemicals. The most interesting development here relates to complex biomolecules rather than small molecules.

7.2.3 Chemical Production of Complex Biomolecules and Highly Potent Chemical Ingredients

DNA and protein synthesis have made significant progress in recent decades. Automated DNA and protein synthesizers have been available for some time (1980 for DNA and 1984 for proteins), and oligomers (even entire genome sequences) can be purchased from specialised companies trading over the internet. More recently, progress has also been made in the automated synthesis of carbohydrates. Automating carbohydrate synthesis is much more challenging than DNA or protein synthesis, involving some 50 different monomers, but it can open up avenues to entirely new categories of products.[4] Carbohydrates are important biomolecules that have interesting structural functions, are used in biological signalling and have many other interesting properties. For example, polysaccharides can be used as building blocks in the manufacturing of engineered nanoparticles, leading to new and less costly vaccines for treating hospital-acquired resistant infections. Clinical trials of such vaccines are expected to start soon. Other applications include new structural materials, microarrays for diagnostics tests, and monoclonal antibodies as diagnostics and treatments.[4]

One of the drivers for these developments is the market. In the past, the industry target for moving a new product from approval to production was between 4 and 6 years. Today, fast track designation, breakthrough therapy designation, accelerated drug approval and priority reviews are increasingly applied to shorten the time to market. In 2014, 2 out of 3 new drugs were designated in one of these fast track categories.[4]

The evolution of targeted delivery methods also has led the pharmaceutical industry to move towards using active ingredients with a much higher biological activity (highly active pharmaceutical ingredients or HAPI). More than a third of all novel drugs fall into this category. This has led to a significant change in the footprint of plants that develop and manufacture such ingredients. These plants are technologically highly complex, operate against short timelines, and require strong managerial, technical, safety and environmental standards. Risk management strategies are applied such as primary and secondary containments, decontamination, regular leakage checks, extensive cleaning procedures, monitoring systems concerning health, safety and the environment, air monitoring, and waste management. Workforce performance and a strong safety culture, compliance with regulatory and company standards and commitment to safety are paramount.[4] Companies control access to critical areas at such manufacturing sites, organise processes in ways that minimise transport, and apply devices for electronic tracking of people and key materials as well as for batch controls. A broad assets base enables them to swiftly scale up production from development to production. Containment standards are similar to what is traditionally used in high-security biotechnology facilities.

These facilities in many respects resemble a larger-scale Schedule 1 facility. But given their production profile they tend to remain under the declaration threshold for DOC facilities and thus are not liable for verification. This poses questions about some of the design assumptions of the CWC verification system in the chemical industry.

7.2.4 Potential Novel Agents

Past State programmes have selected toxic chemicals as chemical warfare agents because of their high acute toxicity, as well as of other properties that affect their utility as weapons materials.[14] Historically, "acute toxicity" meant high lethality, but there also are agents that primarily have an incapacitating effect – blistering agents (mustard agents, lewisite and similar agents), although these agents also have systemic toxic effects and can be deadly; agents causing essentially reversible effects on the respiratory system (lachrymation, coughing, breathing difficulties or vomiting – typical examples are harassing agents or "tear gases"); or agents causing temporary effects on the central nervous system (including BZ and certain glycolates).

Over an entire century, chemical weapons development programmes have screened thousands of small molecules for their utility as chemical weapons – it is fair to assume that with regard to the "traditional" types of CW agents, it is unlikely that research would lead to new types of lethal agents that significantly surpass what has already been achieved by traditional CW programmes.

Perhaps with the exception of certain specific regional contexts (the Korean Peninsula, the Middle East), or in assassination attempts such as the Novichok poisoning of the Skripals in the United Kingdom in March 2018, the general consensus today is that "[W]hilst the threat of 'traditional' chemical warfare with mass casualties has declined significantly since the implementation of the [Chemical Weapons] Convention began, other forms of chemical weapons might appear attractive for their capacity to cause terror, or appear useful in population displacement and social/economic destabilization."[15]

Such novel agents could emerge from a variety of chemical classes, including toxins and bioregulators. What they would have in common is that their primary target would be the central or peripheral nervous system, and that their toxic action (at lower dose levels) would be reversible and not normally life threatening. The CWC does contain provision for the use of riot control agents (RCAs) for law enforcement purposes, and some argue that other types of toxic chemicals may also be used for law enforcement purposes.[16] Concerns have thus been raised over the possibility that States may acquire advanced RCAs or certain incapacitating chemical agents (ICA), ostensibly for law enforcement purposes.

Research that is entirely legitimate but could lead to insights that others could use to design novel incapacitating agents involves several complementary fields: research aimed at a more detailed understanding of regulatory and other physiological processes in the nervous system including the brain, associated with the discovery of new neuropeptides (or receptors) and bioregulators;[17] new methods of drug discovery including predictive modelling to improve extrapolation from *in vitro* to *in vivo* ("systems pharmacology"),[18–20] such as for example a computational suite to estimate the toxic potential of drugs (endocrine and metabolic disruption and certain aspects of carcinogenicity and cardiotoxicology);[21] and more effective means of reaching specific targets in the body (the brain in particular) including by more efficient and selective ways of taking agents across biological barriers and in particular the blood–brain–barrier (BBB).

Neurobiology has seen "an exponential increase of output, improving our understanding of neural network responses associated with behaviours, such as anger and aggression, and physiological conditions such as addiction, fear and narcolepsy. Neural networks can now be manipulated to induce some of these states and work has begun to translate these findings to non-human primate models."[22]

Peptides and proteins play an important role as chemical agents for such manipulations. Their intracellular delivery remains a challenge due to the plasma membrane, and in particular, chemicals acting on the CNS need to overcome or bypass the BBB. This requires techniques for targeted drug delivery, *e.g.* by using modified viruses or engineered nanoparticles as drug carriers. An example for such targeted delivery is a platform that has been derived from the anthrax toxin: the use of a non-toxic PA/LF_N platform (consisting of a protective antigene and non-toxic, N-terminal PA binding domain of the lethal factor derived from anthrax lethal toxin) for the delivery of peptides and proteins into mammalian cells. More than 30 different proteins, polypeptides containing natural and non-natural amino acids, as well as small molecules have been successfully carried into mammalian cells in this research as such "non-native cargo".[23]

Another recent example was the development of a structural model of the protein Mfsd2a. This protein is a transporter at the BBB and thus a potential conduit for drug delivery into the brain. The research allowed insight into the mechanism of how Mfsd2a transports essential lipids across the BBB and could in the future be used for the structure-guided search and design of new scaffolds for drug delivery into the brain.[24]

In addition to research that may lead to the discovery of novel types of chemical agents or ways of transporting them to their target, new tools and techniques are being introduced that may make novel agents more easily accessible. An example from the field of genome editing is CRSPR/Cas 9 – a gene-editing tool that has become truly transformational. This is one of several new gene-editing techniques that—compared to previous methods—is simple, accessible and low cost, and has consequently seen rapid global distribution, accelerating the pace of work in gene-editing. Development has not stopped at CRISPR/Cas 9, and two new systems are attracting interest: Cpf1 and C2c2.

The fear that "deliberate or unintentional use may lead to far-reaching economic and national security implications" has led to the inclusion of genome editing into the 2016 United Sates Worldwide Threat Assessment under risks associated with weapons of mass destruction.[25] An example for gene editing applications that may have prompted this unusual decision are gene drives: these are genetic systems that disrupt, replace or otherwise modify an essential gene and at the same time circumvent the rules of genetic inheritance by greatly increasing the odds that the drive will be passed on to offspring. The drive can potentially spread through an entire population even if it imposes a severe handicap such as infertility. At least in theory, gene drives may open up a way for eradicating an entire species. This is why they have attracted attention for the fight against malaria – as a means of eradicating the vector (the *Anopheles* mosquito). But there is of course potential for misuse. Whether these advances are of any particular relevance to CW arms control (other than that they surely would be a tool in a CW development programme) remains questionable – there are many steps that a weaponeer needs to take from proof-of-concept to a functioning, reliable and predictable weapon. The risks are more likely related to the unpredictable environmental impact of eradicating a species, posing safety, as well as ethical questions.[4]

Nevertheless, gene editing techniques such as CRISPR are beginning to make an impact outside the laboratory, and the technology is being deployed for the development of a broad range of new bio-based products including therapeutics, antimicrobials, animal health products and genetically engineered crops, as well as for research in areas such as cell engineering, phenotypic screening and genomics, and industrial biotechnology (microbial fermentation, therapeutics).[4]

Another example for such enabling tools is biology-oriented synthesis (BIOS) using core structures obtained from natural products as scaffolds of compound collections to access biological diversity around bioactive structures, or deconstructing natural products into fragments and recombining these fragments into biologically active composites.[26]

Toxins, also, have attracted renewed attention. Much progress has been made in understanding the molecular and cellular mechanisms of their action. This includes the identification and characterisation of genetic components associated with toxins, the characterisation of their mechanisms of action, the development of new research tools, as well as the alterations to existing or the discovery of novel forms of toxins.[27]

On the horizon is the development of a new genetic code using non-standard amino acids, thereby expanding the chemical repertoire available for protein synthesis. This could result in novel compounds with profound differences in structure, stability, activity, and binding as compared to the proteins known from nature. In the laboratory, the recoding of genomes by completely replacing the natural system with an artificial codon has already been accomplished. Such experiments are undertaken to test the malleability of the genetic code, to expand the chemistry and functionality of proteins, and to search for new materials with properties different from those accessible by using the 20 natural amino acids. The hope is to fine-tune the properties of these novel materials for new applications as nanostructures, therapeutics, industrial enzymes, sensors, in drug delivery, or to engineer genetic isolation and virus resistance.[4] But of course, the very same techniques could be used to tailor the properties of designer proteins to the requirements of a novel CW agent that, for example, could selectively act on a specific receptor in the CNS.

Finally, one cannot exclude the possibility of chance discoveries of new agents with unexpected properties. An example for research that might lead to such unexpected discoveries is the work on pseudogenes (DNA sequences not naturally used for gene expression). Such research may lead to the discovery of new peptides and proteins that are not normally produced by cells, and that may exhibit unexpected physiological properties.

These and other research directions may be of concern to the CWC as they might lead to the discovery of new types of incapacitating chemical agents or "advanced RCAs". Such non-natural biomolecules are of course covered under the CWC's General Purpose Criterion, and at least in a legal sense are not *per se* a threat to the CWC. But, research involving such compounds may lead to discoveries that some may perceive as viable breakout capabilities. It has often been observed that there is a significant difference between the discovery of a new potential agent, and the introduction of a new chemical (biochemical, biological) weapon. What makes this research potentially problematic is the possibility of an emerging nexus between scientific advancement and the interest of some States in the development of "law enforcement weapons" based on incapacitating chemical agents, or the use of supertoxic chemicals in special operations. In such a context, offensive chemical weapons research and development could be easily concealed in what ostensibly is a legitimate programme to develop new means of law enforcement. This would seriously undermine the existing prohibitions, and challenge the effectiveness of a range of CW countermeasures (detection, physical protection, decontamination, medical countermeasures).[28] There are of course significant drawbacks for such new types of chemical agents (poor stability and difficulty to be produced in large amounts), but it is exactly these issues that a research and development programme directed at weaponising "novel agents" for law enforcement or special operations purposes would attempt to overcome.

7.2.5 Advances in Medical and Other Countermeasures

Throughout the history of chemical warfare, the military utility of chemical weapons was in part determined by the effectiveness of the defences available against them. This included the detection of chemical agents (in the atmosphere, on surfaces, in soil and water, in living organisms), physical protections against exposure to the agents (respirators, protective clothing, filtration systems for vehicles and buildings), medical countermeasures (prophylaxis, treatment of victims, medical aftercare and long-term treatments), and decontamination of personnel, equipment, vehicles, buildings and other aspects of the human environment.

Convergence in chemistry and biology is expected to enhance our ability to diagnose and treat diseases, including acute poisoning. Convergence is also expected to increase the pace of drug discovery and enable targeted drug delivery. Drugs that neutralise the effect of toxins such as ricin or botulinum toxin have been developed, including treatments that confer active ricin immunity and lead to long term protection against the toxin.[41]

The Temporary Working Group on Convergence of the SAB highlighted the emergence of protein and antibody-based drugs as effective medical countermeasures against chemical warfare agents. One example was the use of recombinant human butyrylcholinesterase (BuChE) derived from the milk of transgenic goats or CHO cell expression systems, as a candidate for prophylactic or post-exposure treatment of nerve agent poisoning by agent scavenging. This form of treatment remains experimental because of the large amounts of the drug needed, so medical research is looking for alternatives such as engineered hydrolytic enzymes that could catalytically break up nerve agent molecules rather than merely scavenge them.

Another example is small Open Reading Frames (sORFs) encoded proteins. Experimental methods and bioinformatics approaches in the search for biologically active sORF-encoded peptides have resulted in a three-step process of prediction, analysis and estimation of biological functions. This knowledge has been used for the design of an antibody-based nerve agent scavenger: phage-display libraries ("immunization" and screening *in vitro*) and autoimmune repertoires (*in vivo*) have been combined, leading to a process of reactive immunization and kinetic selection as combinatorial approaches to create novel artificial scavenger/biocatalysts – essentially a "reactibody" that combines screening for bindings with the conversion of scFV into a whole size human antibody.[4]

Another field of interest for improved medical countermeasures is targeted drug delivery (see also the section on nanotechnology below). An example is the use of Antibody–Drug Conjugates (ADCs) to deliver drugs to target organs. This kind of treatment is still rather expensive, but it is highly effective.[4]

Equally important are developments related to the use of enabling technologies, including information and communications systems. Mobile phones are being used as platforms for point of care diagnosis and treatment; crowd-sourcing and information sharing over the internet or mobile phone networks are being used to increase computation power and provide access to reliable databases, in support of diagnoses, as well as epidemiological analysis.

Convergence also affects other fields of protection against chemical weapons, such as detection, physical protection and decontamination.[6] With regard to detection, significant efforts have been made over the years to develop devices that are sensitive, fast responding, show little cross-sensitivity, can detect a range of different agents, can be miniaturised so they are portable and preferably hand-held, and can be ruggedized for use in the field. A range of physical, chemical and biochemical principles have been exploited to this end, including ion traps, ion mobility spectrometry, flame photometry, enzymatic reactions, and mass spectrometry.

Much progress has been made in the development of new types of chemical and biological sensors. Biosensors and instruments that exploit enzyme inhibition such as the inhibition of acetylcholine esterase and butyrylcholine esterase by nerve agents and other anticholinergic agents have been used for many years. What has evolved are techniques used to build these sensing techniques into field instruments, for example by using enzyme immobilisation, and more recently, the use of miniaturised disposable biosensors for agent detection, as well as point of care diagnosis. There is also increasing use of immunoassays in field detection instruments.

With regard to decontamination, progress has been made in the use of enzymes. Enzymes can overcome many of the disadvantages of traditional decontamination solutions, as they are less chemically aggressive. Enzyme based decontaminants can be attractive for the mild decontamination of delicate equipment such as optical or electronic instruments, and of course for skin decontamination. Examples include enzymes that catalyse the hydrolysis of nerve agents. Mutant enzymes can be engineered to show a preference for the more toxic enantiomer of optically isomeric nerve agents such as sarin or soman.

7.3 Nanotechnology

7.3.1 Overview

Nanotechnology and nanoscience are terms that refer to the investigation and use of entities (particles) at the nanoscale, where at least one of the dimensions is somewhere between 1 and 100 nm. Nanoparticles are of course present in natural materials, but given their size they are often large in numbers but small in share of weight or volume. The physico-chemical properties of materials can change significantly when nanoparticles are manufactured that have a uniform particle size distribution in the nanometer range. There is an increasing interest in a wide range of biomedical applications of nanomaterials, including with regard to using nanoparticles to support bioimaging, biosensing, medical therapy and controlled drug release, and gene therapy.[29]

Many of the developments that have already been discussed in the previous section on convergence between chemistry and biology overlap with advances in nanoscience and nanotechnology. At the same time, nanoscience and nanotechnology extend beyond and outside the areas of science that are normally considered in the context of convergence between chemistry and biology. This is why the impact of advances in nanotechnology on CW arms control is discussed here in a separate section.

Three aspects will be looked at in detail: the potential for using nanomaterials as novel toxic agents, their utility as delivery systems for toxic chemicals, and their application in the protection against chemical weapons.

7.3.2 Nanomaterials and Toxicity

The use of nanomaterials has increased steadily over recent years, and so have concerns about exposure and possible health risks. Nanomaterials are used in many aspects of life, including in medicines and cosmetics. Human exposure can happen in many different ways, but the most likely exposure is in the workplace where nanomaterials are produced and handled. In this work environment, they are typically absorbed *via* the respiratory system, where they penetrate deeply into the lungs and are deposited in the gas exchange area.[30]

Nanoparticles are recognised by the body as foreign materials and interact with the host organism. These interactions can be physical, chemical, or immunological. The toxicity of nanomaterials differs from that of "normal" chemicals, whose toxicity follows a dose–response relationship: the toxic effect of a chemical increases with the dose administered to the organism, following an S-shaped curve. The toxic effect is normally expressed as the medium lethal dose or LD_{50} (amount of chemical per kg of body weight at which 50 per cent of the test animals die) or in the case of non-lethal toxic effects as the effective medium dose or ED_{50}. Further increases in the dose above the LD_{50} will increase the percentage of deaths, gradually reaching a plateau until a dose at which all test animals die. For inhalation toxicity, the administered dose depends on the concentration of the chemical in the breathing air and the exposure time, and the toxic dose is expressed as the concentration–time product of the agent at which 50 per cent of the test animals die (LCt_{50}).

There are many biological factors that influence the toxic dose of a chemical, including the species of the test animal, the mode of administration (by inhalation, through the skin, by ingestion, by administration into muscle tissue or the blood stream, *etc.*), age, sex, time of day, and individual predisposition, such as previous illnesses.

In the case of nanomaterials, additional factors need to be considered. These relate to the physico-chemical properties of nanomaterials such as their small participle size, large surface area, other surface characteristics and often-uniform shape (morphology).[31,32] A recent review recalled that given their larger surface area, smaller nanoparticles tend to exhibit higher toxicity, and also that the shape and structure of the nanoparticles contribute to their toxicity.[33] Because of their small size, nanoparticles can cross biological barriers – very small nanoparticles with sizes of less than 10 nm are believed to act similar to a gas and can enter human tissue easily and disrupt the normal cell environment.[34]

In vitro studies have identified a number of mechanisms that underlie the toxicity of nanoparticles. These include oxidative stress, DNA damage caused by physical or chemical interactions, immunogenicity caused by modulation of cell-mediated immune response, cell death (apoptosis) and necrosis caused by the same mechanisms, and befouling caused by association of nanoparticles with proteins. *In vivo* studies have also pointed to translocation of nanoparticles after administration to secondary organs, causing toxic effects. Neurotoxic effects can occur through translocation of nanoparticles and crossing or bypassing the BBB, resulting in inflammatory responses in the brain.[35–38]

Studies in animals as well as human data suggest that nanomaterials can adversely affect the respiratory system, the gastrointestinal tract, the circulatory system and the central nervous system.[39–41]

The knowledge of the toxicity of nanomaterials and the physiological mechanisms involved remains incomplete and most studies over the past 10–15 years were mechanistic rather than toxicological in design, using excessive dose ranges which do not allow the discussion of toxicological end points.[42]

This poses the question of whether there may be a risk of the discovery of new types of chemical agents based on nanomaterials. On balance, this risk appears to be fairly limited and a comparative advantage of nanomaterials over well-established chemical warfare agents is not apparent. This conclusion involves considerations of what is known about their acute toxicity as well as manufacturing costs and other factors that would affect their potential utility. On the other hand, nanomaterials are widely used in human society and there is always a residual risk of unintentional side effects, or a potential for abuse of commercially available materials that could be exploited for hostile purposes. The main concern, however, relates to the use of engineered nanoparticles as delivery vehicles for payloads of toxic chemicals.

7.3.3 Nanomaterials as Delivery Vehicles

Cellular uptake and internalization of drugs are critical steps in the effective delivery of medicines. Uptake of small molecules occurs mainly by endocytosis: this is an active transport process through the lipid bilayer wrapping of cells. With regard to the uptake of nanoparticles, particle size and surface charge are the most important factors. A recent literature review underlined the profound effects of particle size and charge on clearance, enhanced permeability and retention effects, enhanced residence times and accumulation tendency of nanoparticles at tumor sites.[43]

Examples of the types of nanomaterials that have been considered or used as drug delivery systems include polymeric nanoparticles such as polysaccharide chitosan nanoparticles which function due to their ability to facilitate both protein and drug conjugation; poly(lactic-*co*-glycolic acid) as a carrier across the BBB; inorganic ceramic nanoparticles such as silica, titania and alumina that can be used as drug carriers due to their porous nature; metallic nanoparticles including superparamagnetic iron oxide nanoparticles, gold shell nanoparticles and titanium oxide nanoparticles which are routinely used for magnetic resonance imaging contrast enhancement and as drug carriers; silver nanoparticles that are being explored as antibacterial agents; and carbon nanoparticles (fullerenes and nanotubes) which are widely used for drug delivery purposes given the multiple attachment points of the fullerenes and the high electrical conductivity and strength of the nanotubes.[33]

A review published in 2012 highlighted the use of nanocarriers for drug delivery in the form of nanocrystalline drugs, liposomes, polymer–drug conjugates, polymeric micelles, protein nanoparticles, and lipid colloidal dispersions.[44] Solid lipid nanoparticles have also been developed for targeted delivery and controlled release of a wide range of drugs.[45]

There are essentially two types of nanopolymers that have been evaluated as drug delivery systems: nanocapsules (a core of encapsulated drug surrounded by a polymeric membrane or shell) and nanospheres (where the drug is distributed or absorbed throughout a polymeric matrix). A recent literature review concluded that:

"Polymeric NPs are capable to maintain high stability in systemic circulation with enhanced half-life, which can be further optimized by controlling the release of therapeutic agents from the NPs. Moreover, polymeric molecules have various solubility profiles in [a] wide range of solvents. This is advantageous for surface modification or functionalization to achieve different purposes of delivery and targeting. Subsequently, both doses and frequency of administration of therapeutic agents can be reduced due to high payloads into nanocarriers, leading to superior efficacy and minimizing the side effects. Besides, polymeric NPs of desired physicochemical properties are capable of preserving their content from hepatic metabolism, enzymatic degradation and rapid clearance. Specifically, the enormous surface area of polymeric NPs is an attractive feature to control the release kinetics, drug loading capacity and administration route, which can regulate the fate of drug into the body."[43] Another example is the development of celluloses nanoparticles to deliver medicines to specific receptor targets.

Nanoparticles have been engineered to transport drug payloads, including small molecules, therapeutic DNA and peptides, through biological barriers, including the BBB.[46,47] Strategies developed to carry CNS active drugs across the BBB include bypass through intranasal delivery, invasive transport across the BBB, the use of cell-penetrating peptides, passive targeting of diseased tissue, receptor-mediated endocytosis and combinations of these strategies. Nanoparticle brain targeting drug delivery remains, however, still in an experimental stage.[48]

An important property of nanoparticles intended for drug delivery is biodegradability. Biodegradable nanosystems may be effective tools for transporting drug payloads across biological barriers and could be used to deliver chemicals to targets they normally would not reach. Nanomaterials used as drug delivery vehicles also can be designed to protect their chemical payload from biological factors (enzymes in particular) that would otherwise metabolise the compounds, thereby altering (normally reducing) their toxicity and/or promoting their elimination from the body. A recent publication reported the delivery of DNA and RNA into cells using biodegradable silicon nanoneedles.[49]

Another interesting development is the engineering of "smart" nanoparticles. These nanoparticles are designed to respond to environmental stimuli such as changes in temperature, pressure, electric or magnetic fields, light, or chemical parameters such as pH or ionic strength. This creates the possibility of combining their carrier properties with drug release mechanisms that respond to external factors, enabling the targeted delivery of drugs in precise locations in the body when combined with, for example, imaging techniques. Yet another development that could change the way nanoparticles are being exploited for directed delivery are synthetic motors. A first study of such delivery vehicles (an acid powered zinc based micromotor) under *in vivo* conditions has been published recently.[50] Such active delivery strategies are expected to dramatically increase the efficiency compared to traditional passive diffusion of orally administered medicines, and can be combined with engineered features that enable high payload capacity, autonomous payload release and non-toxic self-destruction. Nevertheless, there remain many challenges to the wider practical application of such mechanisms in medical treatment, including cost, performance, controllability, and difficulties in the manufacturing of these smart particles and in scaling up production volumes.[51]

From an arms control perspective, it seems highly unlikely that such smart delivery systems would bring advantages over traditional chemical warfare agents that would justify the significant costs involved, and the need to carefully control the external stimuli to affect agent release as intended would be difficult if not impossible to achieve under field conditions. One should also take into account that under field conditions, agents disseminated over large areas inevitably lead to the loss of vast amounts of agents into the environment without ever reaching any of the intended targets. This favours cheap and easy-to-use types of agents that can be mass-produced and easily disseminated over large areas, over costly and complex agent delivery systems.

Despite these drawbacks, engineered nanoparticles may be of interest for the targeted delivery of chemical agents that act on the CNS. This type of delivery becomes particularly relevant when one considers the development of incapacitating chemical agents based on toxins or bioregulators which are otherwise difficult to deliver and are rapidly metabolised or eliminated. Such types of novel chemical agents and delivery systems would be rather expensive and technically challenging to use, but they might be of interest for special-operations types of scenarios including targetted assassinations.

An area of nanotechnology that may deserve monitoring in the future is the construction of nano-machines. The advances made in this direction have recently been recognised by the award of the Nobel Prize in Chemistry 2016 to Jean-Pierre Sauvage, Sir J. Fraser Stoddart and Bernard L. Feringa, for their design and production of molecular machines. A recent development that may simplify some of the approaches to creating nanoscale tools and eventually machines is DNA origami, which uses single-stranded DNA molecules, adding hundreds of DNA strands for folding, as a template for building nanostructures and patterns. To these templates, functionality can be added by DNA capture strands that enable specific interactions. In this way, researchers have created three-dimensional behaviour of nanoparticles under the influence of external factors such as light or pH changes, including controlled motion, three-dimensional reconfiguration, and distinct optical behaviour. Such systems may eventually find application as bio-sensors, as targeted drug delivery systems, or as nano-surgical tools. This type of work remains, for the time being, at the stage of fundamental research, but the example of CRISPR technology has shown the potential of serendipity. Nano-machines as weapons, however likely or unlikely they may be, would pose some difficult legal challenges to the definitions of the CWC – they would be chemical systems that interfere with life processes, but would their mechanisms qualify as chemical or mechanical actions? The issue may remain futuristic at this stage, but advances should certainly be carefully monitored.

7.3.4 Nanomaterials and Protection Against Chemical Weapons

Nanomaterials are being used to enhance protection against toxic chemicals. These materials can open up new approaches to decontamination, lead to more effective physical protection (new filter materials, clothing that repels or decontaminates toxic chemicals, and self-cleaning surface coatings),[52,53] they can be used as drug delivery systems for antidotes[54] and treatments for poisoning with chemical warfare agents, or they can lead to new types of environmental and biological sensors or detectors.[6,55–59]

Much of the discussion in Chapter 9 that addresses targeted drug delivery is equally relevant for the development of medical countermeasures. This is particularly true for treatments that need to be transported through biological barriers in order to be effective.

Nanobiology has been characterised as an enabler for the detection of disease causing and toxic agents. An example is a nanobiosensor for the detection of ricin that uses aptamer-modified graphene-based catalytical micromotors.[60] Another example is a DNA aptamer for the detection of saxitoxin. This is a surface plasmon resonance sensor that could be used in diagnostic assays.[61]

Nanotechnology also has led to the creation of new materials that reduce the physiological burden for the wearer whilst providing a barrier against CW agent exposure. Examples include the use of carbon layers or fibres integrated into textile materials, or self-decontaminating materials such as the integration of enzymes into clothing materials that can break down chemical warfare agents.

7.4 The Impact of Convergence and Nanotechnology on the Convention

Advances in the life sciences at the interface of chemistry and biology can have a multitude of societal impacts. Millet, in a recent briefing paper on the security impact of biotechnology, points to the benefits in such areas as public health, manufacturing, and food security. But he also identifies a range of potential adversary applications, ranging from bioweapons to alternative production methods for drugs such as heroin, LSD or cannabis, as risks with regard to resource security, energy security, data security and environmental security.[62]

These conclusions were drawn with the regime governing the prohibition of biological and toxin weapons in mind, but similar conclusions can be drawn with regard to the impact that convergence between chemistry and biology and advances in nanotechnology will have on the functioning of the CWC. The effects on the Convention's operations can be beneficial as well as posing new risks, and the nature of the challenges as well as the measures that would be needed to manage them will depend on the assumptions one makes about the nature of future chemical warfare threats and scenarios.

Before discussing these challenges to CW arms control, however, it is important to point out that advances in science and technology are widely expected to have profound beneficial effects on human society in general and also in the context of strengthening the technical basis that underpins chemical weapons arms control and disarmament.

7.4.1 The Benefits

The advances in science and technology, including those that result from convergence at the intersection of chemistry and biology, and progress in nanoscience and nanotechnology are expected to bring about a wide range of benefits.[34,63–66] The areas were these benefits may occur are diverse and relate to many of the essential needs of today's society: in the public, animal and plant health sectors there are hopes for better diagnostic tools, treatments and prophylaxes; there are hopes for an increasing ability to tailor drugs and treatment regimes to the needs of individual patients; with regard to the environment there are expectations for more efficient, environmentally friendly and sustainable means of goods manufacturing and of disposing of or recycling wastes; with regard to food production, more efficient means of farming with stronger resilience to losses by disease or environmental factors, better quality assurance as well as more efficient means of preventing the loss of agricultural produce after harvest are anticipated; with regard to consumer and industrial goods there is hope for a wider variety of materials with properties that are specifically designed to meet particular needs, and for production methods that are more economical than in the past, safer and environmentally sustainable. There are also expectations about how these advances may contribute to stronger resilience against environmental and man-made catastrophes.

These developments in science and technology will change the global environment within which the provisions of Article XI of the CWC can be implemented. The evolving environment of scientific, technological and industrial collaborations and trade is becoming more globally distributed and diverse. This trend is not new, and it is worth recalling what the US

National Academies' Committee on Trends in Science and Technology relevant to the Biological Weapons Convention had to say in 2011: "[T]here has been particularly rapid progress in the power of, and access to, enabling technologies, especially those depending on increased computing power."[67] The Committee then highlighted certain areas where this changing collaborative environment had become most tangible, which included:

- Collaborations between individual researchers and the emergence of global networks of researchers and of "virtual laboratories";
- Increased access globally to sophisticated reagents, commercial kits and services, placing hitherto advanced techniques within reach of less highly trained practitioners as well as expanding the global spread of life science research and its industrial applications;
- Although tacit knowledge remains a critical aspect in first class research, web-based technologies are facilitating the transfer of such knowledge and the formation of global learning communities and partnerships;
- As a consequence, barriers to the global spread and utilisation of science and technologies for legitimate use are being reduced, and conditions for international cooperation in the peaceful application of the life sciences are becoming more favourable;
- At the same time, the risk of abuse remains and mechanisms to militate against these risks of abuse are still needed.

It is this last issue, that the Committee itself observed was "an area that would benefit from more in-depth analysis to gain a more nuanced understanding of the developments and trends and their impact on the norm against biological weapons" – and the same would hold for the norm against chemical weapons.

This chapter focuses on the security and arms control implications of advances in science and technology. Such advances in science and technology and the promises they offer, of course, do not by themselves result in decisions by society about how these new capabilities may be exploited, nor do they alone determine which investments will be made into further advancements of science and technology. Choices in science and technology directions are influenced by funding decisions by governments as well as private industry; they are influenced by expectations of investment returns as well as perceptions of benefits and risks to people and the environment; such decisions also are influenced by the cultural, historical and political context.

It is not science and technology *per se* that an analysis of the risks of abuse must address. In fact, many of the drivers of scientific progress work towards the improvement of the human condition. There is a dark side, and it is important to think about the potential for misuse and the means of prevention and risk mitigation. But, it is also important to bear in mind that the direction science and technology take involves political and societal choices, and to analyse the impact of these advances in their context.

7.4.2 The Risk of the Emergence of Novel Concepts of Chemical Warfare

In the context of the Biological and Toxin Weapons Convention, a group of States from all continents identified two major threats that the BWC ought to address: the spread of biological technology in a way that could enable State and non-State actors to develop and produce a biological weapon, and the emergence of new infectious diseases that have the potential to spread around the world and cause significant damage to individuals and societies.[68] Can the same logic also be applied to the impact of scientific and technological advances at the intersection of chemistry and biology – but with regard to the international norm against chemical weapons?

First, a few words on the emergence of new threats from developments unrelated to weapons programmes. There are of course differences between chemical and biological agents in terms of the size and potential spread of incidents and the ability to control them, and also with regard to the nature of the threats (naturally emerging diseases *versus* manmade chemicals or materials extracted from nature), but nevertheless one can point to certain similarities between the emergence of new infectious diseases and the further diffusion of chemical materials and technology into human society.

Chemicals are an important aspect of daily life. They are ubiquitous and are used widely in such diverse applications as agriculture, pest control, food preservation, water purification, as cleaning products and solvents, as medicines and in many more areas. Some of the chemicals widely available in society are toxic, others can be used as precursors for the manufacture of hazardous chemicals. Access to chemicals can only be controlled to a certain degree, but States, as well as organisations, that use chemicals nevertheless need to take effective measures to limit access to them and prevent their misuse. This underlines the importance of States Parties fully implementing their obligations under Article VII of the CWC, by enacting and enforcing penalties and implementing legislation, as well as administrative measures they require to prohibit and prevent the misuse of toxic chemicals for hostile purposes within their jurisdiction. States Parties also should encourage and promote voluntary measures taken by industry or by other institutions that work with chemicals (for example, universities or research institutions). Such voluntary compliance measures strengthen the CWC norms and promote chemical safety and security. These regulatory and voluntary measures also address the requirement for States Parties, under Article VI of the CWC, to adopt the necessary measures to ensure that toxic chemicals are only used for purposes not prohibited, and that the OPCW's industry inspection and chemicals trade monitoring measures further strengthen the prevention of the misuse of toxic and precursor chemicals for hostile purposes.

Over time, the advances in science and technology may change the pattern of chemical materials present in daily life, which may in fact further reduce the risks involved. As a general trend, one can observe a shift from toxic, aggressive materials to the use of chemicals that are less hazardous to people and the environment, that are biodegradable, and that require smaller amounts to achieve a desired outcome. One could argue, then, that over time, advances in science and technology are likely to reduce the amounts of toxic or otherwise hazardous chemicals present in human society, thereby reducing access to these materials and increasing chemical safety as well as security.

Nevertheless, chemicals will remain part of our daily life and the recent uses in the Syrian conflict of improvised chemical weapons, such as barrel bombs disseminating chlorine gas, indicate that there are forces at work that may undermine the norm against poison gas. These breaches of the norm against chemical warfare, of course, have not been brought about by new science and technology – in fact, the chemical agents and the methods used for their dissemination in Syria and Iraq were both old and improvised. The use of chlorine gas in particular poses the question as to whether the presence of toxic chemicals in our society *by itself* has to be considered a risk factor in the context of armed conflict, and whether the psychological (in contrast to the legal) threshold for using toxic chemicals in armed conflict may have been lowered.

But, is there also a risk that new scientific discovery might lead to a reversal in States thinking about whether or not to acquire and use toxic chemicals as a means of war, or for other hostile purposes?

The broadly held assumption today is that traditional concepts of chemical warfare involving the massive battlefield use of highly toxic (lethal) agents are no longer relevant. This perception holds, despite the continuing use of toxic chemicals in the Middle East and the international condemnation of these recent chemical weapons uses as war crimes.

But the external security environment in which the CWC has to function has changed profoundly. In 2011, the Director-General's Advisory Panel on the future of the OPCW observed that whilst the number of inter-State conflicts had declined, the level of violence had not. It stated that:

> "The borderlines between war, civil war, large-scale violations of human rights, revolutions and uprisings, insurgencies and terrorism as well as organized crime are blurred. In addition to traditional military forces, more non-State actors have appeared on the battlefield, *i.e.* paramilitary groups, warlords and their militias and volunteers, mercenaries and private military companies, terrorists and criminal groups... [T]here are worries, in such types of conflict and with such actors, that the rules of international law applicable in armed conflict, and in particular the principles and rules of international humanitarian law, may be undermined... [D]istinctions between law enforcement, counter-terrorism, counter-insurgency and low-intensity warfare may get blurred, and certain types of chemical weapons such as incapacitants may appear to offer tactical solutions to operational scenarios where civilians and combatants cannot easily be separated or distinguished."[69]

Life science convergence and nano-technology may create expectations that chemically-based non-lethal weapons are feasible, which could lead to the acquisition by States Parties of such types of weapons for law enforcement purposes, including by military forces – along the same lines as riot control agents that have already been authorised by a number of States Parties for use by their armed forces when operating in peacekeeping roles abroad. The risks that emanate from these trends have been highlighted by a number of international meetings organised, respectively, by the ICRC and the government of Switzerland. At the end of 2015, a number of States Parties have warned against the risks associated with aerosolized chemical agents acting on the central nervous system, making an explicit link to the Convention's objective to prevent the re-emergence of chemical weapons.[70]

At the same time, caution should be used not to overstate the risk that new scientific discoveries might lead to new chemical weapons. As has been stated before, there is a long road from proof of concept to an effective weapons system. One issue to consider is tacit knowledge. It is perhaps a small but nevertheless significant hurdle for State programmes. The opportunities that these science and technology advances may create for sub-State actors, on the other hand, are often overstated. It remains unlikely that criminals or terrorists will resort to such high-tech weapons unless the chemical materials and delivery systems already exist (in weapons stockpiles of law enforcement structures, in certain research and development laboratories, or as dual use items that can be used to improvise delivery systems, such as certain agricultural equipment). Opportunistic acquisition and use of such materials and devices and improvisation remain of course possible and measures to prevent access by criminals or terrorists to them will be important.

7.4.3 Stronger Protections Against Toxic Chemicals

Convergence and nanotechnology are expected to contribute significantly to the strengthening of protections against the effects of toxic chemicals on humans as well as the environment. The history of chemical warfare has ample evidence to show that strong protection against chemical weapons has a deterrent effect on their acquisition and use.

It is less clear as to whether the advances in protective capacity and resilience against chemical weapons also inhibit activities in the area of weapons research and development, in particular in those grey areas where it would be difficult to distinguish between research and development aimed at strengthening defences and work that was part of an offensive programme.

Historically, changes in the doctrines and tactics of chemical warfare were often driven by the competition between offensive and defensive means. Advances on the offensive side of chemical warfare can lead to temporary advantages such as new weapons that can overcome existing defences (alarms, filters and other physical protection, means of decontamination, and medical countermeasures). But, such advances would soon be offset by the introduction of new or improved defensive means.

This cycle of course is broken in a context where work towards stronger protections is allowed to continue (as anticipated under Article X of the Convention) whilst offensive CW programmes are closed down (as required under Article I). To rely on this protection, States Parties need to have confidence in the action of all other States Parties, including in the area of protection against chemical weapons and with regard to how they make use of advances in science and technology.

This confidence essentially comes from two sources: transparency in the national steps taken by States Parties to implement the Convention (with regard, in particular, to national implementation measures taken in accordance with Article VII, the submission of complete declarations under Article VI, and the full and timely declaration of their protective programmes as required by Article X), and the effectiveness of the Convention's verification system to detect violations that may lead to the re-acquisition of chemical weapons in whichever form.

7.4.4 Challenges for National Implementation

Convergence is beginning to blur the lines that have hitherto separated the implementation environments of the BWC and the CWC.[3] Science convergence is of course not a new phenomenon as such and the overlap between chemistry and biology has long been recognized. The two Conventions bear testimony to this by inclusion of toxins in both treaties.

But convergence in the science base or overlap of one group of agents is one thing, convergence in the implementation environment (industry and trade) is quite another. This is where current convergence trends will matter most; convergence and nanotechnology are increasingly affecting industrial production of and trade in goods and services. As the industrial application of the life sciences grows and expands to a multitude of uses outside its traditional public health domain (energy, food, consumer products, climate change), the impact of advances in science and technology on CWC implementation are likely to become more profound.

Trends that have been identified in the literature include increased global diffusion of knowledge and technology, global interconnections between research, development and manufacturing organisations, an increase in actual and virtual transfers of goods, services and information, and the emergence of virtual research and development collaborations. The internet has become a platform for research collaborations, as well as a place for trade in chemicals, equipment and technological solutions. At the same time, custom manufacturing of chemical materials, outsourcing, and specialised supply services as well as modular approaches and automation are today characteristic features of the life science industry.

In short, the implementing environment of the CWC has become more fluid, globally interconnected, dependent on increased trade and information sharing, knowledge-based and demand driven. This challenges some of the assumptions that the Convention and national implementation systems have made in such areas as the definitions of facilities and activities subject to declaration and inspection, or the scope of facilities and activities that ought to be opened for international verification measures. National implementation measures and collaborations between States Parties on regulatory and enforcement measures need to become equally flexible and adaptable.

At the same time, there is significant time compression from scientific discovery to wide-spread and large-scale use in industry and society. In contrast, the time it takes States to legislate, regulate and enforce tends to be much longer. The experience of the OPCW has been that adapting guidelines for national implementation systems is difficult and time-consuming at the international level, but it is essential for providing guidance to States Parties on how the provisions of the Convention ought to be implemented and which good practices there are that States can apply.

Given the pace of scientific and technological advances, and increasingly also of their industrial application, one could ask whether regulation in the traditional sense is already coming too late. There have been calls for a broader governance approach to manage the emerging risks and to involve all relevant stakeholders in this process.

The nexus of terrorism, criminal actors and toxic chemicals poses yet another set of challenges to States Parties. The risk that non-State actors might use toxic chemicals to cause terror or coercion reinforces the need for States Parties to enact penal legislation including extending their jurisdiction extraterritorially to their nationals abroad that commit crimes involving chemical weapons, to foster legal cooperation with other States Parties, and to further develop, and where possible, harmonise, the tools they use to enforce prohibitions and punish individuals and organisations that commit crimes involving chemical (and biological) weapons. The recent Russian proposal for an international convention to counter chemical (and perhaps also biological) terrorism,[71] whilst falling short on certain legal requirements and creating potential conflicts with some of the basic norms of the CWC, does indeed address a number of real issues. It aims at increasing practical cooperation among States in preventing and prosecuting acts of chemical terrorism, for example by improving exchanges of information related to such acts and compelling States to apply the principle of "prosecute or extradite". As acknowledged by a critique of this proposal, "[S]tates need to strengthen their capacity to prevent, prohibit, detain, prosecute, and punish such acts, and closer international collaboration and coordination is essential."[72]

Yet another challenge for national implementation relates to what is often referred to as garage biology and amateur science. This is a field of activity that is difficult to regulate, and where regulations are difficult to enforce, given both the dual use nature of these activities and the context within which they are conducted. In 2013, it was estimated that more than 2000 enthusiasts across the world (mostly in the US and Europe), organised in 40 groups, were engaged in amateur science activities in the field of the life sciences.[73] A prominent example for such activities, involving community-based research facilities as well as national and international competitions, is synthetic biology. There have been some concerns about the potential that these activities might create for criminal or terrorist misuse, but there is a broadly held view today that such concerns tend to be over-stated and unjustifiably alarmist. There is clearly a need for adequate regulation and some steps have also been taken towards self-regulation (for example, the 2011 London Congress of the DIY community issued a draft Code of Ethics which includes a pledge that the work undertaken would only be used for peaceful purposes). Furthermore, law enforcement agencies have engaged with the amateur science community to provide advice, increase transparency and promote awareness in these communities about safety and security requirements. But, there remains a concern that governance issues are as yet not adequately addressed.[74]

McLeish and Trapp, during the run-up to the Seventh BWC Review Conference in 2011, expressed the need for a new governance approach with these words: "The life sciences are undergoing a paradigm shift ... From an arms control perspective, these factors translate more and more into a post-proliferation world. In such a world, traditional models of proliferation control are certain to fail, and the traditional top-down government approaches no longer seem appropriate. From a broader regulatory perspective, the role of governments is changing. The state alone is no longer able to control the way that life sciences discoveries are used. The circumstances beg instead for a governance system that brings together all stakeholders—science, industry, government, and the public—and broadens as well as deepens the basis for compliance with the safe and responsible conduct and utilization of science, thus supporting the norm against biological weapons."[75] This conclusion stands even today, and it also applies to chemical weapons disarmament and the broader context of chemical safety and security.

7.4.5 Impact on the CWC Verification System

The impact of advances in science and technology on CWC verification depends, first and foremost, on how broadly and quickly they manifest themselves in the practices of the chemical (biochemical, biotechnological, pharmaceutical, *etc.*) industries. No matter how exciting advances at the laboratory bench are, their practical application at industrial scale will depend not merely on scientific merit, but also on how they can compete with mature technologies in terms of effectiveness, pricing of raw materials and products, investment costs and other factors.

Once a new technology is more broadly used in industry, its primary impact with regard to the CWC will be on industry verification. The CWC industry verification system is designed around the concept of "risk to the object and purpose" of the Convention, taking account of a variety of factors such as, amongst others:

- The properties and amounts of the chemical(s) produced or stored (for example: are they scheduled chemicals or can they easily be converted into scheduled chemicals; what is their toxicity (or in the case of precursor chemicals or feedstock materials which toxic chemicals can they be converted into), do they otherwise have properties that might make them suitable for CW purposes; what are the amounts of these chemical(s) produced, stored, transferred or used);

- The processes, equipment and technologies used (for example: what is the available capacity of the production and processing units; how easy would it be to convert a facility for the purpose of producing scheduled chemicals or other toxic chemicals of relevance to the Convention);
- The nature of the activities carried out at a given industrial site (for example: are there a variety of materials and production or processing units available that in combination could create a capacity to manufacture CW agents; is the facility primarily producing chemicals, or is it largely processing or formulating them, or is it undertaking other activities with these chemicals; are there filling units/lines in place that could be used for filling weapons with toxic chemicals).

As new technologies move into industrial use, key questions that States Parties should ask include whether these new technologies are covered by the existing declaration and verification rules of the CWC (and whether they should be covered if they fall outside the current system); how—in terms of perceived risks to the object and purpose of the CWC—they compare to traditional chemical technologies that are already subject to declaration and verification; whether these new technologies change significantly the pattern of chemicals manufacturing, trade and use and how that might affect the conduct of verification activities; whether new technologies significantly alter characteristic features in chemical production plants that have hitherto been indicators of the risk that a particular production facility poses to the CWC; or whether inspectors need additional expertise and skills to recognise and correctly assess production plants using these new technologies.

Several examples of emerging technologies have been highlighted in the literature that States Parties should consider when reviewing the functioning of the industry verification system. Without attempting to provide a complete list, here are some examples: the coverage of biologically medicated processes under the concept of "production by synthesis" of Part IX of the Verification Annex; how to reflect the capabilities of multipurpose plants in the risk assessment methodology and thus inspection intensity; whether peptides and other bioregulators ought to be given a separate verification subregime given the advances in the manufacturing of such biomolecules and in drug delivery.

The trends towards more extensive production and use of highly active pharmaceutical ingredients (HAPI) challenges some of the assumptions underlying the industry verification system with regard to the risks associated with certain chemical production facilities, or with regard to thresholds for their declaration and inspection. Some HAPI facilities show safety and security features as well as production capabilities that place them squarely into a risk category similar to that of larger Schedule 1 facilities, but more often than not they remain undeclared as they stay under the thresholds of Part IX of the Verification Annex.

These trends are also relevant in light of the discussions about whether or not incapacitating chemical agents are legitimate for use in law enforcement. The issue has been discussed for many years, initially outside the OPCW, but after several years of hesitation, now also among States Parties. In 2015, Australia and other States Parties published a joint national paper in which they expressed serious concerns about the impact on the Convention that the acquisition and use of toxic chemicals would have which target the central nervous system (CNS) for law enforcement purposes. They recommended that States Parties make their positions on the matter known, that the issue be kept under review by the Executive Council, and they expressed their hope for further discussions of this matter among States Parties and with the Technical Secretariat so as to develop "concrete recommendations for how to address CNS acting chemicals in a way that would significantly advance one of the OPCW's priorities—preventing the re-emergence of chemical weapons."[76]

These are examples for new technologies and solutions in the chemical and pharmaceutical industries that could relatively easily be repurposed for CW production purposes, thus "blending" a kind of standby CW production capability into legitimate industrial facilities. This is not a new issue – in fact the question of what to do about "CW capable facilities" had come up at the end of the negotiations and was only temporarily resolved by the inclusion of Part IX of the Verification Annex, on the express understanding that the entire industry verification regime would be reviewed several years after the entry into force of the Convention, based on practical experience.[77] But, whilst certain details pertaining to industry verification have indeed been reviewed and some minor adaptations of implementation practice agreed over the years, the States Parties have failed to do what the Convention asked them to do: to *re-examine* the Conventions provisions dealing with discrete organic chemicals and to undertake an *overall review* of the verification regime for the chemical industry.

States Parties also have shown little inclination to adapt the Schedules of Chemicals to new discoveries. The Convention of course has an in-built assumption that the Schedules are dynamic elements of the verification system, which is why they are specifically subject to the "change" procedure under Article XV of the CWC rather than its normal amendment procedure. This, in the words of the Convention, was done "[I]n order to ensure the viability and the effectiveness of this Convention".[78] However, whilst States Parties have used this change mechanism on several occasions to adapt procedural provisions in the Verification Annex, the Schedules of Chemicals have remained unchanged.

This may not have affected the dependability of the industry verification regime as yet, but as more highly active biomolecules are being produced in larger amounts in industry, and with the distance in time to previous State CW programmes increasing, there will be legitimate questions about whether industry verification is adequately targeting all facilities and chemicals that may pose a risk to the object and purpose of the Convention.

But, the impact on the verification system of advances in science, technology and industry goes beyond routine verification. It also poses challenges to the concepts of challenge inspection and investigations of alleged CW use. These special types of inspections could rely, so far, on what are essentially binary verification questions: the detection of a certain chemical in a certain context was almost certain proof that a chemical weapons related incident had occurred. As more biomolecules are being identified, characterised and manufactured for use in such areas as medicine, agriculture, or in other industries, the CWC may find itself increasingly operating in an environment not dissimilar to that of the Biological Weapons Convention, where the natural background of biological agents complicates the interpretation of investigation findings. The detection of intact sarin or a characteristic degradation product that has no natural background in an environmental or biomedical sample is a strong signal and is almost by itself conclusive evidence of a chemical weapons use. The detection of a neuropeptide or a naturally occurring toxin in the same sample would be an important piece of evidence but by itself might not carry the same weight for concluding whether a chemical weapon had been used, and additional data and contextual analysis might be necessary to rule out natural causes for their presence.

7.5 Conclusions

Advances in science and technology affect the norm enshrined in the Convention as well as its implementation at both the national and international levels. As these advances begin to manifest themselves in industrial production, and in the changing uses of biologically active chemical compounds in different parts of society, it is important that States Parties of the CWC review the impact that these developments may have on the implementation of the treaty, and find ways of adapting treaty implementation to minimise any risks posed by them to the object and purpose of the Convention whilst maximising their societal benefits.

Scientific advances can result in the discovery and acquisition of new chemical agents, or the discovery of new methods of their production or dissemination. However, they can also be utilised to strengthen defences against chemical weapons, for example in the form of more effective, cheaper and more robust methods for detection, protection, medical prophylaxis and treatment, decontamination and recovery.

The most important concept built into the Convention to protect against the detrimental effects of scientific discovery and technological progress is the General Purpose Criterion. The Convention's prohibition covers *all* toxic chemicals and their precursors, and only when toxic or precursor chemicals are intended for peaceful purposes, and only as long as their types and quantities are consistent with such purposes, are they not considered a chemical weapon. In a legal sense, this is a very robust concept indeed. But, applying it in the real world is not always easy.

Identifying the potential risks and benefits of new scientific advances can be difficult when the technology in question is still evolving, and when its potential as well as limitations are not as yet fully understood. Science impact assessment is a complicated process that needs to take into account a range of factors, reaching beyond trying to understand the potential inherent in scientific progress and also assessing the impact of drivers in industry, trade, and elsewhere in society. Many of these issues are multidisciplinary and require a multi-stakeholder approach in order to ensure that the measures that States Parties agree to take are effective as well as accepted by society, with respect to managing the risks, maximising benefits, strengthening the norm of the Convention and ensuring compliance.

The assessment of the impact of advances in science and technology and the development of appropriate responses to them requires effective communications between all stakeholders concerned (governments, industry, the science and technology communities, and civil society), and a governance approach that embeds safety, security, compliance and ethical principles into the culture of the professionals working with chemicals. Only in this way can a construct as complicated as the General Purpose Criterion be brought to life.[4]

References

1. *Organisation for the Prohibition of Chemical Weapons – United Nations Joint Investigative Mechanism, 3rd Report, Document S/2016/738*, 24 August 2016.
2. *Switzerland "Convergence between Biology and Chemistry: Latest Findings of Relevance to the Biological and Toxin Weapons Convention", Working Paper BWC/MSP/2015/MX/WP.8*, 30 July 2015.
3. Spiez Laboratory and CSS ETH Zürich, *Spiez Convergence – Report on the First Workshop 6–9 October 2014*, November 2014, http://www.labor-spiez.ch/en/akt/pdf/Spiez_ Convergence_2014_web.pdf, last accessed September 2016.
4. Spiez Laboratory, *Spiez Convergence – Report on the Second Workshop 5–8 September 2016*, October 2016, https://www.labor-spiez.ch/pdf/en/rue/LaborSpiezConvergence2016_02_FINAL.pdf, last accessed 24 March 2018.
5. Organisation for the Prohibition of Chemical Weapons, *Convergence of Chemistry and Biology Report of the Scientific Advisory Board's Temporary Working Group*, 2014, p. 3.
6. A. Scott, *Chem. Eng. News*, 2016, **94**(19), 25.
7. M. M. Bomgardner, *Chem. Eng. News*, 2016, **94**(18), 26.
8. OPCW, *SAB Presentation to the Meeting of Experts of the BWC*, 2014, http://www.unog.ch/80256EDD006B8954/(httpAssets)/4902A1294789C50DC-1257D2C0055BB22/$file/BWC+MX+2014+-+Presentation+-+OPCW.pdf, last accessed October 2016.

9. M. E. Mangan, Production of medicines from engineered proteins in plants: proteins for a new century, in *Plant Biotechnology*, ed. A. Ricroch, S. Chopra and S. J. Fleischer, Springer, 2014, ch. 20, p. 263.
10. R. Trapp, Synthesis of peptide bioregulators, in *Innovation, Dual Use and Security – Managing the Risks of Emerging Biological and Chemical Technologies*, ed. J. B. Tucker, The MIT Press Cambridge, (MA) and London (England), 2012, ch. 11, pp. 173–185.
11. OPCW Decision, *Meaning of 'production' as Defined in Article II, Subparagraph 12(a), Document C-II/Dec.6*, 5 December 1997.
12. J. R. Riches, R. W. Read, R. M. Black, N. J. Cooper and C. M. Timperley, *J. Anal. Toxicol.*, 2012, **36**, 647.
13. A. E. Smithson, *Bull. At. Sci.*, 2018, https://thebulletin.org/london-attack-saddle-moscow-chemical-weapons-inspections11632.
14. R. Trapp, The role of industry in promoting biosecurity: a case study of the convergence of chemistry and biology, in *Preventing Biological Threats – what You Can Do*, ed. S. Whitby, T. Novossiolova, G. Walther and M. Dando, Bradford Disarmament Reseach Centre, 2015, ch. 9, http://www.brad.ac.uk/social-sciences/peace-studies/research/publications-and-projects/guide-to-biological-security-issues/Guide_final.pdf, last accessed October 2016.
15. OPCW Technical Secretariat, *Report of the Advisory Panel on Future Priorities of the Organisation for the Prohibition of Chemical Weapons, Document S/951/1011*, 2011.
16. W. Krutzsch, E. Myjer and R. Trapp, *CWC a Commentary, Part III, Chapter Article II*, Oxford University Press, 2014.
17. OPCW, *Report of the Scientific Advisory Board on Developments in Science and Technology for the Third Special Session of the Conference of the States Parties to Review the Operation of the Chemical Weapons Convention, Document RC-3/DG.1*, 2012.
18. N. P. Tatonetti, P. P. Ye, R. Daneshjou and R. B. Altman, *Sci. Transl. Med.*, 2012, **4**(125), 125.
19. J. Huang, C. Niu, C. D. Green, L. Yang, H. Mei and J.-D. J. Han, *PLoS Comput. Biol.*, 2013, **9**(3), http://journals.plos.org/ploscompbiol/article?id=10.1371/journal.pcbi.1002998, accessed October 2016.
20. A. Ma'ayan, A. Dd Rouillard, N. R. Clark, Z. Wang, Q. Duan and Y. Kou, *Trends Pharmacol. Sci.*, 2014, **35**(9), 450.
21. A. Verdani, M. Dobler, Z. Hu and M. Smiesko, *Toxicol. Lett.*, 2015, **232**(2), 519.
22. IAP/The global network of science academies, *The Biological and Toxin Weapons Convention – Implications of Advances in Science and Technology, International Workshop on Advances in Science and Technology for the BWC, Conference Report*, 2016.
23. A. E. Rabideau and B. L. Pentelute, *ACS Chem. Biol.*, 2016, **11**(6), 1490.
24. D. Q. Y. Quek, L. N. Nguyen, H. Fan and D. L. Silver, *J. Biol. Chem.*, 2016, **291**, 9383.

25. J. R. Clapper, *Statement for the Record – Worldwide Threat Assessment of the US Intelligence Community, Senate Arms Services Committee*, 9 February 2016, p. 9, https://www.dni.gov/index.php/newsroom/testimonies/217-congressional-testimonies-2016/1313-statement-for-the-record-worldwide-threat-assessment-of-the-u-s-ic-before-the-senate-armed-services-committee-2016, accessed on October 2016.
26. A. Adibekian and P. Stalforth, *ACS Chem. Biol.*, 2016, **11**, 816.
27. IAP/The global network of science academies, *The Biological and Toxin Weapons Convention – Implications of Advances in Science and Technology, Technical Report*, 2016.
28. V. Pitschmann and Z. Hon, *Molecules*, 2016, **21**, 556.
29. G. Chen, I. Roy, C. Yang and P. N. Prasad, *Chem. Rev.*, 2016, **116**, 2826.
30. A. Asbach, *et al.*, *Assessment of Personal Exposure to Airborne Nanoparticles – a Guidance Document NanoIndEx Final Project Report*, 2016, http://www.nanoindex.eu/?page_id=323, last accessed October 2016.
31. V. Raman, S. Suresh, P. A. Savarimuthu, T. Raman, A. M. Tsatsakis, K. S. Goolokhvast and V. K. Vadivel, *Exp. Ther. Med.*, 2016, **11**, 553.
32. H. Bahadar, F. Maqbool, K. Niaz and M. Abdollahi, *Iran. Biomed. J.*, 2016, **20**(1), 1.
33. P. Khanna, C. Ong, B. H. Bay and G. H. Baeg, *Nanomaterials*, 2015, **5**, 1163.
34. H. Bahadar, F. Maqbool, K. Niaz and M. Abdollahi, *Iran. Biomed. J.*, 2016, **20**(1), 1.
35. G. Chen, I. Roy, C. Yang and P. N. Prasad, *Chem. Rev.*, 2016, **116**, 2826.
36. T. Wu, T. Zhang, Y. Chen and M. Tang, *J. Appl. Toxicol.*, 2016, **36**, 345.
37. J. D. Strickland, W. R. Lefew, J. Crooks, D. Hall, J. N. R. Ortenzio, K. Dreher and T. J. Shafer, *Nanotoxicology*, 2016, **10**(5), 619.
38. W. K. Boyes, R. Chen, C. Chen and R. A. Yokel, *Neurotoxicology*, 2012, **33**, 902.
39. P. Khanna, C. Ong, B. H. Bay and G. H. Baeg, *Nanomaterials*, 2015, **5**, 1163.
40. D. C. Shin, J. Y. Yang, J. Y. Yang, S. H. Kim, I. H. Choi, Y. W. Lim, C. S. Kim and G. T. Lee, *NSTI Nanotech.*, 2008, **2**, 142.
41. K. Y. Wi and S. S. Feng, *Biomaterials*, 2005, **26**, 2713.
42. H. F. Krug, *Angew. Chem.*, 2014, **53**(46), 12304.
43. S. M. A. Sadat, S. T. Jahan and A. Haddadi, *J. Biomater. Nanobiotechnol.*, 2016, 7, 91.
44. S. Bamrungsap, Z. Zhao, T. Chen, L. Wang, C. Li, T. Fu and W. Tan, *Nanomedicine*, 2012, 7(8), 1253.
45. N. Yadav, S. Khatak and U. V. S. Sara, *Int. J. Appl. Pharm.*, 2013, **5**(2), 8.
46. M. S. Kalaiarasi, P. Arjun, S. Nandhagopal, J. Brijitta, A. M. Iniyan, S. G. P. Vincdent and R. R. Kanan, *J. Appl. Biomed.*, 2016, **14**(2), 157.
47. Y. Liu, S. An, J. Li, Y. Kuang, X. He, Y. Guo, H. Ma, Y. Zhang, B. Ji and C. Jiang, *Biomaterials*, 2016, **80**, 33.
48. T.-T. Zhang, W. Li, G. Meng, P. Wang and W. Liao, *Biomater. Sci.*, 2016, **4**, 219.

49. A. Chiappini, E. de Rosa, J. O. Martinez, X. Liu, J. Steele, M. M. Stevens and E. Tasciotti, *Nat. Mater.*, 2015, **14**, 532.
50. W. Gao, R. Dong, S. Thamphiwatana, J. Li, W. Gao, L. Zhang and J. Wang, *ACS Nano*, 2015, **9**(1), 117.
51. A. Srivastava, T. Yadav, S. Sharma, N. Nayak, A. Kumari and N. Mishra, *J. Biosci. Med.*, 2016, **4**, 69.
52. G. Metreveli, L. Wågberg, E. Emmoth, S. Belák, M. Strømme and A. Mihranyan, *Adv. Healthcare Mater.*, 2014, **3**(10), 1546.
53. Lawrence Livermore Laboratory, *Nanoworks News*, 2012, http://www.nanowerk.com/news2/newsid=27023.php, last accessed October 2016.
54. J. Yang, L. Fan, F. Wang, Y. Luo, X. Sui, W. Li, X. Zhang and Y. Wang, *Nanoscale*, 2016, **8**, 9537.
55. S. H. Cho, S. S. Kwon, J. Yi and W. Il Park, *Nano Convergence*, 2016, **3**, 14.
56. M. A. Kafi, H.-Y. Cho and J.-W. Choi, *Nano Convergence*, 2016, **3**, 17.
57. T. Y. Jeon, D. J. Kim, S.-G. Park, S.-H. Kim and D.-H. Kim, *Nano Convergence*, 2016, **3**, 18.
58. P. C. Ray, G. K. Darbha and A. Ray, *NSTI Nanotech.*, 2008, **2**, 81.
59. T. C. Dao, T. Q. N. Luong, T. A. Cao, N. M. Kieu and V. V. Le, *Adv. Nat. Sci.: Nanosci. Nanotechnol.*, 2016, **7**, 015007.
60. B. Estaban-Fernández de Ávila, M. A. Lopez-Ramirrez, D. F. Báez, A. Jodra, V. V. Singh, K. Kaufmann and J. Wang, *ACS Sens.*, 2016, **1**(3), 217.
61. S. M. Handy, B. J. Yakes, J. A. DeGrasse, K. Campbeli, C. Telliot, K. M. Kanyuck and S. L. DeGrasse, *Toxicon*, 2013, **61**(99), 30.
62. P. Millet, *The Cipher Brief*, 2016, https://www.thecipherbrief.com/article/security-impact-biotech, last accesses 24 October 2016.
63. European Commission, *Nanotechnologies – Principles, Applications, Implications and Hand-on Activities. A Compendium for Educators, Directorate-general for Research and Innovation, Industrial Technologies (NMP) Unit G.4—Nano Sciences and Nano Technologies, EUR 24957*, Brussels, 2012.
64. National Academies, A Research Strategy for Environmental, Health, and Safety Aspects of Engineered Nanomaterials, in *Committee to Develop a Research Strategy for Environmental, Health, and Safety Aspects of Engineered Nanomaterials*, ed. J. M. Samet, The National Academies Press, Washington DC, 2012.
65. The Royal Society of Chemistry, *Royal Society of Chemistry View on Nanoscience and Nanotechnology*, 2012, http://www.rsc.org/images/Nanotechnology-Position-Consultation_tcm18-223077.pdf, last accessed October 2016.
66. J. F. Sargent Jr, *Nanotechnology: A Policy Primer, CRS Report to Congress 7-5700*, 2012.
67. National Academies, *Life Sciences and Related Fields – Trends Relevant to the Biological Weapons Convention*, The National Academies Press, Washington DC, 2011.
68. *Australia, Canada, Chile, Colombia, Czech Republic, Finland, Germany, Hungary, Japan, Lithuania, Netherlands, Nigeria, Norway, Philippines, Republic of Korea, Slovakia, Sweden, Switzerland "Addressing Modern Threats in the*

Biological Weapons Convention: Follow-up and Recommendations" Paper Presented to the 2015 Meeting of the States Parties to the Convention on the Prohibition of the Development, Production and Stockpiling of Bacteriological (Biological) and Toxin Weapons and on Their Destruction, Document BWC/MSP/2015/WP.6/TRev.1, 14 December 2015.

69. *Note by the Director-General "Report of the Advisory Panel on Future Priorities of the Organisation for the Prohibition of Chemical Weapons", OPCW Technical Secretariat, Document S/951/2011*, 2011.
70. *Australia, Austria, Belgium, Canada, Chile, Cyprus, Germany, Finland, Ireland, Japan, New Zealand, Norway, Poland, Portugal, Republic of Korea, Spain, Switzerland, Turkey United States of America, United Kingdom f Great Britain and Northern Ireland "Aerosolisation of Central Nervous System-acting Chemicals for Law Enforcement Purposes" OPCW Conference of the States Parties Document C-20/Nat/2*, 2015.
71. *Ministry of Foreign Affairs of the Russian Federation "Elements of the Draft International Convention for the Suppression of Acts of Chemical Terrorism"*, 6 April 2016, http://www.mid.ru/en/foreign_policy/international_safety/regprla/-/asset_publisher/YCxLFJnKuD1W/content/id/2211492/pop_up?_101_INSTANCE_YCxLFJnKuD1W_viewMode=tv&_101_INSTANCE_YCxLFJnKuD1W_qrIndex=0, last accessed October 2016.
72. O. Meier and R. Trapp, Russia's chemical terrorism proposal: Red herring or useful tool?, *Bull. At. Sci.*, 2016, http://thebulletin.org/russia's-chemical-terrorism-proposal-red-herring-or-useful-tool9531, last accessed October 2016.
73. A. Jefferson, *The Growth of Amateur Biology: A Dual-use Governance Challenge? Policy Paper 3, Biochemical Security 2030 Project*, 2013, https://biochemsec2030dotorg.files.wordpress.com/2013/08/jefferson-policy-paper-3-for-print.pdf, last accessed October 2016.
74. I. Puscas, *Synthetic Biology and 'Amateur Science': Dual-use and Challenges of Regulation Strategic Security Analysis*, February 2016, p. 2, http://www.gcsp.ch/News-Knowledge/Publications/Synthetic-Biology-and-Amateur-Science-Dual-use-and-Challenges-of-Regulation, last accessed October 2016.
75. C. McLeish and R. Trapp, *Nonproliferation Rev.*, 2011, **18**(3), 527.
76. *Australia, Austria, Belgium, Canada, Chile, Cyprus, Germany, Finland, Ireland, Japan, New Zealand, Norway, Poland, Portugal, Republic of Korea, Spain, Switzerland, Turkey, United States of America, United Kingdom of Great Britain and Northern Ireland "Aerosolisation of Central Nervous System-acting Chemicals for Law Enforcement Purposes" OPCW Document C-20/Nat.12*, 24 November 2015.
77. *Chemical Weapons Convention*, Verification Annex Part IX, paragraph 26.
78. *Chemical Weapons Convention*, Article XV, paragraph 4.

CHAPTER 8

Advances in Understanding Targets in the Central Nervous System (CNS)

MALCOLM R. DANDO

Bradford University, Department of Peace Studies, Richmond Road, Bradford, BD7 1DP, UK
*E-mail: mrdando@bradford.ac.uk

8.1 Introduction: Implications of a Mechanistic Neuroscience

Modern work on lethal chemical agents, particularly on treatments, is covered in other chapters. The focus here is therefore on incapacitants. The classic 1970s Stockholm International Peace Research Institute's (SIPRI) study, *The Problem of Chemical and Biological Warfare*, noted[1] that at the start of the United States' incapacitating agent research programme in the early days of the Cold War, there seemed to be many possible ways in which chemical agents could be used to achieve as the title of a paper[2] by one participant put it "[S]elective malfunctioning of the human machine: New horizons in chemical warfare."

The SIPRI study stated that the US Army Chemical Corps pointed out at least twelve of these possibilities and gave details of some such malfunctions and the chemical agents that might be used. A recent account of the programme quoted one scientific director[3] as stating that the

Preventing Chemical Weapons: Arms Control and Disarmament as the Sciences Converge
Edited by Michael Crowley, Malcolm Dando and Lijun Shang

Published by the Royal Society of Chemistry, www.rsc.org

symptoms of value to the programme included "fits or seizures, dizziness, fear, panic, hysteria, hallucinations, migraines, delirium, extreme depression, notions of hopelessness, lack of initiative...even suicidal mania." A detailed account of the programme by one participant[4] also suggests that these objectives were pursued with enthusiasm as in other countries, but not by current scientific and ethical standards.[5] The SIPRI study certainly demonstrated that the US was not the only major state seeking such objectives at that time.

The strenuous and comprehensive nature of these investigations was made clear in the 1997 version of the US *Textbook of Military Medicine: Medical Aspects of Chemical and Biological Warfare*. The chapter on incapacitating agents stated bluntly that:[6] "[V]irtually every imaginable chemical technique for producing military incapacitation has been tried at some time." The text continued, specifically in regard to the US programme:

> "Between 1953 and 1973, at the predecessor laboratories to what is now the US Army Medical Research Institute of Chemical Defense, many of these were discussed and, when deemed feasible, systematically tested. Chemicals whose predominant effects were in the central nervous system were of primary interest and received the most intensive study."

Yet, as the authors of the chapter concluded "[F]rom this large number of possibilities, chemical compounds in a single group—the 'anticholinergics' —are regarded as most likely to be used as military incapacitating agents." The group, of course, included BZ (3-quinuclidinyl benzilate), which was weaponised (although it was of limited value given its long onset time and unpredictable effects) and never used by the US.

Indeed, the Scientific Advisory Board of the Organisation for the Prevention of Chemical Weapons (OPCW) summarised the outcome of all such investigations, as far as is known from open sources, in its report to the Third Review Conference as follows:[7]

> "The types of chemicals and pharmaceuticals, known to have been considered as incapacitants from open literature sources, were discussed. Most are centrally acting compounds that target specific neuronal pathways in the brain. All of them emerged from drug programmes undertaken from the 1960s to the 1980s, as far as can be judged by the research that has been published."

The difficulty of using chemical incapacitants without killing some of the victims is well known. As the United Kingdom (UK) Royal Society pointed out in addition to side effects of the agents involved:[8]

> "when considered as a complete weapon system in an operational context, uncontrollable variables such as the size, health, age *etc.*, of the target population, secondary injury (*e.g.* airway obstruction), and requirement for medical aftercare introduces further challenges to the development of a safe incapacitating chemical agent."

However, it was clear to scientists at the time of the Cold War in the last century that there was a much more fundamental problem caused by the lack of understanding of the interaction of chemical agents or drugs with the neuronal receptors. As noted in the 1978 second edition of *Principles of Psychopharmacology*:[9]

> "The complexity of the CNS, coupled with inadequate tools for observation, are formidable barriers in gaining information about molecular interactions between CNS drugs and receptor sites."

Yet, as Matthew Meselson[10] argued at the turn of the century, we will be increasingly able to manipulate living systems in the century ahead and some such capabilities could be subject to dual-use. This view is also current amongst some scientists who specialize in the study of chemical and biological weapons.[11] Therefore, the question to be investigated in this chapter is whether the previous situation, in which we could only gain a mechanistic understanding of the neuronal mechanisms underlying behavior in simple or very specialized systems[12] has changed since the late 20th Century, and, in particular, within the advances made is it possible that new and seemingly viable mechanisms of chemical incapacitation could be discovered?

There is certainly a widely held view that major advances can be made in our understanding of the neuronal mechanisms underlying our behavior and that the new technologies will allow further advances.[13] Furthermore, there are large scale national funding initiatives to underpin brain research, for example in the US,[14] Europe,[15] and Japan.[16] This optimism about our increasing ability to investigate the interacting neuronal circuits producing human behavior is clear from the report *BRAIN 2025: A Scientific Vision* for the US National Institutes of Health.[17]

> "In considering these goals and the current state of neuroscience, the working group identified the analysis of circuits of interacting neurons as being particularly rich in opportunity, with potential for revolutionary advances."

The report went on directly to explain:

> "Truly understanding a circuit requires identifying and characterizing the component cells, defining their specific connections with one another, observing their dynamic patterns of activity as the circuit functions *in vivo* during behavior, and perturbing these patterns to test their significance. It also requires an understanding of the algorithms that govern information processing within a circuit and between interacting circuits in the brain as a whole."

Of course, given the distress caused by mental illnesses around the world[18] we would all like to see this kind of work producing useable results.

Yet, the dangers that could arise from a successful search for novel incapacitants through the dual-use of such knowledge has been recognized by some States, and attempts to forestall such malign developments have been led by Switzerland. As the US noted in its generally positive review of the CWC Third Review Conference:[19]

"On the less positive side, we too are disappointed that time ran out before final agreement could be reached on language relating to substances termed 'incapacitating chemical agents'. The United States believes that agreement on language is within reach. We will work closely and intensively with the Swiss and other delegations so that this important discussion can continue."

The US went on to reconfirm that "the United States is not developing, producing, stockpiling, or using incapacitating chemical agents". A number of other States Parties also confirmed that they were not interested in such agents. Germany[20] for example, stated that "Germany, in her implementing legislation has deliberately opted for a narrow interpretation of the Convention." Thus, only standard riot control agents were allowed for law enforcement purposes by military or police forces.

Australia brought such concerns to a head in a 2014 paper which stated that:[21]

"The weaponisation of CNS acting chemicals for law enforcement purposes is of concern to Australia due to the health and safety risks and the possibility of their deliberate misuse, both of which have the potential to undermine the global norm against the use of toxic chemicals for purposes prohibited by the Convention."

This was followed up in 2015 by a paper[22] by Australia and 22 other States Parties titled *Aerosolisation of Central Nervous System-Acting Chemicals for Law Enforcement Purposes* which noted that there have been many calls for the OPCW to increase its focus on "preventing the re-emergence of chemical weapons, including new types of potential chemical agents." It then went on to state that the paper identified this as a particular area of concern and made recommendations for dealing with the problem. The paper was issued again in December 2016 with 36 co-sponsors.[124]

Of course, as the issue of the convergence of aspects of chemistry and biology has become of more concern, the need to carefully analyse the implications of the potential development of novel incapacitants for the Biological and Toxin Weapons Convention (BTWC), as well as for the CWC, has become clear. As the report of the first Spiez Convergence Workshop in 2014 noted:[23]

"In today's and tomorrow's security environment, it is perhaps less the number of possible casualties that matters, more the potential of these weapons to subjugate and coerce." The report went on to explain why this change of focus was needed now: "[N]ew types of war, attempts by sub-state actors to acquire and use chemical or biological weapons, the use of chemical or biological weapons in covert operations, their small-scale use to terrorise and subjugate – all these scenarios define the boundaries for the assessment of how advances in science and technology may affect CBW arms control."

It is, therefore not surprising that the implications of advances in neuroscience has been discussed in meetings of States Parties to the BTWC.

The UK, which has had considerable experience in research on potential incapacitants during the Cold War period,[24,125] suggested the need for more intensive discussion of this issue in a paper titled *The convergence of chemistry and biology: implications of developments in neuroscience* in 2012,[25] and devoted a section of its contribution to the background paper on developments in science and technology for the 2011 Seventh Review Conference of the BTWC to neuroscience.[26] The UK noted that:[27]

> "In this field, technologies for the discovery and development of compounds that act on the central nervous system (CNS) are of particular relevance to the BTWC. These include advances in molecular and genetic neuroscience, drug discovery technology, bioregulators and drug delivery to the brain."

The paper then proceeded to discuss such developments in a series of following paragraphs. Some of this data is illustrated in Table 8.1. Yet, discovery of medicinal drugs or chemical agents of utility for good or ill depends first on knowing what particular synaptic receptor system must be affected in what way in order to disrupt a circuit producing the behavior of interest.

Since the middle of the 20th Century when the study of animal behavior —Ethology —began to become central to modern biology, four types of question have seemed to require answers if a specific behavior is to be properly understood. These questions ask, how did this particular behavior arise in evolution and what function does the behavior have in the animal's life - what are termed 'ultimate' mechanisms? Then, there are the more 'proximate' questions of how does the behavior arise in the individual's development, and finally how is the particular behavior caused when it occurs? It is obviously this last question of the causal mechanism that is of most concern to neuroscientists and it is startling to realize just how recently scientists have begun to understand such causal mechanisms.[28]

Yet, the reason that major States are willing to invest large funds in neuroscience today is because of a clear acceleration in our capabilities to investigate the brain because of advances in technology. The authors of the US report *BRAIN 2025: A Scientific Vision* for the US National Institutes of Health put the situation very clearly when in a section titled "Why Now?" they stated:[29]

> "The reason lies in the technological and conceptual revolution that is underway in modern neuroscience."

Their illustrative examples are summarized in Table 8.2, and in the text they are followed by what seems to be a summary example:

> "Only a short time ago, we were restricted to studying the brain's electrical activity one nerve cell at a time, now we can record from hundreds of nerve cells simultaneously, allowing us to analyse the cooperative activity of nerve cells as they operate in intact circuits: we look towards a future in which we can measure even richer patterns of brain activity, involving millions of nerve cells at any instant."

Table 8.1 Neuroscience Advances that Could be of Concern. Modified from ref. 26.

Possible Concern	UK, 2011
Molecular neuroscience	"Ion channels and neurotransmitter are key components in many cellular processes...and are frequently targeted in the search for new therapeutic drugs..."
Genetic neuroscience	"...Ion channels have been extensively studied at molecular and genetic levels, and hundreds of genes have now been identified that encode their constituent protein subunits.... Mutations in the genes encoding receptor subunits have been shown to play a role in many human diseases, and this has stimulated the search for new drugs that act at these sites..."
Drug discovery technology	"In recent years, advances in drug discovery technology have included screening with high content systems (HCS), which is based on a bioassay approach, using automated digital microscopy and flow cytometry to evaluate multiple biochemical and morphological parameters in cellular systems..."
Bioregulators	"Bioregulators regulate vital cellular processes and a range of important physiological responses through reaction of a peptide ligand with cellular receptors. Peptide bioregulators and their synthetic derivatives continue to be of interest for their therapeutic potential.... The costs for peptide synthesis and production are likely to continue to fall, and be increasingly undertaken by specialist synthesis companies..."
Drug delivery to the brain	"...Large molecules can be re-engineered to cross the BBB[Blood Brain Barrier] as molecular 'Trojan horses' – endogenous peptides, or peptidomimetic monoclonal antibodies which enter the brain from the blood *via* receptor-mediated transport on endogenous BBB transporters. Liposomes and nanoparticles have also been used for drug delivery to the CNS."

Of course, we are not there yet, but it is as well to remember that is where we are heading as we examine a variety of current investigations of neuronal circuits and receptor/transmitter systems.

In short, most of us know that brain imaging techniques allow the identification of areas of the brain that are active when certain tasks are undertaken and that the genomics revolution of the 1990s allowed the identification of many more neuronal transmitter and receptor systems as well as increasing capabilities to create animals with genes 'knocked out' or 'knocked in'. More recently it has become possible to produce animals with particular neuronal cell types made sensitive to light and to turn the activity of such cells on or off in freely moving animals and thus to investigate their functions. It has also become possible to make brain tissues clear so that the anatomical

Table 8.2 The Revolution in Neurosciences. Modified from ref. 17.

Category	BRAIN 2025
Investigation of neuronal cell types	"...New molecular, genetic and cellular tools are generating exquisite insights into remarkably diverse neuronal cell types that exist within our brains, the basic 'parts list' of our neuronal circuits..."
Investigation of interconnections	"...Novel anatomical techniques are producing remarkable new opportunities for tracing the interconnections between brain regions and individual neurons, revealing basic brain circuit maps in unprecedented detail..."
Neural circuit function	"...Innovative electrical and optical recording tools are allowing us to measure the intricate patterns of electrical activity that exist within these circuits across a broad array of behaviors ranging from decision-making to memory to sleep..."

connections between cells can be examined closely.[30] Such advances will continue and increase our capabilities to investigate the neuronal circuits underlying our behavior.

8.2 Selective Malfunctioning of the Human Machine

Seeing the nervous system in isolation from the rest of the body is, of course, an artificial idealization to assist in its detailed investigation, but the nervous system does not exist in isolation from the rest of the body. Most people know that the nervous system is intimately connected to the endocrine (hormonal) system with the pituitary being a master switch connecting the brain to the body's endocrine producing organs and the brain receiving complex feedback as a result of their hormonal output.[31]

Much less well known is the link between the nervous system and the immune (defence) system of the body which responds to threats such as exposure to infections and toxins. As one recent review noted:[32] "[M]echanistically, the systems communicate through a common language of cell-surface G protein and tyrosine kinase receptors that enable them to respond to the signals each produces." With such levels of linkage between three complex systems the question surely arises as to whether 'selective malfunctions' can be generated even now by those with such aims?

8.2.1 Parkinson's Disease

Parkinson's Disease is the second most common neurodegeneration disease in human beings after Alzheimer's Disease.[33] The disease was first described as the 'shaking palsy' in 1817 because of the disruption of the motor control system, but the proximal cause only became clear in the 20th Century.

In the 1960s, Swedish scientists had found a method of tracing the central monoamine containing neurons using fluorescence histochemistry. Their later work revealed the location and tracts of the major dopamine, noradrenaline, adrenaline, 5-hydroxytraptamine and histamine systems in mammalian brains.[34] According to the Swedish investigators' nomenclature cell groups, A1 to A7 contained noradrenaline and cell groups A8 to A16 contained dopamine.

In the 1970s and 1980s a small number of very unusual cases of severe Parkinson's disease came to doctors' attention in California.[35] The cases were unusual as the disease rarely affects people under 50, yet these suffers were much younger. They were also heroin users. Careful work eventually led to the conclusion that a contaminant in the drug that they had injected had led to the destruction of critical dopamine cells in the brain. The contaminant was MPTP (1-methyl-4-phenyl-1,2,3,6-tetrahydropyridine, CAS Number 28289-54-5) which is converted to a toxic metabolite that kills the dopamine cells.

Parkinson's Disease is complex and causes a range of problems, in addition to the disruption of the control of movement, but the cause of the movement disorder is quite specific. One of the groups of dopamine producing cells (A9) lies in the *substantia nigra* of the midbrain and the axons innervate a forebrain structure called the striatum. So, this pathway is called the nigrostriatal tract and large-scale destruction of this input causes Parkinson's Disease because the striatum and other structures, which together are called the basal ganglia, 'gate' the movement commands coming from the motor cortex. Put simply, the dopamine input from the A9 group of dopamine producing cells keep this gate open and after the loss of some 80% of this input control problems are bound to arise.

To an extent, it has proved possible to help people who suffer from the disease by administration of a substitute for the dopamine transmitter, but there are many limitations to this form of treatment and further research is much needed on better means of treatment.[36] There are a small number of cases of Parkinson's disease that have a clear genetic cause, but the cause of most instances of the disease remains unknown. Possibly it is caused in some cases by environmental toxins or perhaps some form of autoimmune response. What is clear is that critical aspects of the disease can be produced by a quite specific induced destruction of one small part of the entire movement control circuit of the brain by a neurotoxin. A happier outcome of the induced Parkinson's Disease of the Californian drug addicts case was that it was followed up by scientists and now there are a sophisticated variety of animal models of the disease that are being used to further our understanding of what goes wrong and what might be done to correct the control mechanism.[†]

[†]See ref. 31 for details of these animal models and their uses.

8.2.2 Manipulation of Mammalian Memory

So, could modern neuroscientists with their novel methodology of investigating the brain by manipulating its activities[37] and using their new toolbox of techniques[38] actually also modify the higher-level functions of the mammalian brain?

The formation of memories in the brain is certainly one of the high-level functions of the brain. Over a hundred years ago it was proposed that memories resulted from the simultaneous activations in the brain that formed lasting connections in what was called an engram. In the late 1940s, Donald Hebb famously suggested that perceptual memory results from the strengthening of connections between neurons that are active at the same time during memory acquisition, and in the 1990s, the phenomenon of activity-induced long-term potentiation (LTP) of central neuron synapses gave weight to this idea. This led to a great deal of informative work on the operation of synapses and widespread agreement that LTP was an important mechanism in memory formation.[39] However, it is only recently that it has been technically possible to modify memories in mammals.

In 2013, the Nobel Laureate Susumu Tonegawa and colleagues reported the experimental creation of a 'false' memory in a mouse.[40] They were able to identify a group of cells activated in the mouse hippocampus when the animal was exploring a container. Later, having put the mouse into a different container, they delivered an electric shock to the animal whilst at the same time optically stimulating the cells that had been activated when the mouse explored the first container. So, an association was formed between the activity in these hippocampal cells and an electric shock which caused the mouse to freeze in fear. When the mouse was put back into the first container and the hippocampal cells became active, it froze in fear even though it had never experienced a shock in that container.

Even more specifically, a report in 2015 again focused on the hippocampal 'place cells'. These cells are activated when a mouse explores a space:[41] Here, the scientists recorded from a particular neuron and noted where the mouse was when the cell was active. When mice sleep, they replay sequences of place cell activity from their active period. In this experiment, a second electrode on the medial forebrain bundle of nerve fibres was automatically activated whenever the identified place cell was active as the animal slept. This stimulation was known to result in the release of dopamine neurotransmitter in a circuit that reinforces pleasurable behavior, such as receiving a food reward. The net result was that when the mice woke up they were much more likely to go to the location linked to the place cell than they were before the treatment. It was as if they expected that a reward was there. In short, the mouse's memory had been modified whilst it slept, by the experimenters. Of course, it is hoped that this type of work will eventually lead to methods of helping people with memory problems such as those with Alzheimer's Disease,[42] but it also clearly illustrates the power of the new techniques available to neuroscientists.

8.3 Mechanisms of Incapacitation

As we all know, since the discovery of chemical agents (drugs) just after the Second World War, that help people with some forms of mental illness, the development of new drugs has been of outstanding importance in medical efforts to deal with such diseases. The fact that in recent decades there has been a slowing of progress[43] will surely provide grounds for greater use of modern neuroscience to help understand the underlying circuits, transmitters and receptor systems involved. As one recent paper concluded:[44]

> "The overview and future directions that we have presented here for pairing of optogenetic circuit analysis with current and next generation pharmacological approaches reveal how novel cellular pathophysiological mechanisms underlying psychiatric disorders can be uncovered."

In short, development of new drugs (pharmacology) will be combined with the new techniques of neuroscience investigation. And, as the paper notes at the end:

> "Current research is still in the very early stages of developing and implementing multidisciplinary approaches for the treatment of complex brain disorders, but unparalleled potential is certainly apparent as revolutionary tools, such as those discussed here, contribute to the future."

Given the past history, noted by the OPCW Director-General's Scientific Advisory Board, that current incapacitants are all based on the civil science of the 1960s through to the 1980s[7] it would be foolish to imagine that these more sophisticated current and future investigations and results could not also be subject to misuse.

Recently, Rachel Wurzman and James Giordano[45] compiled a detailed list of the categories of possible neuropharmacological agents, neurobiological agents and organic neurotoxic agents and they gave examples of agents in each category. Their list included lethal agents which are not relevant here, but, on the other hand, incapacitating toxins such as SEB that act *via* the immune system, and which was in the US biological weapons arsenal, could be added to their list. So, there is a long list of possibilities and it is not possible to review each of these potential agents here. Therefore, the approach taken here is to concentrate on examples of well-known old, current and possible future agents and to look at what the use of modern neuroscience techniques has revealed about the circuits and receptors that they target.

At this time, of course, progress in our knowledge depends on the use of both older and recently available technologies, but, as will become clear, however, this approach does not leave much doubt that our knowledge of targets for incapacitating agents will increase substantially in coming decades.

8.3.1 Old Agents

Many potential agents were investigated by the adversaries in the latter 20th Century Cold War, but most were found to be unsuitable for various reasons, such as being difficult to synthesise, unpredictable in effect, not amenable to effective dispersal and so on. The question of interest here is whether more recent investigations of the mechanisms of action of these 'old' agents could lead to a reconsideration of their utility for the purposes of incapacitation. Two examples are reviewed here to illustrate such possibilities.

8.3.1.1 Psychedelics

Psychedelics are a group of hallucinogens that have a similar indolamine structure to serotonin (5-hydroxytryptamine, 5-HT) and include LSD (lysergic acid diethylamide) and psilocybin.[46] They are known to exert their effects on the brain through action on the 5-HT_{2A} receptors. These effects are dependent on the person, context and dose, but can produce profound alterations in waking consciousness that include:[47] "visual hallucinations, audio-visual synesthesia, and...derealization and depersonalization phenomena." Such substances have a long history of use in human societies, but modern work started with the chance discovery of the effects of LSD by the chemist Albert Hofman in 1943. Following the discovery of serotonin in the brain of mammals a decade later this quickly led to considerable civil research based on the assumption that LSD in some way interfered with the normal function of serotonin in the brain.

There was also, of course, military interest in the possible use of LSD as an incapacitant. The UK, for example, carried out laboratory tests on people briefly in the early 1950s and then more systematically alongside field trials in the early to mid-1960s. However, in 1969, UK officials:[48] "judged LSD unsatisfactory as an operational agent because of the unpredictability of its effects and because it was expensive to produce." That view of the unpredictability of effects was understandable given the limited knowledge of the serotonin system and receptor distribution at the time. And that, basically, remained the situation until the 1990s because societal reaction to the recreational use of the drug led to its very restricted legal classification and thus, to considerable restriction on civil research.

That situation has changed dramatically today. A detailed review of recent research concluded by quoting from a statement made by a scientist in 1980:[49]

> "It does not seem to be an exaggeration to say that psychedelics, used responsibly and with proper caution, would be for psychiatry what the microscope is for biology and medicine and the telescope is for astronomy. These tools make it possible to study important processes that under normal circumstances are not available for direct observation."

Today, there is the additional factor of the revolution in neuroscience technology that allows brain processes to be investigated in other ways that could hardly be imagined in 1980.

Indeed, we now know that the dopamine system discussed previously in regard to the development of Parkinson's Disease (Section 8.2.1) is but one of six long projection systems by which groups of cells in the brain stem with specific transmitters send projections widely through the brain and spinal cord. These major modulatory systems of the brain produce noradrenaline, adrenaline, dopamine, serotonin, acetylcholine and histamine.[50] It is now well understood that these modulatory systems are crucial in the maintenance of sensory awareness, motor responses and the levels of arousal underlying behavior. The serotonin cell groups are located along the midline of the brain stem in the raphe nuclei and send projections down the spinal cord and the ascending projections go to almost all of the forebrain.

There are a large number of different receptor types affected by serotonin. Psychedelics are known to act predominantly as agonists or partial agonists to 5-HT_{2A} receptors in the brain and in order to explain the actions of psychedelics:[51] "[O]ne would expect to find high receptor levels in key brain regions that are responsible for sensory processing and cognition." That is indeed what has been found as this extract continues "[T]he 5-HT_{2A} receptor is in fact expressed in those very areas that include, among others, the neocortex, thalamus, locus coeruleus (LC) and ventral tegmental area (VTA)." But, what have modern investigations discovered about how psychedelics affect the brain? And, what more are they likely to uncover?

In order to answer that question, we need to consider first some important recent work using modern brain imaging technologies. It is well known that brain imaging studies are designed to show the areas of the brain activated when certain tasks are undertaken. In the late 1990s, it began to be noted that a set of brain areas also regularly reduced their activities when such tasks were undertaken. This network has, since the turn of the century, become known as the brain's default mode network. It consists of three major subdivisions:[52] "the ventral medial prefrontal cortex; the dorsal medial prefrontal cortex; and the posterior cingulate cortex and adjacent precuneus plus the lateral parietal cortex." This discovery significantly reinforces the view that rather than being driven by external cues and producing reflex actions, that the brain is intrinsically organized to maintain an operational model of the world and to implement actions through this model. Moreover, the default mode network is seen as an important element within that system that can be disrupted in disease, particularly in regard to the balance between its activities and the activities of other important networks.[53]

There has been an increasing use of brain imaging techniques such as positron emission tomography (PET), functional magnetic resonance imaging (fMRI), and magnetoencephalography (MEG) to investigate the impact of psychedelics such as psilocybin[54] and LSD[55] on the brain. The general conclusion that has been drawn is that:[56]

> "It seems increasingly evident that psychedelics reduce the stability and integrity of well-established brain networks...and simultaneously reduce the degree of separateness or segregation between them...that is they induce network disintegration and desegregation."

It seems probable that this type of research will increase because of the insights it may give as to the causes of debilitating diseases such as schizophrenia and into the nature of consciousness. That is, of course, to be welcomed, but it should also be appreciated that a more detailed and specific understanding of how psychedelics act may open up interest in new routes of incapacitation and misuse.

8.3.1.2 Delta-9-Tetrahydrocannabinol (THC)

THC, the active ingredient of marijuana, was certainly investigated as a potential incapacitant in the 1960s, but it was not found to be sufficiently potent enough to be of interest.[57] However, since that time, our understanding of how the brain system is affected by THC, the function of that system, and much more powerful agents with similar effects to THC, has radically changed.

In 1988, it was possible to identify cannabinoid receptors in the brain that were affected by THC and other cannabinoids. Then, in the early 1990s, it was discovered that there were substances in the brain that had a similar activity to THC. These endocannabinoids include anandamide (arachidonoyl ethanolamide),[58] and a variety of chemicals that affect these receptors were developed in order to further investigate this novel system.

Taking THC through smoking marijuana produces effects of euphoria and exhilaration, and at higher dosages, a relaxed dreamlike state and possibly hallucinations. There are also physiological effects such as increased heart rate and increased hunger. These effects are mediated by two receptor types, CB_1 which is heavily present in the brain, and CB_2 which is more located outside of the brain. The CB_1 receptors are clearly of most interest here and they are present particularly in higher-level structures such as the cortex, basal ganglia, hippocampus and cerebellum.[59]

There has been a high level of social use of THC for many years in the western world and concern about the health effects of such use. These concerns have increased as the potency of illicit cannabis plant material has increased in the last two decades[60] and particularly as the use of synthetic cannabinoids with up to 100 times the potency of THC have become widely available in so-called ‘legal highs’ with brand names like Spice.[61,62] Not surprisingly, the use of these more potent products has increased concerns about the health risks of use and particularly about the possible risk of psychosis being accentuated in vulnerable individuals or induced in healthy people.[63,64]

CB_1 receptors belong to the standard family of G protein metabotrophic receptors, but they have an interesting presynaptic function on inhibitory gamma amino butyric acid (GABA) interneurons. When pyramidal neurons in the hippocampus are active it appears that endocannabinoids are released and diffuse back to the nerve terminals of GABA interneurons to activate the presynaptic CB_1 receptors which act to inhibit the GABA interneurons. This mode of functioning is called retrograde signaling and clearly enhances the activity of the pyramidal cells.

The disruptive function of agents like THC have again been linked to their effects on neural networks in the brain. One recent review noted that:[65]

> "Complex brain functions...are based on distributed processes among multiple cortical and subcortical regions. It has been shown that the neural assemblies involved in these functions engage into transient periods of synchronous oscillatory activity while processing information."

These synchronous neural oscillations, detected electrically, are thought to be mediated mostly by GABA interneurons of the type which are known to be modulated by the release of endocannabinoids. As the review goes on to note:[66]

> "a host of human studies have collaborated these [animal studies] findings, which provide support for the notion that CB_1Rs [receptors] contribute to psychosis *via* disruptions in neural oscillations ... numerous EEG studies have shown that chronic cannabinoids are associated with disrupted theta-band and gamma-band neural oscillations."

Indeed, it appears that the strongest evidence that cannabinoids disrupt neural oscillations comes from a small number of studies of controlled THC administration to human beings.

Clearly there is much work to be done to elucidate how this system functions normally and how it is disrupted by cannabinoids, but, given the possibility of further understanding psychoses and their prevention, it is hardly likely that this work will not be carried out. It may also open up novel avenues for misuse and therefore needs to be carefully monitored by both States and civil society.

8.3.2 Current Agents

Despite the desire to find a wide range of novel incapacitating chemicals, and much effort, during the Cold War, today as the UK Royal Society concluded in 2012:[67]

> "with increasing emphasis on rapid action and short duration of effects, contemporary interest has tended to focus on sedative–hypnotic agents that reduce alertness and as the dose increases, produce sedation, sleep, anaesthesia and death."

A review of these kinds of agents was published in 2000 by the College of Medicine, Applied Research Laboratory at Pennsylvania State University under the title *The Advantages and Limitations of Calmatives for Use as a Non-Lethal Technique.* Perhaps because of the advances made in our understanding of receptors for neurotransmitters in the brain during the 1990s, this report was optimistic about the possibilities noting that:[68]

"the researchers identified many promising new developments that deserve consideration with high potential as prototypical calmatives with availability in the near future."

Clearly, they did not agree with the later widely shared conclusion of the Royal Society that:[69]

"the feasibility of developing an incapacitating chemical agent and delivery system combination that is safe (*i.e.* has a low risk of lethality) is questionable. None of the agents under consideration are free of unintended side effects and current advances in delivery systems do not solve the problem of controlling dose."

This gap in appreciation was perhaps well demonstrated by the title page of the College of Medicine report having the chemical structure of the synthetic opioid fentanyl as the cover illustration. A mixture of two derivatives of fentanyl, of course, was used in an attempt to break the 2002 Moscow theatre siege, and, while many hostages were rescued, over 120 died as a result of the use of these so-called non-lethal incapacitating agents.[70]

The College of Medicine report, however, was particularly enthusiastic about two other classes of chemical agents stating:[68]

"Researchers identified several drug classes (*e.g.* benzodiazepines, alpha-2-adrenoreceptor agonists) and individual drugs (diazepam, dexmedetomidine) found appropriate for immediate consideration as a non-lethal technique."

Benzodiazepines like diazepam target GABA receptors and act by allosteric action on one particular receptor type ($GABA_A$), but there are many sub-types of this receptor. Thus, whilst it has been possible to construct agents such as remimozolam (CNS7056) that act to quickly produce sedation and have a short period of action, this is because it is rapidly metabolized in the body rather than it attacking a specific receptor sub-type with a particular mode of action in the circuits underlying arousal.[71]

8.3.2.1 *Dexmedetomidine*

Our level of awareness is controlled by complex circadian (sleep-wake) and attention (focus) circuits.[72] These circuits can be affected by many drugs, such as the benzediazepines just described, which target the transmitter/receptor systems in such circuits. Another drug investigated by the College of Medicine was dexmedetomidine. This drug affects the noradrenaline neuromodulatory system of the brain. As with the serotonin system previously described, noradrenaline producing neurons in the brain stem innervate different parts of the brain and spinal cord, but the A6 cells of the locus coeruleus are of particular interest, as these cells have extensive projections to the cerebral cortex and are clearly involved in the maintenance of vigilance and responses to unexpected sensory inputs.[50]

When cells of the locus coeruleus are active and producing noradrenaline, this affects inhibitory alpha$_2$-adrenergic receptors on the noradrenaline producing cells themselves as well as their target neurons. Therefore, the output also inhibits the neurons producing the noradrenaline. Dexmedetomidine is an agonist drug which has the same kind of action as noradrenaline. As the noradrenaline modulatory is an important part of the awake/arousal/attention mechanism of the brain, the drug produces sedation effectively.[73] Given the complexity of the mechanisms involved in maintaining the awake attentional state, and the consistency of the response to dexmedetomidine, the College of Medicine report argued for its utility for use in conjunction with other drugs:[74]

> "Used in conjunction with most other sedative agents, the drug markedly (28–90%) reduces the dose requirement for the primary agent, often reducing side effects leading to increased safety of the mixture of pharmaceutical agents."

This possibility is well recognized in medical practice, but the College of Medicine report went further to point out the incapacitating agent possibilities:[74]

> "The concept could be considered that exposure of a group first to the pharmacological agent (leading to mild sedation and sleepiness) can then be added to another direct non-lethal technique."

In short, the difficulties of the operational use of one incapacitating agent might be overcome by using a mixture of agents at a lower (less dangerous) level.

This was exactly the approach taken by Czech scientists in the early years of this century. In an effort that they said was aimed at developing 'pharmacological non-lethal weapons' they reported experiments on various mammals, monkeys and human beings. For example, stating that:[75]

> "The triple combination of *midazolam* at a dose of 0.08 μg kg^{-1}, *dexmedetomidine* 1.5 μg kg^{-1} and *ketamine* 3 μg kg^{-1} i.m. was also tested in 10 nurses." (original emphases)

and that:

> "*Fully reversible analgesic sedation* was also tested in *man*. We tested the triple combination of *dexmedetomidine* 1 μg kg^{-1}, *midazolam* 0.08 mg kg^{-1} along with *fentanyl* 1 μg kg^{-1}. ... The onset of effect following administration ... was very rapid, with appreciable psychomotor sedation seen within 2–3 minutes. Sleep set in within 5–7 minutes." (original emphases)

Fortunately, although the work continued after 2010, the Czech authorities made it clear that the researchers could not be involved in the development of such non-lethal weapons under the law in the Czech Republic.[76]

8.3.2.2 *Future Manipulation of the Locus Coeruleus System*

Until recently there has been a common view among researchers working on the locus coeruleus (LC) system that it functions in a rather homogeneous manner. As one recent review paper noted this understanding:[77]

> "converges on the popular view that the LC is a homogeneous structure responsible for the simultaneous release of NE [noradrenaline] on terminal fields throughout the entire extent of the LC efferent network, primarily the forebrain, cerebellum and spinal cord."

It was, of course, well understood that the LC neurons had various modes of activity. They fired most rapidly when the animal was awake, more slowly during slow wave sleep and not at all during paradoxical (rapid eye movement) sleep. In the awake state, the LC neurons fire at a slow tonic rate or in short high frequency bursts which are associated with focused attention.[78] However the researchers did not think that the whole system was further differentiated.

That simple unified view of the LC, as one system, appears to be breaking down as more sophisticated methods are used to investigate the system in greater detail. In short:[77]

> "Rather than expecting unifying principles of operation it may be more appropriate to acknowledge heterogeneity that predicts the selectivity of action and diversity of function that are required to generate adaptive responses to a multitude of behavioral circumstances."

We will examine the link between the LC and the sleep/wake circadian rhythm in a later section, but this newer viewpoint is particularly interesting because of the variety of neurotransmitter and neuromodulatory influences it implies on the output from and the input to different LC neurons.

The vast majority of LC neurons contain the neuropeptide galanin as well as noradrenaline and it is to be expected that work using modern techniques such as optogenetics will lead to a much greater understanding of the roles played by these interesting neuropeptides in the output functions of diverse LC neurons.[79] However, it may be the huge diversity of inputs to the LC system[80] that provide the most opportunity for manipulation—for good or ill—in the future.

The LC neurons receive input from non-neuropeptides such as glutamate, GABA and serotonin, but also from a wide range of neuropeptides and these neuropeptide inputs come from a wide range of sources within the brain. Corticotrophin releasing factor (CRF) is released onto LC neurons in response to stress and clearly increases their firing rate and thus the impact of noradrenaline release on target neurons. It is thought that this may disrupt attention processes and could be involved in conditions which are related to inappropriate stress responses such as post traumatic stress disorder (PTSD). Orexin input to the LC, which is dealt with in more detail in the next section of this chapter, is clearly involved in increasing the firing rate of LC neurons as a contribution to arousal in moving from sleep to the awake state and in maintaining the awake state, but evidence is also accumulating that demonstrates the role of orexin input to the LC and subsequent noradrenaline output to the hippocampus and amygdala in learning processes. Substance P is also located in axon terminals on LC neurons and related neurokinin-1 tachykinin receptors (NK1) are located on LC neurons. It is likely that this input increases LC neuron discharges in response to pain.

Other neuropeptides act to reduce LC neuron firing rates. Input of endogenous opioids, acting on mu opioid receptors inhibit the activity of LC neurons. Thus:[81]

> "it appears that the opioid system serves to return LC discharge back to basal levels after stressor exposure and serving as a restraint on the LC-NE system during chronic or repeated stress."

Unsurprisingly, there is evidence of comorbidity of opiate abuse with stress disorders such as PTSD.

Melanin concentrating hormone (MCH) cells of the lateral hypothalamus affect MCH1 receptors on neurons of the LC. It is thought that this input is inhibitory and functions to keep the LC neurons silent during rapid eye movement sleep. If this is correct, then any disruption of this input could produce sleep disturbances. Somatostatin (SS) input, like that of the MCH cells, is thought to be inhibitory and to assist in the stability of sleep. Neuropeptide Y (NPY) is widely distributed in the brain and there has been recent evidence that this input from the hypothalamus to the LC is involved in resilience to stress. The mechanisms involved have yet to be clarified but as NPY can be administered intranasally it is being investigated as a possible therapeutic method of dealing with stress-related illnesses.[82]

Clearly then, the regulation of the sleep/wake circadian cycle and levels of arousal through the LC/NE system are complex. However, considerable progress has been made in understanding its mode of operation and there are many more interesting avenues of research which can be expected to elucidate specific aspects of the functional anatomy and physiology of the regulation of diverse behaviours – some of which could well be manipulated by those with hostile purposes in mind.[83] What also needs to be kept in mind, however, is that this is but one illustration of the growing mechanistic understanding of the brain being produced by advances in relevant technologies. So far, we have looked at examples of old and current agents where relevant advances are being made. We turn now to two illustrative examples where interest in misuse is not so evident, but where the progress in our understanding suggests that misuse could be considered in the future.

8.3.3 Novel Agents?

Whilst consideration of incapacitants in the past concentrated on the chemical 'Psycho Agents'[84] it is interesting to note how often over the last two decades consideration has also been given to natural bioregulators and their analogues. For example, a review of the threat from mid-spectrum agents written in 2003 by an informed scientist stated that:[85]

> "The result of exposure to such agents orally or by means of inhalation could have serious consequences including fatigue, fear, physical and mental incapacitation ... Therefore, these agents not only should be considered as lethal agents, but also as non-lethal incapacitating agents."

And, the author continued by giving a series of examples of such agents:

> "relevant bio-regulators are endogenous opioids, which induce analgesia, reduce pain, alter the blood pressure and depress respiration; Substance P, which is the neurotransmitter in pain sensory neurons and may cause irritation and pain; and neuropeptide Y, which alters drinking and feeding behavior, alters blood pressure and body temperature, and influences memory functions."

So, concerns about bioregulatory neuropeptides is not new, but what is perhaps surprising is the speed of the advances in our understanding of the functions of such neuropeptides and the complexity of their actions within sophisticated human behavior.

8.3.3.1 Orexin

In February 2013, the London *Guardian* newspaper's weekly review had an extended article entitled "The narcolepsy family's house of sleep". The article recounted how doctors in Spain had put together an understanding of how one family that had been split up during the Spanish Civil War had been traced through their genetic inheritance of a tendency to narcolepsy and how difficult it was for the people affected "to stay either awake or asleep for a sustained period of time." The doctors involved explained that they hoped that investigating the family's genetics would help in finding a means to deal with narcolepsy, but they also explained that[86] "[T]here is evidence that the origin of the illness is autoimmune in 95% of cases which do not have an obvious genetic cause." This explanation is similar to that which has been put forward to explain the cause of Parkinson's Disease. As the article explains "[S]ufferers seem to produce antibodies that attack the hypocretin neurons, stopping or slowing production...hypocretin (also known as orexin) plays an important role in alertness." The important point here is that orexin was unknown until it was discovered in the brain at the very end of the 1990s.

As the introduction to a themed section on "Orexin Receptors" in a recent edition of the *British Journal of Pharmacology* noted:[87]

> "Compared to other neurotransmitters and neuropeptides, the orexin/hypocretin system represents the new kid on the block. First identified only in 1998, the past fifteen years has seen a steady growth in an understanding of the roles that orexin receptors and their cognate peptide ligands play in the CNS and beyond."

In particular, there is considerable evidence that input from these orexin producing neurons to the locus coeruleus (which was discussed in the previous section of this chapter), is critical in the transition from the sleep to wake state and the maintenance of that state.[78] Clearly, drugs that block the orexin receptors would therefore be of value in the treatment of the widespread problem of insomnia in the modern world. What is striking is how quickly such drugs have been developed.[88] There are two types of orexin neuropeptides and two types of orexin receptors. Further development of drugs that affect these receptors can be expected as the structure of the human OX_2 orexin receptor bound to the drug suvarexant has been determined at a fine resolution and this will facilitate further drug development.[89]

While the control of arousal by orexin neurons[90] is currently of most interest in regard to the possible development of novel incapacitants, there is increasing evidence[91] of involvement of the orexins in other complex brain functions. Clearly there is a lot more to be uncovered about this recently discovered system. Among the scientists involved, there is a widespread consensus about what they intend to achieve. As one detailed paper titled "Understanding how discrete populations of hypothalamic neurons orchestrate complicated behavioural states" which discussed orexin research extensively put it:[92]

> "An important question for future research is whether Hcrt [hypocretin/orexin] neurons can be anatomically and functionally subdivided into multiple, discrete subpopulations, each projecting to a distinct downstream region."

Then, of course, the aim would be to follow the way in which the different modes of activity of these distinct subpopulations affect the downstream targets to produce behavior.

They believe that they already have the necessary technology to carry out these tasks as the paper notes:[93]

> "Optogenetic technology has allowed scientists the ability to stimulate or inhibit genetically-defined populations of neurons with millisecond-precise temporal resolution."

and,

> "Pharmacogenetic technology, such as 'Designer Receptor Exclusively Activated by Designer Drugs' – DREADDS, allows for the ability to hijack a neuron's G-protein signaling cascade to induce or suppress neural activity."

In such circumstances, rapid advances in our understanding of the orexin system seems to be virtually guaranteed.

One point of particular interest in relation to future incapacitating agents is the cause of the disruption of the orexin system in most cases of narcolepsy. There has long been known to be a correlation between a particular human immune system type and the occurrence of narcolepsy and this possible autoimmune cause of narcolepsy seems to have been confirmed by some incidents of narcolepsy that followed the use of a particular vaccine against H1N1 influenza in some countries earlier in this century.[94] However, a recent elegant experiment poses a different possibility. In this experiment, a genetic manipulation of mice removed their adaptive immune system response to infection *via* the nasal inhalation route with H1N1.[95] These mice then developed narcolepsy and disruption of the orexin neuron system suggesting that the virus itself might cause the disease. Thus, the experimenters concluded that:

> "It is therefore possible that several viruses that can invade the brain along the olfactory route, include mutant virus strains, which could more selectively target Orx/Hcrt [orexin] neurons and cause life-long sleep pattern changes in susceptible individuals."

Of course, it is possible that both a direct viral assault and autoimmune responses could operate to cause narcolepsy in susceptible individuals.

8.3.3.2 Oxytocin

In 2013, the US Defense Advanced Research Projects Agency (DARPA) put out a solicitation for work titled *Oxytocin: Improving measurement sensitivity and specificity*.[96] The solicitation explained that "[O]xytocin also affects behaviors relevant to national security." Now oxytocin, unlike orexin, is not a 'new kid on the block' but is well known to biologists as a hormone involved in functions such as reproduction, lactation and pair bonding in rodents.[97] As the commentary on a recent set of review papers pointed out:[98]

> "The oxytocin system has long been the focus of rigorous scientific investigation as a model of neurosecretion from the pituitary and for its role in regulating uterine contractions during labor and milk ejection during nursing."

In this part of the mammalian oxytocin system, cells in the hypothalamus secrete oxytocin (and the related peptide vasopressin) into the peripheral blood stream *via* the pituitary and thus influence body organ systems.[99] However, in evolutionary terms, the genes for oxytocin, vasopressin and related peptides have a very long history and the system has been put to diverse functions in different animal groups and even within related species.[100] Indeed oxytocin producing neurons also function within the brain to affect mammalian behavior and it is this aspect of the function of the oxytocin system—its role as a neurotransmitter within the brain—that the DARPA solicitation addressed.

Earlier this century, experiments suggested that oxytocin delivered intranasally could significantly affect the trusting behavior of human beings.[101] This study initiated a resurgence of interest in oxytocin as a neuromodulator in the brain. As the commentary, previously mentioned, noted in 2016:[98]

> "Oxytocin research has experienced a remarkable resurgence over the past decade, particularly with respect to its role in regulating social cognition and its prospects for treating psychiatric disorders."

So, what was DARPA's interest in this neuropeptide and its sole receptor type in the brain? The solicitation explains as follows:[96]

> "Recent studies have shown the regulation of oxytocin to be a complex process. In particular two forms of oxytocin have been identified. A 10–12 amino acid pro-hormone is first produced, and then, at some point, may be cleaved to a 9 amino acid hormone. This shortened form is the active neuropeptide, oxytocin, known to bind the oxytocin receptor and is credited with oxytocin's behavior altering effects. The biological role, if any of the 12 amino acid pro-hormone is unknown … Additional forms of different lengths or biologically active metabolites of oxytocin may exist, as well."

As the available techniques were not able to distinguish between the different forms of oxytocin, DAPRA was interested in funding the development of new techniques that could do so in order to better understand how 'Narrative Networks' influence peoples' behavior – in short, how we are influenced by stories told by people we trust. Such interest is perhaps understandable given current concerns about the radicalization of young people, but it is important to grasp just how far and fast research on social cognition is taking us and where malign manipulation might occur.

The use of modern neuroscience techniques is beginning to unpick the molecular mechanisms by which oxytocin exerts its effects on behavior in mammals, despite the undoubted complexity of this aspect of the function of the oxytocin system in the brain and its interactions with other neurotransmitter and neuromodulatory systems. Two recent examples of such work illustrate how that understanding is being developed.

A paper in *Nature* in 2015 reported work on maternal behavior in mice. If young pups fall out of the nest they emit ultrasonic distress calls and the mothers (dams) use these calls to locate the pup and bring it back into the nest. Virgin females do not exhibit such behavior, but can be induced to do so by the appropriate use of oxytocin. Virgin females will also pick up the maternal behavior if co-housed with dams and litters, but at a much slower rate. This was shown in an experiment in which three groups of virgin mice co-housed with dams and litters were compared. The first group of virgin mice received systemic oxytocin, the second just saline injections and the third optogenetic enhanced release of oxytocin in the brain. As a result:[102]

> "Within 12 h of being co-housed, virgin females receiving either oxytocin or optical...stimulation began retrieving more than saline injected animals ... Saline-injected virgins generally required at least 2 days of co-housing to express retrieval behavior."

It was also shown that virgin females not co-housed with mothers but only with pups were much faster in retrieving pups if they had been receiving oxytocin.

The paper went on to show that while oxytocin neurons projected to both left and right auditory cortexes, significantly more oxytocin receptors were present in the left primary auditory cortex, and that if the GABA agonist muscimol was infused into the left, but not the right, side retrieval behavior was impaired. Further work studied what happened to the neurons of the left auditory cortex in order to enhance the salience of the pups' vocalisations and led to the general conclusion that the results:[103]

> "...directly demonstrate how oxytocin paired with the pups calls rapidly changes brain state, transforming weaker virgin responses into more robust and temporarily precise maternal responses. The predominant effect of oxytocin is to reduce cortical inhibition within seconds, followed by longer-term modifications over hours ... enhancing spiking and successful maternal care."

This conclusion is in accordance with what appears to be a widespread consensus that:[104] "[O]xytocin increases the salience of social stimuli and promotes parental nurturing and social bonds."

Detailed investigations, it must be stressed, are at an early stage. However, some links to agents that could be of concern are already being revealed. One recent paper was titled "Endocannabinoid signaling mediates oxytocin-driven social reward" and investigated how marijuana-like natural chemicals produce their behavioral effects.[105]

The authors argued that the rewards of social interaction in a group required the induction of long term synaptic plasticity at certain excitatory synapses in the nucleus accumbens and that this was dependent on oxytocin. Then, because the regulation of the reinforcement of some natural stimuli and of neurotransmission in the nucleus accumbens depends also on the endocannabinoid system, they tested the hypothesis that there might be cooperation between the endocannabinoid system and oxytocin to regulate reward for social interaction.

Using mice, they found first that social interaction increases anandamine (the natural endocannabinoid neurotransmitter) in the nucleus accumbens, whereas isolation decreases its concentration. Then they demonstrated that:

> "Pharmacological and genetic experiments show that anandamine mobilization and consequent actuation of CB_1 cannabinoid receptors are necessary and sufficient to express the rewarding properties of social interaction."

And, they further showed that oxytocin was driving the production of anandamine in the nucleaus accumbens. This led them to conclude that "anandamine mediated signaling at CB_1 receptors, driven by oxytocin, controls social reward" and that further that deficits in this mechanism could be a factor in the social impairment underlying autism spectrum disorders.

8.4 Conclusions

The main bearing of this account is that there is a past history, current interest and a probable future for incapacitating chemical agent development based on the misuse of benignly intended advances in civil neuroscience. More significantly, the latter sections of this chapter suggest that the far-reaching aims of the Cold War weapons engineers—and beyond these aims—may soon become available for those with malign aims because of the very rapid advances being made by modern neuroscientists. That is not, of course, to argue that the work of civil scientists should stop, but rather to suggest that care must be taken to erect strong barriers, including legal, oversight, educational and awareness-raising means, against malign misuse of their work.

This argument is in agreement with the more general assessment of the potential misuse of modern biotechnology, made by Professor Matthew Meselson in his 2000 essay on "Averting the Hostile Exploitation of Biotechnology" that was quoted in the introduction to this chapter.[10] As he noted, during this century we will become increasing able to manipulate life and misuse of such capabilities could result in "unprecedented opportunities for violence, coercion, repression or subjugation." Moreover, he went on to point out two particular features of these changes in our capabilities "[U]nlike the technologies of conventional or even nuclear weapons, biotechnology has the potential to place mass destructive capabilities in a multitude of hands." Thus, preventing misuse of such technologies will require a considerable up scaling of current control measures, and particularly in regard to the potential misuse of neuroscience, he continued by noting that these advances will also give us the capability in coming decades "to reach deeply into what we are and how we regard ourselves." Thus, there can be little doubt that the sooner the bulk of neuroscientists become involved in the development of better control and oversight measures the more able the international community will be prepared for these very likely challenges.

In an article in the *Bulletin of the Atomic Scientists* in May 2006,[106] James Giordano suggested that a start could be made now on the task of increasing surveillance of neuroscience research and development around the world as long as care was taken that this did not "lead to a kind of arms race, as nations reacted to new developments by creating counter agents or improving upon one another's discoveries". And, he specifically noted that just this mistake could be made in relation to the development of novel incapacitating agents.

Those who doubt that such intentional deliberate manipulation of human behavior is possible need to consider what the blind forces of evolution have produced in the manipulation of hosts by parasites. In particular, the knowledge we are gaining now of the manipulation of the nervous systems of mammals, including primates, should give cause for concern. A specific example is the protozoan parasite *Toxoplasms gondii* for which there is strong evidence of the manipulation of intermediate hosts like rats, mice and human beings in order to increase its chances of successful transmission to its definitive host—members of the cat family—where it can reproduce.

A recent perspective[107] on a paper examining the mechanism involved was titled "Masterpiece of epigenetic engineering – how *Toxoplasma gondii* reprogrammes host brains to change fear to sexual attraction". Parasites often have more than one mechanism to affect the host's behavior. In this case, one of the mechanisms that changes the rat's natural aversion to cat odour to increase the chances of predation (and this transmission to the cat) includes a direct attack on the amygdala (fear) circuit. As the authors of the related full paper note:[108]

> "infection institutes an epigenetic change in the DNA methylation of the arginine vasopressin promotor in the medial amygdala of the male rat ... Furthermore, we show that loss of fear in the infected animals can be rescued by the systemic hypermethylation and recapitulated by direct hypomethylation in the medial amygdala."

Clearly, however, humans are not now suitable hosts and the effects of infection are generally assumed to be irrelevant to predation. Yet, there is also a possible evolutionary explanation as our nearest relatives, chimpanzees, infected by the parasite lose their aversion to the urine of their natural predator—the leopard—but not to the urine of lions or tigers which are not natural predators of chimpanzees.[109]

There are many forms of host manipulation yet to be investigated in detail,[110] and many unknown potential agents such as venoms[111] and novel toxins[112] that could be found to have military utility.[113] Moreover, it is to be expected that current technologies for the investigation of the impact of such agents will continue to improve substantially. Technologies central to the further investigation of neural networks underlying behavior such as fMRI[114] and optogenetics and phartmacogenetics[115] will certainly be included in those that will substantially improve. It should be noted also that technologies being applied in neuroscience include the CRISPR/Cas system[116] which has been of clear concern to officials involved in national security.[117]

There has also clearly been much use of model organisms such as flies,[118] zebra fish[119] and monkeys[120] to investigate the general properties of nervous systems and the underlying causes of human diseases of the nervous system. Yet, the use of monkeys, in which diseases thought to be similar to autism have been deliberately produced, should surely raise ethical questions about whether there are some experiments we should not undertake.[121] Such questions range well beyond the problem of dual-use, but there are certainly questions of dual-use, if not already with us, then certainly on the horizon. For example, does the development of malodorant incapacitating weapons[122] not signal that we are at a stage beyond peripherally acting riot control agents and are beginning to engage with aspects of the central nervous system controlling more complex behavior than responses to painful peripheral stimulation?

In short, this review of potential novel targets for incapacitating chemical agents in the central nervous system reinforces the UK's conclusion that:[123]

> "Many of the benefits and risks in the neurosciences lie in the future. However, in the development phase it is timely to consider issues related to governance of this dual-use technology area, balancing the obligation to take measures to prohibit and prevent misuse with the need to ensure that the beneficial development of science is not hampered."

Indeed, there would appear to be substantial risks of misuse if this proactive approach is not taken.

References

1. Stockholm International Peace Research Institute, *CB Weapons Today*, Almqvist & Wiksell, Stockholm, 1973, vol. II, ch. 4: Research and development: Implications for the future of CBW, 260–332, pp. 298–299.
2. D. Lindsey, *Mil. Med.*, 1960, **125**, 598.
3. R. Khatchadourian, *The New Yorker*, 2012, December 17, http://www.newyorker.com/magazine/2012/12/17/operation-delirium, last accessed 16/03/2016.
4. J. S. Ketchum, *Chemical Warfare Secrets Almost Forgotten*, Private Publication, Santa Rosa, California, 2006.
5. U. Schmidt, *Secret Science: A Century of Poison Warfare and Human Experiments*, Oxford University Press, Oxford, 2015.
6. J. S. Ketchum, and F. R. Sidell, in *Textbook of Military Medicine: Medical Aspects of Chemical and Biological Warfare*, ed. F. R. Sidell, E. T. Takafuji and D. R. Franz, Office of the Surgeon General, Department of the Army, United States, Washington D. C., 1997, ch. 11: Incapacitating Agents, pp. 291–294.
7. OPCW, *Report of the Scientific Advisory Board on Developments in Science and Technology for the Third Special Session of the Conference of the States Parties to Review the Operation of the Chemical Weapons Convention, RC-3/DG-1*, OPCW, The Hague, 29 October 2012, p. 4.
8. Royal Society, *Neuroscience, Conflict and Security, Brain Waves Module 3*, Royal Society, London, February 2012, p. 52.

9. F. J. Petrasek, in *Principles of Psychopharmacology*, ed. W. G. Clarke and J. del Giudice, Academic Press, New York, 1978, ch. 7: Chemistry and Structure - Activity Relationships of Psychotropic Drugs Part I General Topics, p. 139.
10. M. Meselson, *CBW Conventions Bulletin*, 2000, **48**, 16.
11. V. Pitschmann, *Toxins*, 2014, **6**, 1761–1781.
12. P. Simmons and D. Young, *Nerve Cells and Animal Behaviour*, Cambridge University Press, Cambridge, 3rd edn, 2010.
13. S. Zeki, *Philos. Trans. R. Soc., B*, 2015, **370**, 163.
14. L. A. Jorgenson and W. T. Newsome, *et al.*, *Philos. Trans. R. Soc., B*, 2015, **370**, 164.
15. P. Frackowiak and H. Markham, *Philos. Trans. R. Soc., B*, 2015, **370**, 171.
16. H. Okano, A. Miyawaki and K. Kasai, *Philos. Trans. R. Soc., B*, 2015, **370**, 310.
17. Brain Research through Advancing Innovative Neurotechnologies (BRAIN) Working Group, *Report to the Advisory Committee to the Director, NIH: Brain 2025: A Scientific Vision*, National Institutes of Health, Washington, D. C., 5 June, 2014, p. 5.
18. T. R. Insel, P. Y. Collins and S. E. Hyman, *Foreign Affairs*, January/February, 2015, p. 127.
19. United States, *Statement by Ambassador Robert P. Mikulak, EC-72/NAT.8*, OPCW, The Hague, 6 May, 2013, pp. 1–2.
20. Germany, *Toxic Chemicals for Enforcement, RC-3/NAT.33*, OPCW, The Hague, 16 April, 2013, p. 2.
21. Australia, *Weaponisation of Central Nervous System Acting Chemicals for Law Enforcement Purposes, C-19/NAT.1*, OPCW, The Hague, 19 November, 2014, p. 1.
22. Australia, *et al.*, *Aerosolisation of Central Nervous System-acting Chemicals for Law Enforcement Purposes, C-20/NAT.2*, OPCW, The Hague, 24 November, 2015, p. 3.
23. Spiez Laboratory, *Spiez Convergence: Report on the First Workshop, 6–9 October 2014*, Federal Office for Civil Protection, Spiez, Switzerland, November, 2014, p. 16.
24. Ministry of Defence, *Historical Survey of the Porton Down Service Volunteer Programme 1939-1989, Part IV, Human Studies with Incapacitating Agents, 109-141*, Ministry of Defence, London, June, 2006.
25. UK, *The Convergence of Chemistry and Biology: Implications of Developments in Neurosciences*, BWC/MSP/MX/WP.1, United Nations, Geneva, 12 July 2012.
26. Implementation Support Unit, *New Scientific and Technological Developments Relevant to the Convention: Annex, Submission from States Parties, United Kingdom of Great Britain and Northern Ireland, 21–38, BWC/CONF. VII/INF. 3/Add.1*, United Nations, Geneva, 23 November 2011.
27. Implementation Support Unit, *New Scientific and Technological Developments Relevant to the Convention: Annex, Submission from States Parties, United Kingdom of Great Britain and Northern Ireland, 21–38,*

BWC/CONF.VII/INF. 3/Add.1, United Nations, Geneva, 23 November 2011, p. 30.
28. S. Finger, *Minds Behind the Brain: A History of the Pioneers and Their Discoveries*, Oxford University Press, Oxford, 2000.
29. Brain Research through Advancing Innovative Neurotechnologies (BRAIN) Working Group, *Report to the Advisory Committee to the Director, NIH: Brain 2025: A Scientific Vision*, National Institutes of Health, Washington, D. C., 5 June 2014, p. 14.
30. L. Luis, *Principles of Neurobiology*, Garland Science, New York, 2016, ch. 13: Ways of Exploring, pp. 557–609.
31. M. Wilkinson and R. E. Brown, *An Introduction to Neuroendocrinology*, Cambridge University Press, Cambridge, 2nd edn, 2015.
32. S. Talbot, S. L. Foster and C. J. Woolf, *Annu. Rev. Immunol.*, 2016, **34**(1), 1–27.
33. W. Lee, P. Sayana and J. Jankovic, *Neurotherapeutics*, 2014, **11**, 92.
34. K. Fuxe and T. Hokfelt, in *Principles of Psycopharmacology*, ed. W. G. Clarke and J. del Giudice, Academic Press, New York, 1978, ch. 4: Transmitter-Based Mapping of Central Neurons, pp. 71–96.
35. C. Wallis, *Time*, 24 June 2001, **1**.
36. J. S. Meyer and L. F. Quenzer, *Psychopharmacology: Drugs, the Brain, and Behavior*, Sinauer Associates, Sunderland, Mass., 2005, pp. 124–127.
37. K. Krug, D. C. Salzman and S. Waddell, *Philos. Trans. R. Soc., B*, 2015, **370**, 201.
38. S. S. Hall, *MIT Technology Review*, 2014, 17 June, vol. 1.
39. M. Poo, *et al.*, *BMC Biol.*, 2016, **14**, 40.
40. X. Liu, S. Ramirez and S. Tonegawa, *Philos. Trans. R. Soc., B*, 2014, **369**, 142.
41. G. de Lavilleon, *et al.*, *Nat. Neurosci.*, 2015, **18**, 493.
42. D. S. Roy, *et al.*, *Nature*, 2016, **531**, 508.
43. M. J. Millan, *et al.*, *Eur. Neuropsychopharmacol.*, 2015, **25**, 591.
44. G. D. Stauber and A. O. Mason, *Pharmacol. Rev.*, 2013, **65**, 156.
45. R. Wurzman and J. Giordano, in *Neurotechnology in National Security and Defense*, ed. J. Giordano, CRC Press, Boca Raton, 2015, ch. 7: *"NEURINT" and Neuroweapons: Neurotechnologies in National Intelligence and Defense,* pp. 79–114.
46. D. E. Nichols, *Pharmacol. Rev.*, 2016, **68**, 264.
47. D. E. Nichols, *Pharmacol. Rev.*, 2016, **68**, 264, p. 270.
48. Ministry of Defence, *Historical Survey of the Porton Down Service Volunteer Programme 1939–1989, Part IV, Human Studies with Incapacitating Agents, 109–141*, Ministry of Defence, London, June, 2006, p. 119.
49. D. E. Nichols, *Pharmacol. Rev.*, 2016, **68**, 264, p. 345.
50. C. B. Saper, in *Principles of Neural Science*, ed. E. R. Kandel, J. H. Schwartz and T. M. Jessell, McGraw-Hill, New York, 4th edn, 2000, ch. 45: *Brain Stem Modulation of Sensation, Movement and Consiousness*, pp. 889–900.
51. D. E. Nichols, *Pharmacol. Rev.*, 2016, **68**, 264, p. 299.

52. M. E. Raichle, *Annu. Rev. Neurosci.*, 2015, **38**, 433–447, p. 439.
53. R. L. Buckner, *Dialogues Clin. Neurosci.*, 2013, **15**, 351.
54. R. L. Carhart-Harris, *et al.*, *Proc. Natl. Acad. Sci. U. S. A.*, 2012, **109**, 2138.
55. R. L. Carhart-Harris, *et al.*, *Proc. Natl. Acad. Sci. U. S. A.*, 2016, **113**, 4853.
56. R. L. Carhart-Harris, *et al.*, *Proc. Natl. Acad. Sci. U. S. A.*, 2016, **113**, 4853, p. 4857.
57. J. S. Ketchum, *Chemical Warfare Secrets Almost Forgotten*, Private Publication, Santa Rosa, California, 2006, pp. 35–42.
58. J. S. Meyer and L. F. Quenzer, *Psychopharmacology: Drugs, the Brain, and Behavior*, Sinauer Associates, Sunderland, Mass., 2005, pp. 405–407.
59. H.-C. Lu and K. Mackie, *Biol. Psychiatry*, 2016, **79**, 516.
60. M. A. ElSohly, *et al.*, *Biol. Psychiatry*, 2016, **79**, 613.
61. M. S. Castaneto, *et al.*, *Drug Alcohol Depend.*, 2014, 12.
62. C. Milano, *et al.*, *Front. Neurosci.*, 2016, **10**, 153.
63. M. Sherif, *et al.*, *Biol. Psychiatry*, 2016, **79**, 526.
64. L. Fattore, *Biol. Psychiatry*, 2016, **79**, 539.
65. P. D. Skosnik, J. A. Cortes-Briones and M. Hajós, *Biol. Psychiatry*, 2016, **79**, 568–577, p. 568.
66. P. D. Skosnik, J. A. Cortes-Briones and M. Hajós, *Biol. Psychiatry*, 2016, **79**, 568–577, p. 571.
67. Royal Society, *Neuroscience, Conflict and Security, Brain Waves Module 3*, Royal Society, London, February 2012, vol. 1–64, p. 46.
68. J. M. Lakoski, *et al.*, *The Advantages and Limitations of Calmatives for Use as a Non-lethal Technique*, College of Medicine, Applied Research Laboratory, The Pennsylvania State University, October 3 2000, p. 4.
69. Royal Society, *Neuroscience, Conflict and Security, Brain Waves Module 3*, Royal Society, London, February 2012, p. 52.
70. J. R. Riches, *et al.*, *J. Anal. Toxicol.*, 2012, **36**, 647.
71. M. R. Dando, *Neuroscience and the Future of Chemical-biological Weapons*, Palgrave/Macmillan, Basingstoke, 2015, p. 113.
72. A. M. Rosenwasser, *Brain Res. Rev.*, 2009, **61**, 281.
73. L. E. Nelson, *et al.*, *Anesthesiology*, 2003, **98**, 428.
74. J. M. Lakoski, *et al.*, *The Advantages and Limitations of Calmatives for Use as a Non-lethal Technique,* College of Medicine, Applied Research Laboratory, The Pennsylvania State University, October 3 2000, p. 21.
75. European Working Group on Non-Lethal Weapons, in *Proceedings of the 3rd European Symposium on Non-lethal Weapons*, ed. L. Hess, J. Schreiberova and J. Fusek, Pharmaceutical Non-Lethal Weapons, Pfinztal, Frankfurt ICT, 2005, vol. 23, pp. 7–9.
76. L. Streda and J. Patočka, *Kontakt*, 2014, **16**, 57.
77. B. D. Waterhouse and D. J. Chandler, *Brain Res.*, 2016, **1641(Pt B)**, v–x, p. viii.
78. B. D. Waterhouse and D. J. Chandler, *Brain Res.*, 2016, **1641(Pt B)**, v–x, p. vi.
79. D. Weinshenker and P. V. Holmes, *Brain Res.*, 2016, **1641(Pt B)**, 320.
80. G. A. Zitnik, *Brain Res.*, 2016, **1641(Pt B)**, 338–350.
81. G. A. Zitnik, *Brain Res.*, 2016, **1641(Pt B)**, 338–350, p. 344.
82. E. L. Sabban, L. G. Alaluf and L. I. Serova, *Neuropeptides*, 2015, **56**, 19.

83. S. Terbeck, *et al.*, *Neurosci. Biobehav. Rev.*, 2016, **66**, 54.
84. R. Kirby, *The CBW Conventions Bulletin*, 2006, vol. 71, p. 1.
85. P. Aas, *Prehosp. Disaster Med.*, 2003, **18**, 306–312, p. 307.
86. G. Tremlett, *The Guardian Weekly*, 15 February 2013, pp. 26–27.
87. C. J. Langmead, *Br. J. Pharmacol.*, 2014, **171**, 281–282, p.281.
88. C. J. Winrow and J. J. Renger, *Br. J. Pharmacol.*, 2014, **171**, 283.
89. J. Yin, *et al.*, *Nature*, 2015, **519**, 247.
90. C. Alexandre, M. L. Anderman and J. E. Scanmell, *Curr. Opin. Neurobiol.*, 2013, **23**, 752.
91. J. Li, Z. Hu and L. de Lecea, *Br. J. Pharmacol.*, 2014, **171**, 332.
92. A. K. Graebner, M. Iyer and M. E. Carter, *Front. Syst. Neurosci.*, 2015, **9**, 111, p. 5.
93. A. K. Graebner, M. Iyer and M. E. Carter, *Front. Syst. Neurosci.*, 2015, **9**, 111, p. 2.
94. E. Miller, *et al.*, *BMJ [Br. Med. J.]*, 2013, **346**, f794 .
95. C. Tesoriero, *et al.*, *Proc. Natl. Acad. Sci. U. S. A.*, 2016, **113**, E368-377, p. 376.
96. Department of Defense, *Oxytocin: Improving Measurement Sensitivity and Specificity*, Defense Advanced Research Projects Agency, SBIR Solicitation, 2013.2, 2013, p. 1.
97. E. Choleris, D. W. Plaff and M. Kavaliers, *Oxytocin, Vassopresssin and Related Peptides in the Regulation of Behaviour*, Cambridge University Press, Cambridge, 2013.
98. S. Sharmay-Tsoory and L. J. Young, *Biol. Psychiatry*, 2016, **79**, 150–152, p. 150.
99. D. A. Baribeau and E. Anagnostou, *Front. Neurosci.*, 2015, **9**, 335.
100. R. Feldman, *et al.*, *Biol. Psychiatry*, 2016, **79**, 174.
101. A. Kelle, K. Nixdorff and M. R. Dando, *Preventing a Biochemical Arms Race*, Palgrave/Macmillan, Basingstoke, 2012, pp. 61–87.
102. B. J. Marlin, *et al.*, *Nature*, 2015, **520**, 499–504, p. 499.
103. B. J. Marlin, *et al.*, *Nature*, 2015, **520**, 499–504, p. 503.
104. L. J. Young and C. E. Barrett, *Science*, 2015, **347**, 825–826, p. 825.
105. D. Wei, *et al.*, *Proc. Natl. Acad. Sci. U. S. A.*, 2015, **112** 14084–14089, p. 14084.
106. J. Giordano, *Bulletin of the Atomic Scientists, Analysis: The Neuroweapons Threat*, 21 May 2016.
107. J. Flegr and A. Markoš, *Mol. Ecol.*, 2014, **23**, 5934.
108. S. A. H. Dass and A. Vyas, *Mol. Ecol.*, 2014, **23**, 6114–6122, p. 6114.
109. C. Poirotte, *et al.*, *Curr. Biol.*, 2016, **26**, 98.
110. K. Weinersmith and Z. Faulkes, *Integr. Comp. Biol.*, 2014, **54**, 93.
111. S. J. M. Moreau and S. Asgari, *Toxins*, 2015, **7**, 2385.
112. E. Lemichez and J. T. Barhieri, *Cold Spring Harbor Perspect. Med.*, 2015, **3**, 1.
113. V. Pitschmann and Z. Han, *Molecules*, 2016, **21**, 556.
114. R. A. Poldrack, *Neuroimage*, 2012, **62**, 1216.
115. G. Aston-Jones and K. Deisseroth, *Brain Res.*, 2013, **1511**, 1.
116. M. Heidenreich and F. Zhang, *Nat. Rev. Neurosci.*, 2016, **17**, 36.

117. J. R. Clapper, *Statement for the Record: Worldwide Threat Asssessment of the U. S. Intelligence Community, Senate Armed Services Committee*, U. S. Congress, Washington, D. C., 9 February 2016.
118. D. Owald, S. Lin and S. Waddell, *Philos. Trans. R. Soc., B*, 2015, **370**, 1.
119. F. D. Bene and C. Wyatt, *Dev. Neurobiol.*, 2011, **73**, 404.
120. Z. Liu, *Nature*, 2016, **530**, 98.
121. D. A. Relman, *J. Infect. Dis.*, 2014, **209**, 170.
122. D. Hambling, *New Sci.*, 2012, (2867), Technology News.
123. UK, *The Convergence of Chemistry and Biology: Implications of Developments in Neurosciences*, BWC/MSP/MX/WP.1, United Nations, Geneva, 12 July 2012, p.3.
124. Albania, *et al.*, C-21/NAT.3/Rev.3 at https://www.opcw.org/fileadmin/OPCW/CSP/C-21/national_statements/c21nat03_e_.pdf.
125. J. R. Walker, *The UK and Nuclear, Biological and Chemical Weapons: Arms Control and Programmes 1955-1974*, Ashgate, Farnham, 2012, ch. 3: Inappropriately hilarious, The UK and Incapacitating Agents.

CHAPTER 9

Advances in the Targeted Delivery of Biochemical Agents[†]

K. NIXDORFF

Darmstadt University of Technology, Department of Microbiology and Genetics, Schnittspahnstr. 10, 64287 Darmstadt, Germany
*E-mail: nixdorff@bio.tu-darmstadt.de

9.1 Introduction

Toxins and bioregulators are biochemical agent categories that are most relevant to convergence of biology and chemistry in the context of the Biological and Toxin Weapons Convention (BTWC) and the Chemical Weapons Convention (CWC). The successful application of bioactive materials for either therapeutic or hostile purposes depends to a great extent on how effectively they are delivered to their target. Effective delivery of a bioactive agent is also said to be the most difficult and crucial step in producing an effective bioweapon, and the one that would be most likely to fail.[1] However, new and improved ways of delivering such agents are already at hand and are being developed further at a rapid pace. Improved methods of delivery are most evident in experimental and clinical applications of biochemical therapeutics described in the open scientific literature. The

[†]This chapter is a revision and extension of Policy Paper no. 9 prepared by Kathryn Nixdorff for the Biochemical Security 2030 project: https://biochemsec2030.org/

Preventing Chemical Weapons: Arms Control and Disarmament as the Sciences Converge
Edited by Michael Crowley, Malcolm Dando and Lijun Shang

Published by the Royal Society of Chemistry, www.rsc.org

methods of targeted delivery of therapeutics to treat disease do not exactly mimic those that would be used for the delivery of biochemical agents as weapons. However, by examining these developments it is possible to come to some basic conclusions about the feasibility of delivering toxins and biochemical agents for terrorism or biological warfare purposes.

The two fields of work that have progressed most significantly and appear to be most relevant for both drug delivery and biochemical warfare are aerosol and vector-directed technologies, which are the focus of this study. Significantly, great strides are being made as a result of investments in drug development and delivery as part of cancer treatment, as well as gene and immuno-therapy. Nanotechnology has played a fundamental role in many developments in these areas, particularly through the construction of defined nanoparticles for facilitated uptake through the tissues. In addition, improvements in targeting and gene transfer efficacy of viral and non-viral vectors, along with progress in the protective packaging of these agents in nanoparticles that can also direct them to specific tissues, have made them much more feasible delivery systems. Clinical trials with humans have shown that several of the vectors already developed and armed to deliver specific payloads in cancer and gene therapy have proven to be successful in principle and in some cases in effect. Furthermore, the delivery of viral and non-viral vectors over the aerosol route is increasingly being explored so that this is rapidly becoming a feasible option.

The most sophisticated of these advances in science and technology are certainly not easy to put into practice, requiring extensive expertise (both scientific and tacit knowledge), well-equipped laboratories and substantial funds. Thus, State-supported actors are more likely than terrorists to have such means, although the possibility of hostile use by terrorists can by no means be ruled out. While the potential for misuse is certainly given, it is most difficult to assess just how actual the risk of misuse is in effect. Nevertheless, there still remains a need for proactive approaches to this issue area, which should involve the development and dissemination of criteria to help identify research of potential concern, communicate concerns to relevant communities, and where necessary to amend or expand upon existing governance systems.

9.2 Areas of Convergence of the CWC and the BTWC of Relevance for Targeted Delivery Systems

The BTWC and the CWC are two separate conventions; the conventions themselves are not converging, but rather there is an increasing degree of convergence of biology and chemistry within the two conventions. This convergence of the sciences of biology and chemistry involves both agents and technologies that are covered by both conventions.

9.2.1 Technologies of Relevance for the Targeted Delivery of Biochemicals

Nanotechnology has contributed greatly to improvements in aerosol delivery in this regard. Nanotechnology encompasses a wide range of practices in research and application of materials in the nanoscale dimension. Most relevant for consideration of targeted delivery technologies are nanoparticles ranging in size between 1 nanometer (a billionth of a meter, or around 10 times the size of an atom) and 100 nanometers (the size of large molecules).[2]

Nanoparticles in this size range can penetrate cell membranes and tissues much easier than larger particles. For this purpose, inhalable nanoparticles of defined size, form and surface charge have been developed that greatly facilitate uptake over the mucous membranes of the nasal passage and respiratory tract. A schematic representation of such nanoparticle design[3] is shown in Figure 9.1. With special design, these particles can also be made to cross the blood–brain barrier.[4] In addition, methods have been developed to protect sensitive microorganisms and other biological agents from the detrimental effects of environmental stress elements.[5,6]

These developments are crucial for the targeted delivery of therapeutics to desired locations in the body. However, the same methods can of course be misused to deliver harmful agents.

Systems biology is a further enabling technology for targeted delivery. Systems biology seeks, with the aid of *bioinformatics* to understand how complex physiological systems interact with one another and function as a whole. This is done by the integration of all functional information about biological systems through computer-assisted modelling.[7] For example, systems biology approaches have been used to study host–microbe interaction[8] and more specifically host–parasite interactions[9] to aid in understanding infectious diseases. Systems biology approaches have also been employed to study immune function,[10] and neuroscience.[11] This research yields an enormous amount of information about the regulation of vital physiological processes including, among others, respiration, heartbeat, body temperature, cognition, mood and immune responses. Furthermore, systems biology identifies specific targets in physiological systems for interaction with bioactive substances (bioregulators, see Section 9.2.2) that can generate either positive or negative physiological responses. The main aim of systems biology is to gain knowledge that will enable the steering of disease processes in a beneficial direction, that is, towards health. At the same time this knowledge also provides information about how to manipulate these processes in a detrimental way.

9.2.2 Agents of Relevance for the Targeted Delivery of Biochemicals

The biochemical agents that are representative of the convergence of biology and chemistry within the BTWC and the CWC are bioregulators and toxins.

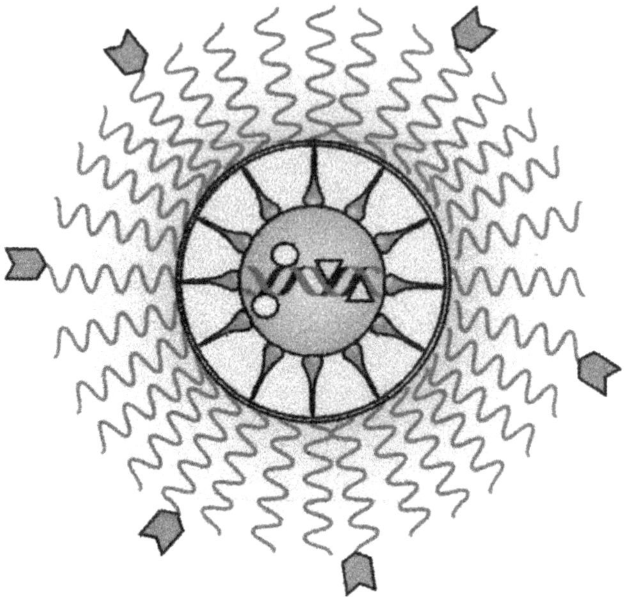

Figure 9.1 Schematic diagram of a nanoparticle designed for the targeted delivery of a bioactive substance. In this example, the bioactive substance in the center of the core (the payload) is composed of a DNA molecule to which peptides have been attached to facilitate transport into the nucleus of the host cell after release in the cytoplasm. Attached to the central core are membrane-active peptides that can destabilize endosomal membranes to facilitate release of the payload inside the host cell. The core has a cationic (positively charged) surface to which polar (negatively charged) flexible peptides are attached (squiggly lines). These serve as a scaffold to which targeting ligands are attached that can interact with host cell surface receptors to trigger uptake of the nanoparticle by receptor-mediated endocytosis. Image reprinted with permission from Macmillan Publishers Ltd: *Nat. Rev. Drug Discovery* (ref. 3), copyright 2006.

Bioregulators are biochemicals produced by the body that steer, regulate or to a great extent control physiological reactions such as those discussed under systems biology (Section 9.2.1). These substances include hormones (endocrine system), cytokines (immune system), neuropeptides and neurotransmitters (nervous system).[12–14] Normally, these substances are produced in optimal amounts to ensure the regulatory balance of physiological functions. However, if they are produced in amounts greater or less than optimal, this causes an imbalance of physiological responses that can be harmful or even lead to death.

A case in point is the famous mousepox experiment.[15] Researchers attempted to suppress a rodent plague in Australia by sterilizing female mice with a vaccine directed against their ova (eggs). To this end, the researchers genetically engineered a mousepox virus as a vector to carry a gene encoding a protein located on the surface of the ova. Mice that were immune to the mousepox virus were infected with the engineered virus vector and were expected to produce antibodies against their ova due to overproduction of the ova protein. As the resulting antibody production was suboptimal, the researchers added another gene to the mousepox vector encoding the cytokine interleukin (IL)-4, which is known to boost antibody responses. However, when mice were infected with this doubly engineered mousepox virus, they unexpectedly died, despite being immune to the wild-type mousepox virus. Apparently, overproduction of IL-4 blocked another arm of the immune system crucial for defence against virus infection, and the mice could no longer control the infection. This demonstrates how disrupting the balance of the production of a bioregulator can have devastatingly harmful effects.

Another class of cytokines that are relevant biological agents are the pro-inflammatory cytokines IL-1, IL-6 and tumour necrosis factor (TNF) alpha. When these cytokines are produced in moderate amounts in response to an infection, they induce mild inflammation reactions and contribute greatly to the healing process by recruiting immune defence cells to the site of infection. If they are produced in particularly large amounts or continually during chronic illnesses, this can lead to various disorders such as coronary insufficiency, thrombus formation, hypoglycemia, and in some cases even to shock and death.[16] It is significant in this respect that IL-1 was reported to be effective in aerosol form in pulmonary absorption studies carried out by the US Army under its medical research programme.[17]

Although the Soviet Union had signed and ratified the BTWC, it ran the largest offensive biological weapons programme of the twentieth century, which employed as many as forty to sixty thousand scientists, doctors, engineers and technicians in dozens of secret weapons facilities spread throughout the USSR.[18,19] Methods employing the newest developments in the fields of genomics, neurobiology, immunology and systems biology were applied in the most sophisticated research and development experiments ranging from attempts to render highly pathogenic bacteria resistant to multiple antibiotics, through to the creation of bacteria–virus and virus–virus chimeras, as well as the development of bioregulators as weapons to disrupt physiological functions. In the latter case, scientists were particularly interested in neuropeptides and one task involved the creation of a vaccinia virus carrying a gene for an endorphin (a peptide opioid produced by the body that has narcotic effects).[20] The aim was possibly to reduce pain or perhaps put a host that received the virus to sleep as a result of overproduction of the endorphin.

Many bioregulators are proteins or peptides, and although they are produced in very small amounts by the body, the methods of genetic engineering and synthetic biology make the production in relatively large quantities possible.[21,22] Bioregulators increase the spectrum of biological agents over that of the classical agents consisting of microorganisms and toxins.[23] The potential of the aerosol route to deliver bioregulators for therapeutic purposes is of great interest, and new technologies, especially nanotechnology, have made dissemination much more feasible.

Toxins are "extremely poisonous products of the metabolism (independently of their nature) of living organisms".[24] The discovery of toxins as pathogenic factors of microorganisms had its origins in the laboratory of Emil von Behring, working together with Shibasaburo Kitasato on diphtheria and tetanus in 1890.[25,26] Later, as toxins from additional living beings were discovered, they were classified according to origin: bacterial toxins, mycotoxins (fungal toxins), phytotoxins (plant toxins), and venoms (toxins from invertebrate or vertebrate animals).[27] There are in addition several toxins that originate from dinoflagellates, small eukaryotic organisms that are not plants, animals or fungi, but are rather placed in the group of the protists, a diverse cluster of eukaryotic organisms that include algae and protozoa. Some of these toxins are called marine toxins as they are produced by organisms found in salt water habitats. Several toxins of particular biological weapons potential are listed in Table 9.1.

Because these substances are toxic but not infectious, they are very similar to toxic chemical agents and are indeed regulated by both the BTWC and the CWC. However, the toxicity of some toxins is much more potent than most chemical weapons agents.[24] For example, botulinum toxin, abrin and

Table 9.1 Some Toxins of Biological Weapons Potential.[a]

Bacterial toxins
Anthrax toxins (*Bacillus anthracis*)
Protective antigen
Lethal factor
Edema factor
Clostridial toxins
Botulinum toxins A–G (*Clostridium botulinum*)
Epsilon toxin (*Clostridium perfringens*)
Cholera toxin (*Vibrio cholerae*)
Diphtheria toxin (*Corynebacterium diphtheriae*)
Shigatoxins/Verotoxins (*Shigella dysenteriae*, *Escherichia coli*)
Staphylococcal enterotoxin B (*Staphylococcus aureus*)
Marine toxins
Anatoxin-A (cyanobacteria)
Anatoxin-A(s) (cyanobacteria)
Microsystins (cyanobacteria)
Saxitoxins (cyanobacteria and dinoflagellates)
Conotoxins (cone snail)
Brevetoxins (dinoflagellate *Karenia brevis*)
Ciguatoxins (dinoflagellate *Gambierdiscus toxicus*)
Tetrodotoxins (puffer fish)
Plant toxins
Abrin (*Abrus precatorius*)
Ricin (*Ricinus communis*)
Mycotoxins (fungal toxins)
Trichothecene
Mycotoxin T-2
Myrotoxin B

[a]Designations in parentheses are the sources of the toxins.

Table 9.2 Comparison of Toxin (TW) and Chemical (CW) Weapons Agents.[a]

Type	Substance	LD_{50}[b] μg kg^{-1} i.p.	Heat stability	Uptake over Inhalation	Uptake over Skin
TW	Botulinum	0.001	±	++	−
	Ricin	3.0	−	++	−
	Abrin	0.7	±	++	−
	Mycotoxin T-2	1200	+	++	+
	Tetrodotoxin	8.0	+	++	+
	Saxitoxin	10	−	++	+
CW	Sarin	168[c]	+	++	±
	VX	20[c]	+	++	+

[a]Values from ref. 24, 28, 29 and 30.
[b]Doses lethal for 50% of those affected (mice after intraperitoneal (i.p.) injection of the toxin).
[c]Doses lethal for 50% of those affected (rats after subcutaneous injection of the toxin).

ricin are considerably more toxic than the most toxic chemical agent VX[28] (Table 9.2).

Some toxins, such as the botulinum toxins abrin and ricin, are protein in nature. In general, the protein toxins are unstable to heat. Exceptions are the relatively heat stable botulinum toxins and abrin. Because these toxins are highly toxic, even after a heavy loss of activity through their dissemination they would potentially still have a large effect. The problem of stabilization of the toxins can in part be solved by new methods of microencapsulation and nanotechnology processes. Other toxins such as the tricothecene toxins mycotoxin T-2 or myrotoxin B and the marine toxins anatoxin A and saxitoxin are not proteins and are therefore relatively heat stable. They are, however, much less toxic than the protein toxins (Table 9.2). The method of action of the toxins is very diverse. Some exercise their effects on the nervous system (anatoxin A, botulinum toxin, saxitoxin) while others inhibit protein biosynthesis (ricin, shigatoxin, thricothecene). Although one talks about toxins in the singular (*e.g.* botulinum toxin), every toxin molecule has in reality several variants (*e.g.* botulinum toxins A–G).[29,30] The toxicity of most toxins is much higher when inhaled than when administered over other routes (oral, intravenous, intramuscular, intradermal, subcutaneous), although aerosol dissemination of some toxins is limited because of their instability when released into the environment.[24] Nevertheless, advances in microencapsulation techniques for the aerosol delivery of drugs[31] show great promise for stabilizing labile toxins as well as other biological agents for effective aerosol dissemination.

9.3 Aerosol Delivery

As mentioned earlier, the aerosol route has always been the classical method of choice for delivering biological agents over wide areas. Aerosols are particles in the form of a liquid or a powder that are suspended in air and can be inhaled. Many infectious microorganisms can enter the body through the mucous membranes lining the nasal and respiratory tracts as well as the intestinal tract. The size of the droplets determines to a great extent where they will be deposited in the airway after inhalation. Particles up to 5 micrometers in diameter can reach deep lung areas (alveoli or air sacs); larger particles will be deposited in more anterior parts of the respiratory tract.[32]

9.3.1 Aerosol Delivery of Microorganisms

The aerosol method of delivery of biological agents has a long history beginning with the development of aerobiology for dissemination of microorganisms and toxins between World Wars I and II.[33] A number of major States developed extensive weapons programmes during those years. Allied nations concentrated mainly on pathogen identification and protection, modes of transmission, infection, and packaging for delivery and dissemination. A wide variety of agents directed against humans, plants and animals were studied, but most offensive work focussed on anthrax, botulinum toxin and methods of preparation and delivery, including the production of anthrax and botulinum bombs.[34] Under the direction of Paul Fildes, an eminent microbiologist in the UK, the British at the Chemical Defence Experimental Station at Porton Down produced five million cattle cakes laced with robust anthrax endospores that were to be dropped from the flare chutes of bomber aircraft over German farming land in order to wipe out cattle and thus deal an economic blow to Germany. Using experience gained through chemical warfare in WWI, Fildes believed that the most effective way of disseminating bacteria would be by aerosol. Work began on an apparatus that could produce clouds of bacteria to be inhaled by experimental animals in order to determine the required doses. It was also shown that a bursting munition could be used to create the aerosol. This was tested against sheep for safety reasons on Gruinard Island off the north-west coast of Scotland, using anthrax endospores as agent. The weapon proved to be more powerful than any known chemical weapon. Gruinard Island was eventually decontaminated from long-lived anthrax endospores in 1986.[35,36] Once again, the Allies did not deploy these weapons in WWII, as they were meant to be used only in retaliation of an Axis attack with biological agents.

The US clearly had an extensive biological weapons programme during and after the Second World War.[34,37,38] In addition to studies concerning protective measures, a wide array of highly infectious microorganisms and extremely poisonous toxins were studied along with several anti-crop and anti-animal agents. The utility of biological agents disseminated *via* aerosols was tested in the laboratory (small scale), in closed chambers (medium scale) and in the open air (large scale). Tests were carried out in the anti-personnel programme both with simulant microbes and with pathogens. With regards to munitions development, the US started with adaptations of the burster-type bombs developed by Britain during the war, but soon extended its programme to include "submunitions, gas expulsion bombs, various types of line spray tanks and highly specialised projectiles and generators as well as insect vectors".[39] The US programme did not, however, weaponise contagious agents.[40]

The UK had largely ended its offensive biological programme by the late 1950s and the US renounced its offensive programme in 1969, but defensive aerosol testing continued.[41] In the early years of the Cold War, the UK and the US performed numerous aerosol field tests on animals with pathogens sprayed from aircrafts, ships and land generators. Although many technical details remain unavailable, the tests were judged to be successful, partially successful, or a failure,[42] success meaning that at least many animals became infected.[41]

A pertinent example[43] of the efficacy of aerosol dissemination of biological agents involved the use of the bacterium *Bacillus thuringiensis* (which produces an insect toxin) to control insurgent populations of the European gypsy moth, which was posing a threat to the lumber industry in and around the Victoria region of British Columbia. This study demonstrates just how effective the aerosol route can be for delivering microorganisms successfully over a large area, even without using sophisticated technology. The report describes the results of aircraft spray application of the biological insecticide Foray 48B (a solution of *B. thuringiensis* endospores) over approximately 30 000 acres including residential and rural areas in the Victoria, B. C. region. The undertaking resulted in greater than 99% mortality of the gypsy moth population in that area. A surprising finding was that enough small (2–7 micrometers) droplets were formed that could penetrate houses and contaminate the nasal passages of residents inside their homes, even though the equipment used was designed to generate droplets of 110–130 micrometers in diameter. While exposure to *B. thuringiensis* should have no detrimental effects on humans, this tells us something about how effective a similar type of operation using highly infectious microorganisms could be.

9.3.2 Aerosol Delivery of Vaccines

A further use of aerosol delivery of biological agents concerns vaccination with microorganisms *via* the aerosol route, which has been practiced for quite some time and is known to be effective. Vaccination *via* aerosol delivery has several advantages over the traditional methods of subcutaneous (beneath the skin) or intramuscular injection using needles and syringes. The drawbacks to these conventional methods include limited acceptance to the use of needles, transmission of diseases through needle pricks and the need for trained healthcare workers.[44] Field trials in Mexico established the effectiveness of mass immunization of children with the measles vaccine virus *via* the aerosol route.[45] The children were exposed to the aerosol output of a classic jet nebulizer driven by an air compressor for a 30 seconds period *via* a paper conical mask held over the mouth and nose. Subsequent tests showed that this type of vaccination compared favourably with that of conventional methods of administration and used a third of the dose normally required. In the former Soviet Union, thousands of people were successfully vaccinated with aerosols of live, attenuated strains of anthrax, plague, tularemia and smallpox agents using tent-exposure systems.[46,47] Although the aerosols were produced in an enclosed environment (direct *via* a mask or in a tent), these practical experiments demonstrated that vaccination against a wide array of biological weapons-relevant microorganisms could be achieved by inhaling aerosol clouds containing the agents.

Aerosol delivery of vaccines has the added advantage that direct delivery to the mucous membranes of nasal and lung surfaces is achieved, where many infections begin. Among the various mucosal sites, nasal delivery is especially attractive for vaccine administration because this is a site of relatively high tissue permeability, low activity of destructive enzymes and the presence of a considerable number of immune response cells.[48] In order to avoid the risk of reversion to the active state of attenuated microorganisms used as vaccines as well as to reduce possible side effects that some whole cell vaccines cause, the use of sub-unit vaccines consisting of only the immunogenic protein components of microorganisms instead of whole cells has gained interest in recent years. Particularly, packaging these components for aerosol delivery into nanoparticles offers improved absorption and uptake. In addition, a more effective antigen recognition and response by immune cells in the mucosa would be achieved.[46] At the same time, these techniques could be misused to deliver biological agents that have the potential to exert detrimental effects on the immune system.

9.3.3 Advances in the Aerosol Delivery of Therapeutics and Biochemical Agents to Specific Sites in the Body

Systems biology is rapidly providing knowledge about the regulation of vital physiological processes through bioregulators such as hormones, peptides, neurotransmitters and cytokines, along with the targets with which they interact. This work is being carried out with the aim of modulating disease processes in the direction of better health. There is, however, increasing concern that such knowledge can be used to pinpoint ways of disrupting the normal functions of vital physiological processes such as respiration, heartbeat and immunity, or affecting mood and cognizance negatively. The potential of aerosols for delivery of drugs is a current area of particular interest, and here nanotechnology plays a central role in improving the aerosol delivery of bioactive substances:

> "A major challenge in nanomedicine is to engineer nanostructures that can efficiently encapsulate drugs at high concentration, cross the cell membrane, and controllably release the cargo at the target site over a prescribed period of time."[49]

Delivery of therapeutics *via* the aerosol route is attractive for a number of reasons. The surface area of the lung is between 80 and 140 square meters. Also, the alveolar (air sac) epithelium (cell lining) in most pulmonary regions is only about 0.1–0.2 micrometers thick, and the distance between the epithelial surface and the blood is much less than it is in the bronchial system, which should facilitate drug uptake.[32] There are, however, a number of absorption barriers in the human lung including the mucus layer, the alveolar lining fluid layer, and competing uptake pathways such as particle engulfment by macrophages (white blood cells that can take up and destroy foreign particles).

While lipophilic (having an affinity for fat) substances are readily absorbable over the nasal mucosa, more polar (hydrophilic, having an affinity for water) compounds such as peptides and proteins, or DNA, are taken up relatively poorly, so that methods aimed at improving their permeability properties have been developed. In this context, packaging drugs into nanoparticles that are coated with cationic (positively charged) substances such as chitosan (a polysaccharide derived from shellfish chitin), polymeric nanocarriers such as poly(lactic acid) or poly(lactic-*co*-glycolic acid), or a combination of these substances, greatly improves uptake. In addition, encapsulation of the particles with polyethylene glycol or polyoxyethylene derivatives can increase their stability. Further improvements have been achieved by cross-linking chitosan with tripolyphosphate in order to increase the release time of encapsulated peptides and proteins or enhance the gene expression of DNA-based vaccines[48] and competing uptake of the particles by phagocytic cells can be reduced by outfitting nanoparticles with self peptides that can specifically inhibit phagocytosis.[50]

As in vaccine administration, the nasal route has emerged as being particularly advantageous for the delivery of drugs. This route also has the added potential of providing direct access of drugs to the brain, and many peptides and proteins (among others: Orexin-A, insulin, leptin, erythropoietin) can be detected in the central nervous system after intranasal delivery.[51] Orexins (or hypocretins) are neuropeptides that stabilise the waking condition, and it has been suggested that:

> "Clearly, if it were possible to interfere with the function of the orexins then there would be available to those with malign intent an impressive means of incapacitation."[52]

Uptake after nasal delivery is achieved by various methods: absorption into olfactory blood vessels and entry into the general circulation; absorption into olfactory lymphatic vessels draining to the deep cervical lymph nodes of the neck; and extracellular diffusion or convection in compartments associated with olfactory nerve bundles and entry into the cranial compartment.[53]

Again, packaging bioactive materials into nanoparticles that have been constructed with absorption-enhancing substances such as chitosan can facilitate uptake *via* the nasal route. Such constructs may also be used to enable biological agents present in the circulation to cross the blood–brain barrier. Normally, the brain is protected from the potentially harmful effects of most substances or cells in the circulation by the extremely tight junctions between the endothelial cells lining blood capillaries. This barrier has been called a double-edged sword:

> "On the one hand, this cellular interface helps to maintain a constant, optimal environment for neuronal function through a combination of barriers and selective transport systems that regulate the passage of wanted and unwanted molecules. But on the other hand, it presents a formidable challenge to medicine because it stops most drugs from passing from the bloodstream to the brain."[54]

The mechanism of absorption enhancement by chitosan and other polycation (positively charged polymers) substances appears to be a combination of adhesion (sticking to tissues) and a transient opening of the junctions in epithelial cell layers lining the mucosal surface of the nasal compartment and respiratory tract.[55] This is also relevant for the delivery of drugs across the blood–brain barrier, as the cancer drug doxorubicin was able to cross the intact blood–brain barrier when attached to nanoparticles coated with polysorbate, another absorption enhancer.[2]

Nanotechnology can be applied in the aerosol delivery of drugs to fight pulmonary infectious diseases such as pneumonia, tuberculosis or fungal infections. For example, nanoscale delivery systems can enhance the absorption of drugs and uptake through the epithelial layer of the mucosa, target the drugs to specific cells/tissues/organs and release them in a controlled manner in response to a specific stimulus.[4] The drugs can be protected from degradation and their release in the body controlled by coating the nanoparticles with substances such as poly(lactic-*co*-glycolic acid) (PLGA) or polyethylene glycol (PEG). Indeed, changing the degree of surface PEGlycolation, of either the bare drug or encapsulating particles can influence the rate of particle degradation and can be manipulated to favor longer release; similar effects can be achieved with liposomes (vesicles of fatty molecules that can enclose bioactive substances).[56]

Several examples of clinical applications and experimental studies have shown that aerosol delivery of drugs and other bioactive biochemicals is feasible not only in principle, but also in effect. Inhaled insulin delivery has been explored for over a decade, and some formulations for inhalation administration have been marketed and found to be effective in the past but have since fallen from the market. Recently, a powdered formulation of the drug, Afrezza ("Technosphere insulin"), has been approved by the US Food and Drug Administration and will be developed and manufactured by MannKind in partnership with Sanofi for the treatment of either Type I or Type II diabetes.[57]

Another example of successful aerosol delivery of a drug concerns the neuropeptide oxytocin, which was reported to increase trusting behaviour in humans given a single dose by nasal spray.[58] It has even been marketed by

Vero Labs as a liquid spray to be used like perfume, "formulated to emphasize its key role in human bonding to improve confidence, enhance relationships and strengthen bonds".[59] A growing number of studies have investigated the effects of oxytocin after nasal administration on human behaviour, cognition and brain activation, particularly in connection with specific disorders such as anxiety, autism and schizophrenia.[60] Churchland and Winkielman have pointed out the difficulties in assigning a specific role to oxytocin in the complex realm of social cognition and that:

> "it may turn out that the best clinical use of intranasal OXT is primarily as an effective (and perhaps non-addictive) anti-anxiety medication".[61]

Other clinical studies on the aerosol administration of drugs have reported further progress in using this route for the treatment of *Pseudomonas* infections in cystic fibrosis patients.[62]

Over the past 20 years, DNA/RNA nanotechnology has revolutionized drug delivery by using "a simple and well-studied molecule as a building block to develop more complex bioactive and versatile systems to control cellular function".[63] In this regard there has been a great deal of interest in using the gene silencing RNA interference (RNAi) system therapeutically. This can be a potent, effective and practical method of interfering with or silencing the expression of unwanted gene activities. Effectors of this method of gene silencing are, among others, short, (21–26 nucleotides) interfering RNA molecules (siRNA). These are recognized by a so-called silencing complex (RISC) which mediates the degradation of gene transcripts (specific messenger RNA expressed by that gene). This essentially ablates, turns off or "knocks-down" the activity of that specific gene since no product can be synthesised from that degraded gene transcript.[64] A multitude of new formulations are being investigated that package siRNA into nanoparticles suitable for uptake by cells, and some studies have used the nasal route for delivery.[65] Recently, a double-blind, placebo-controlled clinical trial of a siRNA-based therapeutic directed against a respiratory syncytial virus and delivered by nasal administration showed that this method has therapeutic activity. It represents the first proof-of-concept efficacy test of the therapeutic effect of RNAi in humans.[66]

9.3.4 Advances in Aerosol Delivery Devices and Techniques

Aerosol delivery of biological agents for experimental and therapeutic clinical applications is increasingly becoming the preferred and most relevant route, but also the least straightforward.[65] Due to the advantages of using the aerosol route for the delivery of biologically active substances, a great deal of research into the development of aerosol delivery devices for therapeutic purposes has been carried out in the last two decades. A huge increase in activity occurred in the 1990s after the Montreal Protocol banned chlorofluorocarbons as propellants. This resulted in the development of hydrofluorocarbon-driven metered dose inhalers (MDIs) as well as dry powder inhalers

(DPIs) for aerosol drug delivery. In the most recent phase, engineered porous powders of low density have been developed that can achieve an increased lung deposition of 40–60% at present, compared with 5–20% in the past. Dry powder inhalers (DPIs) can deliver higher doses and are therefore particularly useful for one-dose delivery programmes. While MDIs can deliver doses in the range of 1 mg or less, DPIs can deliver loads of 50 mg of powder per inhalation.[67]

Nebulizers or atomizers are also used to create aerosols. These are devices that pump air or oxygen through a liquid to produce a mist, which is inhaled through a face mask. A classic jet-type nebulizer driven by an air compressor was used in the successful measles virus aerosol vaccination programme in Mexico discussed in Section 9.3.2. Newer devices use a vibrating mesh or ultrasound systems for generating aerosols of fine particle fractions of a precisely controlled size.[68] They can be applied to deliver a wide variety of drug formulations including highly viscous fluids, proteins, peptides, surfactants and DNA for gene therapy purposes. Indeed, advanced aerosol technologies "continue to expand therapeutic options".[69]

Currently, emerging inhalation therapies applying these delivery devices are being used for the treatment of various illnesses by delivering biologically active substances. These substances include antibiotics for the treatment of respiratory bacterial infections, a neuraminidase inhibitor for influenza, anti-pulmonary arterial hypertension agents, cyclosporine for transplant rejection, cytokines such as interferon beta (IFN-β) for the treatment of asthmatic symptoms and chronic obstructive pulmonary disease (COPD) as well as infectious diseases and alpha 1 antitrypsin for the treatment of cystic fibrosis.[70] Experiments consisting of aerosol delivery of biological agents to animals have employed different types of aerosol chambers depending on the animal type. In the case of mice, nose-only delivery systems (Figure 9.2)[71] are most frequently used,[72] while for larger animals a Henderson-type aerosol chamber (Figure 9.3) is the apparatus of choice.[73] In some cases, such as for non-human primates, a Henderson head-only system is used.[74–76] In this type of apparatus, a non-human primate is anesthetized and placed on a platform in a supine posture, with the head placed through an opening in a type of dental dam covering the round opening on the side of the aerosol chamber.

9.3.5 Feasibility of the Aerosol Delivery of Biochemical Agents as Weapons

The most prominent example of the feasibility of the aerosol delivery of drugs is the incident in which Russian military special forces tried to rescue hostages held at the Moscow Dubrovka Theatre Center by introducing an unidentified "gas" (supposed to have incapacitating effects) into the theatre ventilation system. Of the 800 hostages held in the theatre, 127 died and more than 650 of the survivors required hospitalization in 2003.[77] Many of the patients had classic signs of opioid (narcotic) intoxication, and the Russian Health Minister announced several days later that a derivative of the opioid fentanyl had been used. A subsequent report[78] from British investigators

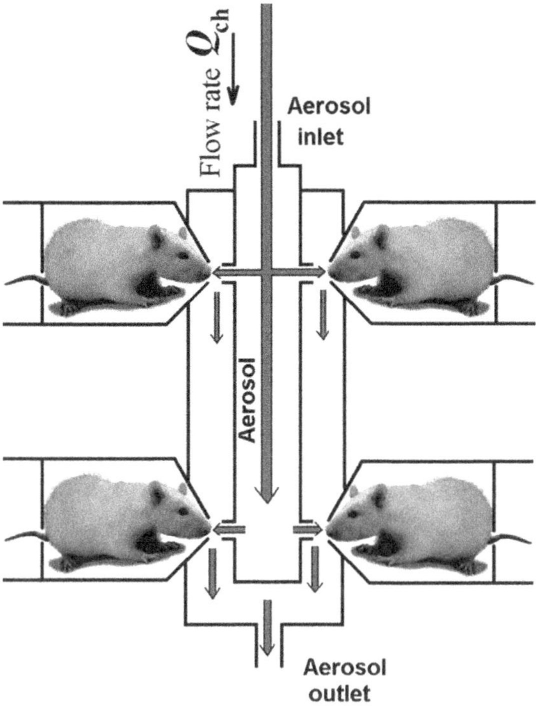

Figure 9.2 Schematic diagram of a nose-only aerosol apparatus. Image reprinted from *J. Aerosol Sci.*, **78**(41), A. A. Onischuk, T. G. Tolstikova, A. M. Baklanov, M. V. Khvostov, I. V. Sorokina, N. A. Zhukova, S. V. Ankov, O. V. Borovkova, G. G. Dultseva, V. V. Boldyrev, V. M. Fomin and G. S. Huang, Generation, inhalation delivery and anti-hypertensive effect of nisoldipine nanoaerosol, 41–54, Copyright 2014, with permission from Elsevier.

provided evidence that a mixture of two anesthetics, carfentanil and remifentanil, were the fentanyl compounds used.

Shoyele and Slowey[79] have produced a list of 15 peptides/proteins that could feasibly be administered over the aerosol route. This list includes among others several cytokines, erythropoietin, calcitonin, insulin, amylin and the growth hormones. Nanotechnology has made aerosol delivery of biochemicals much more feasible. This includes new strategies of nanoparticle design that enable directing the bioactive substances to specific sites of disease as well as protecting and stabilizing sensitive agents to preserve their bioactivity until they have had a chance to exert their effects. Several experimental and clinical applications of the aerosol delivery of therapeutic biochemicals provide proof of principle.

Although the methods of the aerosol delivery of drugs for clinical and experimental purposes are all designed to present the most concentrated agents directly to the subject, and thus do not mimic aerosol delivery in a biological weapons type of scenario, data gained from these experiments can provide a calculated estimate of the feasibility of the delivery of biological

Figure 9.3 An example of a Henderson-type aerosol apparatus. The aerosol generating equipment is contained in a BSL-3 cabinet (right half of the figure) while the equipment for controlling air flow and relative humidity is separate. This image was taken from J. Allen, *Tradeline*, 2007, https://www.tradelineinc.com/reports/2007-5/hpas-precision-aerosol-exposure-and-containment-technology. Photo by the Health Protection Agency (UK), reprinted with kind permission from Tradeline, Inc., Orinda, Calif. (USA).

agents for terrorist or warfare purposes. For example, researchers experienced in the aerosol delivery of ricin to non-primate humans using a Henderson head-only system have estimated that:

> "An aerosol cloud of micron-sized particles containing high concentrations of ricin toxin, *even in the event of significant atmospheric dilution*, could deliver a potentially lethal dose to victims in a military battlefield or civilian terrorist scenario".[74] (Emphasis added.)

In general, the feasibility of delivering toxins *via* aerosols depends on several factors including environmental stability and ease of production, but the most important factor is toxicity.[80] Based on a mathematical model and tested by field trials, it has been estimated that:

> "...a toxin with an aerosol toxicity of 0.025 mg kg^{-1} would require 80 kg of toxin to cover 100 km^2 with an effective cloud exposing individuals to approximately a lethal dose..."[81]

9.4 Viral Vector Delivery

Advances in molecular biology, immunology and tumor genetics have led to the design of novel viral vectors for use in vaccine therapy, cancer, drug and immunotherapy. In general, these viruses act as ferries or vehicles that carry and deliver foreign genes to the body. The strategy of this procedure is that infection with the modified virus would lead to the expression of the foreign genes in the cells of affected tissues. This would result in the synthesis of the selected active substance (the gene product), which can then exert its effect.

In a biomedical and clinical application context these are perfectly acceptable undertakings that can counter disease and promote health, but the dual-use implication here is that these same technologies could be used to arm viruses with a destructive or even deadly payload. The use of viral vectors may seem to be fairly straight-forward in principle, but there are several obstacles to success which have to be overcome and have been the subject of intensive research and trials, including directing the viruses to the right cells, and improving gene transfer efficiency and gene expression. Nevertheless, such progress has been made that it has prompted the prediction that:

> "...in spite of naysayers lacking vision, ... there is now firm hope that gene therapy will soon do for medicine what aeroplanes did for transportation".[82]

Clinical trials with humans have shown that several of the vectors that have already been developed and armed to deliver specific payloads in cancer and gene therapy have proven to be successful in principle, and in some cases in effect. In clinical trials with metastatic melanoma patients, an engineered vaccinia virus armed to deliver an immunostimulatory substance to boost anti-tumor responses showed that the virus could in effect successfully deliver its package to selected tissues.[83] A vaccinia virus enhanced for tumor selectivity has also been armed with a pro-drug activation system, which has been termed "suicide gene therapy".[84] In this case, the virus delivers a gene encoding a non-toxic yeast enzyme that is converted to its highly toxic form when the gene is expressed in tumor cells. The immunogenic capacity of the modified vaccinia virus Ankara strain, as well as its high level of safety and ability to incorporate relatively large foreign gene inserts, has made it an attractive candidate for clinical vaccine development strategies against a variety of infectious diseases.[85] However, the viruses most used in recent years as vectors for drug delivery in experimental and clinical applications have been adenoviruses, adeno-associated viruses and lentiviruses.

9.4.1 Viral Vectors Most Suitable for the Delivery of Biochemical Agents

9.4.1.1 Adenoviruses

It has been said that adenoviruses (AV) have "long been the workhorse of virotherapy".[86] AV can cause mild respiratory (flu-like) and ocular infections, but most persons with an intact functioning immune system recover without any treatment. AV have long been a popular viral vector due to their large host range as well as their ability to achieve efficient transgene expression (activation of the foreign gene that has been transferred by the virus) in both replicating and non-replicating cells, to accommodate relatively large foreign genes, and to deliver their payload without integrating into the host cell genome, which can cause detrimental mutations. Although adenoviruses have a fairly large host range, the range of cell types that adenoviruses can infect has been broadened by engineering the virus to contain new surface proteins. This is what is known as changing the tropism of a virus, so that it can bind to and be taken up by cells that it does not normally infect.[87]

AV have been used primarily in the treatment of cancer, and in this regard, several types of modification of the virus confer specificity in attacking cancerous cells. Most promising has been the delivery of a gene encoding for granulocyte–macrophage colony stimulating factor (GM–CSF), a cytokine that promotes stimulation of anti-tumour immune responses and recruiting of phagocytic cells to the tumour site that can induce strong inflammatory responses that in turn can potentially damage tissues. Indeed, GM–CSF has become the leading cytokine in the virotherapy field.[86] Compared to other vector systems, such as lentiviruses and adeno-associated viruses, the clinical efficacy data has, in general terms, been disappointing. Apparently, this may be due to several reasons:

> "...including longevity of transgene expression using conventional...vectors and the prevalence of preexisting immunity to HAdV-5 [the most frequently used strain of AV] in the population".[88]

However, advanced adenoviral vectors have been developed that have achieved considerably longer transgene expression duration than first-generation adenoviral vectors.[89] One major disadvantage with the use of adenovirus vectors for therapy is the fact that they induce potent immune responses after delivery. On the one hand, this is desirable, as the immune response helps to contain the viral infection. However, on the other hand, this can also be detrimental by limiting the therapeutic efficacy. This problem has been difficult to solve, but advances in strategies to suppress the immune response have shown promise.[86] Nevertheless, adenoviral vectors are still commonly used viral vectors in clinical studies.[90] That AV vectors can successfully deliver their payload when administered over the aerosol route has been demonstrated recently in connection with genome editing technology[91] using the Clustered Regularly Interspaced Short Palindrome Repeats (CRISPR)/CRISPR-associated (Cas) endonuclease (CRISPR/Cas) system. The CRISPR/Cas system basically consists of two components: a guide RNA (gRNA) and a non-specific CRISPR-associated endonuclease (Cas). The gRNA is a combination of a synthetic RNA sequence needed for binding the Cas endonuclease and a targeting sequence homologous to the genomic target to be modified. The genomic target of Cas is thus defined by the targeting sequence present in the gRNA. By varying the strategy, different modifications such as mutations, deletions or insertions can be achieved at the target site. In this regard, researchers aiming to produce a mouse model of malignancy by introducing specific chromosomal rearrangements used an AV vector to deliver CRISPR/Cas9 components to mice *in vivo via* an aerosol (intratracheal instillation). The CRISPR/Cas9 components successfully induced the desired specific chromosomal rearrangements in somatic cells of the animals.[92]

9.4.1.2 Adeno-associated Viruses

Adeno-associated viruses (AAV) are small viruses of the parvovirus family that are found mixed with adenoviruses, as the name implies. This association with adenoviruses is practical, as AAV are defective and need a so-called helper virus (the AV) to supply components necessary for gene expression and replication (reproduction). Because AAV can infect cells but not replicate, they are considered safe vectors as they are non-pathogenic.[93] They also have the ability to transduce (infect host cells and release their payload) both dividing and non-dividing cells. Although AAV vectors have the potential to promote stable long term therapeutic gene expression, a major drawback for gene therapy, that they share with AV vectors, is the development of strong anti-viral, immune responses against the viral capsid proteins (proteins making up the outer covering of the virus).[94] To deal with this problem, many methods are presently being tried, including changing the serotype of the virus, introducing new capsid proteins, modulating the immune response or inducing tolerance.

Although early successes in phase I/II clinical trials for a broad spectrum of AAV vectors have been reported, both pre-clinical and clinical trials with first generation vectors have highlighted the problems of immunogenicity and gene transfer efficacy. However, an array of second generation AAV vectors, engineered through rational and combinatorial approaches to address these problems are now available, and new lead AAV vector candidates are poised for clinical translation.[95] For example, a combination of approaches has been described that involves the modification of particular amino acid residues in the capsid proteins exposed on the surface of the virus along with modifications to the genome designed to achieve high efficiency transduction at lower doses. Apparently, this strategy produced a six-fold increase in transduction efficiency over first generation AAV vectors along with a diminished immune response against one recombinant.[96] Other researchers have also achieved similar increases in the transduction efficiency of AAV vectors by modifying capsid protein residues.[94] Long-term expression of transduced genes at therapeutic levels has been achieved in patients with inherited retinal disorders and haemophilia B, and recent results have raised hopes for the treatment of many other diseases.[97] Indeed:

> "With the many successes of AAV gene therapy in the last 50 years, key discoveries about virus biology, improved methods of vector production, and positive effects seen in long-term clinical trials, it remains clear that there is always room for improvement. Ongoing challenges to evade neutralizing antibody responses to treatment are still being met with novel capsid generation as well as improved vector delivery."[98]

AAV vectors have also been used to deliver CRISPR/Cas components *in vivo* to mice; the *in vivo* route of administration was intravenous injection through the tail vein.[99,100]

Although stable transgene expression primarily results from extrachromosomal AAV vector genomes, there have been reports of integration of AAV into the host genome, so that their safety for gene therapy purposes is a matter of some concern.[101]

9.4.1.3 Lentiviruses

Much work has been invested in the development of lentivirus (the subfamily of retroviruses to which the AIDS virus belongs) vector delivery systems, as these viruses are very efficient in infecting even non-dividing cells and achieving stable expression of the transferred genes in those cells, as well as being only weakly immunogenic.[102] Although lentiviruses have a very narrow host range, this can be broadened or altered by pseudotyping, which involves exchanging the surface proteins of particular strains of viruses during packaging of the virus. Other innovative approaches include outfitting the virus with targeting ligands fused to virus surface proteins to infect specific cells as well as using tissue-specific gene regulating promoters to restrict gene expression to certain target cells and reduce the risk of gene-induced immune responses to new proteins expressed on the surface of targeted cells.[103]

The great promise of lentiviral vector development for clinical use is dampened by the fact that they are retroviruses that integrate randomly into the genome of the host, which could cause detrimental mutations as has been shown in the past occurrence of vector-related leukaemia in a significant number of the children treated to correct severe combined immunodeficiency (SCID).[104] On the other hand, the property of retroviruses to integrate into the host genome has the advantage of potentially long-lived expression of the delivered gene due to its stable insertion. If not integrated into the host genome, the virus genome exists as an episome that frequently becomes lost in the host cell population. Third generation lentivirus vectors have been designed for improved safety as well as performance.[105] For example, integration-deficient lentiviral vector systems with mitotically stable episomes (non-integrated virus genomes that persist in the host cell population) capable of long-term transgene expression have been developed.[106] Although safety concerns are primary for therapeutic use of lentiviruses as vectors, the safety factor would be less relevant for the use of lentivirus vectors as biological weapons. Lentiviruses have also been used to deliver CRISPR/Cas9 genome editing components to human T cells in order to precisely remove HIV-1 (the AIDS virus) that had integrated into the host cell genome. The positive results as well as

> "the complete absence of genomic and off-target functional effects in all assays also provides critical support for the promise of developing this approach for future therapeutic applications."[107]

There is a much interest in developing vectors in combination with RNA interference (RNAi), as described in Section 9.3.3, to silence or block the expression of genes, thus preventing the synthesis of the gene products. Its use in combination with genetically engineered lentivirus, facilitates high efficiencies of small hairpin RNA (shRNA) delivery and/or integration into genomic DNA for stable shRNA expression.[108] Recent studies have shown selective delivery of a lentivirus-based shRNA to macrophages in the lungs of mice by intratracheal administration. The shRNA was specific for targeting a transcription factor that regulates the production of pro-inflammatory cytokines by macrophages. This treatment was successful in modulating the inflammatory response to a significant degree.[109]

9.4.2 Feasibility of Using Viral Vectors for the Delivery of Biochemical Agents as Weapons

In most research and clinical studies viral vectors are administered by injection, in some cases using repeated application, which would not be practical for their delivery as biological weapons. However, some studies have indicated that administration over natural routes such as inhalation is feasible. For example, treatment of cystic fibrosis patients by inhalation of an adeno-associated virus vector engineered with a gene to deliver the transmembrane conductance regulator, which is defective in cystic fibrosis, has resulted in "encouraging trends in improvement in pulmonary function",[110] with new advances in genetic therapies for cystic fibrosis underway.[111] In some 20 clinical trials that have been carried out, use of gene-transfer agents, including adenovirus and adeno-associated virus, have demonstrated "proof of principle for gene transfer to the airway".[112]

It was further shown that lentiviral vectors pseudotyped with the glycoprotein from the Ebola Zaire EboZ filovirus outer surface for specific airway cell targeting could achieve gene transfer in the lungs of mice.[113] The mice were infected by direct instillation of a single dose of the vector, but the potential for infection by inhalation of an aerosol was at least given by the investigation. In another study, a lentivirus vector carrying a foreign gene was administered *via* aerosol to mice by inhalation in a nose-only exposure chamber.[114] The results showed that lentivirus-mediated delivery of the foreign gene *via* aerosol was effective to a significant degree.

Since that time, further advances in the aerosol delivery of viral vectors have been achieved. For example, by targeting immunogenic substances from a pathogen encoded in a viral vector directly to the local, mucous membrane site, protective antibodies and cellular pathogen-specific immune responses can, in principle, be generated at the portal of entry for these pathogens.[115] Furthermore, aerosol delivery of viral vectors has been used in cancer therapy experimental models. In this regard, the protein osteopontin has been involved in cancer metastases, and delivery of lentivirus-based shRNA targeting the osteopontin gene (to silence this gene) in mice with breast cancer *via* a nose-only inhalation system "significantly decreased the expression level of osteopontin and altered the expression of several important metastasis-related proteins".[116] This group of investigators reported later that nose-only aerosol delivery of lentivirus-mediated osteopontin to mice suppressed lung tumorigenesis.[117] Also, aerosol delivery of the cowpox virus to cynomolgus monkeys resulted in severe respiratory disease,[118] and both aerosol inhalation and intranasal administration of influenza A viruses to mice were shown to be equally effective in causing an infection.[119]

Many viruses are quite sensitive to environmental stress, reducing their ability to survive in the atmosphere, a property that would be disadvantageous for their dissemination as biological weapons *via* aerosols. However, an area of intensive investigation involves the development of methods for encapsulating or packaging sensitive substances for controlled drug delivery over the nasal and respiratory routes.[6,120,121] These studies could yield benefits for increasing the resistance of viruses and non-viral agents during aerosol dissemination.

It must be stressed that the requirements for viruses as vectors may be quite different for therapeutic and biological warfare purposes. For therapeutic use, very stringent requirements are placed on viral vectors. For example, safety concerns could prevent the use of viruses that are highly infective and communicable, or that integrate into the host genome. Such viruses might, however, be considerably more effective in delivering drugs and biochemicals for a bioweapons scenario than other viruses that do not have these properties. Also, the requirement for long term expression of the transgene (the gene delivered to the host cell) that would be important for measuring the success in therapeutic use may not be as relevant for terrorist or biological warfare purposes. Again, the methods used for the experimental and clinical administration of viral vectors in aerosols (intratracheal application, nasal instillation, nose-only administration) do not exactly mimic the aerosol dissemination of biological agents as weapons, but they nevertheless demonstrate the principle of concept and provide data that could be used to extrapolate to a weapons type of delivery.

9.5 Artificial Viruses as Vectors for the Delivery of Biochemical Agents

One area of research that is rapidly advancing and needs to be closely monitored is the creation of so-called "artificial viruses" for drug delivery as well as gene and cancer therapy. Artificial viruses, also called non-viral vectors, are being developed in an attempt to overcome the negative aspects of using viruses to deliver genes, such as safety and manufacturing problems, immunogenicity, limited targeting ability and limited transport capacity. Non-viral vectors are a particular type of nanoparticles that are designed to deliver specific payloads to designated cells[122] (see Figure 9.1), thus essentially fulfilling the role of viruses as vectors.

9.5.1 Types of Artificial Vectors: Nanoparticles and Nanorobots

Artificial viruses usually consist of DNA compacted into particles with polycationic substances (positively charged polymers) such as polyethylenimine, oligoethylenimine coupled with short diacrylate linkages, polyaspartylhydrazide and chitosan for enhanced uptake into host cells.[123,124] Shielding molecules such as polyethyleneglycol to protect the DNA cargo and particular surface structures that can target the vectors to specific tissues can be added to these basic nanoparticles. Furthermore, biocompatible and biodegradable polymers like poly(lactic-*co*-glycolic acid) (PLGA), which has been approved by the FDA for use in the delivery of small interfering siRNA, exhibit favorable safety profiles, sustained release properties and improved colloidal stability. A dry powder formulation based on cationic lipid-modified PLGA nanoparticles to deliver siRNA directly to lung tissue for treatment of severe lung disease has been developed and shows promise for clinical application.[125]

The main problem with non-viral vectors is that they have not yet consistently shown gene transfer efficiency comparable to that of viruses, which limits their practical use.[122] Nevertheless, single-stranded siRNA loaded into three-stranded, RNA/DNA hybrid triangular self-assembled nanoparticles "could be efficiently transfected into NCI-H292 cells [human airway epithelial cells] in a dose- and time-dependent manner, resulting in high transfection efficiency".[126] Furthermore, a significant degree of effectiveness in gene delivery to airway cells in mice using a cationic non-viral vector administered through the nasal route has been demonstrated.[127] Also, in a human model for cystic fibrosis (cystic fibrosis bronchial epithelial cells), as well as in mice, the CFTR gene enclosed in a modular, self-assembling nanocomplex consisting of a mixture of cationic liposomes and a receptor-targeting/DNA-binding peptide delivered *via* aerosols in a nebulising chamber demonstrated both safety and effectiveness.[128]

More recently the construction of so-called nanorobots[129] has been gaining attention. Nanorobots have endless applications in medicine. Because of their small size, they can interact directly with cells or even penetrate into them. They can be constructed from DNA or proteins that carry different types of active payloads, with ligands that can attach to specific cell surface structures to direct them to designated targets (Figure 9.4). They can be programmed by specific construction with "logic operations" to reconfigure in response to cell-surface cues to open up and deliver their payloads.[130]

To date only "dumb" nanorobots have been designed, that is, they cannot actively seek out the cells they have been constructed to attack but will reach these cells only by chance. Lenaghan and co-workers[131] have proposed an active nanorobot design for cancer therapy and discussed the challenges involved which include the core design, propulsion, power, sensing and actuation, control, decision making and integration (Figure 9.5). There is great interest in developing these vectors further so that rapid advancement in

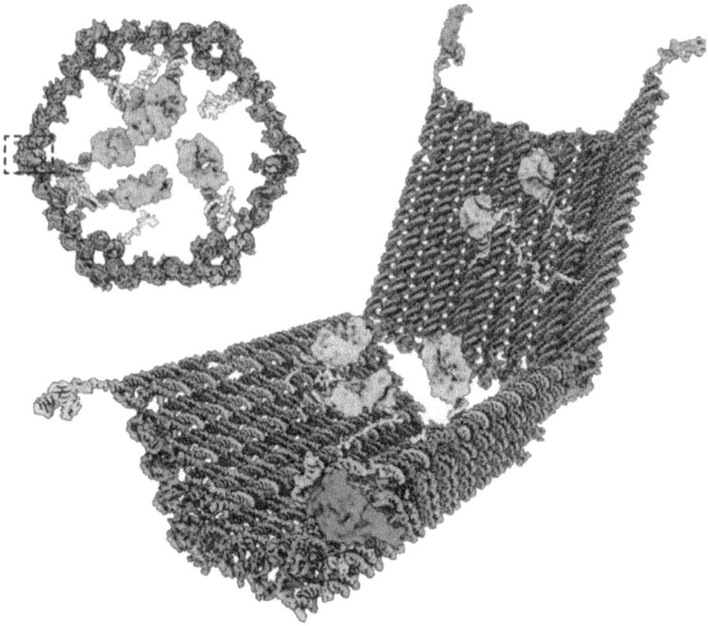

Figure 9.4 Representation of an aptamer-gated DNA nanorobot. Smaller image above, left: view of the closed nanorobot composed of a DNA scaffold (outer structure) with a protein payload (molecules inside). The structure is held closed by aptamer locks consisting of oligonucleotide or protein molecules that can bind to structures on the ends of the nanorobot (boxed area to the left and a similar lock to the right). The large image is a perspective view of the nanorobot opened by protein displacement of aptamer locks. Image taken from S. M. Douglas, I. Bachelet and G. M. Church, *Science*, 2012, **335**, 831, Figure 1, A and C, p. 832. Reprinted with permission from AAAS.

this area can be expected, which could pose a huge potential for misuse in the near future, as these vectors could carry and deliver toxins or bioregulators that could cause the disruption of vital physiological or neurological processes.

9.5.2 Feasibility of Using Artificial Viruses as Vectors for the Delivery of Biochemical Agents as Weapons

Non-viral vectors are being actively developed with the aim of overcoming the negative features of using viruses to deliver payloads, primarily safety considerations. Although efficient gene transfer activity comparable to that achieved by viral vectors remains a problem with non-viral vectors, improvements are being achieved continuously. Although most efforts regarding aerosol gene application *via* non-viral vectors have been limited to *in vitro* and *in vivo* experiments in small animals,[132] a currently ongoing clinical trial in

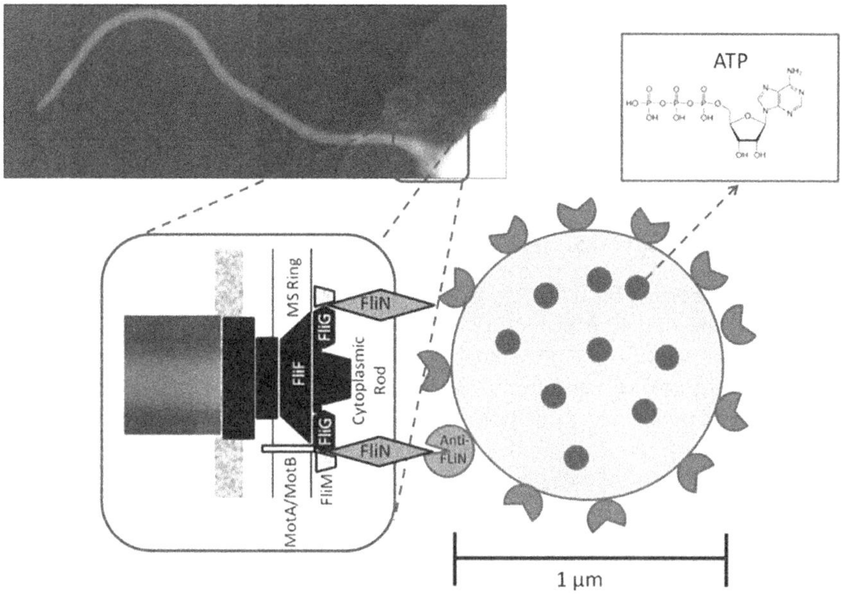

Figure 9.5 Schematic diagram of the envisaged nanorobot. The core of the robot is a polysaccharide-based nanoparticle (right half of the figure), where modular components can be attached using known chemistry. The propulsive system is a fully functional flagellum (left half of the figure) isolated from the bacterium *Escherichia coli*, attached to the core by means of a protein (FliN) on the flagellum and anti-protein receptors (anti-FliN) on the core. To power the propulsive system, adenosine triphosphate (ATP), the universal energy carrier in living organisms, is encapsulated into the nanoparticle core. Image taken from S. C. Lenaghan, Y. Wang, N. Xi, T. Fukuda, T. Tarn, W. R. Hamel and M. Zhang, *IEEE Trans. Biomed. Eng.*, 2013, **60**, 667, Figure 1, p. 668. Reprinted with permission from the Institute of Electrical and Electronics Engineers.

the field of cystic fibrosis therapy was initiated by the UK Cystic Fibrosis Gene Therapy Consortium using an aerosolized non-viral gene transfer agent to deliver the cystic fibrosis transmembrane conductance regulator (CFTR). The consortium has recently reported the results of this clinical trial. The report stated that the first demonstration of a significant change in lung function after CFTR gene therapy in patients using the non-viral vector was encouraging. However, the improvement was modest and the conclusion was that:

> "Although encouraging proof of principle that CFTR gene therapy can lead to changes in a clinically relevant outcome, the modest effect of these to date means that further work is necessary before this becomes a clinical therapy."[133]

More than anything, this study points to the fact that although aerosolized, non-viral vectors for clinical application have shown some degree of efficacy and are of continued great interest, they are still in the early development stages. Thus, due to the problems in the gene transfer efficiency of non-viral vectors at present, they would be less feasible than viral vectors for delivering biological agents for either therapeutic or non-peaceful purposes.

9.6 Conclusions

Progress in science and technology is absolutely essential for fighting disease and improving health and well-being in general. At the same time, the possibility of misuse of these developments for hostile purposes becomes more realistic as the biological agent spectrum increases, due to the rapid accumulation of knowledge about new targets for the interaction of biological agents with vital physiological systems. These discoveries can point to ways in which physiological systems may be manipulated with the aim of improving health, but with the potential to be misused to disrupt vital functions. This is compounded by the equally rapid development and improvement of ways to deliver these agents to their targets. Two methods for the delivery of biochemical agents that have progressed most significantly and appear to be most relevant are aerosol and viral, as well as non-viral vector-directed technologies. Nanotechnology plays a central role in almost all of these advances by improving ways to deliver bioactive substances more effectively.

Great strides are being made in aerosol delivery techniques, particularly in connection with interests in drug development and delivery. In this regard, the production of defined nanoparticles combined with new methods for making substances absorbable through the nasal and respiratory tracts create a potential for greatly improved delivery of bioactive compounds. The addition of substances to nanoparticles to improve absorption and uptake have also offered ways of overcoming the blood–brain barrier in the administration of drugs. Several examples of clinical applications and other studies (insulin, oxytocin, opiates) have shown that aerosol delivery of bioactive biochemicals is feasible in effect.

Considerable improvements have been made in the past ten years in viral-vector technologies for gene therapy or for the delivery of drugs and bioactive substances for therapeutic purposes. Such advances include improvements in targeting and gene transfer efficacy of viral vectors as well as the long-term expression of the delivered transgene. New improvements in methods to protect sensitive viral vectors from inactivation by harsh conditions in the environment add to the feasibility of their aerosol dissemination. Particular interest in developing non-viral vectors for cancer and gene therapy has also brought significant improvements, but the efficacy of these systems has been only modest. However, non-viral vectors are still in the early development phase and further improvements can be expected.

Clearly, aerosol delivery of biochemicals compacted into nanoparticles as well as viral vectors containing genes encoding bioactive proteins have been shown in numerous experimental and clinical applications to be effective to various degrees. Naturally, the methods of targeted delivery of therapeutics to treat disease do not exactly mimic those that would be used for the delivery of biochemical agents as weapons. Nevertheless, by examining these developments it is possible to come to some basic conclusions about the feasibility of delivering toxins and biochemical agents for terrorism or warfare purposes.

For therapeutic use, biochemical agents complexed into nanoparticles or incorporated into viral vectors in the form of genes encoding the active agents are usually administered by injection or by aerosol application. When aerosols are used to deliver a therapeutic agent, clinical application is usually accomplished by using devices such as nebulizers or atomizers to create aerosols that are inhaled through a face mask or by using metered dose inhalers/dry powder inhalers (Section 9.3.4). In experimental situations employing animals, the aerosol is usually administered by intratracheal application, nasal instillation or by nose-only or head-only application in specialized aerosol chambers. All of these therapeutic applications are designed to deliver the agent in a very direct manner, with minimum loss of the biologically active substance before reaching the patient or experimental animal. In a weapons delivery scenario, the agent would most likely be released into the environment in the form of an aerosol cloud, which would be subject to some loss of effect due to possible dilution by wind dispersion or to detrimental effects of environmental stress parameters such as temperature, ultraviolet light, drying out, *etc.* Despite such contingencies, release of microorganisms as well as biochemical agents into the environment in the form of an aerosol cloud can indeed produce significant effects, as documented in the results of numerous tests or actual agent deployment settings reported in Sections 9.3.1 and 9.3.5.

In assessing the feasibility of viral vectors for delivering biochemicals for non-peaceful purposes, it must be stressed that the goals of using armed viruses for gene and cancer therapy are quite different from those of using armed viruses as weapons. For example, the stringent efficacy demands of therapeutic use might not be so crucial in the case of weapons delivery, and the concerns about the safety of highly efficient lentiviral or other viral vectors would presumably matter little to a determined aggressor bent on delivering a biological weapons agent to a chosen target. Certainly, with improvements occurring rapidly and with ever growing experience gained in experimental and clinical applications, delivery *via* viral and non-viral vectors is becoming increasingly feasible. For the final judgement on the feasibility of using viral or non-viral vectors for weapons delivery, questions remain that can only be answered by direct testing of the particular agent in a designated scenario. For example, determinative factors would include: sensitivity of the agent to environmental stress parameters; uptake efficiency of the agent; efficiency of the release of the payload; transgene expression efficiency; amounts of the agent needed for a significant impact for a particular scenario; agent production capabilities.

The most sophisticated of these advances in science and technology are certainly not easy to put into practice, but require extensive scientific, technical expertise as well as tacit knowledge, well-equipped laboratories and substantial funds. While the application of these advances by non-State actors for hostile means can certainly not be ruled out, State-supported actors are more likely than terrorists to have such means. This places a particular responsibility on the States Parties of the CWC and the BTWC to ensure that illicit biochemical warfare programmes using these technologies are not being developed. The BTWC is called upon particularly as the treaty has no verification regime that can test compliance.

Efforts have been made over the past years to strengthen the BTWC, in particular with the review of science and technology developments as a standing agenda item within the 2012–2015 intersessional process (ISP).[134] While some common understandings have indeed been reached during this 2012–2015 ISP, virtually no effective action has been taken on these understandings, despite the improved focus on science and technology developments. States Parties to the BTWC will simply have to find a more dedicated way to review relevant developments in science and technology, assess their implications for the Convention and make recommendations for oversight measures of good practice that will ensure the freedom to pursue responsible science and at the same time provide security against intentional as well as unintentional misuse of the products, technological advances and knowledge resulting from work in the life sciences and related fields. Such measures could serve as a framework or guide for all States Parties to implement at the national level and thus comply with their obligations under the Convention in this regard. This is particularly urgent in light of the very disappointing lack of progress in this area at the Eighth Review Conference of the BTWC in November 2016.[135]

References

1. U.S. Department of Defense, *The Military Critical Technologies List. Part II. Weapons of Mass Destruction Technologies, P. II.3-2*, 1998, available at: http://fas.org/irp/threat/mctl98-2/mctl98-2.pdf, last accessed March 16, 2017.
2. S. S. Suri, H. Fenniri and B. Singh, *J. Occup. Med. Toxicol.*, 2007, **2**, 16.
3. E. Mastrobattista, M. A. E. M. van der Aa, W. E. Hennink and D. J. A. Crommelin, *Nat. Rev. Drug Discovery*, 2006, **5**, 115.
4. F. Andrade, D. Rafel, M. Vidiera, D. Ferreira, A. Sosnik and B. Sarmento, *Adv. Drug Delivery Rev.*, 2013, **65**, 1816.
5. W. E. Rudzinski, A. Palacios, A. Ahmed, M. A. Lane and T. M. Aminabhavi, *Carbohydr. Polym.*, 2016, **147**, 323.
6. Q. Zeng, J. Han, D. Zhao, T. Gong, Z. Zhang and X. Sun, *Int. J. Nanomed.*, 2012, **7**, 985.
7. K. Thiel, *Nat. Biotechnol.*, 2006, **24**, 1055–1057.
8. T. S. Stappenbeck and H. W. Virgin, *Nature*, 2016, **534**, 191.

9. A. Aderem, J. N. Adkins, C. Ansong, J. Galagan, S. Kaiser, M. J. Korth, G. L. Law, J. G. McDermott, S. C. Proll, C. Rosenberger, G. Schoolnik and M. G. Katze, *mBio*, 2011, **2**, 1.
10. D. E. Zak and A. Aderem, *Vaccine*, 2015, **33**, 5241.
11. P. Miller, *F1000Research*, 2016, **5**(F1000 Faculty Rev), 992.
12. M. Dando, *Biotechnol. Res. Int.*, 2011, 973851.
13. R. N. Germain, M. Meier-Schellersheim, A. Nita-Lazar and I. D. C. Fraser, *Annu. Rev. Immunol.*, 2011, **29**, 527.
14. A. Aderem, J. N. Adkins, C. Ansong, J. Galagan, S. Kaiser, M. J. Korth, G. L. Law, J. G. McDermott, S. C. Proll, C. Rosenberger, G. Schoolnik and M. G. Katze, *mBio*, 2011, **2**, 1.
15. R. J. Jackson, A. J. Ramsay, C. D. Christensen, S. Beaton, D. F. Hall and I. A. Ramshaw, *J. Virol.*, 2001, **75**, 1205.
16. E. T. Rietschel and H. Brade, *Sci. Am.*, 1992, **267**, 54.
17. USAMRIID, *Annual Report for Fiscal Year 1987*, United States Army, Fort Detrick, MD, 1987, p. 19, available at: http://www.dtic.mil/dtic/tr/fulltext/u2/a230195.pdf, last accessed March 16, 2017.
18. J. Miller, in *Biowarrior*, I. V. Domaradskij and J. Orent, Prometheus Books, Amhearst, 2003, Forward, pp. 9–12.
19. M. Leitenberg and R. A. Zilinskas, *The Soviet Biological Weapons Program. A History*, Harvard University Press, Cambridge and London, 2012, pp. 79–120.
20. M. Leitenberg and R. A. Zilinskas, *The Soviet Biological Weapons Program. A History*, Harvard University Press, Cambridge and London, 2012, p. 229.
21. V. Pitschmann, *Toxins*, 2014, **6**, 1761.
22. R. Trapp, in *Innovation, Dual Use, and Security*, ed. J. B. Tucker, MIT Press, Cambridge, MA, 2012, ch. 11, pp. 173–185.
23. E. Kagan, *Clin. Lab. Med.*, 2006, **26**, 421.
24. V. Pitschmann and Z. Hon, *Molecules*, 2016, **21**, 556.
25. *Milestones in Microbiology*, ed. T. Brock, American Society for Microbiology, Washington, D.C., 1961, p. 140.
26. J. E. Alouf, in *The Comprehensive Sourcebook of Bacterial Protein Toxins*, ed. J. E. Alouf and M. R. Popoff, Elsevier, San Diego, 3rd edn, 2006, ch. 1, pp. 3–21.
27. M. Salzman, J. M. Madsen and M. I. Greenberg, *Clin. Lab. Med.*, 2006, **26**, 397.
28. L. K. M. Wright, R. B. Lee, N. M. Vincelli, C. E. Whalley and L. A. Lumley, *Toxicol. Lett.*, 2016, **241**, 167.
29. *The Comprehensive Sourcebook of Bacterial Protein Toxins*, ed. J. E. Alouf and M. R. Popoff, Elsevier, San Diego, 3rd edn, 2006.
30. E. Geissler and K. Lohs, in *Biological and Toxin Weapons Today*, ed. E. Geissler, Oxford University Press, 1986, ch. 3, pp. 36–56.
31. M. N. Singh, K. S. Y. Hemant, M. Ram and H. G. Shivakumar, *Results Pharma Sci.*, 2010, **5**, 65.
32. G. Scheuch, M. J. Kohlhaeufl, P. Brand and R. Siekmeier, *Adv. Drug Delivery Rev.*, 2006, **58**, 996.

33. T. Rosebury, *et al.*, *Experimental Air-borne Infection*, The Williams and Wilkins Company, Baltimore, 1947.
34. B. J. Bernstein, *Sci. Am.*, 1987, **256**, 94.
35. G. B. Carter and G. S. Pearson, in *Biological and Toxin Weapons: Research, Development and Use from the Middle Ages to 1945*, ed. E. Geissler and J. E. van Courtland Moon, Oxford University Press, Oxford, 1999, ch. 9, pp. 168–189.
36. *Ad Hoc Meeting of Scientific and Technical Experts from States Parties to the Convention on the Prohibition of the Development, Production and Stockpiling of Bacteriological (Biological) and Toxin Weapons and on Their Destruction, 21 April 1987, Report,* BWC/CONF.II/EX/2, Attachment, United Kingdom Declaration, Geneva, 31 March to 15 April 1987, pp. 15–16.
37. E. A. Kabat, *Annu. Rev. Immunol.*, 1983, **1**, 1.
38. T. Rosebury, *Peace or Pestilence. Biological Warfare and How to Avoid it*, McGraw-Hill Book Company, New York, Toronto, London, 1949, available at: https://archive.org/stream/peaceorpestilenc00rose#page/n9/mode/2up, last accessed March 16, 2017.
39. U.S. Army, *U.S. Army Activity in the U.S. Biological Warfare Programs*, 1977, vol. 1 and 2, available at: http://www2.gwu.edu/~nsarchiv/NSAEBB/NSAEBB58/RNCBW_USABWP.pdf, last accessed September 12, 2016.
40. J. E. van Courtland Moon, in *Biological and Toxin Weapons: Research, Development and Use from the Middle Ages to 1945*, ed. E. Geissler and J. E. van Courtland Moon, Oxford University Press, Oxford, 1999, ch. 11, pp. 215–254.
41. L. Cole, *Health Secur.*, 2016, **14**, 315.
42. B. Balmer, in *Deadly Cultures: Biological Weapons Since 1945*, ed M. Wheelis, L. Rosza and M. Dando, Harvard University Press, Cambridge, MA, 2006, ch. 3, pp. 47–83.
43. D. B. Levin and G. Valadares de Amorim, *Biosecur. Bioterror.*, 2003, **1**, 37.
44. W. F. Tonnis, G. F. Kersten, H. W. Frijlink, W. L. J. Hinrichs, A. H. de Boer and J.-P. Amorij, *J. Aerosol Med. Pulm. Drug Delivery*, 2012, **25**, 249.
45. F. T. Cutts, C. J. Clements and J. V. Bennett, *Biologicals*, 1997, **25**, 323.
46. B. Laube, *Respir. Care*, 2005, **50**, 1161.
47. Y. Roth, J. S. Chapnik and P. Cole, *Ann. Otol., Rhinol., Laryngol.*, 2003, **112**, 264.
48. N. Csaba, M. Garcia-Fuentes and M. J. Alonso, *Adv. Drug Delivery Rev.*, 2009, **61**, 140.
49. J. Liu, A. Stace-Naughton, X. Jiang and C. J. Brinker, *J. Am. Chem. Soc.*, 2009, **131**, 1354.
50. P. L. Rodriguez, T. Harada, D. A. Christian, D. A. Pantano, R. K. Tsai and D. E. Discher, *Science*, 2013, **339**, 971.
51. J. J. Lochhead and R. G. Thorne, *Adv. Drug Delivery Rev.*, 2012, **64**, 614.
52. M. Dando, *Biotechnol. Res. Int.*, 2011, 973851, 9 pages.
53. J. J. Lochhead and R. C. Thorne, *Adv. Drug Delivery Rev.*, 2012, **64**, 614.

54. C. Betsholtz, *Nature*, 2014, **509**, 432.
55. A. M. M. Sadeghi, F. A. Dorkoosh, M. R. Avadi, M. Weinhold, A. Bayat, F. Delie, R. Gurny, B. Larijani, M. Rafiee-Tehrani and H. E. Junginger, *Eur. J. Pharm. Biopharm.*, 2008, **70**, 270.
56. B. K. Rubin and R. W. Williams, *Adv. Drug Delivery Rev.*, 2014, **75**, 141.
57. J. Kling, *Nat. Biotechnol.*, 2014, **32**, 851.
58. M. Kosfeld, M. Heinrichs, P. J. Zak, U. Fischbacher and E. Fehr, *Nature*, 2005, **435**, 673.
59. Vero Labs, http://www.verolabs.com, last accessed March 16, 2017.
60. O. Weisman, O. Zagoory-Sharon and R. Feldman, *Psychoneuroendocrinology*, 2012, **37**, 1582.
61. P. S. Churchland and P. Winkielman, *Horm. Behav.*, 2012, **61**, 392.
62. D. E. Geller, P. A. Flume, D. Staab, R. Fischer, J. S. Loutit and D. J. Conrad, *Am. J. Respir. Crit. Care Med.*, 2011, **183**, 1510.
63. A. Stejskalova, M. T. Kiani and B. D. Almquist, *Exp. Biol. Med.*, 2016, **241**, 1127.
64. P. Sandy, A. Ventura and T. Jacks, *BioTechniques*, 2005, **39**, 215.
65. O. M. Merkel, I. Rubenstein and T. Kissel, *Adv. Drug Delivery Rev.*, 2014, **75**, 112.
66. J. P. DeVincenzo, *Antiviral Ther.*, 2012, **17**, 213.
67. C. L. Leach, *Toxicol. Pathol.*, 2007, **35**, 23.
68. T. Daniels, N. Mills and P. Whitaker, The Cochrane Library, 2013, Issue (4), p. CD007639, DOI: 10.1002/14651858.CD007639.pub2, available at: http://onlinelibrary.wiley.com/doi/10.1002/14651858.CD007639.pub2/pdf, last accessed March 16, 2017.
69. J. C. Waldrep and R. Dhand, *Curr. Drug Delivery*, 2008, **5**, 114.
70. Q. Zhou, P. Tang, S. S. Y. Leung, J. G. Y. Chan and H.-K. Chan, *Adv. Drug Delivery Rev.*, 2014, **75**, 3.
71. A. A. Onischuk, T. G. Tolstikova, A. M. Baklanov, M. V. Khvostov, I. V. Sorokina, N. A. Zhukova, S. V. Ankov, O. V. Borovkova, G. G. Dultseva, V. V. Boldyrev, V. M. Fomin and G. S. Huang, *J. Aerosol Sci.*, 2014, **78**, 41.
72. G. Mainelis, S. Seshadri, O. B. Garbuzenko, T. Han, Z. Wang and T. Minko, *J. Aerosol Med. Pulm. Drug Delivery*, 2013, **26**, 345.
73. J. Allen, *Tradeline*, 2007 available at: https://www.tradelineinc.com/reports/2007-5/hpas-precision-aerosol-exposure-and-containment-technology. last accessed March 16, 2017.
74. C. J. Roy, K. Song, S. K. Sivasubramani, D. J. Gardner and S. H. Pincus, *Curr. Top. Microbiol. Immunol.*, 2012, **357**, 243.
75. A. Nalca, V. A. Livingston, N. L. Garza, E. E. Zumbrun, O. M. Frick, J. L. Chapman and J. M. Hartings, *PLoS One*, 2010, **5**, e12880.
76. P. A. Dabish, J. Kline, C. Lewis, J. Yeager and M. L. M. Pitt, *Inhalation Toxicol.*, 2010, **22**, 224.
77. P. M. Wax, C. E. Becker and S. C. Curry, *Ann. Emerg. Med.*, 2003, **41**, 700.
78. J. R. Riches, R. W. Read, R. M. Black, N. J. Cooper and C. M. Timperley, *J. Anal. Toxicol.*, 2012, **36**, 647.

79. S. A. Shoyele and A. Slowey, *Int. J. Pharm.*, 2006, **314**, 1.
80. M. A. Poli, in *Encyclopedia of Bioterrorism Defense*, ed. R. Katz and R.A. Zilinskas, John A. Wiley & Sons, Inc., Hoboken, New Jersey, 2nd edn, 2011, Toxins, Overview and General Principles, pp. 593–595.
81. D. R. Franz, *Defense against Toxin Weapons*, U.S. Army Medical Research and Materiel Command, Fort Detrick, Maryland, 1997, available at: http://www.usamriid.army.mil/education/defensetox/toxdefbook.pdf, last accessed March 16, 2017.
82. P. Leboulch, *Nature*, 2013, **500**, 280.
83. T. C. Liu, E. Galanis and D. Kirn, *Nat. Clin. Pract. Oncol.*, 2007, **4**, 101.
84. S. Chalikonda, M. H. Kivlen, M. E. O'Malley, X. D. E. Dong, J. A. McCart, M. C. Gorry, X.-Y. Yin, C. K. Brown, H. J. Zeh III, Z. S. Guo and D. L. Bartlett, *Cancer Gene Ther.*, 2008, **15**, 115.
85. M. P. Holgado, J. Falivene, C. Maeto, M. Amigo, M. F. Pascutti, M. B. Vecchione, A. Bruttomesso, G. Calamante, M. P. del Medico-Zajac and M. M. Gherardi, *Viruses*, 2016, **8**, 139.
86. C. Larson, B. Oronsky, J. Scicinski, G. R. Fanger, M. Stirn, A. Oronsky and T. R. Reid, *Oncotarget*, 2015, **6**, 19976.
87. L. Coughlan, H. Uusi-Kerttula, J. Ma, B. P. Degg, A. L. Parker and A. H. Baker, *Hum. Gene Ther.*, 2014, **25**, 385.
88. A. H. Baker, *Hum. Gene Ther.*, 2014, **25**, 255.
89. N. Brunetti-Pierri, T. Ng, D. Iannitti, W. Cioffi, G. Stapleton, M. Law, J. Breinholt, D. Palmer, N. Grove, K. Rice, C. Bauer, M. Finegold, A. Beaudet, C. Mullins and P. Ng, *Hum. Gene Ther.*, 2013, **24**, 761.
90. R. G. Crystal, *Hum. Gene Ther.*, 2014, **25**, 3.
91. J. A. Doudna and E. Charpentier, *Science*, 2014, **346**, 1258096.
92. D. Maddalo, E. Manchado, C. P. Concepcion, C. Bonetti, J. A. Vidigal, Y.-C. Han, P. Ogrodowski, A. Crippa, N. Rekhtman, E. de Stanchina, S. W. Lowe and A. Ventura, *Nature*, 2014, **516**, 423.
93. J. Flint, V. R. Racaniello, G. F. Rall, A. M. Skalka and L. W. Enquist, in *Principles of Virology*, American Society for Microbiology, Washington, D.C., 4th edn, 2015, vol. I, p. 520.
94. C. Li, N. Diprimio, D. E. Bowles, M. L. Hirsch, P. E. Monahan, A. Asokan, J. Rabinowitz, M. Agbandje-McKenna and R. J. Samulski, *J. Virol.*, 2012, **86**, 7752.
95. A. Asokan, D. V. Schaffer and R. J. Samulski, *Mol. Ther.*, 2012, **20**, 699.
96. C. Ling, B. Li, W. Ma and A. Srivastava, *Hum. Gene Ther: Methods.*, 2016, **27**, 143.
97. F. Mingozzi and K. A. High, *Nat. Rev. Genet.*, 2011, **12**, 341.
98. E. Hastie and R. J. Samulski, *Hum. Gene Ther.*, 2015, **26**, 257.
99. E. Senis, C. Fatouros, S. Grosse, E. Wiedtke, D. Niopek, A.-K. Mueller, K. Börner and D. Grimm, *Biotechnol. J.*, 2014, **9**, 1402.
100. F. A. Ran, L. Cong, W. X. Yan, D. A. Scott, J. S. Gootenberg, A. J. Kriz, B. Zetsche, O. Shalem, X. Wu, K. S. Makarova, E. V. Koonin, P. A. Sharp and F. Zhang, *Nature*, 2015, **520**, 186.

101. H. Nakai, E. Montini, S. Fuess, T. A. Storm, M. Grompe and M. A. Kay, *Nat. Genet.*, 2003, **34**, 297.
102. A. Schambach and C. Baum, *Curr. Gene Ther.*, 2008, **8**, 474.
103. C. Frecha, J. Szecsi, F.-L. Crosset and E. Verhoeyen, *Curr. Gene Ther.*, 2008, **8**, 449.
104. D. Escors and K. Breckpot, *Arch. Immunol. Ther. Exp.*, 2010, **58**, 107.
105. J. Matrai, M. K. L. Chuah and T. VandenDriessche, *Mol. Ther.*, 2010, **18**, 477.
106. H. Kymäläinen, J. U. Appelt, F. A. Giordano, A. F. Davies, C. M. Ogilvie, S. G. Ahmed, S. Laufs, M. Schmidt, J. Bode, R. J. Yanez-Munoz and G. Dickson, *Hum. Gene Ther.*, 2014, **25**, 1.
107. R. Kaminski, Y. Chen, T. Fischer, E. Tedaldi, A. Napoli, Y. Zhang, J. Karn, W. Hu and K. Khalili, *Sci. Rep.*, 2016, **6**, 22555.
108. D. B. Stovall, M. Wan, Q. Zhang, P. Dubey and G. Sui, *J. Visualized Exp.*, 2012, (64), e4129.
109. A. A. Wilson, L. W. Kwok, E. L. Porter, J. G. Payne, G. S. McElroy, S. J. Ohle, S. R. Greenhill, M. T. Blahna, K. Yamamoto, J. C. Jean, J. P. Mizgerd and D. N. Kotton, *Mol. Ther.*, 2013, **21**, 825.
110. R. B. Moss, D. Rodman, L. T. Spencer, M. L. Aitken, P. L. Zeitlin, D. Waltz, C. Milla, A. S. Brody, J. P. Clancy, B. Ramsey, N. Hamblett and A. E. Heald, *Chest*, 2004, **125**, 509.
111. P. L. Sinn, R. M. Anthony and P. B. McCray Jr, *Hum. Mol. Genet.*, 2011, **20**, R79.
112. B. Laube, *Respir. Care*, 2005, **50**, 1161.
113. M. F. Medina, G. P. Kobinger, J. Rux, M. Gasmi, D. J. Looney, P. Bates and J. M. Wilson, *Mol. Ther.*, 2003, **8**, 777.
114. S.-K. Hwang, J.-T. Kwon, S.-J. Park, S.-H. Chang, E.-S. Lee, Y.-S. Chung, G. R. Beck Jr, K. H. Lee, L. Piao, J. Park and H.-H. Cho, *Gene Ther.*, 2007, **14**, 1721.
115. C. J. Roy, A. Ault, S. K. Sivasubramani, J. P. Gorres, C.-J. Wei, H. Andersen, J. Gall, M. Roederer and S. S. Rao, *Respir. Res.*, 2011, **12**, 153.
116. K.-N. Yu, A. Minai-Tehrani, S.-H. Chang, S.-K. Hwang, S.-H. Hong, J.-E. Kim, J.-Y. Shin, S.-J. Park, J.-H. Kim, J.-T. Kwon, H.-L. Jiang, B. Kang, D. Kim, C.-H. Chae, K.-H. Lee, T.-J. Yoon, G. R. Beck Jr and M.-H. Cho, *PLoS One*, 2010, **5**, e15623.
117. A. Minai-Tehrani, S.-H. Chang, J.-T. Kwon, S.-K. Hwang, J.-E. Kim, J.-Y. Shin, K.-N. Yu, S.-J. Park, H.-L. Jiang, J.-H. Kim, S.-H. Hong, B. Kang, D. Kim, C.-H. Chae, K.-H. Lee, G. R. Beck Jr and M.-H. Cho, *Cell. Oncol.*, 2013, **36**, 15.
118. R. F. Johnson, D. A. Hammoud, M. G. Lackemeyer, S. Yellayi, J. Solomon, J. K. Bohannon, K. B. Janosko, C. Jett, K. Cooper, J. E. Blaney and P. B. Jahrling, *Virology*, 2015, **481**, 124.
119. J. A. Belser, K. M. Gustin, J. M. Katz, T. R. Maines and T. M. Tumpey, *Virology*, 2015, **481**, 107.
120. B. Nayak, A. Panda, P. Ray and A. R. Ray, *J. Microencapsulation*, 2009, **26**, 154.

121. D. Cipolla, B. S. H. Wu, I. Gonda and H.-K. Chan, *J. Aerosol Med. Pulm. Drug Delivery*, 2015, **26**, 411.
122. K. L. Douglas, *Biotechnol. Prog.*, 2008, **24**, 871.
123. V. Russ, H. Elfberg, C. Thoma, J. Kloeckner, M. Ogris and E. Wagner, *Gene Ther.*, 2008, **15**, 18.
124. M. Ogris, A. K. Kotha, N. Tietze, E. Wagner, F. S. Palumbo, G. Giammona and G. Cavallaro, *Pharm. Res.*, 2007, **24**, 2213.
125. D. K. Jensen, L. B. Jensen, S. Koocheki, L. Bengtson, D. Cun, H. M. Nielsen and C. Foged, *J. Controlled Release*, 2012, **157**, 141.
126. Y. Wang, Z. You, J. Du, H. Li, H. Chen, J. Li, W. Dong, B. He, C. Mao and G. Wang, *J. Controlled Release*, 2016, **233**, 126.
127. T. W. Kim, H. Chung, I. C. Kwon, H. C. Sung, B. C. Shin and S. Y. Jeong, *J. Gene Med.*, 2005, **7**, 749.
128. M. D. I. Manunta, R. J. McAnulty, A. D. Tagalakis, S. E. Bottoms, F. Campbell, H. C. Hailes, A. B. Tabor, G. J. Laurent, C. O'Callaghan and S. L. Hart, *PLoS One*, 2011, **6**, e26768.
129. S. M. Douglas, I. Bachelet and G. M. Church, *Science*, 2012, **335**, 831.
130. J. Elbaz and I. Willner, *Nat. Mater.*, 2012, **11**, 276.
131. S. C. Lenaghan, Y. Wang, N. Xi, T. Fukuda, T. Tarn, W. R. Hamel and M. Zhang, *IEEE Trans. Biomed. Eng.*, 2013, **60**, 667.
132. P. Zarogoulidis, N. K. Karamanos, K. Porpodis, K. Domvri, H. Huang, W. Hohenforst-Schimdt, E. P. Goldberg and K. Zarogoulidis, *Int. J. Mol. Sci.*, 2012, **13**, 10828, available at: http://www.mdpi.com/1422-0067/13/12/17290/s1last accessed March 16, 2017.
133. UK Cystic Fibrosis Gene Therapy Consortium, *Efficacy and Mechanism Evaluation*, 2016, vol. 3, issue (5), DOI: 10.3310/eme03050.
134. United Nations, *Seventh Review Conference to the Convention on the Prohibition of the Development, Production and Stockpiling of Bacteriological (Biological) and Toxin Weapons and on Their Destruction, 5–22 December 2011, Final Document*, BWC/CONF.VII/7, Geneva, 13 January 2012, available at: http://www.unog.ch/80256EE600585943/(httpPages)/F1CD974A1FDE4794C125731A0037D96D?OpenDocument, last accessed March 16, 2017.
135. United Nations, *Eighth Review Conference to the Convention on the Prohibition of the Development, Production and Stockpiling of Bacteriological (Biological) and Toxin Weapons and on Their Destruction, 7–25 November 2016, Final Document*, BWC/CONF.VIII/8, Geneva, 11 January 2017.

CHAPTER 10

The Future of Chemical Weapons: Advances in the Development of Anti-plant Agents

S. WHITBY

University of Bradford, Faculty of Social Sciences, BD7 1DP, Bradford, UK
*E-mail: S.Whitby@Bradford.ac.uk

10.1 Introduction

Published in 1971 and 1973 respectively, comprehensive coverage of worldwide developments in chemical weapons since World War I (WWI) can be found in Volumes I and II of the seminal SIPRI[1] (Stockholm International Peace Research Institute) series of books, *The Problem of Chemical and Biological Warfare*. US aspects of this history are also the subject of discussion by Brophy, Miles, and Cochrane,[†,2] in an official history of the subject published in 1959, titled, *The Chemical Warfare Service: From Laboratory to Field*, the latter covering the period from the First to the Second World Wars. Regarding aspects of UK and US transatlantic collaboration, Carter and Pearson[3]

[†]This was the 2nd of three volumes in a series devoted to the Chemical Warfare Service (CWS) in World War II, later to be known as the Chemical Corps. In accordance with the Preface to this Volume, "... it traces the history of these activities from the World War I period ... until the end of World War II.

Preventing Chemical Weapons: Arms Control and Disarmament as the Sciences Converge
Edited by Michael Crowley, Malcolm Dando and Lijun Shang

Published by the Royal Society of Chemistry, www.rsc.org

give an overview of inter-war collaboration in their 1996 *Journal of Strategic Studies* article titled 'North Atlantic Chemical and Biological Research Collaboration: 1916–1945', and the period of collaboration which also included Canada is highlighted during the period 1937 to 1945 in Bryden's[4] *Deadly Allies*, published in 1989.[1] Developments relating to weapons for targeting plants, are thus viewed in this chapter in the context of historical collaborative UK and US (as well as Canadian) efforts where the UK and US, in particular, exchanged information and expertise and in some senses pooled resources on matters relating to chemical weapons. Insights concerning collaboration in this area are also obtained from existing and recently released primary sources. In this connection, the dynamics of collaboration derive in part from my own work[5] on the history of anti-crop biological warfare programmes, and the latter has been useful in providing insights into collaboration in research and development, and into military precedents set by the UK that ushered in the use of chemicals on a huge scale in Indo-China. The focus in this chapter, therefore, is on developments during World War II (WWII) out of which large-scale anti-plant chemical warfare was conceived. The Malayan 'Emergency' provided the backdrop to UK efforts to move from small-scale use of anti-plant chemicals to large-scale deployment, and to methods that facilitated wide-spread dissemination. Increased public understanding of the use of chemical anti-plant weapons in warfare centred around the question of their use in the Second Indo-China War which began in the early 1960s and ended more than a decade later. One of the most comprehensive accounts of the extraordinary story of operations concerning the use of herbicides, is the seminar work on Vietnam by Paul F. Cecil.[6] Published in 1986, some years after the war had ended, and based significantly on then newly available primary sources as well interviews with those who flew the missions, *Herbicidal Warfare: the Ranch Hand Project in Vietnam*, provides insights into this unprecedented decade-long strategy to deploy huge quantities of commercially-available chemicals in the war against the North Vietnamese and their Chinese backers. As well as providing insights into hitherto undisclosed aspects of this military project, Cecil also traced the scientific and public reaction to use – the latter being an enduring issue in the twenty-first century. Also notable is Arthur H. Westing's *Herbicides in War: the Long-Term Ecological and Human Consequences*.[7] Published in 1984 some two years prior to the work of Cecil, Westing's work approached the issue from the perspective of the environmental consequences of military activity. As an active member of the American Association for the Advancement of Science (AAAS) and participant in the Herbicide Assessment Committee (HAC), Westing had been active—and with others, had enjoyed some success—in trying to bear pressure upon the then US administration to bring about an end to the use of anti-crop agents and herbicides in Vietnam. Also significant is a contribution by Thomas Whiteside titled, *Defoliation: What Are Our Herbicides Doing to Us?*[8] A somewhat concise contribution that was published as early as 1970, this publication also approached the subject from the perspective of the consequences of use for mammals and for the environment.

Alastair Hay's *Chemical Sythe*[9] is particularly noteworthy and like the aforementioned publications, Hay's work provides the backdrop to some of the issues addressed in this chapter. *Inter alia*, all of the above include extensive bibliographical references to sources that provide primary, secondary, qualitative and quantitative data sets on this subject. Further to this, published in 1983, Stanley Karnow's *Vietnam: a History*[10] represents a voluminous account of the country's catalogue of colonial and post-colonial wars that place the Second Indo-China War in Vietnam in historical context. Significantly, insights into contemporary developments in phytobiology, in so far as they relate to auxins, have been provided by Richard Napier, auxins expert at the Napier Laboratory at the University of Warwick, UK.

10.2 Anti-plant Weapons

According to Meselson,[‡,11] anti-plant (chemical weapons) research "may be traced" to original research conducted by Charles Darwin and his son Francis into the way in which plants respond to light and gravity that centred around the isolation and identification of growth promoting substances in oat and canary grass, as reported in their 1880 book: *The Power of Movement in Plants*.[12] As noted by Troyer,[§] debate concerning precisely who was responsible for the 'discovery' of the so-called 'auxins', remains, to this day, a contested issue. Davies[13] speculates that the concept of hormones in plants may have originally derived from work by Sachs who had postulated in the early 1880s that substances move throughout plants. According to Went,[14] Sachs' great achievement was his application to morphology—the study of the form and structure of plants and of the relationship between their structures—of the concept of causality.[¶] However, as Went[15] notes, it would be some time before the precise identification of auxins would take place in that "no direct evidence for the existence of such special substances, as Sachs had postulated, was obtained for nearly 40 years". Assignment of the term 'auxin' ('auxein': meaning to grow) is attributed to the work of Went[16] in 1928. Work on the isolation of plant growth substances lead to its eventual chemical identification in 1931 through the work of Kogl and Haagen-Smit.[17] Subsequently, a

[‡]Meselson, an accomplished award-winning geneticist and microbiologist at Harvard University, had been at the centre of efforts throughout the late 1960s and early 1970s to bring about a change in US policy on the use of anti-crop agents and herbicides in Vietnam. He remains to this day a long-time opponent of the hostile misuse of biology and chemistry in warfare and has worked since the 1960s to uphold the norm of non-use of biological and chemical weapons and to strengthen the international legally binding prohibition regimes that relate to both forms of warfare.

[§]2,4-D or 2,4-dichlorophenoxyacetic acid, 2,4,5-T or 2,4,5-trichlorophenoxyacetic acid, and MCPA or 2-methyl-4-chlorophenoxyacetic acid, are all selective hormone herbicides. For further information, see ref. 49.

[¶]Wend *et al.* conceptualise a hormone as "... essentially that of a correlation carrier, where correlation (as used in regard to plants) is defined as the influence exerted by one part of the plant upon another ... in the sense of causal relationship." For further information, see ref. 50.

single substance would be identified by others as a naturally-occurring plant growth hormone: IAA, indole-3-acetic-acid, the latter being found in small quantities to be a stimulator of growth in plants. In larger doses, however, IAA would result in their death. According to Epp[18] and co-workers, a "herbicide is classified as an auxin if it induces physiological and phenotypic effects similar to those induced by indole-3-acetic acid, a natural plant hormone of the auxin class". Davies[19] notes the need for clarity concerning auxin nomenclature thus: "[T]he term 'plant growth substance' [he notes] is also used for plant hormones but is a rather vague term and does not describe fully what these natural regulators do – growth is only one of the many processes influenced ... While the term plant growth regulator is a little more precise, this term has been mainly used by the agrichemical industry to denote synthetic plant growth regulators as distinct from endogenous growth regulators." As noted by Troyer,[20] work on endogenous growth regulators would lead to the discovery of "the first systemic or hormone herbicides". This followed work by Zimmerman[21] *et al.* who found that synthetic chemicals similar to IAA could inhibit and reduce growth in plants. Thus, research involving endogenous plant growth regulators would result in research into the creation of synthetic plant growth regulators and lead to the adoption for military use of selective synthetic herbicides.

For the purpose of this discussion we focus here on developments at the interface between civil and military work, in particular, that which took place during the course of WWII, on simultaneous, but independent UK and US research and development involving, as noted by Troyer "4 groups of workers in 2 countries".‖ Transatlantic collaboration across this area, together with early UK deployment of chemical anti-plant agents and munitions in the 1950s in Malaya (now Malaysia) as part of its response to the emergency,** was characteristic of the development and early evolution of anti-plant chemical weapons work in the two countries, both during wartime, and in the post WWII era. This work would lead eventually to their large-scale production and deployment by the US in the second Indo-China war in the 1960s and 1970s, in Vietnam. The following paragraphs highlight anti-plant work on both sides of the Atlantic. National domestic arrangements in the UK and US are set out, and, aspects of transatlantic collaboration highlighted. The identification of the now infamous chemical anti-plant agents that saw use in Vietnam is placed in the context of a broad range of chemical agents under consideration.

Objective assessment of the wartime 'effectiveness' of auxin deployment was severely curtailed by a dearth of reliable and verifiable scientific

‖Indeed, Troyer notes that the debate surrounding the discoverer of chemicals remains a contested issue, with this author concluding that credit should be assigned equally to each of the 4 groups in the 2 countries. See Troyer's article for a full explanation.

**New tactics adopted by the British in Malaya in the 1950s recognised the importance of isolating insurgents from their sources of food. Food denial was thus physical separation as well as destruction of plants – a policy that sought also to create a sense of security in populated areas which also sought to separate passive supporters of the insurgency from communist guerrillas.

evidence. With use remaining a highly contentious issue, and this was to become increasingly the case in the mind of the general public and in the minds of service personnel that had been exposed to the chemicals during operational use. For the purpose of this discussion, however, the use of synthetic chemical auxins in Vietnam is thus limited in terms of its scope to the agents in question, the apparent quantities used, and ground covered. End of use is addressed. Auxins as well as other anti-plant chemicals are then discussed in the context of relevant international legal prohibition regimes. Following their deployment in Vietnam, the discussion of 'auxins' in particular concludes by casting the issue in the light of twenty-first century scientific and technological advances in order to understand improvements that may have taken place in regard to rational design, target specificity and effectiveness.

The plant hormone auxin can be categorized as follows: 1. natural auxins and 2. synthetic auxins. Although the distinction between natural and synthetic hormones is significant, our enquiry focuses in on what Troyer[20] referred to as synthetic chemicals that were found during the course of intensive research and development in the UK and US, to have plant inhibiting and growth reducing properties when applied to plants in high concentrations. These would first be regarded as weed-killers for commercial and domestic use; and commercial and wartime secrecy ensured that knowledge of developments of relevance to military matters would remain restricted, particularly during the period of hostilities. Knowledge of their potential for hostile use was thus not widely appreciated by the general public at this time, but such agents would later become associated with their potential to be used for hostile military purposes in crop destruction and in defoliation.

10.2.1 UK Developments

Independent respective 1940s UK and US research and development had taken place on both sides of the Atlantic. Work on the plant growth inhibiting characteristics of chemicals had taken place in the UK at both ICI[20] (by Templeman *et al.*) and at Rothampstead Agricultural Research Station[20] (by Nutman *et al.*), the latter under the auspices of the UK Agricultural Research Service (ARS). The selective properties of two chemicals had been under investigation: one code-named 1313 was identified as isopropyl-*n*-phenyl carbamate (in the carbamate class of herbicides HRAC K2, inhibitors of cell division), and another code-named 1414[22] was identified as calcium-2-methyl-4-chloro-phenoxyacetate (an auxin, HRAC class O). The former was found to be effective against barley, oats, rye, wheat and other cereal crops. The latter was found to be effective against sugar beet and other root crops. According to sources[23] from the UK National Archives, following initial 1940s field trials—the apparent success of which were reported up the UK political hierarchy—Churchill was thought to have given consideration to, but did not act, in connection with their possible use in 1942 against Germany.

10.2.2 US Developments

As detailed by Troyer,[20] consideration of the use of hormone herbicides also arose from near-simultaneous but independent discovery by two groups in the US, the first being attributed to Franklin D. Jones *et al.*, and the second being attributed to a group of researchers lead by Ezra J. Kraus of the University of Chicago. Jones' interest in the use in agriculture and in horticulture of plant-growth-regulating chemicals led to experimentation in 1943 into the effects of hundreds of chemical substances and Jones' work on 2,4-D and 2,4,5-T would result in the submission of first, a Canadian, and then an American commercial patent application in 1944 and in 1945, respectively. The second group involved a collaboration between the University of Chicago and the US Department of Agriculture at Beltsville, MD. Troyer contends[20] that the use of auxins as herbicides in warfare can be attributed to Kraus who had initially suggested the possibility in a letter to colleagues at Beltsville. Both Troyer[20] and Reinhardt[24] are explicit about Kraus' role. In a contention that is broadly aligned with Troyer's understanding of Kraus' role, Reinhardt argues that Kraus received military approval in February 1942 to carry out work, as stated by Reinhardt "for offensive purposes ... developing hormones to destroy crops and forest useful to the enemy". Such approval signalled Kraus *et al.'s* association with US war efforts under the auspices of the Special Project Division charged with the organisation of wartime US CBW activities that would eventually be based at Camp Detrick, MD.

10.2.3 Aspects of Transatlantic Collaboration

Expanding on the topic previously addressed by Brophy, Miles and Cochrane, according to Carter and Pearson,[3] transatlantic collaboration on chemical weapons development had become formalized between the UK and US towards the end of the First World War when in 1918, Anglo-American 'liaison' underpinned the creation of the US Chemical Warfare Service. Thus, formal arrangements between the UK and US would subsequently, during the inter-war period and the Second World War, extend to include Canada and the establishment of a process of tri-partite collaboration in this area – one that continues to this day in respect of the development of defensive measures.

Informal collaboration between the UK and the US on matters relating the chemical warfare thus followed the First World War, and some aspects of formalisation originated in the 1920s with Canadian involvement emerging in the later inter-war years and following the Second World War. Reciprocal arrangements had been established between the UK and US in 1924 where representatives from the US Chemical Warfare Service (CWS) at Edgewood Arsenal attended meetings held by the UK Chemical Warfare Committee and this reciprocal arrangement, with UK Senior Representatives attending meetings in the US, is thought to have endured through the late 1920s, and into, and throughout the 1930s. UK and US collaboration evolved into the integration of technical and administrative arrangements on chemical warfare with the British–US Inter-Services Committee on Chemical Warfare (ISCCW), becoming an integrated UK–US body. Indeed, according to Brophy, Miles and Cochrane,[25] one of the "objectives" of the United States Chemical Warfare Committee (USCWC), an organisation that was set up in the fall of 1942 to "co-ordinate all chemical warfare activities" was to "... insure that all types of chemical warfare material used by the British and Americans would be interchangeable."

Although difficulties regarding World War II CW trans-Atlantic collaboration, communication and information exchange were encountered, and improvements in technical and administrative arrangements would eventually emerge from periodic review, following early CW collaborative arrangements, the 1940s would also see the 'virtual integration'[26] of UK and US chemical programmes with UK staff being stationed in the US, and US staff being stationed in the UK. Likewise, US staff were stationed at the Canadian Experimental Station at Suffield, Alberta, and in other commonwealth countries. UK and Canadian staff had been correspondingly located at Dugway Proving Group, Utah, as well as in Florida, and at the US proving ground at San Jose, in Panama. As well as improving transatlantic collaboration, communication and information exchange, the creation of the Inter-Service Committee on Chemical Warfare in Washington also facilitated high-level administrative and political exchange and this organisation functioned as a conduit for information and exchange at the highest political levels. Following his stationing in Washington in 1943, Lord Stamp, together with his American counterpart in London, played an important part in this process. Subsequent exchanges of information between the UK's Sir Jon Anderson—"an official charged with the organisation of the UKs civilian and economic resources during the war"—explained the significance of UK work on auxins in a 1944 letter, copied to Churchill, that was sent to Vannevar Bush, the then scientific advisor to Roosevelt. This addressed experimental work that had taken place at ICI on the chemicals code-named "1313" and "1414"; and, under a recommendation from Churchill their effectiveness was communicated to the Americans "in case they should wish to use them against the Japanese". The apparent destructive capabilities of such chemicals were relayed[27] to the US by the UK's Sir John Anderson who noted the following that:

> "the group in charge of the experiment estimate that 1lb per acre of either substance would result in almost complete destruction of the vulnerable crops under ideal conditions, and 3lbs per acre would result in almost complete destruction under most conditions that could reasonably be expected."

Bush's reply to Anderson, also copied to Churchill, revealed parallels between UK and US work on the use of chemicals with herbicidal and defoliating properties in that US work had involved: "A project on organic chemical agents as herbicides ... To date [the Report noted] one of the most promising compounds appears to be 2,4-dichlorophenoxyacetic acid". The letter also revealed the existence of a programme involving "defoliation experiments in which large amounts of chemicals [were] used." Indeed, the latter information exchange appears to underplay the extent of US R&D work on chemicals that could be used against plants for hostile purposes, with subsequent reports noting that some 800 chemicals had been the subject of screening in the US for this purpose. Out of this huge number of potential agents, only four were eventually selected – and it is striking to see the similarities between UK and US efforts in this connection. The four agents were described in a Chiefs of Staff Committee Report[28] dated 1945:

(a) "2,4-dichlorophenoxyacetic acid (LN8) or 2,4-D, ... the principal substances employed in the American work and 2-methyl-4-chlorophenoxyacetic acid (LN32) an analogue of equal activity preferred for ease of production in the United Kingdom. These two agents are most active against dicotyledons, although rice is also extremely sensitive. The remaining cereals are however, affected by these agents only at a relatively higher dose level.

(b) 2,4,5-trichlorophenoxyacetic acid (LN14) or 2,4,5-T, chemically a simple extension of LN8, found to be especially effective in reducing the tuber yield of potatoes, a crop relatively less sensitive to LN8 and LN32.

(c) isopropyl-*n*-phenyl carbamate (LN33), especially active against the monocotyledons, *i.e.*, cereals and grasses, but with relatively little effect on broad-leaved plants."

The report by the Inter-Services Sub-Committee also noted the quantities that would be required if a chemical warfare attack was to be mounted against plants on Japan's main islands, with Hokkaido, Honshu, Shikoku, Kyushu and Okinawa consisting of 457, 5895, 368, 1153, and 2, thousands of acres, respectively. The report noted that rice would require the application of some 20000 tons of 2,4-D, corn would require the application of some 10000 tons of isopropyl-*n*-phenyl carbamate, and root crops would require the application of some 1000 tonnes of 2,4-D or 2-methyl-4-chlorophenoxyacetic acid; and large-scale field experimentation with such agents had suggested, as the report notes, that the agents in question might represent "a very promising weapon". Significant recommendations were included in the report regarding the potential difficulties involved in a country organizing defence against attack with such agents, and the potential for such agents to be used by a world security organisation as a possible alternative to the use of nuclear weapons, "as a form of sanction against a recalcitrant nation which would be more speedy than blockade and less repugnant than the atomic bomb". Also considered was their use in the security of the (British) Empire and its territories in the "destruction of food supplies of dissident tribes in order to control an area", and finally the report envisaged their use in the control of vegetation on "ranges" and "airfields". The extent of

US WWII anti-plant chemical warfare preparations, however, remains contested, with few references available to substantiate the assertion contained in Hanson's 1950 account,[29] which suggested that a quantity of chemical anti-plant agents had been shipped in 1945 to a location in the vicinity of Iwo Jima. Although Truman had denied the existence of any Presidential Order actioning this, or approving of the use of such chemicals for hostile purposes, Hanson noted the following, that: "In July and August, 1945, a shipload of US [anti-plant chemical agents] for use in [the] destruction of the Japanese rice crop was en-route to the Marianas [island archipelago, south, south-east of Japan]".

Following its establishment by the Ministry of Supply in 1948, one of the remits of the UK's Crop Committee was to devise "methods of destroying crops and other vegetation and countermeasures", and a Ministry of Supply report of the same year noted the extent of anti-plant warfare developments on both sides of the Atlantic. At the time, the report noted: "the subject of crop destruction is in a very underdeveloped state. During the War some progress was made, principally in the USA in the warlike application of chemical agents, but the quantities [required] raised formidable production and tactical problems ... In this country no work is being done other than research on weed killers for agricultural purposes." However, the six-monthly reports to the UK Advisory Council on Scientific Research and Technical Development showed that UK anti-plant chemical activities had continued between 1948 and 1955 but only on a limited scale, with US anti-plant work being conducted on a much larger scale with some 12000 chemical agents being screened during this period. In spite of apparent limitations regarding UK anti-plant activities, 2,4-D and 2,4,5-T had been used by the UK in testing, along with Endothall (HRAC herbicide class Z, unknown mechanism of action), in Tangyanika and in Kenya.

A precedent regarding use would be set when the *n*-butyl esters of 2,4-D and 2,4,5-T would be deployed by the British in the 14 year war against insurgencies in Malaya where they were used to significant effect in the defoliation of jungle-cover, and as part of a food denial programme in the destruction of food crops – tactics that would be repeated more than a decade later in Vietnam by the US, only on a much larger scale. Indeed, Cecil[30] noted the success of British tactics in Malaya as an approach that was: "later described by the general commanding officer in Malaya as one of the "decisive weapons" in anti-guerilla warfare". As noted in a Report[31] prepared for the Advanced Research Projects Agency (ARPA) by Rand, the tactics adopted by the British in Malaya in the 1950s recognised the importance of isolating insurgents from their sources of food. Food denial was thus physical separation as well as destruction of plants – the former being a policy that aimed to create a sense of security in populated areas which also sought to separate passive supporters of the insurgency from communist guerrillas. In December 1951, Cabinet Memoranda[32] reported on progress with the use of chemicals in anti-crop destruction deployments as follows:

> "Experiments on a small scale have been carried out using two recently discovered hormone weed killers (2,4-D and 2,4,5-T) with sodium trichloroacetate in various combinations. Tests began on the 17th September, 1951, by means of hand sprays, and within ten days the foliage was dead. Three months later no significant regeneration had taken place and the original growths are generally dead. Current observations strongly support the view that there will not be regeneration of any plant growth sufficient to provide cover for the enemy for six months, and probably for one year, after treatment... While these chemicals have been successfully applied by hand the advantage to be derived by this method of control by chemicals can be fully realized only by means of large-scale mechanical methods of application. It is now therefore an engineering problem".

The memorandum was clear about the intended aims of this method of control where use in small-scale clearance operations could be achieved through hand-spraying to reduce the likelihood of ambush. However, in regard to large-scale mechanical methods of application concerning the "... destruction of crops grown by, or for, the bandits in remote jungle areas", it was noted that:

> "...the Federal Government [had] ordered through I.C.I (Malaya) Ltd. two sets of spray rigs for use in Auster aircraft. This is the only purpose yet in mind for application by aircraft. Earliest date of delivery is February/ March, 1952, and despatch by air has been requested."

Regarding future use, Cecil[30] went on to note that, "the positive aspects of British experience in Malaya would form the basis for future American involvement in herbicidal warfare in Southeast Asia."

Although there is little evidence[††] to suggest that the US had deployed anti-plant chemical agents in the 1950 to 1953 war in Korea, Cecil notes that, under a limited initiative to understand the effectiveness of what was referred to as 'environmental warfare', North Korean food crops and production facilities were targeted and attacked by the US through the use of napalm. Significantly, Cecil[33] contends that US research and development with anti-plant agents during the early 1950s would result in the identification of a specific combination of chemicals as follows that, "American scientists at Fort Detrick had determined that the best vegetation control chemicals available were mixtures of the butyl esters of 2,4-D and 2,4,5-T". Such combinations would later be deployed by the US in Vietnam. Corresponding military developments followed the purchase of large quantities of chemicals by the US Air Force in 1953 and testing established the operational utility of 1000 gallon (Hourglass) spray tanks as aerial delivery systems of choice for the dissemination of chemical anti-plant agents allowing, as observed by Cecil,[33]

[††]Cecil notes the case of a civilian witness giving evidence at a Congressional Hearing some 20 years after the event that suggested limited use might have taken place in the closing stages of the war in Korea. For further information, see ref. 51.

B29 and B50 heavy bombers, and C119 transport aircraft to be, "quickly and easily turned into spray planes." Spray tanks for both dry and wet agents would eventually be loaded onto C-123 transport aircraft, the latter being particularly manoeuvrable in low-speed operations of the kind required for the dissemination of chemical spray. The US programme had been significantly curtailed due to budget cuts in 1958, but 1959 saw the first use of this combination of agents at Camp Drum, New York, where Air Force stocks of 2,4-D and 2,4,5-T that had been stored by the Department of Agriculture at Belsville, MD, were mixed in a "one-to-one ratio".[34]

10.2.4 Vietnam

Commercial chemicals used in agriculture in the US were first deployed in Vietnam in testing with 2,4-D and 2,4,5-T in 1961 and were found to be effective against vegetation, and shipments of large quantities of both agents began to arrive at an airbase in South Vietnam in January 1962.[9] During the initial stages of herbicide use in Vietnam, the chemicals were distinguished during shipment in metal drums with blue, green, orange, pink, purple and white stripes designating the particular chemical within. Both water- and oil-based preparations saw initial use and were found to be effective against forest and brushwood vegetation, and against rice and a range of other food crops.

Official statistics regarding area covered were not widely (publicly) available during the war, but the Herbicide Assessment Committee (HAC) included an unofficial estimate in its 1970 report to the Annual Meeting of the AAAS in the same year which stated that by 1969 acres of forest land and crop land sprayed with chemicals could have been as high as 1 221 415, and 65700, respectively.

Although acreage covered was significant, initial concern regarding use, however, was met by a range of responses, some noting the utility of use in relation to its apparent value in saving the lives of US service personnel, and others noting that use was consistent with US domestic application of commercial herbicides in peacetime. However, opposition to use soon began to coalesce around the huge quantities of agents deployed and evident damage to plants and trees and crops, and important questions began to be asked regarding environmental and human health implications. Initial official estimates of use between 1962 and 1964 suggested that 276 000 gallons of 2,4,5-T had been disseminated over Vietnamese forest. Just under 12 million gallons of agent orange were reported to have been deployed between 1965 and 1971. It is estimated that some 20000 herbicide missions were flown under the auspices of Operation Ranch Hand, as noted by Hay,[9] in missions that dispensed between 1962 and 1971 an estimated 19 million gallons of herbicide. Citing a US National Academy of Science Report, Hay contends that "10.3% of Vietnam's inland forests, 36.1% of mangrove forests, 3% of cultivated land, and 5% of 'other' land was affected by the spraying program". Of course, not all chemicals used in Vietnam could be classified as auxins and colloquial chemical names, and active ingredients of selected agents used in

Vietnam included: Orange, a 1 : 1 mixture of the *n*-butyl esters of 2,4-dichlorophenoxyacetic acid (2,4-D) and 2,4,5-trichlorophenoxyacetic acid. Purple was a different combination of chemicals to Orange. White 2,4,-D/Picloram, a 4 : 1 mixture of the tri-iso-propanolamine salts of 2,4-D and 4-amino-3,5,6-trichloro-picolinic acid in water. Blue (not auxin)/cacodylic acid, a 6 : 1 mixture of sodium dimethylarsenate and dimethylarsenic acid in water.

As noted by Meselson, research into the possible "carcinogenic, teratogenic, and mutagenic" effects of a large number of chemicals and pesticides would precipitate first, limitations on use, and subsequently would play an important role in a change in policy that would end the programme in Vietnam. The Bionetics Research Laboratory study, commissioned by the US Cancer Institute, suggested that 2,4,5-T might be dangerous to humans. According to Meselson, the Bionetics Study, "found that 2,4,5-T administered to pregnant mice and rats consistently caused a high frequency of malformations and death in fetuses and newborns and categorised 2,4,5-T as "probably dangerous". However, it transpired a short time later that the results of the Bionetics Study may have arisen from 27 ppm (parts per million) dioxin-contaminated 2,4,5-T, caused, according to Meselson, by 2,3,7,8-tetrachlorodibenzodioxin (TCDD) – the latter being thought to have been an unanticipated but highly-toxic bi-product of an incorrect production process involving synthetic chemical precursors. According to Meselson[35] who cites Young's 2009 work, "Dioxin [was] formed as an impurity... in the synthesis procedure for 2,4,5-trichlorophenol, a precursor of 2,4,5-T, particularly if carried out at too high a temperature." Not all suppliers of agent orange produced contaminated product and the source of the dioxin contamination has not been publically identified. Nevertheless, restrictions on the use of 2,4,5-T thus arose on 29th October 1969, with its use in the destruction of food crops being restricted on 1 January 1970, with all defoliation from fixed-wing aircraft ending in Vietnam by May 1970.

10.3 Prohibition

Until the mid-1970s the US had adopted a stance on the 1925 Geneva Protocol that maintained that herbicides were not covered by the prohibition and in spite of the US depositing its instruments of ratification in respect of the Protocol in 1975, the US would maintain this reservation – a stance that it would later repeat in relation to exemption insisted upon by the US under the Chemical Weapons Convention (CWC) which entered into force in 1997. Those campaigning to bring an end to use in Vietnam applied pressure on the US administration to reverse its position on herbicide use in 1966, to no avail. As noted by Hay,[9] "The government continued its stance that herbicides were not chemical weapons".

The CWC embodies a General Purpose Criterion[36] in Article II.1(a) and Article VI.2., of the Convention. According to Pearson, Article II of the CWC defines chemical weapons as including: "Toxic chemicals and their precursors, *except where intended for purposes not prohibited under this Convention, as long as types and quantities are consistent with such purposes*" As Pearson explains:

> "The phrase in bold (above) embodies the comprehensive scope of the Convention and is generally referred to as the General Purpose Criterion (GPC) which ensures that past, present and future chemical weapons are all prohibited. The implementation of the General Purpose Criterion is placed by Article VI on each State Party which "shall adopt the necessary measures to ensure that toxic chemicals and their precursors are only developed, produced, otherwise acquired, retained, transferred, or used within its territory or in any other place under its jurisdiction or control for purposes not prohibited under this Convention".

The CWC also includes requirements that its States Parties make declarations relating to the existence of chemicals that are listed in Schedules, and specific provisions are contained in the CWC in the form of requirements related to the destruction of stockpiles. However, the CWC was not explicit about what qualified as warfare, and herbicides do not appear in the Schedules and thus are not relevant to declarations under the CWC. According to the wording of the Convention,[37] regarding herbicides, it states that, "the prohibition of their use as a method of warfare is recognised in the CWC Preamble". However, although herbicides are not defined specifically in the Convention and there are no specific declaration or destruction requirements related to them, this does not preclude application of the general purpose criterion to chemicals traditionally considered as herbicides. In other words, as explained by Robinson,[38] "only if the intended purpose of a toxic chemical is not prohibited by the Convention shall the chemical not be considered a chemical weapon". The US Executive Order 11850, titled 'Renunciation of certain uses in war of chemical herbicides and riot control agents', contains the US exemption which amounts to a qualified but limited renunciation of herbicides as a method of warfare – one that permits their deployment in exceptional but specific circumstances. Executive Order 11850 states as follows that: "The United States renounces, as a matter of national policy, first use of herbicides in war except use, under regulations applicable to their domestic use, for control of vegetation within US bases and installations or around their immediate defensive perimeters."[39]

Deliberate damage to the environment became a highly controversial issue and played an important role in the ending of anti-crop and defoliation programmes in Vietnam. Indeed, the CWC reflects concerns raised in related legal instruments regarding the impact of war upon the environment. According to the Preamble[40] of the CWC, this recognises, "the prohibition embodied in the pertinent agreements and relevant principles of international law, of the use of herbicides as a method of warfare." An example of this is the Convention on the Prohibition of Military or Any other Hostile Use of Environmental Modification Techniques[41] (otherwise known as the ENMOD Convention) that was adopted by the UN General Assembly in 1976. In a statement that was reaffirmed in the Final Declaration of its Second Review Conference, according to the so-called ENMOD Convention, which was opened for signature in 1977 and entered into force October 1978 – "Each State Party to this Convention undertakes not to engage in military or any other hostile use of environmental modification techniques having widespread, long-lasting or severe effects as the means of destruction, damage or injury to any other State Party." As noted by the International Committee of the Red Cross (ICRC), deliberate starvation of civilians as well as attacking, destroying, removing or rendering useless objects indispensable to the survival of the civilian population was also prohibited under Article 54 (1&2)[42] of Additional Protocol 1 of the Geneva Convention.

10.4 Advances in Science and Technology

Discussion of 'auxins' now needs to be framed in the context of twentieth- and twenty-first century scientific and technological advances. This requires a brief overview of the history and evolution of synthetic herbicide discovery as well as the potential for rational-design and potential improvements that have emerged, or may in future, with respect to target specificity and effectiveness. By the 1990s, the five main classes of naturally-occurring endogenous plant hormones had been identified, namely the auxins, abscisic acid, cytokinins, ethylene, and gibberellins. In terms of regulating growth and of organised development in plant tissue and organs, as noted by Gaspar *et al.*[43] auxins, cytokinins, and auxin-cytokinin ratios were usually considered to be the most important hormones, whist abscisic acid, ethylene, gibberellins and other hormone-like components were known to have other important regulatory functions in plants. According to Gaspar *et al.*,[43] examples of those included "polyamines, jasmonates, brassinosteroids, oligosaccherins, sterols, salicylic acid, and systemins." In terms of the plant growth regulators, a significant number of synthetic compounds had been discovered up until this period with biological activity which equalled or exceeded that of naturally-occurring hormones. In this connection, extensive developments in commercial herbicides discovery had taken place since the discovery of 2,4-D and other important auxin herbicides, and Epp[44] and co-workers organise the most successful of these that act "*via* an auxinic mode of action" in relation to the following classes, namely: "the aryloxyacetates (2,4-D, MCPA, dichloroprop, mecoprop, triclorpyr, and fluroxypyr), the benzoates (dicamba), the quinoline-2-carboxylates (quinclorac and quinmerac), the pyrimidine-4-carboxylates (aminocyclopyrachlor), and the pyridine-2-carboxylates (picloram, clopyrialid, and aminopyralid)." Picloram and other important picolinate auxin herbicides were discovered in the 1950s and in the 1960s respectively, with 2,4-D, MCPA, the trichlorophenoxys and picloram all being used in Vietnam.

Over time, research and development in phytobiology transitioned from the identification of the different classes of plant growth hormones as listed above to the identification and understanding of hormone receptors as well as the characteristics of binding. Napier[45] considers the implications for future work in this area in terms of advances in understanding ligand binding, and, the following observation is perhaps characteristic of phytobiology transitioning from a descriptive to a predictive science. He notes that: "An understanding of binding provides an entry for molecular modelling and the possibility of knowledge-led design of new and modified ligands, as well as ways of altering sensitivity to ligands. Knowledge of the signalling functions of the receptor will help identify downstream components in the signalling cascade and offer possibilities of manipulating physiological responses to hormones. Progress in this area of plant biology has proved to be slow, although some significant work done in relation to the identification of ligand binding sites and receptors over the last 10 years has led to considerable advances concerning the characterisation of known plant hormone binding sites and receptors. As Napier[45] notes, synthetic analogues of the major classes of plant hormones were described in the phytobiology literature in the 1990s, and that in the prevailing period that the identification of previously unknown binding sites may be derived from research techniques that involve genetic or biochemical screening.

10.5 Conclusions

From a scientific perspective, the discovery of auxins signaled a paradigm shift in phytobiology. In agriculture, auxin herbicides have allowed for the maximization of crop production and the simplification of crop management.[44] Indeed, according to Fryer,[46] such substances have, "transformed agriculture and [are] considered to be amongst the greatest scientific discoveries." Although still an issue of some contention, the phytobiology literature has gone some way to dispel confusion around the discovery[47] of endogenous auxins, and to dispel some of the confusion around the discovery[20] of the first synthetic herbicides. Following application of the latter in military programmes, and the precedent set by the UK regarding use in Malaya, their deployment in the Second Indo-China War had far-reaching implications for human and animal health and was the cause of unprecedented long-term damage to the environment. Given the comprehensiveness of arms control prohibition and related environmental regimes that ban and regulate the use of chemical weapons in warfare, it is difficult—given the acute sensitivity that exists concerning environmental stability and sustainability in the twenty-first century—to imagine a case outside of all-out war between states, where a country might use chemical anti-plant agents on a scale to parallel the use in Vietnam. Outside of terrorism, this leaves open the possibility for use, nevertheless, in the case of anarchic warfare scenarios within states, such as in Syria, for example, where anti-personnel chemicals have been used in a way that is no doubt damaging—as is terrorism—to the norm of non-use that is embodied in the relevant treaties and conventions. However, assuming that sufficient political will exists to move states to act concerning use, alleged state use in the latter context is surely a matter for the Chemical Weapons Convention, the OPCW, and UN Security Council, and use by non-State actors in this context a criminal matter for the state concerned. The use of such chemicals outside of warfare in the context of "peaceful" drug-eradication programmes remains a matter of contention even though such activities may remain faithful to the letter of the general purpose criterion of the Chemical Weapons Convention. In his seminal article titled 'Averting the Hostile Misuse of Biotechnology' in 2000, Meselson lamented that every major technology had been used for hostile purposes. This somewhat downbeat assessment of human propensity to divert its ingenuity to destructive capacity led Meselson to question whether or not biotechnology and genetic engineering would suffer the same fate. However, in so far as the case of auxins are concerned, the answer to Meselson's question remains elusive where progress in the peaceful uses have been, and continue to be, immensely beneficial in agriculture and sustainable food production, so much so that they continue to be very much a work-in-progress requiring considerable intellectual and technological investment.[48] This short survey of the use of chemicals against plants underlines, however, the dual-use quality of science and technology, as well as the importance of organizing science and security in such a way as to ensure that human ingenuity is reserved only for beneficent purposes. Upholding the norm of non-use in relevant prohibition and regulatory regimes is essential. This will assist in strengthening the web of preventative measures required to deter and respond to misuse. Two further initiatives may strengthen the regime of compliance around the non-use of chemicals against plants, namely: raised levels of awareness and education concerning the ethical, legal, and social implications of research and development in phytobiology and in the field of science more generally; and, closer association and collaboration between phytobiologists, scientists, and security communities.

References

1. (a) *The Problem of Chemical and Biological Warfare: The Rise of CB Weapons*, Stockholm International Peace Research Institute, Stockholm Humanities Press, Almquvist & Weksell, New York, 1971, vol. 1; (b) *The Problem of Chemical and Biological Warfare: CB Weapons Today*, Stockholm International Peace Research Institute, Almquvist & Weksell, Stockholm Humanities Press, New York, Paul Elek, London, 1973, vol. 2.
2. L. P. Brophy, W. D. Miles and R. C. Cochrane, *The Chemical Warfare Service: From Laboratory to Field*, Office of the Chief of Military History, Department of the Army, Washington DC, 1959.
3. G. Carter and G. S. Pearson, North Atlantic Chemical and Biological Research Collaboration: 1916-1945, *J. Strateg. Stud.*, 1996, **19**(1), 74–103.
4. J. Bryden, *Deadly Allies*, MacLelland and Stewart, Toronto, 1989.

5. S. Whitby, *Biological Warfare against Crops*, Palgrave, 2001.
6. P. F. Cecil, *Herbicidal Warfare: The Ranch Hand Project in Vietnam*, Praeger, New York, London, 1986.
7. A. H. Westing, *Hebicides in War: The Long-term Ecological and Human Consequences*, SIPRI, Taylor and Francis, London, 1984.
8. T. Whiteside, *What Are Our Herbicides Doing to Us? Defoliation, Friends of the Earth Books*, 1970.
9. A. Hay, Vietnam and 2,4,5-T, in *The Chemical Scythe; Lessons of 2,4,5-T and Dioxin, Disaster Research in Practice*, ed. A. Hay, Springer, USA, 1982.
10. S. Karnow, *Vietnam: A History*, Penguin, 1983.
11. M. Meselson, *From Charles and Francis Darwin to Richard Nixon: The Origin and Termination of Anti-plant Chemical Warfare in Vietnam, Proceedings of the Symposium – 100 Years of Chemical Warfare: Research, Deployment, Consequences*, Berlin, 21–22 April 2015.
12. C. Darwen and F. Darwen, *(1880) the Power of Movement in Plants, Cited in Meselson, M (2016) from Charles Darwen to Richard Nixon: The Origin and Termination of Anti-plant Chemicals in Vietnam, Symposium Proceedings, 100 Years of Chemical Warfare: Research, Deployment and Consequences*, John Murray, Berlin, London, 21–22 April 2015, p. 1.
13. P. J. Davies, The plant hormones: their nature, occurrence, and functions, in *Plant Hormones, Biosynthesis, Signal Transduction, Action!* ed. P. J. Davies, Klewer Academic Publishers, 2004.
14. F. W. Went and K. V. Thimann, *Phytohormones*, Macmillan, 1937, p. 7.
15. F. W. Went and K. V. Thimann, *Phytohormones*, Macmillan, 1937, p. 9.
16. F. W. Went, Wuchsstoff und Wachstum, *Rec. Trav. Bot. Neerl.*, 1928, **25**, 1–116.
17. F. Kogl and A. J. Haagen-Smit, Uber die Chemie des, in *Proc. Sect. Sci.*, ed. W. K. Akad, Wetenschap, Amsterdam, 1931, vol. 34, pp. 1411–1416.
18. J. B. Epp, A. L. Alexander, T. W. Balko, A. M. Buysse, W. K. Brewster, K. Bryan, J. F. Daeuble, S. C. Fields, R. E. Gast, R. A. Green, N. M. Irivine, W. C. Lo, C. T. Lowe, J. M. Rega, J. S. Richburg, J. M. Ruiz, N. M. Satchivi, P. R. Schmitzer, T. L. Siddall, J. D. Webster, M. R. Weimer, G. T. Whiteker and C. N. Yerkes, The Discovery of Arylex TM active and Rinskor TM active: Two Novel Auxin Herbicides, *Bioorg. Med. Chem.*, 2015, **24**, 362–371.
19. P. J. Davies, The Plant Hormones: Their Nature, Occurrence, and Functions, in *Plant Hormones, Biosynthesis, Signal Transduction, Action!* ed. P. J. Davies, Klewer Academic Publishers, 2004, p. 2.
20. J. R. Troyer, *Weed Sci.*, 2001, **49**, 290–297.
21. P. W. Zimmerman and E. Wilcoxon, Several chemical growth substances with case initiation of roots and other responses in plants, *Boyce Thompson Inst.*, 1935, 7, 209–229.
22. S. Whitby, *Biological Warfare against Crops*, Palgrave, 2001, p. 123.
23. *Memorandum from Sir John Anderson to the Prime Minister (Winston Churchill) on Crop Destruction*, 26 June 1944.
24. C. Reinhardt, *Chemical Sciences in the 20th Century: Bridging Boundaries*, Wiley Publishers, 2007, p. 216.

25. L. P. Brophy, W. D. Miles and R. C. Cochrane, *The Chemical Warfare Service: From Laboratory to Field*, Office of the Chief of Military History, Department of the Army, Washington DC, 1959, p. 45.
26. M. Gowing, *Independence and Deterrence: Britain and Atomic Energy: Policy Making*, 1945–1952, vol. 1, p. 94.
27. *Letter from Sir John Anderson to Vannevar Bush*, Director, Office for Emergency Management, Office of Scientific Research and Development (OSRD), 19 April 1944.
28. Note by Joint Secretaries, Potentialities of Weapons of War During the Next Ten years, Chiefs of Staff Committee, Joint Technical Warfare Committee, 12 November 1945. Enclosed Report by the Inter-Services Sub-Committee on Biological Weapons, in *Biological Warfare against Crops*, ed. S. M. Whitby, Palgrave, London, 2001, p. 240.
29. B. W. Hanson, *Great Mistakes of the War*, New York, 1950, pp. 88–107.
30. P. F. Cecil, *Herbicidal Warfare: The Ranch Hand Project in Vietnam*, Praeger, New York, London, 1986, p. 17.
31. R. W. Komer, *The Malayan Emergency in Retrospect: Organization of a Successful Counterinsurgency Effort, a Report Prepared for the Advanced Research Projects Agency, Rand Corporation, R-957-ARPA*, February 1972, ARPA Order No: 189–1.
32. His Britannic Majesty's Government, *Cabinet Memoranda, Cabinet, Malaya: Memorandum by the Secretery of State for the Colonies, Chemical Defoliation of Roadside Jungle, CAB 129/48, Original Reference C 1(51)-62 (51)*, 1951 31 October–29 December 26, pp. 267–268, Appendix VII.
33. P. F. Cecil, *Herbicidal Warfare: The Ranch Hand Project in Vietnam*, Praeger, New York, London, 1986, p. 18.
34. P. F. Cecil, *Herbicidal Warfare: The Ranch Hand Project in Vietnam*, Praeger, New York, London, 1986, p. 19.
35. M. Meselson, *From Charles and Francis Darwin to Richard Nixon: The Origin and Termination of Anti-plant Chemical Warfare in Vietnam, Proceedings of the Symposium - 100 Years of Chemical Warfare: Research, Deployment, Consequences*, Berlin, 21–22 April 2015.
36. The General Purpose Criterion, *Article II, Chemical Weapons Convention*, 29 April 1997.
37. Organisation for the Prohibition of Chemical Weapons, *Brief Description of Chemical Weapons, Chemical Weapon as Defined by the CWC, CW Agent Group, Persistency Rate of Action*, https://www.opcw.org/about-chemical-weapons/what-is-a-chemical-weapon.
38. J. P. Perry Robinson, Private communication to author, July 2017.
39. *Executive Order 11850, 'Renunciation of Certain Uses in War of Chemical Herbicides and Riot Control Agents', the National Archives*, April 8 1975, https://www.archives.gov/federal-register/codification/executive-order/11850.html.
40. *Preamble, Chemical Weapons ConventionL Recognizing the Prohibition, Embodied in the Pertinent Agreements and Relevant Principles of International Law, of the Use of Herbicides as a Method of Warfare*, 1997, https://www.opcw.org/chemical-weapons-convention/preamble/.

41. *Convention on the Prohibition of Military or Any Other Hostile Use of Environmental Modification Techniques*, http://www.un-documents.net/enmod.htm.
42. *Additional Protocol 1, Protection of Objects Indispensable to the Survival of the Civilian Population, Protocol Additional to the Geneva Conventions of 12 August 1949, and Relating to the Protection of Victims of International Armed Conflicts*, 8 June 1977, https://ihl-databases.icrc.org/ihl/INTRO/470.
43. T. Gaspar, C. Kevers, C. Penel, H. Greppin, D. M. Reid and T. A. Thorpe, Plant Hormones and Plant Growth Regulators in Plant Tissue Culture, *In Vitro Cell. Dev. Biol.: Plant*, 1996, **31**, 272–289.
44. J. B. Epp, A. L. Alexander, T. W. Balko, A. M. Buysse, W. K. Brewster, K. Bryan, J. F. Daeuble, S. C. Fields, R. E. Gast, R. A. Green, N. M. Irivine, W. C. Lo, C. T. Lowe, J. M. Rega, J. S. Richburg, J. M. Ruiz, N. M. Satchivi, P. R. Schmitzer, T. L. Siddall, J. D. Webster, M. R. Weimer, G. T. Whiteker and C. N. Yerkes, The Discovery of Arylex TM active and Rinskor TM active: Two Novel Auxin Herbicides, *Bioorg. Med. Chem.*, 2015, **24**, 362–371.
45. R. Napier, Plant Hormone Binding Sites, *Ann. Bot.*, 2004, **93**(3), 227–233.
46. J. D. Fryer, Foreword, in *The Hormone Weedkillers, A Short History of Their Discovery and Development*, ed. C. Kirby, British Crop Protection Council, 1980.
47. T. A. Enders and L. C. Strader, Auxin Activity: Past, Present and Future, *Am. J. Bot.*, 2015, **102**(2), 180–196.
48. R. Napier, *The Napier Laboratory, Interview*, University of Warwick, 14 March 2017.
49. J. R. Troyer, In the Beginning: the Multiple Discovery of the First Hormone Herbicides, *Weed Sci.*, 2001, **49**, 290–297.
50. F. W. Went and K. V. Thimann, *Phytohormones*, Macmillan, 1937, p. 4.
51. P. F. Cecil, *Herbicidal Warfare: The Ranch Hand Project in Vietnam*, Praeger, New York, London, 1986, p. 18.

CHAPTER 11

The Future of Chemical Weapons: Advances in Anti-animal Agents

P. D. MILLETT

University of Oxford, Oxford Martin School, 34 Broad Street, OX1 3BD, Oxford, UK
*E-mail: piers.millett@philosophy.ox.ac.uk

11.1 Weapons Targeting Animals

There is little evidence that traditional chemical warfare agents have been developed or used specifically to target animals. Some animals have been affected by chemical weapons, for example during their use on the Western Front during the First World War. Specialised protective equipment for animals, such as horses, was developed at that time to protect them against the use of chemical weapons. These weapons were likely not developed specifically to target animals, but it may have been a welcome side effect.

Affecting animals as a form of secondary impact is recognised under the Chemical Weapons Convention. The definition of a toxic chemical (which are the agents regulated under the treaty) covers "Any chemical which through its chemical action on life processes can cause death, temporary incapacitation or permanent harm to humans or animals."[1] This important international instrument, therefore, explicitly prohibits the development of anti-animal chemical weapons.

Preventing Chemical Weapons: Arms Control and Disarmament as the Sciences Converge
Edited by Michael Crowley, Malcolm Dando and Lijun Shang

Published by the Royal Society of Chemistry, www.rsc.org

Whilst there is limited historical evidence of anti-animal chemical weapons, there is a much stronger evidence base for the development and use of biological weapons specifically intended to target animals. An overview of this history is provided through a series of case studies in Section 11.2.

This history, in part, led to anti-animal biological weapons also being banned under international law. Sequential review conferences of the Biological Weapons Convention determined:[2]

> "...all naturally or artificially created or altered microbial and other biological agents and toxins, as well as their components, regardless of their origin and method of production and whether they affect humans, animals or plants, of types and in quantities that have no justification for prophylactic, protective or other peaceful purposes, are unequivocally covered."

Based upon the well-established trend towards the convergence of the sciences (in particular chemistry and biology), and the inclusion of toxins as both chemical and biological weapons agents, recent developments in science and technology could alter interest in, or the potential to develop anti-animal biochemical warfare agents. This chapter also provides an overview of recent developments in the life sciences that could increase or decrease the risk of deliberate disease in Section 11.3.

Lessons learned from historical case studies and the insights derived from efforts to consider the implications of developments in science and technology are then combined to consider the future of anti-animal biochemical weapons in Section 11.4.

11.2 Past Offensive Anti-animal Activities

A short history of the use of animal disease as a weapon was included in a recent publication by the World Organization for Animal Health (OIE).[3] This section uses this history to provide an overview of the efforts of States and non-State actors to acquire these weapons. This chapter does not add additional details to this history. It is intended, however, to demonstrate to the reader that not only have these weapons been desired, but that they have actually been deployed on a number of occasions – with questionable efficacy.

11.2.1 German Anti-animal Weapons During World War I

The German biological weapons programme solely targeted animals. It applied recent understanding of germ theory to design and instigate a programme intended to cause deliberate animal disease outbreaks. In particular, the programme targeted militarily-significant animals used "for military logistics, war fighting and as part of the food supply, including horses, donkeys, mules, cattle and reindeer".

The OIE history documents "the widespread use of these weapons. Germany used biological weapons to attack animals in five countries across three continents. Attacks were carried out in Argentina, Norway, Romania, Spain, and the United States".

Details are also provided as to how the agents were produced:

> "In some cases, clandestine forward production facilities were established in target countries. For example, a laboratory in the basement of a house in Washington D.C. produced the agents used in attacks in the United States. In other cases, the agent was produced in scientific institutes in Germany and shipped internationally. Germany made use of spies, diplomatic pouches and even submarines to deliver agents."

The agents and delivery devices used have also been documented:

> "This programme focused on a limited number of diseases, namely glanders and anthrax. It deployed very basic delivery devices. Surviving documentation describes a brush being used to spread agent onto the noses of animals in the United States and sugar cubes with capillary tubes laced with agent being seized in Scandinavia."

Historical records show that German agents reported relatively large losses of animals, in particular those transported from the USA and destined for use in Europe. Allied records confirming such losses have yet to emerge. As a result, the impacts of these efforts are uncorroborated and disputed by experts.[4]

11.2.2 UK Anti-animal Weapons During World War II

Whilst the German anti-animal biological weapons programme during World War I was based upon sabotage, the Allied response during World War II was focused at a more tactical and strategic level. Based upon the earlier German use of biological weapons, Allied forces suspected Axis powers of developing a new weapon making use of the foot-and-mouth disease (FMD) virus.

The UK explored a range of different anti-animal biological weapons, including those intended to cause glanders, swine fever, rinderpest and foot-and-mouth disease. Basic studies on the causative microbes explored how they survived on foodstuffs and grass and research into defensive measures was also undertaken.[5] The UK's most comprehensive response, code named Operation Vegetarian, involved producing:[3]

> "...five million anthrax-laced cattle cakes in less than two years. They were to be distributed over German agricultural land using the flare chutes of long-range bombers. A London-based soap manufacturer produced the empty cattle cakes. Specialist facilities at the chemical and biological

warfare establishment at Porton Down grew *Bacillus anthracis* in converted milk churns. The agent was then processed and the cattle cakes filled. Most of the cakes were destroyed soon after the end of hostilities, and the remainder destroyed in 1972."

Following the end of World War II, the UK began to explore virus-based anti-animal biological weapons, including those using the causative agents of foot-and-mouth disease, rinderpest, Newcastle disease, and fowl plague. At the same time, there was a shift from strategic weapons overtly intended to disable food production, to sabotage-based weapons where it would not be clear that an attack had even taken place. Key parts of the necessary infrastructure were transferred from military to civilian government departments, but additional resources were allocated specifically for continuing both offensive and defensive research and development.

The end of the UK's programme is also documented and the "interest in weapons waned in 1950s and offensive activities were abandoned all together in 1956".[3]

11.2.3 US Anti-animal Weapons During the Cold War

Following the end of World War II, the USA continued to develop strategic level biological weapons for targeting Soviet livestock including cattle, horses and swine. The USA developed weapons based upon six agents split into three categories. The first included the causative agents of FMD and rinderpest; the second the causative agents of classical swine fever, Newcastle disease, and fowl plague; and the third was the causative agent of African swine fever. To deliver these agents, the USA developed:[3]

- "spray tanks to deliver the causative agents of classical swine fever and Newcastle disease;
- balloon bombs for use as a strategic weapon with multiple agents;
- a feather bomb (converted from a leaflet dispersal device) designed to deliver multiple agents; and
- a cluster bomb, still under development during the active phases of the anti-animal programme".

As was the case in the UK, there were specialist facilities dedicated to the development of anti-animal biological weapons. They were slowly transferred from military to civilian command, for example, the anti-animal biological weapons facility run by the US Chemical Corps between 1952 and 1954 as Fort Terry became the Plum Island research facility when it was transferred to the Department of Agriculture in 1954.

Anti-animal biological development was not confined to dedicated facilities but also integrated into the broader offensive weapons programmes. In addition to Fort Terry, there were anti-animal and anti-crop components present within the broader US biological weapons programme, including dedicated sections at the programmes primary research facility at Fort Detrick. Equally, research and development were undertaken in consort with the UK and Canada as part of the broader tri-partite collaboration on biological weapons.

As with the UK, the end of the programme is well documented. Despite having tested and prepared two anti-animal weapons by 1957, the programmes "to develop strategic weapons were abandoned at around this time. Some work on sabotage weapons may have continued until the US unilaterally rejected biological weapons in 1969".[3]

11.2.4 USSR Anti-animal Weapons During the Cold War

Whilst there are comprehensive academic histories of many elements of the Soviet biological weapons programmes, fewer details have been recorded about anti-animal and anti-crop activities. Some details have been included in academic histories and the presence of an anti-animal programme code-named Ecology and ran by the Ministry of Agriculture is widely accepted. Ecology:[3]

> "is reputed to have employed around 10 000 people and applied genetic engineering techniques to enhance the ability of pathogens, including to increase virulence, circumvent or defeat defences (such as vaccines, antibiotics, and detection technologies) as well as to produce novel genetic constructs that cause different symptoms."

The programme is believed to have explored the weaponization of a wide variety of animal diseases, including those that infect pigs, cows, fowl, goats and sheep. Diseases investigated included: African swine fever, Aujeszky's disease, bluetongue, classical swine fever, FMD, fowl pox, goat pox, Newcastle disease, ornithosis, psittacosis, rinderpest, and sheep pox.

As with the UK and the USA, the Soviet Union had facilities dedicated to working on anti-animal biological weapons, including: two research and development facilities – the Scientific Research Agricultural Institute in Kazakhstan and the Scientific Research Institute for Animal Protection in Russia. There was also a production facility, disguised as an animal vaccine facility, at Pokrov in Russia. Anti-animal biological weapons were tested at facilities connected with the anti-personnel bioweapons programme, in particular the Scientific Research Agricultural Institute in Kazakhstan, and VECTOR (the State Research Centre for Virology and Biotechnology) in Russia.

There have also been suggestions that the Soviet Union used some of these weapons, in particular allegations that some of these weapons were used against the horses of the Mujahideen in Afghanistan.[3] There is much less documentation as to the end of the Soviet biological weapons programme than other programmes discussed in this chapter.

11.2.5 Non-State Actors and Anti-animal Weapons

Whilst comparatively rare, there have been examples of non-State actors attempting to acquire anti-animal biological weapons. For example, the Weapons of Mass Destruction (WMD) Terrorism Database maintained by Middlebury Institute of International Studies at Monterey (MIIS), in 2006, identified only 25 out of 852 incidents targeting plants or animals, only 4 of such incidents resulted in animal deaths.

The history published by the OIE contains details of two more recent reviews. It noted that the RAND Database of Worldwide Terrorism Incidents (RDWTI) contained only 13 incidents involving a biological weapon out of 40 126 records and none of those targeted animals. The review also noted that the University of Maryland Global Terrorism Database (GTD) included 35 biological weapon-related incidents (out of 125 087 records) and that none targeted animals, but one was carried out by animal rights activists.

The level of sophistication of these efforts, the agents employed, and the motivation of those seeking these weapons differs in all of the cases documented. For example:

> "They have been used as part of open hostilities. For example, in 1952 in Kenya, the Mau used extracts from African milk bush to poison 33 cattle, killing 8, as part of their efforts to undermine colonial rule... On other occasions, these weapons have been used with more benign aims. For example, in 1997 in New Zealand, farmers smuggled a pathogen into the country and used it an attempt to control rabbit populations... There are also allegations of attacks on economic grounds. For example in 2000, *New Scientist* reported a Brazilian Minister of State asserting that an FMD outbreak in south of country had been started deliberately as economic sabotage".

11.3 New Hazards – The Impact of Scientific and Technical Developments on Anti-animal Weapons

The convergence of chemistry and biology, in particular the ability to use chemistry to make biological agents, and biology to produce chemicals, requires the re-examination of anti-animal weapons across the spectrum of chemistry and biology.[6]

Recent developments in science and technology play a pivotal role in mitigating the risks posed by biochemical weapons, including those targeting animals. For example, a report of a 2015 workshop by the IAP: Global Network of Science Academies noted that over the last five years "our collective capacity to combat disease has markedly improved, regardless of whether the outbreak is naturally occurring or the result of a malevolent act".[7] A summary of recent developments is provided in Section 11.3.1.

The international scientific community, however, has long recognised that the same advances can be misused to cause deliberate harm. For example, in 2004 the US National Academy of Sciences noted that "almost all biotechnology in service of human health can be subverted for misuse by hostile individuals or nations".[8] Some of the possible implications of recent developments are summarized in Section 11.3.2.

A thorough exploration of all recent relevant developments in science and technology is beyond the scope of this chapter. The review conducted by the IAP drew on a comprehensive annotated bibliography and resulted in over 50 pages of conclusions. This section provides an overview of developments pertinent to anti-animal biochemical weapons. Some of the advances discussed were derived from research specifically focused on animals and animal disease, many more have come from other fields, such as research on human diseases, or public health. In other cases, key advances are foundational in nature, perhaps improving our understanding of common disease-causing mechanisms, or allowing the development of tools that facilitate efforts across a broad variety of disciplines. As a result, this chapter considers a broader range of advances than those specifically relating to animal disease. How these advances might impact upon the future of anti-animal biochemical weapons is the subject of the next section.

One typology for considering relevant advances would be to break them down by scientific fields, sub-fields or disciplines (for example, those from microbiology, synthetic biology, neuroscience, *etc.*). Such an approach does not fully illustrate either the cross-cutting nature of many of the key developments (for example, the implication of enabling technologies such as highly portable nanopore-based gene sequencing, or improvements in genome editing tools, which impact many relevant fields); or how these developments interact and feed off one another to result in a collective impact greater than the sum of their parts. For example, the IAP: Global Network of Science Academies noted in 2015 "[I]nterconnections between pathogenicity, transmissibility and drug resistance *e.g.* mutations that can confer additional functional characteristics without degrading others."[7] As a result, this chapter follows the approach they used to identify trends in science and technology pertinent to the Biological Weapons Convention and uses a typology based upon the impact of the advances, rather than their origin.

11.3.1 Developments in Science and Technology Decreasing the Hazards

There have been a wide variety of recent advances that have improved our ability to deal with disease, regardless of cause. This section reviews improvements in: our understanding of detecting, diagnosing and surveillance, as well as preventing, treating and mitigating disease and intoxination (Sections 11.3.1.1–4 respectively).

11.3.1.1 Understanding Disease and Intoxination

Relevant developments have improved our understanding of transmissibility and host range; pathogenicity and virulence; toxins; disease agents, other than bacteria and viruses; immunology and host-pathogen interactions; the role of the microbiome; and the significance of 'biofilms' for pathogen persistence.

There has been notable progress in structural and genetic analysis, production or isolation of pathogens with increased transmissibility or host range, as well as improved tools to investigate them. Relevant examples of specific genetic sequence information connected to transmissibility were identified from amongst recent influenza research. The identification and characterization of important structural elements that affect the host range have been highlighted. In addition to research involving influenza viruses, examples of such research were noted for the dengue fever virus and other pathogenic viruses. Tools that have enabled this research included: "tools to study genetic evolution of viruses and search for mutations known to confer altered transmissibility; synthetic biology approaches; approaches for characterizing host shifts for rabies viruses in bats and Ebola virus between pigs and non-human primates; and model systems that recapitulate *in vivo* viral life cycles".[7]

In terms of pathogenicity and virulence, there have been advances that enable relevant genetic analysis with examples being drawn from research involving bacterial pathogens (such as particularly lethal *escherichia coli*, MRSA, and *mycobacterium tuberculosis*), viral pathogens (such as the SARS coronavirus, influenza viruses, HIV, dengue fever virus, and Ebola Virus), plant pathogens (such as wheat yellow rust), as well as zoonotic pathogens (such as *bacillus anthracis*). Progress in understanding how transcription impacts pathogenicity and virulence has also been highlighted, including relevant research involving influenza viruses. Structural impacts have also been noted, including the role of intracellular nanotubes, secretion systems, capsid components and virus-host structural components. A number of other factors that might impact pathogenicity and virulence have been highlighted, including the physical location of a pathogen in a host, the ability to adapt to different temperatures as well as the manipulation of the host by pathogens, such as by modulating the damage caused by cytokine storms, or inflammatory responses. A host of new tools and approaches have enabled these advances, including those in sequencing genetic material, handling, manipulating and modelling data, studying and building biological systems, and testing clinical impacts.

Toxin research has also benefitted from notable recent advances. This has included "incremental increases in our understanding of the production and action of toxins".[7] This has drawn on improvement in "structural and genetic analysis, characterisation of the mechanisms of action and novel toxins and drugs to neutralise toxins, detection procedures and improved tools to investigate these".[7] This enhanced ability to understand and manipulate the action of toxins has led to an increased interest in using them as research tools, in agriculture or in pharmaceuticals or in the development of novel medical treatments. There has been notable research involving ricin, botulinum toxins, and abrin. A number of newly identified toxins or those with altered or enhanced characteristics were also discussed, as were shortcomings in current detection and diagnostic capacity.[7] Several new drugs to neutralize the effect of toxins were also identified.

There has also been progress in our understanding of disease caused by agents other than bacteria and viruses, including fungi, plasmodia and prions. Relevant advances covered investigations into important historical epidemics, characterization of the genetic elements associated with disease, comparisons with non-diseased states and improvements in screening techniques.

Progress in understanding host-pathogen interactions was noted including in agents potentially relevant to deliberate harm, such as: *yersinia pestis*; flaviviruses such as dengue; influenza; SARs coronaviruses; *brucella* sp.; and close relatives of the smallpox virus. These studies have revealed possibilities for modulating immune responses, opening the door to being able to decrease the clinical impact of a pathogen, for example by reducing inflammatory reactions or preventing cytokine storms. There has also been notable improvement in our understanding of how pathogens evade the immune system. These relate to both innate and acquired immunity – offering opportunities to improve our bodies' abilities to deal with disease and to increase the efficacy of treatments and prophylaxis. This has been complemented by developments in reducing the side-effects of important drugs, such as hypersensitivity caused by certain small molecules. A number of new and improved tools and research methodologies have facilitated these advances, including enhanced data generation, data handling and modelling, better structural analysis, progress in neurobiology as well as imaging techniques.

Research into how pathogens interact with our commensal microbial populations has helped to identify and characterize novel mechanisms connected to antimicrobial resistance – offering opportunities to better optimise the use of these therapeutics. This has drawn upon efforts to map "the structure, function and diversity of the health human microbiome, providing a useful normal data set against which disease states can be compared".[7]

The role of biofilms in pathogenesis has also become better understood in recent years, with improvements of our understanding in how they form, their contribution to pathogenesis and how they enable horizontal gene transfers.

11.3.1.2 Detecting Disease and Intoxination

There has been notable progress in detecting disease and intoxination. The IAP noted "In particular, there have been significant advances in biosensors, biomarkers, mass spectrometry, microscopy and imaging". They also stressed that there were still significant hurdles to overcome, for example "More comprehensive baseline data is needed to assist in comparisons and establish a 'norm' against which to compare unusual events".[7]

In addition, biosensors, "have been improved by increased sensitivity and ability to detect a wider range of agents using a wider range of samples and to detect functional changes in the host."[7] Progress in nanoparticles has been particularly important but a number of other areas have been noted, including synthetic biology and the design, construction and use of engineered organisms, plasmon resonance sensors, planar waveguide devices, DNA scaffolds, quantum dots, and surface enhanced Raman scattering.

We have a greater capacity to identify biological markers for disease (biomarkers). The IAP suggests this is important as it "improves our understanding of disease progression, enables earlier detection of disease, differentiates between different infections, and facilitates triaging of patients and epidemiological investigations".[7] Relevant biomarkers have been located in a wide range of different tissues and locations within the body and have been used to characterise cellular signalling behaviour, identify pre-clinical indicators of infection, differentiate between clinically similar infections, investigate the interactions of drugs, explore how pathogens modulate gene expression, determine likely clinical outcomes of disease in individual patients, and catalogue the infection history of a subject. There have also been notable developments in the ability to use biomarkers to detect rare disease events.

Progress has also continued in more traditional approaches to detecting disease. For example:

- Mass spectrometry established research tools, such as MALDI-TOF "have entered into routine use for detections and diagnosis".[7] Uses of these platforms have included agent identification, disease surveillance, microbial forensics, the detection of complex biological samples, the detection of antibiotic resistance, and differentiation between strains of a pathogen; and
- In microscopy and imaging, recent advances have enabled "direct observation of proteins as they conform and move; 3D imaging on a sub-cellular level in live cells...; tracking the course of pathogens during infection of a living host; and building all-atom models of viral capsids".[7] Ultra-cheap and repurposed equipment has also become available in recent years.

These developments have also enabled more sophisticated capabilities for differentiating between deliberate and natural outbreaks, which is key for determining that an unusual disease event has an unnatural origin. Of particular note were advances in "genomics, PCR and mass spectrometry and the emerging discipline of 'microbial forensics' [that] can help establish attribution if a malevolent deployment is suspected".[7]

11.3.1.3 Diagnosing and Disease Surveillance and Intoxination

There have been advances that help to "identify causative agents more rapidly, expediting the selection of optimal treatment options and preventing transmission".[7] These have included:[7]

> "...rapid diagnosis of unknown pathogens; sequencing; PCR diagnostics; distributed diagnostics and point-of-care devices; centralisation of certain types of laboratory capacity; genetic and molecular epidemiology; as well as the use of cheap and disposable equipment. Substantial improvements in diagnostic speed and accuracy were noted... Advances in metagenomics have enabled the identification of previously unculturable microorganisms, previously unidentified pathogens, and the identifications of pathogens from within mixed environmental samples."

Of particular relevance to this chapter is the progress made in microbial forensics drawing on advances in: whole genome sequencing; comparative genomics; bioinformatics; metagenomics; mass spectrometry; and applications of nanotechnology. In some cases, these tools "have been validated for use in bioforensics, such as for high-throughput genome sequencing platforms".[7]

They noted a number of new capabilities that have come on line in recent years, including:[7]

- "linking specific phenotypic qualities of an agent with sequence data;
- using sequence data to differentiate an organism of interest from background microbes;
- cataloguing genomic diversity and mapping circulating mutations, providing better reference data sets;
- use of molecular epidemiology, to trace the evolution of an agent of interest more accurately;
- distinguishing between samples of a pathogen indistinguishable using traditional tests;
- to detect any disconnect from a natural chain of infection;
- linking whole genome information to existing datasets of genetic fingerprint, previously used for pathogen identification, increasing the available reference data;
- giving 'insights into production methods, such as particle sizes, associated signatures, geographical origins, and intelligence information';
- characterizing mixed environmental samples;
- establishing the relationship between a novel agent and known pathogens; and
- determining evolutionary dynamics".

Advances in sequencing platforms and capabilities have resulted in increases in the amount and range of data available. This has been especially true for pathogens involved in disease and has led to notable improvements in the amount of benchmark and background data available. Recent advances have "enabled a wider set of samples to be used in diagnostics, decreased time for confirming a diagnosis, and enabled in-country diagnostic capacities".[7] Of particular note, has been the advent of molecular epidemiology – using genetic or other biological data to assist in reconstructing and tracking the transmission of a pathogen, subsequently, "have been increasingly used for diagnosis and surveillance during disease outbreaks".[7]

The IAP noted important progress in PCR-based diagnostics. They highlighted their comparative benefits over traditional serological approaches, for example, fewer reagents are needed and it produces results that are more easily interpreted. There has also been notable progress in developing high-throughput versions and "[A]utomated, rapid, and sensitive PCR-platforms capable of identifying over a thousand pathogens from clinical samples now exist".[7] PCR diagnostic platforms have also become more user-friendly and "[C]artridge-based diagnostic PCR platforms intended for use by non-specialists have been developed. Some are intended for in-field use, including detection of the potential use of biological weapons".[7]

More broadly, the IAP noted "a trend 'to make sophisticated tests, or assays, more easily performed with less training, leading to decentralization and diagnosis closer to the point-of-care'."[7] This has important implications for combatting highly-infectious diseases where the turnaround time for diagnosis and subsequent implications for clinical management are particularly important. There has also been increasing integration of nanotechnology, resulting in "improved accuracy, increasing speed and reductions in costs of point-of-care diagnostics". There is also a trend towards a more distributed approach to diagnostics, with the "integration with modern information technology platforms, such as smart phones".

The IAP also noted changing trends in laboratory capacity, with the "consolidation of testing in larger labs, driven by economies of scale, availability of expertise, and the development of newer diagnostic approaches that have substantial complexity and costs". This was the result of "more advanced (and likely more expensive) diagnostic tests [that] may require specialized equipment, resources and laboratories (*e.g.* reference laboratories)".

11.3.1.4 Preventing, Treating and Mitigating Disease and Intoxination

There have been a number of notable trends in efforts towards the capacity deployed to prevent, treat and mitigate disease. The IAP noted "[I]t is increasingly simpler, faster and cheaper to industrialize production processes".[7] Some advances are connected to changing industrial production capacities, for example, increased use of outsourcing of production to multipurpose facilities decreasing the need for dedicated vaccine production capacity. There is also a more distributed production infrastructure, which decreases the distance products have to travel before they are used. Other developments are connected to the platforms used in such facilities, for example, "[S]ingle-use equipment and modular production technologies shorten turn-around times."

It is not only the production capacity that has changed considerably. The IAP highlighted important developments in the design, development and improvement of drugs and vaccines. For example, "[H]igh-throughput platforms and 'big-data' approaches continue to reveal a wealth of potential new targets and candidates for drugs and vaccines". The increasing 'digitalization' of biology has also proven important for design, testing and optimisation processes, and has been supported by parallel advances in "better computational technologies, improved capacity for rational design, integration of synthetic biology approaches, more sophisticated modelling tools, enhanced synthesis technologies and a wider array of platform technologies".[7]

Important advances highlighted by the IAP in recent years covered:

- *Drug development*, including the continued development of high-throughput systems, the application of computational and bioinformatics approaches, and synthetic biology;
- *Novel drugs*, including new drug classes, continued improvement of existing drugs, better ways of combining drugs, off-label uses of licensed drugs, antibody therapies and drugs targeting important disease vectors;
- *Vaccine design*, including through the application of new tools and bioinformatics platforms, the advent of modularized vaccines with common backbones, the use of synthetic biology approaches, uptake of the use of virus-like particles, and a more sophisticated ability to improve the impact and efficacy of vaccines;
- *Vaccine and drug production*, including the reduced size of production facilities, opportunities for quicker and simpler production, changing costs associated with production, greater use of bio-based production and synthesis; and the use of recombinant plants and animals for production;
- *Vaccine and drug delivery*, including the ability to increase the environmental stability of vaccines and drugs, enhanced drug-delivery vehicles, in particular nanoparticle-based systems, as well as improvements in trans-dermal drug delivery and cellular delivery;
- *Antimicrobial resistance*, including understanding the mechanisms involved, the identification of novel targets for antimicrobials, and progress in developing novel antimicrobials; and
- *Recovering from disease*, including improvements in generating community buy-in for responders, medical management and infection control, protective equipment, and decontamination.

11.3.2 Developments in Science and Technology Increasing the Hazards

Many of the same advances and developments that are reducing the risks of deliberate disease could potentially be applied in increasing those same risks. Reviews of such a potential suggest that recent advances in science and technology could help those set on acquiring weapons agents, improving their efficacy, producing and stockpiling them, as well as targeting, delivering and dispersing them (Sections 11.3.2.1–4).

When combined, recent advances have had a notable impact on the risk posed by deliberate disease. They could "facilitate almost every step of a biological weapons programme and technological barriers to acquiring and using a biological weapon have been conspicuously eroded".[7]

The IAP noted, however, that whilst such uses were possible, they were far from straightforward:[7]

> "Many of these developments are at the leading edge of current capabilities. They are expensive and complicated to acquire and deploy successfully. Making use of them for prohibited purposes would probably require the resources of a State. This situation may change in the future, reinforcing the need for on-going efforts to review relevant developments in science and technology."

11.3.2.1 Acquiring Weapons Agents

The IAP identified a number of ways that such agents might be acquired, including:

- "*Acquiring agents from nature*; including an expanded range of possible agents and locations in which they are to be found, as well as tools for characterising previously 'unculturable' microbes;
- *Synthesising existing agents*; non-specialists can now compile gene 'cassettes' coding for, for example, virulence factors and 'reboot' some viruses; pathogens responsible for historical epidemics can now be synthesised or reactivated; and many small peptides, bioregulators and toxins can now be produced by chemical synthesis;
- *Designing and synthesising novel agents*, this is now easier through using genome engineering platforms, 'cloud-labs', 'biofabs', and more sophisticated tools, and the availability of standards for designing, manipulating and compiling microbes, their parts, and proteins. This allows novel pathogens to be produced".

11.3.2.2 Improving the Efficacy of Weapons Agents

The advances in our understanding of disease discussed in Section 11.3.1.1 were reviewed for potential improvement of a weapon agent. The IAP concluded that recent advances did offer opportunities to alter agents to make them better weapons, including by enhancing pathogenicity and transmissibility or expanding the host range, circumventing host immunity, adding antimicrobial or drug resistance, or improving environmental stability.

Given that they are mid-spectrum agents and are explicitly covered by both the Biological Weapons Convention and the Chemical Weapons Convention, possible malign applications of advances related to toxins are of particular relevance to this chapter. There has been notable progress in characterizing these agents, including that "[T]he genetic components and mechanisms of action of toxins are increasingly well characterized and their characteristics can be manipulated". These advances have benefitted from "[M]ore sophisticated tools for researching and manipulating toxins". Progress in understanding and manipulating toxins, when combined with advances in biological engineering, have led to an increased capacity to produce them through metabolic pathway engineering. Such production pathways may be of particular note "in the case of those toxins that are awkward to extract in large quantities from natural sources".[7] Other advances, have "made it easier to produce toxins on a large-scale, which has commercial applications but also means it complicates efforts to prevent the use of toxins as bioweapons".

11.3.2.3 Producing and Stockpiling Weapons Agents

The IAP identified a wide range of ways that the increasing digitization and industrialization of biotechnology could enable those attempting to produce or stockpile a weapon agent, including:

- "*Concealing prohibited activities*. Changes to production signatures and a shift towards the use of multiple smaller reactors compromises efforts to identify sites of biological weapons production;
- *Industrialising biological production processes*. Less space and time are now required for scale up, narrowing windows for interdiction. The process can also be simplified using new technologies, though at significant cost;
- *Producing biological agents*. The increased use of biosynthesis and bio-based production, scaffolds, and 'biopharming' accelerates the speed and yield. This also applies to vaccine production;
- *Switching production from permitted to prohibited activities*. The use of single-use, disposable and modular production equipment offers possibilities for faster technological breakout;
- *Acquiring relevant equipment*. Critical laboratory materials such as reaction vessels (including those currently covered by control lists) could potentially be fabricated using 3D printing technology. If this is possible, it might reduce costs and further lower barriers to prohibited activities. Once again, this complicates efforts to enforce non-proliferation measures.

- *Distributed production*. The decoupling of design and manufacture has led to the growth of stand-alone fabrication and production facilities. Whilst limited in number and geographic distribution at the moment, the potential for the growth of such facilities and their impact in changing the footprint of prohibited activities might warrant closer attention over the coming years;
- *Outsourcing biological production*. Multipurpose biological production facilities suitable for varying-scale production of biological agents as well as for the synthesis of genetic material and other synthetic genomics techniques are now commonplace. The existence of many 'virtual' biotech companies demonstrates the potential in this space;
- *Storing biological agents*. Increasing the environmental stability of biologics, together with the use of other approaches removes the requirements for cold-chain storage and its associated infrastructure. Improvements in production techniques have reduced the need for 'stockpiling' whilst the proliferation of freeze-drying capabilities enables this should it prove desirable."

The developments that had enabled these changes were drawn from: the changing footprint of production capacities; improvements in the industrial scale-up of biotechnology; increasing reliance on bio-based production and biosynthesis; molecular scaffold technology; biopharming; use of outsourced production and modularization of bio-based production capacity; the trend towards the use of disposable, synthesized and repurposed equipment; advances in the modification of agents for storage, freeze-drying and non-cold chain storage; and microencapsulation and smart particles.

11.3.2.4 Targeting, Delivery and Dispersal of Weapons Agents

The delivery of a biological weapon has been simplified by advances in a number of key areas, including:

- "*Nanotechnology*. A wider range of nanoparticles, of different sizes, can more efficiently deliver complex payloads to diverse targets. Nanoparticles can now target previously inaccessible physiological sites and cell types (*e.g.* by crossing the blood–brain barrier). Nanoparticles also add other desirable characteristics to agents, such as increased persistence in the body and immune avoidance. Nanoparticles suitable for aerosol release have been developed since the Seventh Review Conference;
- *Aerobiology*. Offers powerful tools for modelling the release of bioweapons, including both environmental and indoor dispersal patterns, helping optimise the release of an agent. There have also been advances in equipment for generating and modelling aerosol dispersion;

- *Use of chemical co-factors to increase the uptake of biological agents.* These have been identified for use with biologically active proteins; and
- *Increasing capacity to deliver biological weapons via the alimentary route.* The use of sophisticated formulations can improve absorption from the gastro-intestinal tract."

11.4 Future Risks for Biochemical Anti-animal Weapons

The short history of anti-animal biological weapons programmes published by the World Organization for Animal Health concluded:

> "Biological weapons to target animals have been developed and used by both State and non-State actors for a variety of reasons. Significant capabilities can be compiled using low technology approaches and good laboratory practice. More sophisticated approaches have also been attempted. To date, efforts to acquire these weapons have been infrequent; their use is even scarcer."

Based upon these historical precedents, there may be grounds to assume that future development and use of anti-animal weapons would also be comparatively rare. It is also likely that there might be different types of weapons – ranging from low-technology, improvised attacks, through to more sophisticated efforts. Whilst there have been historical examples of strategic level programmes intended to cripple enemy food production, many of the programmes that began looking at such weapons eventually switched to explore capabilities to initiate animal disease in a more covert manner.

There are those, such as Ban Ki Moon, the UN Secretary General in 2015, that argue the norm against biological weapons is so strong that no State openly identifies itself as possessing them.[9] Perhaps it is the strength of this norm that has driven States from exploring the overt use of biochemical weapons targeting animals to consider sabotage approaches that offer greater plausible deniability.

It is worth noting that animals are still important for military purposes in parts of the world. This suggests that weapons targeting animals for tactical purposes remain a possibility. However, such a State would once again run into the strength of the norm against the use of biochemical weapons. It is difficult to imagine a State deciding that the use of tactical anti-animal weapons would be worth risking international condemnation and possible response to their use. It may be possible if there was plausible deniability – but once again this may be more likely using wild-type agents, minimizing the utility of applying many of the recent developments in science and technology discussed in this chapter. Furthermore, historical biological weapons programmes failed to produce a militarily reliable weapon – that operating in an environment in which they have been used is too complicated and that the additional burden placed on military logistics, doctrine and training is untenable.

Whilst the impact of sabotage style anti-animal agents has been hard to document, it is certainly true that animal disease outbreaks continue to have a major impact on our societies. For example, according to the UK National Audit Office, the 2001 outbreak of Foot-and-Mouth Disease "had a devastating effect on the rural economy and parts of the tourist industry" and "cost the public sector over £3 billion and the private sector more than £5 billion".[10] Such an impact might speak to a continued vulnerability to, and hence potential desirability of a capacity to instigate, deliberate animal disease.

The findings of the 2015 IAP review of recent relevant developments in science and technology provide a stark insight into what might be possible today. They noted that many of the barriers to producing biochemical weapons have been undermined. They also highlighted a wide variety of ways in which natural agents might be enhanced, modified or in other ways improved to make better weapons. However, they also highlighted that doing so today is at the very limit of what is possible. It is appropriate, therefore, to ask both who might be able to accomplish such a feat, and why those able to do so would wish to.

It would seem that capabilities to develop advanced anti-animal biochemical weapons might be beyond all those without access to State level resources today. Noting the lack of historical interest by non-State actors in these weapons and well documented trends for terrorists to employ weapons with which they are familiar (as opposed to something as exotic as an advanced anti-animal biochemical weapon), there seems grounds to question whether such groups would currently be interested in acquiring and using such weapons, even if they could get them.[11]

Should a State be interested in developing biochemical weapons they would still be confronted with the strength of the norm against them. Some of the advances discussed in this chapter would potentially help such a State – complicating efforts to detect that an agent has been altered, reducing the window to interdict their development efforts, and altering the footprint of their production platforms. However, the use of the weapon would still need to be concealed. As a result, in the short to medium term it may be that advanced tactical and strategic anti-animal biochemical weapons remain undesirable.

Perhaps the most worrying scenario might be that a State decides to invest heavily in applying recent developments in science and technology to produce an advanced anti-animal biochemical weapon for sabotage use. Such approaches have been attempted in the past as the resources dedicated to similar efforts by Germany during World War I demonstrate. A modern advanced weapon might be harder to detect, prevent or mitigate.

It might be produced in a programme that is harder to detect, interdict or counter. It might be readily concealed as an unusual, yet natural disease event. However, given current shortcomings in effective animal disease surveillance, mitigation and response, and vulnerability to traditional animal disease, it is unclear what the advantage such an advanced weapon would provide. Why invest the resources needed when simply moving an existing disease from one place to another could have almost the same impact?

It is also important to note all the recent developments in dealing with disease regardless of cause or origin. In other words, a host of new defences are coming on line and we are unlocking unprecedented capabilities to deal with unusual animal disease outbreaks. It is unclear how, why or when the technical developments discussed in this chapter might be applied to anti-animal biochemical weapons. It is more certain that they are already being applied to dealing with animal disease. It would seem that, with respect to weapons targeting animals, we might be moving into a period where there is a net reduction of the risk of deliberate animal disease.

11.5 Conclusions

This chapter provided a brief overview of important historical examples of the acquisition and use of biochemical agents targeting animals. It explored a number of ways in which recent developments in science and technology could alternatively increase or help reduce the associated risks. It identified a series of challenges to applying these advances so as to cause harm – from the technical, through to the political to a very basic comparison of costs and benefits. It argued that the future risk posed by these weapons might actually decrease over the coming decades as these advances are applied to reducing risk, but might not be applied so readily by those intent on causing deliberate harm.

In conclusion, the current threat posed by anti-animal biochemical weapons is real – it has precedent, and is likely impacted by recent developments in science and technology. Measures to combat this threat, however, are also well developed.

There appears to be a strong norm against these weapons. Both the Biological Weapons Convention and Chemical Weapons Convention explicitly prohibit anti-animal biochemical weapons. The Chemical Weapons Convention's definition of toxic chemicals, which underpin the ban, specifically prohibits certain uses of chemicals that cause death, temporary incapacitation or permanent harm to animals. Successive review conferences of the Biological Weapons Convention have repeatedly stressed that the ban covers weapons targeting humans, animals and plants. The next step will be ensuring that these bans keep pace with developments in science and technology so that these regimes remain well placed to deal with the anti-animal weapons threats of tomorrow.

References

1. Chemical Weapons Convention, *Convention on the Prohibition of the Development, Production, Stockpiling and Use of Chemical Weapons and on Their Destruction, Organization for the Prohibition of Chemical Weapons*, 31 August 1994, https://www.opcw.org/fileadmin/OPCW/CWC/CWC_en.pdf.
2. Biological Weapons Convention, *Final Document of the Seventh Review Conference, BWC/CONF.VII/7*, United Nations Office at Geneva, Switzerland, 13 January 2012, http://www.unog.ch/80256EDD006B8954/(httpAssets)/C2BAA955E58E15C7C1257D01005260D3/$file/BWC_CONF.VII_07.pdf.
3. P. Millett, *Animal Disease as a Weapon: A Short History, Bulletin of the World Organization for Animal Health,* No. 2015-3, http://www.oie.int/fileadmin/Home/eng/Publications_%26_Documentation/docs/pdf/bulletin/Bull_2015-3-ENG_new.pdf.
4. *Biological and Toxin Weapons: Research, Development and Use from the Middle Ages to 1945, SIPRI Chemical & Biological Warfare Studies No. 18*, ed. E. Geissler and J. E. van Courtland Moon, Oxford University Press, 1999.
5. *United Kingdom, Confidence Building Measure Return for 2017 (Covering Data for 2016) for the Convention on the Prohibition of the Development, Production and Stockpiling of Bacteriological (Biological) and Toxin Weapons and Their Destruction*, 31 March 2017, p. 42, http://www.unog.ch/80256EDD006B8954/(httpAssets)/65F5D8D533B96C47C12580F400571598/$file/BWC_CBM_2017_United+Kingdom.pdf.
6. Scientific Advisory Board of the Organization for the Prohibition of Chemical Weapons, *Convergence of Chemistry and Biology: Report of the Scientific Advisory Board's Temporary Working Group, SAB/REP/1/14, Organization for the Prohibition of Chemical Weapons*, June 2014, https://www.opcw.org/fileadmin/OPCW/SAB/en/TWG_Scientific_Advsiory_Group_Final_Report.pdf.
7. IAP: Global Network of Science Academies, *The Biological and Toxin Weapons Convention: Implications of Advances in Science and Technology – Technical Report*, IAP, 2015, https://royalsociety.org/topics-policy/projects/biological-toxin-weapons-convention/.
8. U.S. National Academies, *Biological Research in the Age of Terrorism*, National Academies Press, 2004, http://www.nap.edu/catalog/10827/biotechnology-research-in-an-age-of-terrorism.
9. United Nations, *We Must Remain Vigilant,' Ban Urges as Biological Weapons Convention Turns 40*, United Nations News Centre, 26 March 2015, http://www.un.org/apps/news/story.asp?NewsID=50430#.V-2A5ZMrKu4.
10. UK National Audit Office, *The 2001 Outbreak of Foot and Mouth Disease, Comptroller and Auditor General, HC 939 Session 2001-2002*, 21 June 2002, https://www.nao.org.uk/report/the-2001-outbreak-of-foot-and-mouth-disease/.
11. US Office of Technology Assessment, *Technology against Terrorism: The Federal Effort*, Congress of the United States, July 1991.

CHAPTER 12

Development and Hostile Use of Toxic Chemical Means of Delivery and Dispersal

MICHAEL CROWLEY

University of Bradford, Bradford, West Yorkshire, BD7 1DP, UK
*E-mail: m.j.a.crowley1@bradford.ac.uk

12.1 Introduction

Under the Chemical Weapons Convention (the CWC), a "chemical weapon" is defined as comprising:

(a) toxic chemicals or their precursors, except where intended for purposes not prohibited by the Convention, as long as the types and quantities are consistent with such purposes;
(b) *munitions and devices specifically designed* to cause death or other harm through the toxic properties of those toxic chemicals specified in subparagraph (a), which would be released as a result of the employment of such munitions and devices;
(c) *any equipment specifically designed* for use directly in connection with the employment of the munitions and devices referred to in (b)."[1]

In addition to "toxic chemicals", this definition clearly recognises and underlines the critical importance of "munitions and devices" and related

Preventing Chemical Weapons: Arms Control and Disarmament as the Sciences Converge
Edited by Michael Crowley, Malcolm Dando and Lijun Shang

Published by the Royal Society of Chemistry, www.rsc.org

"equipment" as intrinsic components of chemical weapons. Consequently, under the CWC all such components when "specifically designed" for use in chemical weapons would automatically be prohibited under the Convention even if they were not filled with toxic chemicals; all stockpiles of such "specifically designed" munitions, devices and related equipment would need to be declared to the OPCW by possessor States and verifiably destroyed.

In contrast, according to the legal *Commentary* on the CWC by Krutzsch, Myjer and Trapp, this definition "*excludes*, for example, dual-use delivery systems even if they can be or in fact have been used to deliver chemical weapons, such as artillery pieces, bomber planes or missiles".[2] This ensures that the vast stocks of conventional munitions and associated delivery systems that could potentially be, but are not actually being, and are not intended to be, employed as chemical weapons, are not caught by accident within the Convention's definition of a chemical weapon; and do not have to be declared and destroyed. However, if any such conventional munitions and associated delivery systems (or indeed any other means of delivery) are actually being developed, stockpiled or utilised for the dissemination of toxic chemicals for hostile purposes, those specific munitions and devices would be considered to be chemical weapons, and would need to be declared and destroyed.

Reflecting these discussions, this Chapter will attempt to provide an overview of the wide variety of contemporary mechanisms that have been developed and/or employed by State and non-State actors to deliver toxic chemicals to specific targets or disperse toxic chemicals over wide areas, for hostile purposes. Consequently, in addition to examining the development and employment of munitions and delivery devices solely intended for use as chemical weapons, this Chapter will also explore the adaptation and employment of conventional munitions as chemical weapons, and also the malign application of dual-use technologies in the development of improvised chemical weapons. Where possible, this Chapter will employ "real life" illustrative case studies of such activities that have occurred since the negotiation of the Chemical Weapons Convention. It will illustrate how continuing technological advances are facilitating the ongoing development and promotion of certain means of delivery and dispersal, with greater payload capacity, wider dispersal area and extended strike range; and have led to greater accessibility of such technologies to a broader range of actors, with consequent implications for the effective regulation of such technologies.

12.2 Improvised Chemical Delivery Mechanisms

12.2.1 Introduction

The following sections will explore the contemporary development and use by both State and non-State actors of a wide variety of chemical weapons employing improvised means of delivering toxic chemicals to specific targets or dispersing toxic chemicals over a wide area in which the targets are situated. Firstly, the development and use of ground-based chemical improvised explosive devices (IEDs) and aerial chemical IEDs will be examined. Although there are diverse types of such devices, they are all made, fully or in part, from commercially available products not originally intended by the manufacturers for such malign purposes. This will then be followed by an exploration of the potential malign application, as chemical weapon delivery mechanisms, of commercially available general utility aerosol dispersal devices and aerial vehicles.

The examination of such improvised chemical delivery mechanisms raises difficult questions regarding the appropriate regulation of relevant dual-use or rather multi-purpose technologies – *i.e.* those technologies that can be utilised for both hostile and peaceful ends, with no, or only minor, modifications. Critical to such considerations is the recognition of the importance of intent in determining whether a particular dual-use technology is so employed. The hostile use of a specific dual-use technology does not arise automatically from the inherent properties of that technology, but requires the active intervention of relevant actors. Consequently those seeking to develop and implement controls of relevant dual-use or multi-purpose technologies must wrestle with how best to ensure that such technologies are effectively regulated to prevent potential misuse by the few without unduly limiting access to and use of these technologies for legitimate purposes by the many. The balance between limitation and access to the relevant technology will depend on the specific characteristics of the technology under consideration, particularly its susceptibility to, and the potential consequences of its misuse, as well as the societal benefits of its legitimate use. Consequently, an important element in the process to develop appropriate responses is an examination of previous successful and unsuccessful attempts to employ such technologies and goods for malign purposes.

12.2.2 Chemical Improvised Explosive Devices

An improvised explosive device (IED) is a bomb fabricated in an *ad hoc* manner incorporating destructive, lethal, noxious, pyrotechnic or incendiary chemicals that is designed to destroy, incapacitate, harass or distract. IEDs may incorporate military, commercially sourced or homemade explosives or other commercial components that are used in a manner other than intended by the manufacturer.[3] IEDs have been widely employed by non-State actors including criminals, terrorists, and armed opposition groups and also by State forces. Normally an IED is comprised of: a switch, an initiator, container, an explosive charge, and a power source. Anti-personnel IEDs typically also contain "enhancements" such as nails, glass or metal fragments designed to increase the amount of shrapnel propelled by the explosion.[4] In a chemical IED, these "enhancements" are replaced by a toxic chemical or chemicals, normally toxic industrial chemicals (TICs), homemade chemical agents or sometimes chemical warfare agents. In comparison to non-chemical IEDs which are filled with enough explosives to ensure significant destructive kinetic and mechanical effect, chemical IEDs contain far lower amounts of explosive charge, normally sufficient only to rupture chemical cylinders and crack open the outer shell of the container so as to release the toxic chemical(s). Indeed, if more explosives are employed there is a concordant risk that the chemical agent will be destroyed in the consequent heat and kinetic energy of the explosion, and thereby reduce the effectiveness of the device as a chemical weapon.

12.2.2.1 Ground-based Chemical IEDs

Ground-based chemical IEDs are normally static devices placed at particular sites including buildings and other enclosed spaces, or by roadsides; alternatively they can be placed in a variety of cars, trucks, *etc.* – so-called vehicle borne improvised explosive devices (VBIEDs). Ground-based chemical IEDs can be detonated remotely, by timer, by suicide bomber or by victim contact. Contemporary use of ground-based chemical IEDs has been restricted to a small number of terrorist, insurgent or armed opposition groups, most notably those operating in Iraq and Syria. To date, there have been no confirmed contemporary instances of State forces developing or employing such devices.

12.2.2.1.1 Development and Use of Ground-based Chemical IEDs: Al-Qaeda, Islamic State and Related Organisations. Since the late 1990s, Al-Qaeda and subsequently Islamic State (IS) and allied groups, have shown a strong and continuing interest in developing ground-based chemical IEDs. Initially their technical capabilities for production and delivery were rudimentary, but subsequently appear to have grown. In February 2003, Saudi Arabian authorities arrested a five member Jihadist cell loosely affiliated with Al-Qaeda which had been planning a subsequently aborted chemical attack on the New York subway system. The investigators found a data file on one of the suspects containing plans for a home-made chemical dispersal device called a Mubtakkar. However, when Central Intelligence Agency (CIA) chemists reconstructed the chemical dispersal device from the plans in the confiscated computer file, they determined that it would not have worked effectively. The components would have reacted violently, causing the device to blow apart and abort the generation of the lethal gas.[5]

In April 2004, Jordanian authorities announced they had halted a major chemical attack by Al-Qaeda in Amman, which they estimated saved the lives of between 20 000–80 000 people.[6,169] The plot reportedly involved 20 tons of chemicals (including pesticides, potassium cyanide, hydrogen peroxide, glycerin, and acetone among others)[7] and explosives in three specially configured trucks to target the headquarters of the General Intelligence Directorate (GID), the Prime Minister's office, and the US embassy.[8] Two of the trucks were reportedly to be converted into large bombs while the third was to carry containers of toxic chemicals in an effort to produce a toxic chemical "gas cloud".[6,9,10] According to *Jane's Intelligence Review*, "The chemicals were designed to have two distinct effects: the first specifically sulfuric acid would be to chemically enhance the explosive power of the IEDs; the second and more deadly additions nitric acid and acetone, among others would result in the creation, after the explosions, of a cloud of toxins [sic] that would disperse around the GID compound and out into the city, inducing mass casualties."[8] Following conflicting statements by Al-Qaeda leaders it is unclear and contested whether this was an attempted chemical weapons attack or instead the employment of chemicals as conventional explosives.[6,10,11]

From late 2006 to mid-2007 Al-Qaeda in Iraq/Islamic State launched a series of chemical attacks combining chlorine with vehicle-borne improvised explosive devices (VBIED).[6,12–14] In the first documented attack which occurred in October 2006 in Ramadi, Al Anbar Governorate, the VBIED was an armoured dump truck containing two cylinders of chlorine and 120 mm mortar rounds; it was detonated by the suicide driver and resulted in four casualties.[15] Collectively, dozens of people died over the course of this eight month chemical bomb campaign, however the majority, if not all, of the deaths appear to have resulted from the physical effects of the explosions. It appears that the detonations of the IEDs destroyed much of the toxic chemical agent rather than dispersing it, and consequently, although the chlorine gas clouds resulted (over the course of the campaign) in hundreds of people suffering vomiting, breathing problems and other toxic effects, the amounts of toxic chemical released were not sufficient to cause many (if any) fatalities. In an effort to enhance the toxic effects of the bombs, the insurgents experimented with different proportions of chlorine and explosive before finally abandoning their efforts.[16] The attacks reportedly stopped partly because Islamic State failed to kill as many people as they intended, and also because US Armed Forces in Iraq "hunted down the bomb makers responsible for the campaign, while simultaneously clamping down on the availability of chlorine."[17]

Quillen has argued that these chlorine attacks were "targets of opportunity", as Islamic State had an existing manufacturing capability for VBIEDs, and chlorine was readily available at least until the attacks started, so "combining the two took little effort or imagination."[18] Hoffman has contended that the pairing of IEDS with toxic chemicals was the "most novel and innovative use of such devices" and that "much as the Iraq conflict has served as a proving ground for other terrorist weapons and tactics, it has also served this purpose with chemical weapons."[13] Tucker presciently noted in 2011 that "their repeated use in Iraq may have crossed a psychological threshold that could make a return to such tactics more likely".[16]

In October 2014, following significant gains in territory and resources, Islamic State recommenced the use of chemical IEDs, with targets including Iraqi security services, Kurdish forces and the civilian population.[19]

This time, however, the most common method of attack involved roadside IEDs instead of large VBIEDs, with the Iraqi Bomb Disposal Team claiming to have defused dozens of these home-made devices filled with chlorine.[20] Whilst these smaller devices are less likely to result in large numbers of fatalities, they appear more successful in dispersing (rather than destroying) the toxic chemical and thereby creating fear and injury amongst the target population. However, on occasion Islamic State has reverted to its original design. In January 2015, Kurdish Peshmerga fighters claimed to have identified a large truck VBIED travelling on the road between Mosul, Iraq and the Syrian border. They reported that the VBIED contained about 20 cylinders of chlorine and the initiation system was rigged to be operated by a suicide bomber. The VBIED exploded when the Peshmerga forces opened fire on it.[15]

Subsequently, there have been indications that individuals linked to Islamic State have attempted to develop and employ ground based chemical IEDs beyond Iraq. In February 2015 an IED with a chlorine enhancement was discovered in a shopping mall in Jakarta, Indonesia. Fortunately, the device did not detonate and consequently no casualties were inflicted. Indonesian terrorists believed to have returned from fighting with Islamic State forces in Syria were suspected of being behind the event.[21]

12.2.2.2 Chemical Improvised Rocket-assisted Munitions

An improvised rocket-assisted munition (IRAM) is an IED comprised of a large metal canister, often an acetylene, oxygen or propane tank, placed on top of a conventional rocket munition. Anti-personnel IRAMs filled with explosives, scrap metal, and ball bearings have been employed by armed opposition groups in Iraq since 2007.[170] Because IRAMs employ pressurized gas canisters, these devices can also potentially be used to carry and disperse toxic chemicals. To date, there has been reported isolated small-scale use of chemical IRAMs by IS in Qayyarah, Iraq, in September and October 2016.[22] In contrast, Syrian Government forces have been accused of employing chemical IRAMs in several parts of Syria on a larger scale.

12.2.2.2.1 Use of Chemical IRAMs by Syrian Armed Forces. In its report covering July 2016–February 2017, the UN Commission of Inquiry (COI) determined that "Government and pro-Government forces continued to target civilian areas deliberately with improvised chlorine munitions, both air-dropped and through the use of IRAMs in the suburbs of Damascus and Idlib Governorate."[23] Similarly, Human Rights Watch presented witness accounts, video and photographic evidence of two attacks (and eyewitness accounts of a further four attacks) by Syrian Government armed forces utilising chlorine-filled IRAMs upon territory near Damascus held by armed opposition groups, injuring both civilians and combatants.[24] The UN COI report highlighted an attack on 30 January 2016 when "an IRAM with an improvised chlorine payload struck near a frontline (armed opposition group) position in Sultan al-Marj village injuring 11 men ... photographs provided by witnesses show a 107 mm rocket remnant that reportedly delivered the munition."[25] Analysing photographs of munition remnants from the attack, an online civil society weapons monitor, Hady al-Khatib, identified them as Iranian-made 107 mm rockets, typically launched from a Type-63 multiple rocket launcher, a system that both Syrian Government forces and armed opposition groups have used in Syria.[26] The design of the weapon indicated that it would have had a relatively short range and would have been highly inaccurate.

12.2.2.3 Aerial Chemical Improvised Explosive Devices

There are a range of aerial improvised explosive devices often called barrel bombs, lob bombs or flying IEDs, that comprise a container (*e.g.* an oil drum or gas cylinder) dropped from an aircraft and filled with explosive, incendiary or other substances and often additional materials to increase fragment projection. Given the requirement for aerial delivery, their use appears to be restricted to State forces, which have reportedly included the US in Vietnam in the 1960s;[27] Sri Lanka[28] and Sudan[29] in the 1990s; and Iraq[30] and Syria during the ongoing armed conflicts. A variation of such aerial IEDs, specifically designed to disperse toxic chemical agents, has been manufactured and employed by Syrian armed forces against both civilians and armed opposition groups in that country.

12.2.2.3.1 The Development and Use of Aerial Chemical IEDs by Syrian Armed Forces. The OPCW Fact-Finding Mission (FFM) mandated to "establish the facts surrounding allegations of the use of toxic chemicals ... for hostile purposes" in Syria documented a series of chemical attacks employing aerial chlorine IEDs in 2014 and 2015 killing at least 13 people and injuring between 350 and 500.[171–173]

From its examination of images of unexploded devices and munition remnants in the Idlib Governorate, the OPCW FFM concluded that certain Syrian aerial chemical IEDs were variations on a common structure, comprising a number of inner cylinders or canisters, apparently containing chlorine, within an outer metal jacket. The devices could also contain a detonation cord or an explosive mixture and an ordinary time fuse consisting of a blasting cap and a time-calculated fuse wick. It appeared that in some designs (see Figure 12.1) the cylinders contained hydrochlorofluorocarbons (HCFCs) instead of chlorine and in such cases the munitions also held smaller bottles, apparently of potassium permanganate, which acted as an oxidising agent. Following aerial delivery, the detonation cord or explosive mixture in the device would be triggered consequently rupturing the gas cylinders, (and where relevant, containers of potassium permanganate), and the outer shell, thereby releasing the toxic chemical compound(s). Analysis of aerial

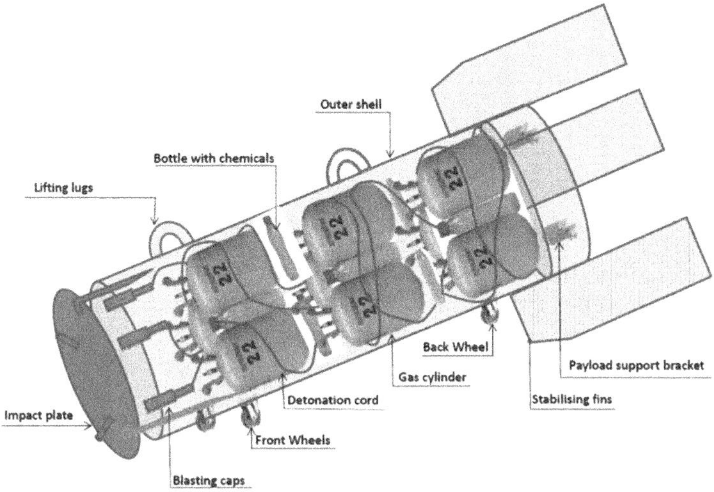

Figure 12.1 Depiction of an improvised chemical munition. Image taken from the Report of the OPCW Fact-Finding Mission in Syria regarding alleged incidents in the Idlib Governorate of the Syrian Arab Republic between 16 March and 20 May 2015.

chemical IED designs observed through social media demonstrated an evolution of their manufacture, likely driven by trial and error.[31] As they are improvised munitions, their size and weight have varied, but some have been estimated to weigh up to 400 kg.[32]

The OPCW–UN Joint Investigative Mechanism (JIM) subsequently examined the OPCW FFM findings to determine responsibility for such activities. Although the Syrian Government claimed that aerial chemical IEDs had been employed by armed opposition groups potentially using UAVs, this has been discounted by the OPCW–UN JIM which concluded that owing to their weight, such aerial chemical IEDs can be deployed only from helicopters.[33] The OPCW–UN JIM concluded that Syrian Government armed forces were the perpetrators of three chlorine weapon attacks on Talmenes (21 April 2014), Qmenas (16 March 2015) and Sarmin (16 March 2015). The JIM determined that helicopters from Syria's 63rd Helicopter Brigade, operating from Hama and Hmeimim airbases, had carried out the attacks dropping chlorine aerial IEDs in all three cases.[33,34]

Reports of additional and continuing attacks by the Syrian Government armed forces in 2016, 2017 and 2018 utilising improvised aerial delivered chlorine weapons have been presented by the UN Commission of Inquiry,[35] civil society monitors and international human rights organisations.[36] Such attacks have reportedly resulted in further fatalities and hundreds of casualties. There have been no confirmed attacks by the armed opposition forces utilising such improvised aerial chemical munitions. In addition to the aerial chlorine IEDs described previously, Syrian Government force helicopters have also reportedly simply dropped gas cylinders filled with chlorine onto opposition held areas. For example, according to Human Rights Watch, such attacks by Government armed forces took place in Allepo from 17 November–13 December 2016 during the battle for control of that city, and affected hundreds of civilians.[37]

12.2.3 Commercial Aerosol/Smoke/Fog Generators and Dispersal Devices

There are a range of commercially available devices which generate and disperse particulate smoke, aerosols or dense vapour fogs. Such devices often employ either an inert gas or pump to propel mineral oil, glycol, or a glycerine and water mixture into a heat exchanger, where the solution is vaporized, the vapour then mixing with the cooler external air at a controlled rate to condense to a fog or mist with a controlled droplet size. Such devices are commonly and legitimately employed in the entertainment industry, fire-fighter/emergency services training and by the military for creating smokescreens to obscure troop movement. However, such devices could potentially be employed for the weaponised dispersal of toxic chemicals. Indeed, certain commercial manufacturers of multipurpose smoke devices highlight their potential use as RCA delivery mechanisms (see 12.2.3.1).

Since the coming into force of the CWC, there has been no confirmed large scale State or non-State actor use of aerosol/smoke/fog generators and dispersal devices to disseminate weaponised toxic chemical agents, for hostile purposes. To date, one non-State actor, the Japanese apocalyptic cult, Aum Shrinynko, has carried out chemical attacks employing air borne dissemination, (see 12.2.3.2) and as described below, it employed improvised evaporation methods rather than off-the-shelf aerosol generators. However, the increasing capacity and range of commercially available devices, together with their ability to be employed on diverse ground, marine or aerial vehicles, raise concerns of potential future misuse by both State and non-State actors.

12.2.3.1 Afterburner Multipurpose Aerosol Smoke System

The US manufacturer, MSI Delivery Systems Inc., has developed and promoted the Afterburner 2000, which it claims to be "The Fastest, Fully Self-Contained, Portable, Multi-Mission Aerosol Delivery System in the World ... [a] ... multi-purpose smoke generator capable of rapidly blanketing large areas with dense smoke."[38] The Afterburner 2000 can reportedly release over 1500 cubic feet of smoke with a range greater than 100 feet in one second;[39] the "dependent version with high-capacity backpack expelling 320 000 cubic feet of smoke on a single charger".[39] According to the company, the applications for the basic smoke generator include: "First Responder Training (fire, police, EMS, civil defense)... Realistic training on ships, aircraft, factories, tunnels, *etc*...Urban Warfare/Battlefield Obscuration."[39]

In addition, the company states that the "non-toxic camouflage smoke (training smoke) can be mixed with irritants such as OC, CS, Pepper". This "upgrades mission capabilities" to include "Crowd Control & Civil Unrest; SWAT Teams & Tactical Incursions; Prison Systems for Riot Control/Prisoner Extraction; Non-Lethal Terrorist Suppression/Repel Pirates; Military Operations (MOUT/COIN)."[†,39]

12.2.3.2 *Air-borne Sarin Attack Employing* Ad hoc *Evaporation – Aum Shrinynko*

On 27 June 1994, Aum Shrinynko staged a nerve gas attack in a residential area of Matsumoto, Japan. The primary targets were three judges about to reach a verdict against the cult in a fraud case. Aum Shrinynko members drove to Matsumoto in a truck that was equipped with an electric hot plate, which they used to vaporize drops of sarin, and a fan and nozzle to vent the toxic fumes. Over a period of roughly 25 minutes, the team vaporized three litres of sarin, generating a toxic cloud that drifted down wind over a residential neighbourhood. The Matsumoto attack killed seven and injured 144 others, including the three targeted judges.[40] Subsequently, the cult conducted a larger scale chemical attack on the Tokyo subway on 20 March 1995. This attack was carried out at very short notice as a diversionary tactic in response to information that the cult's headquarters was to be raided by police. Consequently, the cult synthesised 24 litres of sarin quickly without distilling it, thereby reducing its purity to 25%. The agent was also diluted with acetonitrile to reduce its volatility and thus give the perpetrators time to escape. During the rush hour, cult members boarded five different underground cars carrying sarin filled plastic bags, which they subsequently punctured with sharpened umbrella tips, and quickly left the trains. The punctured bags released puddles of liquid sarin that slowly evaporated, generating toxic fumes. Seventeen people were in a critical condition requiring intensive care and twelve of them subsequently died; 37 suffered from severe symptoms and a further 984 displayed less severe symptoms.[41] In both cases, the means of aerial dissemination was very basic and inefficient; there may well have been significantly more deaths and injuries if the sarin had been dispersed more effectively.

12.2.4 Wide Area Spray Aircraft

A range of commercially available aircraft have either been specially developed or converted for agricultural use including aerial application of pesticides (crop dusting), fertilizer or seeds. These range from small fixed wing single piston engine planes to larger more powerful aircraft, including turboprop, twin engine or even jet planes. They normally have

---

[†]The company states that "Irritants [are] available through authorized 3rd-party vendors".

spraying systems attached to the trailing ends of their wings, and pumping machinery driven by air turbines linked to the engine or an auxiliary power source. Standard fixed wing crop dusters can carry between 300 to 600 gallons of pesticide or other liquid payloads, however some such as the AT-802A AirTractor can carry 800 gallons of liquid payload.[174] Aircraft of similar design and payload capacity have been specifically developed for use in aerial firefighting,[175] whilst others such as the Antonov An-2 biplane have been developed as general utility aircraft and modified for multiple purposes. Helicopters have also been commercially promoted for agricultural spraying,[‡] aerial firefighting or as general utility aircraft, some having significant lifting capacity such as the Erickson S-64 Air-Crane which has a 2600 gallon storage tank[42] and the Mi-26 TP which can carry 4000 gallons.[43]

Previously, certain States have employed spray devices attached to fixed wing aircraft for wide area dispersal of herbicides and other toxic chemicals as an integral part of their military tactics. For example, during its war in Vietnam, the US Air Force from 1962 conducted a nine year herbicide spraying programme utilising C-123 aircraft loaded with 1000 gallon spray tanks. According to an official US Air Force history of this programme, *Operation Ranch Hand*, some eighteen million gallons of chemicals were sprayed by this means on an estimated twenty percent of South Vietnam's jungles.[44] Following the coming into force of the CWC, there has been no confirmed contemporary use by any State of equivalent weaponised aerial chemical spray mechanisms in armed conflict.

Concerns have regularly been raised regarding the risks of non-State actors, notably terrorist organisations, employing commercially available aerial spray devices for malign purposes. These grew following reports that Al-Qaeda had explored aerial delivery of biological and chemical agents *via* crop dusters in the US.[176–180] In their 2012 Report to the Third CWC Review Conference the OPCW Scientific Advisory Board noted that: "While the few instances of the release of toxic chemicals by non-State actors have used crude devices, spray and fogging devices developed by the pesticide industry or developed for veterinary treatment of large-scale animal farms are of concern."[45]

The true level of the risk and consequences of such employment are contested. Although there are a range of commercially available spray/dispersal aircrafts that could potentially be utilised to deliver significant amounts of toxic chemical agents over wide areas, there are important logistical and technical barriers that any non-State actor would have to overcome to effectively employ such aircraft as viable chemical weapon delivery mechanisms. In addition to successfully obtaining these aircraft (and relevant

---

[‡]A variety of helicopter models have been adapted for agricultural spraying with the agent tanks placed on or outside the helicopter body, while a spray rig, extending outward to the sides, is attached well below the helicopter's main rotor blades.

toxic chemical agents) without alerting the security and intelligence services, the non-State actor would need to acquire the relevant tacit knowledge in their use. As Smithson explains "pilots of crop dusters are required to have a 1 year apprenticeship to learn how to operate and fly the aircraft safely. Someone with limited piloting experience in light aircraft would be considerably challenged to translate those basic skills into an ability to get a loaded crop duster successfully off the ground."[46] To date, an open source review by the author could discover only one successful confirmed contemporary chemical attack by a non-State actor employing agricultural aircraft.

12.2.4.1 Paraguay: Weaponised Use of a Commercial Agricultural Spray Aircraft

The Paraguayan Government, national and international media and human rights organisations have highlighted an apparent chemical attack upon five Avá Guaraní Indigenous communities in Itakyry district, eastern Paraguay, in November 2009. The Guarani had been in dispute over their ancestral lands; their ownership challenged by soya farmers from over the nearby border with Brazil.[47] On 6 November 2009, following a failed attempt by the soya farmers to forcibly remove the indigenous people from their land, an agricultural aircraft sprayed an unidentified toxic chemical onto the homes of the Guarani communities. According to the Paraguayan Health Minster, Esperanza Martinez: "a low-level flight by airplane [took place] where the people were working in their fields, in their community orchards, with the spraying of a liquid over the people who were there, who later presented with massive symptoms."[48] Although the identity of the toxic chemical(s) employed has never been made public, it was reportedly claimed to have been a pesticide normally employed on soya crops. It appears to have been a planned attack rather than accidental discharge, as the plane was not circulating over any areas where soya crops were being grown.[49]

According to the Health Minister, "For us, it's very clear that this is an acute community-wide intoxication caused in a premeditated manner by an unknown substance."[48] Approximately 220 people were subsequently affected, suffering vomiting, diarrhoea, fainting and breathing difficulties, among other symptoms. At least seven people were taken to hospital and one person was reported to be in a serious condition. According to press reports, clean drinking water was subsequently supplied to the communities as their own water supply had been contaminated in the chemical attack.[50] On 10 November 2009, Paraguayan President Fernando Lugo ordered an investigation into the use of toxic chemicals against Indigenous Peoples in Itakyry. However, according to Amnesty International, although investigations were opened, they were not "duly pursued" and no prosecutions were made.[51]

12.2.5 Unmanned Aerial Vehicles

Unmanned aerial vehicles (UAVs) have been defined as "powered aerial vehicles that do not carry a human operator, use aerodynamic forces to provide vehicle lift, can fly autonomously or be piloted remotely, can be expendable or recoverable, and can carry a lethal or nonlethal payload".[52] Over the last decade, there has been a dramatic increase in the research, development and possession of UAVs by State military and security forces. According to an October 2014 analysis by Horowitz and Fuhrmann, 87 States had some UAV capability, at least 24 possessing advanced unarmed UAV models and nearly 30 having armed UAV development programs.[53] According to a 2016 analysis by the New America Foundation, over a dozen States possessed armed UAVs, and at least nine—Azerbaijan, Iran, Iraq, Israel, Nigeria, Pakistan, Turkey, UK and the US—had employed them in armed combat.[54] To date, there have been no reports of State development, possession or employment of UAVs with chemical weapons payloads. However, concerns have been raised regarding the development of UAVs promoted for wide area dispersal of riot control agents in law enforcement operations (as discussed in Section 12.3).

Similarly, in the civilian sphere, there has been contemporary development and increased online accessibility of a growing range of small hobbyist and mid-size commercially available UAVs capable of delivering several kilogram payloads. In addition to such multipurpose UAVs, there has been the associated development of a variety of specialized UAVs (notably agricultural and firefighting drones) intended to carry (sometimes significant) liquid payloads. For example, the Lockheed Martin Corporation and Kaman Aerospace Corporation have transformed the K-Max helicopter into a semi-autonomous UAV. It is promoted commercially as a general utility lift aircraft capable of performing a variety of functions including aerial firefighting, battlefield supply or humanitarian aid delivery, crop spraying and oil/gas drilling operations. The K-Max has been designed to lift and deliver a payload of over 2700 kg or approximately 680 gallons of water or other liquid, which is comparable to the capacity of a standard fixed-wing crop sprayer aircraft.[§]

The advancing capabilities and greater commercial availability of UAVs has led to increased interest in their employment by non-State actors, with a number of terrorist organisations—including Hamas, Islamic State, and Jabhat Fateh al-Sham—establishing dedicated UAV programmes employing repurposed commercial models, whilst Hezbollah, have a fleet of military grade UAVs reportedly provided by Iran. Non-State armed opposition groups or terrorist organisations have employed such systems for a range of functions including intelligence, surveillance, target acquisition and reconnaissance, information and propaganda operations and increasingly

---

[§]For details of K-MAX and potential missions, see the Lockheed Martin company website: http://www.lockheedmartin.co.uk/us/products/kmax.html.

offensive operations.[55,56] UAVs have reportedly been used by non-State actors to deliver improvised explosive devices in Colombia, Gaza, Iraq, Syria and the US.[57,58]

To date, there has been no successful employment of UAVs by terrorist organisations or other non-State actors to deliver chemical IEDs or otherwise deliver or disperse toxic chemical agents, although there have been a small number of failed bids to do so. The first reported attempt was by Aum Shinrikyo who acquired two "remote control helicopters of the kind used in Japan for aerosol spraying of crops."[59] When the group was asked by the vendor whether or not they wanted the spray tank attachments, they said, 'No we already have our own, thank you.'[60] In late 1993 and early 1994, they experimented with, and considered employing, these helicopters with an attached aerosol dispersal device to disseminate sarin in an assassination attempt against the leader of a rival Buddhist group.[61] The plan reportedly failed as "the helicopters crashed during testing".[62]

In June 2013, Iraqi authorities arrested five men allegedly linked to Al-Qaeda who were reportedly planning to use small remote-control helicopters to distribute toxic chemical agents as part of an attack against unspecified targets in Iraq, North America and Europe.[63] According to the Iraqi Defence Ministry, three terrorist workshops for manufacturing the toxic chemical agents, including sarin and mustard gas, were uncovered and remote-controlled helicopters were also seized at the sites. A Defence Ministry spokesperson stated that the remote-control devices were to have been used to release the chemical agents over the target from a "safe" distance of 1.5 km.[63]

Despite the potential utility of UAVs as means for chemical agent dissemination, the current level of technology commercially available is somewhat limited, specifically with regards to range, endurance, and payload (albeit with certain exceptions, as noted above). However, security analysts have warned that over the course of the next decade, significant advances will be made to the hardware and other core factors that currently limit the threat potential of the non-State actor misuse of commercially available UAVs. For example, it is predictable that future commercially available UAVs will be able to carry heavier payloads, fly and loiter longer, venture farther from their controller, survive in more difficult weather conditions and be able to do so *via* more secure communications links that are less susceptible to disruption.[64] In addition, further dual-use technological developments have resulted in increased autonomous capability and inter-operability between UAVs. Such advances provide so-called "swarming" capability, whereby numerous autonomous aircraft, networked together, can be employed towards common goals. Whilst such swarming technology has benign application in agriculture or for search and rescue missions, it also has a potential offensive application, whereby a UAV swarm could be tasked with overwhelming a target or targets through weight of numbers and avenues of attack.[4] In certain posited scenarios, large swarms of relatively small UAVs could be armed with chemical or biological agent payloads, collectively delivering significantly large amounts of agent onto or dispersing such agents over the target populations.

12.3 Wide Area Riot Control Agent Means of Delivery

12.3.1 Definition and Concerns

When utilized in accordance with manufacturers' instructions and in line with international human rights law and standards, riot control agents can provide an important alternative to other applications of force more likely to result in injury or death, such as firearms. Law enforcement officials throughout the world regularly use them to disperse violent crowds or to subdue dangerous individuals. Yet, they also are open to misuse. A study by the author of reports from international non-governmental human rights organizations, United Nations and regional human rights bodies, identified human rights violations committed by law enforcement officials utilizing riot control agents in at least 95 countries or territories from 2009 to 2013.[65]

This situation could dramatically worsen as a result of the ongoing development, marketing, and potential acquisition and subsequent deployment of a range of systems capable of delivering far greater amounts of riot control agent over wider areas or more extended distances than is currently possible with standard law enforcement RCA dispersal mechanisms, such as handheld sprays, grenades, and single-projectile launchers. In addition to their potential misuse for collective ill treatment or punishment of crowds, such wide-area RCA delivery mechanisms could be employed as force multipliers in conjunction with firearms, making lethal force even more deadly on a large scale. Although nominally developed or acquired for law enforcement, they may also be incorporated into military arsenals, potentially resulting in their subsequent use in armed conflict. Certain RCA delivery mechanisms, particularly large-calibre mortar or artillery shells, designated as RCA munitions, could also be used to disperse other toxic chemicals, such as classic chemical warfare agents. In addition, current weak trade controls in this area could result in the acquisition of wide area RCA delivery mechanisms and misuse by non-State actors, including armed opposition forces, unregulated private military and security companies, criminal groups, or terrorist organizations.

12.3.2 The Development and Marketing of "Wide Area" RCA Means of Delivery

Research undertaken by the Bradford Non-Lethal Weapons Research Project (BNLWRP) and the Omega Research Foundation (ORF) has detailed the development, production, possession or promotion by State or commercial

Figure 12.2 Image of 2 12-barrel IronFist 38 mm weapon system modules photographed at Milipol 2015, Paris, France © Robin Ballantyne/ORF.

entities in 16 countries of "wide area" RCA means of delivery since the coming into force of the CWC in 1997. The following case studies, illustrate some of the types of delivery systems of potential concern.[¶]

12.3.2.1 *Multi-barrel and Automatic Launchers*

Companies in Australia, China, France, India, Israel, South Korea, the Russian Federation and the US have developed multi-barrel launchers potentially capable of firing large salvoes of RCA projectiles, whilst Chinese and Russian companies have marketed automatic grenade launchers that can fire large numbers of RCA projectiles from a single barrel. Both launcher types potentially enable users to indiscriminately deluge substantial numbers of people with RCAs in a very short period.

12.3.2.1.1 Iron Fist [United States]. The US manufacturer, NonLethal Technologies Inc., has developed and as of November 2017 has marketed the IronFist "less lethal 38 mm weapon system with up to 36 barrels ... [intended] to rapidly deploy a blanket of less lethal munitions into, or over, a hostile crowd."[66] When the IronFist 36 barrel system (Figure 12.2) is loaded with high capacity 10 inch CS rounds (each with 10 mini-grenades) "a tactical officer can deploy up to 360 CS mini-grenades into a riotous crowd as needed from a single vehicle station..."[66]

12.3.2.1.2 Vehicle Mounted 38 mm and 64 mm Automatic Grenade Launchers [China]. In 2012, the China Ordnance Equipment Research Institute began to promote a vehicle mounted 38 mm automatic grenade

¶For a further discussion of these case studies and other RCA wide area means of delivery of potential concern including correspondence with relevant Governments and companies, see ref. 65 (chapters 4 and 12), and 181–183.

Figure 12.3 Image of a vehicle mounted 38 mm automatic grenade launcher photographed at China Police 2014 © Robin Ballantyne/ORF.

launcher (Figure 12.3) and a 64 mm automatic riot grenade launcher, capable of firing tear gas grenades.[67] These two weapon systems were designed to "cope with (large-scale) mass events quickly and effectively".[67] The 38 mm launcher has a maximum firing rate of 200 grenades per minute and a maximum effective range of at least 300 m. The 68 mm launcher has a maximum firing rate of 60 shots/minute and a maximum effective range of 600 metres.[68]

12.3.2.1.3 30 mm Grenade Round and Automatic Launcher [Russian Federation]. According to the 2009 English language version of the 2006 "*Ordnance and Munitions*" volume of *Russia's Arms and Technologies*,[69] a Russian company developed a 30 mm grenade round filled with irritant action pyrotechnic composition designed for the AGS-17 automatic grenade launcher. Employing such munitions, the AGS-17 has a maximum firing range of 1700 metres and a maximum rate of fire of between 350 and 400 rounds per minute. Reportedly, it can be used to "temporarily incapacitate armed lawbreakers on the open or rough terrain and those hiding in buildings, various structures and vehicles".[69]

12.3.2.2 Mortar Munitions and Other Large Calibre Projectiles

Companies or State research entities in China, India, the Russian Federation, Serbia, Turkey and the US have developed and/or promoted a disturbing range of RCA mortar munitions or other large calibre RCA projectiles capable of delivering significant quantities of RCA over wide areas and/or extended ranges. They vary in terms of their calibre, weight, design, material construction, potential payloads, area coverage and range, as well as the purposes for which they have been promoted.

12.3.2.2.1 120 mm Mortar Shell [Turkey]. In November 2003, *Jane's Defence Weekly* reported that the Turkish (State-owned) arms manufacturer, Makina ve Kimya Endustrisi Kurumu (MKEK) had developed a 120 mm mortar round—the CS MKE MOD 251—filled with CS.[70] The CS MKE MOD 251 mortar round weighed 17.34 kg and had a maximum range of 8132 metres.[71] It was promoted by MKEK on their website,[72] and at international security exhibitions including the Africa Aerospace and Defence (AAD) exhibition held in Cape Town, South Africa, in September 2010. Following the discovery of these activities, BNLWRP, ORF and the Institute for Security Studies (ISS) sought information from the company and the Turkish Government. In subsequent correspondence,[73] the Turkish Government stated that 1000 CS MKE MOD 251 munitions had been produced in 1996, prior to Turkey's ratification of the CWC, with roughly 150 used for testing purposes during the initial R&D phase in 1997. The facility for their production was subsequently discontinued after 1997. At the time of CWC ratification, there remained 850 pieces of CS MKE MOD 251 type munitions in the inventory of the Turkish Armed Forces. From that time and until 2011, Turkey stated that none of the remaining 850 munitions were used, but were stored at the Turkish Armed Forces ammunition destruction facility awaiting disposal. In July 2011 Turkey reported that such destruction had been completed.[73]

12.3.2.2.2 Mortar Shells, Heliborne-dispensed RCA Munitions and 500 kg Cluster Munitions [Russian Federation]. A Russian company reportedly developed 82 mm and 120 mm mortar shells filled with irritant-action pyrotechnic composition. The 82 mm shell which could be fired from a variety of mortars, weighed 3.5 kg, and had a maximum range of 2670 metres. The 120 mm shell weighed 16 kilograms and had a maximum range of 6.8 kilometres when mortar fired and 6.6 kilometres when fired from artillery pieces.[69] The company reportedly developed a heliborne dispenser of packages of sub-munitions filled with irritant-action pyrotechnic composition, which could be released from helicopters at an altitude of 50–300 m, in hovering mode, or at a flying speed of 150–300 km h^{-1}.[69] The company also reportedly developed a 500 kg cluster bomb packed with sub-munitions charged with irritant-action pyrotechnic composition. Reportedly, it could be "dropped from a fixed-wing or rotary-wing aircraft in an altitude span of 100–12 000 metres at a speed of up to 1200 km h^{-1}. The bomb permits high concentrations of an irritant agent to be attained within a short time."[‖]

12.3.2.2.3 XM1063 155 mm Malodorant Projectile [United States]. General Dynamics Ordnance and Tactical Systems worked under the direction of the US Army's Armament Research, Development and Engineering

‖Information taken from ref. 69. Details of Russian RCA munitions described in this section were last made publicly available in 2009; it is not known whether production continues or stocks remain.

Center (ARDEC) to develop a 155 mm artillery projectile, the XM1063.[74] The XM1063 was designed to carry out three interrelated functions, to "separate combatants from non-combatants; suppress, disperse or engage personnel [and] deny personnel access to, use of, or movement through a particular area, point or facility".[75] The XM1063 was intended to have a range of at least 20 km, and potentially up to 28 km.[76] The multiple sub-munitions would be released above the target area and then fall to the ground and disperse their payloads.[76] Estimates of the area covered vary up to a reported maximum of 10 000 m^2.[77] The proposed payload was described as a "non-lethal personnel suppression agent"[78] and a legal review conducted in 2007 indicated that malodorant agents were considered for this munition.[79] Testing of the XM1063 was reportedly completed successfully in 2007 and it was due for low-rate production from 2009. However, production was subsequently put "on hold".[77] No further information regarding the current status of the XM1086 R&D programme has been made public by the US government.

12.3.2.3 Unmanned Aerial and Ground Vehicles

A significant contemporary technological and market trend has been the development and/or promotion by companies or State research bodies in Austria, China, France, Israel, Spain, South Africa, Turkey, the United Arab Emirates and the US, of unmanned ground vehicles or drones capable of carrying RCA spraying devices or RCA projectile launchers.

12.3.2.3.1 Modular Advanced Armed Robotic System (MAARS) [United States]. QinetiQ North America, Inc. (QNA) developed the Modular Advanced Armed Robotic System (MAARS) through "partnership with various agencies in the [US] Department of Defense."[80] It is remotely controlled by an operator who can be over 1 kilometre away.[81] As well as a M240B medium machine gun, MAARS incorporates a 4 barrelled 40 mm grenade launcher that has the capability to utilise either 40 mm high-explosive grenades or a range of less-lethal ammunition, including 40 mm tear gas grenades.[82] According to a November 2010 *New York Times* article, US Army Special Forces had bought six MAARS "for classified missions", and the National Guard had requested "dozens more to serve as sentries on bases in Iraq and Afghanistan."[83] Although there have been no reports of its field deployment by the US military, MAARS has reportedly continued to be trialed and studied, for example in July 2016 by the US Marine Corps Warfighting Laboratory,[184] and is currently promoted by QNA on its website and at trade fairs.

12.3.2.3.2 Skunk Riot Control Copter [South Africa]. The South African company, Desert Wolf, has developed the Skunk riot control copter to "control unruly crowds without endangering the lives of the protestors or the security staff".[84] According to company promotional material, the UAV is equipped with four high-capacity paintball barrels which can collectively fire

"80 Pepper balls per second stopping any crowd in its tracks";[84] its "current hopper capacity of 4000 balls ... allows for real stopping power".[84] The UAV can be "operated in formation by a single operator [employing] the Desert Wolf Pangolin ground control station".[84]

In a July 2014 interview with *BBC News*, Desert Wolf's managing director Hennie Kieser stated that the company "received an order for 25 units" from "an international mining house".[85] Mr Kieser claimed other potential customers included "Some mines in South Africa, some security companies in South Africa and outside South Africa, some police units outside South Africa and a number of other industrial customers."[85]

In October 2015, Desert Wolf had reportedly entered into an agreement with an un-named South American manufacturer who would supply a range of "non-lethal" grenades including tear gas for use in an upgraded Skunk model – the Skunk GL UAV. It was reported that each UAV would be capable of carrying "up to 48 grenades" and that "the grenades will be dropped over crowds and descend *via* parachute, minimising collateral damage on the ground and allowing the smoke to spread from above."[86]

12.3.2.3.3 AVS 100/150 Drones with Multiple Purpose Payload Launcher [France]. In November 2015, at the Milipol (Military & Police) 2015 exhibition in Paris, the French UAV manufacturer, Aero Surveillance first publicly displayed its AS-100 and AS-150 drones with a new Multi-purpose Payload Launcher (the MPL 30) which can deploy a range of payloads from an altitude of several hundred metres, including up to 18 tear gas grenades. According to Aero Surveillance, which continues to promote the MPL 30 on its website and at arms exhibitions (Figure 12.4), the system is designed for "law enforcement and homeland security" operations.[87,88]

Figure 12.4 Image of an unmanned aerial vehicle with attached Multi-purpose launcher/aero cougar photographed at Eurosatory 2016, Paris, France, © Robin Ballantyne/ORF.

Figure 12.5 Images of an Eagle Eye Drone photographed at Eurosatory 2016, Paris, France, © Robin Ballantyne/ORF.

12.3.2.3.4 Eagle Eye Drone [Turkey]. The Turkish company Yavascalar (YAVEX) developed the Eagle Eye Anti Riot Drone (Figure 12.5), which was promoted at the Eurosatory 2016 arms and security fair in Paris, France, in June 2016. According to the company's marketing materials, it has an effective operating range of 5 km^2 and a maximum operative altitude of 3000 feet, and can carry a payload of up to 5.5 kg. It is fitted with an RCA projectile delivery mechanism which can release nine CS smoke projectiles which drop free-fall onto the target crowd below.[89] According to *MSI*, the *Turkish Defence Review Magazine*, Eagle Eye can be effective over an area of 700–800 square metres with the projectiles it carries, depending on environmental conditions.[90] The project has currently been suspended following the acquisition of Yavascalar by the Turkish company, Sarsilmaz.

12.4 Military Toxic Chemical Means of Delivery

12.4.1 Introduction

This Section explores the contemporary development and/or employment of a range of toxic chemical delivery and dispersal mechanisms solely, or primarily, intended for use in armed conflict. In addition to munitions and delivery mechanisms that have been specifically manufactured to deliver or disperse chemical warfare agents, this Section will also explore a range of conventional munitions and other conventional weapons that have, subsequent to manufacture, been adapted for use as chemical weapons. To illustrate the range of potentially applicable delivery and dispersal mechanisms, the scale of certain reported development programmes and diversity of those seeking to develop and employ such chemical weapons, three case studies are examined. The first two explore contemporary reported State military chemical weapons development or employment in Syria and reported development in North Korea. In contrast, the last explores the attempts of Islamic State to develop and employ a range of chemical munitions in the area then under its control.

12.4.2 Syrian Chemical Weapons Munitions

12.4.2.1 Introduction

To date, only limited information has been made public concerning the size, nature and origin of Syria's past and current chemical weapons stockpile. Although the Syrian Government, on accession to the CWC in October 2013, provided information to the OPCW on quantities and types of chemical precursors, chemical agents, and munitions it possessed, specific details of the Syrian declaration were never disclosed by the OPCW. However, evidence gathered during subsequent UN, OPCW–UN and OPCW investigations prior to and after the Syrian CWC accession, together with research by civil society organisations, the media, and through the release of intelligence reports and analysis by certain States,[185–187] provide indications of the types of chemical munitions and agent held at some stage by the Syrian Government in its chemical weapons stockpiles, and subsequently employed in the ongoing civil war. Whilst there have been numerous reported chemical weapons attacks—some apparently employing sarin in binary form deployed in a variety of munitions**—the information in the majority of instances could not be verified or was inconclusive. In the following three cases, however, enough substantive evidence has been gathered on the agent and munitions to determine that they originated from Syrian chemical weapon stockpiles.

12.4.2.2 Indigenously Factory Manufactured Sarin Rocket

On 21 August 2013, two separate opposition-controlled districts in Damascus—the Zamalka neighbourhood of Eastern Ghouta and the Moadamyiya neighbourhood of Western Ghouta—were attacked by multiple sarin filled rockets, injuring over 3600 people and killing over 800.[91] On 28–29 August 2013, a UN Secretary-General's Mechanism (UNSGM) investigation team visited munition impact sites in Eastern Ghouta. At the Zamalka site they photographed and later removed the remnants of a munition which had landed in earthy, relative soft, ground where the shaft/engine of the projectile remained dug in, and was undisturbed until investigated.[92] From its analysis, the UNSGM investigation team concluded that the munition was a 330 mm surface-to-surface rocket (Figure 12.6), capable of carrying a chemical payload. The approximate capacity of liquid in the warhead was between 50–62

**This issue is covered in detail in ref. 188.

Figure 12.6 Images of a 330 mm surface-to-surface rocket, photographs taken from Report of the United Nations Mission to Investigate Allegations of the Use of Chemical Weapons in the Syrian Arab Republic on the alleged use of chemical weapons in the Ghouta area of Damascus on 21 August 2013.

litres.[93] Analysis of the rocket remnants and environment confirmed that it had contained sarin.[94] The UNSGM Mission did not establish the origin of the munition or the entity which had employed it.

The UN Commission of Inquiry (UN COI) also investigated the Ghouta attacks. Although it did not determine the perpetrators of the attacks, it did conclude that: "The evidence available concerning the nature of the quality and quantity of the agents used indicated that the perpetrators likely had access to the chemical weapons stockpile of the Syrian military, as well as the expertise and equipment necessary to safely manipulate large amounts of chemical agents."[95]

According to Human Rights Watch, the dimensions of the Zamalka rocket were compatible with the Iranian-produced 333 mm Falaq-2 rocket launcher, or close copies and derivatives thereof.[189] Arms Research Services (ARES) have documented the Syrian Government forces' possession and use of the Falaq-2 launching system with unidentified 330 mm rockets.[190] This rocket type had a non-aerodynamic design indicating that it would have a relatively short range and would not be capable of accurate targeting. While Human Rights Watch could not establish where such rockets were manufactured, their basic design and unique size matching the Iranian rocket launching system led the organisation to suggest a Syrian industrial origin.[96]

12.4.2.3 Foreign Factory Manufactured Sarin Rocket

According to eyewitness accounts, the Moadamiya neighbourhood of Western Ghouta was struck by at least seven rockets at about 5 a.m. on 21 August 2013.[97] On 26 August 2013, the UNSGM investigation team visited two of the munition impact sites, where they photographed and later removed remnants of a rocket (Figure 12.7).[98] From the observed and measured

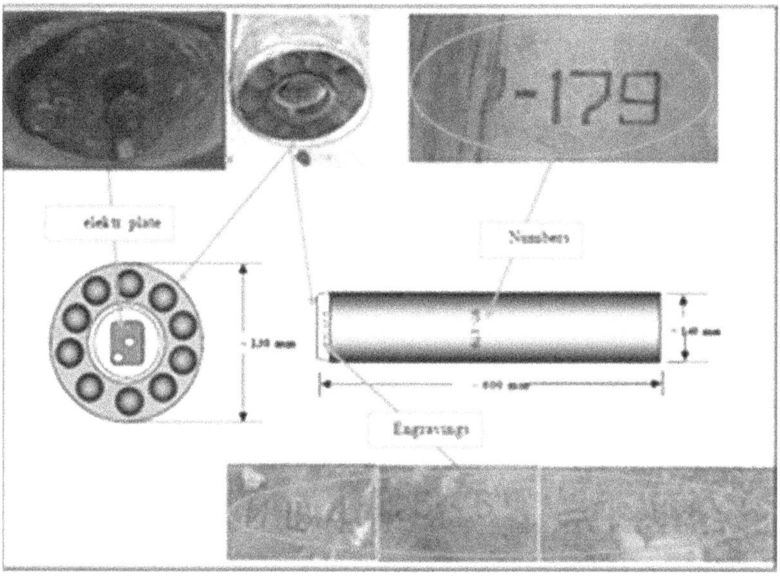

Figure 12.7 Remnants of an artillery rocket. Photographs taken from Report of the United Nations Mission to Investigate Allegations of the Use of Chemical Weapons in the Syrian Arab Republic on the alleged use of chemical weapons in the Ghouta area of Damascus on 21 August 2013.

characteristics, the UNSGM investigation team concluded that the munition was a 140 mm surface-to-surface rocket, capable of carrying a chemical payload. Analysis of the rocket remnants and environment confirmed it had contained sarin.[99] Although, the UNSGM Mission did not establish the origin of the munition or the entity which had employed it, it stated that the munition "indicatively matched one of the variants of the M-14 artillery rocket, with either an original or an improvised warhead (not observed at the impact site)." According to the UNSGM investigation team, the relative positions of the two impact sites were "fully congruent with the dispersion pattern commonly associated with rockets launched from a single, multi-barrel, launcher."[100]

According to *Janes Ammunition Handbook*, the M-14 was a Soviet Union era artillery rocket, produced in three versions including a 140 mm chemical rocket carrying an 18.3 kg warhead containing 2.2 kg of liquid sarin to be spread by a TNT burster charge controlled by a point detonating fuze.[101] Cyrillic text recorded by both the UNSGM investigation team and Human Rights Watch appears to indicate that the munition was manufactured at "Factory 179" in Novosibirsk, which was one of the largest producers of artillery and rockets during the Soviet period, and a known manufacturer of the M-14.[102] According to information held on SIPRI's Arms Transfers Database, the Soviet Union transferred 200 BM-14 launchers "second hand" to Syria in 1967–1969; the BM-14 is the most common launcher for Soviet 140 mm rockets.[103] However there is no publicly available information concerning the quantities and types of any associated 140 mm M-14 rockets transferred to, or in the possession of, Syria.

12.4.2.4 Factory Manufactured Sarin Bomb

In the early morning of 4 April 2017, widespread public reports emerged that air strikes had released sarin upon the town of Khan Sheikhoun. Interviewees and early warning reports gathered by the UN COI indicated that a Sukhoi 22 aircraft conducted four air strikes upon Khan Shaykhun at around 6:45 am on that day; and that only Syrian forces operated such aircraft. The Commission "identified three conventional bombs... and one chemical bomb used in the attack".[104] According to the UN COI "The chemical bomb killed at least 83 persons, including 28 children and 23 women, and injured another 293 persons, including 103 children."[105] The UN COI stated that "photographs of weapon remnants depict a chemical aerial bomb of a type manufactured in the former Soviet Union."[106] It concluded that there were "reasonable grounds to believe that Syrian forces attacked Khan Shaykhun with a sarin bomb."[107] Independent research and analysis by Human Rights Watch led the organisation to conclude that "the Syrian warplane dropped a factory-made sarin bomb", which Human Rights Watch tentatively identified as a Soviet Union produced KhAB-250 – a "Chemical Aviation Bomb" weighing 250 kg and containing 44.1 kilograms of sarin.[108]

From reviewing the evidence gathered by the OPCW FFM and conducting its own research, the OPCW–UN JIM stated that it was "confident" that Syria was "responsible for the release of sarin at Khan Shaykhun".[109] According to the OPCW–UN JIM: "The sarin identified in the samples taken from Khan Shaykhun was found to have most likely been made with a precursor [methylphosphonyl difluoride] DF from the original stockpile of the Syrian Arab Republic."[110] Furthermore "*the results confirm that sarin was produced by the binary route*, in which DF is combined with isopropanol (iPrOH) in the presence of hexamine".[111] Following analysis of the impact crater, an external expert commissioned by the OPCW–UN JIM concluded that the "most probable type of munition to have caused the crater would be a relatively large bomb, with a mass of 300–450 kg... dropped from a medium or high altitude, of between approximately 4000 and 10 000 m."[112] In addition, "information obtained by the OPCW–UN JIM", regarding images showing a filler cap from a chemical munition found at the site, "with two closure plugs" was "uniquely consistent with Syrian chemical aerial bombs".[113] Although there are differences in the characterisation of the chemical munition by the UN COI, OPCW–UN JIM and Human Rights Watch, the three investigations provide evidence that Syrian forces employed a factory-made sarin aerial bomb at Khan Shaykhun.

12.4.3 North Korea – a State Chemical Weapons Programme Outside the OPCW

12.4.3.1 Chemical Weapons Programme Overview

The international community has recognised the threat to international peace and security from North Korea's attempts to develop nuclear, chemical and biological weapons and has sought to address this through UN Security Council sanctions and other measures.[191] However, due to the secrecy and isolation of the North Korean regime, public information on its chemical weapons programme is limited and unverified. In 2007, the Director of the US Defense Intelligence Agency asserted that "North Korea has had a longstanding chemical weapons stockpile of nerve, blister, blood, and choking agents".[114,192] In 2016, the South Korean Ministry of National Defense estimated this stockpile as between 2500 and 5000 metric tons of chemical weapons.[115] It is claimed that at maximum capacity, North Korea has the ability to produce up to 12 000 tons of chemical warfare agents, including sarin and VX.[116] North Korea is thought to be potentially capable of weaponizing such chemical agents through employment of ballistic missiles and a range of conventional munitions and associated means of delivery.[193]

12.4.3.2 Conventional Munitions

Although estimates vary, North Korea's conventional arsenal is believed to be sizeable; in 2011, the US and South Korea concluded that North Korea had more than 13 000 artillery and multiple rocket launcher (MRL) systems.[117] Analysts believe that there are a thousand of these systems (or more) that fall into the "long-range" category, and they consist of long-range 170 mm self-propelled guns, augmented by long-range 240 mm multiple rocket launchers.[118] According to a 2016 CSIS report, "thousands of artillery of various calibres" reportedly including "artillery pieces with calibres larger than 80 mm" are "capable of being fitted with chemical warheads." In addition, according to CSIS, a 2012 South Korean report stated that North Korea was able to equip its rocket launchers, mortars, and field artilleries with biological and toxin weapons and assessed that botulinum toxins, (as well as smallpox, and anthrax) were the most likely to be weaponized.[119]

12.4.3.3 Ballistic Missile Programme

North Korea has produced, tested and fielded short and medium range missiles including Scud-B and Scud-C variants with ranges of 300 km and 500 km, respectively; the Nodong missile with a range of 1300 km, and the Musudan missile with a minimum range of 3000 km.[121,194] According to the 2016 South Korean Defense White Paper, "through these successive additions to its missile inventory, North Korea has gained direct strike capabilities against

Table 12.1 Selected North Korean Ballistic Missiles.[126,127,195]

Name	Range (km)	Payload (kg)	Status
KN-02	100–120	250–500	Operational
Hwasong-5	300–320	1000	Operational
Hwasong-6	500	700	Operational
Hwasong-7	700–1000	400–750	Operational
KN-11	1000	Unknown	Testing
Nodong	1300–1500	250–700	Operational
Musudan	3200–4000	1200	Testing
Taepodong-1	2500–5500	100–200	Operational
Taepodong-2	6000–9000	100–500	Operational
Unha-3	10 000+	100–1000	Operational
Hwasong-12	4500	500–650	Testing
Hwasong-14	10 000+	300–700	Testing
KN-08	11 500	300–700	Testing

[South Korea] and the surrounding countries of the Korean Peninsula."[122] In the 1990s, North Korea initiated the research, development and flight testing of longer range missiles including the Taepodong-1 with a maximum range of 5000 km, the Taepodong-2 with a maximum range of 9000 km, the Unha-3 rocket (ostensibly for its space programme) and the Hwasong-14, both with an estimated range of over 10 000 km (See Table 12.1).[123–125] The development of such intermediate and long-range missiles potentially allows North Korea to strike Japan, China, parts of Europe and Continental US.

Although US Government sources have given no figures of North Korean missile holdings, they have estimated that North Korea has deployed over 600 Scud missile variants, about 200 Nodong (Rodong) missiles, and fewer than 50 Musudan and Taepodong missiles.[128] Information concerning the numbers of chemical missiles stockpiles are similarly scarce. A 2016 CSIS report stated that North Korea had deployed "hundreds" of short-range Hwasong-5/-6 missiles and Frog-5/-7 missiles "capable of being fitted with chemical warheads". In addition, the CSIS stated that "according to defector accounts, the DPRK's long-range missiles such as the Nodong and other ballistic rockets...are capable of delivering CW agents"[129] and that the South Korean Government had previously estimated that half of North Korea's long-range missiles were able to deliver biological or chemical weapons.[130] In April 2017, Japanese Prime Minister Shinzo Abe told the Japanese Parliament that "There is a possibility that North Korea already has a capability to deliver missiles with sarin as warheads."[131]

12.4.3.4 Concerns Grow but Potential Chemical Weapons Holdings Remain Unverified

Concerns regarding North Korea's potential chemical weapons capabilities and intent have been underpinned by the regime's long-standing military doctrine promoting such weapons;[132] its forward deployment of an estimated 8–10 000 artillery and multiple rocket launchers, some of which are potentially capable of firing chemical munitions;[133] and its previous large scale attack in 2010 upon South Korea's Yeonpyeong Island, employing 100 conventional artillery shells and rockets.[134] Recently, concerns have been further heightened by North Korea's increased efforts throughout 2017 to improve its existing ballistic missiles and augment their capabilities;[135–137] and its suspected involvement in the assassination in February 2017 of Kim Jong-nam, the half-brother of North Korea's leader Kim Jong-un, who was killed at Kuala Lumpur airport in Malaysia through the administration of the nerve agent VX.[138,139]

Whilst the preceding reports of North Korean activities are deeply concerning, the claims regarding its chemical weapons holdings have not been verified, and the information to date indicates the *potential* of North Korea to arm its conventional munitions and ballistic missiles with chemical warheads. Whilst North Korea apparently has significant chemical agent stockpiles, a wide range of conventional munitions and means of delivery, and an ever growing missile capability, it is currently unknown whether it has filled *any* of its conventional munitions or equipped *any* of its ballistic missiles with reliable and effective chemical warheads. To date, there have been no confirmed test flights in North Korea of missiles with chemical warheads. Consequently, the functionality and reliability of these weapons—if they exist—is therefore unknown.[140]

12.4.4 Islamic State Military Chemical Weapon Munitions

12.4.4.1 Introduction

Historically, chemical weapons programmes employing delivery and dispersal mechanisms capable of delivering significant quantities of toxic chemicals over wide areas or extended ranges have been undertaken by States. However, recently one non-State actor—Islamic State (IS) acquired the resources and capability to produce and employ such chemical munitions and delivery systems, albeit to a limited degree. The following section examines potential routes by which IS developed and/or acquired a range of conventional munitions, some of which it filled with toxic chemicals—including chlorine and sulfur mustard—and subsequently employed in attacks against combatants and civilian populations in Iraq and Syria.

12.4.4.2 Indigenous Industrial Manufacture of Conventional Munitions

Islamic State forces indigenously manufactured a range of mortars, rockets and other projectiles; although the primary purpose has been for employment as conventional munitions, some could potentially be filled with toxic chemicals. In February 2015, the NGO Conflict Armament Research (CAR) detailed IS production in Kobane, Iraq, of significant quantities of improvised conventional munitions, most based on mortar rounds of various calibres and 105 mm rockets.[141] Subsequently, the CAR investigations highlighted increasing sophistication in IS forces' indigenous munition production. In December 2016, following field investigations in Mosul, Iraq, the CAR documented IS's coordinated specialised weapons manufacturing plants indigenously producing 120 mm mortars, and 110 mm and 115 mm rockets.[142] Mortar production, at least, was on a relatively large scale with the CAR estimating that IS forces probably manufactured tens of thousands of 120 mm mortar rounds.[143] The CAR concluded that "Although production facilities employ[ed] a range of non-standard materials and chemical explosive precursors, the degree of organisation, quality control, and inventory management, *indicate[d] a complex, centrally controlled industrial production system.*"[144]

12.4.4.3 Acquisition of Conventional Munitions

Islamic State (IS) reportedly acquired a substantial arsenal of foreign factory produced conventional weapons and associated munitions originally "designed or manufactured in more than 25 countries".[145] These included rocket propelled grenade (RPG) launchers, 60 mm and 82 mm mortars, 106 mm recoilless rifles, 122 mm and 130 mm field guns, 122 mm, 152 mm and 155 mm howitzers, multiple rocket launchers, man portable air defence systems (MANPADS), surface-to-surface missiles, tanks and armoured fighting vehicles, together with associated conventional munitions.[146–149] Although the provenance of specific weapons and munitions is often uncertain, it appears that some were captured from Iraqi or Syrian combatants or stockpiles, whilst others may have been acquired through illicit transfers.

Open source information indicates that Islamic State attempted to repurpose certain conventional munitions as chemical weapons. For example, in January 2017 the *Associated Press* reported the discovery by Iraqi Special Forces of a chemical weapons facility in Mosul, Iraq, containing a tank of sulfur mustard and a warehouse of more than a dozen surface-to-surface rockets bearing Russian inscriptions.[149] According to Brigadier Fadhil from the Iraqi Special Forces, the types of rockets found at the site suggested that IS was attempting to weaponize the chemical agent.[149]

12.4.4.4 Islamic State Attacks Employing Conventional Munitions Filled with Toxic Chemicals

Islamic State has since 2007 successfully made and employed a variety of simple chemical IEDs, predominately in Iraq. However, since 2015, there have been a growing number of reported chemical attacks by IS against both combatants and civilian populations, apparently employing a range of conventional munitions capable of delivering significant amounts of chemical agent over extended distances. Initially, such attacks were predominately in Iraq, but subsequently were increasingly reported in Syria as well. In many of these attacks the specific nature of the means of delivery employed could not be properly confirmed. In certain cases however, as illustrated below, evidence has confirmed the IS use of conventional mortars, rockets or artillery shells filled with either chlorine or low grade sulfur mustard.[††] IS did not appear to have gained any military or strategic benefits from the development of such chemical munitions, and their use dramatically decreased from early 2017 onwards, following the military defeats of the organisation and recapture of much of its territory in Iraq and Syria.

12.4.4.4.1 IS Chemical Attack Employing Mortars. According to information gathered by Human Rights Watch,[150] the Syrian Government, the Russian Federation Government, the OPCW FFM[151] and the OPCW–UN JIM,[109] the village of Um-Housh (or Umm Hawsh) was attacked on the evening of 15 September 2016 and early morning of 16 September 2016, by mortar shells fired by IS forces. Two female casualties who sought medical treatment exhibited symptoms of exposure to sulfur mustard, which was subsequently confirmed from blood sample analysis.[152] According to the Syrian Government investigation, "locally made mortar shells of several calibres landed", with some failing to function as designed.[153] A 217 mm mortar shell was recorded and removed by Russian Federation investigators; its contents were subsequently subject to non-destructive evaluation and sampling by the OPCW FFM and were found to contain a "liquid sulfur mustard fill".[154] A defence research institute, a forensic institute and an expert consultant, all engaged by the OPCW–UN JIM, observed that the mortar shell was of "poor production quality"[155] whilst experts in energetic materials noted that the mortar shell had a "similar appearance to those made by makeshift large-scale production methods."[156] According to the JIM "mortars fitting 217 mm calibre mortar shells are easily produced from available high-quality steel tubing. The range of such munitions is from 1 to 2 km... Forensic institutes and experts... noted that the range and accuracy of [such] homemade mortar shells are imprecise and depend on a number of variables, including the amount of explosive powder used."[157] According to information provided to the OPCW–UN JIM, Islamic State had "developed the capability to produce sulfur mustard as of 2015" and "had ample access to industrial zones, including oil and gas fields to produce the delivery systems, munitions and chemicals."[158]

12.4.4.4.2 IS Chemical Attack Employing Rockets and Mortars. According to international media reports, IS forces carried out a chemical attack upon

[††]It appears that from 2015, IS indigenously manufactured sulfur mustard through the Levinstein chemical reaction process, which does not yield high quality agent, but is a "relatively uncomplicated method of producing sulfur mustard to which non-State actors might turn." For further information see ref. 196.

the village of Taza near Kirkuk in Northern Iraq on 8 March 2016. A Peshmerga Forces commander, interviewed by *Reuters*, claimed that a total of 24 mortar shells and Katyusha‡‡ rockets filled with "poisonous substances" were fired into Taza from the nearby Bashir area.[159] A local Shia Muslim defence unit member, interviewed by the *BBC*, stated that "the rockets didn't explode, but when they landed a chemical agent seeped out."[160] The *BBC* reported that three children were killed and some 600 people were injured, 35 of them critically. Doctors in the local hospital stated that the symptoms that the victims exhibited after the attack, including burns and respiratory problems, were consistent with exposure to chemical agents.[160–162] The OPCW Director-General expressed "serious concerns" about the alleged chemical attacks in Taza and elsewhere and offered technical assistance to Iraq in its investigations.[163] Although neither the OPCW nor Iraq has formally identified the chemical agents used; the BBC reported that some rockets "were loaded with what has been identified locally as sulfur mustard ... and chlorine."[160]

12.4.4.4.3 IS Chemical Attack Employing Artillery Shells. On the morning of 21 August 2015, the Syrian town of Marea was the target of a chemical attack. A subsequent OPCW FFM investigation established "with the utmost confidence that at least two people were exposed to sulfur mustard ... and ... it is very likely that the effects of sulfur mustard resulted in the death of an infant."[164] A further 85 people also reportedly sought medical assistance with injuries and symptoms related to sulfur mustard exposure over the course of the next four days.[165] Subsequent investigations by the OPCW–UN JIM determined that IS were responsible for the attack and provided more information about the munitions employed. According to the OPCW–UN JIM Report, "Several witnesses, as well as a number of other sources provided information that Marea was bombarded by around 50 artillery shells, several of which were filled with sulfur mustard, from the east or south-east, an area under the control of [IS]."[166] Furthermore, "sources stated that the munitions used in all these locations were 130 mm artillery shells ... [and] ... open source information shows pictures of [IS] having 130[mm] artillery shells and towed-field guns."[167] According to the OPCW–UN JIM, the 130 mm artillery shells were "thought to be easily repurposed and filled with different payloads."[168]

12.5 Conclusions

Analysis of contemporary cases has shown that both State and non-State actors have developed and used a variety of chemical weapons employing improvised means of delivering toxic chemicals to specific targets or dispersing toxic chemicals over a wide area. An examination of ground-based

‡‡The term "Katyusha" is often used by Iraqi observers to refer to indigenously produced makeshift Iraqi rockets rather than the original Russian models.

chemical improvised explosive devices (IEDs) has shown that development and use of such weapons has to date been restricted to a small number of terrorist, insurgent and armed opposition groups, most notably those associated with Al-Qaeda and Islamic State, operating in Iraq and Syria. Whilst there are no recently confirmed instances of State forces developing or employing ground-based chemical IEDs, Syrian Government forces in 2016, 2017 and 2018 reportedly employed improvised rocket assisted munitions filled with chlorine against civilians and combatants. Of greater concern, given the scale of injury and death, has been the development and widespread use in many parts of Syria since 2014 of aerial chemical IEDs, most commonly containing chlorine, as documented by the OPCW FFM. Given the requirement for aerial delivery, their use appears to be restricted to State forces. In at least three cases the OPCW–UN JIM has determined that such chemical weapons have been employed by the Syrian Air Force against civilians. Reports of subsequent chemical aerial IED attacks against both combatants and civilians by Syrian Government forces were reported in 2016, 2017 and 2018 by the UN COI and human rights NGOs.

Although there are diverse of types of chemical IEDs, they are all made fully or in part from commercially available products not originally intended by the manufacturers for such malign purposes. Likewise, there are a range of commercial aerosol generators and wide area spray aircrafts that could potentially be employed as improvised chemical agent delivery mechanisms. Although concerns have regularly been raised regarding potential terrorist use of commercial aerial dissemination devices, particularly crop dusters, there appear to be significant logistical and technical barriers to such use, and only one successful contemporary non-State actor attack by such means has been discovered by the author. One area where there does appear to be interest by certain State and non-State actors is the potential application of UAVs for the dissemination of toxic chemicals. Although limited information is available concerning State programmes, it appears that non-State actor employment of commercially available UAVs may well be constrained in the near future due to existing technology limitations notably in the range, endurance and payload of such devices. However, significant advances in these areas are predicted in the next decade, so technology monitoring needs to be undertaken and threat assessments in this area may well need to be regularly revised.

Investigations conducted by Bradford Non-Lethal Weapons Research Project and the Omega Research Foundation have uncovered the development, production and commercial promotion of a range of munitions and delivery systems intended to disseminate significant amounts of RCA over wide areas and/or extended distances, including large smoke generators, multi-barrel projectile launchers, automatic grenade launchers, rocket propelled grenades, mortar munitions, artillery shells, heliborne munition dispensers, aerial bombs and cluster munitions, and delivery mechanisms on unmanned ground vehicles and UAVs. The employment of certain wide-area RCA delivery mechanisms may be justifiable, in particular in large-scale law enforcement situations provided that they comply with the CWC provision on "types and quantities" and are used in strict conformity with human rights standards. However, other forms of wide-area RCA delivery mechanisms that have been developed, such as artillery shells, mortar shells, aerial bombs and cluster munitions, appear to be completely inappropriate for any form of legitimate law enforcement, having possible utility instead in large-scale human rights abuses or armed conflict. They should be considered as chemical weapons and treated accordingly (*i.e.* they should be formally declared to the OPCW and verifiably destroyed). The development and promotion of new forms of wide area RCA means of delivery is continuing apace, in part reflecting advances in science and technology as well as market opportunities; it is an area that should be closely monitored by the OPCW and relevant human rights bodies.

An examination of Syria's continuing employment of chemical weapons has highlighted both the difficulties in ensuring that States abide by their obligations under the CWC and also the importance of identifying chemical munitions used in such attacks. Evidence gathered during UN, OPCW–UN and OPCW investigations prior to and after the Syrian CWC accession, together with research by civil society organisations and intelligence reports released by certain States, provide some indications of the types of chemical munitions and agent held at some stage by the Syrian Government in its chemical weapons stockpiles, and subsequently employed in the ongoing civil war. These munitions include unknown quantities of Syrian manufactured 330 mm sarin rockets and associated launchers, apparently of Iranian origin; Soviet manufactured 140 mm sarin rockets and associated launchers; and large sarin aerial bombs apparently of Soviet origin that can be dropped from Syrian warplanes. Although all declared Syrian chemical weapons stockpiles were subsequently verifiably destroyed under OPCW–UN oversight, concerns emerged that the Syrian Government did not provide full and accurate details to the OPCW of its chemical weapons programme and holdings. Consequently, the work of the OPCW Declaration Assessment Team established to clarify these inaccuracies must continue until all uncertainties are fully resolved, and this must be followed by the complete verified removal and destruction of any remaining stockpiles of chemical weapons agents, filled or unfilled chemical munitions and other means of delivery. In addition, the OPCW FFM's investigations into allegations of the use in Syria of sarin and improvised chemical weapons employing chlorine and other toxic chemicals must likewise continue. Where evidence of the development of such chemical weapons is uncovered, the OPCW must act swiftly to halt such activities. All evidence of such activities must also be secured so that the perpetrators of such chemical weapons attacks can be identified through a UN-mandated investigation or other appropriate mechanism.

Of the four States yet to accede to or ratify the CWC, the international community has recognised that the greatest current potential threat to international peace and security comes from North Korea. North Korea reportedly has significant chemical agent stockpiles; large numbers of conventional munitions and associated weapons, and an ever-growing missile capability - much of it *potentially* capable of being utilised for chemical delivery. Whilst currently available information is deeply concerning, it indicates *potential capability*; it is currently unknown whether North Korea has filled *any* of its conventional munitions or equipped *any* of its ballistic missiles with reliable and effective chemical warheads. However, concerns regarding North Korea's chemical weapons capabilities and intent have been exacerbated by its long-standing military doctrine promoting such weapons; its forward deployment of large numbers of CW compatible conventional weapons near South Korea's borders and its apparent use of a chemical weapon in the assassination of high profile North Korean opponent. It is clearly vital that the OPCW and the wider international community continue to engage with North Korea (and the other non-CWC States) to promote CWC accession and the consequent declaration and verifiable destruction of all chemical weapon stockpiles and production facilities. In the case of North Korea, such engagement needs to be accompanied by continuance of UN arms embargoes and associated measures intended to combat nuclear, chemical and biological weapons production and proliferation to other State or non-State actors.

Historically, chemical weapons programmes that employed munitions capable of delivering significant quantities of toxic chemicals over wide areas or extended ranges were undertaken by States. Recently, one non-State actor has been successful in such endeavours, albeit to a limited degree. Following its acquisition and retention of large tracts of Iraq and Syria, Islamic State apparently had the material resources, know-how and territorial security to indigenously manufacture, and/or capture or otherwise acquire, a range of conventional munitions, including mortars, rockets and artillery shells, some of which it filled with toxic chemical payloads including chlorine and sulfur mustard. These chemical munitions were subsequently employed by IS, utilising captured or acquired conventional weaponry, from 2015 against civilians and armed forces in Iraq and Syria. IS did not appear to have gained any strategic or military advantage from such chemical attacks, and they dramatically decreased from early 2017 onwards, apparently in response to the military defeats of IS and the recapture of much of its territory in Iraq and Syria, the bombing of its chemical weapons facilities and killing or capture of certain chemical weapon developers. In terms of strategies for preventing future non-State actor chemical weapons programmes, this case illustrates the importance of denying non-State actors the material resources and "safe space" to indigenously manufacture or acquire the toxic chemicals and conventional munitions needed for weaponization. Consequently, all States should introduce and enforce national measures fully implementing their obligations under the CWC and UNSCR 1540 to combat both State and non-State CW development programmes; prevent illicit trafficking of relevant conventional weapons and munitions; ensure relevant conventional weapons and munitions stockpiles are secure; and ensure that all former chemical weapons stockpiles and production facilities in their territories are verifiably destroyed.

References

1. OPCW, *Chemical Weapons Convention*, 1993, Article II.1.
2. W. Krutzsch, E. Myjer and R. Trapp *The Chemical Weapons Convention, a Commentary*, Oxford University Press, 2014, p. 80.
3. Adapted from: United Nations, *International Ammunition Technical Guidelines*, 2011, available from: https://www.un.org/disarmament/un-saferguard/guide-lines/, accessed 14 December 2017; NATO, IED, *NATO Glossary of Terms and Definitions, (AAP-6)*, 2013, pp. 2-1-2; US Department of Homeland Security, *Developing an Empirical Understanding of Improvised Explosive Devices*, 28 August 2009.
4. US Department of Homeland Security, *IED Attack Improvised Explosive Devices, a Fact Sheet from the National Academies and the Department of Homeland Security*, undated.
5. J. Tucker, *The future of chemical weapons*, The New Atalantis, 26, Fall 2009/Winter 2010, pp. 3–29, pp. 17–18.
6. C. Quillen, The Islamic State's Evolving Chemical Arsenal, *Stud. Conflict Terrorism*, 2016, **39**(11), 1019–1030.
7. J. Warrick, *Black Flags: The Rise of ISIS*, Doubleday, New York, 2015, p. 140.
8. *Jordan Was Chemical Bomb Target*, BBC News, 17 April 2004; S. Gorka and R. Sullivan, Jordanian Counterterrorist Unit Thwarts Chemical Bomb Attack, *Jane's Intell. Rev.*, 2004, 22–23.
9. J. Warrick, *Black Flags: The Rise of ISIS*, Doubleday, New York, 2015, p. 139.
10. Al Qaeda's Poison Gas, *Wall St. J.*, 2004, A16.
11. News Chronology, February through April 2004, *The CBW Conventions Bulletin*, June 2004, vol. 64, p. 47.
12. D. Cave and A. Fadam Iraqi, *Militants Use Chlorine in 3 Bombings*, The New York Times, 21 February 2007.
13. B. Hoffman, Low-tech Terrorism, *National Interest*, March–April 2014, p. 70.
14. B. Roggio, *Al Qaeda's Chlorine Attacks: The Dirty War in Anbar*, Long War Journal, 17 March 2007, https://www.longwarjournal.org/archives/2007/03/al_qaedas_chlorine_a.php.
15. Counter Improvised Explosive Devices Centre of Excellence, *Chemical IEDs Current Status and Trend Threat*, November 2016, p. 10.
16. J. Tucker, The future of chemical weapons, *The New Atalantis*, 26, Fall 2009/Winter 2010, pp. 3–29, p. 11.
17. P. Bergen, Reevaluating Al-Qa'ida's Weapons of Mass Destruction Capabilities, *CTC Sentinel*, 2010, 3(9), 2.

18. C. Quillen, The Islamic State's Evolving Chemical Arsenal, *Stud. Conflict Terrorism*, 2016, **39**(11), 1023.
19. C. Quillen, The Islamic State's Evolving Chemical Arsenal, *Stud. Conflict Terrorism*, 2016, **39**(11), 1025.
20. Islamic State, *'Using Chlorine Gas' in Iraq Roadside Bombs*, BBC News, 12 March 2015.
21. Counter Improvised Explosive Devices Centre of Excellence, *Chemical IEDs Current Status and Trend Threat*, November 2016, p. 12.
22. Human Rights Watch, *Iraq: Fleeing ISIS Forces Fired Toxic Chemicals*, 11 November 2016.
23. UN, Human Rights Council, *Human Rights Abuses and International Humanitarian Law Violations in the Syrian Arab Republic, 21 July 2016 – 28 February 2017, Independent International Commission of Inquiry on the Syrian Arab Republic, A/HRC/34/CRP.3*, 10 March 2017.
24. Human Rights Watch, Death by Chemicals, *The Syrian Government's Widespread and Systematic Use of Chemical Weapons*, 1 May 2017, pp. 40–46.
25. UN, Human Rights Council, *Human Rights Abuses and International Humanitarian Law Violations in the Syrian Arab Republic, 21 July 2016 – 28 February 2017*, Independent International Commission of Inquiry on the Syrian Arab Republic, A/HRC/34/CRP.3, 10 March 2017, paragraph 51.
26. H. Al-Khatib, *New Visual Evidence about Chlorine Gas Attacks in Eastern Ghouta*, Bellingcat, 14 February 2017, https://www.bellingcat.com/news/mena/2017/02/14/new-visual-evidence-chlorine-gas-attacks-eastern-ghouta/, accessed 20 November 2017.
27. Operation Inferno, *Operation Bannish Beach*, Globalsecurity.org, undated, https://www.globalsecurity.org/military/ops/banish-beach.htm, accessed 15 December 2017.
28. UN Human Rights Commission, *Special Rapporteur on the Implementation of the Declaration on the Elimination of All Forms of Intolerance and of Discrimination Based on Religion or Belief, E/CN.4/1993/62*, 6 January 1993.
29. Human Rights Watch, *Under Siege: Indiscriminate Bombing and Abuses in Sudan's Southern Kordofan and Blue Nile States*, 11 December 2012.
30. Human Rights Watch, *Iraq: Civilian Toll of Government Airstrikes*, 23 July 2014.
31. OPCW, *Report of the OPCW Fact-finding Mission in Syria Regarding Alleged Incidents in the Idlib Governorate of the Syrian Arab Republic between 16 March and 20 May 2015, S/1319/2015*, Annex 2, pp. 81–82.
32. UN, Security Council, *Third Report of the Organization for the Prohibition of Chemical Weapons-united Nations Joint Investigative Mechanism, S/2016/738*, 24 August 2016, p. 11.
33. UN, Security Council, *Third Report of the Organization for the Prohibition of Chemical Weapons-united Nations Joint Investigative Mechanism, S/2016/738*, 24 August 2016, paragraph 44.

34. UN, Security Council, *Fourth Report of the Organization for the Prohibition of Chemical Weapons – United Nations Joint Investigative Mechanism, S/2016/888*, 21 October 2016, paragraphs 27–33.
35. UN General Assembly, *Commission of Inquiry, Human Rights Council, A/HRC/34/64*, 2 February 2017, paragraphs 52–56.
36. Human Rights Watch, Death by Chemicals, *The Syrian Government's Widespread and Systematic Use of Chemical Weapons*, 1 May 2017, pp. 15–16; Amnesty International, *Syria: Witness testimony reveals details of illegal chemical attack on Saraqeb*, 6 February 2018; Human Rights Watch, *Syria: A Year on, Chemical Weapons Attacks Persist*, 4 April 2018; Bellingcat Investigation Team, *Open Source Survey of Alleged Chemical Attacks in Douma on 7th April 2018*, Bellingcat, 11 April 2018, available at https://www.bellingcat.com.
37. Human Rights Watch, Death by Chemicals, *The Syrian Government's Widespread and Systematic Use of Chemical Weapons*, 1 May 2017, p. 15.
38. MSI Delivery Systems.Inc., *MMADS Afterburner 2000*, available at: http://wpsitetest.msi-deliverysystems.com/products/, accessed 14 December 2017.
39. MSI Delivery Systems.Inc., *The AB2K, Multi-Mission Aerosol Delivery System*, available at: http://www.msi-deliverysystems.com/Products/page9/page9.html, last accessed 14th February 2013, now removed.
40. J. Tucker, *Chemical Terrorism in: High-impact Terrorism, Proceedings of a Russian-American Workshop*, US National Academy of Sciences, 2002, p. 119.
41. J. Tucker, *Chemical Terrorism in: High-impact Terrorism, Proceedings of a Russian-American Workshop*, US National Academy of Sciences, 2002, p. 120; Monterey Institute of International Studies, *Chronology of Aum Shinrikyo's CBW Activities*, 2001.
42. *S-64 Erickson Air-Crane*, available at: http://ericksoninc.com/wp-content/uploads/2015/05/Erickson-MRO-S-64.pdf, accessed 20 November 2017.
43. *Mi-26 Halo Heavy Lift Helicopter*, Russia, available at: http://www.airforce-technology.com/projects/milmi-26heavylifthel/, accessed 20 November 2017.
44. W. Buckingham, Operation Ranch Hand, *The Air Force and Herbicides in Southeast Asia 1961-1971. Office of Air Force History*, US Air Force, Washington DC, 1982.
45. OPCW, *Report of the Scientific Advisory Board on Developments in Science and Technology for the Third Special Session of the Conference of the States Parties to Review the Operation of the Chemical Weapons Convention (Third Review Conference), RC-3/DG.1*, 29 October 2012, paragraph 56.
46. National Agricultural Aviation Association, *Agricultural Aviation Security*, undated, http://www.pnwaaa.org/Security.pdf, accessed 14 December 2017.
47. Amnesty International, *Paraguay: Amnesty International Submission to the UN Universal Periodic Review, AMR 45/003/2010*, 12 July 2010, Amnesty International, *Paraguay indigenous Community Threatened by Illegal Eviction and Pesticide Attack*, 10 November 2009.

48. CNN, *More than 200 Paraguay Villagers Thought Sprayed with Pesticide*, 11 November 2009, http://edition.cnn.com/2009/WORLD/americas/11/10/paraguay.pesticide.attack/, accessed 20 November 2017.
49. National, Minister confirms that they sprayed uncultivated land in Itakyry, 8 November 2009.
50. Amnesty International, *Further Information on UA 300/09 – Indigenous Communities Face Reprisals, AMR 45/049/2009*, 9 November 2009.
51. Amnesty International, *Committee on the Elimination of Racial Discrimination, Briefing on Paraguay, AMR 45/001/2011*, July 2011.
52. E. Bone and C. Bolkcom, *Unmanned Aerial Vehicles: Background and Issues for Congress, Report for Congress, RL31872*, 25 April 2003.
53. M. Horowitzand and M. Fuhrmann, *Droning on: Explaining the Proliferation of Unmanned Aerial Vehicles,* 1 October 2015, Last Revised 26 January 2016, http://papers.ssrn.com/sol3/papers.cfm?abstract_id=2514339, accessed 14 December 2017.
54. New American Foundation, *World of Drones*, undated, https://www.newamerica.org/in-depth/world-of-drones/, accessed 20 November 2017.
55. K. Sayler, *A World of Proliferated Drones: A Technology Primer*, Center for a New American Security, June 2015, p. 15, http://www.cnas.org/sites/default/files/publications-pdf/CNAS%20World%20of%20Drones_052115.pdf, accessed 20 November 2017.
56. C. Abbott, M. Clarke, S. Hathom, and S. Hickie, *Hostile Drones: The Hostile Use of Drones by Non-state Actors against British Targets*, Remote Control Project Oxford Research Group and Open Briefing, January 2016, pp. 1–19, http://remotecontrolproject.org/wp-content/uploads/2016/01/Hostile-use-of-drones-report_open-briefing.pdf, accessed 14 December 2017.
57. L. Friese, N. Jenzen-Jones and M. Smallwood, *Emerging Unmanned Threats: The Use of Commercially-available UAVs by Armed Non-state Actors*, Armament Research Services, 2016, p. 35. See also: D. Rassler, *Remotely Piloted Innovation: Terrorism, Drones and Supportive Technology, Combating Terrorism Center at West Point*, United States Military Academy, October 2016, p. 51.
58. D. Rassler, *Remotely Piloted Innovation: Terrorism, Drones and Supportive Technology, Combating Terrorism Center at West Point*, United States Military Academy, October 2016, p. 51.
59. K. Olson, *Overview: Recent Incidents and Responder Implications Responding to the Consequences of Chemical and Biological Terrorism Seminar*, 11-13 July 1995, pp. 2–42, cited by D. Rassler; C. Quillen, The Islamic State's Evolving Chemical Arsenal, *Stud. Conflict Terrorism*, 2016, **39**(11), 8.
60. K. Olson, Overview: Recent Incidents and Responder Implications, presentation and remarks given at the Responding to the Consequences of Chemical and Biological Terrorism Seminar, 11–13 July 1995, pp. 2–42.
61. D. Rassler, Remotely Piloted Innovation: Terrorism, Drones and Supportive Technology, *Combating Terrorism Center at West Point*, United States Military Academy, October 2016, pp. 13–14.

62. M. Gips, *A Remote Threat. Security Management*, October 2002, pp. 14–16, cited by D. Rassler.
63. *Iraq uncovers Al-Qaeda 'chemical Weapons Plot'*, BBC World News website, 1 June 2013; D. Blair, *Iraq Foils Suspected Al-Qaeda Plot to Launch Remote-controlled Helicopters Carrying Chemical Weapons*, Daily Telegraph, 3 June 2013.
64. D. Rassler, Remotely piloted innovation, terrorism, drones and supportive technology, *Combating Terrorism Center at West Point*, US Military Academy, October 2016, p. 60.
65. M. Crowley, *Chemical Control: Regulation of Incapacitating Chemical Agent Weapons, Riot Control Agents and Their Means of Delivery*, Palgrave Macmillan, Basingstoke, Hampshire, November 2015, pp. 50–70.
66. The IronFist, *NonLethal Technologies*, 2016, see also Product Catalogue, *NonLethal Technologies*, 2017, both available from company website: http://nonlethaltechnologies.com/, accessed 20 November 2017.
67. China Ordnance Equipment Research Institute, *Undated Catalogue, Distributed at Asia Pacific China Police Expo*, 2012, pp. 8–13.
68. China Ordnance Equipment Research Institute, *Undated Catalogue, Distributed at Asia Pacific China Police Expo*, 2012, p. 13.
69. Non-Lethal Munitions Section, *Volume 12 Ordnance and Munitions (English Language Version), Russia's Arms and Technologies. The XXI Century Encyclopedia, Version 2006.1eng*, Arms and Technologies Publishing House, 5th May 2009.
70. C. Foss, *Turkey Details 120 mm Automatic Mortar*, Jane's Defence Weekly, 12 November 2003.
71. *Turkish Defence Industry Catalogue*, 2007, previously available at: http://www2.ssm.gov.tr/katalog2007/data/24509/9/uruning/uruning34.html, last accessed 30 September 2010, subsequently removed.
72. MKEK was promoting the CS mortar round till at least mid-November 2009 on its website, previously available at: http://mkekexport.com/ammunition.htm, last accessed 16 November 2009, subsequently removed.
73. Correspondence from Ambassador Dogan, *Permanent Representative of Turkey to the OPCW, to BNWLRP, ORF and ISS*, 25th February 2011; Correspondence from Mr Utkan, Counsellor, *Permanent Representation of Turkey to the OPCW, to BNLWRP, ORF and ISS*, 8 July 2011.
74. US Army, Picatinny Centre, *Non-lethal Artillery Structural Firing (FYO4) Purchase Order Contract in Support of the FY04 155 mm Non-lethal Artillery Projectile Program, Contract Number W15QKN-04-m-0328*, 14 September 2004.
75. J. McCormick, *Presentation on 155 mm XM1063 Non-lethal Personnel Suppression Projectile, General Dynamics OTS, National Defense Industrial Association, 42nd Annual Armament Systems: Gun and Missile Systems Conference and Exhibition*, Charlotte, North Carolina, 23rd–26th April 2007, p. 4.

76. J. McCormick, 155 mm XM1063 Non-Lethal Personnel Suppression Projectile, *Presentation to the 41st Annual Armament Systems: Guns and Missile Systems, Conference & Exhibition*, National Defense Industrial Association (NDIA), Sacramento, US, 27–30 March 2006, p. 12.
77. D. Hambling, *US Weapons Research Is Raising a Stink*, The Guardian, 10 July 2008.
78. J. McCormick,155 mm XM1063 Non-Lethal Personnel Suppression Projectile, *Presentation to the 41st Annual Armament Systems: Guns and Missile Systems, Conference & Exhibition*, National Defense Industrial Association (NDIA), Sacramento, US, 27–30 March 2006, p. 7.
79. Joint Non-Lethal Weapons Directorate, *Non-lethal Weapons (NLW) Reference Book*, 30 June 2011, available at: http://publicintelligence.net/dod-non-lethal-weapons-2011/, accessed 16 October 2017, p. viii.
80. QinetiQ North America, *MAARS Advanced Armed Roboticsystem*, http://www.qinetiq-na.com/products/unmanned-systems/maars/, accessed 16 October 2017.
81. QinetiQ North America, *MAARS Advanced Armed Roboticsystem*, http://www.qinetiq-na.com/products/unmanned-systems/maars/, accessed 16 October 2017, p. 1.
82. QinetiQ North America, *MAARS Datasheet*, https://www.qinetiq-na.com/wp-content/uploads/MAARS_DataSheet_LR.pdf, accessed 16 October 2017.
83. J. Markoff, *War Machines: Recruiting Robots for Combat*, The New York Times, 27 November 2010.
84. Skunk Riot Control Copter, *Desert Wolf Website*, available at: http://www.desert-wolf.com/dw/products/unmanned-aerial-systems/skunk-riot-control-copter.html, accessed 16 October 2017.
85. L. Kelion, *African Firm Is Selling Pepper-Spray Bullet Firing Drones*, BBC News, 18 June 2014, available at: http://www.bbc.co.uk/news/technology-27902634, accessed 16 October 2017.
86. G. Martin, *Desert Wolf Adding Grenades to Skunk Riot Control UAV*, Defenceweb, 7 October 2015, http://www.defenceweb.co.za/index.php?option=com_content&view=article&id=40961:desert-wolf-adding-grenades-to-skunk-riot-control-uav&catid=35:Aerospace&Itemid=107, accessed 16 October 2017.
87. *Aero Surveillance Introduces Payload Launcher*, 18 November 2015, Shepherdmedia.com, https://www.shephardmedia.com/news/uv-online/aero-surveillance-introduces-payload-launcher/, accessed 16 October 2017.
88. MILIPOL 2015 *Air Security: Aero Surveillance Introduces a New Multi Purpose Payload Launcher*, 17 November 2015, Emmergency-Live.com, available at: www.emergency-live.com/en/equipment/milipol-2015-air-securityaero-surveillance-introduces-a-new-multi-purpose-payload-launcher/, accessed 16 October 2017.
89. Eagle Eye Anti Riot Drone, Yavascalar Catalogue, undated, pp. 48–49.
90. MSI Turkish *Defence Review Magazine*, Number 26, August 2016.

91. Médecins Sans Frontières, *Syria: Thousands Suffering from Neurotoxic Symptoms Treated in Hospitals Supported by MSF*, 24 August 2013; Human Rights Watch, *Attacks on Ghouta Analysis of Alleged Use of Chemical Weapons in Syria*, September 2013.
92. United Nations, *General Assembly/Security Council, Report of the United Nations Mission to Investigate Allegations of the Use of Chemical Weapons in the Syrian Arab Republic on the Alleged Use of Chemical Weapons in the Ghouta Area of Damascus on 21 August 2013*, Note by the Secretary-General, A/67/997–S/2013/553, 16 September 2013, p. 26.
93. United Nations, *General Assembly/Security Council, Report of the United Nations Mission to Investigate Allegations of the Use of Chemical Weapons in the Syrian Arab Republic on the Alleged Use of Chemical Weapons in the Ghouta Area of Damascus on 21 August 2013*, Note by the Secretary-General, A/67/997–S/2013/553, 16 September 2013, p. 24.
94. United Nations, *General Assembly/Security Council, Report of the United Nations Mission to Investigate Allegations of the Use of Chemical Weapons in the Syrian Arab Republic on the Alleged Use of Chemical Weapons in the Ghouta Area of Damascus on 21 August 2013*, Note by the Secretary-General, A/67/997–S/2013/553, 16 September 2013, paragraph 28.
95. UN, Human Rights Council, *Report of the Independent International Commission of Inquiry on the Syrian Arab Republic, A/HRC/28/69*, 5 February 2015, paragraph 44.
96. Human Rights Watch, *Attacks on Ghouta Analysis of Alleged Use of Chemical Weapons in Syria*, September 2013, ref. 120 (Human Rights Watch), p. 12.
97. Human Rights Watch, *Attacks on Ghouta Analysis of Alleged Use of Chemical Weapons in Syria*, September 2013, pp. 4–5.
98. United Nations, *General Assembly/Security Council, Report of the United Nations Mission to Investigate Allegations of the Use of Chemical Weapons in the Syrian Arab Republic on the Alleged Use of Chemical Weapons in the Ghouta Area of Damascus on 21 August 2013*, Note by the Secretary-General, A/67/997–S/2013/553, 16 September 2013, p. 6.
99. United Nations, *General Assembly/Security Council, Report of the United Nations Mission to Investigate Allegations of the Use of Chemical Weapons in the Syrian Arab Republic on the alleged use of chemical weapons in the Ghouta area of Damascus on 21 August 2013*, Note by the Secretary-General, A/67/997–S/2013/553, 16 September 2013, paragraph 28.
100. United Nations, *General Assembly/Security Council, Report of the United Nations Mission to Investigate Allegations of the Use of Chemical Weapons in the Syrian Arab Republic on the alleged use of chemical weapons in the Ghouta area of Damascus on 21 August 2013*, Note by the Secretary-General, A/67/997–S/2013/553, 16 September 2013, p. 26.
101. *Janes Ammunition Handbook Sixth Edition, 1997-98*, ed. T. Gander, Janes Information Group, Coulsdon, Surrey, 1998, pp. 544–545.
102. Human Rights Watch, *Attacks on Ghouta Analysis of Alleged Use of Chemical Weapons in Syria*, September 2013, p. 5.

103. Stockholm International Peace Research Institute, *Arms Transfers Database*, http://armstrade.sipri.org/armstrade/page/trade_register.php, Recipient report for Syria for the period 1960–2016.
104. UN, Human Rights Council, *Report of the Independent International Commission of Inquiry on the Syrian Arab Republic, A/HRC/36/55*, 8 August 2017, paragraphs 72–74.
105. UN, Human Rights Council, *Report of the Independent International Commission of Inquiry on the Syrian Arab Republic, A/HRC/36/55*, 8 August 2017, paragraph 75.
106. UN, Human Rights Council, *Report of the Independent International Commission of Inquiry on the Syrian Arab Republic, A/HRC/36/55*, 8 August 2017, paragraph 74.
107. UN, Human Rights Council, *Report of the Independent International Commission of Inquiry on the Syrian Arab Republic, A/HRC/36/55*, 8 August 2017, paragraph 77.
108. Human Rights Watch, Death by Chemicals, *The Syrian Government's Widespread and Systematic Use of Chemical Weapons*, 1 May 2017, pp. 29–30, and appendix II, pp. 53–56.
109. United Nations, *Seventh Report of the Organisation for the Prohibition of Chemical Weapons-united Nations Joint Investigative Mechanism, S/2017/904*, 26 October 2017, paragraph 93.
110. United Nations, *Seventh Report of the Organisation for the Prohibition of Chemical Weapons-united Nations Joint Investigative Mechanism, S/2017/904*, 26 October 2017, paragraph 93 (h).
111. United Nations, *Seventh Report of the Organisation for the Prohibition of Chemical Weapons-united Nations Joint Investigative Mechanism, S/2017/904*, 26 October 2017, paragraph 84.
112. United Nations, *Seventh Report of the Organisation for the Prohibition of Chemical Weapons-united Nations Joint Investigative Mechanism, S/2017/904*, 26 October 2017, paragraph 53.
113. United Nations, *Seventh Report of the Organisation for the Prohibition of Chemical Weapons-united Nations Joint Investigative Mechanism, S/2017/904*, 26 October 2017, paragraph 56.
114. M. Maples, *Current and Projected National Security Threats to the United States, Statement before US Senate Select Committee on Intelligence*, 11 January 2007, p. 13.
115. Republic of Korea, *Defense White Paper*, Ministry of National Defense, December 2016, p. 34.
116. L. Yoon-Geol, *North Korea has 5,000 Tons of Chemical Weapons as Scary as Nuclear Weapons, Sisa Journal*, number 1121, 13 April 2011.
117. J. Geramone, *Thurman Wants to Bolster Ties with South Korea*, Armed Forces Press Service, 30 June 30 2011, URL: http://8tharmy.korea.army.mil/20110630thurman-garamone.asp, accessed 14 December 2017.
118. B. Bechtol, Maintaining a Rogue Military: North Korea's Military Capabilities and Strategy at the End of the Kim Jong-il Era, International Journal of Korean Studies, Spring 2012, Vol. XVI No. 1.pp. 160–191, p. 164.

119. A. Cordesman, with C. Ayers, and A. Lin, North Korean Nuclear Forces and the Threat of Weapons of Mass Destruction in Northeast Asia., Working Draft: July 25 2016, CSIS, p. 15. citing NTI and the James Martin Center for Nonproliferation Studies, "North Korea Biological Chronology", August 2012.
120. Republic of Korea, *Defense White Paper*, Ministry of National Defense, December 2016, p. 32, available at: http://www.nti.org/learn/countries/north-korea/delivery-systems/.
121. NTI, North Korea, Missile, last updated: July, 2017, and Design Characteristics of North Korea's Ballistic and Cruise Missiles, both available at: http://www.nti.org/learn/countries/north-korea/delivery-systems/.
122. Republic of Korea, *Defense White Paper*, Ministry of National Defense, December 2016, p. 32.
123. NTI, North Korea, Missile, last updated: July, 2017, citing Wright, D. North Korea Appears to Launch Missile with 6,700 km Range, All Things Nuclear Blog: Union of Concerned Scientists, 3 July 2017, www.allthingsnuclear.org, accessed 14 December 2017.
124. Schilling, J. North Korea Finally Tests an ICBM, 38 North, 5 July 2017, www.38north.org, accessed 14 December 2017.
125. Hwasong-14 (KN-20), *Missile Defense Project, CSIS*, https://missilethreat.csis.org/missile/hwasong-14/, accessed 14 December 2017.
126. NTI, *North Korean Ballistic Missile Models*, last updated: March 2016, http://www.nti.org/analysis/articles/north-korean-ballistic-missile-models/, accessed 14 December 2017.
127. NTI, *North Korea, Missile*, last updated: July, 2017, *The CNS North Korea Missile Test Database*, last updated: August 2017, http://www.nti.org/analysis/articles/cns-north-korea-missile-test-database/, accessed 14 December 2017.
128. NTI, *North Korea, Missile*, last updated: July, 2017, citing Strategic Weapon System, *Korea, North, Jane's Sentinel Security Assessment*, 20 January 2011.
129. A. Cordesman, C. Ayers and A. Lin, *North Korean Nuclear Forces and the Threat of Weapons of Mass Destruction in Northeast Asia*, CSIS, Working Draft: July 25 2016, p. 6; NTI, *North Korea: Chemical*, updated December 2015.
130. A. Cordesman, C. Ayers and A. Lin, *North Korean Nuclear Forces and the Threat of Weapons of Mass Destruction in Northeast Asia*, CSIS, Working Draft: July 25 2016, p. 15, citing NTI, *North Korea – Biological*, updated http://www.nti.org/country-profiles/northkorea/biological/December 2015,.
131. Reuters, *North Korea may be capable of sarin-tipped missiles: Japan PM*, 13 April 2017.
132. Federation of American Scientists, Doctrine, last updated: 3 March 2000, https://fas.org/nuke/guide/dprk/doctrine/, accessed 14 December 2017.
133. International Crisis Group, *North Korea's chemical and biological weapons programs, Asia Report No 167*, 18 June 2009, p. 8.

134. *Report: N. Korea fires on S. Korea, injuring at least 17*, CNN, 23 November 2010, at: http://edition.cnn.com/2010/WORLD/asiapcf/11/23/nkorea.skorea.military.fire/index.html?hpt=T1&iref=BN1, accessed 14 December 2017.
135. NTI, North Korea, Missile, last updated: July, 2017, citing Wright, D. North Korea Appears to Launch Missile with 6,700 km Range, All Things Nuclear Blog: Union of Concerned Scientists, 3 July 2017, www.allthingsnuclear.org.
136. J. Schilling, *North Korea Finally Tests an ICBM, 38 North*, 5 July 2017, www.38north.org.
137. *North Korea fires Missile over Japan in 'unprecedented Threat'*, BBC World News, 29 August 2017.
138. R. Paddock, C. Sang-Hun, K. Jong-nam Was Killed by VX Nerve Agent, Malaysians Say, The New York Times, 23 February 2017.
139. D. Bradley, *VX Nerve Agent in North Korean's Murder: How Does It Work?*, Scientific American, 24 February 2017.
140. A. Cordesman, C. Ayers and A. Lin, *North Korean Nuclear Forces and the Threat of Weapons of Mass Destruction in Northeast Asia*, CSIS, Working Draft: July 25 2016, p. 5.
141. Conflict Armament Research, *Islamic State Weapons in Koban*, April 2015.
142. Conflict Armament Research, *Standardisation and Quality Control in Islamic State's Military Production*, December 2016.
143. Conflict Armament Research, *Standardisation and Quality Control in Islamic State's Military Production*, December 2016. pp. 7, 11, 12 & 18.
144. Conflict Armament Research, *Standardisation and Quality Control in Islamic State's Military Production*, December 2016, p. 4.
145. Amnesty International, *Taking Stock: The Arming of Islamic State, MDE 14/2812/2015*, December 2015, p. 5.
146. Amnesty International, *Taking Stock: The Arming of Islamic State, MDE 14/2812/2015*, December 2015, pp. 14–17.
147. *Islamic State Captures Masses of Iranian-supplied Weaponry Near Khanasir*, 16 April 2016, https://www.bellingcat.com/news/mena/2016/04/16/islamic-state-captures-masses-of-iranian-supplied-weaponry-near-khanasir/.
148. J. Bender & A. Rosen, *As ISIS Continues to Gain Ground, Here's What the Militants Have in Their Arsenal*, Business Insider, 14 November 2014, http://uk.businessinsider.com/isis-military-equipment-arsenal-2014-11?op=1/#g-7s-14.
149. Associated Press/Guardian, *Chemical weapons found in Mosul in Isis lab, say Iraqi forces*, 29 January 2017.
150. Human Rights Watch, *Syria: New Deadly Chemical Attacks*, 28 September 2016, https://www.hrw.org/news/2016/09/28/syria-new-deadly-chemical-attacks, accessed 14 December 2017.
151. UN Security Council, Letter dated 4 May 2017 from the President of the Security Council addressed to the Secretary-General, S/2017/400, 5 May 2017, containing Report of the OPCW FFM regarding the incident of 16 September 2016 as reported in the Note Verbale of the Syrian Arab Republic number 113 dated 29 November 2016.

152. UN Security Council, Letter dated 4 May 2017 from the President of the Security Council addressed to the Secretary-General, S/2017/400, 5 May 2017, containing Report of the OPCW FFM regarding the incident of 16 September 2016 as reported in the Note Verbale of the Syrian Arab Republic number 113 dated 29 November 2016, paragraph 6.3.
153. UN Security Council, Letter Dated 4 May 2017 from the President of the Security Council Addressed to the Secretary-general, S/2017/400, 5 May 2017, containing Report of the OPCW FFM regarding the incident of 16 September 2016 as reported in the Note Verbale of the Syrian Arab Republic number 113 dated 29 November 2016, paragraph 5.16.
154. UN Security Council, Letter dated 4 May 2017 from the President of the Security Council addressed to the Secretary-General, S/2017/400, 5 May 2017, containing Report of the OPCW FFM regarding the incident of 16 September 2016 as reported in the Note Verbale of the Syrian Arab Republic number 113 dated 29 November 2016, paragraph 5.34.
155. United Nations, *Seventh Report of the Organisation for the Prohibition of Chemical Weapons-united Nations Joint Investigative Mechanism, S/2017/904*, 26 October 2017, paragraph 10.
156. United Nations, *Seventh Report of the Organisation for the Prohibition of Chemical Weapons-united Nations Joint Investigative Mechanism, S/2017/904*, 26 October 2017, paragraph 11.
157. United Nations, *Seventh Report of the Organisation for the Prohibition of Chemical Weapons-united Nations Joint Investigative Mechanism, S/2017/904*, 26 October 2017, paragraphs 11 & 13.
158. United Nations, *Seventh Report of the Organisation for the Prohibition of Chemical Weapons-united Nations Joint Investigative Mechanism, S/2017/904*, 26 October 2017, paragraph 22.
159. Reuters, *Islamic State Used 'poisonous Substances' in Village Shelling: Officials*, 9 March 2016, https://www.reuters.com/article/us-mideast-crisis-iraq/islamic-state-used-poisonous-substances-in-village-shelling-officials-idUSKCN0WB21H, accessed 14 December 2017.
160. N. Kohnavard, *Iraqi Town Taza 'hit in IS Chemical Attack' Appeals for Help*, BBC News, 25 March 2016, http://www.bbc.co.uk/news/world-middle-east-35898990.
161. Associated Press/The Guardian, *Isis Launches Two Chemical Attacks in Northern Iraq*, 12 March 2016.
162. C. MacDiarmid, *Inside Taza, the Iraqi Town Gassed by the Islamic State*, VICE News, 16 March 2016.
163. OPCW, *Director-general Expresses Concern over Alleged Recent Chemical Attacks in Iraq*, 23 March 2016, https://www.opcw.org/news/article/director-general-expresses-concern-over-alleged-recent-chemical-attacks-in-iraq/.
164. United Nations, Security Council, *Report of the OPCW Fact-finding Mission in Syria Regarding Alleged Incidents in Marea, Syrian Arab Republic, August 2015, S/2015/908*, 24 November 2015, paragraph 4.6.

165. UN, Security Council, *Third Report of the Organization for the Prohibition of Chemical Weapons-united Nations Joint Investigative Mechanism, S/2016/738*, 24 August 2016, paragraph 31.
166. UN, Security Council, *Third Report of the Organization for the Prohibition of Chemical Weapons-united Nations Joint Investigative Mechanism, S/2016/738*, 24 August 2016, paragraph 42.
167. UN, Security Council, *Third Report of the Organization for the Prohibition of Chemical Weapons-united Nations Joint Investigative Mechanism, S/2016/738*, 24 August 2016, paragraphs 21, 23 and 42.
168. UN, Security Council, *Third Report of the Organization for the Prohibition of Chemical Weapons-united Nations Joint Investigative Mechanism, S/2016/738*, 24 August 2016, paragraph 21.
169. R. Pita, Assessing al-Qaeda's Chemical Threat, *Int. J. Intell. Counterintell.*, 2007, **20**(3), 480–511.
170. C. Chivers, *The Gun Blog, More Data on Arms to Share: The IRAMs of Iraq and Syria*, 2013, http://cjchivers.com/post/33870206289/more-data-on-arms-to-share-the-irams-of-iraq-and, accessed 14 December 2017.
171. OPCW, *Technical Secretariat, Office of the Director-general*, Note by the Technical Secretariat, Second report of the OPCW fact-finding mission in Syria, S/1212/2014, 10 September 2014.
172. OPCW, *Technical Secretariat, Office of the Director-general, S/1230/2014*, Note by the Technical Secretariat, Third report of the OPCW fact-finding mission in Syria, 18 December 2014.
173. OPCW, *Technical Secretariat, S/1319/2015*, 29 October 2015, Note by the Technical Secretariat, Report of the OPCW fact-finding missing in Syria regarding alleged incidents in the Idlib Governorate of the Syrian Arab Republic between 16 March and 20 May 2015.
174. Air Tractor, Inc, available at: https://airtractor.com/aircraft/at-802a/, accessed 20 November 2017.
175. *AT-802F Fire Boss*, AirTractor, Inc., available at: https://airtractor.com/aircraft/at-802f-fire-boss/, accessed 20 November 2017.
176. US Office of Technology Assessment, *Technologies Underlying Weapons of Mass Destruction, US Office of Technology Assessment*, US Congress, Washington, D.C., OTA-BP-ISC-115, December 1993.
177. J. Blum and D. Eggen, *Airborne Poison via Crop Duster?* International Herald Tribune, 25 September 2001.
178. I. Sample, *Airborne Biological Weapon Attacks Are Serious Concern*, New Scientist, 24 September 2001.
179. L. Klotz and E. Sylvester, *Crying Wolf: The Terrorist Crop-duster*, The Huffington Post, 25 May 2011, http://www.huffingtonpost.com/dr-lynn-c-klotz/crying-wolf-the-terrorist_b_538204.html, accessed 20 November 2017.
180. B. Graham and J. Talent, *et al.*, *World at Risk: The Report of the Commission on the Prevention of WMD Proliferation and Terrorism*, Washington DC, December 2008.

181. M. Crowley, *Tear gassing by remote control: The development and promotion of remotely operated means of delivering or dispersing riot control agents, Remote Control Project*, December 2015.
182. M. Crowley, *Drawing the Line: Regulation of "wide Area" Riot Control Agent Delivery Mechanisms under the Chemical Weapons Convention*, Omega Research Foundation/University of Bradford, April 2013.
183. Omega Research Foundation, *Manufacture, trade and use of 'tools of torture' in the Council of Europe*, January 2018.
184. Marine Corps Warfighting Laboratory MAGTAF Integrated Experiment (MCWL), Defense Imagery Distribution System, 9 July 2016, https://www.dvidshub.net/image/2729263/marine-corps-warfighting-laboratory-magtaf-integrated-experiment-mcwl#.V8mPkBmM0o8.facebook, accessed 16 October 2017.
185. *Syria/Syrian chemical programme – National executive summary of declassified Intelligence (English Translation)*, French Defence Ministry, Paris, 3 September 2013, https://www.diplomatie.gouv.fr/IMG/pdf/Syrian_Chemical_Programme.pdf, accessed 4 December 2017.
186. *Syria: Reported chemical weapons use, Letter from Chairman of the Joint Intelligence Committee to UK Prime Minister*, Ref: Jp 115, 29 August 2013, https://www.gov.uk/government/uploads/system/uploads/attachment_data/file/235094/Jp_115_JD_PM_Syria_Reported_Chemical_Weapon_Use_with_annex.pdf, accessed 4 December 2017.
187. M. Gebauer, Germany offers clue in search for truth in Syria, Der Spiegel, 3 September 2013, Newly Declassified CIA Report on Syria's Chemical Weapons, Matthewaid, August 29, 2013, http://www.matthewaid.com/post/59722308253/newly-declassified-cia-report-on-syrias-chemical, accessed 4 December 2017.
188. J. P. Robinson, Observing Chemical-Weapons Proliferation, Why did Syria invest in binary nerve gas? Harvard Sussex Program, Science Policy Research Unit, forthcoming (advance copy circulated to author).
189. Human Rights Watch, Attacks on Ghouta Analysis of Alleged Use of Chemical Weapons in Syria, September 2013, p. 9. citing a brochure extract from Iran's Defense Industries Organization on the Falaq-2 launcher and its FL2-A rocket, available at: Nic Jenzen-Jones, "Alleged CW Munitions in Syria Fired from Iranian Falaq-2 Type Launchers," Post to "The Rogue Adventurer" (Blog), August 29, 2013, http://rogueadventurer.com/2013/08/29/allegedcw-munitions-in-syria-fired-from-iranian-falaq-2-type-launchers/, accessed September 6, 2013.
190. N. R. Jenzen-Jones, Y. Lyamin and G. Wright, Iranian Falaq-1 and Falaq-2 Rockets in Syria, Research Report No.2, ARES; N. Jenzen-Jones, Alleged CW Munitions in Syria Fired From Iranian Falaq-2 Type Launchers, http://rogueadventurer.com/2013/08/29/alleged-cw-munitions-in-syria-fired-from-iranianfalaq-2-type-launchers/, accessed 14 December 2017.

191. United Nations Security Council, *Resolution 1718 (2006), S/RES/1718(2006), 14 October 2006, and Subsequent UNSCR Resolutions Including: UNSCR 2270 (2016) and UNSCR 2371 (2017).*
192. US Central Intelligence Agency, *Unclassified Report to Congress on the Acquisition of Technology Relating to Weapons of Mass Destruction and Advanced Conventional Munitions*, 1 July through 31 December 2006.
193. B. B. Bell, statement before the House Armed Services Committee, 9 March 2006, cited in: International Crisis Group, *North Korea's chemical and biological weapons programs, Asia Report No 167*, 18 June 2009, p. 7.
194. *The CNS North Korea Missile Test Database*, available at: http://www.nti.org/analysis/articles/cns-north-korea-missile-test-database/.
195. NTI, *Design Characteristics of North Korea's Ballistic and Cruise Missiles*, last updated July 2014, http://www.nti.org/media/pdfs/design_characteristics_of_north_koreas_ballistic_and_cruise_missiles.pdf?_=1406744044, accessed 14 December 2017.
196. United Nations, *Seventh Report of the Organisation for the Prohibition of Chemical Weapons-united Nations Joint Investigative Mechanism, S/2017/904*, 26 October 2017, paragraphs 21–22.

CHAPTER 13

Advances in Poisoning Mechanisms, Biomarkers, and Drugs for the Prevention and Treatment of Traditional Chemical Weapons

XINBO ZHOU[a], WU ZHONG*[a], PENGBIN BU[b] AND LIJUN SHANG[c]

[a]Beijing Institute of Pharmacology and Toxicology, No.27, Taiping Road, Haidian Distinct, P.R. China, 100850; [b]National Health and Family Planning Commission of the People's Republic of China, National Health and Family Planning Commission, No.1 Xizhimengwainan Road, Xicheng Distinct, Beijing, P. R. China, 100044; [c]School of Chemistry and Biosciences, Faculty of Life Sciences, University of Bradford, Bradford, UK, BD7 1DP
*E-mail: zhongwu@bmi.ac.cn

13.1 Introduction

Traditional chemical weapons, such as mustard gas and nerve agents, remain the biggest global threat; their pathogenesis is complex, particularly mustard gas and soman poisoning, and thus there has been no suitable means of prevention and treatment developed to date. Mustard gas has been employed in the Syrian civil war and in Iraq, thereby indicating that research

Preventing Chemical Weapons: Arms Control and Disarmament as the Sciences Converge
Edited by Michael Crowley, Malcolm Dando and Lijun Shang

Published by the Royal Society of Chemistry, www.rsc.org

on its underlying molecular mechanism of toxicity is warranted. In the past 10 years, significant research progress on the mechanisms, biomarkers, and drugs for the prevention and treatment of exposure to traditional chemical weapons, such as mustard gas and soman poisoning, has been made. This progress demonstrates that the convergence of biology and chemistry has had a beneficial impact on our understanding of the mechanisms underlying the actions of these agents on living systems and has opened up new possibilities for effective prevention and treatment. This chapter presents the major research breakthroughs in this field (internationally and in China), as well as providing a reference for the development of an integrated means of control and rapid diagnosis of exposure to these toxicants.

13.2 Research Progress on the Mechanism of Mustard Gas and Nerve Agent (NA) Poisoning

13.2.1 Research Progress on the Mechanism of Mustard Gas Poisoning

Mustard gas (sulfur mustard, SM) is an erosive chemical agent that has been used in the past two world wars and in the Iran–Iraq war and now, by Daesh. The respiratory tract, cornea, and skin are the most common injury sites of SM.[1,2]

(1) The lung is the main target organ of SM injury. Acute respiratory distress syndrome and pulmonary infection are the main early causes of death due to SM poisoning, and pulmonary fibrosis and cancer are the main later causes of death due to SM poisoning.
(2) SM skin damage includes erythema and blister formation, accompanied by symptoms of pain, itching, and sensitivity to touch. High doses of SM cause a rapid development of erythema accompanied by varying degrees of skin blistering and tissue necrosis, often requiring 2 to 3 months for the skin to fully heal. In general, the healing time of SM injuries is longer than that of ordinary scalds, is generally difficult to cure, and highly prone to relapse; therefore, research studies in chemical defense medicine have focused on the elucidation of its molecular mechanism.
(3) The eye is more sensitive to SM injury than the skin and respiratory tract mostly because SM is generally applied as a vapor or mist, and a small number of SM droplets can directly splash into the eyes. SM eye injuries are classified as mild, moderate, and severe based on the length of the incubation period, clinical symptoms, and signs. SM corneal injury is generally not serious and mainly involves the corneal epithelium. Initial repair after injury is generally simple, although this may also lead to persistent or recurrent episodes of localized corneal epithelial defects, as indicated by a longer period of corneal epithelial

punctate fluorescein staining. Corneal epithelium damage makes the cornea lose its natural barrier, and once infected, may impair visual function. In the acute phase, within a few hours after SM exposure, symptoms begin to appear, and their severity is generally related to the dose, concentration, and exposure time.[3,4] Ghasemi *et al.*[5] reported that SM eye injuries are highlighted by photophobia and eye surface discomfort, and that recovery may take several years. Suffering from delayed onset SM keratitis due to an initial SM contact approximately 12, 20, or even 40 years ago, indicates that its subclinical process lasts for a long time.[6]

The chemical structure of SM is simple and it is easy to produce, store, and use. To date, no effective prevention and treatment of SM damage has been established because its underlying mechanism of SM toxicity has not yet been fully elucidated.[7] A large number of experimental and clinical studies have shown that the mechanism of SM damage to organs mainly involves inflammation, apoptosis, DNA damage, and oxidative stress.[8–11] In this chapter, the progress of these studies is reviewed.

13.2.2 Inflammatory Response to Tissue Injury Induced by SM

13.2.2.1 Inflammatory Response to SM-induced Lung Injury

Acute lung injury induced by SM inhalation involves inflammatory cells and cytokines, and erosion and inflammation are two of its major pathological features. During acute injury to the lung, the pro-inflammatory activity of macrophages increases, and the tumor necrosis factor-α (TNF-α) induces pulmonary inflammation. The broncho alveolar lavage fluid (BALF) of rats exposed to SM at 0.7–1.4 mg kg^{-1} for 16, 24, 48 hours and 7 days shows an increase in the BALF protein levels, as well as inflammatory markers such as cyclo-oxygenase-2 (COX-2), TNF-α, inducible nitric oxide synthase (iNOS) and matrix metalloproteinases (MMPs). 2-chloroethyl ethyl sulfide (CEES)-induced pulmonary toxicity is also significantly reduced in TNFR1-deficient mice. This suggests that CEES may mediate a heightened signaling by TNF-α through the TNF receptor 1 pathway, thereby inducing inflammatory reactions, leading to pulmonary toxicity.[12]

MMPs can degrade extracellular matrix components and participate in SM-induced lung injury processes, which mainly increase vascular permeability and inflammatory cell migration. During SM-induced lung injury in rats, MMP-9 protein expression is upregulated in a dose-dependent manner, and the activities of the MMP-9 protein and gelatinase are increased. Clinical data have shown that patients with SM lung injury for 20 years have elevated levels of sputum and serum TNF-α and interleukin-1β (IL-1β) and decreased levels of IL-1 receptor antagonist (IL-1Ra) and fibrinogen levels, which are indicative of the persistence of SM-induced airway inflammation.[13]

Yazdani *et al.*[14] found that production of NF-κB1 and NF-κBp65 was upregulated in patients with SM-induced lung injury, thereby suggesting that these nuclear factors play a role in the induction of an inflammatory reaction by SM. Research studies have shown that the airway response value, the number of eosinophils, monocytes, and lymphocytes, and the levels of IL-4 and interferon gamma (IFN-γ) in guinea pigs that inhaled SM were significantly higher than those in a control group that inhaled a neutral solution. After treatment with vitamin E and dexamethasone, a decrease in airway response and the number of eosinophils, monocytes, and neutrophils were observed, whereas the levels of serum IL-4 and IFN-γ increased.[15,16]

13.2.2.2 Inflammatory Responses to SM-induced Skin Injury

Inflammatory mediators are the early molecular signals of SM injury, and skin that comes into contact with SM is the first to secrete pro-inflammatory cytokines. Previous studies involving the observation of clinical manifestations and histopathological features of SM injury have indicated that a large number of inflammatory complexes are present in the lesion. Therefore, early suppression of inflammatory cytokine expression can alleviate further damage that may be caused by SM.

Different kinds of animal skin and *in vitro* tissue and cell experiments have shown that upon SM injury, various inflammatory mediators such as interleukin TNF-α, (IL)-1α, IL-1β, IL-6, IL-8, as well as monocytes, TGF-β, leukotriene B4, monocyte chemoattractant protein-1, and other inflammatory factors, are upregulated.[17] The SM-induced free radical-mediated activation of TNF-α production, which in turn leads to the production of further inflammatory mediators, may be the main pathway that leads to pulmonary inflammation.[18] When human keratin cells were treated with SM at concentrations of 50–300 μmol L^{-1}, TNF-α expression and NF-κB activity increased. In TNF-α-overexpressing human keratinocytes, this was manifested as an increase in IL-6 and IL-8 release, which in turn leads to increased cellular sensitivity to SM and increased cell death rate.

Studies have shown that SM induces inflammatory cytokine production mainly through the p38-mitogen-activated protein kinase (MAPK) pathway. SM triggers phosphorylation of p38-MAPK in keratinocytes, and p38-MAPK inhibitors inhibit the release of SM-induced inflammatory cytokines.[19] Phosphorylation of p38-MAPK occurs within 5 min of exposure to 200 μmol L^{-1} of SM.[20]

Dillman *et al.*[21] investigated the role of p38-MAPK in normal human epidermal keratinocytes and found that SM induces p38 phosphorylation in a dose-dependent manner within 15 min of exposure. Furthermore, the upstream kinase activators of p38-MAPK (MKK3 and MKK6) are also rapidly activated, and with continuous p38 phosphorylation, several p38 molecules are activated, thereby increasing the production of inflammatory cytokines. When epidermal keratinocytes were treated with a p38 inhibitor (*e.g.*, SB203580), SM-induced inflammatory cytokine levels of IL-6 and IL-8 significantly decreased, that of TNF-α decreased by 90%, and that of IL-1β decreased by 50%, thereby suggesting that inhibition of inflammatory cytokine induction is one way of mitigating SM damage.

SM-induced blister formation is associated with the release of proteases, particularly MMP-9 and MMP-2, which degrade type IV collagen and other basement membranes, including that separating the epidermis from the dermis. Ray *et al.*[22] further confirmed that SM toxicity is related to the stimulation of trypsin/chymotrypsin-like serine proteases and depends on the synthesis of Ca^{2+} and new proteins.

In addition, Anderson *et al.*[23] determined that inhibiting protease activity can effectively reduce the degree of damage induced by SM. Protease inhibitors have a certain therapeutic effect on SM-induced damage due to skin exposure. The US Institute of Chemical Defense Research showed that the protease inhibitor ethyl *p*-guanidinobenzoate hydrochloride (ICD 1579) can attenuate the SM-induced pathological damage in mouse auricular skin. Cowan *et al.*[24] exposed a HEK cell model to 200 $\mu mol\ L^{-1}$ of SM and found that two protease inhibitors, ICD 1579 and *N*-tosyl-L-lysine chloromethyl ketone (TLCK), applied one hour after SM application suppressed the production of SM-induced inflammatory cytokine IL-8 in a dose-dependent manner.

13.2.3 SM-induced Apoptosis of Tissues

13.2.3.1 Apoptosis of the Lung Due to SM Exposure

Several studies have shown that apoptosis (a form of cell death) is a major mechanism of SM lung injury. Apoptosis can be elicited *via* the Fas pathway.[25] The Fas receptor (FasR) is a transmembrane protein belonging to the TNF receptor (TNFR) superfamily. A Fas-associated death domain (FADD) is associated with the cytoplasmic part of the receptor. When the Fas ligand (FasL), a transmembrane protein in the TNF cytokine family, engages the FasR, Fas signaling recruits a death inducing signaling complex (DISC) of intracellular proteins to the FADD and this leads to the activation of the enzyme caspase 8. Caspase 8 then activates caspase 3, triggering an activation cascade (caspases 3/6/7), leading ultimately to death of the cell. Caspase 3, which is the crucial executioner of the death reaction, can be activated through alternative pathway reactions involving mitochondrial products and caspase 9.

Apoptosis can theoretically be elicited through TNF *via* a pathway similar to the Fas pathway but initiated through TNF engagement of TNFR.[26] However, in the majority of TNF-responsive cells, the apoptosis pathway is apparently blocked so that this signaling does not undergo apoptosis, but rather elicits pro-inflammatory cytokine responses as a major role of TNF signaling, as discussed above.[27]

It has been suggested that the Fas apoptotic pathway is indeed involved in SM-induced cell damage. In normal human bronchial epithelium (NHBE) cells, SM as well as a Fas agonist antibody CH11 (an antibody that can mimic the action of FasL) both induced caspase 3 activation. This activation was inhibited by a small interfering RNA (siRNA) designed to block the expression of FasR as well as an antibody directed against FasR (ZB4), which indicated that SM-induced apoptosis occurred *via* the Fas response pathway. SM-induced caspase 3 activation could be inhibited by FasR siRNA when added to the NHBE cultures up to 8 hours after SM application.

This suggests that SM promotes Fas–FasL interaction, which in turn induces apoptosis. Pirzad *et al.*[28] examined the BALF of patients approximately 20 years after SM exposure, which showed significantly higher Fas–FasL levels than that of control subjects who did not have any history of SM exposure; however, no significant difference in caspase-3 activity was observed between the two cohorts. Immunohistochemical staining of the airway and submucosal epithelia of rats intratracheally injected with SM (2 mg kg^{-1}) showed significantly higher caspase-3 and caspase-9 expression levels compared to that of the control group, suggesting that SM, possibly through the intrinsic mitochondrial-caspase 9 pathway, induces apoptosis of tracheal parietal cells.[29]

On the other hand, Ray *et al.*[30] reported that the exposure of respiratory epithelial basal cells to SM for 16 hours induces an increase in the activities of caspases 3 and 8, indicating that caspase 8 is the hallmark of the extrinsic Fas death receptor (DR) pathway. The results of other studies have also suggested that SM can induce airway epithelial cell apoptosis through the DR pathway. In this regard, SM enters the respiratory tract and initially interacts with airway epithelial cells, which in turn induces apoptosis *via* caspase-3 cascade activation. SM stimulated NHBE and small airway epithelial cells (SAEC) *in vitro*, with the caspase 8 and caspase 3 DR pathways being significantly activated.[31] Thus, both Fas death receptor and mitochondrial apoptotic pathway reactions are involved in SM-induced cell death.

13.2.3.2 Apoptosis of Skin Cells Due to SM Exposure

SM-induced cell responses are relatively complex, including protein expression upregulation and modification. Human skin keratinocytes treated with SM (300 μmol L^{-1}) at pH 8.5 induce a significant increase in the expression of nuclear glyceraldehyde-3-phosphate dehydrogenase, thereby suggesting that this specific enzyme is related to apoptosis, DNA structural changes, DNA repair, regulation of chromosome integrity, and telomere structure.

Rosenthal *et al.*[32] reported that SM can induce the upregulation of Fas, its ligand FasL, and the caspase precursors procaspase-3, -7, -8, and -9, with caspase-3 acting as the main executive molecule of apoptosis. Fas knockdown in keratinocytes prevents an increase in the activity of caspase-3 as well as DNA strand breakage, that were induced by SM.

Several studies have confirmed, however, that after application of high concentrations of SM (1 mmol L^{-1} or higher), Poly(ADP-Ribose) Polymerase (PARP) overexpression enhances the consumption of its substrate, β-nicotinamide adenine dinucleotide (NAD^{+}), which then induces adenosine triphosphate (ATP) deficiency, ultimately leading to cell necrosis, a form of cell death different from that of apoptosis.[33] Treatment of human normal keratinocyte HaCaT cells with lower concentrations (1–1000 μmol L^{-1}) of SM results in cell apoptosis, which is characterized by minor PARP cleavage by caspase-3 without ATP depletion. DNA fragmentation increases in a time- and concentration-dependent manner. The role of PARP depends, however, on the cell type and degree of cell damage. For example, human dermal fibroblasts show a PARP-dependent, SM-induced necrosis, whereas keratinocytes display markers of an apoptotic death. The death of PARP-deficient keratinocytes is unaffected.[34] However, it should be recalled that in general, when a dose of SM greater than 1 mmol L^{-1} is applied, PARP activation results in intracellular NAD^+ depletion, which in turn induces necrotic cell death that subsequently leads to the formation of blisters in SM injury.

Studies on other DNA damaging agents support this conclusion. Although PARP can mediate cell death signals (necrosis) that are caused by the depletion of NAD^+ in cells, this may not be the only mechanism by which PARP mediates cell death. A previous study has shown that PARP can form poly(ADP-ribose), PAR, which is an apoptotic signal, and that the depletion of PAR mediates cell death and a decrease in PARP levels.[35] However, the investigators reported that PARP inhibitors have provided only a minor degree of protection of cells in culture, regardless of the mechanism of PARP-induced cell death (necrosis *via* NAD^+ depletion or apoptosis *via* PAR production).

13.2.4 DNA Cross-linking in SM-induced Tissue Injury

SM is a cytotoxic bifunctional alkylating agent that causes DNA base alkylation, intra-chain crosslinking, and double-strand breakage. SM is lipophilic and diffuses into internal organs in multiple ways. DNA is the most sensitive molecular target of SM that responds by forming mono- and di-adducts, thereby generating a modified DNA structure. Acute SM poisoning results in DNA damage, and long-term damage can cause cancer. SM induces DNA damage and triggers PARP activation. A high concentration of SM promotes PARP over activation, as well as NAD and ATP depletion, which consequently leads to cell necrosis. Exposure of human epidermal cells to SM for 6 hours results in an increase in PARP-1 activity, coupled with a 22% reduction in intracellular ATP levels. Compared to wild-type ($PARP^{+/+}$) mice fibroblasts, gene knockout ($PARP^{-/-}$) mouse fibroblasts exposed to SM incur nucleosome DNA breakages, degradation and activation of caspase-3, and upregulation of annexin V. Cultured human skin keratinocytes exposed to SM and CEES, respectively, showed a significant increase in cellular DNA adducts.

Behravan *et al.*[36] detected DNA fragmentation by single cell gel electrophoresis in patients approximately 23 to 27 years after SM exposure, as well as a significant increase in the percentage of lymphocyte DNA damage, thereby suggesting that SM-induced DNA damage persists years after SM exposure. Homologous recombination of genes related to the DNA repair pathway can promote cell survival.

Jowsey *et al.*[37] reported that homologous recombination is the main repair pathway of SM injury, and that nucleotide excision repair and non-homologous end-ligation also contribute to cell survival. In addition, exposure of human TK6 lymphoid stem cells to SM results in the activation of the DNA damage signaling pathway, which follows a concentration- and time-dependent manner. The alkylating effect of SM triggers changes in various cellular components, thereby resulting in DNA single adducts and cross-linking, which in turn causes DNA double-strand breaks, as well as a disruption of the homologous recombination DNA repair pathway.[38]

Moser *et al.*[39] reported that exposure of human peripheral blood lymphocytes to SM (10–100 μmol L^{-1}) resulted in a dose- and time-dependent increase in DNA damage. Blaha *et al.*[40] exposed human epithelial cells to CEES for 2 hours, and electron microscopy revealed cytoplasmic extravasation, nuclear chromatin condensation, nuclear membrane and cytoplasmic structure rupture, and cytoplasmic vacuolization, suggesting that CEES induces cell apoptosis that is characterized by terminal differentiation.

13.2.5 Oxidative Stress in SM-induced Tissue Injury

The main molecular mechanisms underlying SM poisoning include inflammation and polymerase overactivation-induced DNA alkylation, commonly resulting in oxidative stress.[41] The mechanism underlying SM-induced lung injury involves oxidative stress, apoptosis, and systemic disruption of the balance among various proteolytic and anti-proteolytic enzymes, inflammation, and DNA damage. Extensive evidence indicates that oxidative stress and apoptosis are two key elements.[42,43]

Jafari *et al.*[44] reported a significant increase in the activities of superoxide dismutase (SOD), catalase, and glutathione peroxidase in the BALF and plasma red blood cells of 54 patients with SM lung injury, which may be related to the depletion of glutathione (GSH) and an increase in malondialdehyde (MDA), thus suggesting that SM in lung cells may cause oxidative stress by disrupting the antioxidant defense system and inducing lipid peroxidation. Shohrati *et al.*[45] described a decrease in serum GSH levels and an increase in serum MDA levels in 250 SM lung injury patients, which indicated that SM-induced lung injury can lead to an imbalance of the oxidative-antioxidant system. BALF collected from guinea pigs at 7 days after being intratracheally injected with CEES (2 mg kg^{-1}) showed a significant increase in SOD-1 activity (35%) and a marked decline in SOD-3 content (62%), which suggest that oxidative stress is one of the major mechanisms involving SM poisoning.

Glutathione S-transfer (GST) is one of the major antioxidants in the human airway that can reduce oxidative stress in various types of environment. In SM-induced chronic obstructive pulmonary disease (COPD), GSTP1 and GSTA1 are upregulated in the bronchial epithelia, whereas GSTP1 immunoreactivity in the bronchial walls is weakened, thereby suggesting that GST plays an important role in the cytoprotection of anti-oxidative stresses.[46]

Male mice exposed to CEES (1068 mg kg^{-1}) for 24 hours, 7 days, and 14 days and examined for oxidative stress showed that GSH and glutathione-bound enzyme (GSHE) were depleted, whereas lipid peroxidation was enhanced. Male mice that prophylactically received oral S-2(ω-aminoalkylamino) alkylaryl sulfide (DRDE-07) (249 mg kg^{-1}) 30 minutes before SM exposure *via* the skin showed an effective reduction in SM-induced oxidative stress levels. The results showed that DRDE-07 plays a beneficial role in alleviating the toxic effects of CEES and may thus be potentially utilized in the treatment of SM/CEES poisoning.[47]

Recent studies have indicated that paraoxonase-1 (PON1) has antioxidant activity under oxidative stress and thus can prevent lipoprotein oxidation. Evaluation of patients with SM lung injury indicated a decrease in serum PON and aryl esterase activity, together with the presence of the ApoAl protein only in the BALF.[48,49] Chang *et al.*[50] suggested that stress involving the endoplasmic reticulum triggers the cell survival pathway. SM can also concomitantly induce protein misfolding in the endoplasmic reticulum, which may lead to the degradation of target proteins.

Recent investigations have focused on the molecular toxicology of SM, which showed that treatment using 3 μmol L^{-1} of SM induces a 35- and 6–10-fold increase in the expression of microRNA183 and microRNA21, respectively. MicroRNA183 plays an important role in cell migration and stromal–epithelial interactions, whereas microRNA21 regulates monocyte differentiation, oxidative stress, and inflammation. MicroRNAs have recently been described as a core factor that play an important role in the maintenance of normal skin morphology and wound healing, and may thus be potentially utilized as a treatment regimen for refractory wounds, as well as a molecular target for the treatment of SM-induced skin injury.

13.2.6 Conclusions

The mechanisms underlying SM-induced tissue and cellular damage are relatively more complex than previously conceived and mainly involve the process of alkylation. Main reactions include the induction of inflammatory responses, apoptosis, necrosis, DNA fragmentation and activation of matrix metalloproteases.

SM-induced free radical-mediated activation of TNF-α signaling *via* the TNFR, which in turn leads to the production of further inflammatory mediators such as the pro-inflammatory cytokines TNF-α, IL-1α, IL-1β, IL-6 and IL-8, may be the main pathway that leads to pulmonary inflammation.

SM can induce cell death in the form of apoptosis *via* the FasL–Fas pathway. Apoptosis proceeds through both the death receptor and the mitochondrial sub-pathway reactions. SM can also elicit cell death by necrosis, which is characterised by cell autolysis and blistering of the skin. Whether apoptosis or necrosis is induced upon SM treatment of cells depends on such factors as dose, duration of treatment and cell type. Matrix metalloproteases, which degrade type IV collagen and other basement membranes, including that separating the epidermis from the dermis, may contribute to the blistering effects seen as a result of necrosis.

Although a variety of molecular signals related to SM injury have been elucidated, novel treatment methods still remain in the theoretical experimental stage and have not been applied in clinical settings. Furthermore, interactions involving various signal molecules in SM injury remain elusive, such as the relationship between p53, NF-κB, and apoptosis, the reaction after the activation of p53 signaling molecules, and the activation of various signal molecules and the development of SM injury. Further research on these issues may elucidate the mechanism underlying SM-induced tissue damage, which in turn may facilitate in the identification of effective targeted treatment methods for SM injury.

The mechanism involving SM lung injury is very complex and thus there is a need to investigate the molecular biological characteristics (transcriptional regulation, gene expression) of pulmonary epithelial cells by using stable and accurate dose–response and time-dependent animal models of SM toxicity. In addition, the identification of SM lung injury-specific biomarkers and therapeutic targets as novel strategies for the prevention and treatment of SM-induced tissue damage are also warranted in the area of anti-chemical medicine, locally and internationally.

13.2.7 Research Progress on the Mechanism of NA Poisoning

NAs are a class of strong cholinesterase inhibitors and organic phosphorus pesticides (such as dimethoate and dichlorvos) that belong to the same family of organic phosphate compounds and thus have similar physical and chemical properties and toxic effects. Common NAs include soman, sarin, and VX. To date, the use of NAs in terrorist-related activities remains a worldwide threat. Sarin was used by AS in 1995 in the Tokyo subway. Organophosphate poisoning remains an important health problem in the West Pacific and South Asia regions of Sri Lanka. Poisoned patients there generally present symptoms such as tremors and convulsions of the salivary muscle that often require tracheal intubation, whereas some develop OP-induced delayed neuropathy (OPIDN) symptoms (such as myasthenia gravis). Furthermore, those who survive organophosphate poisoning are at a higher risk for encephalitis and brain cell death, whereas individuals who experience low-dose poisoning may eventually develop long-term progressive neuritis and neurodegenerative diseases.

The mechanism of NA poisoning involves its irreversible or reversible binding to acetylcholinesterase (AChE), which prevents it hydrolyzing acetylcholine (ACh), thereby leading to the excessive accumulation of ACh. NAs then continue to act on the cholinergic receptor, leading to cholinergic nervous system hyper function that often manifests as salivation, muscle tremors, convulsions, difficulty breathing, and coma, and eventual death. Previous studies have shown that a dysfunctional cholinergic system cannot fully explain all the changes that are observed when the central nervous system is exposed to poisons. Animal experiments have indicated inconsistencies between the extent of brain damage that is caused by NAs and AChE inhibition, which in turn suggests that the disruption of the cholinergic system is not the only factor that is involved in NA-induced brain injury, and that a non-cholinergic mechanism may also be involved in this process. A variety of non-cholinergic mechanisms play an important role in NA-induced brain injury, which are specifically related to cell excitotoxicity, calcium overload, oxidative stress, inflammatory response, and changes in the blood–brain barrier permeability. Previous studies have suggested that soman intoxication can be divided into three stages: (1) the early cholinergic stage that includes seizures within 5 min of poisoning; (2) the middle stage that consists of a combination of cholinergic and non-cholinergic symptoms (seizures within 40 minutes of poisoning); (3) the late non-cholinergic stage that comprises seizures at around 40 minutes after poisoning and plays a critical role in the development of brain injury. At present, prevention and first-aid measures of NA poisoning have been the focus of research studies both in China and internationally, and have made substantial progress. New treatment methods for NAs that target the non-cholinergic mechanism may thus become an effective adjuvant therapy.

13.3 Research Progress on SM Biomarkers and NA Poisoning

13.3.1 Research Progress on Related Biomarkers in SM-exposed Human Populations

SM is an alkylating agent that may interfere with various biochemical components, and the analytical detection of its hydrolysates, adducts, and other biomarkers are of great significance for the rapid diagnosis of SM poisoning, assessment of degree of poisoning, and medical treatment. *In vivo* biological proinflammatory cytokines may be utilized as indicators for the clinical treatment and diagnosis of SM poisoning. In this study, we review recent advances in the *in vivo* study of SM injury-related markers from two aspects of various adducts and inflammatory markers.

13.3.1.1 In vivo *Adducts*

Thiodiglycol (TDG) and Thiodiglycol sulfoxide (TDGO) Currently, urine is the main sample used for the determination of SM exposure, based on the presence of TDG and TDGO. Riches[51] used isotope dilution gas chromatography-ion trap tandem mass spectrometry to analyze the SM hydrolysis products thiodiglycol and thiodiglycol sulfoxide in urine samples; however, due to a high background, these could not be easily detected. Sulfur diethylene glycol sulfoxide is reduced by titanium trichloride, which is then indirectly analyzed for the simultaneous detection of hydrolysates and cleavage products. However, thiodiglycol and thiodiglycol sulfoxide are non-specific indicators of human poisoning as trace amounts of these molecules (<10 ng) can also be detected in urine samples from unexposed individuals.

SM-glutathione adducts SM and its oxidation products bind to glutathione; furthermore, through β-lyase or acetyl transferase, SM generates small β-cleavage products that are utilized as specific biomarkers of SM injury. Read *et al.*[52] used liquid chromatography-mass spectrometry/mass spectrometry (LC-MS/MS) to analyze β-lysates after SM exposure, with the lowest detection limit within the range of 0.1–0.5 ng mL^{-1}, which is lower than that of previous detection methods. Furthermore, the process is relatively simple.

SM–DNA adducts SM–DNA adducts are the main molecules that have been investigated in relation to SM poisoning, and *N*-(2-hydroxyethylthio)-deoxyguanine (HETEG) is the most important SM–DNA adduct. Jowsey *et al.*[53] suggested that the activation of the DNA repair pathway in homologous chromosomes after SM exposure is essential for cell survival, and proposed that the dosing and timing of SM exposure affects cell damage in TK6 lymphatic stem cells.

SM–protein adducts Compared to SM–DNA adducts, the retention time of protein adducts *in vivo* is longer, even lasting up to a few months. Therefore, the detection of SM protein adducts to confirm SM poisoning serves as a more reliable approach. SM–protein adducts are usually detected by gas chromatography-mass spectrometry (GC-MS) or liquid chromatography-mass spectrometry (LC-MS). Edman degradation of SM–hemoglobin adducts generates N-terminal valine adducts, which are then detected by GC-MS.

13.3.1.2 Inflammatory Response-related Biomarkers Interleukins (ILs)

When human skin comes into contact with SM, it immediately enters the acute phase of epidermal basal keratinocyte necrosis, which is characterized by the development of erythema and blisters. The acute phase is then followed by a series of inflammatory reactions that include epidermal cells incurring damage, the release of IL-1 to induce an inflammatory response, and a significant increase in IL-1 levels. Yaraee *et al.*[54] investigated the levels of the proinflammatory cytokines (TNF, IL-1α, IL-1β) as well as the IL-1 receptor antagonist (IL-1Ra) in a population that was exposed to SM approximately 20 years ago. The levels of TNF, IL-1α, IL-1β, and IL-1Ra in the control group were 23.79, 1.89, 1.91, and 32.9 pg mL^{-1}, respectively, whereas in the SM-exposed population, on the other hand, these were 11.11, 0.81, 1.73, and 26.7 pg mL^{-1}, respectively. The levels of pro-inflammatory cytokines in the mustard-exposed group were significantly lower than those in the normal control group ($P < 0.01$). These findings indicate that the long-term effects of exposure were in part due to the regulation of cytokine levels, whereas the downregulation of pro-inflammatory factors may facilitate the development of a treatment scheme for clinical complications.

Attaran *et al.*[55] reported that veterans with SM-induced chronic lung diseases have an IL-6 level of 15.01 ± 0.61 pg dL^{-1}, compared to the 4.59 ± 3.40 pg dL^{-1} of the healthy control group. These findings indicate that IL-6 is strongly associated with chronic obstructive pulmonary disease.

The eye is the most vulnerable target tissue that is affected by SM. Delayed keratitis is the most dangerous eye complication that is caused by SM exposure. Ghasemi *et al.*[56] determined that due to local and systemic immune downregulation, the serum IL-8 levels of people exposed to SM 20 years ago were significantly lower than those of the control group.

Tumor necrosis factor (TNF) The TNF-α level in the animal model of SM exposure was significantly higher, which in turn stimulated cells to secrete IL-8 and IFN-γ. These findings indicate that TNF-α plays an important role in lymphoid tissue hyperplasia with respect to SM exposure.

In 2012, Panahi *et al.*[57] reported that the serum TNF-α, TGF-β, and IFN-γ levels of patients previously exposed to SM were significantly lower than those without a history of SM exposure. The treatment of SM-induced dermatitis and pruritus with TNF-α inhibitors and recombinant IFN-γ has achieved good therapeutic effect.

Granulocyte–macrophage colony-stimulating factor (GM–CSF) Amiri *et al.*[58] showed that patients exposed to SM about 25 years earlier have serum GM–CSF levels that were similar to those of healthy controls. In the SM-exposed group, the GM–CSF serum levels were positively correlated with the proportion of eosinophils, IL-1α, IL-1β, TNF-α, and IL-6, but were not significantly correlated with IL-8 and clinical complications. The target and specific mechanism involving the hematopoietic and immune systems of patients exposed to SM 25 years ago thus requires further study. Emad *et al.*[59] found that the GM–CSF level in the alveolar lavage fluid of the veteran group with SM-induced pulmonary fibrosis differed significantly from that of the normal control group. Furthermore, the number of leukocytes, eosinophil percentage, and levels of IL-8, and GM–CSF in the alveolar lavage fluid were significantly correlated with SM-induced pulmonary fibrosis.

Interferon (IFN) and transforming growth factor (TGF) The most common adverse reaction to SM exposure includes chronic skin pruritus. However, the pathogenesis of SM-induced pruritus is not clear. In 2013, Panahi *et al.*[60] proposed that the assessment of SM-induced cutaneous pruritus could be monitored based on the levels of IFN-γ and TGF-β, which were significantly higher in the normal control group compared to the SM-exposed group. On the basis of this study, it has been proposed that drugs regulating serum IFN-γ and TGF-β levels be used in the treatment of pruritus to potentially improve the quality of life of these people.

13.3.1.3 Conclusions

For toxicological traceability analysis, relevant *in vivo* biomarkers for SM injury are generally used in analysis. *In vivo* adducts are highly specific markers of SM injury; however, these are often difficult to detect, and the assay required to do so is relatively expensive. Inflammatory factors in the body can directly reflect the real-time exposure levels of poison in the blood, and its detection is simple and inexpensive, yet its specificity is relatively low. Therefore, the combination of the two biomarkers may be beneficial to research studies and detection of relevant biomarkers in mustard-exposed populations. Methods in selecting the appropriate detection technology, improving the specificity and sensitivity of the method, and making it suitable for the detection of poisoned samples are still important issues that need to be addressed in future research studies.

13.3.2 Advances in Adduct Markers of NA Poisoning and Their Analytical Methods

NAs are highly effective and rapid killing chemical warfare agents that impart their toxicological effects mainly by inhibiting the biological activity of acetylcholinesterase. NAs enter the body and are easily distributed in various organs; the main forms of NAs include poisonous prototypes, hydrolysis products, and macromolecular adducts. The detection of these compounds in the body confirms NA exposure and the extent of poisoning using this NA, as well as assists in monitoring a patient during treatment. The NA and its associated biomacromolecular adducts are superior biomarkers because these are easily detected and highly specific, and the application of modern analytical instruments can further increase sensitivity in detecting trace amounts of neuropathic poisons. Upon entry into the body, NAs mainly interact with two types of biological macromolecules: cholinesterase and albumin. NAs and albumin adducts are highly sensitive, aging-resistant, and are unaffected by rejuvenating drug treatments, along with other prominent advantages. This has become the focus of recent NA biomarker research.

13.3.2.1 Analytical Detection of in vivo *Diagnostic Biomarkers for NA Poisoning*

Analytical detection of toxicant prototypes. The analytical detection of toxicant prototypes can directly determine the type and structure of the target molecule. Analysts have adopted gas chromatography (GC) for the analytical detection of five kinds of poisoning agents, namely, GB, GD, GF, VX, and Russia VX (RVX). However, because some NAs have a relatively rapid degradation rate (*e.g.*, the half-life of GA at pH 7 is only 8.5 hours), Munro *et al.*[61] have indicated that these methods do not meet the requirements for the detection of trace amounts of NA after exposure. Because of the limited sensitivity of this method, cases involving a small amount of a toxicant prototype in the body, or older samples, may generate false-negative detection results. Koskela *et al.*[62] reported that liquid chromatography-solid phase extraction-nuclear magnetic resonance (LC-SPE-NMR) was used to detect a toxicant prototype in off-line mode.

Analytical detection of degradation products of toxicants. Toxicant prototypes are very unstable *in vivo*, and their associated toxic protein adducts undergo natural aging, thereby easily producing the corresponding degradation products, alkyl methylphosphonic acids (AMPAs) and methylphosphonic acid (MPA). Several analytical detection methods for the degradation products of toxicants in various biomedical samples have been established.

Analytical detection of toxicant degradation products in urine. Analysis of urine toxicant degradation products is the simplest way of identifying casualties due to NA exposure. Most of the metabolites are polar molecules with relatively small molecular weight that can be analyzed directly by LC-MS, gas chromatography (GC), gas GC-MS, or gas chromatography-mass spectrometry/mass spectrometry (GC-MS/MS) after derivatization.[63–66] Hui *et al.*[64] performed gas chromatography-flame ionization detector (GC-FID) detection using urine samples from Tokyo metro victims after derivatization with trimethylchlorosilane, and reported a high fluoride ion concentration of 27.7 ± 1.83 txmoL g^{-1}, which was significantly higher ($P < 0.01$) than that in normal urine (15.5 ± 1.98 moL g^{-1}). The limitation of this method is that the traceable time is short, because the wounded can excrete more than 90% of the metabolites within the initial 48–72 h after exposure.

Analytical detection of toxicant degradation products in blood samples. To detect toxicant degradation products in blood, the collected samples are initially centrifuged, subjected to protein removal and solid phase extraction (SPE), followed by direct detection by using capillary electrophoresis (CE) and LC-MS or they are analyzed by GC-MS after derivatization. Tsuchihashi *et al.*[67] have extensively utilized the separation and detection methods of electron microprobe analysis (EMPA) for a hydrolysate of VX in human sera. In selected ion mode (SIM), the LOD was up to 3 μg L^{-1}.

Detection of toxicant-protein adducts. Nerve agents are electrophilic chemical substances and can quickly form phosphorylase, after entering into the body, with oxygen atoms of serine hydroxyl in the acetylcholinesterase (AChE) center. They can also combine with plasma butyrylcholinesterase (BuChE) and carboxylesterase (CaE), which further interact with albumin *in vivo* to form a variety of toxicant–protein adducts. AChE and BuChE adduct detection methods mostly use biological mass spectrometry in their analysis. The biological half-life of BuChE *in vivo* is 5–16 d, and the concentration of BuChE in plasma may reach up to 80 nmol L^{-1}, which is 26-fold higher than that of AChE. Therefore, BuChE is the best biomarker for confirming exposure to NAs.

Over the past 10 years, adducts of albumin (with tyrosine at position 411) and toxicants have also been extensively investigated.[68,69] Although the binding constant of albumin and toxin is much smaller than that of cholinesterase and toxin, the concentration of albumin in organisms is 100 times that of the concentration of cholinesterase, whereas albumin–toxin adducts are not found in aged samples. Therefore, the development of a method for detecting albumin–toxin adducts as a biomarker for toxic agents is warranted.

13.3.2.2 Conclusions and Prospects

Toxic biological samples are readily traced in various body tissues within a few days or weeks. However, most NAs have a small median lethal dose and thus, the established detection method must be highly sensitive and specific. Chromatography and GC-MS have become an important research method for the *in vivo* diagnostic marker analysis of NAs. For rapid screening and diagnostic purposes, a number of *in situ*, real-time, high selectivity, and high specificity detection methods for toxicant prototypes and their related compounds, such as fluorescence, electrochemistry, and immunological techniques will be extensively utilized for the diagnosis of NA poisoning.

13.4 Research Progress on Drugs Used in the Prevention and Treatment of SM and NA Poisoning

13.4.1 Advances in SM Scavenger (SM Disinfection Technology)

To date, SM remains one of the major poisons of foreign military equipment. SM is a typical persistent agent that is highly stable and can remain in water and soil substrates for an extended period of time. The Japanese invaders left large amounts of SM poison in China's northeastern region, which resulted in several poisoning incidents, and continues to threaten China's ecological environment and public health. SM disinfection technology has become an interesting field of research that circumvents the limitations of traditional disinfection technologies such as low efficiency, toxicity, and environmental effects.

13.4.2 Oxidation Disinfection Technology

Potassium ferrate disinfection technology Potassium ferrate is a non-toxic and highly oxidizing compound that not only decomposes organic toxicants, but also disinfects heavy metal ions, and thus may be potentially applied to SM disinfection. Although potassium ferrate can effectively disinfect SM, it is relatively unstable in aqueous solutions, undergoes rapid decomposition under normal temperature condition, and the associated cost for the preparation of solid potassium ferrate is significantly high. Therefore, the application of potassium ferrate for the disinfection of chemical warfare agents is limited.

Peroxide disinfection technology Compared to traditional disinfection technologies, peroxide disinfection technology is of low toxicity, minimally corrosive, has a broad spectrum of reactants, and is widely used in the decontamination of chemical warfare agents. This method has been mainly employed in disinfecting SM through oxidation reactions, wherein SM is selectively oxidized into non-erosive sulfoxide, and with prolonged reaction times, may be further oxidized into erosive sulfone. Peroxide water-based disinfection is the most commonly used method for peroxide disinfection; one of the most extensively studied systems is the bicarbonate/hydrogen peroxide (BAP) system, which has been sparingly applied in clinical settings. The American Chemical and Biological Center developed the disinfectant DECON GREEN system using peroxide disinfection technology and this has been utilized by the United States Army. The disinfectant system consists of hydrogen peroxide, Triton X-100, the surfactant potassium carbonate, and potassium molybdate, which generate non-toxic and non-corrosive sulfoxide products. The DECON GREEN system has been further modified by integrating water-in-oil microemulsion technology to increase the SM disinfection rate by enhancing the SM solubility. In this system, molybdate mainly acts as a catalyst for hydrogen peroxide to produce an oxidizing group that diffuses into the micro-emulsion and then reacts with the SM molecules.

In addition, US Lawrence Livermore National Laboratory researchers mixed potassium persulfate (OXONE) with silica white carbon black, with Cu^+ as a catalyst to generate an acidic silica gel disinfectant that effectively disinfects SM. A silica gel disinfectant composed of H_2O_2–OXONE–$FeSO_4$ and gas phase silica gel, its sterilization time of SM simulant 2-GEES was less than 3 min, and disinfection potency reached 10 g L^{-1}.

Nano-metal oxide self-disinfection technology Nano-metal oxides are good disinfectants of SM and its simulants and can be used for the coating of combat uniforms, weapons, and equipment surfaces. It has been reported that nano-alumina has the best disinfection effect on CEES, which mainly yields hydrolysates and dechlorination reaction products.

13.4.3 Photocatalytic Disinfection Technology

The photocatalytic method is a photocatalyst TiO_2-based method of catalytic disinfection that utilizes visible light or ultraviolet (UV-A) radiation at room temperature. It can completely oxidize and decompose organic matter, reduce heavy metal ions, is non-toxic, harmless, and non-corrosive, thus allowing its application for SM disinfection. It was found that TiO_2, which has a relatively large specific surface area, showed a weaker light intensity, and the addition of trace water vapor catalyzed the photocatalytic sterilization reaction under gas phase reaction conditions. However, because the reaction process generated products that are difficult to convert into volatile substances that accumulate on the catalyst surface, this readily results in the deactivation of TiO_2. Under gas phase conditions, due to the impact of composition and structure, the mechanisms of disinfection and degradation vary with different simulants.

Ramacharyulu *et al.*[70] studied SM disinfection by using nano-TiO_2 and by comparing the results of UV-A light exposure and no-light conditions, indicated that the photocatalytic efficiency of TiO_2 decreased with an increase in nanoparticle size. TiO_2 molecules with a particle size of 11 nm showed excellent disinfection ability under light irradiation conditions, achieved complete sterilization of SM within 6 hours, and generated the disinfection products, acetaldehyde, thiol, and sulfoxide. One disadvantage of using the TiO_2 catalyst is that it requires the use of UV light at a wavelength of 388 nm to excite the catalytic reaction, and poor reactions occur in the presence of sunlight. The metals that were used to impregnate the TiO_2 catalyst mainly included zirconium, vanadium, and iron.[71]

The current SM disinfection technology has several pros and cons, thereby requiring additional research to identify factors that are involved in the disinfection mechanism as well as to improve disinfection performance because the disinfecting rate of the existing peroxide disinfection system on SM is relatively slow, the product also contains a toxic product, sulfone, and a need to eliminate the occurrence of corrosion of the material. Novel peroxide activator multifunctional surfactant and disinfectant solutions for SM peroxide disinfection need to be established.

13.4.4 Drug Research on SM-induced Injury

Oxidative stress (OS) is one of the initiating and key linkages of SM toxicity.[72,73] The nucleophilicity of the sulfur atoms of SM molecules produces ortho groups that participate in the reaction and facilitate in the cleavage of the chlorine ion, thereby forming a strong electrophilic (reagent) sulfonium ion, which is the root cause of oxidative stress in the body. The production of reactive oxygen species (ROS) may be mainly related to respiratory bursts after SM-induced neutrophil activation, immune cell-associated oxidative stress, and lack of anti-oxidants due to the overactivation of polyadenosine diphosphate ribose polymerase, thereby resulting in insufficient cellular energy supplies.[74] Therefore, SM eradication and anti-oxidative stress treatment are important anti-SM toxicity strategies.

13.4.4.1 Scavengers and Anti-oxidative Stress Drugs

Sulfhydryl/sulfur compounds have been one of the main areas of research in SM control, which include thiosulfate ($Na_2S_2O_3$). In recent years, long-chain compounds such as cysteine mercaptopurine and ethyl benzene sulfide have been examined in terms of their role in scavenging free radicals and DNA protection. Cysteine (Cys) and GSH are both sulfhydryl-containing compounds that are capable of binding directly to free SM molecules. However, due to poor lipid solubility, the resulting increase in intracellular Cys and GSH concentrations was not significant. Studies have shown that short-chain cysteine ester CysNAC and cystine dimethyl ester, regardless of concentration in lung tissues, do not impart any protective effect, whereas low concentrations of long-chain cysteine isopropyl ester renders complete protection to lung tissues.[75] Methods of improving the lipid solubility of these molecules as well as increasing intracellular GSH levels are thus important research directions to improve their antitoxic activity.

Diphosphine purine (2,6-dithiopurine, DTP) is a nucleophilic scavenger that can directly bind with SM to prevent DNA damage. DTP enters the cells and reacts with the SM analogue CEES and 2-chloroethyl methyl sulfide (CEMS) to reduce cell toxicity.[76] Blue/white colony screening showed that DTP could completely eliminate CEES and CEMS-induced mutations in human skin cells. Furthermore, mutagenic screening of a mouse model (big blue mice) indicated that DTP treatment after 1 hour of exposure could completely repair gene mutations.[77] However, whether these mutations have a long-term protective effect has yet to be determined.

The ethyl benzene sulfide compound amifostine is an FDA-approved cytoprotective drug for the treatment of radiation-induced injuries and has been mainly used in the treatment of cancer. Its protective mechanism involves the elimination of free radicals, which indirectly results in DNA protection.[78]

Amifostine (WR-1065) is a prodrug that is metabolized *via* membrane-bound alkaline phosphatase (ALP) dephosphorylation. Higher local alkaline phosphatase activity, the abundance of blood vessels, and higher pH has been correlated with higher rates of amifostinephosphorylation. Previous studies have shown that the preventive administration of amifostine may also result in anti-SM toxicity, whereas no significant therapeutic effect was observed after exposure,[79] which may be related to SM-induced oxidative stress that occurs during early stages of injury.

DRDE-07 is an analogue of amifostine that is more lipophilic than amifostine. DRDE-07 administered 30 minutes before SM exposure significantly increased the median lethal dose for mice, whereas no significant effect was observed 2 hours after SM exposure. DRDE-07 is also less toxic than amifostine, and its starting dose is significantly lower than that of amifostine.[80]

13.4.4.2 Metal Porphyrin Compounds

Catalytic metalloporphyrins such as AEOL-10150 are novel small molecule antioxidants that possess significant SOD-like and catalase-like activities, as well as the capacity to scavenge for peroxynitrite and lipid peroxides. The Aeol company is currently developing a preclinical study to develop AEOL-10150 as a new anti-chemical and radiation damage drug with funding from the US government. In addition, the study has also determined that rats subjected to continuous nasal inhalation of 5% CEES for 15 min followed by AEOL-10150 (5 mg kg^{-1}) twice a day at 1 hour and 9 hours after exposure were capable of counteracting CEES-induced pulmonary edema and showed reduced inflammatory cell infiltration.

13.4.4.3 Polyphenolic Compounds

Polyphenols mainly include flavonoids and phenolic compounds that possess antioxidant activity due to the presence of an ortho-phenolic hydroxyl group that can be easily oxidized into a quinone structure. Polyphenols have a strong ability to capture free radicals such as active oxygen molecules and thus may be utilized as strong antioxidants and free radical scavengers. Currently, the compounds reported to have antitoxic effects include equol,[81–83] tocopherols and their analogues,[84,85] Indian cottonseed,[86] and silibinin (SB).

In addition, SM-related studies have shown that melatonin can increase the production of low-molecular weight antioxidants and reduce oxidative stress, such as lipid peroxidation, in rats exposed to SM. The effect of melatonin on a variety of oxidative stress molecules, especially the significant inhibition of peroxynitrite production and its induced cytotoxic effects, suggests that melatonin may also have a protective impact on the residual effects of SM.

13.4.4.4 Inflammation-based Drugs

The tissue injury induced by SM is closely related to inflammatory responses, and a large number of neutrophils, monocytes and macrophages that have accumulated in the area of injury can release myeloperoxidase and inflammatory cytokines such as IL-1β, IL-6, TNF-α, IL-8, and MCP-1 (a chemokine that attracts macrophages), as well as induce the release of substance P, cathepsin, plasminogen activator, and histamine, thereby leading to tissue redness, blisters, necrosis, and other pathologic damage. The inflammatory response induced by SM involves multiple signaling pathways, including the arachidonic acid metabolic pathway, mitogen-activated protein kinase (MAPK) signaling pathway, and the nuclear transcription factor-κB (NF-κB) signaling pathways. Arachidonic acid can be catalyzed by cyclooxygenase and lipooxygenase to produce prostaglandins and leukotrienes. The MAPK and NF-κB pathways can regulate the transcription of a variety of inflammatory mediators. Extensive research studies on therapeutic drugs targeting the key signal molecules of the above signaling pathways have been performed for the prevention and treatment of SM injury. The mechanism of SM injury is complex, and the efficacy of a single drug is limited; therefore, there is a need to develop a combination therapy and comprehensive treatment measures to improve therapeutic outcomes.

13.4.4.5 SM Corneal Injury Treatment

The eye is the most sensitive organ to SM, and 75–90% of the population exposed to SM show varying degrees of acute and delayed eye involvement, of which SM keratopathy (corneal disease) is the most common. In recent years, with the awareness and attention on mustard gas keratopathy, various experts have attempted to resolve the technical problems associated with SM prevention and treatment. A rabbit SM cornea injury model has been established, of which the expression of matrix metalloproteinase-9 by immunohistochemistry, basement membrane electron microscopy, and immunofluorescence has been examined. The results showed that oxytetracycline or hydrogel accelerates the corneal healing rate of SM-induced damaged tissues,[87] reduces the expression of matrix metalloproteinase-9, recovers corneal thickness, and inhibits angiogenesis, thereby significantly improving wound healing. Previous studies have also shown that thymosin β4 accelerates corneal wound healing, as well as contributes to epidermal cell regeneration and anti-inflammatory mediator regulation, and thus can be used in the treatment of MGK and ocular trauma.[88]

In vitro studies have shown that thymosin β4 stimulates conjunctival epithelial cell migration, thereby leading to focal adhesion formation and extracellular laminin-5 deposition, and that increased cell migration may be associated with the production of fibronectin-5.[89] The most effective treatment of MGK to date is lamellar keratoplasty and limbal stem cell transplantation.[90] The respiratory tract, eyes, and skin are the main entry routes for SM absorption, and the pathophysiological mechanism of corneal injury is not clear, and its prevention and treatment remain major issues that are currently difficult to overcome.

13.4.4.6 Conclusions

In summary, SM is an erosive chemical agent that has no specific antidote, and thus remains a research topic of interest in the area of chemical defense in various countries. Because SM can rapidly undergo alkylation with various kinds of biomolecules *in vivo*, it is generally difficult to dissociate it from the bound state. Therefore, methods of removing free SM and blocking these binding sites to prevent further damage are key to the prevention and treatment of SM-induced oxidative stress. Antioxidant stress treatment may play a pivotal role in SM treatment, and an in-depth study of related drug(s) may facilitate in the development of SM antidotes.

13.4.5 Advances in Prophylactic Drugs and Treatment of NA Damage

13.4.5.1 Prophylactic Drugs

Currently, NA prophylactic drugs mainly include chemical drugs such as reversible AChE inhibitors, anticholinergic drugs, and reactivators. The use of reversible AChE inhibitors can protect AChE from being phosphorylated by organophosphorus compounds within a short time, which in turn may be targeted in the development of novel preventative schemes.[91] Biological drugs, namely biological scavengers, generate blood organic phosphorus molecules that lose activity prior to binding with AChE, and thus can be used as prophylactic agents.

13.4.5.1.1 AChE Inhibitors. Carbamate compounds can cause carbamylation of AChE active sites within a short period of time, thereby protecting it from inhibition by organophosphorus compounds. In the past, pyridostigmine bromide was considered as a NA poisoning preventive drug; however, it is incapable of traversing through the blood–brain barrier and thus cannot protect AChE in the CNS. To improve neuroprotective efficacy, pyridostigmine is often combined with other chemical drugs. Studies have shown that pyridostigmine may have some adverse reactions; for example, soldiers involved in the Gulf War developed adverse reactions that may be related to exposure to low-dose NA pesticides or the use of the prophylactic drug pyridostigmine.[92] Huperzine A (HupA) selectively and reversibly inhibits AChE and plays a major role in the prevention of NA poisoning. The combination of HupA and benzodiazepine drugs can enhance the protection of the central nervous system.

In addition, HupA does not impair cognitive function, thereby not resulting in any behavioral side effects. Galantamine is an AChE competitive inhibitor that acts as an allosteric modulator of nicotinic receptors, and it is less toxic than HupA.[93] Animal experiments have shown that it imparts a preventive effect on soman or sarin poisoned guinea pigs. However, whether HupA or galantamine and other reversible AChE inhibitors can be used as a universal protective agent against various types and modes of exposure to organophosphorus compounds remains unclear. Galantamine is used in the treatment of Alzheimer's disease. Previous studies have determined that galantamine could be used alone or in combination with atropine to neutralize immediate or delayed toxicity after the neurotoxic injury of guinea pigs of different ages and sexes. Male adolescent guinea pigs that received galantamine and atropine before and after exposure to lethal doses of neurotoxin or organophosphate pesticides were able to effectively neutralize the associated toxic effects. Administration of galantamine before and atropine after exposure to organophosphorus pesticide resulted in a two-fold decrease in the LD_{50}, with a survival rate of up to 100%.[94]

Prevention of NA poisoning may also be achieved by using a cholinesterase re-activator. Transdermal patches containing oxime HI-6 have been successfully tested in the field, and the Czech Republic and Slovak army has been equipped with cholinesterase activators and Benactyzine tablets for protection against major toxins.[95]

13.4.5.1.2 Biological Macromolecules. Biological macromolecules, such as most enzymes and proteins, have a certain role to play against organophosphorus compounds. Skin tissue and blood biological scavengers can play a role in detoxifying organophosphorus compounds in the body, even when the content of detoxification enzymes is equivalent to that of albumin (about 0.6 mmol L^{-1} albumin in the plasma and lymph fluid). Therefore, to increase the protective effect on the human body, an endogenous biological scavenger is required, and the ideal biological scavenger should be able to counteract five times that of the median lethal dose (LD_{50}), and all organophosphorus compounds are clearly antagonistic. Unlike drug prophylaxis, the iatrogenic toxicity of biological scavengers cannot be ignored. A large number of active enzymes may disrupt certain metabolic processes, and repeated administration of non-human enzymes may lead to adverse immunological reactions. The use of nanomaterial encapsulation or polyethylene glycol (PEG) modifications to reduce or prevent immune responses may thus be necessary.

13.4.5.1.3 BuChE. Animal experiments have shown that high doses of human BuChE can antagonize up to 5.5 times that of the LD_{50} of soman or 8 times that of the LD_{50} of VX,[96] however, the dose required may be too large. Early reports have shown that for protection against twice the LD_{50} of soman, 200 mg of BuChE is needed.[97] Human BuChE may be produced *in vivo* for a short time by using gene delivery methods. An adenovirus system can also be employed for the expression of human BuChE in mice, and mice with five times the LD_{50} for VX are capable of eliciting an excellent antagonistic effect.[98] However, before the actual application of gene therapy, there is a need to build a safe carrier or a vector that does not produce toxic viral proteins nor induce a detrimental immune response. Because excessive organic phosphorus can also inhibit AChE, the current combination of oxime compounds and atropine is not ideal for the treatment of high concentrations of organophosphate poisoning.

BuChE as an antidote has many advantages, including *in vivo* wide distribution, rapid response to NAs and organophosphate insecticides, long retention time *in vivo*, and the absence of an immune response despite repeated treatment; therefore, it can be used as a preventive and/or therapeutic drug,[99,100] mainly as a biological scavenger.

Recombinant HBuChE polyethylene glycol (Protexia, PEG–BuChE) can neutralize organophosphates and become an effective treatment after exposure to a lethal dose of neurotoxin. Protexta (Protexia, PEG–BuChE) can neutralize organophosphates and can be an effective treatment for exposures to lethal doses of a neurotoxin. Through a pre-loaded automatic injector, biologically active Protexia can be directly injected into the body, and in the case of a large number of wounded individuals, is highly suitable as a first aid treatment.[101] The therapeutic effects of Protexia have been initially verified on guinea pigs exposed to VX, and the results showed that 2 hours after the administration of Protexia, these animals were all still alive, whereas the control group showed 100% mortality. The treatment group also showed mild neurobehavioral lesions, thereby suggesting that Protexia has a large-scale protective effect against neurotoxins.[102]

13.4.5.1.4 Paraoxonase-1 (PON-1). Organophosphorus compounds (OPs) are the substrates for catalytic biological scavengers; therefore, the protective effect of small-dose catalytic biological scavengers is more effective than that of high-dose chemical-equivalent biological scavengers. Attempts have been made to study an enzyme that mutates human cholinesterase into degradable organophosphorus compounds. Researchers have constructed more than 60 mutants of BuChE and AChE.[103] The reported human enzymes include plasma paraoxonase-1 (PON-1), erythrocyte pro-peptidase, and human liver senescence marker (SMP-30).

There are very limited data on the enzymatic activity of SMP-30 hydrolyzing organophosphorus compounds. Human propeptidase exhibits a similar structure and catalytic properties as those of organophosphate dehydratase (OPAA)/proline. Although human propeptidase is easily expressed in *Escherichia coli* and that G-type NA[104] could be hydrolyzed at a high rate, studies involving this enzyme are still at a primary stage.

PON-1 is an effective catalytic biological scavenger that can be hydrolyzed at a very high rate. PON-1 levels of most organophosphorus compounds have a negative correlation with the poisoning sensitivity of organophosphorus compounds. For example, larvae expressing human PON-1 can protect >100 times against the LD_{50} exposure to chloramphenicol. In contrast, KO-mice expressing PON-1 are much more sensitive to organophosphorus compounds than wild-type mice. Although PON-1 hydrolase preferentially hydrolyzes low-toxicity isomers of soman, *in-vitro* studies using the AChE reverse titration method have demonstrated that PON-hydrolytic enzymes can effectively hydrolyze tabun. In addition, studies of the intravenous injection of purified human PON-1 into guinea pigs have shown that it is effective against sarin and soman poisoning.[105] Studies have shown that a 10–100-fold increase in the catalytic potency of PON-1 is sufficient for protection against NAs.

In addition, it has been reported that PON-1 can be obtained by gene transfer and combined with high-throughput screening, may after 5 passages generate a mutant that is capable of hydrolyzing most cyclosporine analogues, as well as having a 10 million-fold increase in catalytic efficiency, thereby protecting mice against the $1LD_{100}$ cyclosarin analogue. Purified plasma-derived PON-1 is difficult to obtain, and its stability depends on the presence of similar substances and a hydrophobic environment. Therefore, mass production of PON-1 from expired plasma is required to solve several technical problems. To overcome problems related to production stability and bioactive drug pharmacokinetics, gene delivery experiments using PON-1 have been conducted, which showed that mice that are capable of endogenously producing high levels of PON-1 can counteract 4-fold the LD_{50} doses of diazoxon (an organophosphate).[106] Therefore, it is expected that PON-1 mutants formed by safe gene delivery techniques against NAs may be promising catalytic biological scavengers.

13.4.5.1.5 AChE-induced/Stabilized Peptide. AChE, which disrupts cholinergic signaling, is a target of organophosphate neurotoxic agents, and its inactivation can cause rapid collapse or death. Recently, it has been found that amino-terminal peptides based on non-AChE catalytic subunits can stabilize newly synthesized AChE molecules and protect these against degradation in the skeletal muscle. When AChE-induced/stabilized peptides are injected into mice, this induces a significant increase in AChE activity, particularly of that between neuromuscular synapses,[107] thereby rapidly increasing the protective effect on twice the LD_{50} DFP-challenged animals. The above experiments have led to a new approach for enhancing the *in vivo* expression of AChE against organophosphate poisoning and accelerating healing.

13.4.5.2 Drug Treatment

For acute organophosphorus poisoning, routine first aid measures include the administration of anticholinergic and re-activator agents. At present, the most commonly used NA poisoning first-aid drugs include pyridine oxime compounds, which are represented by AChE reactivator, anticholinergic drugs represented by atropine, as well as benzodiazepine anticonvulsants such as diazepam. In addition, based on the need for a treatment for NA poisoning, the abovementioned drugs are usually placed in a carrier for automatic intramuscular injection. US military equipment for anti-NA automatic injection needles includes two syringes, one of the automatic syringes contains 600 mg of chlorpyrifos and the other contains 2 mg of atropine.[108] Among the various treatment drugs for the prevention and treatment of NA poisoning, anticholinergic drugs are mainly used for first aid and symptomatic treatment, cholinesterase protection agents are used for the prevention of poisoning, and re-activators are used in the treatment of NA poisoning. To date, most re-activators consist of a quaternary ammonium salt, which does not penetrate the blood–brain barrier, and thus is not useful for application to aged AChE.

13.4.5.2.1 Anticholinergic Drugs. Anticholinergic drugs are highly diverse and can be selected based on the type of poisoning and its progression. The more commonly used anticholinergic drugs include atropine sulfate, scopolamine, benzene trihexyphenidyl, carbamoxifen procyclidine, and pentenoyl chloride hydrochloride (Changtuoning). Atropine can antagonize peripheral muscarinic receptor-like symptoms of poisoning, but its antagonistic role in central muscarinic, nicotine receptors (M, N receptors) is weak. On the other hand, the central anticholinergic drug benzathine has a strong antagonistic role in the central MN receptor, whereas its antagonistic effect of peripheral M-like symptoms is weak. A combination of the two can produce the best anticholinergic effects. Scopolamine has been successfully used for the anticholinergic treatment of soman intoxication. However, the anticonvulsant effects of atropine and scopolamine are limited to early stages of seizures (0–40 min after the onset of seizures). The CNS anticholinergics trihexyphenidyl, scopolamine, cycloprofen bazatropine (benzatropine), and piperidantin inhibit AChE in the CNS and can cause memory and cognitive decline; therefore, there is a need to combine drugs to reduce the adverse reactions. Benthiactzine is a diphenylacetic acid derivative that plays an antagonistic role in both the M and N receptors. The soman–tabun VX-induced circulatory failure model in dogs, dichlorvos of parathion-induced circulatory failure in rats, and other experiments have confirmed that benthiactzine is a rapid and effective treatment for cholinesterase inhibitor (ChEI) poisoning-induced circulatory failure.

13.4.5.2.2 Cholinesterase Re-activator. Oxime cholinesterase re-activators consist of three generations. The first generation consists of pyridine formaldehyde oxime, and its representative drug is pralidoxime. The second generation includes bis-quaternary ammonium pyridine aldehyde oxime, and its representative drug is obidoxime. The third generation includes the H series and the HGG and BDB classes, and the representative drugs are HI-6 and HLö-7. Chlorpheniramine (PAM-Cl) has strong activity and low toxicity and is currently the preferred re-activator in China. Methotrexate is a phosphate solution of methanesulfonate with poor water solubility that is mainly used in the United Kingdom. Obidoxime (DMO4) has strong re-activation effects, is more toxic, and is commonly used in European countries.

Trimedoxime (TMB4) has the strongest re-activation, but it is also highly hepatotoxic and is still used in the United States and other NATO countries. HI-6 is a bipyridine quaternary ammonium salt re-activator that is more effective, with lower toxicity, and highly water soluble. However, its aqueous solution is very unstable and thus it has not been widely used in clinical settings. Compared to pralidoxime and obidoxime, it has obvious re-activation advantages on sarin, cyclosarin, soman, VX, and other NA poisons.

In vitro re-activation assays have shown that the re-activation k_r^2 value of HI-6 on AChE was 1.003 L (mmol min)$^{-1}$ in the brain of sarin-poisoned rats, whereas k_r of cyclosarin poisoning was as high as 29.167 L (mmol min)$^{-1}$.[109] Meanwhile, the k_r values of HI-6 on human AChE that are inhibited by sarin, VX, cyclosarin, and tabun are 13.5, 20.9, 27.5 and 0 L (mmol min)$^{-1}$, respectively. HI-6 has a significantly higher re-activation k_r^2 value in human AChE that is inhibited by cyclosporin than that of obidoxime (0.42 L mmol^{-1} min),[110] and re-activation of soman-inhibited human erythrocyte AChE was as high as (31.61 ± 1.02)%. However, HI-6 did not re-activate the AChE that was inhibited by tabun.[111] However, *in vitro* experiments have shown that HI-6 and obidoxime impart a combined re-activation effect on tabun-inhibited AChE that was similar to that of obidoxime alone.[112] In addition, methyl oxime (MMB-4), K series oxime, O-oxime, and HLö-7 are in the developmental stage of testing.

13.4.5.2.3 Re-activators that Penetrate the BBB. Oxime compounds (such as HI-6) can effectively re-activate AChE, but these have a very low CNS penetration rate and can hardly penetrate the BBB, and are thus incapable of effectively re-activating AChE in the case of CNS poisoning. How the efficiency of oxime compound penetration through the BBB can be improved has been a research topic of interest in the field of drug discovery for the treatment of NA damage. Oxime compounds are specific substrates of the P-glycoprotein (P-gp) efflux pump, which is located along the BBB. Joosen *et al.*[113] intravenously injected a non-competitive specific substrate of P-gp, tariquidar, into rats, and by microdialysis technology quantitatively determined that the brain HI-6 levels increased by two-fold, whereas those in the plasma and the control group did not differ significantly. In addition, the administration of HI-6 and atropine to rats pretreated with tariquidar almost completely prevented the occurrence of soman-induced seizures. Recent studies on three oximes, isopropyl nitroacetone (MINA), diacetylmonooxime (DAM), and diethylaminoketoxime, have shown that these can re-activate AChE in the CNS to prevent convulsions in NA-infected animals, improve animal survival, and prevent behavioral reactions after poisoning.

Broad-spectrum non-bis-quaternary amine oxime research has also made progress, wherein chloramphenicol precursors have been shown to pass through the BBB, as well as rapidly change into chlorpyrifos in the brain, thereby preventing sarin- and VX-induced animal seizures. Researchers have synthesized new glyoxal derivatives of pyridinoxime, whose ability to pass through the BBB *via* glucose transporters requires further evaluation.[114] The newly synthesized derivatives of 4-alkoxybenzaldehyde oximes were connected to neutral peripheral ligands *via* ethylene glycol linkers, and AChE amidine-oxime, which could be reactivated by NA, and was able to permeate the BBB. *In vivo* experiments have shown that these compounds have neuroprotective effects on soman and sarin-challenged mice.

In terms of the ability to reactivate human AChE inhibited by NAs and insecticides, uncharged derivatives of newly synthesized phenyl-tetraisoquinoline are similar or superior to pyridyloxime-based compounds. All of these new activators are expected to become new drugs that can penetrate the BBB.

13.4.5.2.4 Other Re-activator Drugs. Pyridine-based re-activators have been shown to re-activate phosphorylated AChE, but cannot effectively re-activate BChE, which is similarly inhibited. TAB_2OH was screened from a modified compound based on edrophonium chloride, and its activation *in vitro* on inhibited hBuChE was found to be 3–5 times higher than that of pralidoxime chloride. TAB_2OH and the chemical equivalent of hBuChE can increase the protection index by 30–60%, as well as reduce the dose of BuChE. Among the oxime- type re-activators synthesized based on imidazoles, the quaternary imidazole oximes have a faster re-activation rate than the tertiary amine analogs, and even faster than previously reported TAB_2OH. A VX-BuChE re-activating compound was screened from such compounds; however, its activation of OP-hAChE is relatively slow.

13.4.5.2.5 Anticonvulsants and Neuroprotective Agents. Previous studies have confirmed that although benzodiazepine drugs have a clear anticonvulsant effect, these also have limitations such as recurrence and poor neuroprotection after the convulsions have been controlled. In recent years, research teams from around the world have utilized glutamate receptor antagonists to antagonize NA-induced seizures. These studies have included the use of glutamate receptor antagonists, as well as drugs with dual roles as cholinergic receptor and glutamate receptor to antagonize soman-induced seizures that may result in brain damage.

13.4.5.3 Prospects

Despite the variety of existing NA prevention and treatment drugs that can control the effects of chemical poisons, several other issues remain unresolved. The traditional treatment of organophosphate poisoning focuses on the acute signs of AChE accumulation in neural synapses after poisoning, but it has not yet considered the involvement of other signaling pathways. Evidence suggests that after AChE inhibition, non-cholinergic signals are recruited and play a role in long-term neurological deficits. Early accumulation of other neurotransmitters such as serotonin in the hippocampus after organophosphate poisoning may lead to subsequent long-term changes in the serotonergic signals, which in turn leads to permanent neurobehavioral disturbances and cognitive changes in mood. Nallapaneni *et al.*[115] reported that acute organophosphate exposure initially activates excess hippocampal serotonergic signals, which in turn leads to persistent neurochemical and behavioral changes. Early endocannabinoid(eCB) signal amplification can be blocked by eCB hydrolase, which in turn reduces acute and long-term toxic effects. eCBs are endogenous neurotransmitters expressed by depolarizing neurons that reverse the release of various neurotransmitters from presynaptic terminals. Preliminary studies have demonstrated that eCB-degrading enzyme inhibitors have a significant protective effect in acute DFP poisoning. Future NA drug research studies should focus on the following:

(1) Better use of existing drugs, including the application of new administration routes such as transdermal and nasal administration, to achieve rapid drug administration at poisoning sites; develop a more targeted treatment regime to achieve the rationalization of drug combination;
(2) Based on a deep understanding of the AChE spatial structure, using computer simulation technology to optimize the synthesis of stronger cholinesterase re-activators; study a new drug delivery system targeting the CNS to address the problem of re-activating poisoned enzymes in the CNS;

(3) Based on the latest advances in biotechnology research, conduct highly effective biological scavenger research studies; explore the application of gene therapy and cell therapy in the recovery of injured neurons.

13.5 Conclusions

The convergence between the life sciences and chemistry is often viewed predominantly as a concern because it may lead to the development of novel forms of chemical weapons agents. However, as this chapter has demonstrated, there are also many beneficial implications of the convergence, for example in the development of new means of prevention and treatment of the damage caused by chemical weapons based on our growing understanding of the mechanisms of action of these agents, and of how the damage they cause to living systems can be prevented and moderated.

References

1. Q. Q. Wang, R. A. Begum, V. W. Day and K. Bowman-James, Sulfur, oxygen, and nitrogen mustards: stability and reactivity, *Org. Biomol. Chem.*, 2012, **10**(44), 8786.
2. K. Ghabili, P. S. Agutter and M. Ghanei, *et al.*, Sulfur mustard toxicity: history, chemistry, harmacokinetics, and pharmacodynamics, *Crit. Rev. Toxicol.*, 2011, **41**(5), 384–403.
3. R. N. Saladi, E. Smith and A. N. Persaud, Mustard: a potential agent of chemical warfare and terrorism, *Clin. Exp. Dermatol.*, 2006, **31**(1), 1–5.
4. A. Baradaran-Rafii, M. A. Javadi and M. Rezaei Kanavi, *et al.*, Limbal stem cell deficiency in chronic and delayed-onset mustard gas keratopathy, *Ophthalmology*, 2010, **117**(2), 246–252.
5. H. Ghasemi, T. Ghazanfari and M. Babaei, *et al.*, Long-term ocular complications of sulfur mustard in the civilian victims of Sardasht, Iran, *Cutaneous Ocul. Toxicol.*, 2008, **27**(4), 317–326.
6. N. Lagali and P. Fagerholm, Delayed mustard gas keratitis: clinical course and *in vivo* confocal microscopy findings, *Cornea*, 2009, **28**(4), 458–462.
7. Y. H. Jan, D. E. Heck, R. Malaviya, R. P. Casillas, D. L. Laskin and J. D. Laskin, Cross-linking of thioredox in reductase by the sulfur mustard analogue mechlorethamine (methylbis (2- chloroethyl) amine) in human lung epithelial cells and rat lung:selective inhibition of disulfide reduction but not redox cycling, *Chem. Res. Toxicol.*, 2014, **27**(1), 61–75.
8. L. B. Joseph, D. E. Heck, J. A. Cervelli, G. M. Composto, M. C. Babin and R. P. Casillas, *et al.*, Structural changes in hair follicles and sebaceous

glands of hairless mice following exposure to sulfur mustard, *Exp. Mol. Pathol.*, 2014, **96**(3), 316–327.

9. A. Shiyovich, Y. Rosman, A. Krivoy, L. Statlender, M. Kassirer and S. Shrot, Long-term complications of sulfur mustard exposure: a therapeutic update, *Harefuah*, 2014, **153**(3–4), 199–205.
10. M. Batal, I. Boudry, S. Mouret, C. Cléry-Barraud, J. Wartelle and I. Bérard, *et al.*, DNA Damage in internal organs after cutaneous exposure to sulphur mustard, *Toxicol. Appl. Pharmacol.*, 2014, **278**(1), 39–44.
11. M. Asgher, N. A. Khan, M. I. Khan, M. Fatma and A. Masood, Ethylene production is associated with alleviation of cadmium-induced oxidative stress by sulfur in mustard types differing in ethylene sensitivity, *Ecotoxicol. Environ. Saf.*, 2014, **106**, 54–61.
12. V. R. Sunil, K. Patel-Vayas, J. Shen, A. J. Gow, J. D. Laskin and D. L. Laskin, Role of TNFR1 in lung injury and altered lung function induced by the model sulfur mustard vesicant, 2-chloroethyl ethyl sulfide, *Toxicol. Appl. Pharmacol.*, 2011, **250**(3), 245–255.
13. R. Yaraee, Z. M. Hassan, S. Pourfarzam, A. Rezaei, S. Faghihzadeh and M. Ebtekar, *et al.*, Fibrinogen and inflammatory cytokines in spontaneous sputum of sulfur-mustard-exposed civilians-Sardasht-Iran Cohort Study, *Int. Immunopharmacol.*, 2013, **17**(3), 968–973.
14. S. Yazdani, M. H. Karimfar, A. A. Imani Fooladi, L. Mirbagheri, M. Ebrahimi and M. Ghanei, *et al.*, Nuclear factor κB1/RelA mediates the inflammation and/or survival of human airway exposed to sulfur mustard, *J. Recept. Signal Transduction Res.*, 2011, **31**(5), 367–373.
15. M. H. Boskabady, N. Vahedi, S. Amery and M. R. Khakzad, The effect of Nigella sativa alone, and in combination with dexamethasone, on tracheal muscle responsiveness and lung inflammation in sulfur mustard exposed guinea pigs, *J. Ethnopharmacol.*, 2011, **137**(2), 1028–1034.
16. M. H. Boskabady, S. Amery, N. Vahedi and M. R. Khakzad, The effect of vitamin E on tracheal responsiveness and lung inflammation in sulfur mustard exposed guinea pigs, *Inhalation Toxicol.*, 2011, **23**(3), 157–165.
17. A. Emad and Y. Emad, Increased granulocytecolony stimulating factor (G-CSF) and granulocyte-macrophage colony stimulating factor (GM-CSF) levels in BAL fluid from patients with sulfur mustard gas-induced pulmonary fibrosis, *J. Aerosol Med.*, 2007, **20**(3), 352–360.
18. R. Malaviya, V. R. Sunil and J. Cervelli, *et al.*, Inflammatory effects of inhaled sulfur mustard in rat lung, *Toxicol. Appl. Pharmacol.*, 2010, **248**(2), 89–99.
19. B. Rebholz, K. Kehe and T. Ruzicka, *et al.*, Role of NF-kappaB/RelA and MAPK pathways in keratinocytes in response to sulfur mustard, *J. Invest. Dermatol.*, 2008, **128**(7), 1626–1632.
20. H. Y. Ham, C. W. Hong and S. N. Lee, *et al.*, Sulfur mustard primes human neutrophils for increased degranulation and stimulates cytokine release *via* TRPM2/p38 MAPK signaling, *Toxicol. Appl. Pharmacol.*, 2012, **258**(1), 82–88.

21. J. F. Dillman, K. L. McGary and J. J. Schlager, An inhibitor of p38 MAP kinase downregulates cytokine release induced by sulfur mustard exposure in human epidermal keratinocytes, *Toxicol. In Vitro*, 2004, **18**(5), 593–599.
22. R. Ray, B. Keyser and B. Benton, *et al.*, Sulfur mustard induces apoptosis in cultured normal human airway epithelial cells: evidence of a dominant caspase-8-mediated pathway and differential cellular responses, *Drug Chem. Toxicol.*, 2008, **31**(1), 137–148.
23. D. R. Anderson, S. L. Taylor and D. P. Fetterer, *et al.*, Evaluation of protease inhibitors and an antioxidant for treatment of sulfur mustard-induced toxic lung injury, *Toxicology*, 2009, **263**(1), 41–46.
24. F. M. Cowan, C. A. Broomfield and W. J. Smith, Suppression of sulfur mustard-increased IL-8 in human keratinocyte cell cultures by serine protease inhibitors: implications for toxicity and medical countermeasures, *Cell Biol. Toxicol.*, 2002, **18**(3), 175–180.
25. B. M. Keyser, D. K. Andres, E. Nealley, W. W. Holmes, B. Benton and D. Paradiso, *et al.*, Postexposure application of Fas receptor small-interfering RNA to suppress sulfur mustard-induced apoptosis in human airway epithelial cells: implication for a therapeutic approach, *J. Pharmacol. Exp. Ther.*, 2013, **344**(1), 308–316.
26. C. Gu, J. Zhang and Y. Chen, *et al.*, A trigger model of apoptosis induced by tumor necrosis factor signaling, *BMC Syst. Biol.*, 2011, **5**(suppl. 1), S13.
27. C. J. Kearney, S. P. Cullen and G. A. Tynan, *et al.*, Necroptosis suppresses inflammation *via* termination of TNF- or LPS-induced cytokine and chemokine production, *Cell Death Differ.*, 2015, **22**(8), 1313–1327.
28. G. Pirzad, M. Jafari, S. Tavana, H. Sadrayee, S. Ghavami and A. Shajiei, *et al.*, The role of Fas-FasL signaling pathway in induction of apoptosis in patients with sulfur mustard-induced chronic bronchiolitis, *J. Toxicol.*, 2010, **2010**, 373612.
29. X. J. Zhu, R. Xu, X. Meng, H. B. Chu, C. Zhao and C. J. Lian, *et al.*, Mechanistic insights of sulfur mustard-induced acute tracheal injury in rats, *Int. J. Toxicol.*, 2014, **33**(5), 382–392.
30. R. Ray, B. Keyser, B. Benton, A. Daher, C. M. Simbulan-Rosenthal and D. S. Rosenthal, Sulfur mustard induces apoptosis in cultured normal human air way epithelial cells: evidence of a dominant caspase-8-mediated pathway and differential cellular responses, *Drug Chem. Toxicol.*, 2008, **31**(1), 137–148.
31. R. Ray, C. M. Simbulan-Rosenthal, B. M. Keyser, B. Benton, D. Anderson and W. Holmes, *et al.*, Sulfur mustard induces apoptosis in lung epithelial cells *via* a caspase amplification loop, *Toxicology*, 2010, **271**(3), 94–99.
32. D. S. Rosenthal, A. Velena and F. P. Chou, *et al.*, Expression of dominant-negative Fas-associated death domain blocks human keratinocyte apoptosis and vesication induced by sulfur mustard, *J. Biol. Chem.*, 2003, **278**(10), 8531–8540.

33. K. Kehe, K. Raithel and H. Kreppel, *et al.*, Inhibition of poly(ADP-ribose) polymerase (PARP) influences the mode of sulfur mustard (SM)-induced cell death in HaCaT cells, *Arch. Toxicol.*, 2008, **82**, 461–470.
34. U. Wormser, B. Brodsky and E. Proscura, *et al.*, Involvement of tumor necrosis factor-alpha in sulfur mustard-induced skin lesion; effect of topical iodine, *Arch. Toxicol.*, 2005, **79**(11), 660–670; K. Kehe, K. Raithel, H. Kreppel, M. Jochum, F. Worek and H. Thiermann, *Arch. Toxicol.*, 2008, **82**, 461.
35. A. L. Ruff and J. F. Dillman, Signaling molecules in sulfur mustard-induced cutaneous injury, *Eplasty*, 2007, **8**, e2.
36. E. Behravan, S. A. Moallem, S. Khateri, E. Maraghi, P. Jowsey and P. G. Blain, *et al.*, Deoxyribonucleic acid damage in Iranian veterans 25 years after war time exposure to sulfur mustard, *J. Res. Med. Sci.*, 2013, **18**(3), 239–244.
37. P. A. Jowsey, F. M. Williams and P. G. Blain, D. N. A. Damage responses in cells exposed to sulphur mustard, *Toxicol. Lett.*, 2012, **209**(1), 1–10.
38. I. Khaheshi, S. Keshavarz, A. A. Imani Fooladi, M. Ebrahimi, S. Yazdani and Y. Panahi, *et al.*, Loss of expression of TGF-βs and their receptors in chronic skin lesions induced by sulfur mustard as compared with chronic contact dermatitis patients, *BMC Dermatol.*, 2011, **11**, 2.
39. J. Moser, C. F. Levine, D. R. Thomas-Dunmeyer and W. J. Smith, Cross-linking interferes with assessing sulfur mustard-induced DNA damage in human peripheral blood lymphocytes using the comet assay, *Toxicol. Mech. Methods*, 2004, **14**(3), 195–202.
40. M. Blaha, J. Kohl, D. DuBose, W. Bowers Jr and J. Walker, Ultrastructural and histological effects of exposure to CEES or heat in a human epidermal mode, *In Vitro Mol. Toxicol.*, 2001, **14**(1), 15–23.
41. M. Pohanka, Antioxidants countermeasures against sulfur mustard, *Mini-Rev. Med. Chem.*, 2012, **12**(8), 742–748.
42. M. R. Nourani, M. Ebrahimi, M. H. Roudkenar, E. Vahedi, M. Ghanei and A. A. Imani Fooladi, Sulfur mustard induces expression of metallothionein-1A in human airway epithelial cells, *Int. J. Gen. Med.*, 2011, **4**, 413–419.
43. M. Ghanei and A. A. Harandi, Molecular and cellular mechanism of lung injuries due to exposure to sulfur mustard: a review, *Inhalation Toxicol.*, 2011, **20**(7), 363–371.
44. M. Jafari and M. Ghanei, Evaluation of plasma, erythrocytes, and bronchoalveolar lavage fluid antioxidant defense system in sulfur mustard-injured patients, *Clin. Toxicol.*, 2010, **48**(3), 184–192.
45. M. Shohrati, M. Ghanei, N. Shamspour, F. Babaei, M. N. Abadi and M. Jafari, *et al.*, Glutathione and malondialdehyde levels in late pulmonary complications of sulfur mustard intoxication, *Lung*, 2010, **188**(1), 77–83.
46. M. R. Nourani, S. Azimzadeh, M. Ghanei and A. A. Imani Fooladi, Expression of glutathione S-transferase variants in human airway wall after long-term response to sulfur mustard, *J. Recept. Signal Transduction Res.*, 2014, **34**(2), 125–130.

47. S. D. Sawale, P. D. Ambhore, P. P. Pawar, U. Pathak, U. Deb and R. M. Satpute, Ameliorating effect of S-2(ω-aminoalkylamino)alkylaryl sulfide(DRDE-07) on sulfur mustard analogue, 2-chloroethyl ethyl sulfide-induced oxidative stress and inflammation, *Toxicol. Mech. Methods*, 2013, **23**(9), 702–710.
48. L. Golmanesh, F. Bahrami, F. Pourali, E. Vahedi, H. Wahhabaghai and H. Mehrani, *et al.*, Assessing the relationship of paraoxonase-1 Q192R polymor phisms and the severity of lung disease in SM-exposed patients, *Immunopharmacol. Immunotoxicol.*, 2013, **35**(3), 419–425.
49. A. Taravati, S. K. Ardestani, A. A. Ziaee, A. Ghorbani, M. R. Soroush and S. Faghihzadeh, *et al.*, Effects of paraoxonase 1 activity and gene polymorphisms on long-term pulmonary complications of sulfur mustard-exposed veterans, *Int. Immunopharmacol.*, 2013, **17**(3), 974–979.
50. Y. C. Chang, J. D. Wang, K. K. Svoboda, R. P. Casillas, J. D. Laskin and M. K. Gordon, *et al.*, Sulfur mustard induces an endoplasmic reticulum stress response in the mouse ear vesicant mode, *Toxicol. Appl. Pharmacol.*, 2013, **268**(2), 178–187.
51. J. Riches, R. W. Read and R. M. Black, Analysis of the Sulphur mustard metabolites thiodiglycol and thiodiglycol sulphoxide in urine using isotope-dilution gas chromatography-ion trap tandem mass spectrometry, *J. Chromatogr. B*, 2007, **845**(1), 114–120.
52. R. W. Read and R. M. Black, Analysis of betalyase metabolites of sulfur mustard in urine by electrospray liquid chromatography-tandem mass spectrometry,*J. Anal. Toxicol.*, 2004, **28**(5), 346–351.
53. P. A. Jowsey, F. M. Williams and P. G. Blain, DNA damage responses in cells exposed to sulphur mustard, *Toxicol. Lett.*, 2012, **209**(1), 1–10.
54. R. Yaraee, T. Ghazanfari and M. Ebtekar, *et al.*, Alterations in serum levels of inflammatory cytokines(TNF, IL-1alpha, IL-1beta and IL-1Ra)20 years after sulfur mustard exposure:Sardasht-Iran cohort study, *Int. Immunopharmacol.*, 2009, **9**(13–14), 1466–1470.
55. D. Attaran, S. M. Lari and M. Towhidi, *et al.*, Interleukin-6andairflow limitation in chemical warfare patients with chronicobstructive pulmonary disease, *Int. J. Chronic Obstruct. Pulm. Dis.*, 2010, **5**, 335–340.
56. H. Ghasemi, T. Ghazanfari and R. Yaraee, *et al.*, Evaluation of thetear and serum levels of IL-8 in sulfur mustard intoxicated patients 20 years after exposure, *Cutaneous Ocul. Toxicol.*, 2012, **31**(2), 132–137.
57. Y. Panahi, S. M. Davoudi and F. Beiraghdar, *et al.*, Relationship between levels of IFNγ, TNFα, and TGFβ and pruritus in sulfur mustard-exposed veterans,*J. Immunotoxicol.*, 2013, **10**(2), 173–177.
58. S. Amiri, T. Ghazanfari and R. Yaraee, *et al.*, Serum levels of GMCSF 20years after sulfur mustard exposure : Sardasht-Iran Cohort Study, *Int. Immunopharmacol.*, 2009, **9**(13–14), 1499–1503.
59. A. Emad and Y. Emad, Increased granulocyte-colony stimulatingfactor (G-CSF) and granulocyte-macrophage colony stimulating factor (GM-CSF) levels in BAL fluid from patients with sulfur mustard gas-induced pulmonary fibrosis,*J. Aerosol Med.*, 2007, **20**(3), 352–360.

60. Y. Panahi, S. M. Davoudi and F. Beiraghdar, *et al.*, Relationship between levels of IFNγ, TNFα, and TGFβ and pruritus in sulfur mustard-exposed veterans, *J. Immunotoxicol.*, 2013, **10**(2), 173–177.
61. N. B. Munro, S. S. Talmage and G. D. Griffin, *et al.*, The sources, fate, and toxicity of chemical warfare agent degradation products, *Environ. Health Perspect.*, 1999, **107**, 933–974.
62. H. Koskela, M. Ervasti, H. Björk and P. Vanninen, On-flow pulsed field gradient heteronuclear correlation spectrometry in off-line LC-SPE-NMR analysis of chemicals related to the chemical weapons convention, *Anal. Chem.*, 2009, **81**(3), 1262–1269.
63. Y. R. Shu, A. K. Su, J. T. Liu and C. H. Lin, Screening of nerve agent degradation products by MALDI-TOFMS, *Anal. Chem.*, 2006, **78**(13), 4697–4701.
64. D. M. Hui and M. Minami, Monitoring of fluorine in urine samples of patients involved in the Tokyo sarin disaster, in connection with the detection of other decomposition products of sarin and the by-products generated during sarin synthesis, *Clin. Chim. Acta*, 2000, **302**(1/2), 171–188.
65. M. Kataoka and Y. Seto, Discriminative determination of alkyl methylphosphonates and methylphosphonate in blood plasma and urine by gas chromatography-mass spectrometry after *tert*-butyldimethylsilylation, *J. Chromatogr. B*, 2003, **795**(1), 123–132.
66. J. Riches, I. Morton, R. W. Read and R. M. Black, The trace analysis of alkyl alkylphosphonic acids in urine using gas chromatography-ion trap negative ion tandem mass spectrometry, *J. Chromatogr. B*, 2005, **816**(1/2), 251–258.
67. H. Tsuchihashi, M. Katagi, M. Nishikawa and M. Tatsuno, Identification of Metabolites of Nerve Agent VX in Serum Collected from a Victim, *J. Anal. Toxicol.*, 1998, **22**(5), 383–388.
68. N. H. Williams, J. M. Harrison, R. W. Read and R. M. Black, Phosphylated tyrosine in albumin as a biomarker of exposure to organophosphorus nerve agents, *Arch. Toxicol.*, 2007, **81**(9), 627–639.
69. R. W. Read, J. R. Riches, J. A. Stevens, S. J. Stubbs and R. M. Black, Biomarkers of organophosphorus nerve agent exposure: comparison of phosphylated butyrylcholinesterase and phosphylated albumin after oxime therapy, *Arch. Toxicol.*, 2010, **84**(1), 25–36.
70. P. V. R. K. Ramacharyulu, G. K. Prasad and K. Guanesan, *et al.*, Photocatalytic decontamination of surlfur mustard using titania nanomaterials, *J. Mol. Catal. A: Chem.*, 2012, **353–354**, 132–137.
71. P. V. R. K. Ramacharyulu, J. P. Kumar and G. K. Prasad, *et al.*, Sunlight of assisted photocatalytic detoxification of sulfur mustard on Vanadium ion doped titania nanocatalysts, *J. Mol. Catal. A: Chem.*, 2014, **387**, 38–44.
72. A. Korkmaz, Z. I. Kunak and S. D. Paredes, *et al.*, The use of melatonin to combat mustard toxicity, *Neuro Endocrinol. Lett.*, 2008, **29**(5), 614–619.

73. M. Pohanka, R. Stetina and H. Svobodova, *et al.*, Sulfur mustard causes oxidative stress and depletion of antioxidants in muscles, livers, and kidneys of Wistar rats, *Drug Chem. Toxicol.*, 2013, **36**(3), 270–276.
74. M. Pohanka, J. Sobotka and M. Jilkova, *et al.*, Oxidative stress after sulfur mustard intoxication and its reduction by melatonin: efficacy of antioxidant therapy during serious intoxication, *Drug Chem. Toxicol.*, 2011, **34**(1), 85–91.
75. P. E. Wilde and D. G. Upshall, Cysteine esters protect cultured rodent lung slices from sulphur mustard, *Hum. Exp. Toxicol.*, 1994, **13**(11), 743.
76. K. L. Powell, S. Boulware and H. Thames, *et al.*, 2,6 -Dithiopurine blocks toxicity and mutagenesis in human skin cells exposed to sulfur mustard analogues, 2-chloroethyl ethyl sulfide and 2-chloroethyl methyl sulfide, *Chem. Res. Toxicol.*, 2010, **23**(3), 497–503.
77. S. Boulware, T. Fields and E. McIvor, *et al.*, 2, 6-Dithiopurine, a nucleophilic scavenger, protects against mutagenesis in mouse skin treated *in vivo* with 2-(chloroethyl) ethyl sulfide, a mustard gas analog, *Toxicol. Appl. Pharmacol.*, 2012, **263**(2), 203–209.
78. N. P. Praetorius and T. K. Mandal, Alternate delivery route for amifostine as a radio-/chemo-protecting agent, *J. Pharm. Pharmacol.*, 2008, **60**(7), 809–815.
79. P. Kumar, A. Gautam and C. J. Prakash, *et al.*, Ameliorative effect of DRDE 07 and its analogues on the systemic toxicity of sulphur mustard and nitrogen mustard in rabbit, *Hum. Exp. Toxicol.*, 2010, **29**(9), 747–755.
80. A. Gautam and R. Vijayaraghavan, DRDE-07: a possible antidote for sulphur mustard toxicity, *Cell. Mol. Biol. (Noisy-le-grand)*, 2010, **56**(suppl.), OL1334–OL1340.
81. R. C. Siow, F. Y. Li and D. J. Rowlands, *et al.*, Cardiovascular targets for estrogens and phytoestrogens: transcriptional regulation of nitric oxide synthase and antioxidant defense genes, *Free Radical Biol. Med.*, 2007, **42**(7), 909–925.
82. G. E. Mann, B. Bonacasa and T. Ishii, *et al.*, Targeting the redox sensitive Nrf 2-Keap1 defense pathway in cardiovascular disease: protection afforded by dietary isoflavones, *Curr. Opin. Pharmacol.*, 2009, **9**(2), 139–145.
83. R. Vijayaraghavan, A. Gautam and M. Sharma, *et al.*, Comparative evaluation of some flavonoids and tocopherol acetate against the systemic toxicity induced by sulphur mustard, *Indian J. Pharmacol.*, 2008, **40**(3), 114–120.
84. M. H. Boskabady, S. Amery and N. Vahedi, *et al.*, The effect of vitamin E on tracheal responsiveness and lung inflammation in sulfur mustard exposed guinea pigs, *Inhalation Toxicol.*, 2011, **23**(3), 157–165.
85. R. Bhattacharya, R. K. Tulsawani and R. Vijayaraghavan, Effect of trolox and quercetin on sulfur mustard-induced cytotoxicity in human peripheral blood lymphocytes, *Indian J. Pharmacol.*, 2006, **38**(1), 38–42.

86. A. Gautam and R. Vijayaraghavan, Prophylactic effect of gossypin against percutaneously administered sulfur mustard, *Biomed. Environ. Sci.*, 2007, **20**(3), 250–259.
87. S. S. Anumolu, A. S. DeSantis and A. R. Menjoge, *et al.*, Doxycycline loaded poly (ethylene glycol) hydrogels for healing vesicant-induced ocular wounds[ADJ], *Biomaterials*, 2010, **31**(5), 964–974.
88. M. Badamchian, A. A. Damavandy and H. Damavandy, *et al.*, Identification and quantification of thymosin beta 4 in human saliva and tears, *Ann. N. Y. Acad. Sci.*, 2007, **111**(2), 458–465.
89. G. Sosne, P. Qiu and M. Kurpakus-Wheater, Thymosin beta 4: a novel corneal wound healing and anti -inflammatory agent, *Clin. Ophthalmol.*, 2007, **1**(3), 201–207.
90. M. R. Jafarinasab, S. Feizi and M. A. Javadi, *et al.*, Lamellar keratoplasty and keratolimbal allograft for mustard gas keratitis, *Am. J. Ophthalmol.*, 2011, **152**(6), 925–932.
91. S. Eckert, P. Eyer and F. Worek, Reversible inhibition of acetylcholinesterase by carbamates or hyperzine A increases residual activity of the enzyme upon soman challenge, *Toxicology*, 2007, **233**(1–3), 180–186.
92. B. A. Golomb, Acetylcholinesterase inhibitors and Gulf War illnesses, *Proc. Natl. Acad. Sci. U. S. A.*, 2008, **105**(1), 4295–4300.
93. E. F. Pereira, Y. Aracava and M. Alkondon, *et al.*, Molecular and cellular actions of galantamine: clinical implications for treatment of organophosphorus poisoning, *Mol. Neurosci.*, 2010, **40**(1–2), 196–203.
94. R. P. Gullapalli, Y. Aracava and J. Zhuo, *et al.*, Magnetic resonance imaging reveals that galantamine prevents structural brain damage induced by an acute exposure of guinea pigs to soman, *Neurotoxicology*, 2010, **31**(1), 67–76.
95. J. Bajgar, J. Fusek and J. Kassa, *et al.*, Chemical aspects of pharmacological prophylaxis against nerve agent poisoning, *Curr. Med. Chem.*, 2009, **16**(23), 2977–2986.
96. A. Saxena, W. Sun and J. M. Fedorko, *et al.*, Prophylaxis with human serum butyrylcholinesterase protects guinea pigs exposed to multiple lethal doses of soman or VX, *Biochem. Pharmacol.*, 2011, **81**(1), 164–169.
97. K. Parikh, E. G. Duysen and B. Snow, *et al.*, Genedelivery butyrylcholinesterase is prophylactic against the toxicity of chemical warfare nerve agents and organophosphorus compounds, *Pharmacol. Exp. Ther.*, 2011, **337**(1), 92–101.
98. C. Kronman, O. Cohen and O. Mazor, *et al.*, Next generation of OP-bioscavengers: a circulatory long-lived 4-PEG-hypolysine mutant of F338A-huAChE with optimal pharmacokinetics and pseudocatalytic activity, *Chem. Biol. Interact.*, 2010, **187**(1–3), 253–258.
99. M. Dwyer, S. Javor and D. A. Ryan, *et al.*, Novel human butyrylcholinesterase variants: toward organophosphonate detoxication, *Biochemistry*, 2014, **53**(27), 4476–4487.

100. D. M. Cerasoli, E. M. Griffiths and B. P. Doctor, *et al., In vitro* and *in vivo* characterization of recombinant human butyrylcholinesterase (Protexia) as a potential nerve agent bioscavenger, *Chem. Biol. Interact.*, 2005, **157–158**, 363–365.
101. Y. J. Huang, Y. Huang and H. Baldassarre, *et al.*, Recombinant human butyrylcholinesterase from milk of transgenic animals to protect against organophosphate poisoning, *Proc. Natl. Acad. Sci. U. S. A.*, 2007, **104**(34), 13603–13608.
102. H. Mumford and J. K. Troyer, Postexposure therapy with recombinant human BuChE following percutaneous VX challenge in guinea-pigs, *Toxicol. Lett.*, 2011, **206**(1), 29–34.
103. P. Masson and O. Lockridge, Butyrylcholinesterase for protection from organophosphorus poisons: catalytic complexity and hysteretic behavior, *Arch. Biochem. Biophys.*, 2010, **494**(2), 107–120.
104. R. C. DiTargiani, L. Chandrasekaran and T. Belinskaya, *et al.*, In search of a catalytic bioscavenger for the prophylaxis of nerve agent toxicity, *Chem. Biol. Interact.*, 2010, **187**(1–3), 349–354.
105. M. Valiyaveettil, Y. Alamneh and P. Rezk, *et al.*, Protective efficacy of catalytic bioscavenger, paraoxonase 1 against sarin and soman exposure in guinea pigs, *Biochem. Pharmacol.*, 2011, **81**(6), 800–809.
106. R. L. Rotundo, S. G. Rossi and L. M. Kimbell, *et al.*, Targeting acetylcho-linesterase to the neuromuscular synapse, *Chem. Biol. Interact.*, 2005, **157–158**, 15–21.
107. E. Marrero, S. G. Rossi and A. Darr, *et al.*, Translational regulation of acetylcholinesterase by the rNA-binding protein Pumilio-2 at the neuromuscular synapse, *J. Biol. Chem.*, 2011, **286**(42), 36492–36499.
108. K. Cannard, The acute treatment of nerve agent exposure, *J. Neuro. Sci.*, 2006, **249**(1), 86–94.
109. K. Kuca, J. Cabal and D. Jun, *et al., In vitro* reactivation potency of some acetylcholinesterase reactivators against sarin and cyclosarin-induced inhibitions, *J. Appl. Toxicol.*, 2005, **25**(4), 296–300.
110. F. Worek, H. Thiermann and L. Szinicz, *et al.*, Kinetic analysis of interactions between human acetylcholinesterase, structurally different organophosphorus compounds and oximes, *Biochem. Pharmacol.*, 2004, **68**(11), 2237–2248.
111. F. Worek, R. Widmann and O. Knopff, *et al.*, Reactivating potency of obidoxime, pralidoxime, HI-6 and HLö 7 in human erythrocyte acetylcholinesterase inhibited by highly toxic organophosphorus compounds, *Arch. Toxicol.*, 1998, **72**(4), 237–243.
112. F. Worek, N. Aurbek and H. Thiermann, Reactivation of organophosphate-inhibited human AChE by combinations of obidoxime and HI-6 in vitro, *J. Appl. Toxicol.*, 2007, **27**(6), 582–588.
113. M. J. Joosen, M. J. van der Schans and H. M. Wortelboer, *et al.*, Increasing oxime efficacy by blood brain barrier modulation, *Toxicol. Lett.*, 2011, **206**(1), 67–71.

114. G. E. Garcia, A. J. Campbell and J. Olson, *et al.*, Novel oximes as blood brain barrier penetrating cholinesterase reactivators, *Chem. Biol. Interact.*, 2010, **187**(1–3), 199–206.
115. A. Nallapaneni, J. Liu and S. Karanth, *et al.*, Pharmacological enhancement of endocannabinoid signaling reduces the cholinergic toxicity of diisopropylfluorophosphate, *Neurotoxicology*, 2008, **29**(6), 1037–1043.

CHAPTER 14

Current Challenges to Export Controls Related to Chemical Warfare Interdiction

PHILIPPE MAUGER AND RAYMOND A. ZILINSKAS*

James Martin Center for Nonproliferation Studies, Middlebury Institute of International Studies at Monterey, 460 Pierce Street, Monterey, CA, 93940, USA
*E-mail: rzilinskas@miis.edu

14.1 Introduction

The international export control regime that seeks to prevent would-be State and non-State proliferators from acquiring chemical weapons must adapt to three current trends. First, the worldwide growth of chemical-related industries has led to a globalization of the production of dual-use equipment, software, chemicals, and information. By the term "dual-use," we mean an item or a piece of information that has civilian applications but can also be misused by a would-be proliferant in their attempt to acquire chemical weapons. Second, scientific advances and the so-called "convergence" between biology and chemistry have led to the development of new dual-use equipment and chemicals that have not been traditionally covered by chemical warfare (CW) related export controls. Third, the internet has profoundly changed how dual-use equipment, software, chemicals, and information are sourced. The first two trends expand the scope of items that should be subject to CW-related export controls; the third

Preventing Chemical Weapons: Arms Control and Disarmament as the Sciences Converge
Edited by Michael Crowley, Malcolm Dando and Lijun Shang

Published by the Royal Society of Chemistry, www.rsc.org

has revolutionized procurement processes in ways that enable illicit purchasers. The advent of the internet has facilitated malicious actor access to some types of items and information applicable to chemical weapons. Extant export control measures at the international and national levels must be rapidly enhanced to meet these challenges.

In this chapter we highlight the challenges posed to export controls on dual-use chemical equipment, software, chemicals, and information, with an emphasis on the disruptive role of the internet. In view of non-State groups recently having acquired and deployed chemical weapons,[1,2] it is imperative that all States improve methods designed to prevent dual-use items and information from reaching would-be State and non-State proliferators. In order for the international chemical arms control community to take such action, we explain in the ten sections that follow which problems must be overcome and how it might be done.

Following this introduction, the second section surveys the current export control regime. In the third section, we describe and discuss recent State and non-State illicit CW-related procurement attempts. In the fourth section we develop a typology of actors potentially interested in misusing products controlled by CW interdiction efforts. The fifth section discusses the proliferation of dual-use equipment. In the sixth section, we discuss dual-use software. The seventh section examines problems pertaining to dual-use chemical agents, while the eighth section addresses the very difficult problem related to the control of sensitive information. In the ninth and final section we present our thoughts on what can be done by the CW arms control community to prevent dual-use items and information from reaching would-be proliferants, including dealing with the difficulties generated by the internet.

We find that the advent of the internet has increased malicious actor access to proliferation-sensitive equipment, software, chemicals, and information that can be misused to obtain laboratory quantities of CW agents. While industrial-scale production of CW agents remains highly difficult, the capability requirements for mounting a small-scale attack have significantly diminished. The critical capacity that a State or non-State proliferator must acquire in order to establish and operate a large-scale CW program depends on their ability to hire or coerce chemical engineering expertise.

14.2 Current Export Control Measures for CW Interdiction

Four major international accords govern export controls of relevance to CW interdiction, namely the Australia Group (AG), the Chemical Weapons Convention (CWC), the 1996 Wassenaar Arrangement (WA), and the 1987 Missile Technology Control Regime (MTCR). Their individual contributions to the existing CW export control regime are presented in this chapter, as is a general analysis of current trends in CW export controls subsequent to United Nations Security Council Resolution (UNSCR) 1540.

In May 1985, as a consequence of Iraq having used chemical weapons in the Iran–Iraq War (1980–1988), 15 industrialized countries and the EU formed an informal organization with the aim of discouraging and impeding chemical weapons proliferation through the harmonization of national export controls on CW materials. Australia proposed the first meeting and that is the reason for the group's name. In the early 1990s, the Australia Group's (AG) scope was widened to include items that could be used to acquire biological weapons.[3] The AG has, since its inception, expanded in scope and in membership. At the time of writing, 42 States, including all Members of the European Union (EU) and the EU Commission, are AG participants. A number of other countries use the information and have suggested regulations generated by the AG as a basis for their export controls. The AG's main objective is to standardize laws and regulations for controlling exports of chemical and biological equipment, related software and know-how, and chemical and biological agents.[4] Specifically, AG members have agreed to five Common Control Lists that regulate the export of:

- Precursor chemicals to chemical warfare agents;
- dual-use chemical manufacturing facilities and equipment, and related technology and software;
- dual-use biological equipment and related technology and software;
- human and animal pathogens and toxins;
- plant pathogens.

With regard to equipment, the AG participants recognized the ubiquity of many items in the civilian chemical industry and therefore did not attempt to institute blanket controls. As such, the control lists do not attempt to control all equipment of a particular type, nor all chemical exports. For example, not every chemical reactor vessel requires licensing under AG standards, but rather only vessels with a volume greater than 0.1 m^3 and less than 20 m^3, and where the surfaces in direct contact with the chemical(s) being processed are made of specific corrosion-resistant materials.[5] As such, a proliferator can attempt to obtain equipment that is just below the control threshold specified on the control lists, find "loophole" items that are technically not controlled, but that provide similar performance to items that are controlled, or seek to obtain chemicals not covered by the list for CW purposes. To address these concerns, AG participants agreed in 2002 to supplement the control lists through the inclusion of a "catch-all" provision in their respective national export control regulations.[6] A "catch-all" clause provides a mechanism by which a State can block such circumvention attempts targeting unlisted items if either the State or the would-be exporter has cause to believe that the item is destined for CW use. If either the exporter becomes aware that an unlisted item sought for purchase "may be intended ... for use in conjunction with chemical or biological weapons activities," or if the government makes the exporter aware of this risk, the would-be exporter must seek an export license.[7] AG participants have increasingly emphasized the importance of this "catch-all" mechanism.[8]

The CWC (opened for signatures in January 1993 and entered into force in April 1997) is a treaty banning the development, production, stockpiling and use of chemical weapons.[9] It currently has 192 States Parties and one signatory (Israel) that has signed but not yet ratified the Convention.[10] Out of all UN recognized States, Egypt, Israel, North Korea, and South Sudan are the four States that remain outside of the CWC, which means that they are not subject to the verification measures as implemented by the CWC's Organization for the Prohibition of Chemical Weapons (OPCW).[11] Each CWC State Party must adopt "necessary measures to ensure that toxic chemicals and their precursors are only developed, produced, otherwise acquired, retained, transferred, or used within its territory or in any other place under its jurisdiction or control for purposes not prohibited under this Convention."[12] The specific requirements of relevance to export controls are as follows. CWC controls are divided into three lists of chemicals called schedules. The three schedules are sub-divided into two parts, A and B, with the frank CW agents listed under Part A and their main precursors generally listed under Part B (this explanation is a simplification, as certain precursors can be listed on lower Schedules than the warfare agents they can be used to produce).[13] Schedule 1 chemicals are those that have "high potential" for use as chemical weapon agents and have "little or no use for purposes not prohibited" by the CWC.[13] Schedule 2 chemicals pose "significant risk" of misuse, but are known to have some civilian use.[13] Schedule 3 chemicals are those that "pose a risk" of misuse, but are used in large quantities in civilian activities.[13] The production and transfer of Schedule 1 chemicals between nations are strictly controlled and limited to small quantities.[14] As for Schedule 2 and 3 chemicals, CWC States Parties must provide national data on, *inter alia*, imports and exports, broken down by country involved.[15–17] In addition, States Parties cannot transfer Schedule 2 chemicals to States that do not belong to the CWC, and States Parties must provide annual import and export data for all Schedule 2 production or consumption facilities above a given production or use threshold.[18] The CWC has an extensive verification component implemented by the OPCW designed to verify the compliance of States Parties to its provisions, including the validity of State declarations regarding imports and exports.

The 41 WA participants have agreed to place export controls on a wide range of military and dual-use equipment.[19] They do so through two control lists, the Munitions List and the List of Dual-Use Goods and Technologies. The WA participants have agreed to control, among many other item categories, the export of certain military vehicles and vessels, aircraft and unmanned aerial vehicles (UAVs), artillery pieces and their ordnance, missiles, submunitions, mines, and smoke canisters.[20] Although not the sole objective of the WA, these controls restrict the sale of delivery systems and ordnance that could be employed to deliver chemical weapons, as well as military vehicles and vessels with CW protection systems that a proliferator would need to capitalize on CW use.[21] The Munitions List contains the following entries with direct CW applicability:

- CW chemicals (including "riot control" agents) and key precursors, as well as certain chemical defoliants;[22]
- Equipment "specially designed or modified for military use" to disseminate such agents;[22]
- Protective, detection, and decontamination equipment and related supplies (such as biopolymers for detection purposes and biocatalysts for decontamination purposes);[22]
- Software designed to operate or maintain equipment controlled under the Munitions List (such as equipment listed in the bullet points above);[22]
- Software "for determining the effects of conventional, nuclear, chemical or biological weapons."[22]

In addition, WA participants must control the export of dual-use protective and detection equipment and components that could be misused for CW purposes as noted in the List of Dual-Use Goods and Technologies.[23]

The last of the four accords constituting the CW control regime is the Missile Technology Control Regime (MTCR). Its 34 members have agreed to exercise restraint on the transfer of missiles and UAVs capable of delivering a chemical, biological, or nuclear payload, as well as on specialized equipment and parts used in the manufacture of said systems.[24] Items controlled by the MTCR are divided into two categories.[25] Category I items include "complete rocket systems (including ballistic missiles, space launch vehicles and sounding rockets) and unmanned air vehicle systems (including cruise missiles systems, target and reconnaissance drones) with capabilities exceeding a 300 km/500 kg range/payload threshold; production facilities for such systems; and major sub-systems including rocket stages, re-entry vehicles, rocket engines, guidance systems and warhead mechanisms."[25] Category II items are "complete rocket systems (including ballistic missiles systems, space launch vehicles and sounding rockets) and unmanned air vehicles (including cruise missile systems, target drones, and reconnaissance drones) not covered in item I, capable of a maximum range equal to or greater than, 300 km," as well as related equipment, material, and technologies.[25] MTCR members commit themselves not to export Category I production facilities, promise to export Category I systems only in "rare occasions" and only after having obtained binding bilateral safeguards that the systems will not be misused, and to show restraint on exports of Category II systems.[25] As with many of the WA clauses, the MTCR helps to restrict exports of systems particularly suited for the delivery of chemical weapons.

The interplay between the AG, CWC, and WA controls on chemicals warrants explanation. All AG and all WA participants are also CWC States Parties, and so AG and WA requirements are in addition to those stated in the CWC.[26] However, both the AG and the WA lists only partly overlap with the CWC Schedules. The WA controls all of the CWC's Schedule 1 CW agents and precursors.[27] The munitions list itself is vague with regards to the toxins on the CWC Schedules (saxitoxin and ricin), but the explanatory text found within the dual-use list makes clear that toxins fall within the "biological agents" controls, and therefore are controlled.[28] With the exception of the incapacitating agent BZ, the WA list does not include Schedule 2 and 3 chemicals.[29] However, unlike the CWC Schedules, it includes a number of "riot control agents" and defoliants.[29]

The AG list contains only precursors, not proper CW agents. The list was created "on the basis of actual attempts to procure [chemicals] for chemical weapons purposes" and lists additional dual-use chemicals not found on the CWC Schedules.[30] For instance, the AG list controls sodium fluoride and sodium cyanide, two non-Schedule chemicals that were sought by Iraq for the production of the nerve agents sarin and tabun.[30] However, the AG list does not include all precursors named in the CWC Schedules. For example, chlorosarin (*o*-isopropyl methylphosphonochloridate) and chlorosoman (*o*-pinacolyl methylphosphonochloridate) are CWC Schedule 1 precursors, which also are on the WA control list, but are not listed on the AG Chemical Precursors control list.[31,32] These omissions probably occurred because the latter considers these chemicals to be CW agents in their own right, and thus not solely precursors. As such, although the AG advertises itself as a means of meeting the CWC's "necessary measures" clause, membership does not by itself satisfy all of the CWC's export-control related requirements. At the same time, since precursor chemicals solely controlled by AG members are less controlled globally, they may be particularly attractive to proliferators. To cite a recent acquisition attempt of such a chemical for unknown end-use, in 2013, Taiwanese customs interdicted a North Korean attempt to import phosphorus pentasulfide, an AG-listed but non-Schedule chemical used in the manufacture of insecticides, fuels, and nerve agents.[31,33,34] The source of the chemical was not revealed in open sources, but before interdiction, the ship had last stopped in Bangkok, Thailand – a non-AG State.[35] As such, North Korea had tried to acquire a non-Schedule but AG-controlled chemical by orchestrating the shipment in such a way that it would transit through at least two non-AG countries before reaching its final destination.[35]

While the international CW control regime as created by these four accords has continued to grow in size, the majority of United Nations (UN) member States are not party to the AG, WA, or MTCR. By contrast, the CWC has near-universal membership. The AG in particular, despite its crucial role in ensuring export controls over production equipment, is almost entirely composed of Western States and long-time US allies such as South Korea and Japan, and does not include major industrial nations such as Brazil, China, India, and Russia. The AG's membership demographic reflects the Cold War divide that was in place at the time of its creation and is unlikely to change given the security and political realities borne out of the AG intelligence sharing clause. Given the great expansion in chemical industries worldwide, the AG's membership increasingly fails to encompass States with proliferation-sensitive chemical equipment exports. In a similar vein, the WA and the MTCR are increasingly challenged by the proliferation of military production capabilities to non-member States. As a result, the importance of these agreements increasingly lies in the precedent they have set for export controls over particular items. States outside of the AG, WA, or MTCR can incorporate the export control lists generated by these agreements as is, or modify them as they see fit. For example, China's export control lists are mostly but not entirely in line with AG standards, and China has pledged to respect the MTCR. In summary, the continued efficacy of the CW control regime is increasingly tied to the universalization of national CW-related export controls.

This reality is well-understood. When the UN Security Council adopted UNSCR 1540 in 2004, the constituent members created a legal obligation for all UN member States to "establish domestic controls" to prevent the spread of all weapons of mass destruction (WMDs).[36] UNSCR 1540 further stipulated that all member States would "adopt and enforce appropriate effective laws that prohibit any non-State actor to manufacture, acquire, possess, develop, transport, transfer or use nuclear, chemical or biological weapons and their means of delivery."[36] However, the UNSC's decision to impose a legally-binding obligation on all member States in this fashion has not gone unchallenged. For instance, several States raised the legal defense that they were under no obligation to enact new national export control laws to meet their international obligations under UNSCR 1540 because no items subject to controls were present within their borders.[37]

Despite this opposition, a reading of the four major review documents on the implementation of UNSCR 1540 reveals worldwide improvements in export control laws and regulations since its inception.[38–41] By 2006, roughly a quarter of the 193 UN Member States had assembled packages of national chemical accounting measures combined with CW-related export controls.[42,43] The number of States with legislation to account for the internal use and the export of chemicals roughly doubled in the 2006–2011 period, and now approximately half of all States in the world have some such measures on their books.

The world-wide growth in national legislations establishing CW-related export controls has been encouraging. However, internal accounting and export control legislation means little if stakeholders are not aware of their existence and if law enforcement lacks the capability to enforce them.

Japanese strategic trade control data demonstrates the importance of outreach initiatives to potential exporters to ensure that they are cognizant of existing controls. Of all prosecuted import–export violations during 2010–2014, 27% had been committed because the culprit was not aware of the relevant laws and regulations and another 13% because they had improperly interpreted the rules or had made operational errors in their application, *versus* some 13% of willful or gross negligence violations.[44] That is, it was almost three times more likely that a violation was the cause of a lack of knowledge or of a mistake in implementation than the result of a deliberate decision to break controls for profit. No breakdown was provided for just CW-related export control violations, but the general rule that a violation is more likely to be unwitting than deliberate is expected to hold for this subset of cases. Only certain subsets of dual-use chemical equipment are controlled for export. However, it can be difficult even for the seller, let alone customs officials inspecting cargo, to tell the difference between controlled and uncontrolled variants within the same equipment category. As a result, chemical equipment and associated software exporters may not always understand that the products they are dealing with are controlled for export. To return to the chemical reactor vessel example introduced in the AG section above, the difference between an AG-grade and a non-controlled reactor vessel depends on the internal composition of the reactor, and as such, may not be discernable without the equipment's associated documentation. However, equipment re-sellers may not have the product documentation necessary to make this determination, for instance when the equipment is discontinued. The problem of knowing whether an item is controlled or not is then compounded when the re-seller has no background in chemical equipment, for example company liquidators with no specialization in chemical equipment sales. In cases like these, not only will the vendor have trouble making the determination regarding the item, but they may even lack sufficient awareness of CW issues to know that they are supposed to make this determination in the first place. This situation is fundamentally unlike those governing exports of products, such as night-vision goggles or holographic rifle scopes, with obvious military applicability. For this reason, governmental outreach to sensitize companies involved with CW-relevant products is critical.

In parallel to such outreach initiatives, governments must provide their law enforcement and border control forces with the mandate and the capability to enforce regulations. In this regard, the 2006 1540 Committee report flagged that "a large number of States might not have either the technical capacity or the equipment to implement the full range of border controls called for under UNSCR 1540 or might not appropriately interpret their existing border control authority to extend to items and technologies related to weapons of mass destruction."[45] The same identification issue outlined above for exporters also presents itself to law enforcement agents: how can law enforcement distinguish between controlled and uncontrolled variants of a particular item, for instance a chemical reactor vessel? And how can they verify that a chemical stock readied for export really is a non-Scheduled chemical? Because CW-relevant cases are rare compared to the panoply of other border crimes, law enforcement knowledge of CW issues cannot be taken for granted. With regards to chemical precursors, "not even all chemists, let alone customs service officers, know what [thionyl chloride and thiodiglycol] can be used for."[46]

As for equipment identification, contrasting the multi-slide contents from a training tool designed to teach officers to recognize AG-grade equipment with the short and highly technical AG control list definitions confirms that the latter were not written with the non-specialist in mind.[47] These identification problems are particularly acute for law enforcement agents because of the enormous volume of shipments to inspect, the tradeoff between impeding trade and verifying cargo, and because the exporter may be malicious and can falsify documentation regarding their would-be export. Specialized training on the verification of would-be exports of CW-relevant products is therefore necessary. In the case of chemical agent exports, as with other potentially dangerous products, field agents must also be provided with protective equipment and associated training.[48]

Several multilateral agencies assist States seeking to conduct company outreach and law enforcement training, and new programs have been designed since the UNSCR 1540 came into force. The 1540 Committee maintains a list of requests for assistance by Member States, in an effort to match States or international organizations with additional resources to recipients with clear projects.[49] A total of 16 international organizations have offered assistance; of these, the organizations offering assistance directly related to CW interdiction include the OPCW, the MTCR, the United Nations Office of Drugs and Crime, the World Customs Organization, the European Union, and the International Criminal Police Organization (INTERPOL).[50]

CW-relevant export control measures have improved in the last decade, but it remains to be seen whether the aforementioned initiatives are sufficient to match increased CW-relevant commerce and the evolution of proliferator networks. In the next section we provide an overview of the illicit CW procurement networks these export-control assistance programs are meant to combat. The following historical and recent cases demonstrate both the impetus for, and the positive outcomes of, increased export controls.

14.3 State and Non-State Illicit Procurement Attempts

14.3.1 State Events

State-level CW-related procurement prior to the CWC's negotiation involved some direct weapons transfers, although no direct weapons transfers are known to have taken place since the CWC entered into force. To cite a confirmed example of such a transfer dating back to the Cold War, the Chinese provided 16.6 metric tons of chemical weapons agents to Enver Hoxha's regime in Albania.[51–53] Yet even when some States were willing to transfer chemical weapons directly for geopolitical gain during the Cold War, most proliferators sought to establish indigenous production capabilities to reduce their dependence on their foreign patrons. For this reason, transfers of CW chemicals, equipment, and know-how are not a new phenomenon. The way these transfers have been orchestrated, however, has drastically changed since AG export control measures were instituted.

Before the AG's inception, purchases were often carried out directly by the actual end-user with little obfuscation, and foreign assistance included everything from plant design to equipment installation ("turn-key" facility purchases) as well as the direct provision of key precursors. We present some historical examples from Yugoslavia and Iraq below to show how procurement networks have evolved.

Yugoslavia made most of the CW equipment purchases needed to construct its Mostar CW plant in 1958, purchases that included at least one chemical reactor from France.[54] Although the Yugoslavs thoroughly maintained a civilian cover story for the Mostar plant and apparently did not allow foreigners to have access to the site, some business transactions took place in the nearby city of Mostar.[54] When interviewed by CIA officials in the mid-1990s, knowledgeable Yugoslavs expressed their conviction that "Western European countries" must have known of their CW program in part because of what they apparently saw as the insufficient obfuscation of the equipment purchase transactions.[54]

The pre-1985 Iraqi procurement program was (even) less careful, and involved the purchase of turn-key factories from West German providers.[55] The Iraq Compendium authored by the United Nations Monitoring, Verification and Inspection Commission (UNMOVIC) details the evolution of Iraq's program in response to the development and application of CW export controls. Iraq's reliance on foreign supplies can hardly be overstated. In UNMOVIC's judgment, "the vast majority of the chemical processing equipment came from foreign suppliers, and about 95 percent of all precursor chemicals used for the production of CW agents was procured outside [of] Iraq."[56] The advent of multilateral CW-relevant export controls in 1985 forced this massive procurement program underground, and in so doing increased costs and brought about technical complications.

Iraq's CW program was established at the Al Hazen Ibn Al Haitham Institute (1974–1978) near Samarra, which directly and quickly procured chemical synthesis laboratory equipment, analytical instruments, and kilogram quantities of major CW precursors.[57] This combination of large purchases issued by a single institute, combined with the "specific layout, design, and chemical process equipment" of its facilities under construction, all pointed to a potential military end-use and should have raised red flags, had Western exporters at the time been inclined to halt sensitive shipments to Iraq.[57]

Similarly, little effort was spent obscuring the large scale-up in Iraq's CW program between 1978 and the mid-1980s. In August 1981, Iraq's Revolutionary Command Council (RCC) issued a decree that ordered the establishing of the State Establishment for Pesticide Production (SEPP), which was to serve as a cover for a vastly increased CW program.[†] In order to enable increased Iraqi chemical attacks against Iranian forces in the ongoing Iran–Iraq War, SEPP acted quickly to refurbish and reactivate four chemical production plants at the Muthanna State Establishment (MSE) that had been formerly operated by the Al Hazen Institute. Thus, little effort was made to distance SEPP's affiliation with already-established CW sites.[58]

Rather than use front companies to distance the end-user from the foreign supplier, "SEPP was ... authorized to import and purchase locally all necessary materials, substances and equipment, [and] to interact with governmental, local and foreign agencies and entities both inside and outside Iraq."[59] Iraq hired foreign companies to design, construct, and equip additional production complexes at the site in that timeframe.[60] Simultaneously, Iraq imported several CW precursors for use at these facilities.[61] Most blatantly, in the same 1981–1983 period when foreign contractors were on-site to assist with the construction projects, Iraq built large reinforced concrete bunkers in an annex to the production facilities.[62] As such, the complex's military end-use was barely hidden.[‡]

The advent of international CW-related export controls in the post-1985 period saw Iraq's procurement activities become increasingly complicated. The establishment of the AG and the development of its chemical control list complicated Iraq's attempts at importing precursors.[63] In response, during the 1988–1990 period, Iraq attempted to acquire the technology to indigenously produce all precursors for its chemical weapons and hence become self-reliant.[64] These attempts were also unsuccessful, thanks in part to new AG controls on turn-key facilities and on production equipment.[64] In response, Iraq was forced to improve the cover of its institutes and diversify its CW sites. Starting in February 1989, Iraq began tying the old SEPP (which by then had changed its name to MSE) institutes to its pharmaceutical industry to establish more credible cover stories for existing sites, and began establishing new satellite sites to distance new construction from its prior CW sites.[65] Iraq used known companies that had legitimate civilian needs for process equipment, such as the State Organisation for Oil Refineries and Gas Industry (SORGI), to serve as fronts to procure dual-use equipment.[66] It also increasingly turned to creating new intermediary companies, banking accounts, and transportation companies to obscure actual end-users and facilitate transfers.[66] In addition, Iraq

---

[†]The RCC was established in 1979 and soon evolved to become Iraq's ultimate decision-making organ. Its chairman was Saddam Hussein.

[‡]However, to fool enemy aircraft, the Iraqis built many dummy bunkers that looked like the real bunkers but contained nothing of value.

shifted its procurement activities toward countries that had weaker export controls and did not belong to the AG.[66]

There are contrasts between the historical pre-1985 Iraqi case and the several recent Syrian attempts to procure CW protective equipment from North Korea as the Syrian civil war escalated. These new cases are instructive for three reasons. First, they demonstrate the level of obfuscation employed by modern proliferators—and these were botched attempts—as well as the resilience of their networks to interdiction efforts. Second, the cases demonstrate the need for strong export controls governing trans-shipments as well as at-sea interdictions, given that key proliferators like North Korea have no intention of being bound by any domestic export controls and frequently shield shipments by taking advantage of unclear flagging and ownership laws governing maritime cargo inspections.[§] And third, they reveal that even a State like Syria, that had a large and long-running CW program at the time, still had to depend on foreign suppliers for crucial CW-relevant products, such as protective equipment.

In October 2009, South Korean authorities found North Korean dual-use chemical protective suits destined for Syria onboard the ship *MSC Rachelle* during an inspection in the South Korean port of Busan.[67] Subsequent investigation revealed that the products had first been shipped from the North Korean port of Nampo to the Chinese port of Dalian, where they were loaded on board the *MSC Rachelle*.[67] To further conceal the operation's North Korean origin, the ship flew the Panama flag and was technically owned by a Swiss company.[67] One month later, Greece interdicted another shipment.[68] Some 13 000 chemical protective suits identical to those seized from the *MSC Rachelle*, North Korean gas masks, and 23 600 gas indicator ampoules used to detect specific chemical substances were discovered in another ship bound for Syria.[69,70] As with the MSC *Rachelle*, the equipment was manufactured in North Korea, shipped out of Nampo, and trans-shipped through Dalian.[69] In addition, the ship made subsequent trans-shipments through Jeddah in Saudi Arabia and through unstated "other ports" to obfuscate the final Syrian destination.[69] A UN Panel of Experts members posited the existence of at least one additional shipment, since the November shipment lacked any protective boots.[69] In one last published naval case, on April 3, 2013, Turkish authorities found North Korean gas masks alongside small arms and ammunition in the *El Entisar*, once again destined for Syria.[71] This time, the ship was Liberian flagged.[71] Turkish authorities intercepted the ship as it transited through the Dardanelles Straits as part of a circuitous route probably designed to obfuscate the final destination point.[75]

Overall, these were all large shipments that required the use of North Korean vessels for transport. The latter attempted to hide the illicit shipments within legitimate civilian naval traffic. Interdicting naval traffic poses international law quandaries, and failed interdictions can have serious

---

[§]On the legal difficulties posed by maritime interdictions, see ref. 247.

political ramifications, as happened when the US strong-armed an inspection of a Chinese cargo ship called the *Yinhe*, mistakenly thought to be carrying CW precursors, in 1993.[72,73] The maritime "shell game" strategy used by North Korea is therefore more difficult to counter than these successful interdictions may lead one to believe.

For smaller shipments that are logistically easier to orchestrate covertly, trans-shipping from a front company is the commonly-used method. In these cases, the supplier does not have to be a co-conspirator, although co-conspirators can and have used front companies to obscure the end-user of their shipments. Rather, the supplier may simply be negligent in its export control verifications, or otherwise fall prey to a resourceful proliferator. The general front company technique is as follows. The proliferator sets up a front company that appears to have a legitimate civilian need for the desired dual-use equipment, as Iraq did when it used its pharmaceutical companies in the post-1989 period as cover for CW equipment purchases. The front company can be located in a different country, for instance in a busy commercial port city such as Singapore, to bypass country-specific enhanced controls designed to impede known proliferator countries. The front company then attempts to purchase the controlled equipment. Since the front company has an apparent civilian need for the equipment and since its ties to the proliferator country are hard to detect, the supplier is likely to receive export approval even if it actually respects the law and seeks export control approval. Once the equipment is received at the front company, a trans-shipper can then be used to ferry the equipment to a CW facility in the proliferator country. One such front operation operating on behalf of Syrian agents was recently dismantled. The conspirators allegedly "prepared false invoices which undervalued and mislabeled the goods being purchased and listed false information as to the identity and geographic location of the purchasers of the goods" to arrange the illicit shipment of CW agent production and detection equipment to Syria.[74]

Aside from obfuscating shipping records and hiding actual end-users, modern proliferators must hide the underlying financial transactions, which are coming under increasing scrutiny. A front company is also advantageous to a proliferator for this purpose, since the front can be used to launder money and thereby obfuscate the initial source of the funds. Of note, in the last Syrian network example given above, the perpetrators were also charged with wire fraud and money laundering.[74]

14.3.2 Non-State Events

Few terrorist groups have shown an interest in CW, but the groups that have done so have had considerable resources to work with.[¶] In the following subsection, we review the known cases that involved exports.

[¶]See also the historical cases presented in the edited chapters of ref. 248.

The millenarian terrorist group Aum Shinrikyo produced sarin, soman, and VX nerve agents, the blood agent hydrogen cyanide, as well as chlorine, explosives, and a range of narcotics.[75] In drawing lessons from the Aum case, it is important to note that the group operated before the CWC entered into force. A modern non-State proliferator existing within a CWC State Party (like Japan) would now need to evade their national authority's industry registration and licensing requirements to avoid being inspected by the OPCW. In theory, then, a well-resourced non-State group will have more trouble today in avoiding detection when using a front chemical industry company to procure equipment in a CWC Member State.[76] However, as is evident from large-scale illicit drug production facilities, a well-resourced group can still obtain industrial quantities of chemical equipment and operate effectively underground even within CWC Member States.[77]

The current authoritative account of Aum Shinrikyo's CW program is found in the second edition of a report by Danzig *et al.*, partly based on extensive interviews with key convicted Aum Shinrikyo members.[78] The CW program was run by Masami Tsuchiya.[79] The group's first chemical laboratory was established by Tsuchiya's superior, Hideo Murai, in November 1992.[79] Upon his return from a trip to Russia in June 1993, Murai told Tsuchiya that the cult was threatened by CW and tasked the latter with a review of "analytical methods so if chemical weapons are used we can know what they are."[79] Tsuchiya used the Tsukuba University's library to identify potential CW agents and reported back to Murai that sarin appeared to be the most cost-effective agent.[79] Under further tasking by Murai, Tsuchiya ordered chemistry equipment necessary to produce small amounts of sarin directly through Aum Shinrikyo's official company, apparently without obfuscating the end use since Tsuchiya did not think the project to be illegal.[79] By mid-1993, the group was producing 10–20 grams of sarin.[79] Aum then began construction of a facility, Satyan 7, in order to scale up production to approximately 70 tons.[79] This time, two dummy companies were created under the already-existing Aum shell company Hasegawa Chemical to order the necessary equipment and chemicals.[80] Work at the facility failed to produce sarin due to managerial incompetence and Aum's technical inability to modify production pathways from laboratory-scale to industrial-scale production.[81] However, the group's acquisition efforts were highly successful; the equipment inside the company facility was estimated to have been worth some $30 million.[80] Although a detailed inventory of the equipment has not been published, it appears as if the equipment and the precursor chemicals came from Japanese industrial companies. In one case, Aum even rented a reactor vessel from a Japanese chemical company.[82] Danzig *et al.* found no evidence that the group obtained restricted CW know-how or chemical equipment from Russia, concluding that "although Russian information may have aided planning the factory production system, we see no evidence of a Russian contribution to the cult's method of sarin production."[83] Therefore, the only known import case for Aum's CW program, aside from information, involved three US-origin molecular modeling software suites used to model organic chemical synthesis reactions.[84]

Contrary to Al-Qaeda Central's BW program, which involved equipment and pathogen export attempts to Afghanistan,[85,86] its CW program appears to have relied on indigenous materials. Indeed, military photographs of alleged Al-Qaeda-linked "CW laboratories" have shown only rudimentary equipment and laboratory-quantities of chemicals.[87] The only known CW-related export sought by Al-Qaeda Central was CW information.[||] A plea for such information was presented in a document captured in Bin Laden's compound during the raid that killed him and subsequently declassified and released by US authorities, titled "Terror Franchise: The Unstoppable Assassin, TECHS Vital role for its success."[88] The document was signed by Al-Qaeda's then senior planner Abu-Salih al Somali, and if he did write the text, it must have been prepared shortly before his death in August 2009.[89] In garbled English, the author calls for experts ("techs") to prepare tutorials relevant to a long information wish list arranged in order of importance by the author. These experts were exhorted to upload their "(long waited for) work to web sharing sites, Jihadi discussion forums, and send emails advertising... the material[s]" or to send technical information by physically mailing DVDs.[88] After an initial list of requests related to explosives, the author called for information on the following topics under a section titled "Toxicants: Immediately needed":[88]

(1) "Actual improvised production and testing of cyanides, ricin. (immediately needed).
(2) Preparation and testing (rabbits is ok) of any lethal (delayed and immediate) ingested toxicants.
(3) On Camera production of any of war gases (Phosgene, VX, *etc.*). Look at NbK file and scientific principles of improvised home warfare. (onsite production apparatus also.)
(4) Actual production and testing of biological toxicants (Anthrax, Botulism, clostridium, endotoxins, Exotoxins *etc.*).
(5) Production of (HCl) or whatever is needed in the production of toxicants.
(6) Any other options that can be used as toxicant... plants, *etc*... detailed, local names pictures, incidents, cultivation...Insects...*etc*... read scientific principles of improvised home warfare volume 2, 5, 6.
(7) Bacteria based weapons... how? Detail. Any other practical options.
(8) Airborne substance that when sprayed in small quantity or mixed (tablet form) with water, tranquilizes the entire inhabitants of a hall or plane. And its antidote."[88]

[||]For instance, the following chronology makes no mention of CW export/import activities by the group in ref. 249.

The passage demonstrates both Al-Qaeda's desire for CW knowledge, as well as its apparently minimal technical capacity given the wide range of requested information as late as mid-2009. The author's passing remark that technical information is "long waited for" implies that Al-Qaeda Central's soldiers had so far received minimal assistance from external experts through online platforms, despite the author's clear enthusiasm for this information pathway.

Compared to Al-Qaeda's small scale acquisition effort, the so-called Islamic State in the Levant (ISIL) has relied on its extensive local resources and a procurement network spanning 20 countries for acquiring chemicals used to make explosives, munitions, and associated supplies such as blasting cord.[90] When special materials are needed for WMD-related projects in-region, it contracts out acquisition efforts to smugglers; this setup was demonstrated during a recent "red mercury" scam.[91] Narrowing the scope to their CW program, ISIL appears to rely on sources already present in their main operational areas in Syria and Iraq; spare shells and explosives, stocks of industrial chemicals including chlorine, and even old Iraqi mustard munitions.[92] ISIL has tried to recruit foreign fighters with scientific expertise, such as medical doctors and chemists, but given their need for medical personnel and explosives manufacturers, this does not necessarily indicate a desire to import CW know-how.[**,93] Indeed, its CW program appears to rely on individuals previously involved in the Iraqi CW program.[94,95]

As tentatively seen in the above three cases, non-State groups will not necessarily increase foreign acquisition attempts even as their material capability to conduct such attempts increases. Rather, non-State groups have so far primarily been drawn to sources of chemical equipment and chemicals in close proximity to their areas of operation. Domestic controls rather than export controls therefore appear to be the priority when attempting to deny non-State actors access to CW-relevant items. Know-how appears to be the one exception, and know-how transfer has been sought in a variety of forms; from human contact both in person and online, from reviewing openly published information, and from the acquisition and use of specialized chemical reaction prediction software.

14.4 A Typology of Malicious Export Violators

This section introduces a typology of actors potentially interested in misusing some or all of the products that CW interdiction efforts seek to control. Different proliferators are likely to seek different tangible and intangible products to produce different quantities and types of CW agents, and are therefore affected differently by export controls. The goal of the typology is

**Scholar Tom Bielefeld pointed out on 9 March 2016 on social media that the following German-language article claimed one ISIL personnel vetting file included the comment: "Important. Has chemical knowledge." See ref. 250 for further information.

to capture differences in proliferator motive and in capability. The typology we use is as follows:

1. Individuals
2. Organized crime groups
3. Terrorist groups
4. Weak States
5. Strong States

We include organized crime groups in our typology because a sizeable illicit trade in chemicals and chemical equipment exists to feed the demands of producers of illegal drugs, counterfeit medicine and cosmetics, and doping compounds for athletes.[96–103] These illicit channels present challenges to export controls aimed at CW interdiction. Although much of the equipment and compounds currently trafficked do not meet existing criteria for CW control, some exceptions exist. For instance, fentanyl (a potent CW agent if aerosolized) has a long history of illegal use as a heroin substitute in Estonia, and a rising history of use in the United States.[99] Carfentanil, a chemical related to fentanyl but roughly 100 times more toxic, is sold by criminal drug dealers in major American cities.[104] The issue of illicit botulinum toxin production and sale in the form of counterfeit Botox is another example.[105] While our focus remains solely on misuse for CW activities, we draw on published cases of individual and organized crime events to discuss what is already available for misuse.

There is a spectrum between weak and strong States. The weakest States lack the capability to produce precursor chemicals and production equipment and are therefore entirely dependent on imports. The strongest States have the capability to produce all precursors and all associated equipment. They can still be interested in, but are not dependent on, high-quality and specialized imports.

Since capability increases as one moves down the typology list, the scale and capability of CW program capabilities of proliferators likewise increase. However, even States can content themselves with small-scale programs, such as those designed solely for assassinations. For instance, apartheid South Africa's Project Coast joint CW and BW program was focused mainly on assassinations.[106] As was the case in the former Soviet Union, States can also have a CW program for assassination purposes directed by an intelligence agency in parallel to a military-operated CW program.[107] Since such programs are independently run by an agency with different priorities and a separate budget line than the regular military, they may have their own separate procurement networks. As such, although the potential program scale increases as one goes down the typology list, the existence of small-scale but potentially technically-advanced State programs are a demonstrated possibility.

Two distinct paths are available to the five typologies of actors to bypass CW-related export controls. The path likely to be chosen by the actor depends on their motive and capability.

Individuals, criminals, terrorist groups, and weak States interested in using chemicals for violent acts are likely to seek out toxic industrial chemicals (TICs) not controlled by the AG or the CWC.[108,109] TICs are hazardous chemicals routinely used in the chemical industry for peaceful purposes. There are far more TICs than controlled chemicals. Over 1700 TICs have been evaluated for their military diversion potential, although fewer than 100 chemicals are currently publicly identified as posing high, medium, or low hazards.[110] Their relative availability compared to most of the controlled chemicals on the CWC schedules makes them attractive to low-capability proliferators or those seeking poisons for assassinations. At the individual level, homicides, including mass homicides, committed through poisoning typically depend on common chemicals such as pesticides.[111] ISIL has used TICs, more precisely chlorine and tentatively phosphine, delivered during attacks in Syria and Iraq through improvised chemical munitions and improvised explosive devices.[2] As an example of a TIC used by a weak State, military forces under Bashar al-Assad's control in Syria have dropped chlorine barrel bombs from helicopters subsequent to the OPCW's dismantlement of Assad's chemical weapons arsenal.[112–114] Apartheid South Africa is another example, having provided toxic pesticides to Rhodesian special forces.[115] These were used in mass poisoning operations where military clothes designed to fall into guerilla hands were swabbed with poison beforehand.[115] TICs are already readily available in most countries and little can be done at the international export control level to mitigate their threat aside from encouraging the phase-out of such chemicals in favor of less-toxic alternatives whenever possible, and perhaps better accounting and tracking processes. For this reason, we do not consider TICs any further in this chapter.

Strong States pose a different challenge to export controls in that they have the capability to design novel CW agents, including such agents and some precursors that are not listed in the CWC's schedules. This eventuality was raised in the early 1990s by Vil Mirzayanov, a former Soviet weapons chemist, who believed that doing so was one of the Soviet *Foliant* program's objectives.[116,117] Strong States may be attracted to biologically-derived compounds, such as toxins and peptides, as a source of novel agents: these are the so-called "mid-spectrum" agents positioned between traditional chemical and biological weapons agents.[118–120] This possibility has been repeatedly raised by Czech and Croatian chemical defense experts,[120] as well as by scholar Malcolm Dando and others, and is further addressed in Chapters 6, 7 & 8 of the current volume. There has already been attempts at weaponizing biologically-derived substances since the advent of genetic engineering; the Soviet Union's KGB ran at least one such program, which was focused on developing weapons whose payloads were peptides and other bioactive substances for use in assassinations.[121,122] Of particular importance for this chapter is the fact that the production of such agents can be carried out using biological rather than chemical processing equipment, which greatly expands the scope of CW-relevant controlled items. We detail these export control circumvention problems posed by the novel agents, in particular bioactive substances, in the subsequent sections.

14.5 Proliferation-sensitive Equipment

14.5.1 Production Equipment

Although the control criteria specifics vary by item category, AG-grade production equipment items are essentially controlled for their corrosion resistance and their ability to process chemicals in quantities appropriate for a CW program. The production of many CW agents, and in particular the organophosphate nerve agents, involves strong acids and highly exothermic reactions. As a result, corrosion-resistant equipment is necessary to prevent leaks. The AG chemical equipment control list contains certain reaction vessels, storage tanks, heat exchangers or condensers, distillation or absorption columns, filling equipment, and agent or CW munition incinerators, as well as the valves, multi-walled piping, and pumps necessary for their operation.[5]

Four trends pose acute problems for export control officials over such products: the general growth in the trade of such equipment, the growth of AG-grade equipment production in non-AG Member States, the growth in small and medium equipment suppliers, and the disruptive effects of the revolution in e-commerce. The alarming impact of these trends for CW-related export controls were detailed during a 2014 US National Academy of Sciences (NAS) workshop.[123] We discuss each trend in turn below.

To give a sense of the scale of the modern chemical industry, chemical production facilities declared to the OPCW in 2008 by States Parties were very roughly an order of magnitude more than the 500 OPCW facility inspections carried out that year.[124] More specifically, there were some 4500 to 5000 "Other Chemical Production Facilities" declared to the OPCW. These are facilities that do not currently produce scheduled chemicals (but, in many cases, could be adapted to such purposes). The global commerce in chemicals has more than doubled in value in the last decade, and sales of production equipment have expanded to meet demand.[125] Most of the growing demand for chemicals has been in developing countries, and has been met by increased chemical production in these countries.[126] The traditional European AG Member States have so far not been well-positioned for these emerging markets, which now represent roughly half of the worldwide chemicals market.[126,127] Indeed, European chemical production has stagnated in recent years, while production in China and other emerging markets has had high growth rates.[128] As a result, chemical production has increasingly shifted from AG Member States toward non-member countries. To cite the most impressive non-AG example, China has roughly tripled its market share: the country went from controlling a little under 3% of the total chemical export market share in 1995 to approximately 9% in 2012.[129,130]

Chemical equipment has increasingly been manufactured within these emerging markets. It was comparatively easy to keep dual-use equipment under export controls when the major producers were all under the AG's aegis. That categorization is not accurate today: India and China, to name two economic powerhouses outside of the AG, now produce and market plenty of such equipment, including corrosion-resistant AG-grade equipment. For instance, about 3000 Chinese firms are capable of building chemical production equipment in general, and some 30 factories make glass-lined corrosion-resistant equipment.[131] Overall, according to industry experts present at the aforementioned NAS workshop, the chemical equipment manufacturing industry has "a low barrier to entry."[132] Given this lack of steep technical barriers to production, a growing number of small and medium companies in the developing world are expected to become chemical equipment producers in the coming years.[133] This is a particularly problematic trend for CW-relevant export controls, as small firms are generally less visible to law enforcement, are typically more dependent on service providers to move their products, and hence have a greater equipment tracking challenge, and are less likely to have resources for zealous export compliance measures.[251††]

Compounding the first three trends, the way in which these items are traded has been revolutionized by the advent of the internet and the meteoric rise in e-commerce.[134] In the past, obtaining items from foreign suppliers typically required a would-be proliferator to visit in-country to discover potential suppliers and arrange a deal, or otherwise physically invite potential suppliers. These physical visits and queries placed a proliferator at risk of discovery, as we saw in the Yugoslavian case recounted above in Section 14.3.1. In addition, a proliferator unfamiliar with foreign firms would have difficulty finding the small and less visible suppliers that could not afford to maintain large export control departments. Online trading has profoundly changed this dynamic by de-materializing the trading process. Hundreds of suppliers from all over the world can list their goods on high-profile e-commerce websites, thereby gaining visibility. Queries are sent and answered and purchases are arranged electronically through the e-commerce website itself, obviating the need for physical travel.

We have previously studied AG-grade biological equipment offered for sale online and provided market research evidence that partially supports these hypotheses regarding the disruptive effects of the rise in e-commerce on export controls.[135] The positive global development of civilian biopharmacy and bioindustry depends on a lively trade in dual-use biotechnology products, and we stress here that we do not allege that the vendors whose products are cited in this prior report (or in the current chapter) are involved in wrongdoing. Nevertheless, e-commerce in such products does pose growing export control difficulties. For instance, we found that online vendors routinely offered certain payment options known to be vulnerable to abuse by unscrupulous buyers seeking to obfuscate their actual location and identity. This prior study also has direct implications for CW export controls, as some of these biotechnology items can be used to produce toxins or other biologically-derived chemical compounds with potential military applicability.

[††]The equipment tracking challenge is detailed in ref. 250

It confirms the growing availability of such equipment from small and medium enterprises, defined as employing 50 individuals or less, based in non-AG countries.

14.5.2 Protective Equipment

Based on the repeated Syrian procurement efforts described in Section 14.3.1, chemical protective equipment seems to be of particular interest to modern proliferators. As explained in Section 14.2, although chemical protective equipment is not controlled by the AG common control lists, they are controlled by the WA control lists. It was therefore disquieting to us to discover that new and military surplus CW-relevant gas masks and NBC protective sets are widely advertised for sale online, from vendors located in WA Member States and non-Member States alike.

Unlike dual-use production equipment, chemical protective equipment models are extremely easy to identify online. Numerous vendors can be found through simple keyword searches. Since CWC Schedule 2 and 3 chemicals are used in industry, a plethora of sellers offer to export gas masks advertised to defend against chemical agents such as hydrogen cyanide, phosgene, and lewisite, as well as a range of TICs such as phosphine and organophosphorus pesticides. Some gas mask producers sold multiple industrial filter canisters designed to fit general-purpose masks for civilian industrial use. NBC protective suit vendors and several gas mask vendors explicitly advertised their products for defensive military use.[252–254]

Whether the gas masks and the NBC protective suits available for sale are truly capable of protecting users against nerve agents remains unknown, given that minute quantities of nerve agent are sufficient to kill. The aforementioned CWC Schedule 2 and 3 agents advertised as covered by the protective equipment, however, would be perfect for the type of less-resourced proliferators seeking small-scale attacks that would be interested in using online vendors to acquire protective equipment.

Some vendor networks were found to be complex, and appeared to involve WA controlled goods trans-shipped to non-WA Member States for resale. For instance, one company listed as based in Kenya but as shipping out of Shanghai advertises NBC protective suits manufactured in Spain.[254] Although Spain is a WA Member State, Kenya and China are not.

As was the case for the biotechnology equipment vendors discussed in the previous subsection, the vendors' accepted payment options included hard-to-trace services.

As with the online dual-use equipment vendors discussed above, better product sales tracking and company outreach by local or national authorities is necessary to prevent export violations. Many of the companies openly advertise the military applicability of their products. As such, they can hardly plead ignorance of the sensitivity of the equipment they are selling. Outreach efforts could focus on detecting "red flag" indicators of front companies, and on building up a relationship between the authorities and the company to facilitate reporting suspicious would-be buyer requests.

14.5.3 Munitions and Delivery Systems

The current rate of proliferation of CW-relevant munitions and other delivery systems is alarming. Only the long-range, high-payload MTCR Category I systems are still outside the reach of most States. Since the flow of advanced delivery systems seemingly cannot be staunched, greater attention must be paid to the specialized sub-systems necessary for CW agent delivery.

Conventional munitions, including free-fall bombs, artillery shells, and artillery rockets, are so poorly controlled on the export market that any country with the will to do so can acquire large stocks. States seeking CW capabilities for warfighting need a large and stable source of munitions, and systems available on the export market are readily adaptable to CW purposes. The Iraqi CW program relied heavily on modified foreign conventional munitions obtained on the export market.[136] One unnamed "foreign company" even provided assistance with testing potential CW adaptations to artillery and rocket projectiles outside of Iraq.[137,138] More recently, the two types of sarin-filled projectiles used by Assad's forces in Syria appear to have been foreign-designed or modified from foreign conventional munitions.[‡‡,139] One of the munitions appeared to be a Soviet-era M14 artillery rocket; the recovered rocket engine bottom rings analyzed by the UN investigative team had Cyrillic lettering, pointing to the foreign origin of at least one component of the original munition.[140] The origin of the other munition type is more tentative, but the large rocket system seems to be a modification of a known 333 mm Iranian conventional rocket and appears to have been launched from an Iranian launcher in Syrian service or a derivative thereof.[141,142] The Syria case therefore appears to parallel the Iraqi CW program in its use of modified foreign munitions obtained on the export market.

States that are unable to tap into the regular arms export market, for instance because of strongly-enforced embargoes, can turn to illicit suppliers. During the Iran–Iraq War and despite sanctions on Iraq, the latter was still able to find foreign suppliers offering 122 mm rocket warheads suitable for chemical fill or conventional rockets that could be cannibalized for parts to help produce domestic CW variants.[143] As documented in the UNSC Panel of Experts reports on North Korean sanctions, the latter is active today in provisioning other States in artillery rocket parts, and more generally has

‡‡The UN report on the 21 August 2013 chemical attack did not fully identify the origins of the munitions, but noted that one munition "indicatively matches one of the variants of the M14 artillery rocket" and another "indicatively matches a 330 mm caliber, artillery rocket." See ref. 255 for further information.

supplied turn-key ammunition production facilities.[144–147] The collapse of State control in Libya, Syria, and parts of Iraq has opened up enormous stocks of such weapons for use even by non-State actors.[148]

Munitions that disseminate aerosols (such as smoke munitions) or other non-explosive payloads (illumination rounds) pose an increased risk of diversion to CW programs. The UNMOVIC Iraq Compendium provides several examples. For instance, Iraq's 155 mm smoke shells, when filled with mustard, "were the most reliable chemical munitions ever produced by Iraq's CW programme."[149] Similarly, Iraq's BR-250 and BR-500 smoke bombs "required only minor modifications" to accommodate a CW fill.[150] In yet another instance of converted non-explosive munitions, the Iraqis designed an experimental binary VX artillery shell using a foreign 152 mm illumination projectile as a starting point.[151]

A subset of such aerosol-generating systems, large area coverage "riot control" munitions, can be directly used for some military missions and hence pose the greatest threat of diversion. These have included long-range systems, such as artillery and air-delivered projectiles, with highly dubious applications to "riot control."[152] For instance, a 2006 weapons catalog meant to promote Russian ordnance and munitions detailed a 500 kg cluster bomb designed to generate "high concentrations of an irritant agent [...] within a short time."[153] The numerous issues surrounding the development of these munitions are detailed by Crowley in Chapter 12; it suffices to note here that some such weapons have been developed by private companies, marketed domestically, and in some cases even made available for export.

Other systems enabling the dissemination of CW agents, such as UAVs, aircraft, helicopters, submarines, and ballistic and cruise missiles continue to proliferate. UAVs are particularly complex to control given the incredible expansion in existing dual-use models, only some of which are suitable for CW purposes. The US Government Accountability Office estimated that some 41 countries had UAVs in 2004 and at least 76 had UAVs in 2012.[154] According to a recent count, approximately 87 countries have some type of UAV capability, with 23 States possessing more advanced UAV models.[155] Small-scale "hobbyist" UAVs capable of delivering a few kilograms of payload are now widely sold online, and terrorist groups have already acquired and used such systems.[156,157] Aside from greater licensing requirements, one suggestion designed to redress this alarming trend that merits further study is requiring manufacturers to code no-fly zones into their UAV's firmware to be monitored by the GPS unit, so that the drone would desist from flying in non-civilian airspace or refuse to take off from war-torn countries.[158]

In our biotechnology equipment e-commerce report, we noted that while we had not yet identified AG-grade UAV-mounted spray systems for sale on e-commerce websites, such as Alibaba, we had found existing dual-use systems that were nearing the minimal AG control standards.[159] In addition, we found that purchases of other potential dissemination systems such as individual ground aerosol dispensers, and even large dissemination platforms, such as crop duster airplanes and mini-submarines that could accommodate spray systems, were being offered on e-commerce websites.[159] The latter delivery systems would presumably, by their very nature, be difficult to rent, sell, or divert to those lacking requisite licenses. Ground-based aerosol dispensers, for their part, deserve greater attention by export control practitioners. Current AG and MTCR standards are focused exclusively on sprayers that can be mated with airborne platforms, but the new generation of ground-based small-particle aerosol dispensers for medical and pest control purposes is of significant dual-use concern.[160] Although these systems might appear rudimentary in an age of UAVs, ground projectors were widely employed for CW attacks in WWI.

Missile system proliferation has continued apace. By one recent count, over 80 countries have anti-ship cruise missiles—and hence at least a rudimentary cruise missile capability—and some 22 countries currently possess manufacturing capabilities for such weapons.[161] Ballistic missiles have likewise continued to spread globally since the surge in demand in such systems unleashed by the Iran–Iraq War in the late 1980s.[162] To give one relevant example, Syria and North Korea have recently demonstrated their possession of a highly-accurate solid-fuel Soviet-era ballistic missile system, the Tochka (NATO reporting name SS-21 Scarab).[163,164] According to specifications provided by the UNSC Panel of Experts on North Korean sanctions, the Soviet version has a 482 kg warhead capable of dispensing 50 explosive sub-munitions, which in turn can inflict effective combat effects on an area 3 hectares in size.[165] To put the potential effectiveness of the weapon into perspective, it would disseminate explosive sub-munitions onto an area roughly the size of three rugby fields. Although not officially confirmed by the Soviet Union or Russia, military experts believe that the Soviet Union maintained a chemical cluster munition warhead variant for the Tochka; this belief is so widespread that even a Russian nuclear, biological, and chemical (NBC) troop training manual mentions it as an example of a delivery system.[166]

While we have no knowledge as to whether North Korea or Syria possess CW cluster munition warheads for the Tochka, this case should serve as the impetus for much closer attention to exports of specialized sub-systems, such as cluster sub-munitions, that can be adapted for CW agent delivery. Monitoring initiatives should pay close attention to weapons used in past CBW programs. The link between conventional sub-munitions and CW was made explicit in the UNMOVIC Iraq Compendium. According to "experts' opinion, the technology that lends itself to the successful design and manufacture of conventional cluster munitions, may by virtue of its very design contribute to the efficient dissemination of chemical or biological agents,"[167] and "the nature of cluster munitions is such" that once a design is "proven and tested with conventional sub-munitions, it easily can be applied to chemical or biological sub-munitions."[168] A 2014 RAND study noted that controls on sub-munition warheads for ballistic missiles could be reinforced by explicitly including them as part of MTCR-controlled sub-systems.[169] Aside from the sub-munitions themselves, warhead sub-munition dispensing systems and the associated sensing and fuzing systems and software necessary to ensure optimal sub-munition release for CW are products that require closer scrutiny by the arms control community. In particular, CWC States Parties should promote a development and export ban on "riot control" chemical-dispensing cluster munitions, given the obvious inapplicability of the weapons to actual "riot control" situations.

14.6 Proliferation-sensitive Production Software

Production software comes in two forms: individual equipment software that facilitates the use of a specific piece of equipment, and facility-wide software that is used to monitor the operations of a chemical plant as a whole.

The AG has an unclear "carve-out" clause enabling software exports. The "controls on 'software' transfer only apply where specifically indicated in [the AG chemical equipment control list] sections I and II above," but the only mentioned software in the stated sections is dedicated software for toxic gas monitoring systems.[5] While equipment software could be considered under the broad "related technology" clause in section III of the list, this section is left out by the software clause cited above. We do not know if this was an intentional decision or simply an oversight, because the discussion on intangible technology transfers included by the authors of the AG common control list handbook volume on chemical technology makes it sound as if AG controls are meant to encompass more than just this single software type.[170] In the United States, AG chemical process equipment and entire chemical process plant exports are both classed under entry 2B350 of the Bureau of Industry and Security (BIS) Commerce Control List.[171,172] Toxic gas monitoring system software is listed under the subsequent, separate, entry 2B351. As we write this, no software control entries are listed as relevant to 2B350. In contrast, toxic gas monitoring systems have a controlled software entry (2D351). Thus, production software by itself may not be currently controlled, perhaps because the AG list was drafted before highly-computerized chemical process equipment became a common occurrence. In any case, production software is proliferation-sensitive and is at a minimum subject to "catch-all" controls.

14.6.1 Equipment-specific Control Software

By equipment-specific control software, we mean the software that facilitates and converts data from mechanical sensors in the equipment into digital form for output to the user, as well as software that enables equipment control panels designed to facilitate user handling of processes carried out by the equipment. Before the advent of the internet, equipment-specific control software either came with the equipment itself or could be purchased separately, but in either case the software had to be physically transferred from one country to another (for instance on a CD-ROM). Since one can now transfer software over the internet, a modern proliferator can obtain or re-export controlled software without carrying out a physical transfer.

We emphasize the importance of controlling equipment-specific software independently of the underlying piece of equipment because proliferators have in the past acquired software independently of the underlying hardware to facilitate illicit procurement.[173] A future proliferator may also seek software updates for an existing piece of equipment, or may seek equipment software to refurbish a second-hand piece of equipment.

Chemical equipment facilities are increasingly relying on individual equipment sensors for monitoring purposes. To facilitate the transfer of such information to a control center running facility-wide control software, growing numbers of equipment-specific control software are now internet-enabled.[§§,174,175] Such control sensors can assist companies leasing or selling controlled equipment to destinations deemed at risk, where best practices dictate that they insist on a right of inspection as part of their commercial contract. In some past sensitive cases, embassy personnel would from time to time visit in-country workshops to ensure that the controlled equipment sold by companies from the embassy's country had not been diverted.[¶¶] Since these visits were infrequent, a proliferator could nevertheless misuse the equipment in the interim period. A sensor-equipped device, however, could be designed to constantly beam its status to the embassy's personnel over the internet or otherwise store this information in memory for later retrieval. A temporary diversion would become easier to catch. Therefore, a proliferator's equipment provision prospects are complicated by the growing connectivity of chemical equipment. However, there is a growing realization within the industry that enabling local (intranet) and especially internet connectivity on critical chemical equipment poses sizeable security risks.[176] Thus, it is unclear whether the trend toward connected equipment will continue.

14.6.2 Facility-wide Control Software

For laboratory-scale CW programs, facility-wide control software is not necessary. However, facility monitoring software is extremely important in ensuring the continuous and safe operation of an industrial CW plant. This is doubly true for plants that produce lethal CW agents rather than less-lethal incapacitants, since the industrial-scale production processes for most lethal CW agents involves highly exothermic reactions and corrosive and toxic precursors. Such reactions can occasion leaks of lethal products and even explosions if improperly controlled. The inevitability of human errors

---

[§§]By contrast, manual retrieval of sensor data is typically performed using an acquisition card and a laptop.

[¶¶]Such an arrangement is described in ref. 256.

and equipment breakdowns compound these issues. Even proliferators with no regard for their workers' health are likely to seek such software, since leaks or malfunctions would interrupt production and risk detection by foreign intelligence agencies.

Enforcing export controls on such software, however, is extremely challenging. Aside from leak detection software, the AG control list does not explicitly place controls on facility-wide control software. As such, facility-wide control software exports are principally covered by "catch-all" controls or by additional national regulations. We hope to demonstrate the weak nature of these measures through a case study of the chemical process control software used at the Hungnam Fertilizer Factory located in Hungnam, North Korea. This case study demonstrates the myriad ways by which CW-relevant software may be obtained.

The Hungnam Fertilizer Factory produces fertilizer for civilian use.[177,178] It also probably supplies the ammonium nitrate necessary for the production of military explosives at a nearby site in the same city.[179] Satellite imagery reveals that the factory has a rail line connection to stores with sizeable berms meant to prevent chain detonations from spreading across several depots in case of an accident.[‖‖] A once-highly-classified Soviet General Staff city plan of Hungnam in our possession confirms that this second site is an ammunition storage ground.[180] Hungnam city is widely cited as housing some chemical weapons relevant infrastructure, with the Hungnam Fertilizer Factory routinely listed as one of the sites allegedly involved in CW agent production.[181–184] South Korean defense white papers published in 1996 and 1998 emphasized the existence of CW infrastructure in or near Hungnam.[185] Any attempted chemical industry exports to the Hungnam Fertilizer Factory, as for any other chemical facility within Hungnam, would therefore almost certainly trigger a catch-all clause for export licensing.

North Korean-published footage of the factory's control room shows its staff using a chemical process control program.[186] Joseph Bermudez Jr, an expert on the North Korean military, noted that the facility had reportedly been upgraded and computerized in the early 2000s alongside other critical chemical complexes.[187] How this software ended up at the facility remains a mystery. Although North Korea claims to produce process control software, we remain suspicious that this particular version is foreign-designed.[188] First, the facility control computers were running on Windows rather than on North Korea's Red Star Linux-based operating system. Had North Korea indigenously designed the software, one might expect it to have been designed for their own operating system. Second, some footage showed three-dimensional displays, which are uncommon in the industry

‖‖The Hungnam Fertilizer Plant entrance is at: 39°50′31″N, 127°37′38″E. The explosive storage is at: 39°49′28″N, 127°35′05″E. To avoid potential confusion, we also provide the coordinates of the oft-mentioned February 8 Vinalon Complex, which is also in Hungnam but is a separate site: 39°51′31″N, 127°34′10″E.

and are needlessly complicated for a first design. Third, in some propaganda footage, North Korea had blurred a small portion of the screen to obscure the software's name but not the rest of the display.[189] Fourth and most importantly, North Korea attempted to import industrial control software suites in the same timeframe, indicating a lack of domestic capability in this regard.[190,191]

When we attempted to determine the software's origin, we were surprised to discover that the facility had been named as a potential candidate to receive foreign investments in South Korean and Japanese technical publications at around the time the facility was reportedly upgraded.[192] These must be placed in their historical context: up until the 2003 breakdown of a diplomatic agreement known as the Agreed Framework, foreign relations with North Korea were in a state of détente. This apparent willingness to invest in the plant is nevertheless troubling given the aforementioned allegations surrounding the site.

Perhaps the software was actually domestically designed. Else, perhaps the software was provided by a foreign company that did not discover and was not informed of the facility's likely military dimensions, or whose country did not have catch-all CW controls in place. It is even possible that such a hypothetical company could have requested and received an export license given the political context at the time. Other, more complex, potential explanations cannot be discounted. A company may have exported chemical control software to another chemical facility in North Korea that did not have a negative reputation, and North Korean engineers could have subsequently illicitly transferred the software and equipment setup to Hungnam. Or perhaps the software was reverse-engineered by North Korean engineers themselves, or even possibly by a third party which then sold it to North Korea (with or without installation assistance).

We spoke on background with a former procurement officers that handled (*inter alia*) export-control compliant software procurement for a large multinational company to ascertain the feasibility of each of these routes. They made clear that such software was extremely difficult to install and run without assistance from the manufacturer. They explained that even their former employee's large and highly-trained engineering teams stayed in daily contact with their software suppliers for troubleshooting assistance. Based on this personal experience, they believed it most likely that the software (whatever it may be) had been both sourced and installed by its manufacturer. A second, but in their opinion, less likely explanation was that a third-party counterfeit group had reverse-engineered an existing piece of software and had provided installation assistance. The former procurement officers stressed that an unassisted installation and especially the outright reverse-engineering of chemical control software would be a major feat of software and hardware engineering. They emphasized that any group using counterfeit software were bound to run into technical issues. They felt that the reverse-engineering of such software was a large investment for a relatively small and risky payoff and was implausible outside of the most desperate and well-resourced of

State programs. However, they did not rule out this possibility in the case of a North Korean high-priority project.

None of these options are reassuring from a CW-relevant software export control point of view. If this facility was deemed a permissible export location by a company or a government despite its well-known affiliation with conventional military activity and widespread rumors of illicit CW activities, the bar for control appears in our view impermissibly low. In all other cases, one known proliferator of CW-relevant equipment has significantly more capability in facility-wide chemical control software procurement than previously known.

14.7 Proliferation-sensitive Chemical Agents

As mentioned in the production equipment discussion in Section 14.5.1, the number of facilities capable of producing Scheduled chemicals have greatly increased and their locations have become globalized. Because Scheduled chemicals fall under the purview of the OPCW, preventing the diversion of such chemicals toward CW programs reduces to the problem of strengthening national implementation of the CWC's provisions.

In contrast, biosynthesis developments pose a significant challenge to CW-relevant control measures.[193] These advances have made the cheap mass production of a growing range of biologically-derived compounds possible, including certain CW-relevant toxins. In the past, militarily-useable quantities of toxins or other biologically-derived compounds had to be synthesized chemically if a suitable method could be found, or else harvested from natural sources at enormous expense. The biotechnology revolution and the resultant "convergence" between biology and chemistry has enabled the biosynthesis of chemicals, whereby bacteria are genetically modified to produce a desired compound. Growing civilian pharmaceutical demand has buoyed the biosynthesis of novel toxins and other biologically-derived compounds with potential CW interest. To cite one example out of many, biosynthesis capabilities for certain conotoxins have been developed by civilian researchers in China.[194] Conotoxins have medical potential, but CW defense experts have also flagged their potential for misuse.[195] The range of potential CW agents will continue to grow as biosynthesis capabilities expand to compounds used as, or as precursors for, the manufacture of medication and commercial products.[196] The following four trends, all directly or indirectly brought up by experts within the framework of the Biological Weapons Convention's (BWC) intersessional meetings, presage a growth in supplier capabilities and thus increased export control challenges:

- The growth in commercial services for the synthesis of small peptides, toxins, and bioregulators;[197,198]
- The future mass-production of peptides by such suppliers, whereby production yields of up to several hundred kilograms per year per plant are on the horizon;[199]

- The increasing capability for the discovery of novel biologically-derived compounds with harmful effects as a result of improved screening techniques;***
- The emerging potential for the design of novel toxins by harnessing effective protein engineering techniques and improved knowledge enabling better predictions of protein structures.[200]

Current controls on the commercial production and export of biologically-derived compounds are poor for two reasons. First, biological facilities rather than chemical facilities can be used to produce these compounds, and OPCW declaration and verification measures are concerned with the latter. As a result, it is far easier to conceal the production of biologically-derived compounds for warfare use from the OPCW compared to the production of traditional CW agents. And second, since many biologically-derived compounds are not CWC Scheduled chemicals, the trade in such compounds does not have to be declared by Member States.

In addition to issues posed by the growth in the legal trade, the illegal market for chemicals is of growing concern with regard to proliferator access to small quantities of CW agents. Potential CW agents are available from existing illicit drug, doping, and counterfeiting markets.[200–202] The fact that a drug syndicate's laboratory was using an obscure drug synthesis pathway published in 1943 employing a radioactive catalyst demonstrates that the services of some remarkably skilled chemists are available to some well-resourced non-State groups.[203]

"Convergence" issues likewise apply to illicit production, and toxins and other biologically-derived substances are already available from black market suppliers. For instance, doping laboratories have turned to producing peptides in an effort to avoid detection and therefore have the technical capability to produce a range of potential CW agents at least in small quantities.[202] Looking forward to the next few years, the demonstrated capability to produce opioids in yeast poses obvious illicit drug production risks.[204–206] These same compounds could be used as lethal or incapacitant CW agents. As legitimate biosynthesis production expands to encompass precursor chemicals used to produce medicines, counterfeit medication production laboratories are likely to follow and pursue these pathways themselves. Overall, illicit laboratories are therefore expected to follow the trend set by their licit counterparts in having the ability to biosynthesize small quantities of a growing range of potential CW agents. Other non-State actors could then acquire small quantities of such agents on the black market for CW use.

Issues of small-scale black market supply are compounded by the emergence of the new internet-enabled proliferation pathway for these substances. A growing proportion of illicit drug and doping product sales are now organized online, providing users with easier access to illicit

***Not directly stated, but implied, in ref. 257.

suppliers worldwide.[†††,207–209] A few individuals have already been caught through FBI "sting" operations trying to acquire ricin through online marketplaces either out of curiosity, or to aid and abet murder.[210,211] As additional illicit actors obtain the means of producing biologically-derived CW agents, a growth in actual suppliers might occur to meet this apparent demand.

14.8 Proliferation-sensitive Information

Proliferation-sensitive information can be shared in three different ways: through in-person transfer, through the exchange of physical objects such as reports, and through interactions on the internet. These pose different challenges to export controls.

Direct information transfer through in-person training has been used by proliferators such as Iraq as a means of obtaining sensitive CW information, and these events remain a dissemination pathway to monitor.[212] One example of such a collection effort occurred in 1981 with the worldwide travels of then Brigadier Wouter Basson to elicit sensitive CBW information for use by apartheid South Africa.[213] Advances in biometrics and data analysis software used to identify, monitor, and prevent the entry and exit of individuals with known foreign military ties might improve the ability of States to prevent this type of short-term attendance. However, past proliferators have also been willing to send their specialists for long-term education and work abroad as a way of eventually obtaining sensitive knowledge, and these types of arrangements are increasingly difficult to prevent. The advent of multinational companies hiring from a global workforce translates into a massive growth in potential transfers of sensitive information that must be monitored. The possible security concern over foreign nationals working in sensitive industries is often in conflict with the ability of national firms to compete globally but also with broader values of scientific openness and with the basic human right of freedom of speech.[214,215] Partly in response to this proliferator practice, the US has "deemed exports" rules that allow the prosecution for export violation as soon as controlled information is transferred to a foreign national. As such, the information does not have to leave the country for an export violation to occur. But enforcing "deemed exports" rules means being able to prove that sensitive information has been transferred, and this has mostly required tying the information to a specific controlled item. To cite a recent example, a US-origin multinational company settled an export control violation case with the US Department of Commerce's Bureau of Industry and Security (BIS) in part over allegations that a Russian national working at its California-based plant had repeatedly been allowed to access export controlled blueprints for controlled electronics.[216,217] It would be extremely difficult to enforce export controls in such a manner over large swathes of

---

[†††]Not directly stated, but implied in ref. 257

advanced chemistry training relevant to CW, such as aerosol toxicology work, given that this training would not be tied to the manufacturing or installation of controlled equipment.[258]

Physical objects such as sensitive reports, manuals, articles, and books are other traditional routes by which sensitive information can be shared. The advent of internet trading platforms where physical copies can be sought and bought has enhanced a modern proliferator's ability to obtain information in this manner. For instance, military manuals have been found for sale on e-commerce platforms that allow book sales, raising fears that sensitive copies not meant for public release could be exported abroad through such networks.[218,219]

Aside from enhancing the trade in physical objects containing sensitive information, the advent of the internet has opened a new and virtually uncontrolled channel for the diffusion of proliferation-sensitive information. The aforementioned controlled electronics case demonstrates how transfers of sensitive information can now be carried out without physically transferring objects. Rather, foreign national employees were provided login and passwords to access the company's US-based mainframe, which housed computer files containing sensitive information.[220] Since these non-US employees now had the ability to remotely access and copy the export-controlled blueprints over the internet, the BIS invoked US "deemed exports" regulations.[220] Furthermore, because one non-US individual employed in a company subsidiary located outside of the US allegedly used their access to open one such file (and hence transferred information across the internet from the US mainframe to their workstation in a foreign country), the BIS also invoked regular export control regulations.[220] Beyond its ability for non-physical and near instantaneous transfers, the internet enables global access to information. A document that has been improperly declassified or leaked can now rapidly be uploaded and indexed and hence become easily retrieved online by anyone in the world. The rise of free automated translation software exacerbates this problem, since a proliferator can leverage data released or published by many different countries. Information classified in one country may not be classified in another. In addition, declassification mistakes have occurred. A skilled proliferator can therefore leverage translation software to conduct thorough literature searches on the former CW producing States to leverage any differences in classification and more readily identify and obtain sensitive information.

Having reviewed the ways in which transfers of CW-relevant information can and have been orchestrated, we now turn to a discussion over what constitutes *sensitive* CW-relevant information. Proliferation-sensitive information of relevance is divided here into two categories: information on the manufacture and effects of CW-relevant chemical compounds, and information of relevance for proper CW agent dissemination. Each of these categories are analyzed in turn in the subsections that follow.

In general, CW-relevant information can be generated by scientists not affiliated with a government laboratory and published in academic journals, as well as information from government scientists published in journals or contained in non-classified, declassified, or leaked government documents. The lion's share of scholarly attention to the issue of information export controls has been placed on issues surrounding academic publications, notably those detailing the synthesis of novel CW-relevant chemicals.[221–225] As we argue here, however, the real threat comes from documents revealing information of relevance for proper CW agent pilot or industrial production and dissemination, which are unlikely to be penned outside of government CW defense institutions.

14.8.1 Information on the Manufacture of CW-relevant Chemical Compounds

The Iraqi and Aum Shinrikyo case studies demonstrate that, even before the widespread development of the internet, open-source research could uncover synthetic pathways for CW agents. UNMOVIC concludes that:[212]

> "Iraqi scientists involved in the CW programme obtained much of their information on basic CW production technology from training courses held in foreign institutions, open source publications, foreign patents, international conferences and international forums. This information obtained from open source material was first tested by Iraq at the laboratory level to identify and adjust unknown parameters of the synthesis of chemical warfare agents that could not be found in open sources, such as kinetics of chemical reactions, combinations of catalysts, equipment specifications and scale-up procedures."

Assuredly, pathways for developing and producing nerve agents are available in open sources. It is telling that the synthetic route chosen by Iraq for the production of sarin was "well known from open scientific literature."[226] One Iraqi interviewee told UNMOVIC inspectors: "the literature was the source of so much data that we did not need to test or check so much."[226] Similarly, open literature information "presented many routes for the production of V-agents."[227] However, based on the above UNMOVIC paragraph, the key parameters for pilot or industrial-scale production are probably not available from open sources, and as such the open literature lacks "many important technical details" in this regard.[228] Known proliferator cases are consistent with this conclusion. In Aum Shinrikyo's case, although Masami Tsuchiya rapidly learned how to prepare laboratory-scale amounts of sarin from open sources, the group floundered when faced with industrial-scale production problems.[229] Libya similarly failed in its efforts to go from laboratory-scale to the industrial-scale production of sarin and soman.[230] In addition to a lack of key parameter data, the reliability of existing publicly available information cannot be taken for granted by a proliferator. For instance, the published stoichiometry for certain synthesis pathways of CW-relevance are known to be wrong.[231] This would complicate a proliferator's attempt at industrial scale-up if they did not correct for these errors, since incorrect ratios of feed chemicals would be ordered and used. At both the industrial and the laboratory scales, such mistakes would yield a product of inferior quality.[231]

The security issue surrounding publishing research on novel chemicals with a potential for misuse is not new. Chemists are already all too accustomed to the dilemma of publishing findings regarding psychoactive compounds that may then be abused by organized crime for the production of synthetic illicit drugs, as well as findings on novel anesthetic or poisonous compounds that could be misused to commit crimes. Since novel chemicals with a potential for misuse, particularly novel psychoactive compounds, have both an organized crime and a proliferation component, it makes little sense to advocate for a new, separate, publication control mechanism solely for proliferation threats. A rational approach to the publication problem should not see the rising threat of terrorist misuse greatly upset the existing balance between the need for free academic publishing and the need for security, given the existing and already staggering security problems posed by the illicit drug trade. Any control on sensitive information must be carefully weighed against the potential chilling effect on fundamental research they could have (Table 14.1).

Aside from certain industrial biosynthesis production methods, relatively few academic publications on CW-relevant chemicals would be of interest to a proliferator beyond what has already been published. The publication risk for work on potential novel agents crucially depends on the relative complexity of the synthetic pathway discovered and its potential for industrial scale-up. The scientific literature abounds with compounds known to be more toxic than the nerve agents. However, proliferators have historically not weaponized many of these agents in part because their synthesis was even more difficult than that of the nerve agents, which were understood to be sufficiently toxic to be militarily effective. A modern proliferator would mostly be after agents of similar effectiveness but of lower difficulty of manufacture than the traditional nerve agents or that could be produced using less controlled precursors. Publications detailing scalable production pathways through biosynthesis and filtration of biologically-derived compounds known to be potential CW agents fall within this category and are therefore cause for concern. When it comes to purely chemical pathways, however, it is far less clear what constitutes an improvement—let alone an improvement necessitating publication controls—over what has already been made public.

Indeed, academics not affiliated with a government CW defense program are unlikely to have the tools to assess CW-relevant publication risk, since as discussed above, industrial process details on synthetic pathways for nerve agents and on additive effectiveness are not (and should not be) easy to find in open sources. Aside from certain industrial biosynthesis production methods, the risk posed by academic publications therefore seems to reside mostly in accidental discoveries; research findings that are discovered to be

Table 14.1 Publication risk assessment for findings of potential relevance to CW agent production.

Risk assessment factors	Comments on the risk assessment
Is it easy to understand the finding's applicability to CW?	High risk for government publications or for publications by military or government-affiliated researchers. High risk if a known CW agent, precursor, or CW defense site is mentioned. Increased risk if publications cited are themselves by military or government-affiliated researchers or mention one of these topics
To what extent are the findings novel?	Risk increases if similar techniques or findings have not been published, or if similar work is only found in hard-to-access sources (ex. rare monographs in a foreign language that are not available online)
The publication presents a...	
...new synthesis pathway for a known CW agent or precursor	Risk depends on agent effectiveness and current availability: high for CWC Schedule 1 chemicals and precursors. Risk increases when the new precursors are less-controlled
...synthesis pathway using different equipment (ex. toxin biosynthesis *vs.* chemical synthesis)	As above, risk depends on agent effectiveness and current availability. High risk if the new equipment is less-controlled. High risk if the new equipment allows for illicit production outside of chemical facilities declared to the CWC
...novel lethal agent	Risk depends on comparative efficiency and difficulty of synthetic pathway compared to closest known analog
...novel incapacitant	Risk depends on comparative efficiency and difficulty of synthetic pathway compared to closest known analog
...novel additive	High risk when the associated CW agent(s) is/are explicitly identified, when the additive's comparative effectiveness is equal or higher when compared to known CW additives, and when the publication's overall level of detail is greater than prior publications on known additives. Extreme risk if the additive also confers mask-breaking or skin penetrability functionality to the final mixture when added to an agent that does not have such functionality[247]
...production technique that enhances delivery or broadens delivery options (ex. microencapsulation)	Risk increases with the availability of the underlying production and dissemination equipment. Risk also depends on the technical sophistication necessary to operate such equipment

of CW relevance only after their publication or that are not known to be of CW relevance by civilians. No proposed control schemes can be effective in such cases. At the same time, these publications will not be easy to associate with CW, decreasing the chances that the findings become utilized by proliferators.

14.8.2 Dissemination Information: Models and Munitions

Information of relevance for proper CW agent dissemination includes aerosol generation and propagation data, as well as data on munitions particularly well-adapted for delivering aerosols over large areas. In the past, technical publications on munitions were vetted mostly in the context of highly capable State-level Cold War adversaries. Information on designs deemed obsolete were allowed to be published. Today, non-State groups have demonstrated interest in the design of munitions. Examples include the design of improvised chemical mortar shells by ISIL in Syria and the modification of surface-to-air missiles into improvised surface-to-surface missiles by rebel groups in Libya.[232–234] Designs deemed obsolete by a State could be of great interest to these groups. In particular, ISIL appears to have repeatedly failed in modifying their chemical munitions so that they release an aerosol. Publicly-known attacks have so far seen their munitions spill liquid contents on impact.[235,236] Munitions knowledge is of course only part of the problem in adapting existing munitions for CW use, and engineering and parts supply no doubt pose difficult challenges to non-State actors. One could nevertheless hope for greater controls over such sensitive details, at least in what is made available online.

Two case studies are presented here to make this point. First, we recall how US CBW munitions patents that had been made public during the Cold War were finally pulled from circulation in the early 2000s due to their potential online availability.[237,238] Second, we note that a North Korean publication on designing a fuel-air explosive munition cited one Naval Weapons Laboratory (NWL) report and two Sandia National Laboratory reports.[239] These US reports were marked unclassified, distribution unlimited, although the original NWL report had initially been restricted to US government agencies and contractors. As of this review, all three reports were publicly available online (we contacted the US government websites hosting these papers prior to publication). Although we cannot prove that the North Koreans acquired these reports online, this appears to be the simplest explanation. A look at other North Korean publications shows that Western articles only available online are sometimes cited, confirming that their researchers actively seek out such publications online.[‡‡‡,§§§] The cruel irony that US national lab research has indirectly facilitated the

‡‡‡For example, ref. 259 cites "C. Blundo *et al.*, cryptology ePrint archive report, 2006/170, 2006".

§§§Another likely example is ref. 260 that cites "IEEE P1363a/D9 (Draft Version 9)," which is a resource almost always retrieved online.

development of a North Korean weapon system designed to kill US troops and their allies should serve as a wake-up call for greater controls over public online access to detailed munitions design information, especially regarding those with potential CW applicability.

14.9 Concluding Remarks and Recommendations

We find that malicious actors now have increased access to dual-use equipment, software, chemicals, and sensitive information that can be misused to obtain laboratory quantities of CW agents. While industrial-scale production of CW agents remains highly difficult and preconditioned on advanced technical training, the capability requirements for mounting a small-scale attack are demonstrably diminishing.

Our specific recommendations are as follows:

- Assistance to States seeking improved chemical export controls and export tracking procedures should continue, for instance as part of UNSC 1540 assistance and as part of peer activities through the WCO. These initiatives must be adapted to help national law enforcement establish partnerships with e-commerce website owners based in-country. E-commerce website owners should in turn inform the myriad small and medium enterprises operating through their service on export control requirements.
- Given significant proliferator interest in such gear, law enforcement should conduct urgent outreach activities toward online sellers of CW protective equipment.
- Additional efforts to prevent the spread of dissemination systems appear necessary. Export controls on high-capacity ground aerosol dispensers should be considered and, if found to be practicable, implemented through the AG. Dual-use UAVs should have stricter licensing requirements enabling better export tracking, and the GPS-locking of such systems should be considered; these requirements could be set through national controls or through the WA. Better controls on cluster sub-munitions and their associated sensor and fuzing subsystems should be implemented in part by adding them to the MTCR controlled technology list, even for the few systems that are excluded under the Convention on Cluster Munitions.[240] So-called "riot-control" chemical-dispensing cluster munitions should be declared illegal under the CWC. Other "large-area" chemical-dispensing munitions should be, at a minimum, banned for export (see Chapter 12 by Crowley).
- Outreach activities by law enforcement should urgently include companies offering the commercial synthesis of peptides, toxins, and bioregulators if they have not done so already. National authorities as well as leading companies should encourage self-imposed security best

practices at the industry consortia level as was done by gene synthesis service providers.[241,242]

- AG control list language on software must be clarified. The AG list authors should explicitly state what type of software aside from leak detection software is controlled, if any. AG participants should consult with chemical process and software engineers to determine if equipment-specific software currently sold in association with controlled equipment is sufficiently specific to the controlled equipment to justify export controls.
- With regard to sensitive information controls, the highest risk is posed by potential government (mis)-declassification rather than new non-governmental research. Namely, any further declassification of technical data on the industrial scale-up of CW production lines would pose an extreme proliferation risk. More resources need to be allocated to often overwhelmed declassification staff to ensure that such data does not slip through controls. In addition, data on munitions with a high risk of conversion to CW munitions should be more strictly controlled. As for academic publications, work describing the industrial production through biosynthesis methods of known CW toxins or other bioactive molecules with known CW potential is one area that poses significant risk of misuse and probably warrants additional oversight.

In a recent review of the field, one long-time export control expert noted that their application appeared to be shifting "toward reinforcing nonproliferation awareness rather than 'controlling' trade."[243] Given the rising volume and the rapidly evolving nature of CW-relevant commerce, the effectiveness of export controls in this field is increasingly dependent on buy-in from companies. Today, small businesses can easily sell and re-sell an increasingly varied range of CW-relevant products to global buyers through e-commerce platforms. States must do more to make the application of export controls feasible for such firms, which do not have the luxury of retaining lawyers to interpret opaque control language and do not necessarily have the sophisticated parameter data all too often necessary to determine whether the item their livelihood depends on is controlled for export. Security practitioners have encouraged firms to incorporate nonproliferation concerns into their corporate sustainability plans and industry standards so as to build awareness of export controls and develop best practices for their application.[244,245] Certainly, the advent of the internet and the resultant explosion in available data provides opportunities for companies to conduct thorough export control verifications in novel ways, ranging from the use of satellite imagery to remotely "inspect" a shipment's stated end destination, to using freely-accessible Automatic Identification System portals to verify in real time that ships and aircraft do not stop or re-transit to at-risk destinations while a controlled shipment is being made. While such developments are laudable, national authorities have a responsibility to maintain export control regulations in a responsive state so as to ensure a common and effective minimum security standard.

References

1. Organisation for the Prohibition of Chemical Weapons (OPCW), *Director-general Circulates OPCW FFM Reports to States Parties*, November 6 2015, https://www.opcw.org/news/article/director-general-circulates-opcw-ffm-reports-to-states-parties/.
2. C. J. Chivers, *ISIS Has Fired Chemical Mortar Shells, Evidence Indicates*, New York Times, July 17 2015, http://www.nytimes.com/2015/07/18/world/middleeast/islamic-state-isis-chemical-weapons-iraq-syria.html.
3. Australia Group, *The Origins of the Australia Group*, http://www.australiagroup.net/en/origins.html.
4. Australian Group, *Australian Group Membership*, http://www.australiagroup.net/en/membership.html.
5. Australia Group, *Control List of Dual-use Chemical Manufacturing Facilities and Equipment and Related Technology and Software*, September 2014, http://www.australiagroup.net/en/dual_chemicals.html.
6. A. Kelle, CBW export controls: towards regime integration? in *Non-proliferation Export Controls: Origins, Challenges, and Proposals for Strengthening*, ed. D. Joyner, Hampshire: Ashgate Publishing Limited, 2006, p. 104.
7. Australia Group, *Guidelines for Transfers of Sensitive Chemical or Biological Items*, June 2015, clauses 7a,b, http://www.australiagroup.net/en/guidelines.html.
8. Australia Group, *Statement by the Chair of the 2014 Australia Group Plenary*, June 6 2014, http://www.australiagroup.net/en/media_june2014.html.
9. Organisation for the Prohibition of Chemical Weapons (OPCW), Convention on the prohibition of the development, production, stockpiling and use of chemical weapons and on their destruction, *Article I. General Obligations*, https://www.opcw.org/fileadmin/OPCW/CWC/CWC_en.pdf.
10. OPCW Technical Secretariat, *Note by the Technical Secretariat: Status of Participation in the Chemical Weapons Convention as at 17 October 2015*, October 19 2015, pp. 1–6, S/1315/2015, https://www.opcw.org/fileadmin/OPCW/S_series/2015/en/s-1315-2015_e_.pdf.
11. OPCW, *Angola Joins Chemical Weapons Convention*, September 21 2015, https://www.opcw.org/news/article/angola-joins-chemical-weapons-convention/.
12. OPCW, Convention on the prohibition of the development, production, stockpiling and use of chemical weapons and on their destruction, *Article VI. Activities Not Prohibited Under this Convention*, **vol. 3**, p. 16.

13. OPCW, Convention on the prohibition of the development, production, stockpiling and use of chemical weapons and on their destruction, *Annex on Chemicals, A. Guidelines for Schedules of Chemicals*, pp. 49–50.
14. OPCW, Convention on the prohibition of the development, production, stockpiling and use of chemical weapons and on their destruction, part VI, *Activities Not Prohibited Under the Convention in Accordance with Article VI, Regime for Schedule 1 Chemicals and Chemical Facilities Related to Such Chemicals*, p. 122.
15. OPCW, Convention on the prohibition of the development, production, stockpiling and use of chemical weapons and on their destruction, part VII, *Activities Not Prohibited Under the Convention in Accordance with Article VI, Regime for Schedule 2 Chemicals and Facilities Related to Such Chemicals, C. Transfers to States Not Party to This Convention*, p. 133.
16. OPCW, Convention on the prohibition of the development, production, stockpiling and use of chemical weapons and on their destruction, part VII, *Activities Not Prohibited Under the Convention in Accordance with Article VI, Regime for Schedule 2 Chemicals and Facilities Related to Such Chemicals, A. Declarations, Declarations of Aggregate National Data*, p. 127.
17. OPCW, Convention on the prohibition of the development, production, stockpiling and use of chemical weapons and on their destruction, part VIII, *Activities Not Prohibited Under the Convention in Accordance with Article VI, Regime for Schedule 3 Chemicals and Facilities Related to Such Chemicals, A. Declarations, Declarations of Aggregate National Data*, p. 134.
18. OPCW, Convention on the prohibition of the development, production, stockpiling and use of chemical weapons and on their destruction, *A. Declarations, Declarations of Plant Sites Producing, Processing or Consuming Schedule 2 Chemicals*, pp. 127–129.
19. Wassenaar Arrangement, about us, http://www.wassenaar.org/about-us/.
20. Wassenaar arrangement on export controls for conventional arms and dual-use goods and technologies, *List of Dual-use Goods and Technologies and Munitions List, WA-LIST (15) 1*, December 3 2015, pp. 169–171, sections 9.A.12, ML2, ML3, ML4, ML10, http://www.wassenaar.org/wp-content/uploads/2015/08/WA-LIST-15-1-2015-List-of-DU-Goods-and-Technologies-and-Munitions-List.pdf.
21. Wassenaar arrangement on export controls for conventional arms and dual-use goods and technologies, *List of Dual-use Goods and Technologies and Munitions List, WA-LIST (15) 1*, December 3 2015, section ML9.
22. Wassenaar arrangement on export controls for conventional arms and dual-use goods and technologies, *List of Dual-use Goods and Technologies and Munitions List, WA-LIST (15) 1*, December 3 2015, p. 197, section ML7.

23. Wassenaar arrangement on export controls for conventional arms and dual-use goods and technologies, *List of Dual-use Goods and Technologies and Munitions List, WA-LIST (15) 1*, December 3 2015, p. 5, section 1.A.4.
24. Missile Technology Control Regime, *Equipment, Software and Technology Annex, MTCR/TEM/2015/Annex*, October 8 2015, pp. 1–80, http://www.mtcr.info/english/MTCR-TEM-Technical_Annex_2015-10-08_Current_version.pdf.
25. Missile Technology Control Regime, *MTCR Guidelines and the Equipment, Software and Technology Annex*, http://www.mtcr.info/english/guidelines.html.
26. Australia Group, Australia group participants, *OPCW Technical Secretariat, Note by the Technical Secretariat: Status of Participation in the Chemical Weapons Convention*, 17 October 2015, pp. 1–6, http://www.australiagroup.net/en/participants.html.
27. Wassenaar arrangement on export controls for conventional arms and dual-use goods and technologies, *List of Dual-use Goods and Technologies and Munitions List*, pp. 174–175, section ML7.
28. Wassenaar arrangement on export controls for conventional arms and dual-use goods and technologies, *List of Dual-use Goods and Technologies and Munitions List*, pp. 6–174, sections 1.A.4, ML7.
29. Wassenaar arrangement on export controls for conventional arms and dual-use goods and technologies, *List of Dual-use Goods and Technologies and Munitions List*, pp. 174–176, ML7.
30. Australia Group, *Australia Group Booklet*, July 2007, p. 10, http://www.australiagroup.net/en/agb_july2007.pdf.
31. Australia Group, *Export Control List: Chemical Weapons Precursors*, April 2015, http://www.australiagroup.net/en/precursors.html.
32. OPCW, *Schedule 1: B. Precursors*, https://www.opcw.org/chemical-weapons-convention/annexes/annex-on-chemicals/schedule-1/.
33. *US Lauds Taiwan for Confiscating Perilous Chemicals*, Taipei Times, August 14 2003, http://www.taipeitimes.com/News/taiwan/archives/2003/08/14/2003063677.
34. F. K. Kawahara and B. L. Mickel, Jet fuel composition, Patent US 3049871A, publication date August 21, 1962, filing date May 12, 1958, priority date May 12, 1958, http://www.google.com/patents/US3049871.
35. C. Cooper, Taiwan seizes chemicals from north korea ship, *Wall St. J.*, 2003, http://www.wsj.com/articles/SB106073362340201900.
36. United Nations Security Council, Resolution 1540 (2004), *Adopted by the Security Council at its 4956th Meeting*, 28 April 2004, pp. 2–3, S/RES/1540(2004), para. 2-3, http://www.un.org/en/ga/search/view_doc.asp?symbol=S/RES/1540%20%282004%29.
37. A. Tzanakopoulos, *Disobeying the Security Council: Countermeasures against Wrongful Sanctions*, Oxford Academic Press, Oxford, 2011, p. 118.
38. United Nations Security Council, *UNSCR 1540 Committee, Annex: Report of the Committee Established Pursuant to Security Council Resolution 1540*

(2004), p. 41, S/2006/257, Annex IX, http://www.un.org/en/ga/search/view_doc.asp?symbol=S/2006/257.

39. United Nations Security Council, *UNSCR 1540 Committee, Annex: Report of the Committee Established Pursuant to Security Council Resolution 1540 (2004)*, pp. 14–15, S/2008/493, para. 60, Annex VIII.B p. 43, Annex XV.B p. 56, http://www.un.org/en/ga/search/view_doc.asp?symbol=S/2008/493.
40. United Nations Security Council, *UNSCR 1540 Committee, Annex: Report of the Committee Established Pursuant to Security Council Resolution 1540 (2004)*, S/2011/579, Annex IX p. 43, Annex XI p. 45, Annex XIV p. 48, http://www.un.org/en/ga/search/view_doc.asp?symbol=S/2011/579.
41. United Nations Security Council, *UNSCR 1540 Committee, Annex: Report of the Committee Established Pursuant to Security Council Resolution 1540 (2004)*, S/2016/1038, Annex XI p. 64, Annex XIII p. 66, Annex XVI p. 69, http://www.un.org/en/ga/search/view_doc.asp?symbol=S/2016/1038.
42. United Nations, *Growth in United Nations Membership*, 1945, http://www.un.org/en/members/growth.shtml.
43. United Nations Security Council, *UNSCR 1540 Committee*, S/2006/257, p. 21 para. 97, p. 22 para. 99–100, p. 22 para. 102–103, Annex IX p. 41.
44. Japanese Ministry of Economy, *Trade and Industry (METI), Trade and Economic Cooperation Bureau, Trade Management Section, Trade Control Division, Trends in Post-examination of Cases*, p. 2, http://www.meti.go.jp/policy/external_economy/trade_control/boekikanri/download/jigoshinsa/2014/20140411_129_jg.pdf.
45. United Nations Security Council, *Letter Dated 25 April 2006 from the Chairman of the Security Council Committee Established Pursuant to Resolution 1540 (2004), Addressed to the President of the Security Council*, pp. 20–21, para. 91.
46. H. Di, *U.S. Caught in its Own Game Playing on Arms Control, Scholar Says*, Stanford University Press Release, February 18 1994, http://news.stanford.edu/pr/94/940218Arc4432.html.
47. (a) T. Colgan, *Advanced Inspection Tools and Training to Prevent Illicit Shipments of WMD-related Strategic Commodities, Presentation Given at the WCO Technology and Innovation Forum*, March 5–9, 2012, p. 10, http://www.wcoomd.org/en/events/event-history/2012/~/~/media/A6DB10B887824951A5E03F6B84442E55.ashx; (b) T. Perry, *Technology and Innovation for Preventing Illicit Shipments of WMD-related Strategic Commodities, Presentation at the WCO Technology and Innovation Forum*, November 2–4 2010, p. 6.
48. Å. Samuelson, *Swedish Customs, Do We Need Better Protection? Presentation Given at the 2010 World Customs Organization Technology Forum*, November 20 2012, http://www.wcoomd.org/~/media/WCO/Public/Global/PDF/Events/2010/Technology%20Forum/Presentations/Ake_Samuelsson.ashx?db=web.
49. *1540 Committee, Summary Requests for Assistance from Member States*, 2010, http://www.un.org/en/sc/1540/assistance/requests-for-assistance-from-states.shtml.

50. *1540 Committee, Assistance Programmes and Offers from International, Regional and Subregional Organizations and Other Arrangements*, http://www.un.org/en/sc/1540/assistance/offers-of-assistance/assistance-programmes.shtml.
51. J. Warrick, *Albania's Chemical Cache Raises Fears about Others, Long-forgotten Arms Had Little or No Security*, Washington Post, January 10 2005, http://www.washingtonpost.com/wp-dyn/articles/A61698-2005Jan9.html.
52. J. Warrick, *U.S. To Aid Albania in Destroying Chemicals*, Washington Post, October 22 2004, http://www.washingtonpost.com/wp-dyn/articles/A52695-2004Oct21.html.
53. R. P. Mikulak and A. C. Weber, *Statements to the Fourteenth Session of the Conference of the States Parties of the Organization for the Prohibition of Chemical Weapons*, The Hague, Netherlands, November 30 2009, http://www.state.gov/t/isn/rls/rm/133099.htm.
54. *Uncovering a Chemical Weapons Program in the Balkans, Studies in Intelligence*, approved for release, http://www.foia.cia.gov/document/0006122213, September 10 2014, p. 5.
55. T. Marshall, *3 Germans Sentenced in Libya Poison Gas Case*, Los Angeles Times, October 10 1991, http://articles.latimes.com/1991-10-10/news/mn-336_1_poison-gas.
56. United Nations Monitoring, *Verification and Inspection Commission (UNMOVIC), Compendium of Iraq's Proscribed Weapons Programmes in the Chemical, Biological and Missile Areas*, ch. VI, p. 1035.
57. United Nations Monitoring, *Verification and Inspection Commission (UNMOVIC), Compendium of Iraq's Proscribed Weapons Programmes in the Chemical, Biological and Missile Areas*, ch. III, pp. 53–54.
58. United Nations Monitoring, *Verification and Inspection Commission (UNMOVIC)*, ch. III, pp. 51–56, fn. 2, 58–59.
59. United Nations Monitoring, *Verification and Inspection Commission (UNMOVIC)*, ch. III, p. 57.
60. United Nations Monitoring, *Verification and Inspection Commission (UNMOVIC)*, ch. III, pp. 61–64.
61. United Nations Monitoring, *Verification and Inspection Commission (UNMOVIC)*, ch. III, pp. 62–67.
62. United Nations Monitoring, *Verification and Inspection Commission (UNMOVIC)*, ch. III, p. 65.
63. United Nations Monitoring, *Verification and Inspection Commission (UNMOVIC)*, ch. III, pp. 142–143.
64. United Nations Monitoring, *Verification and Inspection Commission (UNMOVIC)*, ch. III, p. 143.
65. United Nations Monitoring, *Verification and Inspection Commission (UNMOVIC)*, ch. III, pp. 145–146.
66. United Nations Monitoring, *Verification and Inspection Commission (UNMOVIC)*, ch. III, p. 1034.

67. United Nations Security Council, *UNSCR 1874 Panel of Experts, Report of the Panel of Experts Established Pursuant to Resolution 1874*, 2009, p. 25, S/2010/571, para. 62.
68. AFP, *Greece seizes N. Korea Chemical Weapons Suits: Diplomats*, Korea Herald, November 17 2011, http://www.koreaherald.com/view.php?ud=20111117000214.
69. United Nations Security Council, *Note by the President of the Security Council, Report of the Panel of Experts Established Pursuant to Resolution 1874(2009)*, June 14 2012, pp. 27–29, S/2012/422, para. 65–66.
70. J. S. Bermudez Jr, *North Korea's Chemical Warfare Capabilities, 38 North*, October 10 2013, http://38north.org/2013/10/jbermudez101013/.
71. B. Demick, *North Korea tried to Ship Gas Masks to Syria, Report Says*, Los Angeles Times, August 27 2013, http://articles.latimes.com/2013/aug/27/world/la-fg-wn-north-korea-syria-gas-masks-20130827.
72. P. E. Tyler, *No Chemical Arms Abroad China Ship*, New York Times, September 6 1993, http://www.nytimes.com/1993/09/06/world/no-chemical-arms-aboard-china-ship.html.
73. *U.S. Caught in its Own Game Playing on Arms Control, Scholar Says*, Stanford University Press Release, February 18 1994, http://news.stanford.edu/pr/94/940218Arc4432.html.
74. *U.S. Immigration and Customs Enforcement, Counter Proliferation Investigation Unit, Extradited British Resident Pleads Guilty to Conspiracy to Illegally Export Restricted Laboratory Equipment to Syria*, May 26 2016, https://www.ice.gov/news/releases/extradited-british-resident-pleads-guilty-conspiracy-illegally-export-restricted.
75. R. Danzig, M. Sageman, T. Leighton, L. Hough, H. Yuki, R. Kotani and Z. M. Hosford, *Aum Shinrikyo: Insights into How Terrorists Develop Biological and Chemical Weapons*, Center for a New American Security, 2nd edn, 2012, pp. 20–48, http://www.cnas.org/sites/default/files/publications-pdf/CNAS_AumShinrikyo_SecondEdition_English.pdf.
76. R. Danzig, M. Sageman, T. Leighton, L. Hough, H. Yuki, R. Kotani and Z. M. Hosford, *Aum Shinrikyo: Insights into How Terrorists Develop Biological and Chemical Weapons*, Center for a New American Security, 2nd edn, December 2012, p. 62, fn. 156.
77. M. Faul, *Nigeria arrests Mexicans at Super-meth Lab*, Associated Press, March 14 2016, http://bigstory.ap.org/article/263eb182df5644ec8e78c6a4ebe853f4/nigeria-arrests-mexicans-meth-super-lab.
78. R. Danzig, M. Sageman, T. Leighton, L. Hough, H. Yuki, R. Kotani and Z. M. Hosford, *Aum Shinrikyo: Insights into How Terrorists Develop Biological and Chemical Weapons*, Center for a New American Security, 2nd edn, December 2012, pp. 20–33.
79. R. Danzig, M. Sageman, T. Leighton, L. Hough, H. Yuki, R. Kotani and Z. M. Hosford, *Aum Shinrikyo: Insights into How Terrorists Develop Biological and Chemical Weapons*, Center for a New American Security, 2nd edn, December 2012, p. 29.

80. R. Danzig, M. Sageman, T. Leighton, L. Hough, H. Yuki, R. Kotani and Z. M. Hosford, *Aum Shinrikyo: Insights into How Terrorists Develop Biological and Chemical Weapons*, Center for a New American Security, 2nd edn, December 2012, p. 30.
81. R. Danzig, M. Sageman, T. Leighton, L. Hough, H. Yuki, R. Kotani and Z. M. Hosford, *Aum Shinrikyo: Insights into How Terrorists Develop Biological and Chemical Weapons*, Center for a New American Security, 2nd edn, December 2012, pp. 7–30.
82. R. Danzig, M. Sageman, T. Leighton, L. Hough, H. Yuki, R. Kotani and Z. M. Hosford, *Aum Shinrikyo: Insights into How Terrorists Develop Biological and Chemical Weapons*, Center for a New American Security, 2nd edn, December 2012, p. 31.
83. R. Danzig, M. Sageman, T. Leighton, L. Hough, H. Yuki, R. Kotani and Z. M. Hosford, *Aum Shinrikyo: Insights into How Terrorists Develop Biological and Chemical Weapons*, Center for a New American Security, 2nd edn, December 2012, pp. 29–35.
84. M. Leitenberg, Aum shinrikyo's efforts to produce biological weapons: a case study in the serial propagation of misinformation, *Terrorism Polit. Violence*, 1999, **11**(4), 150–154.
85. R. Pita and R. Gunaratna, *Revisiting Al-Qàida's Anthrax Program*, CTC Sentinel, 15 May 2009, vol. 2, (5), pp. 1–4, https://www.ctc.usma.edu/posts/revisiting-al-qaida%E2%80%99s-anthrax-program.
86. J. B. Petro and D. A. Relman, Understanding threats to scientific openness, *Science*, 2003, **302**(5652), 1893.
87. R. Pita, *Al-Qàida and the Chemical Threat, ASA Newsletter 05-3, 108*, June 9 2005, http://www.asanltr.com/newsletter/05-3/articles/053a.htm.
88. A.-S. Al Somali, Terror franchise: the unstoppable assassin, TECHS vital role for its success, *Bin Laden's Bookshelf*, Office of the Director of National Intelligence, http://www.dni.gov/files/documents/ubl/english/Terror%20Franchise.pdf.
89. D. Francis, *Al Qaeda's Blueprint for How to Start a Homegrown Terror Franchise, Foreign Policy*, May 20 2015, http://foreignpolicy.com/2015/05/20/al-qaedas-blueprint-for-how-to-start-a-homegrown-terror-franchise/.
90. Conflict Armament Research, *Tracing the Supply of Components Used in Islamic State IEDs: Evidence from a 20 month Investigation in Iraq and Syria*, February 2016, pp. 1–107, http://www.conflictarm.com/wp-content/uploads/2016/02/Tracing_The_Supply_of_Components_Used_in_Islamic_State_IEDs.pdf.
91. C. J. Chivers, *The Doomsday Scam*, New York Times, November 19 2015, http://www.nytimes.com/2015/11/22/magazine/the-doomsday-scam.html?_r=0.
92. A. Tilghman, *U.S. Confirms Islamic State Use of Chemical Weapons*, Military Times, August 21 2015, http://www.militarytimes.com/story/military/2015/08/21/isis-used-mustard-gas-makhmour-against-kurds/32116637/.

93. A. Amarasingam, *Families of ISIS Fighters Look for Closure in Leaked Documents*, CBC, March 24 2016, http://www.cbc.ca/firsthand/features/families-of-isis-fighters-look-for-closure-in-leaked-documents.
94. H. Cooper and E. Schmitt, *ISIS Detainee's Information Led to 2 U.S. Airstrikes, Officials Say*, New York Times, March 9 2016, http://www.nytimes.com/2016/03/10/world/middleeast/isis-detainee-mustard-gas.html.
95. Q. Abdul-Zahra and S. George, *Iraqi Officials: US Captured Top IS Chemical Arms Engineer*, Associated Press, March 9 2016, http://bigstory.ap.org/article/82a9e40e92124661ab0321f3027eb646/iraqi-officials-us-captured-top-chemical-arms-engineer.
96. D. Levin, *In China, Illegal Drugs Are Sold Online in an Unbridled Market*, New York Times, June 21 2015, http://www.nytimes.com/2015/06/22/world/asia/in-china-illegal-drugs-are-sold-online-in-an-unbridled-market.html.
97. D. Ovalle and J. Weaver, *South Florida's Source for Synthetic Drugs: The China Pipeline*, Miami Herald, September 16 2015, http://www.miamiherald.com/news/local/crime/article35417625.html.
98. D. Ovalle and J. Weaver, *South Florida's Source for Synthetic Drugs: The China Pipeline*, Miami Herald, September 19 2015, http://www.miamiherald.com/news/local/crime/article35867541.html.
99. D. Ovalle and J. Weaver, *The China Pipeline, Part 3: The Deadly Toll of Synthetic Drugs in South Florida*, Miami Herald, September 26 2015, http://www.miamiherald.com/news/local/crime/article36723141.html.
100. Office of National Drug Control Policy, The white house, *Synthetic Drugs (a.k.a. K2, Spice, Bath Salts, etc.)*, no date, https://www.whitehouse.gov/ondcp/ondcp-fact-sheets/synthetic-drugs-k2-spice-bath-salts.
101. U.S. Drug Enforcement Agency (DEA), *151 Arrested in DEA-led Investigation of Synthetic Drug Rings*, October 15 2015, http://www.dea.gov/divisions/hq/2015/hq101515.shtml.
102. T. Thompson, *Police, Prosecutors Say They're Missing Critical Weapon in Battle against Synthetic Drugs*, NBC Washington, November 25 2015, http://www.nbcwashington.com/investigations/Police-Prosecutors-Missing-Critical-Weapon-in-Battle-Against-Synthetic-Drugs-354033881.html.
103. World Anti-Doping Agency, *The Independent Commission Report #1, Final Report*, November 9 2015, pp. 90–309, https://www.wada-ama.org/en/resources/world-anti-doping-program/independent-commission-report-1.
104. European Monitoring Centre for Drugs and Drug Addiction, Fentanyl drug profile, http://www.emcdda.europa.eu/publications/drug-profiles/fentanyl.
105. K. Coleman and R. A. Zilinskas, Fake botox, real threat, *Sci. Am.*, 2010, **302**(6), 84–89.

106. C. Gould and A. Hay, in *The South African Biological Weapons Program, Deadly Cultures: Biological Weapons since 1945*, ed. M. Wheelis, L. Rózsa and M. Dando, Harvard University Press, Cambridge, 2006, pp. 191–212.
107. V. J. Birstein, *The Perversion of Knowledge: The True Story of Soviet Science*, Basic Books, New York, 2004.
108. J. B. Tucker, The role of the chemical weapons convention in countering chemical terrorism, *Terrorism Polit. Violence*, 2012, **24**, 107.
109. W. M. Alley and J. L. Jones, An analysis of the threat of malicious chemical use by nonstate actors: questioning the state-based approach to chemical nonproliferation, *Nonproliferation Rev.*, 2015, **22**(3–4), 301–319.
110. V. D. Hauschild and G. M. Bratt, Prioritizing industrial chemical hazards, *J. Toxicol. Environ. Health, Part A*, 2005, **68**(11–12), 857–876.
111. A. Ramzy, *Two Children Die in Mass Poisoning at Chinese Kindergarten*, New York Times Blog - Sinosphere, March 21 2014, http://sinosphere.blogs.nytimes.com/2014/03/21/two-children-die-in-mass-poisoning-at-chinese-kindergarten/?_r=0.
112. Human Rights Watch, *Syria: Strong Evidence Government Used Chemicals as a Weapon, Civilian Casualties in Barrel Bomb Attacks on 3 Towns*, May 13 2014, https://www.hrw.org/news/2014/05/13/syria-strong-evidence-government-used-chemicals-weapon.
113. *China arms Giant Says Never Exported Chlorine Gas to Syria*, Reuters, May 16 2014, http://www.reuters.com/article/2014/05/16/us-syria-crisis-china-idUSBREA4F08A20140516.
114. Violation Documentation Center, *Allegations of the Use of Internationally Prohibited Gases in Kafer Zeta – Hama Province*, April 14 2014, http://www.vdc-sy.info/index.php/en/reports/1397487934#.U0wGXfldVDN.
115. M. Burger and C. Gould, *Secrets and Lies: Wouter Basson and South Africa's Chemical and Biological Warfare Programme*, Struik Publishers, Cape Town, 2003.
116. W. Englund, *Ex-soviet Scientist Says Gorbachev's Regime Created New Nerve Gas in '91*, Baltimore Sun, September 16 1992, http://articles.baltimoresun.com/1992-09-16/news/1992260123_1_chemical-weapons-new-nerve-gas-developing-chemical.
117. J. B. Tucker, *War of Nerves: Chemical Warfare from World War I to Al-qaeda*, Anchor Books, New York, 2007, pp. 299–324.
118. S. Bokan, Terorizam I Biološko I Toksinsko Oružje, *Biological and Toxin Weapons – Bioterrorism Arh Hig Rada Toksikol*, 2003, vol. 54, pp. 29–43.
119. S. Bokan, The toxicology of bioregulators as potential agents of bioterrorism, *Arh. Hig. Rada. Toksikol.*, 2005, **56**, 205–211.
120. J. Patočka and L. Středa, Protein biotoxins of military significance, *Acta Med.*, 2006, **49**(1), 3–11.
121. I. V. Domaradskij and W. Orent, *Biowarrior: Inside the Soviet/Russian Biological War Machine*, Promethus Books, New York, 2003, p. 109.
122. K. Alibek and S. Handelman, *Biohazard*, Dell Publishing, New York, 2000, pp. 41–172.

123. K. Hughes and J. Alper, *The Global Movement and Tracking of Chemical Manufacturing Equipment: A Workshop Summary*, National Academies Press, Washington, 2014.
124. K. Hughes and J. Alper, *The Global Movement and Tracking of Chemical Manufacturing Equipment: A Workshop Summary*, National Academies Press, Washington, 2014, p. 2.
125. American Chemistry Council, *Global Business of Chemistry*, https://www.americanchemistry.com/Jobs/EconomicStatistics/Industry-Profile/Global-Business-of-Chemistry.
126. O. Schulz, T. Rings, R. Forrest, J. v. Hoyningen-Huene, *Chemical Industry Vision 2030: A European Perspective*, A. T. Kearney, pp. 4–9, https://www.atkearney.de/documents/856314/1214628/BIP_Chemical_Industry_Vision_2030.pdf/554a744e-4897-452c-a25c-66fc12f00b24.
127. A. Gocke and N. Lang, *How the Global Chemical Industry Has Become a Battle of Equals, BCG Perspectives*, June 24 2015, https://www.bcgperspectives.com/content/articles/globalization-automotive-how-global-chemical-industry-become-battle-equals/.
128. BASF The Chemical Company, *Trends in the Chemical Industry in 2013, BASF Online Report*, 2013, https://report.basf.com/2013/en/managements-report/the-business-year-at-basf/economic-environment/chemical-industry.html.
129. Oxford Economics, *Evolution of Competitiveness in the German Chemical Industry: Historical Trends and Future Prospects, a Report for VCI*, September 2014, p. 5, https://www.vci.de/vci/downloads-vci/publikation/vci-oxford-economics-report-evolution-of-competitiveness-in-german-chemical-industry.pdf.
130. K. Hughes and J. Alper, *The Global Movement and Tracking of Chemical Manufacturing Equipment: A Workshop Summary*, National Academies Press, Washington, 2014, p. 3.
131. K. Hughes and J. Alper, *The Global Movement and Tracking of Chemical Manufacturing Equipment: A Workshop Summary*, National Academies Press, Washington, 2014, pp. 9–13.
132. K. Hughes and J. Alper, *The Global Movement and Tracking of Chemical Manufacturing Equipment: A Workshop Summary*, National Academies Press, Washington, 2014, pp. 9–11.
133. K. Hughes and J. Alper, *The Global Movement and Tracking of Chemical Manufacturing Equipment: A Workshop Summary*, National Academies Press, Washington, 2014, p. 9.
134. K. Hughes and J. Alper, *The Global Movement and Tracking of Chemical Manufacturing Equipment: A Workshop Summary*, National Academies Press, Washington, 2014, pp. 23–25.
135. R. A. Zilinskas and P. Mauger, *Biotechnology E-commerce: A Disruptive Challenge to Biological Arms Control, CNS Occasional Paper No. 21*, May 2015, http://www.nonproliferation.org/biotechnology-e-commerce-a-disruptive-challenge-to-biological-arms-control/.
136. United Nations Monitoring, *Verification and Inspection Commission (UNMOVIC)*, ch. III, pp. 139, 147–177, 186, 189–190.

137. United Nations Monitoring, *Verification and Inspection Commission (UNMOVIC)*, ch. III, p. 149.
138. United Nations Monitoring, *Verification and Inspection Commission (UNMOVIC)*, ch. VI, p. 1036.
139. Human Rights Watch, *Attacks on Ghouta: Analysis of Alleged Use of Chemical Weapons in Syria*, September 2013, pp. 4–20, https://www.hrw.org/sites/default/files/reports/syria_cw0913_web_1.pdf.
140. United Nations mission to investigate allegations of the use of chemical weapons in the syrian arab republic, *Report on the Alleged Use of Chemical Weapons in the Ghouta Area of Damascus*, 21 August 2013, pp. 18–21, Appendix V.
141. N. R. Jenzen-Jones, *Update: Alleged CW Munitions in Syria, the Rogue Adventurer*, August 31 2013, http://rogueadventurer.com/2013/08/31/update-alleged-cw-munitions-in-syria/.
142. N. R. Jenzen-Jones, *Alleged CW Munitions in Syria Fired from Iranian Falaq-2 Type Launchers, the Rogue Adventurer*, August 29 2013, http://rogueadventurer.com/2013/08/29/alleged-cw-munitions-in-syria-fired-from-iranian-falaq-2-type-launchers/.
143. United Nations Monitoring, *Verification and Inspection Commission (UNMOVIC)*, ch. III, p. 140.
144. United Nations Security Council, *UNSCR 1874 Panel of Experts, Report of the Panel of Experts Established Pursuant to Resolution 1874*, 2009, S/2010/571, p. 25, para. 60.
145. United Nations Security Council, *UNSCR 1874 Panel of Experts, Report of the Panel of Experts Established Pursuant to Resolution 1874*, 2009, S/2013/337, pp. 30–34, 36–37, 39, para. 67–79, 82, 86–89, 94.
146. United Nations Security Council, *UNSCR 1874 Panel of Experts, Report of the Panel of Experts Established Pursuant to Resolution 1874*, 2009, S/2014/147, pp. 25–26, 32–33, 38–39, para. 64–68, 90–93, 108–111.
147. United Nations Security Council, *UNSCR 1874 Panel of Experts, Report of the Panel of Experts Established Pursuant to Resolution 1874*, 2009, pp. 40–41, S/2016/157, para. 101–106.
148. A. Taylor, *DIY Weapons of the Libyan Rebels*, Atlantic, June 14 2011, http://www.theatlantic.com/photo/2011/06/diy-weapons-of-the-libyan-rebels/100086/.
149. United Nations Monitoring, *Verification and Inspection Commission (UNMOVIC)*, ch. III, pp. 149–150.
150. United Nations Monitoring, *Verification and Inspection Commission (UNMOVIC)*, ch. III, p. 155.
151. United Nations Monitoring, *Verification and Inspection Commission (UNMOVIC)*, ch. III, p. 168.
152. M. Crowley, *Drawing the Line: Regulation of Wide Area Riot Control Agent Delivery Mechanisms under the Chemical Weapons Convention*, Bradford Non-lethal Weapons Project & Omega Research Foundation, April 2013, pp. 9–52.

153. M. Crowley, *Drawing the Line: Regulation of Wide Area Riot Control Agent Delivery Mechanisms under the Chemical Weapons Convention*, Bradford Non-lethal Weapons Project & Omega Research Foundation, April 2013, p. 37.
154. United States Government Accountability Office, *Nonproliferation: Agencies Could Improve Information Sharing and End-use Monitoring on Unmanned Aerial Vehicle Exports*, p. 9, http://www.gao.gov/assets/600/593131.pdf.
155. M. C. Horowitz and M. Fuhrmann, *Droning on: Explaining the Proliferation of Unmanned Aerial Vehicles*, October 1 2015, last revised January 26 2016, http://papers.ssrn.com/sol3/papers.cfm?abstract_id=2514339.
156. K. Sayler, *A World of Proliferated Drones: A Technology Primer*, Center for a New American Security, June 2015, p. 15, http://www.cnas.org/sites/default/files/publications-pdf/CNAS%20World%20of%20Drones_052115.pdf.
157. C. Abbott, M. Clarke, S. Hathom and S. Hickie, *The Hostile Drones: The Hostile Use of Drones by Non-state Actors against British Targets, Control Project Oxford Research Group and Open Briefing*, January 2016, pp. 1–19, http://remotecontrolproject.org/wp-content/uploads/2016/01/Hostile-use-of-drones-report_open-briefing.pdf.
158. C. Abbott, M. Clarke, S. Hathom and S. Hickie, *Hostile Drones: The Hostile Use of Drones by Non-state Actors against British Targets, Control Project Oxford Research Group and Open Briefing*, January 2016, p. 19.
159. R. A. Zilinskas and P. Mauger, *Biotechnology E-commerce: A Disruptive Challenge to Biological Arms Control, CNS Occasional Paper No. 21*, May 2015, pp. 26–44, http://www.nonproliferation.org/biotechnology-e-commerce-a-disruptive-challenge-to-biological-arms-control/.
160. R. A. Zilinskas and H. Alramini, in *Aerosol Vaccines, in Innovation, Dual Use, and Security: Managing the Risks of Emerging Biological and Chemical Technologies*, ed. J. B. Tucker, MIT Press, Cambridge, 2012, pp. 261–270.
161. N. Fahrenkopf, Anti-ship missiles: a dangerous gateway, *Bull. At. Sci.*, February 9, 2016, http://thebulletin.org/anti-ship-missiles-dangerous-gateway9154.
162. A. Karp, in *The Maturation of Ballistic Missile Proliferation, the International Missile Bazaar: The New Suppliers' Network*, ed. W. C. Potter and H. W. Jencks, Westview Press, Boulder, 1994, pp. 4–7, 10–15.
163. S/2016/157, pp. 15–16, pp. 86, 87, para. 29-21 and Annex 6.
164. N. R. Jenzen-Jones, *9K79 Tochka Tactical Ballistic Missile Launchers in Syria, the Rogue Adventurer*, http://rogueadventurer.com/2013/01/05/9k79-tochka-tactical-ballistic-missiles-in-syria/.
165. S/2016/157, pp. 86–87, Annex 6.
166. *Radiation, Chemical, and Biological Protection – Tutorial [In Russian]*, ed. A. B. Pokryshkin and E. N. Starshinova, SUSU Publishing House, Chelyabinsk, 2007, pp. 22–23.
167. *Radiation, Chemical, and Biological Protection – Tutorial [In Russian]*, ed. A. B. Pokryshkin and E. N. Starshinova, SUSU Publishing House, Chelyabinsk, 2007, p. 173.

168. *Radiation, Chemical, and Biological Protection – Tutorial [In Russian]*, ed. A. B. Pokryshkin and E. N. Starshinova, SUSU Publishing House, Chelyabinsk, 2007, p. 175.
169. R. H. Speier, K. Scott McMahon and G. Nacouzi, *Penaid Nonproliferation: Hindering the Spread of Countermeasures against Ballistic Missile Defenses*, RAND Corporation, 2014, pp. 53–54, http://www.rand.org/pubs/research_reports/RR378.html.
170. Australia Group, *Australia Group Common Control List Handbook: Chemical Weapons-related Common Control Lists*, vol. 1, p. 175, http://www.australiagroup.net/en/controllisthandbooks.html.
171. Bureau of Industry and Security, *Commerce Control List: Category 2 – Materials Processing*, U.S. Department of Commerce, https://www.bis.doc.gov/index.php/forms-documents/doc_view/734-ccl2.
172. Bureau of Industry and Security, *Chemical and Biological Controls*, U.S. Department of Commerce, January 2014, https://www.bis.doc.gov/index.php/policy-guidance/product-guidance/chemical-and-biological-controls.
173. J. Adams, *Bull's Eye: The Assassination and Life of Supergun Inventor Gerald Bull*, New York, Times Books, 1992, p. 236.
174. *CEO Perspective: The Internet of Things for Chemicals, Top Priorities to Build a Successful Strategy*, http://www.iotsworldcongress.com/wp-content/uploads/2016/01/document.pdf.
175. M. Rothwell, *Web-enabling Your PLC, Automation*, http://www.automation.com/library/articles-white-papers/programmable-control-plc-pac/web-enabling-your-plc.
176. E. Stoye, Security experts warn chemical plants are vulnerable to cyber-attacks, *Chemistry World*, 2015.
177. R. Dormels, *Profiles of the Cities of DPR Korea – Hamhung*, 2014, pp. 339–340, http://www.univie.ac.at/koreanologie/fileadmin/user_upload/DigitalNK/Forschungsarbeit/Hamhung.pdf.
178. *AP Photos: North Koreans Hunker Down in Loyalty Campaign*, Associated Press, KTAR News, March 25 2016, http://ktar.com/story/982557/ap-photos-north-koreans-hunker-down-in-loyalty-campaign/.
179. *Hŭngnam Chemical Fertilizer Complex*, Nuclear Threat Initiative, http://www.nti.org/learn/facilities/605/, accessed April 1 2016.
180. *Soviet General Staff City Plan*, Hamhung and Hungnam, 1980, quadrants J-52-3,4, scale 1:25000.
181. J. S. Bermudez Jr, *The Armed Forces of North Korea*, I.B. Tauris & Co Ltd., London, 2001, pp. 223–227.
182. Federation of American Scientists, *DPRK – Chemical Weapons*, March 1 2000, http://fas.org/nuke/guide/dprk/facility/cw.htm.
183. Federation of American Scientists, *DPRK – Hungnam*, November 28 1998, http://fas.org/nuke/guide/dprk/facility/hungnam.htm.
184. L. Waldenström, L. Norlander and G. Puu, *North Korea's Chemical and Biological Weapons Programmes in 2005: Real or Outdated Threats?* Swedish Defence Research Agency (FOI), October 2005, pp. 38–42, FOI-R—1679—SE.

185. Federation of American Scientists, *Chemical Weapons*, March 1 2000, http://fas.org/nuke/guide/dprk/facility/cw.htm, accessed August 26, 2016.
186. Uriminzokkiri [an official DPRK state media outlet], *The Flames of Production Are Rising: Local Broadcast Shows Fertilizer Production Increase at the Hungnam Fertilizer Complex Enterprises (In Korean)*, February 5 2013, pp. 2–37, https://www.youtube.com/watch?v=_IApBfzzBXM.
187. J. S. Bermudez Jr, *North Korea's Ponghwa Chemical Factory, 38 North*, July 10 2015, http://38north.org/2015/07/jbermudez071015.
188. *Software and it Exhibition Opened, North Korea Leadership Watch*, October 18 2012, https://nkleadershipwatch.wordpress.com/2012/10/18/software-and-it-exhibition-opened/.
189. Uriminzokkiri [an official DPRK state media outlet], *Seething with Created Miracles – Hungnam Fertilizer Complex Enterprises (In Korean)*, April 7, 2016, pp. 1–42, https://www.youtube.com/watch?v=49QpOeRXDqs.
190. ABB Group, *ABB Signs Cooperation Agreement with Democratic People's Republic of Korea*, December 1 2000, http://www.abb.com/cawp/seitp202/c1256c290031524bc12569a800354534.aspx, accessed August 23, 2016.
191. ABB Group, *ABB Opens Office in Democratic People's Republic of Korea*, June 28 2001, http://www.abb.com/cawp/seitp202/c1256c290031524bc1256a790027c137.aspx, accessed August 23, 2016.
192. Korea Research Institute for Human Settlements, *International Cooperation for Industrial Development in North Korea-based on the Experience of the Former GDR and Poland (In Korean)*, 2004, p. 37.
193. R. Trapp, Advances in science and technology and the chemical weapons convention, *Arms Control Today*, March 1 2008.
194. *CAS Scientists Clone Purified Conotoxin*, May 16 2006, http://english.cas.cn/resources/archive/news_archive/nu2006/201502/t20150215_137864.shtml.
195. J. Patočka and L. Středa, Protein biotoxins of military significance, *Acta Med.*, 2006, **49**(1), 3–5.
196. M. Peplow, Malaria drug made in yeast causes market ferment, *Nature*, February 13 2013, http://www.nature.com/news/malaria-drug-made-in-yeast-causes-market-ferment-1.12417.
197. United Kingdom of Great Britain and Northern Ireland, *The Convergence of Chemistry and Biology: Implications of Developments in Neurosciences*, p. 3, BWC/MSP/2012/MX/WP.1, para. 11, http://www.unog.ch/80256EDD006B8954/(httpAssets)/23877DDF04285912C1257A370050CB00/$file/Adv-BWC_MSP_2012_MX_WP.1.pdf.
198. Australia, *The Convergence of Chemistry and Biology: Implications for the Review of Developments in the Field of Science Related to the Convention*, July 20 2012, p. 3, BWC/MSP/2012/MX/WP.16, para. 10, http://www.unog.ch/80256EDD006B8954/(httpAssets)/EE75CF7A28029463C1257F3800504434/$file/BWC_MSP_2012_MX_WP.16.pdf.
199. United Kingdom of Great Britain and Northern Ireland, BWC/MSP/2012/MX/WP.1, p. 3 para. 10.

200. A. Pitt, *Potential Advances in Technologies in the Life Sciences, Presentation Given at the BWC Meeting of Experts*, Geneva, Switzerland, July 2012, pp. 6–7, http://www.unog.ch/80256EDD006B8954/(httpAssets)/50131D7240BAA16AC1257A40002AF1F7/$file/Pitt+1.pdf.
201. B. Fincoeur and K. van de Ven, The symbiotic evolution of anti-doping and supply chains of doping substances: how criminal networks may benefit from anti-doping policy, *Trends Organize. Crime*, 2015, **18**, 229–250.
202. World Anti-Doping Agency, *The Independent Commission Report #1, Final Report*, November 9, 2015, pp. 90, 142–143, 146, 158, 306, 308–309, https://www.wada-ama.org/en/resources/world-anti-doping-program/independent-commission-report-1.
203. K. J. Moody, I. D. Hutcheon and P. M. Grant, *Nuclear Forensic Analysis*, CRC Press, Boca Raton, 1st edn, 2005, pp. 461–473.
204. R. Ehrenberg, Engineered yeast paves way for home-brew heroin, *Nature*, 2015, http://www.nature.com/news/engineered-yeast-paves-way-for-home-brew-heroin-1.17566.
205. K. A. Oye, J. C. H. Lawson and T. Bubela, Drugs: regulate 'home-brew' opiates, *Nature*, 2015, http://www.nature.com/news/drugs-regulate-home-brew-opiates-1.17563.
206. T. Wogan, *Chemistry World*, 2014, http://www.rsc.org/chemistry-world/2014/08/yeast-turned-morphine-opioid-biotechnology-factories.
207. A. Lavorgna, Internet-mediated drug trafficking: towards a better understanding of new criminal dynamics, *Trends Organize. Crime*, 2014, 250–270.
208. N. Leontiadis and A. Hutchings, Scripting the crime commission process in the illicit online prescription drug trade, *J. Cybersecur.*, 2015, **1**(1), 81–92.
209. B. Fincoeur and K. van de Ven, The symbiotic evolution of anti-doping and supply chains of doping substances: how criminal networks may benefit from anti-doping policy, *Trends Organize. Crime*, 2015, **18**, 240.
210. Department of Justice, *U.S. Attorney's Office, Southern District of New York, New York Man Convicted in Manhattan Federal Court for Attempting to Acquire Deadly Toxin, Ricin*, August 27 2015, https://www.justice.gov/usao-sdny/pr/new-york-man-convicted-manhattan-federal-court-attempting-acquire-deadly-toxin-ricin.
211. J. Halliday, *Breaking Bad Fan Found Guilty of Ordering Ricin Delivery from FBI Agent*, Guardian, July 29 2015, http://www.theguardian.com/uk-news/2015/jul/29/liverpool-man-who-ordered-breaking-bad-style-ricin-delivery-found-guilty.
212. United Nations Monitoring, *Verification and Inspection Commission (UNMOVIC)*, ch. III, p. 77.
213. C. Gould and P. I. Folb, The South African chemical and biological warfare program: an overview, *Nonproliferation Rev.*, 2000, **7**(3), 12.
214. J. Krige, National security and academia: regulating the international circulation of knowledge, *Bull. At. Sci.*, 2014, **70**(2), 42–52.

215. R. Shaw, Export controls and the life sciences: controversy or opportunity? *EMBO Rep.*, 2016, **17**(4), 474–480.
216. H. L. Clark and W. Clark McFadden II, *U.S. Government Penalizes Company for Alleged Deemed Exportand Other Technology*, Orrick, March 7 2014, https://www.orrick.com/Events-and-Publications/Documents/International-Trade-and-Compliance-Advisory-2014-03-07.html.
217. Bureau of Industry and Security, U.S. Department of Commerce, Order Relating to INTEVAC, Inc., retrieved as document E2365 at, http://efoia.bis.doc.gov/index.php/electronic-foia/index-of-documents/7-electronic-foia/227-export-violations.
218. K. Corbin, *eBay Bans Military Manuals Including Historical Items, ECommerceBytes*, June 17 2013, http://www.ecommercebytes.com/cab/abn/y13/m06/i17/s01, accessed August 12, 2016.
219. eBay, *Military Items Policy*, http://pages.ebay.com/help/policies/military-items.html, accessed August 12 2016.
220. Bureau of Industry and Security, U.S. Department of Commerce, Order RelatingtoINTEVAC,Inc.,retrievedasdocumentE2365at,http://efoia.bis.doc.gov/index.php/electronic-foia/index-of-documents/7-electronic-foia/227-export-violations, p. 2.
221. P. S. Kampas, Should everyone know everything? The freedom of information act *vs.* governmental and national security, *J. Biosecur. Biosaf. Biodefense Law*, 2014, **5**(1), 129–152.
222. S. B. Ouagrham-Gormley, *Barriers to Bioweapons: The Challenges of Expertise and Organization for Weapons Development*, Cornell University Press, Ithaca, 2014.
223. J. Krige, National security and academia: regulating the international circulation of knowledge, *Bull. At. Sci.*, 2014, **70**(2), 42–52.
224. J. Azarani, A dangerous curiosity: examining constitutional protection of scientific research, *J. Biosecur. Biosaf. Biodefense Law*, 2014, **4**(1), 85–108.
225. B. Schwarzentraub, Mandatory laboratory regulations: is the United States over regulating select agent research? *J. Biosecur. Biosaf. Biodefense Law*, 2013, **4**(1), 207–228.
226. United Nations Monitoring, *Verification and Inspection Commission (UNMOVIC)*, ch. III, pp. 82–110.
227. United Nations Monitoring, *Verification and Inspection Commission (UNMOVIC)*, ch. III, p. 86.
228. United Nations Monitoring, *Verification and Inspection Commission (UNMOVIC)*, ch. III, pp. 136–143.
229. United Nations Monitoring, *Verification and Inspection Commission (UNMOVIC)*, ch. III, p. 29.
230. J. B. Tucker, The Rollback of Libya's Chemical Weapons Program, *Nonproliferation Rev.*, 2009, **16**(3), 376.
231. United Nations Monitoring, *Verification and Inspection Commission (UNMOVIC)*, ch. III, pp. 130–131.
232. *Kh-29 Air-to-surface Missiles Used as Unguided Rockets in Libya*, Oryx Blog, August 18 2014, http://spioenkop.blogspot.com/2014/08/kh-29-air-to-surface-missiles-being.html.

233. *Libya Dawn Going DIY: S-125 SAMs Used as Surface-to-surface Missiles*, Oryx Blog, April 25 2015, http://spioenkop.blogspot.de/2015/04/libya-dawn-going-diy-s-125-sams-used-as.html.
234. *Libya Dawn Going DIY: 2K12 SAMs Used as Surface-to-surface Missiles*, Oryx Blog, April 27 2015, http://spioenkop.blogspot.com/2015/04/libya-dawn-going-diy-2k12-sams-used-as.html.
235. M. Vickery, *Eyewitness Account: ISIL Steps up Chemical Weapons Attacks on Kurds in Iraq*, USA Today, March 10 2016, http://www.usatoday.com/story/news/world/2016/03/10/eyewitness-account-isil-steps-up-mustard-gas-attacks-kurds-iraq/81579492/.
236. M. Stephens, *The Ugly Threat of an ISIL with Chemical Weapons*, Al Jazeera, August 20 2015, http://www.aljazeera.com/indepth/opinion/2015/08/ugly-threat-isil-chemical-weapons-syria-iraq-kurds-150819124716038.html.
237. S. Shane, *U.S. Reclassifies Many Documents in Secret Review*, New York Times, February 21 2006, pp. 1–2, http://www.nytimes.com/2006/02/21/politics/21reclassify.html?hp&ex=1140584400&en=aefb4d8fc1e315bc&ei=5094&partner=homepage.
238. M. Leitenberg and R. A. Zilinskas, *The Soviet Biological Weapons Program: A History*, Harvard University Press, Cambridge, 2012, p. 302.
239. R. R. Jun, *Formation of Fuel Air Mixed Cloud by Explosive Dispersal of Liquid Fuel and Detonation of it (Official English Title Reproduced as Given), Chemistry and Chemical Engineering (In Korean)*, 2008, pp. 1–8.
240. *Convention on Cluster Munitions*, May 30 2008, 2a, b and c, http://www.clusterconvention.org/files/2011/01/Convention-ENG.pdf.
241. J. B. Tucker, Double-edged DNA: Preventing the Misuse of Gene Synthesis, *Iss. Sci. Tech.*, 2010, **26**(3), http://issues.org/26-3/tucker-2/.
242. International Gene Synthesis Consortium (IGSC), *The Promotion of Biosecurity*, http://www.genesynthesisconsortium.org/index.php, accessed August 29 2016.
243. R. Shaw, Export controls and the life sciences: controversy or opportunity? *EMBO Rep.*, 2016, **17**(4), 480.
244. A. Kurzrok and G. Hund, Beyond compliance: integrating nonproliferation into corporate sustainability, *Bull. At. Sci.*, 2013, **69**(3), 31–42.
245. S. M. Maurer and S. v. Engelhardt, Industry self-governance: a new way to manage dangerous technologies, *Bull. At. Sci.*, 2013, **69**(3), 53–62.
246. Stockholm International Peace Research Institute, The problem of chemical and biological warfare, *CB Weapons Today*, Humanities Press Inc., New York, 1973, vol. 2, p. 294.
247. D. Guilfoyle, maritime interdiction of weapons of mass destruction, *J. Conflict Secur. Law.*, 2007, **12**(1), 1–36.
248. *Toxic Terror: Assessing Terrorist Use of Chemical and Biological Weapons*, ed. J. B. Tucker, Belfer Center for Science and International Affairs, Cambridge, 2000.
249. R. Mowatt-Larssen, *Al Qaeda Weapons of Mass Destruction Threat: Hype or Reality?* Harvard Kennedy School Belfer Center for Science and

International Affairs, January 2010, vol. 13–19, p. 6, http://belfercenter.ksg.harvard.edu/files/al-qaeda-wmd-threat.pdf.

250. V. G. Heil, V. Kabisch and G. Mascolo, *German Federal Criminal Police Office Is in Possession of Secret IS Papers, [in German] Süddeutsche Zeitung*, March 7 2016, http://www.sueddeutsche.de/politik/terrorgefahr-bundeskriminalamt-im-besitz-geheimer-is-papiere-1.2895848.
251. K. Hughes and J. Alper, *The Global Movement and Tracking of Chemical Manufacturing Equipment: A Workshop Summary*, National Academies Press, Washington, 2014, p. 12.
252. Fushun Huateng Safety, *China Protective Nbc Gas Mask*, Alibaba, http://www.alibaba.com/product-detail/china-protective-nbc-gas-mask_1863887792.html?spm=a2700.7724838.0.0.Z1wKuK, accessed August 22, 2016.
253. China Xinxing Import and Export Corporation, *Chemical Military/Police Gas Mask*, Alibaba, http://www.alibaba.com/product-detail/Chemical-Military-police-Gas-Mask_60008346362.html?spm=a2700.7724838.0.0.Z1wKuK, accessed August 22 2016.
254. Earthlab Limited, *Military NBC Protection Suit*, Alibaba, http://www.alibaba.com/product-detail/Military-NBC-Protection-Suit_50010576719.html?spm=a2700.7724838.0.0.Z1wKuK, accessed August 22 2016.
255. United Nations Mission to Investigate Allegations of the Use of Chemical Weapons in the Syrian Arab Republic, *Report on the Alleged Use of Chemical Weapons in the Ghouta Area of Damascus*, 21 August 2013, pp. 18–21, Appendix V, http://www.un.org/disarmament/content/slideshow/Secretary_General_Report_of_CW_Investigation.pdf.
256. *Nuke Expert Says Germany Naïve about Burma Weapons Exports*, Mizzima, March 29 2011, http://archive-2.mizzima.com/news/world/5075-nuke-expert-says-germany-naive-about-burma-weapons-exports.html.
257. Australia, *BWC/MSP/2012/MX/WP.16*, p. 2, para. 7.
258. Bureau of Industry and Security, U.S. department of commerce, *Deemed Exports and Fundamental Research for Biological Items, Undated*, https://www.bis.doc.gov/index.php/policy-guidance/product-guidance/chemical-and-biological-controls/14-policy-guidance/deemed-exports/111-deemed-export-and-fundamental-research-for-biological-items.
259. K. J. Chol and C. K. Hyok, *New Lower Bound on Pixel Expansion of (2,n)-visual Cryptography Schemes (Official English Title Reproduced as Given), Electronics (In Korean)*, 2009, pp. 14–15.
260. R. T. Chol and K. H. Hwang, *A Method for Efficient Implementation of Elliptic Curve Cryptosystem and Digital Signature (Official English Title Reproduced as Given), Research Notification (In Korean)*, 2005, pp. 18–20.

Section IV

Implications for Arms Control and Disarmament

CHAPTER 15

The Future Chemical Weapons Convention – Out-dated Model or CWC 2.0

R. TRAPP

Independent Disarmament Consultant, 116 Route de la Contamine, Chessenaz-74207, France
*E-mail: ralf.trapp@gmail.com

15.1 Key Challenges

The Chemical Weapons Convention (the CWC) is a cornerstone arms control agreement prohibiting an entire category of weaponry. It is of unlimited duration and has achieved nearly universal adherence. As the elimination of most of the past chemical weapons programmes nears completion, efforts to maintain and strengthen the comprehensive ban on chemical weapons must traverse several parallel avenues:

- Increase efforts to bring the remaining few States not party into the regime;
- Complete the tasks in eliminating past chemical weapons (CW) programmes (including the elimination of any CW programmes of the remaining States not party once they join, the completion of the outstanding destruction/conversion activities of the declared CW stockpiles and production facilities, and the resolution of concerns regarding

Preventing Chemical Weapons: Arms Control and Disarmament as the Sciences Converge
Edited by Michael Crowley, Malcolm Dando and Lijun Shang

Published by the Royal Society of Chemistry, www.rsc.org

possible undeclared stockpiles, production facilities and CW development establishments);

- Ensure that the Convention's implementation and compliance assurance mechanisms remain effective and adapt to change, in order to prevent the re-emergence of chemical weapons in whichever form (enhanced national implementation; creative verification that adapts to change, is truly risk-based and holistic in approach, and that makes best use of opportunities offered by advances in science and technology; effective compliance assurance and problem resolution mechanisms);
- Take full advantage of the opportunities for improving international cooperation and assistance for peaceful purposes among States Parties—developed in synergy with the compliance assurance mechanism of the Convention—that flow from advances in science and technology and the changes in the global landscape of chemicals manufacturing and trade as well as the changing environment of science and industry collaborations;
- Strengthen the intellectual, legal, institutional, scientific and technical capacity of the Convention's institutions (the different constituents and advisory boards of the Organisation for the Prevention of Chemical Weapons (OPCW), National Authorities), and further enhance the engagement with other stakeholders (non-governmental organisations, chemical industry, educational and scientific institutions and civil society), to ensure continuing compliance and full and effective implementation of all CWC provisions.

Achieving full universality of the Convention will require appreciation of the specific historical, political and security context that has prevented the remaining few States not party from joining the Convention. It seems that the OPCW has achieved as much as was possible by engaging with States that have yet to become a party, within the context of its own political activities, attempts to create dialogue, and efforts of persuasion. Already, the accessions of some countries in recent years could not have been achieved by the OPCW acting on its own, but have required concerted efforts and political conversations within a broader strategic framework, or the independent action of other key actors (key Member States, the United Nations, regional actors), or a link (at least implicitly) between CWC adherence and other security and strategic issues.

To make further progress on universalization will therefore require the OPCW to understand and design its actions within a broader policy and security framework, and to coordinate actions with other actors and initiatives even if they are not directly/predominantly related to chemical weapons. Success with such effort is difficult to predict, but as past experience has shown, what matters is that the OPCW and its Technical Secretariat are capable of responding swiftly and effectively to the requirements prompted by accession (or ratification) of any of the remaining States not party. The OPCW will need to maintain this capacity, including with innovative systems of ramping up human capacity on short (or no) notice and by maintaining critical expertise and competencies even when these are rarely called upon in its routine business.

Preventing the re-emergence of chemical weapons (including the emergence of novel types of agents or concepts of chemical warfare) will also call for a broader synchronisation of actions in different, overlapping domains. These include the strengthening of the legal power of the General Purpose Criterion (which ensures that the Convention is 'future proof' against any new scientific and technological developments), the adaptation of national implementation systems and international verification measures to changes in the implementing environment, improving the mechanisms of monitoring and (more importantly) evaluating the impact of advances in science, technology, industry and trade, and the building of governance structures that bring together the different governmental and non-governmental actors that have a stake in the field of chemical weapons disarmament. All this calls for an overarching vision of the future of the CWC and the OPCW – shared by its Member States, underpinned by a common strategy pursued by the policy-making organs and implemented by the Technical Secretariat.

Significant progress towards such a shared vision has already been made, promoted by activities of the Director-General and the Technical Secretariat,[1] supported by the Executive Council and its Open-ended Working Group on Future OPCW Priorities, and also underpinned by the work of the SAB and the new Advisory Board on Education and Outreach; the Fourth Review Conference will provide a further opportunity to make progress in this regard.

Furthermore, the OPCW's claim as the global repository of knowledge and expertise on chemical weapons and their disarmament will remain credible only if the OPCW maintains and strengthens its knowledge base. It will also have to find a way of expanding its activity range in an appropriate manner in areas of safety and security emanating from toxic chemicals beyond a narrowly-understood mandate of verifying and eliminating chemical weapons stockpiles and CW production capabilities. The OPCW's relevance and credibility will increasingly be measured not merely by its ability to oversee and verify issues related to chemical weapons disarmament, but also in terms of how it responds to other chemical security (and related safety) challenges, from issues emanating from the legacy of past CW programmes (old and abandoned CW that remain buried at this stage, sea-dumped CW, environmental contamination related to past CW programmes and locations involved in development, testing, production, stockpiling or use of chemical weapons), to new challenges related to the misuse of toxic chemicals by terrorist or criminal actors, to safety and security risks emanating from the presence or the use of toxic chemicals in society in general.

This signals a shift in emphasis from the Convention's provisions directly related to the elimination of past CW programmes (Articles III, IV and V), to the provisions that deal with ensuring full treaty implementation and compliance (Articles VI, VII and IX) as well as those dealing with enhanced cooperation among States Parties and developing their capacity in peaceful chemical activities (Articles X and XI).

In the following sections, some of these challenges will be further elaborated, leading to the formulation of possible recommendations that the Fourth Review Conference may wish to address.

15.2 Preserving the Integrity of the Prohibitions and Preventing the Re-emergence of CW

15.2.1 Challenges to Maintaining the Comprehensive Nature of the CW Prohibition

The previous chapters of this book have set out in considerable detail how the advances in science and technology, the convergence of science and engineering disciplines at the intersection of chemistry and biology, and trends in the chemical and biotechnological industry have begun changing (and may well affect in future) the implementation environment within which the CWC is operating. They have illustrated the benefits that these advances are expected to bring for human society, including with regard to better protections against the effects of toxic chemicals – whether in the form of chemical weapons or toxic hazards emanating from human activity or natural causes. But, they have also highlighted potential risks and indicated possibilities of how these developments might be misused for hostile purposes.

As the world is moving away from the chemical warfare concepts and weapons types associated with the past century and the Cold War era, the risks associated with toxic chemicals and their potential misuse as weapons is becoming more diffuse, and less well defined in terms of chemical compounds or dissemination methods of concern. The CWC's Schedules are less and less reliable as guidance to the types of chemicals that are of concern, both from a prohibition and a verification perspective. The absence of dedicated CW development programmes in States Parties means that (at least those publicly accessible) evaluations of risks associated with new science and technology are based on the *potential* created by such advances, rather than on evidence from actual hostile uses and real-world chemical warfare scenarios. As for monitoring advances in science and technology and assessing their potential impact on the regime, this has two consequences: it may lead to evaluations of the impact of science and technology on the CWC that overstate risks (not everything that is theoretically possible is also practically feasible), but it may also result in evaluations that miss the potential risk inherent in certain advances, in particular when science and technology are making non-linear progress as is the case with convergence, but also when legacies of past CW programmes re-surface that have not been built into the routine verification system.

Previous chapters of this book have drawn the attention of the reader to how research in such areas as the neurosciences, brain research, nanotechnology, synthetic and systems biology, toxinology and bioregulators are deepening the understanding of the functioning of the "human machine" and are helping to design new agents for the treatment of disease, but at the same time, also increase the capacity and knowledge base that would be available should States decide to re-open or embark anew on a chemical weapons programme. Advances in targeted drug delivery, including by progress made in nanotechnology, aerobiology, or the use of natural and artificial viruses as drug delivery vehicles, will be important for medical treatments with much higher specificity and fewer side effects; but they may also offer tools for using types of agents that in the past were ruled out because they were too difficult to deliver to the target organs in the human body. Advances in biotechnology, catalysis, new types of equipment and manufacturing processes will increase the flexibility in the chemical industry to respond to market demands and tailor solutions to changing requirements and demands, but they would also change the footprint and distinguishing features of a chemical weapons programme should States decide to re-open such a programme.

This all underlines the fundamental importance of the General Purpose Criterion (GPC) as a safeguard for the Convention's prohibition of chemical weapons. Every previous Review Conference has emphasised the comprehensive nature of that prohibition, and the GPC has been hailed as the legal construct to ensure this all-encompassing and lasting character of the chemical weapons ban (for a detailed discussion see, for example, Krutzsch' and Trapp's discussion of the relevant provisions of Article II of the Convention).[2]

However, there have been challenges to the GPC. The provisions on riot control agents and the inclusion of "law enforcement including domestic riot control" have led to discussions of the legality of using certain incapacitating chemical agents for law enforcement purposes (and thus potentially the development, production and stockpiling of weapons systems for their dissemination for law enforcement purposes, which may include operations to enforce law and order by military forces).[3] There have also been concerns about whether certain weapons systems for the dissemination of riot control agents over large areas were consistent with the prohibition of their use in warfare.[4]

It is important to note that these challenges did not result from advances in science and technology *per se*. They resulted instead from concepts and activities that exploit certain margins in the legal constructs of the Convention—exemptions from the total ban that were necessary to ensure the continuation of what States perceived as legitimate activities (the use of riot control agents in law enforcement, for example, or certain work in the area of chemical protection that could conceivably pose questions about its intent). Some of these activities could lead to capabilities (even if they were not intended for CW purposes) that may be hard to distinguish from activities related to "real" chemical weapons. This is nothing entirely new, as the use of chlorine in the armed conflict in Syria has again underlined. Chemistry and chemical technology are dual-use and the risks of the misuse of toxic chemicals for weapons purposes do not entirely disappear when military CW programmes are closed down.

What is new is the pace at which science and technological innovation are advancing. Some of the barriers that traditionally stood between new discoveries of toxic chemicals with potential CW relevance and the introduction of a new chemical weapon (such as high production price, lack of control over the dose, difficulty disseminating the agent under field conditions, or instability in storage or delivery), may well fall or reduce as the pace of science and innovation increases. The mere promise that scientific advances or new technological solutions might overcome some of the limitations for novel types of chemical agents, might create uncertainties about the robustness of the regime.

This is why the initiative of a group of States Parties to make State positions known on the use for law enforcement of aerosolised chemicals that act on the CNS, and to initiate a discussion in the OPCW context about the use of such substances in law enforcement, was so important. The paper published by this diverse group of States Parties invited other States Parties to join this discussion, and it made an explicit link to the objective to prevent the re-emergence of chemical weapons.[5] It can only be hoped that the Fourth Review Conference will provide a platform for continuing and updating this discussion; a process that should eventually lead to clear guidelines being designed that prevent the research, development and production activities intended to support law enforcement that would undermine the very basis of the prohibitions of the CWC. However, in any event, it will be important for the Review Conference to again stress the comprehensive and lasting nature of the prohibitions enshrined in the Convention.

15.2.2 Challenges to and Opportunities for National Implementation and Verification

The theme of preventing the re-emergence of chemical weapons should guide the discussions and decisions at the Fourth Review Conference in the areas of national implementation and verification.

In the case of national implementation, significant progress has been made with the establishment of competent and fully empowered National Authorities, and the enactment of penal and other implementing legislation. Progress has also been made on further harmonising the way in which different States Parties are implementing the requirements of the Convention. Nevertheless, significant problems remain – examples include discrepancies between the transfer declarations of States Parties for scheduled chemicals; concerns about the accuracy and completeness of industry (and in some cases also CW) declarations; or questions over the application of the transfer provisions of the Convention to States not party.

Convergence and the advances in science and technology will have several effects on national implementation as well as verification, which the Fourth

Review Conference should address. Firstly, there is the issue of what to do about biological and biologically mediated processes for the production of discrete organic chemicals – an issue that has already been discussed at the end of the Preparatory Commission in the mid-1990s. The SAB has consistently recommended that these production processes should be covered under the Convention, but States Parties have yet to agree on how to treat these types of production plants. Whilst biologically mediated chemicals manufacturing remains a niche business, it can no longer be described as something that *might* become relevant in the future – bio-manufacturing of chemical products is part of today's industrial reality. It is also clear that in terms of risk potential for the Convention, some of the products and processes used by this industry are as relevant to the CWC as those used by other OCPF facilities, including the ones that the Technical Secretariat considers to pose significant risks; other such products and processes are not CWC-relevant at all. A discussion of how to deal with these types of production plants therefore needs to be held with a certain degree of granularity in order to distinguish between the relevance of different types of plants for CWC verification purposes.

Secondly, as convergence trends increasingly manifest themselves in industrial manufacturing, the landscape of the global chemical industry will change, and so will the relevance of different industrial processes and products/intermediates for the Convention. The Schedules of Chemicals will remain important triggers of verification measures for traditional CW agents known from past State programmes. However, new pathways to old agents may become feasible as a result of technological advances, using different intermediates not covered under the Schedules today, or new agent types may become relevant involving intermediates and products not listed in any of the Schedules. The industry verification regime as well as the analytical methods and databases available for challenge inspection need to be adapted to these new technological and chemical realities.

Thirdly, and more generally speaking, this adaptation requires a better appreciation of the concepts underlying the "risk to the object and purpose of the Convention" of chemical products, intermediates and processes. Risk perceptions at the beginning of the operation of the Convention were associated with the experience of past CW programmes, and the routine verification system consequently focussed on activities and facilities involved with scheduled chemicals. Increasingly, in today's security environment, other types of toxic chemicals are also (and sometimes more) relevant than these chemicals. From a perspective of risks associated with State actors, convergence and advances in enabling technologies have increased the technical feasibility of developing chemical agents that are less lethal, more targeted in their physiological effect, and perceived to be more suitable for the emerging operational context of military operations in populated areas. From a perspective of risks associated with non-State actors, the types of chemicals that may pose significant risks may be quite different from those listed in the Schedules, and choices of potential chemical agents are more likely driven by accessibility than by factors that a military programme would take into account. Terrorist threats involving homemade toxic agents or toxic industrial chemicals are well recognised today. Criminals, too, are known to use toxic chemicals, such as in acid attacks. Whilst one can debate whether the latter qualifies as an offence that should be pursued under the Chemical Weapons Convention, or whether it should be considered in a verification context, some of the practical implementation tools of the Convention (penalisation and prosecution of criminal acts involving the use of toxic chemicals as weapons, domestic control measures for materials of particular concern, preventive measures developed in partnership of governments and industry, educational efforts) may well be of interest.[6]

If the risk to the object and purpose of the Convention is to remain a key design factor of the verification regime, then the emphasis in the conduct of industry inspections should gradually shift from verifying the accuracy of declared data and checking on the absence of undeclared scheduled chemicals (with particular emphasis on Schedule 1 chemicals and their production), and place more weight on the general verification aims set out in the Convention (*i.e.*, that activities at an inspected industry facility are consistent with the obligations undertaken under the Convention).

Fourthly, such a shift to a more qualitative verification methodology would also call for a more holistic approach. In 2015, the SAB endorsed the report of its Temporary Working Group on Verification, which put forward a number of recommendations on how the verification regime of the Convention should evolve to stay relevant. The SAB called for the adoption of improved methodologies and new technologies for verification purposes, the availability of adequate human resources and equipment, and the maintenance of competent staff and an attractive work environment at the Secretariat. Highly qualified staff members with initiative, it argued, are required to ensure that verification can adapt to changing circumstances.[7] These recommendations were mirrored by the Technical Secretariat's vision paper on "The OPCW in 2025: Ensuring a world free of chemical weapons"[1] which sets out a series of specific action areas with regard to verification, including the enhancement of the Secretariat's analytical capabilities to make them more robustly geared towards gathering, validating and evaluating information befitting an independent and more holistic assessment of Convention implementation, and enhanced capabilities to monitor the full spectrum of relevant toxic chemicals falling under the scope of the Convention (including, as recent events have shown, with regard to toxic chemicals that were part of past CW development programmes but had not been procured as part of military stockpiles). However, enhanced verification alone will not be able to address the changing risk spectrum and implementing environment. Many new risks will be primarily susceptible to national action rather than international verification, and consequently, the interplay between national and international controls is likely to become much more important. For the OPCW and its Technical Secretariat, that could mean that it ought to pursue several action areas:

- A closer association between verification evaluation results and assessments of the effectives of as well as of weaknesses in national implementation systems;

- Augmented support for National Authorities for implementation measures that reach beyond the scheduled chemicals and address risks related to other toxic chemicals;
- A more interactive system of monitoring and evaluating advances in science, technology and industrial applications, involving the Technical Secretariat, the SAB, States Parties, industry and other non-governmental experts to create a more broadly-based and continuous mechanism for evaluating changes that the verification and national implementation systems ought to adapt to, which would enable the Conference of the States Parties to provide timely guidance on such adaptations to the Technical Secretariat and to States Parties.

Finally, convergence and scientific advances as well as progress in enabling technologies will result in new tools for verification as well as national implementation, from advances in information technology to new analytical methods and instruments that can be used in national controls as well as international verification measures, or advanced methods of data collection and processing (for example with regard to satellite imagery for tracking chemical signatures, or the monitoring and analysis of social media). Convergence will also provide solutions that can augment inspector training. As a result, both national implementation and verification measures could make use of new and more effective tools and techniques, increase the quality of their respective technical systems, and reduce costs (including the costs of verification and compliance with national regulations affected on industry). Article VIII of the Convention in its paragraph 6 calls for the OPCW to make use of advances in science and technology for undertaking its verification measures – the Fourth Review Conference is yet another opportunity for the States Parties to embrace this requirement and generate the political support to make full use of the opportunities created by science and technology.

15.2.3 Challenges to Preventing and Responding to Hostile Uses of Toxic Chemicals

As the orientation of the OPCW has begun to shift from the verified elimination of past CW programmes to the prevention of the re-emergence of chemical weapons, its *raison d'être* will increasingly be defined in terms of ensuring chemical security and mitigating the risks of hostile uses of toxic chemicals, rather than in terms of CW demilitarisation by States. The OPCW will need to work together with a broader range of actors and consider a wider range of scenarios that are relevant in this context, define its role *vis-à-vis* other actors on the international scene, and more clearly define the interface of its own evolving range of activities and programmes with other international actions, such as in chemical safety and chemicals management in general. This process of developing partnerships is well under way, involving UN entities such as the United Nations Office for Disarmament Affairs (UNODA), the United Nations Office for the Coordination of Humanitarian Affairs (UNOCHA), the World Health Organisation (WHO), the International Criminal Police

Organization (INTERPOL) and other organisations. This range of partners also includes industry, non-governmental organisations, academia and civil society. As the Fourth Review Conference is approaching, the OPCW has intensified the discussion of its contribution to global anti-terrorism efforts, in the framework of its Open-Ended Working Group on Terrorism and its Sub-Working Group on Non-State Actors. The Technical Secretariat supports these discussions and at the same time has enhanced information sharing and collaboration with its international partners, including by hosting a table-top exercise involving 15 international organisations and agencies from across the UN system to improve interagency coordination and response to terrorist incidents involving toxic chemicals.

The OPCW experiences of its operations in Syria and Libya have demonstrated the challenges—but also opportunities—that it faces when it operates in such a broader context. It has also prompted decisions about the development of a stronger capacity within the Technical Secretariat to respond to contingencies such as missions to prevent the involvement of chemical weapons in conflict situations (such as in Syria in 2013/2014, and Libya in 2016), or the delivery of assistance to victims of chemical weapons use (whether involving the use of military chemical weapons or improvised devices that disseminate toxic industrial chemicals). A recent step in this direction was the establishment of a Rapid Response and Assistance Mission (RRAM) by the Technical Secretariat.

At the same time, and operationally connected with preparing for contingency operations, the OPCW is further improving its operational and technical capacity to conduct investigations of the alleged use of chemical weapons as well as challenge inspections. Significant progress has been made on developing a network of designated laboratories for the analysis of biomedical/biological samples, and the OPCW's SAB has begun studying problems and opportunities in the area of chemical forensics. This signals a strive towards a more general adoption of a forensic approach by OPCW investigation teams. The advances in science and technology are providing new tools and methods to this end; and, as the Skripal affair in March 2018 has underlined, it will be equally important to compile, expand and properly curate the databases of reference spectra, and collections of reference materials, that will be needed for forensic analyses.

Complementary to its efforts to further develop its investigation and response capabilities, the OPCW can also contribute significantly to strengthening the resilience of States Parties against hostile uses of toxic chemicals (as well as against natural and accidental releases of toxic chemicals). Its programme portfolio in the area of assistance and protection has steadily increased over time, and has adopted a strong regional approach towards capacity building in the areas of prevention, detection, preparedness, and response. The Technical Secretariat's vision paper highlights as specific goals, amongst others, the OPCW certification ("recognition") of regional training centres for targeted and tailor-made capacity development programmes; the establishment of regional capacities for an effective response to the use of chemical weapons or the accidental release of toxic chemicals with OPCW-trained experts and supported by the regions or sub-regions with regard to equipment resources; an expansion of the number of States Parties that have a viable national protective programme, supported by OPCW training and capacity building where needed; an established capacity of the Technical Secretariat to engage in major international responses to the hostile use of toxic chemicals, in concert with other international organisations, compliant with inter-operability requirements and tested in regular exercises and collaborative training including with States Parties; and a laboratory twinning programme to facilitate exchanges between developed and developing countries. The vision paper explicitly sets the OPCW in a wider context and depicts it as a "leading platform for facilitating the exchange of best practices, catalysing international partnerships and standards, promoting national capabilities in the area of chemical security, and establishing a chemical security culture."[8]

This vision places the OPCW on a par with several other global initiatives and mechanisms, such as the efforts of the UN Security Council's 1540 Committee, the EU's CBRN Centres of Excellence Initiative and other measures implemented under the EU's Instrument contributing to Stability and Peace, the outreach programmes of the United States and some other countries, and the chemical security working group of the Global Partnership. The Fourth Review Conference will provide an opportunity to reach out to these partners and to foster the relationship with each of them; this will not only be important to avoid undesirable duplication of efforts or activities that are not sufficiently coherent and connected, but it will also help the partner countries at the receiving end of capacity building support to more effectively absorb these measures and build the results back into their own systems and procedures to ensure sustainability of results.

The convergence in the life sciences will create new opportunities for strengthening the resilience of societies against hostile uses of toxic chemicals, as well as incidents such as natural catastrophes or accidents that may lead to the release of hazardous chemicals. These may include new diagnostic and detection techniques and instruments, better medical treatments, better physical protections, less aggressive decontamination methods and the like. The OPCW's assistance and protection programme and its industrial connections offer a useful platform for exchanges among States Parties on such enhanced protections and good practices.

But, convergence in the life sciences and their industrial application can also help in reducing the day-to-day risks associated with the presence of toxic chemicals in industry, agriculture and elsewhere in society. For many years, there has been a drive towards reducing the use of hazardous chemicals in society, of moving industrial processes to intrinsically safer solutions, and to improving chemicals management, thus improving chemical safety and security. These advances have the potential of reducing chemical risks and at the same time promoting the peaceful application of chemistry for purposes beneficial to society – objectives that are in line with the provisions of Article XI of the Convention.

15.3 Challenges and Opportunities in International Cooperation for Peaceful Uses of Chemistry

Capacity building and the promotion of peaceful collaborations among States Parties in the chemical field are an important aspect of the implementation of the CWC. Not only is this a reflection of the general concept that disarmament and the promotion of economic and technological development are two sides of the same coin, but that international cooperation, assistance and capacity building programmes are also important factors to engage with the States Parties, to attract new Member States, and to marshal sustainable support for the Convention's objectives on a global scale (in society as well as political leaderships).

The OPCW implements its international cooperation programmes in a global political and economic environment that is constantly changing, creating both new risks and opportunities for the economic and technological development of the States Parties. Within this evolving environment, there are tensions between globalisation and protectionism, between the removing of trade and other barriers to promote the diffusion of applications of science and technology, and the erection of new such barriers. At the same time, there have been significant migrations of chemical production capacity from traditional chemical manufacturing regions in North America, Western Europe and Japan to countries in Asia (China in particular), the Middle East, Eastern Europe and finally, it seems, also Africa. These shifts in production location come with changes in global trade patterns in chemicals and chemical technologies. The global implementation environment is changing, resulting in challenges for national implementation and verification but also creating opportunities for further developing international cooperation among States Parties in the field of the peaceful application of chemistry.

Convergence in the life sciences is expected to accelerate the global diffusion of new scientific techniques and knowledge as well as their worldwide application in industry. The emerging technology environment is promoting exchange on a more global scale, the sharing of information, and the work of "virtual teams" that bring together researchers across borders and continents utilising the internet as a collaboration platform as well as research tool. What is happening in the life sciences today will surely also manifest itself in industrial manufacturing, trade, and the application of chemical products and solutions tomorrow.

International collaboration is an intrinsic quality of convergence in the life sciences. The OPCW should exploit the opportunities that emerge from this changing science, technology and industry environment and adapt its programme design in the field of international cooperation accordingly. This will require the support of governments to facilitate international cooperation for the peaceful application of chemistry, but at the same time it will depend crucially on how effective the OPCW can partner with industry, academia and civil society. The Fourth Review Conference will be an opportunity to discuss how the OPCW can make best use of convergence as a motor to promote collaborations among States Parties – not merely for building capacity and promoting scientific collaborations at the laboratory bench, but increasingly by working together with and through partners in industry to foster collaborations and at the same time promote responsible and ethical conduct in the production and use of chemical products and technologies.

The Fourth Review Conference will also be an opportunity to further improve the OPCW's partnership with other organisations and mechanisms that are important actors in international co-operation in the chemical field, most notably the Strategic Approach to International Chemicals Management (SAICM) and the associated environmental chemical safety regimes including the Basel, Stockholm and Rotterdam conventions.

15.4 Maintaining Organisational Strength, Leadership, Competence

15.4.1 Leadership

The operations of the OPCW in Syria and Libya have underscored the importance of leadership in the OPCW's policy-making organs (the Executive Council in particular), and at the helm of the Technical Secretariat. Without strong leadership, it would not have been possible to create the coalition of States Parties that supported the elimination of the Syrian chemical weapons programme, or that enabled the removal of the Libyan chemical weapons materials from the conflict zone for destruction outside of the country.

On the other hand, the discussions, disputes and votes in the Council prompted by the findings of the Joint Investigation Mechanism addressing the attribution of responsibility for the uses of chemical weapons in Syria were a reminder of the importance of political cohesion and support for effective CWC implementation in difficult circumstances, and the severe tensions that a lack of such unity can cause. The OPCW has mastered several crises in the past, and it has managed to maintain (or re-establish) a common sense of purpose after each of them, by basing itself firmly on the objectives of the Convention. This has required careful and quiet negotiations and (to the degree possible) consensus building to broaden the political support for crucial decisions. In an environment that is increasingly characterised by suspicions and mistrust between some of the key actors, it will be important that the EC find ways to act again as a constructive framework for problem resolution and confidence building among its members. The preparation of the Fourth Review Conference by itself is not a suitable context for such confidence building or the resolution of specific compliance concerns. However, what the preparatory process should allow is a re-affirmation of the common principles and mechanisms of the Convention that States Parties should be using to resolve such concerns. The preparatory process for the Fourth Review Conference may thus provide an opportunity to strengthen the common approach of the Council to deal with compliance issues, if its members are willing to take this opportunity.

Providing leadership and promoting unity in complicated circumstances (within the Council but equally within the Technical Secretariat) falls to the Director-General – this is a crucial task. The selection process for the successor to Ambassador Üzümcü at the helm of the Technical Secretariat has been completed. Ambassador Fernando Arias of Spain will take this important assignment and become the next OPCW Director-General on 25 July 2018; he will face considerable challenges in steering the OPCW through its transition and at the same time directing the OPCW's actions in response to any challenges that no doubt will come its way.

15.4.2 The OPCW as a Learning Organisation and Fit for Purpose

The transition of OPCW priorities from CW disarmament to the prevention of the re-emergence of chemical weapons can be conceptualised as a transition of an institution that defines itself primarily by reference to explicit mandates of the Convention, to one that can be described as competence-based. This will be particularly important for the programme areas that are not developed beyond general objectives and principles in the Convention itself. Whilst the requirements for national implementation and verification in the chemical weapons and industry areas are spelled out in much detail in Articles IV, V, VI and IX of the Convention and the related Parts of the Verification Annex, the OPCW's roles and programmatic directions under Articles VII, X and XI are less detailed and much more open-ended. This has created uncertainties and diverging views among States Parties about the scope and nature of OPCW programmes in international cooperation, assistance and implementation support in the past, but this vagueness has also allowed for a degree of flexibility and innovation in designing effective and attractive OPCW programmes.

As the mandates of the OPCW shift towards the prevention of the re-emergence of chemical weapons, and towards strengthening chemical security and helping build human, technical and institutional capacity to this end in States Parties, the knowledge base and competence of the Technical Secretariat will become even more important than during the chemical weapons disarmament period. This will require effective knowledge management, sharing best practices across the organisation, and adopting a personnel policy that balances the maintenance of institutional memory and competence with the need to rotate staff and bring into the Secretariat fresh ideas and new knowledge and experiences. As stated in the Secretariat's vision paper, this must entail to "develop further essential management and governance tools in a number of areas including administration, human resources, and information and computing technology (ICT). The resulting efficiencies and flexibility will increase the Organisation's ability to serve as a platform for States Parties and relevant stakeholders to promote the objectives and purposes of the Convention."[9]

15.5 Conclusions – the Way Forward and Opportunities for Action by the Review Conference

Several practical measures and approaches will be important to pursue in this context, and the Fourth Review Conference will provide a forum for States Parties to discuss and consider recommendations towards these ends. In no particular order, here are a number of issues and action areas that States Parties may find useful to consider.

Be ready for the unknown. The implementation environment (security contexts, industrial capacities and relationships, science and technology) will continue to evolve. This may include evolutionary changes, but also certain sudden developments that cannot be foreseen. As Syria and Libya have shown, the OPCW may have to take action in circumstances it has not planned for or expected, and that will require it to quickly adapt its decision-making processes, planning approaches and practical implementation measures to scenarios it had not anticipated or provided funding for. Similarly, convergence in the chemical and life sciences is often characterised as revolutionary change – that means that progress will not always be incremental, but may also happen suddenly and unpredictably. The Technical Secretariat therefore needs to further develop its systems of monitoring relevant developments outside the OPCW to detect early possible challenges and upcoming tasks and engagements that it needs to prepare for. At the same time, it must maintain a state of preparedness and technical, operational as well as political readiness to respond quickly should it be called upon to act in contingency situations.

To be able to do so, the Technical Secretariat must identify the critical competencies and skills it needs to respond to non-standard situations, and maintain its competence basis including with realistic planning, training and exercises. It also must more clearly define the interfaces between its structures and processes for such contingency operations as well as special procedures such as challenge inspection and investigations of alleged use of chemical weapons, and its capacities and processes dedicated to the routine business of implementing the Convention (routine verification, international cooperation and assistance programmes, implementation support and the like). Contingency operations, by their very nature, are an all-TS effort and will place demands not only on those directly involved in special missions, but on all other units of the Technical Secretariat as well. They will require other staff to take on extra duties; they will require reach-back to knowledge and skills that reside in parts of the Secretariat that are not directly involved in the operation; they will require ramping-up capacity for contingency missions from across the Secretariat or from the outside of the OPWC as well as back-filling of units that have made significant numbers of staff available to the contingency mission. At the technical level, the Secretariat will need to maintain or acquire the tools it needs for these special missions. The Syria missions have underlined the importance of having at the Secretariat's disposal equipment and systems in such areas as security equipment for field use, health and safety equipment, tools and procedures for the collection and interpretation of satellite imagery, secure data and voice communications, and there will be other types of equipment that an OPCW mission in such scenarios may need that are not among the standard set of tools which the Technical Secretariat has at its disposal.

Maintaining adequate capacity and effective management systems. Already at the First Review Conference, the OPCW decided to move towards results-based budgeting. Since then, significant progress has been made to develop management systems towards managing by objectives and planning for outcomes and impact. Several further improvements are possible in this respect, examples include: (a) the development of key performance indicators to assess outcomes as well as the impact and sustainability of results, and the integration of these indicators into planning, results monitoring, evaluation and reporting systems; (b) use of a mid to long-term planning horizon (mid-term 5 years, strategic level 10 years or longer) that commits both the Technical Secretariat and the States Parties to longer-term objectives and priorities, and that ties in with the terms of office of the Director-General; (c) a move to a longer (*e.g.*, 2 year) programme and budget cycle to increase financial stability and predictability of programme implementation; (d) more effective and integrative use of information technology in support of all management processes (budgeting and programming, budget implementation and measurement of how agreed objectives are being met, personnel policy, confidentiality and security, health and safety, procurement, outsourcing/contracting of services, and other areas of management) as well as to enhance the analytical capabilities of the Technical Secretariat with regard to the evaluation of CWC implementation and verification of compliance, including with regard to a more holistic approach to the analysis and reporting of verification data and results.

Enhanced mechanisms to interact with other stakeholders. The OPCW needs to nurture and further develop its relationship with other CWC stakeholders in three essentially overlapping directions: (a) in areas where it interacts and collaborates closely with partners to fulfil its mandate, such as in the conduct of operations in the field where OPCW personnel need to work together with personnel of other agencies such as UN entities, other international organisations and NGOs, or in areas of national implementation such as the monitoring of the trade in relevant chemicals through national customs systems, for example; (b) in areas where other stakeholders support the full and effective implementation of CWC requirements with their own activities, for example through voluntary compliance measures, awareness raising and training of staff in the chemical industry and in academic and research organisations dealing with chemicals; and (c) in areas where the OPCW benefits from information exchanges and advice received from competent organisations in such areas as the monitoring and evaluation of advances in science and technology, or in industry.

The OPCW has, of course, already put in place advisory bodies and mechanisms to this end, in particular, its Scientific Advisory Board and its Education and Outreach Advisory Board. It has also developed a productive and stable relationship with chemical industry associations, and in particular, the International Council of Chemical Associations (ICCA) and regional as well as national chemical industry associations. These stakeholder relationships form part of the wider governance structure of the OPCW. Similar relationships have evolved with the international science union in chemistry—the International Union of Pure and Applied Chemistry (IUPAC)—and civil society organisations that have come together in the CWC Coalition.

These engagements will be gaining in relevance for the OPCW as it defines itself more broadly as an international centre of excellence in achieving and maintaining a chemical weapons free world, and one that responds, and is seen to respond effectively, to the evolution of its strategic context. This broader understanding of its mandate will on the one hand require that the OPCW further enhance its verification system and maintain its core capabilities in the different implementation areas, and on the other hand that it respond to the needs of the States Parties and the evolving challenges at the intersection of chemical weapons arms control, chemical security, and chemical safety.

References

1. OPCW, *Note by the Technical Secretariat The OPCW in 2015: Ensuring a World Free of Chemical Weapons, Document S/1252/2015*, 2015.
2. W. Krutzsch and R. Trapp, in *The Chemical Weapons Convention – a Commentary*, ed. W. Krutzsch, E. Myjers and R. Trapp, Oxford University Press, 2014, pp. 76–82.
3. A. M. Pearson, M. I. Chevrier and M. Wheelis, *Incapacitating Chemical Weapons – Promise or Peril*, Lexington Books, 2007.
4. M. Crowley, *Drawing the Line: Regulation of "wide Area" Riot Control Agent Delivery Mechanisms under the Chemical Weapons Convention*, Omega Research Foundation and Bradford University, 2013.
5. OPCW, *Joint Paper by Australia, Austria, Belgium, Canada, Chile, Cyprus, Germany, Finland, Ireland, Japan, New Zealand, Norway, Poland, Portugal, Republic of Korea, Spain, Switzerland, Turkey, United States of America, United Kingdom of Great Britain and Northern Ireland, Aerosolisation of Central Nervous System-acting Chemicals for Law Enforcement Purposes, Document C-20/NAT/2*, 2015.

6. J. Revill, C. McLeish and A. Ghionis, Could our approach to chemical weapons help reduce the threat of acid attacks? *The Guardian*, 17 July 2017.
7. OPCW, *Scientific Advisory Board's Temporary Working Group Verification Report, Document SAB/REP/1/15*, 2015.
8. OPCW, *Note by the Technical Secretariat the OPCW in 2015: Ensuring a World Free of Chemical Weapons, Document S/1252/2015*, 2015, paragraph 30(h).
9. OPCW, *Note by the Technical Secretariat the OPCW in 2015: Ensuring a World Free of Chemical Weapons, Document S/1252/2015*, 2015, paragraph 5.

CHAPTER 16

Implications for the Biological and Toxin Weapons Convention

JEZ LITTLEWOOD

Norman Paterson School of International Affairs (NPSIA), Carleton University, Ottawa, Ontario, K1S 5B6, Canada
*E-mail: Jeremy.Littlewood@carleton.ca

16.1 Introduction

The outcome of the Eighth Review Conference of the Biological and Toxin Weapons Convention (BTWC) has been described as "minimal", "disappointing" and "less than expected", by a variety of States Parties.[1] Among its underwhelming achievements was the minor updating of the additional understandings that flow from the Final Declaration of each review conference; the Final Declaration relates to the Article-by-Article review of the Convention, which is the central purpose of the meeting itself. In 2016, only Article VII (on assistance in the event of use) and the Solemn Declaration were updated in any substantive form. All other articles, including Articles I (prohibition), III (non-proliferation), IV (national implementation), V (consultations), VI (investigations), and X (peaceful cooperation) were updated only in an administrative sense, removing or amending specific date or procedural references tied to the last cycle of the Intersessional Programme of work between 2012 and 2015. States Parties to the BTWC have, therefore, failed to fulfill their basic procedural and administrative responsibilities in the conduct of the Article-by-Article review. In addition, they failed to agree on a new Intersessional Process.

Preventing Chemical Weapons: Arms Control and Disarmament as the Sciences Converge
Edited by Michael Crowley, Malcolm Dando and Lijun Shang

Published by the Royal Society of Chemistry, www.rsc.org

The scheduled meeting of the BTWC States Parties in December 2017 was tasked to "seek to make progress on issues of substance and process for the period before the next Review Conference,"[2] tantamount to a procedural last minute save that prevented the collapse of a process. This diplomatic fumble may not have long-lasting consequences: a compromise emerged in December 2017 but it involves a carefully constructed and modest programme of work before the Ninth Review Conference that will steer the Convention back into the doldrums[3] while maintaining a regularized series of meetings and work intended, albeit modestly, to strengthen the Convention.

How States Parties recover from 2016 will determine whether or not the BTWC will be a leading forum for consideration of and response to scientific and technological developments that the 2015 report of the InterAcademy Partnership (IAP) noted "give cause for concern."[4] A failure to recover will likely result in a Convention that remains reactive to such developments. These separate processes—scientific and technological developments and the political-diplomatic developments in the BTWC—will influence each other, but the extent to which they determine outcomes in each process in the near and mid-term is complex. It cannot be predicted, but a central question is the verification of the BTWC. Differing opinions over the feasibility of verifying biological disarmament are at the heart of the current dispute. As such, the issue of verification is the subject of this chapter. The chapter begins with an overview of the current context of the verification issue and outlines four options for thinking about the future up to the end of the Tenth Review Conference in 2026. It then explores each option in turn before considering the likely evolution of the Convention in the next five to ten years vis-à-vis verification and the issue of convergence in science and technology.

16.2 The Existing Context

Fundamental differences between States Parties on the issues of compliance and verification continue to exist. No resolution of these differences is in sight and those differences will continue to have a direct effect on how the BTWC evolves up to 2026. These differences, in turn, have a direct effect on how States Parties consider developments in science and technology relevant to the Convention and any appropriate response to such developments. An important issue in this regard is that developments in science and technology, including convergence, will not result in a decision by States Parties to begin a renewed effort to negotiate or agree upon verification measures. Although convergence is of fundamental importance to the future of the Convention, it is politics, rather than science, that drives decision-making in the Convention. As other chapters in this volume attest, convergence in chemistry and biology, as well as other scientific, technological, and engineering disciplines is indisputable, but the potential for misuse alone is insufficient to alter diplomatic and political fundamentals in a forty-year treaty regime.

Actual misuse on a scale that affects the priorities of key States would likely have to occur before a major reorientation of the BTWC towards verification will emerge. This stark assessment is based on two foundations. First, the practice of States Parties in the BTWC over the last four decades which has tended to incremental change because States Parties cannot agree upon more fundamental and far-reaching revisions to the structure and modalities of the BTWC. Second, because science and technology are now both a global undertaking and at the base of expected economic development: both of these aspects are revealed in the current practice of science and in the emergence of the bioeconomy.[5] Science, therefore, is within a larger framework of political, economic, safety, and security activity rather than separate to or above such considerations. As Evans notes: "Knowledge is always produced within a social context, and security is only one of many goals a society is striving for. From that starting place, we can begin to build a contextual approach to governance that is appropriate for the complex practice of real-world science."[6]

Despite the claim that scientific developments will not result in a decision to begin verification debates in the BTWC, these developments and convergence will shape and influence inevitable discussions about verification. Notwithstanding the disappointment of 2016, it does appear inevitable that a discussion on compliance and verification will emerge formally in the next few years for three reasons. First, Australia and others revivified the debate, in their assertion that *we need to talk about compliance*.[7] Second, the Non-Aligned Movement (NAM), Russia, and other States have remained committed to verification of the BTWC since 2001 when the BTWC Protocol negotiations collapsed. Third, and linked to both the preceding reasons, the differences between States Parties on verification explain why the Eighth Review Conference produced a disappointing outcome. Iran's opposition to expanding the mandate of the Intersessional Process and increasing the resources of the Implementation Support Unit blocked further incremental progress and constitutes an attempt to force back on to the agenda of the BTWC a legally binding agreement.[8] Taken together, they signal that a formal debate on compliance and verification will be difficult to avoid. The road to recovery from 2016 will almost certainly have to include verification as a subject for debate, otherwise Iran, and probably the NAM, will stymie any outcome.

Although the verification debate is much broader than the issue of science and technology, perceptions about and awareness of convergence and its implications will shape attitudes to compliance and verification issues. How States Parties identify which aspects of convergence are relevant to the BTWC and what mechanisms should be put in place to assure themselves that militarization of the life sciences is not occurring within States, is far from simple. As a result, although the meeting in 2017 did reach a decision on a future programme of work related to verification and compliance issues, the process will almost certainly not entail the development of a verification or compliance architecture between 2018 and 2020. At best, an agreement may be reached on exploring options related to compliance, reassurance of implementation, and the question of verification. Real decisions have been put off until the Ninth Review Conference in 2021. Even this optimistic possibility of a meaningful discussion about verification is far from assured.

In this vein, it is important to explore options for verification and compliance in the coming years, rather than prescribe mechanisms that draw upon arms control as practiced in the second half of the 20th century. Four options can be identified to explore the issue. First, no change to be made to the *status quo* of the BTWC. Among other things, this risks convergence eroding the BTWC as the Convention becomes less relevant to on-going challenges posed by developments in this area.[9,10] Second, absorption of the BTWC into the institutional framework of the CWC whereby the BTWC is absorbed into the OPCW under the guise of synergistic efficiency and effectiveness of the separate-but-connected anti-chemical and anti-biological weapons regimes. Absorption is not a merger, but a realignment of separate legal agreements and implementation organs under one entity. It is the least likely outcome of the four options explored in this chapter, but remains valid as a means to explore issues and open up conceptual thinking about the future of the BTWC. Third, mimicry, whereby the BTWC States Parties adopt practices and procedures comparable to those of the CWC and OPCW but do so independently and under a stand-alone agreement. This is likely to be closely aligned with a return to negotiations on a verification agreement. And, fourth, innovation, whereby the BTWC States Parties establish, or adapt, and adopt mechanisms that form a more dynamic and resilient network of legal and normative practices that are responsive to challenges but place differential requirements on various actors. Such innovation may be more likely outside the BTWC, but by some like-minded States Parties, rather than within the Convention and by all States Parties. As a consequence, innovation is likely to result in the continued evolution of a complex and layered network regime against biological and toxin weapons.

As outlined below, option two (absorption) and option three (mimicry) are likely to be unacceptable in the next five years. Option one (*status quo*) is the most likely outcome in the next few years, but the *status quo* reflects the process of incremental evolution, not stasis. Thus, modification to understandings and practices are likely to emerge, but may not be universally accepted or adopted because decisions reached in the BTWC Meetings of States Parties are not considered legally binding. This differentiated approach gives further credence to option four (innovation) as the most likely medium term trajectory of BTWC evolution and implementation. It builds on the existing notions of a 'web of deterrence', 'holistic arms control', 'multilevel governance', 'preventive security governance' and a 'disaggregated' approach to strengthening the Convention.[11–15] In summary, a complex regime arrangement.[16]

Each of these options is outlined in further detail in the ensuing pages. This framework is a means to explore how convergence may affect the verification debate. It is a heuristic rather than prescriptive approach, and the options are not mutually exclusive they are simply a means to identify different approaches to responding to the challenges facing the BTWC, of which one singly important aspect is science, technology, and convergence. The chapter builds on existing work in this volume and does not repeat the existing debates related to strengthening the BTWC in the run-up to the Eighth Review Conference.

16.3 Status Quo

The BTWC has charted its own path of incremental evolution over the last four decades.[17] During its life, the Convention has certainly experienced the vicissitudes of a treaty in force, as Sims documented during its first decade of existence,[18] but a significant crisis of confidence has not emerged for three reasons. First, as Carus' history notes, use of biological weapons since 1975 has been remarkably low-level and ineffective and only by non-State actors.[19] Second, although offensive biological warfare programmes existed after 1975 in the Soviet Union, Iraq and South Africa, changing perceptions about the utility of biological weapons, the normative constraints on poison weapons, and the prohibition regime that has evolved over time have all had a positive effect in reinforcing biological disarmament.[20] No State Party has withdrawn from the Convention and no State openly admits to biological weapons. Third, substantive concerns over biological and toxin weapons programmes have been addressed by additional mechanisms that do not involve all States Parties. This has not always been effective or without political implications. For example, the Soviet Union's offensive programme and the Trilateral Process[21] and the Iraqi offensive programme.[22] South Africa's decision was a result of a national undertaking at the end of the Apartheid era.

Moreover, BTWC States Parties have developed processes that work, albeit not as efficiently, effectively, or with as much support as many States Parties and 'Friends of the Convention' would prefer. The *status quo* works, and incentives to adopt new procedures, absorb additional costs, and expand upon institutional mechanisms and processes for the BTWC have repeatedly failed to attract sufficient support over time. The result is partial evolution and small incremental changes to the BTWC.

In this regard, the *status quo* model of the Convention is not one of stasis, or ossification, whereby the Convention and its implementation remain unchanged. As the additional understandings related to the Convention and mechanisms and processes that have developed over time illustrate, implementation of the BTWC does evolve. Two issues are of central importance in considering the *status quo* model and the issue of convergence. First, the relevance of the BTWC to science and technology in practice. Second, the ability of States Parties to act in a manner that adapts implementation of the Convention to ensure that convergence and other developments do not circumvent its prohibitions.

As histories of the Convention confirm,[15,23] since its First Review Conference in 1980, the additional understandings developed and implemented by States Parties have led to a Convention that has evolved in a constructive direction. Scientific and technological developments have repeatedly been taken into account and review conferences have developed understandings that make it clear, as in 2011, "that Article I applies to all scientific and technological developments in the life sciences and in other fields of science relevant to the Convention" (see Chapter 4). From the documentary record it is evident that States Parties (and professional associations) are aware of the implications of science and technology on the BTWC and the additional understandings are a proactive response to these developments, proactive in the sense that reaffirmation of the scope of the BTWC and consensus agreement recorded in successive Final Declarations repeatedly confirm that such developments do not fall outside the Convention and its prohibitions. These words are not without significance. They signal States Parties confirmation that scientific and technological developments do not circumvent the General Purpose Criterion. Awareness of science and technology is not the problem for the BTWC, as the history of review conferences demonstrate. Rather than awareness, the problem is agreement on how to respond to these developments in terms of the implementation of Articles III (non-proliferation), IV (national implementation), VII (assistance) and X (peaceful cooperation). It is the lack of action in these areas that give rise to a justified skeptical assessment that the words in successive final declarations are an insufficient bulwark against the reemergence of biological weapons.

While incrementalism is not ideal, it has been sufficient. Moreover, ideas take time to ripen and be taken up by States Parties. Two proposals in 2016, the establishment of a steering committee of key States Parties—a kind of Executive Council of the Convention *in lieu* of a formal organization or institution—and decision-making powers for annual Meetings of States Parties from 2017, illustrate this. Both are recognizable from previous review conferences. A steering committee from Sims' proposal for "serious administration"[18] and decision making outside the review conference was mooted in 2006. This sedate form of action may be less than ideal, but, coupled with activity outside the BTWC framework in areas such as the Australia Group, United Nations Security Council Resolution 1540 and efforts to assist States in their implementation of the Convention, it has worked.

Whether or not incrementalism remains sufficient is a significant question. Prior to the Eighth Review Conference, a number of perspectives identified concerns over the sustainability of the Convention over time and a coalition of non-governmental groups openly suggested that the BTWC was "not fit for purpose."[24] Without further and more rapid adaptation, the Convention risked being irrelevant to real world science given the pace of change and implications of developments.[9] Awareness of the implications of science beyond States Parties is a central issue of concern, as Dando and Rappert illustrated that lack of awareness among scientists may result in irrelevance at the laboratory level.[25] Maintaining the relevance of the Convention has been at the heart of efforts by some States since the mid-2000s, notably articulated in the US *National Strategy for Countering Biological Threats*[26] and successive Common Positions of the European Union with respect to the Convention, and the efforts by States Parties themselves since 2001. In addition, the work of the Inter Academy Partnership (IAP) has also resulted in an increasing recognition of the risks posed by developments in science and technology.

The central concern of 2016 was that a quinquennial review of developments in science was no longer sufficient to address the pace of developments. As the IAP noted, advances in a range of scientific disciplines "could also facilitate almost every step of a biological weapons programme and technological barriers to acquiring and using biological weapons have been conspicuously eroded since the Seventh Review Conference."[27] Calls for the establishment of a Scientific Advisory Panel for the Convention similar to that in existence for the Chemical Weapons Convention were encapsulated in the IAP report: "a technique for systematically assessing the risk of new scientific developments would be of enormous help in assessing the threat of future trends."[28] Such a mechanism would need to be flexible, inclusive, agile, responsive, and able to foster engagement from a variety of actors, as well as be transparent.

These attributes are not in existence in the BTWC, suggesting that either some would have to be sacrificed, or that such a mechanism would have to remain outside the Convention. Both formal, from States Parties themselves, and informal and independent scientific and technical input already exist. The existing, *status quo*, approach to considering scientific and technological developments of relevance to the BTWC has three components. First, it involves States in their individual capacity submitting reports in preparation for a review conference. Second, States Parties collectively within the Convention at review conferences and under the last intersessional programme of work have debated certain aspects of science and technology. And, third, civil society and professional associations such as the IAP remain engaged on the issue. In terms of technical advice to States Parties this may suffice. Whether such a mechanism is within the Convention or not, however, still leaves a fundamental question over how to respond to such developments and address such implications within Articles I, III, and IV of the BTWC. Critics of the BTWC are correct to note that awareness raising and advice is not enough: action is still required, because global society remains "a long way from grasping the opportunity to put in place effective policies to protect benignly intended work...from hostile misuse."[29]

The *status quo* approach may, therefore, be sufficient to react to convergence and its challenges, providing incremental adaptations to practice among States Parties emerge in the next few years. The problem with this option is, however, two-fold. First, incremental adaptation and agreement among States Parties on binding changes to practice have been absent since

2001. As Pearson, Dando and Sims observed, while three Intersessional Processes demonstrated the benefits of regular contact and discussion among States Parties they produced "little if any effective action or results."[30] The second challenge is with the acceptance of 'sufficient' as a bulwark against developments that are lowering the barriers to bioweapons.

16.4 Absorption

In 2009 the convergence of the BTWC and the CWC was explored by Sims, even though he noted that a merger of the two Conventions in the short to medium term was inconceivable.[31] Absorption is a not a merger of the Conventions. Rather, absorption reflects the possibility that implementation and institutional support within a single institution or international organization focused on chemical and biological weapons may emerge over time. Absorption is not the BTWC being subsumed wholesale into the CWC; or the adoption and application of the CWC's verification annex and related provisions into the BTWC. Absorption is a limited mode that allows for differences and distinctions between the two Conventions to be accommodated, but similarities to be exploited, synergies to be developed, and gains in both efficiency and effectiveness brought to bear.

Although absorption is a remote possibility, it is worth exploring for two reasons. First, simply to identify synergies and efficiencies related to implementation that will enhance the effectiveness of both Conventions. Second, because expanding the conceptual space for thinking about the BTWC and its future will be necessary to overcome the binary obstacle of verification or no verification. No serious consideration of the issue can emerge until around 2020–2022 when the OPCW will likely report on the complete destruction of known chemical weapons stockpiles and cost savings will begin to loom over the OPCW. By this point in time the BTWC may also have experienced another underwhelming Intersessional Programme of work as it contemplates the Ninth Review Conference. The early 2020s might, therefore, engender fundamental reassessment of the CWC and the BTWC, although it should be noted that the OPCW and CWC States Parties are already considering the issues.

It is important to note that absorption is as likely to be driven by the needs of the CWC and the OPCW as it is by the BTWC as States Parties in both Conventions begin to realize the implications of two quite different Convention regimes having to address a spectrum of chemical and biological weapons that are no longer easily discernible. Moreover, as Dando observed[32] "it seems likely that the CWC, even with greater capabilities than the BTWC, will continue to have difficulty adjusting effectively to the rapid state of scientific and technological change in areas such as neuroscience." Trapp, equally, notes an environment that suggests a scientific revolution is in the making may require that the systems or regimes against chemical and biological weapons operate much closer with each other and together.[33] Neither Dando nor Trapp are alone in their concerns, as the

IAP report underlines, and the Advisory Board of the OPCW itself notes: "Convergence in the sciences does not in itself lead to convergence of the regimes, but exchanges of experience and joint technical reviews could be helpful to understand how it affects the implementation of both treaties at the interface between chemistry and biology. That is particularly pertinent as there is an overlap between the two treaties with regard to the prohibition of toxin weapons."[34]

Convergence is occurring within a broader future context where the OPCW will be seeking to reaffirm its existence and funding upon completion of the destruction of declared chemical weapons stockpiles.[35] Maintaining a world free of chemical weapons when budgets are under pressure suggests that the OPCW could become a victim of its own success.[36] Taken together, convergence in myriad scientific disciplines, continued pressure for economic efficiencies in multilateral arrangements, and challenges related to the OPCW following completion of destruction and the challenges of a BTWC that lacks implementation organs, financial support, and an architecture for implementation, may lead to new thinking about both Conventions.

The arc of BTWC–CWC convergence, integration, or at least formal cooperation between the OPCW and the ISU can be found in history through the evolution of the normative constraints on poison weapons, the 1925 Geneva Protocol that prohibited the use of chemical and biological weapons, and the separation of chemical and biological weapons following the UK proposal of 1968.[37] As Walker's history notes, the UK "was not blind to the limitations inherent in its proposal [to separate chemical and biological weapons], but Ministers and officials ultimately recognized that there comes a time when risks must be taken if progress is to be made in multilateral disarmament."[38] While biological and chemical weapons were duly separated in key areas, they were equally conjoint in others: toxins remained under the prohibition and purview of both the BTWC and the later CWC; the Geneva Protocol remained extant in its prohibition on the use of either type of weapon and the United Nations Secretary General's Investigation Mechanism equally applied to chemical and biological weapons. At the national level, shared institutions to address chemical and biological weapons remain in defence and protection establishments, in diplomatic functions, among experts, and within law.[31] Indeed, post 2001, significant efforts by States and civil society actors to ease the burden of national implementation *via* synergies, assistance, and common legislation are evident.[39] Convergence and its realities aside, chemistry and biology have been long been bedfellows in defence, diplomatic and legal efforts within states.

Absorption of the BTWC into the CWC institutional framework would, potentially, permit synergies between two Conventions to deliver efficiency and effectiveness benefits that would enhance both the CWC and the BTWC. At face value, co-location in The Hague of the OPCW and ISU may offer easier and enhanced formal cooperation and collaboration in scientific and technical assessments, training of investigators for deployment under the Secretary-Generals' mechanism, common access to equipment, and shared services (translation, interpretation, information technology and communication support, administrative support), physical infrastructure, and co-location of diplomatic and expert communities that would over time benefit from covering both the CWC and the BTWC. Nevertheless, two treaty regimes have existed side-by-side for nearly twenty years and prior to 1997 the BTWC existed alone for more than two decades. Both substance in law and opportunity in politics must be taken into account: the two Conventions do not have the same membership; the obligations in one are different to the other; and, the verification regimes are entirely different in that the CWC has verification and the BTWC does not. The obstacles to absorption are very real and its prospect is an extremely long-way off.

However, greater cooperation between the CWC and BTWC is likely due to necessity in future years. The period between 2021 and 2025 may be another moment in time that requires that risks be taken if progress is to be made. A merger may be inconceivable, but a confluence of factors may make absorption of common CWC and BTWC functions attractive in financial, political, and technical realms. By the second half of the 2020s, an Organisation for the Prohibition of Chemical and Biological Weapons (OPCBW) may no longer be science fiction.

16.5 Mimicry

Mimicry as developed here is most closely aligned with an additional protocol or legally binding agreement that would follow the AHG process or more closely resemble the CWC–OPCW mechanism: declarations, inspections, investigation procedures, an implementation organization, and the gamut of obligations and measures that are standard in legally binding arms control agreements. Three routes to a legally binding agreement are identifiable. First, a reversal of position by the United States, and a decision or policy that supports multilateral negotiations. This is highly unlikely, but would be embraced by other States. Why a reversal of the US position would occur is an important consideration in this regard. Demonstrated non-compliance by a State actor or a series of biological events by a non-state actor may induce a change of policy in the US, but an embrace of verification is not inevitable if a major incident does occur. The second route would involve like-minded states agreeing upon a negotiation process outside the BTWC, probably *via* the United Nations General Assembly where consensus is not a requirement. The decision to begin negotiations on a nuclear weapons Ban Treaty were initiated by a 2016 General Assembly resolution for a "legally binding instrument to prohibit nuclear weapons, leading to their total elimination"[2] and the negotiation of the Arms Trade Treaty indicates that this is possible. This idea is not detectable among the thinking of States Parties at present but cannot be ruled out in future years. A third road to verification is *via* a step-by-step model. A 'modular' approach[40] would resemble Russia's proposal to focus on less contentious issues while balancing both peaceful cooperation

(Article X) and disarmament (Article I) obligations *via* legally binding declarations and other means.[41] This Russian approach failed to gain support among Western States, not least given concerns about Russian biodefence activities and remaining questions related to its inherited (Soviet) offensive biological weapons program.[42] However, the breakdown in 2016 at the Review Conference indicates that Iran, and others, may consider that the route back to verification can be achieved by denying consensus and blocking progress on a new Intersessional Process. As noted above, the route to a new Intersessional Process will probably have to include a track that includes verification or some focused discussion on compliance options and a small step may be all that is necessary in 2017 to kick start a process that at least explores mimicry as a future option.

Mimicry is, therefore, possible, but unlikely in the medium term of the next five to ten years. The US is unlikely to alter its position on verification before 2020. Thus, the next decision point would be in 2021 at the Ninth Review Conference and even allowing for a decision to begin negotiations, agreement would not be reached until at least the Tenth Review Conference. Opting for a multilaterally negotiated legally binding agreement to address compliance and verification issues therefore poses two significant challenges directly relevant to the convergence dilemma. The first issue is that of time; the second concerns the nature of the agreement and its sustainability as a response to continued scientific progress.

With regard to time, negotiation of a verification agreement would take at least five years and likely require an additional two to four years before ratification and entry into force; moreover, it would be a further five years before many States Parties would have in place the necessary national implementation legislation for any future agreement. Realistically, therefore, any new mechanism would not take effect for a decade (2026) and perhaps longer, or may suffer the fate of the Comprehensive Nuclear Test Ban Treaty (CTBT) and remain not in force for over two decades. Concerns about the impact of convergence in the contemporary period suggest that a push for formal verification and legally binding arrangement(s) as a solution to these concerns entails a willingness to ignore the impact of developments over the next decade on the assumption that a legally binding agreement can be agreed and will enter into force.

The second challenge relates to the nature of the agreement. As an instrument of technology governance and control treaties "once concluded, are often too inflexible to keep pace with rapid technological change."[43] Dando's observation related to the challenges that the CWC faces in dealing with scientific developments affirms this and Tucker's own assessment of dual use technologies across disciplines "suggests that a one-size-size-fits-all approach to managing the risk of misuse is neither possible nor appropriate."[44] In the context of the BTWC, a single legally binding agreement will impose similar obligations and mechanisms on all States that will subsequently become outdated, require assessment of both efficacy and effectiveness, and entail a further round of additional understandings and non-legal agreements similar to how the BTWC has evolved over the last four decades. Furthermore, avoiding a 'one size fits all' approach has been a mantra of some States Parties to the BTWC for over a decade. A return to verification negotiations may be appealing, but the history of the BTWC itself and the experience of other agreements suggest any decision to mimic the CWC will be fraught with risks of failure and, even if successful, pose different challenges in future years.

16.6 Innovation

The logical step is to explore flexible agreements, some of which might be legally binding, whereas others may be voluntary, like-minded, or flow from decisions in other fora, such as the United Nations Security Council or the OPCW. This would entail a change of mindset within States Parties and require embracing new approaches to the challenges that science and technology pose to the future of biological disarmament. Innovation is a catch-all term intended to encompass developments that do not easily fit within the three options above: *status quo*, absorption, mimicry. As noted, the four options are not mutually exclusive and innovation is arguably the area where a blending of the status quo with mimicry and absorption will manifest itself. Innovation is likely to emerge from groups of like-minded States Parties, civil society, and other actors outside formal meetings of the BTWC. It already exists within Millett's "evolved network" that emerged "out of necessity"[45] and is explicit in Koblentz's 'biosecurity reconsidered' that "will depend in part on increasing coordination among national governments, international organizations, the private sector, and nongovernment organizations."[46] In political science it is referred to as 'international regime complexity' which is "the presence of nested, partially overlapping, and parallel international regimes that are not hierarchically ordered."[47] Crowley elucidates this approach with respect to ICAs, RCAs and their means of delivery under the holistic arms control model that provides analytical framework that "seeks to actively explore the utility of employing a multiplicity of mechanisms and to facilitate the active engagement of a variety of relevant actors in the regulation of the weapons or weapons-related technology."[48]

Although Crowley's approach is focused on incapacitating and riot control agents, it underlines what others have argued over the last decade and a half; namely that it is necessary to think beyond the BTWC. Lennane[49] provides the most recent articulation of the emergent model, but similar concepts can be discerned in Koblentz and Lentzos,[50] Tucker and Bansak,[43] Smithson,[22] Millett,[9] and Littlewood.[15] As Smithson noted in 2011, biorisks (broadly conceived by Koblentz and others) require an "elevated problem-solving process,"[51] but the process can itself draw upon "a rich, practical array of options...because virtually all the constituent tools already exist in some form." The central premise of both Smithson's and Crowley's approach is that there are sufficient institutions and experience in existence to develop additional and innovative mechanisms to address the challenges that convergence poses. However, it requires an acceleration of multi-level governance and acceptance of the BTWC as a foundation for action rather than an all-encompassing institution.

In this regard, innovation in the BTWC will likely mirror Crowley's holistic arms control where the Convention is the most appropriate forum for consideration of the issues, but must be supplemented by alternative mechanisms. This gives rise to, or recognition of, parallel regimes, overlapping regimes, and nested regimes. Parallel regimes exist "where there is no formal or direct substantive overlap" whereas overlapping regimes are "where multiple institutions have authority over an issue, but are not mutually exclusive or subsidiary to another" and nested regimes entail "institutions that are embedded within each other in concentric circles, like Russian dolls."[52] As regime complexity scholars observe, "[M]ost new cooperative endeavors and most efforts to include new actors will need to be located within a structure that already has a lot of actors, interests and hardened beliefs."[53] The most salient aspect of this to the BTWC beyond 2016 is that of overlapping regimes: the BTWC, CWC, the UNSCR1540 Committee, and the United Nations Secretary General's investigation mechanism are the readily identifiable multiple entities with authority over discrete aspects of the biological weapons challenge. These intergovernmental arrangements are not the sole entities: equally it has been professional associations like the IAP and like-minded States acting in concert with each other as in the Australia Group, the Global Partnership Against Weapons of Mass Destruction, the Global Health Security Agenda (GHSA), and other cooperative threat reduction programmes, that have had a role in the biological weapons regime, but not always formally within the BTWC.

A new model for the strengthening the Convention would then embrace the legal foundation of the BTWC and the best aspects of its approach to biological disarmament and move beyond the Convention in a manner that "incorporates the various *ad hoc* innovations that states and other actors have devised to work around the limitations" of the BTWC and its methods of working.[54] Innovation would, therefore, resemble, a "loosely-linked regime complex rather than a single international regime" that would involve discrete components of the regime "created at different times, and by different groups of countries...crafted in a context of diverse interests, high uncertainty, and shifting linkages ... [and] ... not integrated, comprehensive, or arranged in a clear hierarchy."[55] Arguably, this loose regime complex is the existing reality in which the BTWC resides, but States Parties to the Convention have been unable to further develop the model and strengthen the Convention in this manner because a large number of States refuse to abandon traditional forms of arms control.[56]

16.7 Conclusion: Practical Responses to Convergence

The challenges that convergence poses to the BTWC will not determine the future form of the Convention, but they will certainly shape it and, perhaps, influence the regime complex that exists and continues to evolve around biological and toxin weapons. The Eighth Review Conference laid bare the central cleavage between States Parties over the question of verification and the process of how to strengthen the Convention. These differences of view are not based on scientific developments or divergence over the implications of convergence. Rather, the cleavage is political. It can be overcome, sidestepped, or bridged only by a political accommodation and agreement.

Neither opponents nor proponents of verification are likely to concede: but both may step back from the diplomatic brink and the compromise work programme to 2020 or 2021 should assist that development. Thus, the most likely outcome of the 2018 to 2020 meetings of States Parties will be similar to previous review conferences: incremental, sub-optimal, processes that are neither binding nor embedded in the practices of all States Parties. This outcome will produce some benefits, but is unlikely to result in binding agreements.

Such an outcome is highly problematic for those concerned about the ability of the BTWC to remain salient in the face of an on-going scientific revolution that already "facilitate[s] almost every step of a biological weapons programme."[27] Waiting until 2021 for another opportunity to propose a science advisory mechanism in the hope that an advisory process will be both adopted and subsequently result in changes in practice among States Parties is optimistic. The IAP's own report noted that any advisory process would not be "the sole source of science advice" but would be complemented by the roles of States Parties to the BTWC and the international scientific community.[57] The failure of the Eighth Review Conference to agree on such an advisory process underlines the necessity of existing modes of advice and communication and exploring new models to respond to the challenges of convergence.

The technology governance response required to address the challenges of convergence entail a mix of available hard law, soft law, and informal measures that will be negotiated by a variety of different actors: State, intergovernmental, non-state, public, and, private. These actors have different perceptions of the risk that science and technology pose, the benefits of such technology, the costs of regulation, and the need to balance domestic political interests, industry buy-in for additional measures, and capacity to implement any decision requiring action.[58] If all stakeholders await the outcome of a consensus decision by States Parties to the BTWC in 2017 or beyond, technology governance will almost certainly remain undeveloped over the next five years. It is much more likely that informal collaboration and cooperation involving various actors will prevail, as in the existing IAP role on science and technology. Certain recommendations may be adopted and formalized by groups of like-minded States Parties who will, in turn, attempt to engage and persuade other States Parties to adopt similar mechanisms under the ambit of good practice.

Compounding this delay is the fact that "emerging technologies are 'moving targets' that undergo a continual process of evolution and refinement,"[59] and flexibility in response may be as, or more, important than rigidity in law. In a future that may entail neuroweapons, inhibition approaches would have to be "constantly vigilant in the face of changing technologies and fuzzy distinctions between medical and military uses, all while navigating the complexities of the health-care industry, political and military ethics, and international law."[60] The BTWC is a treaty and part of international law. As such it is an essential component of the regime, but as Lenanne, and others, note, the Convention may now be transitioning to a foundation for action rather than a forum for activity in order to address new challenges and embrace Evans' observation that security is but one issue in the complex practice of real world science.

These challenges are evident to anyone who has had to deal with science and technology in a political-diplomatic context[61,62] and has observed the accommodation of national interests and "pernicious ignorance" that Robinson noted had affected the trajectory and effectiveness of the CWC.[63] Moreover, they are not new challenges. Pearson's call for a 'web of deterrence' required "a sustained effort by states to integrate various policy approaches with the goal of a strong web of deterrence against the production, possession, and the use of these weapons."[64] Twenty-plus years on, his vision of an effective arms control regime has not materialized. A layered web or complex regime has evolved, but remains a work in progress. The regime itself should already be understood as an extant system of overlapping and interlocking mechanisms where discrete components are not dependent upon each other but work in complementary and supplementary ways and are driven forward not by all States Parties to the BTWC, but by certain groupings of them at specific points of time: a regime complex system already 'overlapping'. Convergence poses significant challenges to the BTWC and the CWC, but the repeated inability of States Parties to the BTWC to agree upon binding commitments that respond to scientific and technological developments suggest that the Convention may be at the limits of its evolution. To date, the international scientific community has proven itself able to bring to the attention of States the potential dangers of science and technology. Where States Parties have not acted under the BTWC, groups of like-minded states have created arrangements such as the Australia Group or acted under the United Nations Security Council. A practical response to convergence therefore entails a continuation of the evolution of regime complexes, with multiple actors engaged in a variety of activities and mechanisms. The objective would be in line with the central purpose of the BTWC—a world free from biological weapons—but would not manifest itself solely under the Convention.

References

1. R. Guthrie, *The Eighth BWC Review Conference: A Minimal Outcome, Bio-Weapons Prevention Project, RevCon Report 16*, December 1 2016, http://www.cbw-events.org.uk/RC16-16.pdf.
2. United Nations, *Taking Forward Multilateral Nuclear Disarmament Negotiations, United Nations General Assembly*, First Committee, A/C.1/71/L.41, United Nations, New York, October 14 2016.
3. N. A. Sims, *Disarmament Diplomacy*, 2003, p. 70, http://www.acronym.org.uk/old/archive/dd/dd70/70op2.htm.
4. Interacademy Partnership, *The Biological and Toxin Weapons Convention: Implications of Advances in Science and Technology, Conference Report*, DES4023_2 (December), 2015, p. 6.
5. OECD, *International Futures Project. The Bioeconomy to 2030: Designing a Policy Agenda*, OECD, Paris, 2009.
6. S. W. Evans, *2016 Issues in Science and Technology*, Fall, 2016, vol. 84–88, p. 88.
7. United Nations, *We Need to Talk about Compliance, Submitted by Australia, Canada, Japan, New Zealand and Switzerland*, BWC/MSP/2012/WP.11, United Nations, Geneva, 12 December 2012.
8. Islamic Republic of Iran, Statement by H. E. Mohsen Naziri Asl, Ambassador Extraordinary and Plenipotentiary of the Islamic Republic of Iran to the United Nations Office at Geneva, At the Adoption of the Final Outcome Document of the 8th Review Conference of the BWC, United Nations, Geneva, 25th of November 2016.
9. P. Millett, *J. Conflict Secur. Law.*, 2010, **15**(1), 25–43.
10. C. A. McLeish and R. Trapp, *Nonproliferation Rev.*, 2011, **18**(3), 527–543.
11. G. S. Pearson, *Weapons Proliferation in the 1990s*, ed. B. Roberts, The MIT Press, Cambridge, 1995, Prospects for Chemical and Biological Arms Control: The Web of Deterrence, pp. 287–304.
12. M. Crowley. *Chemical Control*, Palgrave Macmillan, Basingstoke, 2016.
13. J. B. Tucker, *Innovation, Dual Use and Security*, ed. J. B. Tucker, The MIT Press, Cambridge, 2012, Review of the Literature on Dual Use, pp. 19–44.
14. D. Garcia, *Eur. J. Int. Secur.*, 2016, **1**(1), 94–111.
15. J. Littlewood, *The Biological Weapons Convention*, Ashgate, Aldershot, 2005.
16. K. J. Alter and S. Menuier, *Perspect. Polit.*, 2009, **7**(1), 13–24.
17. N. A. Sims and J. Littlewood, *Nonproliferation Rev.*, 2011, **18**, 499–511.
18. N. A. Sims, *The Diplomacy of Biological Disarmament: Vicissitudes of a Treaty in Force, 1975–1988*, St. Martin's Press, New York, 1988, pp. 288–309.
19. S. W. Carus, *Health secur.*, 2015, **13**(4), 1–37.
20. *Deadly Cultures Biological Weapons since 1945*, ed. M. Wheelis, L. Rózsa and M. Dando, Harvard University Press, Cambridge, 2006.
21. Vertic, *The Verification Yearbook 2002*, ed. D. C. Kelly, Russell Press, Nottingham, 2002, The Trilateral Agreement: lessons for biological weapons verification, pp. 93–109.

22. A. E. Smithson, *Germ Gambits*, Stanford University Press, Stanford, 2011.
23. M. R. Dando, *Preventing Biological Warfare*, Palgrave Macmillan, Basingstoke, 2002; N. A. Sims, *The Evolution of Biological Disarmament*, Oxford University Press, Oxford, 2002.
24. K. Millett, *Statement to the UN General Assembly First Committee on Disarmament and International Security*, United Nations, New York, October 12 2016.
25. M. R. Dando and B. Rappert, *Codes of Conduct for the Life Sciences: Some Insights from UK Academia, Bradford Briefing Paper No. 16 (2nd Series)*, University of Bradford, May 2005.
26. United States, *National Strategy for Countering Biological Threats*, National Security Council, Washington D.C., 2009.
27. Interacademy Partnership, *The Biological and Toxin Weapons Convention: Implications of Advances in Science and Technology, Conference Report*, DES4023_2 (December), 2015, p. 15.
28. Interacademy Partnership, *The Biological and Toxin Weapons Convention: Considerations for a Science Advisory Mechanism*, Meeting Summary, DES4395 (July), 2016, p. 8.
29. M. R. Dando, *Neuroscience and the Future of Chemical-Biological Weapons*, Palgrave Macmillan, Basingstoke, 2015, p. 170.
30. G. S. Pearson, N. A. Sims and M. R. Dando, *Key Points for the Eighth Review Conference*, ed. G. S. Pearson, N. A. Sims and M. R. Dando, University of Bradford, Bradford, 2016, Time for structural changes to make the BTWC more effective, pp. 3–17.
31. N. A. Sims, *The Future of Biological Disarmament*, Routledge, London, 2009, p. 245.
32. M. R. Dando, *Neuroscience and the Future of Chemical-Biological Weapons*, Palgrave Macmillan, Basingstoke, 2015, p. 152.
33. R. Trapp, *Convergence at the Intersection of Chemistry and Biology – Implications for the Regime Prohibiting Chemical and Biological Weapons, Policy Paper 6 Biochemical Security 2030*, University of Bath, Bath, 2014, pp. 27–30.
34. Organisation for the Prohibition of Chemical Weapons, *Note by the Director General. Report of the Advisory Panel on Future Priorities of the Organisation for the Prohibition of Chemical Weapons*, S/951/2011, (25 July), OPCW, The Hague, 2011, p. 20.
35. Organisation for the Prohibition of Chemical Weapons, *The OPCW in 2025: Ensuring a World Free of Chemical Weapons Note by the Technical Secretariat*, S/1252/2015 March 6, OPCW, The Hague, 2015.
36. R. Guthrie, in *The Future of the CWC in the Postdestruction Phase*, ed. J. P. Zanders, Report No. 15, EU Institute for Security Studies, France, Post-destruction era compliance under the CWC, 2013, vol. 28–39, p. 39.
37. J. P. Zanders, *J. Conflict Secur. Law.*, 2003, **8**(2), 391–410.
38. J. Walker, *Britain and Disarmament*, Ashgate, Aldershot, 2012, p. 71.
39. Vertic, *Legislative Guide to National Implementation of UN Security Council Resolution 1540 (2004)*, Vertic, London, 2014.

40. T. Findlay and A. Woodward, *Enhancing BWC Implementation: A Modular Approach*, Weapons of Mass Destruction Commission, Paper No. 23, 2004.
41. SIPRI, *SIPRI Yearbook 2015: Armaments, Disarmament and International Security*, ed. J. Hart and P. Clevestig, Oxford University Press, Oxford, 2015, Chemical and Biological Security Threats, pp. 557–585.
42. R. A. Zilinskas, *Russian Biodefensive Efforts and Apparent Concerns*, July 7, 2016, http://www.nonproliferation.org/wp-content/uploads/2016/08/160829_toward-the-8th-bwc-review-conference__Russian_biodefense_7_2016.pdf.
43. K. C. Bansak and J. B. Tucker, *Innovation, Dual Use, and Security*, ed. J. B. Tucker, The MIT Press, Cambridge, 2012, Governance of Emerging Dual-Use Technologies, pp. 306–329.
44. J. B. Tucker, *Innovation, Dual Use and Security*, ed. J. B. Tucker, The MIT Press, Cambridge, 2012, Review of the Literature on Dual Use, vol. 19–44, p. 39.
45. P. Millett, *J. Conflict Secur. Law.*, 2010, **15**(1), 25–43, p. 9.
46. G. Koblentz, *Int. Secur.*, 2010, **34**(4), 96–132, p. 124.
47. K. J. Alter and S. Menuier, *Perspect. Polit.*, 2009, **7**(1), 13–24, p. 13.
48. M. Crowley, *Chemical Control*, Palgrave Macmillan, Basingstoke, 2016, p. 268.
49. R. Lennane, *Divide and Delegate: the future of the BWC*, BWC Review Conference Series Paper No.1, International Law and Policy Institute, Helsinki, 2016.
50. G. Koblentz and F. Lentzos, Risks, Trade-Offs & Responsible Science BWC, *Review Conference Series Paper No.3*, International Law and Policy Institute, Helsinki, 2016.
51. A. E. Smithson, *Germ Gambits*, Stanford University Press, Stanford, 2011, pp. 248–252.
52. K. J. Alter and S. Menuier, *Perspect. Polit.*, 2009, **7**(1), 13–24, p. 15.
53. K. J. Alter and S. Menuier, *Perspect. Polit.*, 2009, **7**(1), 13–24, p. 21.
54. R. Lennane, *Divide and Delegate: the future of the BWC*, BWC Review Conference Series Paper No.1, International Law and Policy Institute, Helsinki, 2016, p. 4.
55. R. O. Keohane and D. G. Victor, *Perspect. Polit.*, 2011, **9**(1), 7–23, p. 19.
56. O. Meier, *Governance or Arms Control? The Future of the Biological and Toxin Weapons Convention*, Council of Councils, October 26 2016, http://www.cfr.org/councilofcouncils/global_memos/p38432.
57. Interacademy Partnership, *The Biological and Toxin Weapons Convention: Considerations for a Science Advisory Mechanism*, Meeting Summary, DES4395 (July), 2016, p. 6.
58. J. B. Tucker, *Innovation, Dual Use and Security*, ed. J. B. Tucker, The MIT Press, Cambridge, 2012, The Decision Framework, vol. 67–83, pp. 67–83.
59. J. B. Tucker, *Innovation, Dual Use and Security*, ed. J. B. Tucker, The MIT Press, Cambridge, 2012, The Decision Framework, vol. 67–83, p. 82.

60. J. Giordano, *2016 Bulletin of the Atomic Scientists May 31*, 2016, http://thebulletin.org/neuroweapons-threat9494?platform=hootsuite.
61. R. Quirion, A. Carty, P. Dufour and R. Jabr, *Palgrave Communications*, 2016, vol. 2, p. 16071, DOI: 10.1057/palcomms.2016.71, Published online 13 September 2016.
62. A. Moodie, Reforming the Biological and Toxin Weapons Convention's S&T Review Process, Guest Blog Post on the Bifurcated Needle 2016 May 16, http://www.bifurcatedneedle.com/new-blog/2016/5/16/reforming-the-biological-and-toxin-weapons-conventions-st-review-process, September 30, 2016.
63. J. P. Perry Robinson, *Int. Aff.*, 2008, **84**(2), 223–239, p. 234.
64. G. S. Pearson, *Weapons Proliferation in the 1990s*, ed. B. Roberts, The MIT Press, Cambridge, 1995, Prospects for Chemical and Biological Arms Control: The Web of Deterrence, vol. 287–304, p. 303.

Section V

The Role of Civil Society

CHAPTER 17

Role of the Non-governmental Chemical Science Community in Combatting the Development, Proliferation, and Use of Chemical Weapons

K. W. BOWMAN*[a], A. W. M. HAY[b] AND J. L. HUSBANDS[a]

[a]Board on Life Sciences, National Academies of Sciences, Engineering, and Medicine, 500 5th Street N.W., Washington, DC, 20001, USA; [b]University of Leeds, LIGHT Laboratories, Claredon Way, Leeds LS2 9JT, UK
*E-mail: kbowman@nas.edu

17.1 Introduction

A 2015 report from the Organization for the Prohibition of Chemical Weapons (OPCW) argues that:[1]

> "Preventing the re-emergence of chemical weapons comprises a broader set of activities and results to be achieved than the current core objective of non-proliferation suggests. In fact, preventing the re-emergence of chemical weapons cuts across all traditional results areas of the Organisation, and contains elements of verification, capacity development, improved national implementation, effective national protective programmes,

Preventing Chemical Weapons: Arms Control and Disarmament as the Sciences Converge
Edited by Michael Crowley, Malcolm Dando and Lijun Shang

Published by the Royal Society of Chemistry, www.rsc.org

engagement with relevant stakeholders through education and outreach activities, promotion of responsible use of science, the continued pursuit of universality, as well as the monitoring of relevant new developments in science and technology."

Achieving the effective chemical security policy and practice described above will require active engagement by many stakeholders, including an array of non-governmental organizations (NGOs) and individuals. The non-governmental chemical science community has an important role to play in this engagement. In particular, scientific organizations can provide opportunities for scientists to contribute *as scientists* to the effective implementation of the Chemical Weapons Convention (CWC) and other disarmament and nonproliferation efforts.

This chapter focuses on the relationship that the International Union for Pure and Applied Chemistry (IUPAC; see Box 17.1) has built with the OPCW in two major areas:

- Providing independent input to the CWC about the implications of developments in science and technology for potential risks that could undermine the Convention and for potential opportunities to improve its implementation; and

Box 17.1 What is IUPAC?

The non-governmental chemical science community encompasses a sometimes confusing variety of national and international organizations. In addition, researchers have formed a growing global array of formal and informal collaborations to address particular scientific problems or to support the development of a field through such tasks as agreement on common standards, or the creation of necessary infrastructure. IUPAC, formed in 1911, is an example of an international scientific union. Along with its activities to promote strong and productive chemical sciences, IUPAC engages in policy activities such as those it has undertaken with the OPCW over the years. According to its website:[a]

> "IUPAC members form a Global Chemistry Network. Our members, associates, and affiliates – chemical societies, national academies of science, industrial companies, research and development institutions, universities, laboratories, and individuals from around the world represent most of the world's chemistry research and industry. IUPAC also has formal and informal ties to international organizations that touch on every aspect of society. Our Membership Network currently comprises more than 50 national member countries that are actively engaged in the governance of the Union; approximately 31 associated organizations that share common goals and interests; about 58 company associates interested in supporting or participating in IUPAC activities through the Committee on Chemistry and Industry; nearly 1500 individual volunteers engaged in committees and projects; and almost 2000 individuals who are affiliate members, and 2160 fellows."

[a]https://iupac.org/who-we-are/member-organizations/, last accessed 17 September 2016.

- Raising awareness and knowledge *via* education and outreach about the risks connected with the potential misuse of developments in chemical science and the contributions that chemistry practitioners can make in reducing them.

As "the only independent, nongovernmental, international organization devoted to chemistry and the chemical sciences and their application in research, industry, and society", IUPAC is a particularly strong partner for the OPCW.[2] This chapter provides a brief history of the activities that IUPAC has undertaken and also discusses the increasing connections between the chemical and life sciences communities as a reflection of the growing convergence of chemistry and biology.

17.2 Developments in Science and Technology (S&T)

Earlier chapters describe the rapid developments in science and technology that are challenging the future of the CWC as well as providing opportunities to improve its implementation. One message that emerges is the difficulty that governments or international treaties face in their efforts to stay abreast of S&T advances. No government or treaty organization, however well endowed, can track or evaluate all the trends that one would ideally want to understand. The growing overlap between cutting edge developments in chemistry and biology opens whole new arenas that need to be followed.

As illustrated in this section, the non-governmental chemical science community can assist with *monitoring* what is happening. Fostering and sustaining contacts and networks among those in the research community and technical experts in government, who may also be doing research, is one obvious function. The community can also help with *assessment*, since the ultimate goal is to provide decision makers with the information and insights they need to prepare for or take action. This includes finding ways to translate the technical material and insights for diplomats and officials who frequently lack technical training. By enlisting scientists to use their expertise to support the disarmament and nonproliferation enterprise, these activities have the particular virtue of *making scientists part of the solution, not part of the problem*.

IUPAC first engaged with the OPCW through an independent project on assessing options for the destruction of chemical weapons stockpiles, the first major technical and political challenge facing the CWC.[3] Our story begins in 2000, as preparations for the treaty's first review conference got underway. Even though the CWC had a formal Scientific Advisory Board (SAB), its members quickly recognized the benefits of engaging a wider array of expertise to provide input about relevant developments in S&T. This led to a formal request to IUPAC to organize a workshop that could provide information and analysis to the SAB. The SAB would then use the results of the IUPAC meeting to prepare its own report to the OPCW Director General for the States Parties, which also made it easier for IUPAC to provide conclusions and recommendations. The SAB asked for advice about trends that could pose threats to the CWC, as well as new developments that could aid the OPCW in fulfilling its missions. This gave the workshop a mix of both cutting edge science and practical applications.

To the best of our knowledge, the SAB's request to IUPAC was a first; no other nonproliferation or disarmament treaty review conference had ever sought independent input from an international scientific organization. IUPAC's standing as an international NGO also provided advantages in dealing directly with another international body, because relevant national affiliates and adhering bodies were able to engage in the discussions under IUPAC's international umbrella. For example, since IUPAC's small staff did not have the capacity to support the international committee that planned the workshop, the national adhering organization in the country where its headquarters was located, the US National Academy of Sciences (NAS), provided the staff to organize the meeting.

The IUPAC workshop, which was held in Bergen, Norway in 2002, assembled almost 80 participants from over 30 countries. It also became a model for how such events could be effectively organized. The participants were a mix of researchers from academia and industry whose work had potential implications, either positive or negative, for the operation of the treaty, as well as OPCW and government technical experts. Policy experts with scientific backgrounds or affinities were also included. The format of the workshop allowed researchers to present their work, followed by discussions in smaller breakout sessions to bring out the implications for the CWC. The insights from the small group discussions were then fed back into the plenary sessions for general consideration. This approach often made it easier to recruit "pure" researchers since it enabled them to begin by giving a standard technical presentation and then to engage with others in thinking about potentially new implications of their work. The discussions of cutting edge research often brought emerging leaders into the process, reflecting the degree to which younger scientists can make breakthroughs when a field is advancing rapidly. And, by asking researchers to think explicitly about key technical and conceptual hurdles in the path of their work as a way to help the OPCW consider potential risks and benefits, the discussions produced an element of modesty about the immediate prospects for the research that may be missing in the hyper-competitive environment of current science. Finally, the results of the workshop were made available through official channels from the SAB to the OPCW as well as to a broader audience *via* publication in IUPAC's journal.[4]

The success of the first meeting led the OPCW to ask IUPAC to organize another workshop for the 2nd CWC Review Conference.[5] The US NAS again provided staff support to IUPAC and the international organizing committee. This time, the meeting was held a full year before the review conference, in Zagreb, Croatia in 2007, reflecting lessons learned on the most effective timing to provide input into the review conference preparations. The meeting was approximately the same size – just under 70 participants from almost

Figure 17.1 Participants in the Spiez Workshop. Photograph courtesy of Patrick John Y. Lim.

30 countries. It is worth noting that there was a significant contingent of representatives from government departments and laboratories and from National Authorities and 11 members of the SAB. Together, these constituted over 40 percent of the participants. Since a number of the staff involved in organizing the Zagreb meeting had also taken part in planning a meeting on S&T trends for the Biological Weapons Convention in the previous year (see below), a number of themes and issues reflecting the growing convergence between chemistry and biology were addressed in both events.

The IUPAC workshop for the 3rd CWC Review Conference was held at the Swiss national CBRN laboratory in Spiez in February 2012, where the chair of the SAB headed the chemistry division.[6] This workshop also convened 80 participants from approximately 25 countries and international organizations, including a number of OPCW SAB members (see Figure 17.1). It was structured along similar lines to the prior two meetings, mixing plenary and small group sessions and incorporating perspectives from scientific practitioners in academia, industry and government along with chemical security and chemical sciences policy experts.

One advantage provided by the series of IUPAC workshops was continuity, which enabled analysis and reflection about trends over time. As the report of the Spiez workshop noted:[7]

> "Some of the S&T developments discussed in Spiez 2012, Zagreb 2007, and Bergen 2002 include biological or biomediated processes for chemical synthesis; potential applications of nanotechnology to the CWC; microreactors and other novel chemical production methods (*e.g.*, microwave, novel catalysis, photochemistry, *etc.*); development of new toxic chemicals; and advances in methods for CW detection (*e.g.*, "lab on a chip", DNA arrays, microbial sensors, *etc.*)."

More generally, all three IUPAC workshops concluded that "the pace of S&T development is accelerating in the area of chemical and biological convergence."[8]

As mentioned above and discussed in Chapter 18, the success of the IUPAC workshops inspired a similar effort for the Biological and Toxin Weapons Convention (BTWC). Over the last decade, the Biosecurity Working Group of the InterAcademy Partnership (IAP), in cooperation with international scientific partners, has organized assessments of trends in S&T for the 6th, 7th, and 8th BTWC Review Conferences, as well as focused examinations of topics such as research needs for microbial forensics.[†] In contrast to chemistry, where IUPAC is clearly the dominant international union, there are perhaps a dozen international unions in the life sciences, reflecting the diversity and fragmentation of the field.[‡] Like the IUPAC workshops, these projects have informed discussions on whether and how developments in S&T could impact upon the scope and implementation of the BTWC, because the Convention itself does not currently have a systematic infrastructure for examining such scientific questions.[9] They have essentially become an informal "science advisory system" for the BTWC.[§] In addition, the chair of the organizing committee for the 6th and 7th BTWC Review Conference workshops became a member of a temporary working group on the convergence of chemistry and biology organized by the OPCW SAB, which is discussed in Chapter 6.[10]

In the last few years, the SAB has become much more active in examining a broader portfolio of scientific issues. The temporary working group (TWG) on convergence just mentioned is a primary example, as are another TWG on verification whose report was released in June 2015, and a new TWG on investigative science and technology that began its work in early 2018.[11] Rather than a general review of S&T trends, for the 4th CWC

[†]The InterAcademy Partnership (IAP), formally launched in South Africa in March 2016, brings together established networks of academies of science, medicine and engineering, namely the IAP, the global network of science academies, the InterAcademy Medical Panel (IAMP) and the InterAcademy Council (IAC). Under the new InterAcademy Partnership, more than 130 national and regional members/academies work together to support the special role of science and its efforts to seek solutions to address the world's most challenging problems. For further information see: http://www.interacademies.org/ (last accessed 22 September 2016). The Biosecurity Working Group, created in 2003, currently includes the national academies of Australia, China, Cuba, Egypt, India, Nigeria, Pakistan, Poland, Russia, the United Kingdom, and the United States.

[‡]The International Union of Toxicology and the International Union of Biochemistry and Molecular Biology (IUBMB) also represent important elements of chemistry. Given its clear crossover with both chemistry and biology, the IUBMB has been a partner in several of the BTWC-related activities, such as the workshop for the 7th review conference and meetings on broader biosecurity policy and education.

[§]The executive summary of the report prepared to inform the 7th BTWC Review Conference was actually used by the BTWC Implementation Support Unit as part of the official background material on relevant trends in S&T. The document may be found at http://daccess-dds-ny.un.org/doc/UNDOC/GEN/G11/640/10/PDF/G1164010.pdf?OpenElement (last accessed 26 September 2016).

Review Conference IUPAC organized a more focused workshop to permit the exploration of a key topic in depth as part of a series of such events. The "International Workshop on Innovative Technologies for Chemical Security," organized in cooperation with the Brazilian and US academies of sciences, was held in Rio de Janeiro in July 2017. The meeting explored how advances in technology in other sectors, such as industrial and environmental monitoring and precision agriculture, could be applied to support the OPCW's missions. There are some trade-offs with this approach between depth and continuity, but IUPAC continues to be a valuable partner for the OPCW. In addition to its work on S&T, the next section discusses the longstanding IUPAC connections in education and outreach.

17.3 Education and Outreach: Part One[12]

17.3.1 Background: the General Challenge

Increasing awareness of the CWC through education faces many of the same challenges as any effort to introduce new material into a curriculum. The lessons learned from attempts in one field can inform similar ones in others, in particular from the life sciences (see Chapter 18). For example, in 2009, the IAP, in partnership with several life sciences unions, convened a workshop at the request of the US State Department to:[13]

- Survey strategies and resources available internationally for education on dual use issues and identify gaps,
- Consider ideas for filling the gaps, including development of new educational materials and implementation of effective teaching methods, and
- Discuss approaches for including education on dual use issues in the training of life scientists.

The workshop, hosted in Warsaw by the Polish Academy of Sciences, gathered researchers from the life sciences and experts in bioethics and biosecurity. One of the meeting's innovations was to include specialists in the science of how people learn and the most effective methods for teaching. Among the challenges, that are as relevant for the CWC as the BTWC, are intense competition for space in an already crowded curriculum, pressures on students, especially in graduate school, to focus on their research, and the relatively low value placed on teaching relative to research in many academic settings. A key recommendation was that, in most cases, scientists should be introduced to [bio]security issues as one part of their broader training in responsible conduct and the social responsibilities of science. Standalone courses to provide greater depth and expertise could follow for those whose research or work required it. The workshop also highlighted the importance of "champions" in gaining and sustaining attention to an issue. To support the champions, another important set of findings underscored:

- The importance of networks to build knowledge and then support and sustain those endeavoring to introduce new topics and new approaches;
- The need for relevant materials, especially in languages other than English, to reach the widest possible range of students; and
- The power of active learning methods to enhance engagement and encourage genuine learning.

This essential combination of materials, methods, and networks provided the basis for a series of activities relevant to the BTWC by the NAS that are described in Box 17.2 and complement those described in Chapter 18. Three individuals who have been involved in the NAS activities (see below) are inaugural members of the OPCW's Board on Education and Outreach. The various educational activities undertaken in recent years by the OPCW, on

Box 17.2 Educational Institutes on Responsible Science.

Working in partnership with scientists and scientific organizations in the Middle East/North Africa (MENA) and South and Southeast Asia, the US National Academy of Sciences (NAS) has successfully adapted a model originally developed to transform the way biology is taught in US colleges and universities to help scientists and students engage deeply with issues of Responsible Science, including preventing the misuse of research to cause deliberate harm. The Educational Institutes on Responsible Science are a vehicle to create an international network of researchers and educators in the life sciences who can (1) use methods of adult teaching/pedagogy (*e.g.*, the substantial body of research about how people learn) that promote improved [student] learning and (2) teach about responsible science and responsible conduct of research.

The Educational Institutes, overseen by committees of subject matter and pedagogy experts and supported by a team of NAS staff, feature:

- *Week-long residential programs* to provide education about both content and pedagogy;
- Intensive and varied *hands-on learning* experiences, including teamwork and group presentations;
- Emphasis on *active learning*, with many options for use in varied educational settings;
- *Follow-up* to promote sustainability, including small implementation grants and reunion meetings to share challenges, experiences and opportunities.

To date, the NAS has implemented a number of Educational Institutes:

- A pilot Institute for the MENA region was held in Aqaba, Jordan, in September 2012 in partnership with the Bibliotheca Alexandrina and The World Academy of Sciences (TWAS) with faculty from *Algeria, Egypt, Jordan, Libya, and Yemen.* Small, competitively awarded grants enabled faculty to implement what they learned in their own classrooms and report their experiences at a

Box 17.2 (*continued*)

reunion meeting in Spring 2013. A report of the Institute and lessons learned is available as a free PDF, along with a summary in both English and Arabic.[a]

- The first regional Institute in Asia was held in Kuala Lumpur, Malaysia, in August 2013 in partnership with the Academy of Sciences Malaysia for faculty from India, Malaysia, and Pakistan. Competitive grants were awarded and the reunion was held in Spring 2014.
- A second regional Institute for the MENA was held in Trieste, Italy, in May 2014, in collaboration with TWAS with faculty from Egypt, Morocco, Lebanon, and Yemen, as well as observers from Iraq. The reunion meeting for successful grantees was held in Spring 2015.
- A third regional MENA Institute was held in Istanbul, Turkey in May 2015 with faculty from Iraq, Libya, Morocco and Turkey. The reunion meeting was held in Spring 2016.

In addition, the NAS has supported its Egyptian colleagues to develop their own institutes in a partnership with Ain Shams University and Egyptian faculty-alumni of the MENA regional Institutes. Several key resources for teaching about responsible science have been translated into Arabic, and the NAS is working with scientists-alumni of the MENA Institutes in Algeria on the development of a national bioethics curriculum that combines higher education-specific content with essential pedagogical elements for effective education. One four-day long program led by alumni of the Asian Institute was held in Mangalore, India in April 2015 in collaboration with Nitte University. Several one and half day workshops showcasing the teaching of responsible science using active learning were held in India, Indonesia and Malaysia in late 2015. The project maintains a website with information about all its activities that may be found at http://nas-sites.org/responsiblescience/.

[a]National Research Council, *Developing Capacities for Teaching Responsible Science in the MENA Region: Refashioning Scientific Dialogue*, National Academies Press, Washington, DC, 2013. Available at: https://www.nap.edu/catalog/18356/developing-capacities-for-teaching-responsible-science-in-the-mena-region, last accessed 27 September 2016. National Research Council, *Summary: Developing Capacities for Teaching Responsible Science in the MENA Region: Refashioning Scientific Dialogue* (in English and Arabic), National Academies Press, Washington, DC, 2013. Available at: http://dels.nas.edu/Materials/Special-Products/MENA, last accessed 27 September 2016.

its own or in cooperation with IUPAC, reflect similar insights into the most effective approaches to education for the CWC.

17.3.2 The Relationship Between IUPAC and the OPCW in E&O

The IUPAC–OPCW relationship on issues related to education and outreach (E&O) grew out of the IUPAC workshop in Bergen in 2002. OPCW had already undertaken an Ethics Project in 2001, focused on raising awareness among those who would feel a direct impact from the implementation of the CWC. Awareness was considered an essential component of effective implementation, which encompasses more than just legal and administrative measures.

The report of the Bergen workshop included recommendations about E&O and expressed the hope that "In due course, chemical weapon prohibition and non-proliferation considerations may even be incorporated into university and school curricula as part of chemistry education in a similar way to that in which environmental issues, ethics of genetics, and similar issues have been incorporated into chemistry and biology education in the recent past."[14] The SAB took up the recommendations in its report to the first review conference and E&O became part of IUPAC's relationship with the SAB and OPCW from then on.

The first joint OPCW–IUPAC project on E&O was a workshop on "The Chemical Weapons Convention, Chemistry Education and the Professional Conduct of Chemists," which was held at Oxford University in July 2005.[¶] The workshop led to two IUPAC projects, one to produce educational materials under the leadership of Alastair Hay and Peter Mahaffey and a second to pursue an IUPAC code of conduct *via* a working group headed by Graham Pearson.

17.3.2.1 The Multiple Uses Project

A number of those attending the Oxford meeting formed the team to develop educational material, which introduced chemists to the objectives of the Chemical Weapons Convention. The funds provided by IUPAC were to be used to develop a website containing material which introduced students, educators, and policymakers to the topic of the multiple uses of chemicals. Eschewing any label of "dual-use," the project became known as *Multiple Uses of Chemicals* to reflect the fact that chemicals have many uses and that it is the choice of the individual practitioner as to whether the use is for beneficial or harmful purposes. The objective of the education material was to enable discussion about the control of chemicals, their safe use, and how to prevent misuse. Visitors to the site were introduced to various case studies including ones on illegal drugs and chemical weapons. The intention behind this approach was to help initiate a discussion on how to prevent illegal drug manufacture and the controls needed on the chemicals which might be used for this. When the discussion moved on to discuss chemical weapons, participants soon realise that control measures for chemicals to prevent illegal drug production are very similar to those needed to stop chemical weapons being produced. Refinement of the site followed several pilot workshops in Leeds (UK), Moscow (Russia), Seoul (S. Korea) and Bologna (Italy).

Thousands visited the site each year, but the advances in interactive, web-based education required changes to it and the site was updated in 2013/14 with funds from the European Union through the OPCW. Undergraduate students and faculty at the King's Centre for Visualization in Science, directed

---

[¶]It is worth noting that this was also the year that the intersessional topic for the Biological Weapons Convention was "the content, promulgation, and adoption of codes of conduct for scientists."

by Peter Mahaffy and Brian Martin made the changes, in consultation with scientists and educational specialists from both IUPAC and the OPCW.

The expanded *Multiple Uses of Chemicals* website is highly interactive and uses numerous case studies and role-playing scenarios to communicate information to a range of audiences.‖ The site has three separate portals: Brief Overview, Students, and Educators and Policymakers. Major features of the resource, heavily condensed, are contained in the Brief Overview. The Students portal is for secondary and post-secondary chemistry students. Here the aim is to enable discussion of scientific responsibility and integrity around various concepts in general and organic chemistry; the material is also useful for courses on ethics. Ideas on implementing the resource in presentations or classroom discussions, learning outcomes for topics, worksheets for students, and links to other useful websites are contained in the third section for Educators and Policymakers. This covers seven major topics including Multi-Use Chemicals, Responsible Choices in Chemistry, Convergence of Chemistry and Biology, Codes of Conduct, Chemical Weapons Convention, Biological and Toxin Weapons Convention, and the Australia Group. The first two topics review problems associated with the regulation and distribution of multiple use chemicals.

Like its predecessor, the updated material was piloted at several IUPAC meetings in 2013 and 2014 before a final launch. However, the general content of the site and its emphasis on the responsible use of chemicals has proved useful since 2005 in many workshops around the world, including in Central and South America, North and Southern Africa, numerous Mediterranean countries, China, and venues in Europe. By adopting a highly interactive approach, the content can be used to great effect to facilitate intense, but very enjoyable and informative discussions around the control of chemicals and their safe use.

17.3.2.2 Codes of Conduct

IUPAC conducts its work through projects overseen by international committees of volunteers. This was the approach taken for the workshops on S&T and also in considering whether to develop a code of conduct for IUPAC after the Oxford workshop. The group spent several years on the endeavor and concluded that, rather than producing a formal code, it would be more effective to develop principles that should be considered in any code of conduct. This reflects the view that such codes are most likely to be effective if they are developed by those to whom they will apply, fostering a sense of "ownership" among practitioners.** This approach also allows flexibility in tailoring the

‖Available at www.iupac.org/multiple-uses-of-chemicals and www.opcw.org/special-sections/education/multiple-uses-of-chemicals/ (last accessed 17 September 2016).

**This is also the approach ultimately taken by the IAP Biosecurity Working Group in developing the IAP Statement on Biosecurity (see http://www.interacademies.net/Activities/Projects/17806.aspx, last accessed 17 September 2016).

application of core principles to the specific circumstances or group of practitioners for whom the code is intended. The IUPAC working group produced a set of such principles that, although never formally adopted by IUPAC, played a role in the development of The Hague Ethical Guidelines discussed later in the chapter.[15]

17.4 Education and Outreach: Part Two

17.4.1 E&O and the Future of the CWC

The progress made in destroying declared stocks of chemical weapons sparked renewed interest in education and outreach on the importance of preventing the "re-emergence" of chemical weapons and toxic chemicals as instruments of international conflict became evident. The appointment of Ambassador Üzümcü, a strong E&O advocate, as the OPCW Director General in 2009, set the stage for increased activity. The importance of support for E&O from the SAB and the Third Review Conference is described in Chapter 6. The OPCW this time created a temporary working group of the SAB to address E&O. It was established in early 2012 with Djafer Benachour, an SAB member from Algeria, as the chair. Several members of this working group had been involved in the earlier IUPAC efforts on education and codes of conduct.[16]

The working group provided an opportunity for the SAB and OPCW to engage IUPAC in a variety of ways. In addition to preparing recommendations for future action, the group was able to take advantage of existing projects or new ideas to support the development of several E&O activities as a way to showcase what such efforts could yield. A project on education in Argentina (see Box 17.3) illustrates the organizational advantages

Box 17.3 Argentina's Activities on Education.

In 2010, the Argentine National Authority initiated several national activities as part of the next stage in the implementation of the Chemical Weapons Convention in Argentina. These activities included the establishment of a Working Group to study outreach and dissemination of the obligations under the Convention and the national legislation that implements it. Significantly, the Working Group considered that the subject was too important to be circumscribed to only providing a better understanding of the obligations of the Convention to future managers of declarable facilities. As described in an Argentine national report submitted to the Eighteenth Session of the Conference of the States Parties: "[T]he Group therefore proposed a broader scope for the project, in order to promote a culture, among all professionals in the chemical fields, of the responsible use of technical and scientific knowledge, in order to be aware of the potential danger and to prevent all misuse and abuse of chemicals."[a] It was proposed that the project should also include practicing chemists, research scientists and university laboratory professionals and the group also considered that, at a second stage, the project should be targeted at secondary level students.

Given the federal nature of the Argentine university system, the Argentine National Authority, based in the Ministry of Foreign Affairs, sought the support

Box 17.3 (*continued*)

of the Ministry of Education. Both ministries agreed to work together and signed a Memorandum of Strategic Cooperation in August 2013, which established the goals and actions to implement a "[N]ational Project on Education on the Responsible and Secure Use of the Chemical Sciences and Technologies for the Scientific, Economical and Social Development of the Argentine Republic". This partnership between the National Authority and the Ministry of Education has been key to the success of the Argentine national project.

The project consists of four main elements:[b]

1. An overarching "Network of Networks" coordinated by the National Authority and the Ministry of Education which holds annual meetings;
2. A "Train-the-Trainers" Programme Coordinated by the National University of Rosario and the Southern National University (Bahía Blanca). A first workshop was held in Rosario in June 2013 and a second in Bahía Blanca in November 2014;[c]
3. A Virtual Classroom coordinated by the Kennedy University with the National University of Lomas de Zamora. Approved on-line content is available to all participating universities;
4. A Travelling Class coordinated by the National University of Córdoba with the intention of conducting outreach to faculties where there are no educational tools or trained personnel yet.

In December 2013, the paper submitted by the Argentine National Authority reported "great interest" in the project from professors and students. The same paper also said that another report would be submitted to the Nineteenth Session of the Conference of the States Parties in December 2014, measuring the implementation of the project against a number of well-chosen indicators.

[a]C-18/NAT.3, dated 2 December 2013. Available at: https://www.opcw.org/index.php?eID=dam_frontend_push&docID=16910, last accessed 17 September 2016.
[b]For a comprehensive description of the national project, see the presentation by the Executive Chairman of the Argentine National Authority at http://www.opcw.org/index.php?eID=dam_frontend_push&docID=17818, last accessed 17 September 2016.
[c]For more information on the June 2013 workshop in Rosario, see A. Suarez and R. Spanevello, *OPCW Today*, 2013, 2(5), 27.

Source: This is an edited excerpt from *Education and Engagement: Promoting a Culture of Responsible Chemistry,* SAB/REP/2/14, 2014. Available at: https://www.opcw.org/fileadmin/OPCW/SAB/en/Education_and_Engagement-v2.pdf, last accessed 17 September 2016.

enjoyed by the CWC through its structure of National Authorities and provides an example of what a well designed "top down" approach can accomplish.

More specifically, as discussed above, the OPCW was able to provide funding to enable the updating of the Multiple Uses of Chemicals website. Along with the *Fires* project, a series of short documentary films for use with a general audience as well in education programs,[††] these activities in particular supported the Director General's call in his Nobel Peace Prize acceptance speech to think about "how [organizations] can employ new

[††]More information about the Fires project is available at http://www.thefiresproject.com/.

communication tools to raise awareness of the need to practice responsible science, to instill the highest ethical standards in our future scientists and researchers."[17]

In addition to the further development of the *Multiple Uses of Chemicals* website, there have been an increasing number of interactions between the OPCW and IUPAC about ways to collaborate to advance chemical safety and security. Staff members from the OPCW Technical Secretariat have taken part in a number of IUPAC meetings, including conferences dedicated to education and the wider General Assemblies and World Congresses, as well as more specialized Green Chemistry and CHEMRAWN (Chemical Research Applied to World Needs) conferences. The Director General made a video presentation to the 2013 IUPAC World Chemistry Congress. In late 2014, an agreement was reached to have a staff member from the OPCW serve as an advisory member for IUPAC's standing Committee on Chemistry Education and that IUPAC has permanent observer status on the OPCW's Advisory Board on Education and Outreach (see below). The OPCW has relationships with other professional societies in chemistry, such as the American Chemical Society, as well as with industry organizations, but its ties to IUPAC are particularly strong.

A major recommendation of the TWG was the creation of a permanent advisory group for the OPCW to deal specifically with education and outreach. A new advisory board has now been formed and had its inaugural meeting was held in April 2016.[‡‡] The board has been charged with developing both short and long-term strategies and considering how education and outreach can be used to support the CWC. Strategies developed by the board will consider ways of promoting the CWC in the widest possible sense and not just within the chemistry and disarmament communities.

17.4.2 IUPAC and the Hague Ethical Guidelines Project

The renewed interest in E&O has extended to a further exploration of how codes of conduct, which are often seen as fundamental tools for fostering and sustaining a culture of responsibility within professions and industry, could support efforts to prevent re-emergence of chemicals as weapons. Yet, even though chemical weapons have been used in at least two major conflicts in the last four decades, there are relatively few codes within the non-governmental chemistry community and the chemical industry that explicitly include provisions related to the CWC or chemical weapons.[§§] A web search and analysis in 2015 of chemistry-relevant codes in the English language from the chemistry community or industry yielded 7 out of over 140 codes from 46 countries that mention chemical weapons, and 3 that mention the CWC.[18]

[‡‡]More information about the Board may be found at https://www.opcw.org/documents-reports/subsidiary-bodies/advisory-board-on-education-and-outreach/.

[§§]See, for example, the background papers prepared for the 2005 BTWC discussions http://www.unog.ch/__80256ee600585943.nsf/(httpPages)/da292636ae31f1cbc125718600361e55?OpenDocument&ExpandSection=2#_Section2. These papers make a distinction between codes that explicitly refer to the BTWC and those that do not but are nonetheless relevant.

In December 2014, the German representative to the CWC's Conference of States Parties introduced a proposal for a "Hippocratic Oath" for chemists. The Conference ultimately:[19]

> "Welcomed the initiative for a text of ethical guidelines for chemical professionals related to the Convention and invited the Secretariat to inform the Council of its efforts for the advancement of the initiative and its objectives in close collaboration with relevant professional and chemical industry organizations. The Conference encouraged States Parties to discuss the matter further."

The German government provided special funds for two workshops in 2015 to develop the text for ethical guidelines.[20] The project was notable because, although funded and organized *via* the OPCW, the work was done by an international group of 35 scientists from academia and the chemical industry in 24 countries from all regions of the world. Several members of the TWG with close ties to IUPAC, as well as its current and immediate past presidents, and its president elect, took part in the workshops.

The results of the workshops, the *Hague Ethical Guidelines*, are a set of principles that can be used to support the development of new codes, but also as the basis for reviewing existing codes to explore whether they are in line with the goals of the CWC. This reflects an important conclusion in the Guidelines: "A code need not mention chemical weapons or the CWC to support its goals."[¶¶] Workshop participants endorsed the Guidelines, which are available in all six official languages of the OPCW, as well as German.[‖‖] The OPCW welcomes additional endorsements such as the following when, in April 2016, IUPAC sent a letter to the Director General providing its official endorsement of the Guidelines. The letter notes that:[21]

> "A recent IUPAC project on recommendations for codes of conduct for chemists follows a similar model for promulgation of codes, encouraging its members to consider the guidelines in updating and preparing codes of conduct for their organizations. The Core Element of the Hague Ethical Guidelines, which states, "Achievements in the field of chemistry should be used to benefit humankind and protect the environment," is fully consistent with the IUPAC strategic mission and vision."

The Guidelines have been disseminated widely to professional societies and industry organizations. The American Chemical Society, which was represented in the workshops, has since produced what it refers to as a global code of ethics for chemists, drawing on the Guidelines.[22]

[¶¶] See https://www.opcw.org/fileadmin/OPCW/Science_Technology/Hague_Ethical_Guidelines_Brochure.pdf, where the statement is available in all of the official languages.

[‖‖] The German version is available at http://www.den-haag-cw.diplo.de/Vertretung/den-haag-cw/de/04_20Aktuelles/German_20Initiative_20Ethical_20Guidelines.html (last accessed 17 September 2016).

17.4.3 Condemning the Use of Chlorine as a Weapon

On occasion, IUPAC undertakes outreach in the realm of policy. Beginning in the fall of 2016, chemical societies and chemical industry associations began voicing their concern about the use of chlorine as a chemical weapon in the Syrian conflict. As of late spring 2017, statements condemning the use of chlorine as a weapon have so far been made by all 32 societies (including Turkey and Israel) represented by the European Chemical Societies' organisation (EuCheMS), as well as the European Chemical Industry Council, the American Chemical Society and American Chemistry Council issued a joint statement, and they were joined by the Federation of African Chemical Societies, as well as chemical societies from individual countries in Africa, as well as societies of chemists in Mexico and Japan. More recently, EuCheMS has condemned the 4 April (2017) use of sarin in Syria as has the International Council of Chemical Associations (ICCA), a body representing members who account for over 90% of the global sales of chemicals.*** IUPAC has also been active and in a recent letter underscored the "concerns of IUPAC regarding recent reports of the use of chemicals, particularly chlorine, as weapons in Syria and other locations."††† These actions are in keeping with the determination, in the words of The Hague Ethical Guidelines that "Achievements in the field of chemistry should be used to benefit humankind and protect the environment."

17.5 Conclusions

This chapter, through the story of the relationship between the OPCW and IUPAC, has presented a picture of how one organization from the nongovernmental chemistry community has contributed to efforts to promote chemical nonproliferation and disarmament and more recently to prevent the re-emergence of chemical weapons in international conflict. As Chapter 18 on the contributions of the life sciences community underscores, the existence of a formal treaty organization with institutional and financial resources to support E&O is a major advantage in the pursuit of chemical safety and security. In addition, this chapter has sought to underscore two key messages:

- As recognized by the OPCW and the CWC States Parties, effective chemical security policy and practice will require active commitment by scientists, who are part of many relevant sectors (academia, industry, and government) and the scientific community has an important role to

***The EuCheMS statement may be found at http://www.euchems.eu/syrian-conflict-euchems-condemns-use-chemical-weapons/ and the ICCA statement at https://www.icca-chem.org/blog/ (both last accessed 8 April 2017).

†††The IUPAC letter and other statements may be found on the OPCW website at https://www.opcw.org/news/article/chemical-industry-councils-and-scientific-societies-condemn-use-of-chlorine-as-weapon/ (last accessed 15 May 2017).

play. The experience of IUPAC illustrates just how scientific organizations can provide opportunities for scientists to contribute to effective CWC implementation and other nonproliferation and disarmament initiatives.

- Putting the emphasis on Responsible Science as the introduction to chemical security and dual use issues *makes scientists part of the solution, not part of the problem.* This framing provides important opportunities to build upon the existing culture of responsibility in chemical science (and science more generally). The *Multiple Use of Chemical Project* and The Hague Ethical Guidelines are two examples of how this is being put into practice by the OPCW.

Looking ahead, two of the greatest challenges facing chemical safety and security, especially in education, are *scaling up* from the many successful smaller projects to comprehensive programs that reach a broad swath of the community and *sustain* the efforts over time. E&O is a never-ending task that must be renewed and revived for every generation of students and practitioners. The lessons from the IUPAC experience can offer ideas and approaches for the CWC and beyond.

References

1. OPCW, *The OPCW in 2025: Ensuring a World Free of Chemical Weapons, S/1252/2015*, 2015, available at: https://www.opcw.org/fileadmin/OPCW/S_series/2015/en/s-1252-2015_e_.pdf, last accessed 22 September 2016.
2. L. Sydnes, *Chem. Int.*, 2013, **35**(4), 4.
3. (a) IUPAC created its Task Force on Scientific Aspects of the Destruction of Chemical Warfare Agents while the CWC was being negotiated. See *Chem. Int.*, 1993, **15**(4), 121; (b) G. S. Pearson and R. S. Magee, *Pure Appl. Chem.*, 2002, **74**(2), 187.
4. The results of the first workshop were published as a special issue of the IUPAC journal. See *Pure Appl. Chem.*, 2002, **74**(12), 2229–2352.
5. M. Balali-Mood, P. S. Steyn, L. K. Sydnes and R. Trapp, *Pure Appl. Chem.*, 2008, **80**(1), 175.
6. K. Smallwood, R. Trapp, R. Mathews, B. Schmidt and L. K. Sydnes, *Pure Appl. Chem.*, 2013, **85**(4), 851.
7. K. Smallwood, R. Trapp, R. Mathews, B. Schmidt and L. K. Sydnes, *Pure Appl. Chem.*, 2013, **85**(4), 853.
8. K. Smallwood, R. Trapp, R. Mathews, B. Schmidt and L. K. Sydnes, *Pure Appl. Chem.*, 2013, **85**(4), 858.
9. (a) Royal Society, *Report of the RS-IAP-ICSU International Workshop on Science and Technology Developments Relevant to the Biological and Toxin Weapons Convention*, RS Policy Document no. 38(06), 2006; (b) National Research Council, *Life Sciences and Related Fields: Trends Relevant to the Biological Weapons Convention*, National Academies Press, Washington,

DC, 2011, The reports of the workshop for the 8th review conference and a follow-up meeting on potential S&T review processes for the Convention may be found at https://royalsociety.org/topics-policy/projects/biological-toxin-weapons-convention/, last accessed 22 September 2016.
10. OPCW Scientific Advisory Board, *Convergence of Chemistry and Biology: Report of the Scientific Advisory Board's Temporary Working Group, SAB/REP/1/14*, 2014, available at: http://www.opcw.org/index.php?eID=dam_frontend_push&docID=17438, last accessed 27 September 2016.
11. (a) OPCW Scientific Advisory Board, *Verification Report of the Scientific Advisory Board's Temporary Working Group, SAB/REP/1/15*, 2015, available at: https://www.opcw.org/fileadmin/OPCW/SAB/en/Final_Report_of_SAB_TWG_on_Verification_-_as_presented_to_SAB.pdf, last accessed 27 September 2016; (b) OPCW Scientific Advisory Board, *Report of the Scientific Advisory Board at Its Twenty-Fifth Session, SAB-25/1**, 2017, available at: https://www.opcw.org/fileadmin/OPCW/SAB/en/sab2501_e_.pdf, last accessed 15 May 2017.
12. SAB's Temporary Working Group on Education and Outreach, *Education and Engagement: Promoting a Culture of Responsible Chemistry, SAB/REP/2/14,* Released in Late 2014, available at: https://www.opcw.org/fileadmin/OPCW/SAB/en/Education_and_Engagement-v2.pdf, last accessed 17 September 2016.
13. National Research Council, *Challenges and Opportunities for Education about Dual Use Issues in the Life Sciences*, National Academies Press, Washington, DC, 2011.
14. G. W. Parshall, G. S. Pearson, T. D. Inch and E. D. Becker, *Pure Appl. Chem.*, 2002, **74**(12), 2323.
15. G. S. Pearson, E. D. Becker and L. K. Sydnes, *Chem. Int.*, 2011, **33**(6), online edition, available at: http://www.iupac.org/publications/ci/2011/3306/2_pearson.html, last accessed 27 September 2016.
16. OPCW, *Education and Engagement: Promoting a Culture of Responsible Chemistry,* SAB/REP/2/14, 2014, pp. 14–15, available at https://www.opcw.org/fileadmin/OPCW/SAB/en/Education_and_Engagement-v2.pdf, last accessed 17 September 2016.
17. OPCW, *OPCW Today*, 2013, vol. 2, (5), p. 6.
18. OPCW, *Report of the Second Workshop on Ethical Guidelines for the Practice of Chemistry under the Norms of the Chemical Weapons Convention*, September 17–18, 2015, pp. 40–44, available at: https://www.opcw.org/fileadmin/OPCW/SAB/en/Hague_Ethical_Guidelines_2nd_Workshop_Report.pdf, last accessed 17 September 2016.
19. OPCW, *Report of the Nineteenth Session of the Conference of the States Parties 1-5 December 2014*, C-19/5, paragraph 23.3, available at: https://www.opcw.org/fileadmin/OPCW/CSP/C-19/en/c1905_e_.pdf, last accessed 26 March 2018.
20. (a) OPCW, *Report of the Workshop on Guidelines for the Practice of Chemistry under the Norms of the Chemical Weapons Convention*, March 2015, available at https://www.opcw.org/fileadmin/OPCW/SAB/en/March_2015_Ethical_Codes_Workshop-Report.pdf, last accessed 17 September 2016;

(b) OPCW, *Report of the Second Workshop on Ethical Guidelines for the Practice of Chemistry under the Norms of the Chemical Weapons Convention*, September 2015, available at: https://www.opcw.org/fileadmin/OPCW/SAB/en/Hague_Ethical_Guidelines_2nd_Workshop_Report.pdf, last accessed 17 September 2016.
21. *IUPAC letter to OPCW*, 22 April 2016, available at: https://www.opcw.org/fileadmin/OPCW/Science_Technology/IUPAC_Letter_Hague_Ethical_Guidelines.pdf, last accessed 17 September 2016.
22. E. Stoye, *Chem. World*, 23 May 2016, available at https://www.chemistryworld.com/news/global-code-of-ethics-planned-for-chemists/1010209.article, last accessed 17 September 2016.

CHAPTER 18

Role of the Non-governmental Life Science Community in Combatting the Development, Proliferation and Use of Chemical Weapons

T. A. NOVOSSIOLOVA

Center for the Study of Democracy, Sofia, Bulgaria. Formerly Division of Peace Studies, University of Bradford, Bradford UK.
*E-mail: tatyana.novossiolova@csd.bg

18.1 Introduction

The comprehensive and effective implementation of the Biological and Toxin Weapons Convention (BTWC) requires that a robust norm of biosecurity is embedded and sustained in the life sciences, whereby appropriate regulations, policies, codified and voluntary controls, and basic professional practices are integrated in a unified and coordinated approach, that is underpinned by a strong commitment to the goal of minimising the risks that the life sciences may be misused accidentally to cause harm or deliberately for hostile purposes. The essential role of sustained and wide-ranging engagement with the life science community has been widely

Preventing Chemical Weapons: Arms Control and Disarmament as the Sciences Converge
Edited by Michael Crowley, Malcolm Dando and Lijun Shang

Published by the Royal Society of Chemistry, www.rsc.org

recognised and extensively acknowledged within the framework of the Convention. Indeed, when considering Article IV of the Convention at the Seventh Review Conference in 2011,[1] States Parties noted the value of measures to:

(a) "implement voluntary management standards on biosafety and biosecurity;

(b) encourage the consideration of development of appropriate arrangements to promote awareness among relevant professionals in the private and public sectors and throughout relevant scientific and administrative activities;

(c) promote amongst those working in the biological sciences awareness of the obligations of States Parties under the Convention, as well as relevant national legislation and guidelines;

(d) promote the development of training and education programmes for those granted access to biological agents and toxins relevant to the Convention and for those with the knowledge or capacity to modify such agents and toxins;

(e) encourage the promotion of a culture of responsibility amongst relevant national professionals and the voluntary development, adoption and promulgation of codes of conduct;"

In addition, the States Parties further agreed that the Intersessional Process in 2012–2015 should address three Standing Agenda Items each of which included consideration of biosecurity.[1] The Standing Agenda Item on cooperation and assistance includes the topics:

(e) "education, training, exchange and twinning programmes and other means of developing human resources in the biological sciences and technology relevant to the implementation of the Convention, particularly in developing countries;

(f) capacity-building, through international cooperation, in biosafety and biosecurity, and for detecting, reporting, and responding to outbreaks of infectious disease or biological weapons attacks, including in the areas of preparedness, response, and crisis management and mitigation;"

The Standing Agenda Item on scientific and technological developments includes the topic:

(e) "education and awareness-raising about risks and benefits of life sciences and biotechnology."

And, the Standing Agenda Item on strengthening national implementation includes the topic:

> "(d) national, regional and international measures to improve laboratory biosafety and security of pathogens and toxins;"

It is therefore evident that education and measures to strengthen biosafety and biosecurity and promote engagement with the life science community have a key factor in the consideration of the three Standing Agenda Items in the Intersessional Process.

Likewise, the States Parties to the Chemical Weapons Convention have recognised the importance of education and outreach. At the 17th meeting of the Scientific Advisory Board (SAB) of the CWC in November 2011, the Director-General[2] in a session considering the future priorities of the OPCW, underlined the need to refocus the OPCW's activities through the deepening of some of its programmes and through the establishment of a new balance between activities to prevent the re-emergence and misuse of toxic chemicals on the one hand, and international cooperation and assistance on the other. The Director-General went on to say:

> "14.2 In the context of prevention, the Director-General referred, *inter alia* to the need, through educational outreach, to raise awareness among the academic community and industry of the dual-use risks associated with toxic chemicals. To this end, he expected that the SAB would provide important inputs. With regard to chemical safety and security, both training and using the Organisation as a platform for identifying and disseminating best practices would be important elements in this process."

At that same SAB meeting:

> 16.5 "The SAB noted the role that education and outreach could play in chemical safety and security.
>
> 16.6 The SAB recommended the establishment of a temporary working group on education and outreach, which would build upon earlier work done by the SAB."

The Final Document of the Third Review Conference of the CWC[3] included the following language on education and e-learning:

> "National implementation measures
>
> 9.96 The Third Review Conference reaffirmed that full, effective, and nondiscriminatory implementation of Article VII is essential for the realisation of the object and purpose of the Convention.

9.101 The Third Review Conference acknowledged the role of education, outreach and awareness-raising as a relevant activity for the national implementation of the Convention, including awareness among academia and relevant scientific communities of the provisions of the Convention, the domestic laws and regulations relevant to the Convention. Accordingly, the Third Review Conference welcomed the establishment of the SAB temporary working group on education and outreach.

9.103 The Third Review Conference, having reviewed national implementation measures:

(a) Called upon all States Parties to adopt, in accordance with constitutional processes, the necessary measures to fully implement their obligations under the Convention ...

(d) Called upon all States Parties to keep the effectiveness of national implementation measures under review so as to ensure at all times that the provisions of the Convention are implemented within their territory or in any other place under their respective jurisdiction;

(e) Encouraged the Secretariat, in concert with the SAB temporary working group on education and outreach, to assist States Parties, upon request, in implementing education and outreach activities, including by disseminating materials, conducting workshops and regional meetings;

(i) Encouraged the Secretariat to expand the use of contemporary technological developments to assist and promote its training methods, including further development of e-learning modules;"

A subsequent section on assistance and protection against chemical weapons includes the language:

"9.118 The Third Review Conference, having reviewed the assistance and protection against chemical weapons:

(b) Encouraged the Secretariat to engage in more active cooperation with relevant regional and subregional organisations as well as international organizations that have mandates relevant to assistance and protection against chemical weapons. Such cooperation could include joint exercises and training including by the use of e-learning modules;"

And the section on economic and technological development includes the language:

"9.131 The Third Review Conference, having reviewed economic and technological development:

(p) Encouraged the Secretariat to expand the use of e-learning as one of the means for capacity-building and outreach activities;"

In addition, a subsequent section on the general functioning of the OPCW states:

"The general functioning of the Organisation for the Prohibition of Chemical Weapons

9.155 The Third Review Conference, having reviewed the general functioning of the OPCW:

(d) Called upon States Parties and the Secretariat, as part of efforts to promote the ethical norms of the Convention, to encourage and promote efforts by the appropriate national and international professional bodies to inculcate awareness amongst scientists and engineers at an early stage in their training that the knowledge and technologies used for beneficial purposes should only be used for purposes not prohibited under this Convention;"

In November 2014, the SAB Temporary Working Group on Education and Outreach issued its Final Report where the Group recommended that:

"3. An ongoing expert advisory group on education and outreach with respect to the responsible use of science, particularly as it is relevant to the CWC, should be established to help OPCW fulfil its mandate for education and outreach;"

A year later, in 2015, the Conference of the States Parties to the CWC decided at its twentieth session on the establishment of an Advisory Board on Education and Outreach that the primary goal of the Board is "to ensure that the Organisation's education and outreach activities, and those of States Parties, are effective, sustainable, cost-effective, and benefit from the latest advances in education and outreach theory or practice."[4]

When comparing the experience with scientist engagement of the BTWC and CWC, one significant difference needs to be noted. Whereas in the context of the CWC education and outreach have been internationally institutionalised under the auspices of the OPCW, efforts to promote awareness of biosecurity under the BTWC remain largely limited to non-governmental organisations, such as think tanks and academia. That said, the importance of the work that has been carried out within the life science community over the past decade can hardly be overstated, not least because many of those activities have some relevance to the efforts of promoting awareness of the CWC among those in the chemical sciences. The purpose of this chapter therefore is to look into the initiatives which have been undertaken in the life sciences and their implications for strengthening the prohibition of chemical weapons.

18.2 Fostering a Norm of Biosecurity

The present section looks into the various projects and initiatives carried out in support of the BTWC with the goal of promoting a culture of biosecurity in the life sciences.

18.2.1 Biosecurity Education: Development of Content and Building Capacity

During the past decade, significant efforts have been dedicated to the development of content for biosecurity education. Work that has been carried out by the University of Bradford (UK), the US National Academy of Sciences, and the Landau Network-Fondazione Volta (Italy), among others, has made a substantial contribution to the production of training material addressing issues of relevance to the Convention, biosecurity, and dual use for those in the life sciences. Table 18.1 provides an indicative summary of relevant projects and initiatives designed to develop training content for biosecurity education, as well as promote the fostering of networks among universities and institutes at national, regional, and international level.

Another significant event in regard to biosecurity education was the launch of two new books at a side event at the Meeting of States Parties (MSP) in December 2015. The books were jointly funded by the Governments of Canada and the UK under the Global Partnership and are freely available online. The books are designed to help life scientists to engage with the issues of biosecurity and dual use.[5] The first book *Guide to Biological Security Issues, Preventing Biological Threats: What You Can Do* is an edited volume featuring contributing chapters from a wide range of prominent disarmament experts, practitioners, and academics. Key contributors to the book include the University of Bradford, the US National Academy of Sciences (NAS), the Royal Scientific Society – Jordan, the International Federation of Biosafety Associations (IFBA), the Centre for Biosecurity and Preparedness – Denmark (CBB), the Public Health Agency of Canada (PHAC), the US Federal Bureau of Investigation (FBI), the Jordanian Armed Forces, the Royal Medical Society – Jordan, Ministry of Health – Jordan, and INTERPOL. The Guide covers a broad spectrum of issues related to biological security, looking into the role of different stakeholders, including governments, law enforcement agencies, academia, industry, and professional associations in promoting new and strengthening already existing measures, policies, and practices designed to ensure that the life sciences do not cause harm. The Guide is available in different languages, including French, Russian, and Ukrainian.[†]

The accompanying book, *Biological Security Education Handbook: The Power of Team-based Learning* shows the value of, and provides guidance on,

[†]For further details see the Resource Repository pages of the National Implementation section of the BTWC website (ref. 7).

Table 18.1 Projects on biosecurity education content development, outreach and network-building.

Organisation	Project description	Date	Details
University of Bradford, UK, National Defense Medical College, Japan, and the Landau Network-Centro (Fondazione) Volta, Italy	Development of an Educational Module Resource in Biosecurity Funded by the British Council	2008	As part of this project, online teaching material on biosecurity and dual-use bioethics in English, Japanese, Russian, French, Spanish, Urdu, Romanian/Moldovan, Georgian, and Polish was produced.
National Academy of Sciences, USA	*Challenges and Opportunities for Education About Dual Use Issues in the Life Sciences*	2010	The report presents a workshop that was held to engage the life sciences community on the particular security issues related to research with dual use potential. The workshop sought to identify a baseline about (1) the extent to which dual use issues are currently being included in postsecondary education in the life sciences; (2) in what contexts that education is occurring; and (3) what online educational materials addressing research in the life sciences with dual use potential already exist. The report is available at https://www.nap.edu/catalog/12958/challenges-and-opportunities-for-education-about-dual-use-issues-in-the-life-sciences.
University of Bradford, UK	30-Credit Masters-level Online Distance-learning Train-the-trainer Programme in Applied Dual-use Biosecurity Funded by the US Biological Engagement Programme	2010–2012	An accredited Masters-level module featuring lectures and interactive group seminars delivered in a fully-equipped virtual classroom. Participants from the following countries were trained: Afghanistan, Argentina, Brazil, Cameroon, Egypt, Ghana, Georgia, Indonesia, Iraq, Jordan, Kenya, Libya, Malaysia, Mali, Morocco, Nigeria, Pakistan, The Philippines, Qatar, Russian Federation, South Africa, Uganda, United Arab Emirates, and Yemen
Landau Network-Centro (Fondazione) Volta, Italy	European Biosecurity Awareness Raising Network (EUBARnet) Funded by the Prevention of and Fight against Crime Programme (ISEC) of the European Commission–Directorate–General Home Affairs	2011–2012	The project aimed to connect universities and organizations in the EU that are working on raising the awareness of life scientists on biosecurity through the preparation of interdisciplinary informational materials and organization of events and seminars on biosecurity. The project website is available at http://landaunetwork.org/index.php/2012/12/the-european-biosecurity-awareness-raising-network/.

University of Bradford, UK	Development of National Series in Biosecurity Funded by the UK Biological Engagement Programme	2011–2012	As part of this project, online lectures and facilitation notes on biosecurity for the following countries were developed: Algeria, Armenia, Azerbaijan, Egypt, Georgia, Iraq, Jordan, Kazakhstan, Kyrgyzstan, Morocco, Saudi Arabia, Tajikistan, Tunisia, Ukraine, and Uzbekistan. Implementation tests were carried out in Georgia, Tajikistan, and Ukraine
National Academy of Sciences, USA	*Research in the Life Sciences with Dual Use Potential: An International Faculty Development Project on Education about the Responsible Conduct of Science*	2012	The report describes the outcomes of the planning meeting for a two-year project to develop a network of faculty who will be able to teach the challenges of research in the life sciences with dual use potential. This report is intended to provide guidelines for that effort and to be applicable to any country wishing to adopt this educational model that combines the principles of active learning and training with attention to norms of responsible science. The report is available at https://www.nap.edu/catalog/13270/research-in-the-life-sciences-with-dual-use-potential-an.
National Academy of Sciences, USA	*Developing Capacities for Teaching Responsible Science in the MENA Region: Refashioning Scientific Dialogue*	2013	This report describes efforts to address dual use issues by developing institutes around the world that will help life sciences faculty learn to teach about the responsible conduct of science. Based on the successful National Academies Summer Institute for Undergraduate Biology Education, the report's authoring committee designed a general framework for the faculty institutes and chose the Middle East-North Africa (MENA) region to test a prototype faculty institute. The report is available at http://dels.nas.edu/Report/Developing-Capacities-Teaching-Responsible-Science/18356.

(*continued*)

Table 18.1 (*continued*)

Organisation	Project description	Date	Details
University of Bradford, UK and University of Manchester, UK	Development of a Neuroethics Educational Module with a biosecurity component Funded by the Art and Humanities Research Council, UK	2012–2014	A 12-unit open-access online educational resource designed to assist educators and facilitators in the development of modules and courses on neuroethics and responsible conduct in science. Units 8–12 cover issues related to biosecurity, including the BTWC, the Chemical weapons Convention (CWC) dual use, and the history of biological and chemical weapons. The resource was used to develop lectures as part of the ISIS–Tempus Euro–Mediterranean Masters in Neuroscience and Biotechnology, 2013–2014 . Information about the project on the development of an Interdisciplinary Network on Teaching Ethics to Neuroscientists is available at http://gtr.rcuk.ac.uk/projects?ref=AH/J005533/1.
Insubria Centre for International Security (ICIS), University of Insubria, Italy	EU CBRN centres of excellence project 3, knowledge development and transfer of best practice on Biosafety/Biosecurity/Biorisk management Funded by the EU Commission in cooperation with the United Nations Interregional Crime and Justice Research Institute (UNICRI)	2013–2014	The project aimed to promote sustainable knowledge development on biosafety, biosecurity and biorisk management and transfer of best practice through a "train the trainers" model and to harmonise international biosafety and biosecurity standards among the participating countries and strengthen regional and international collaboration and cooperation through knowledge sharing, networking and awareness raising on bio-related issues
Landau Network-Centro (Fondazione) Volta, Italy	EU CBRN centres of excellence project 18, *International Network of Universities and Institutes for Raising Awareness on Dual-use Concerns in Bio-technology* Funded by the EU Commission in cooperation with the United Nations Interregional Crime and Justice Research Institute (UNICRI)	2013–2014	The mission of the project was that of modernizing and internationalizing biotechnology education; improving cooperation through the sharing of international standards and good laboratory practices; and improving the bio-safety of society and environment, with special attention on the "next generation of scientists"

the use of active learning methods for teaching biosecurity. In particular, the Handbook provides a set of active learning exercises for each substantive chapter of the Guide based on a Team-based Learning format.[6] It also provides detailed information on how to prepare, organise, and hold interactive seminars, lectures, and courses in biosecurity for different audiences. It has been demonstrated that the application of active learning methods significantly enhances the quality, effectiveness, and efficiency of biosecurity education, not least because such methods encourage reflection, critical thinking, and self-assessment. The Team-based Learning format has been specifically selected, as it is a special form of collaborative learning which shifts the focus from the teacher to the learners, allowing the latter to move from simply memorising concepts to the application of concepts by means of team-work. This learner-centred approach has the in-built advantage of being structured on a set sequence of activities, which in turn enables easy replication, thus maximising its effectiveness and efficiency. As a result, the Team-based Learning format is suitable for the development and implementation of train-the-trainer courses at various levels of instruction and in different educational and training contexts. The Handbook has been translated into Arabic, French, Russian and Ukrainian.[7]

Taken together, the two books constitute an important training resource in the area of biosecurity, as they combine strategy and content, offering expert-level open-source material in a user-friendly and easily accessible form and in different languages. The books are also an important reference material for the International Certification Programme in Biosecurity, that was launched by the International Federation of Biosafety Associations (IFBA) in October 2016.[8]

18.2.2 Implementation of Biosecurity Education and Outreach

Over the past several years, significant efforts have been made to reach out to life scientists and raise awareness about the potential risks that could arise from their work. Besides the examples outlined in the preceding section, it is noted that several states have taken steps to promote the responsible conduct of science and biosecurity. For instance, in Canada, Denmark, and the Netherlands, specific government agencies—the Public Health Agency of Canada,[9] the Centre for Biosecurity and Biopreparedness[10] in Denmark and the Biosecurity Office[11] in the Netherlands—have been tasked with the development and promulgation of biosecurity-related policies, including measures for fostering capacity-building and training in biosecurity and addressing dual use. Similarly, Tunisia is currently working towards the establishment of a National Authority on Biosafety.[12] According to the draft decree, the Tunisian National Authority is to fulfil a range of functions, including proposing national biosafety and biosecurity strategies; preparing the final draft of the national legal framework on biosafety; maintaining a list with relevant biosafety and biosecurity experts; and facilitating the exchange of scientific, technical, and legal expertise in the field of biosafety and biosecurity. In the

USA, the Federal Bureau of Investigation (FBI) has been running an outreach programme for several years, that is designed to inform life scientists and 'amateur' biologists—such as the 'Do-It-Yourself' biology movement—about the possible risks posed by novel biotechnology advances.[13] The FBI also participates on a regular basis in the International Genetically Engineered Machine (iGEM) competition—an annual event that brings together students from various disciplines to develop projects using synthetic biology—by giving talks and elaborating on the biosecurity implications of modern life sciences. More recently, the US Federal Experts Security Advisor Panel has produced a series of policy documents on the importance of biosafety and biosecurity culture, as well as tools for strengthening relevant national and institutional policies.[14]

There is some evidence that biosecurity is gradually being embedded into courses at university level. Specialised courses are already being taught at the National Agrarian University, Georgia, the University of Mohammed V-Agdal, Morocco, and Quaid-i-Azam University, Pakistan.

In 2014, the Ukrainian Biosafety Association and the Palladin Institute of Biochemistry in Kiev, Ukraine have launched project P633, *Biosafety and Biosecurity Education in Ukraine*.[15] The project was funded by the UK Ministry of Defence and aimed to develop a nationwide network for biosafety, biosecurity, and bioethics education in Ukraine. To this end, a series of international conferences and regional seminars were held within the framework of the project. The conferences sought to bring together government officials, international experts, lecturers, and practitioners both from inside and outside Ukraine to facilitate the exchange of ideas, best practices and lessons learned. The seminars were specifically designed as interactive events, aimed at fostering expertise in biosafety and biosecurity teaching, as well as promoting the development of recommendations and action plans on the implementation of biosecurity and biosafety education.

As part of Project P633, a train-the-trainer course with lecturers and educators from different Ukrainian universities and institutes was held at the Palladin Institute of Biochemistry in Kiev on 17–20 May 2016. The primary objectives of the course were to: (1) launch the Ukrainian versions of the *Guide to Biological Security Issues* and the *Biological Security Education Handbook* described above; (2) demonstrate the use of active learning and specifically, the Team-based Learning format; and (3) encourage broader engagement with and reflection upon how the presented materials can be used for the development of biosecurity training courses and awareness-raising programmes. The course was based upon the Team-based Learning format, whereby participants are required to work in teams for the entire duration of the course. As a next step, each participant was required to plan, prepare, and hold at least one lecture or seminar using the Guide and the Handbook in their teaching and share their experience with fellow trainees *via* the project forum.

Biosecurity has also begun to feature in the continued professional education of life scientists. For instance, the EU CBRN Centres of Excellence Project 3, *Knowledge Development and Transfer of Best Practice on Biosafety/Biosecurity/Biorisk Management*, included in Table 18.1, sought to foster biosafety and biosecurity competence among practising life scientists. This project combined online teaching *via* an interactive web-based platform and in-presence (face-to-face) training designed to facilitate in-depth understanding of substantive matter and concepts, as well as promoting regional collaboration among participating countries. In addition, a number of high-profile international conferences have included specialised training courses on biosecurity in their pre-conference programmes. Examples include:

- *Bioethics and Responsible Conduct of Life Science*, pre-conference course at the 5th Annual Conference of the Biosafety Association of Central Asia and the Caucasus (BACAC), March 2014, Tbilisi, Georgia.
- *Dual-use Bioethics and Biosecurity Training Course*, pre-conference course at the International Workshop, 'Education and Awareness-raising in Dual-use Sciences', June 2015, Sarob, Tajikistan.
- *Dual-use Research and Responsible Conduct of Science*, a pre-conference course at the 58th Annual Conference of the American Biological Safety Association (ABSA), October 2015, Providence, RI, USA.
- *Bioethics and Biosecurity Education Tools*, a pre-conference course at the 11th Annual Conference of the Asia-Pacific Biosafety Association (APBA), May/June 2016, Siem Reap, Cambodia.

Another notable development is that some specialised training courses in laboratory biosafety have been expanded to include elements and components of biosecurity as related to dual use and the BTWC. Particular examples in this regard are:

- *Laboratory Biosafety and Biorisk Management Training Course*, EU/UNICRI Project B1, 'Establishment of Regional Training and Resource Centre in Biosafety, Biosecurity and Laboratory Management in the South Caucasus', 14–25 August 2015, NCDC/Lugar Centre, Tbilisi, Georgia.
- *Laboratory Biosafety Training Course*, 23–27 May 2016, Luigi Sacco University Hospital, Milan, Italy.

A further option for promoting biosecurity awareness among those in the life sciences, that has been largely overlooked over the years, is the incorporation of relevant seminars and courses at high school level. Reaching out to life scientists with regard to biosecurity at an early stage of their careers is important, in order to enable them to develop a basic understanding of the potential risks arising from biotechnology. In June 2016, an interactive seminar, *Dual-use Bioethics: Responsible Science in the 21st Century*, was held with high school students at the First English Language School, Sofia, Bulgaria. The primary objective of the seminar was to pilot the *Guide to Biological Security Issues* and the *Biological Security Education Handbook* with prospective life scientists, in order to determine whether (1) the content can be adapted for courses and seminars with high school students; and that (2) the format of Team-based Learning can be readily adopted for teaching at high school level. The total number of participants in the seminar was 32. Based on the outcomes of the seminar and the highly positive feedback that participants provided, it was evident that the content of the Guide, *Preventing Biological Threats: What You Can Do* can readily be adapted for holding seminars with high school students and that the format of Team-based Learning as described in the *Biological Security Education Handbook: The Power of Team-based Learning* is suitable for use at high school level (Figures 18.1 and 18.2).

18.2.3 Promoting Competence: International Professional Certification in Biosecurity

The International Federation of Biosafety Associations (IFBA) has launched a new certification programme for biorisk management, biosafety, and biosecurity professionals worldwide.[16] This programme is in accordance with the IFBA's chief mission of achieving "*safe, secure, and responsible work with biological materials*" by advancing the body of knowledge and increasing the competency in the field of biosafety, biosecurity, and biorisk management. Certified individuals bring increased value to their employers by demonstrating competence to carry out their responsibilities and by achieving high standards of excellence, professionalism, and continuous learning. By earning certifications from the IFBA, individuals reap the benefits of safer workplaces, career advancement, and international recognition among colleagues.

The Professional Certification on Biosecurity (PC) was launched in October 2016. It is important to note that the exam which practitioners will be

Figure 18.1 High school students' poster on biosecurity and responsible science (1).

Figure 18.2 High school students' poster on biosecurity and responsible science (2).

required to take in order to become certified biosecurity professionals covers a broad scope of issues which extends beyond laboratory biosecurity. The Certification identifies individuals with demonstrated competencies in the fundamental principles and practices of working securely with biological materials and sensitive scientific information in a laboratory, healthcare facility, field response to a disease outbreak and other environments where biological materials are handled.[17] Candidates applying for this certification must first successfully complete the PC in Biorisk Management before they are eligible for examination. The PC in Biosecurity is suited to a wide range of professionals working with and around biological materials such as biorisk management and biosafety officers, scientists, laboratory technicians, researchers, disease outbreak response personnel, facility operations and maintenance personnel, security personnel, educators, consultants and policy makers. Individuals holding this certification possess the knowledge and skills in sufficient degree to securely manage biological risks in the laboratory, healthcare facilities, field response to disease outbreaks and other settings where biological materials are handled. The exam in Biosecurity covers the following domains:

- Domain A: Biosecurity Conventions, Guidelines, and Standards
- Domain B: Biosecurity Risk Assessment and Programme Management
- Domain C: Physical Biosecurity Measures
- Domain D: Pathogen Accountability
- Domain E: Personnel Reliability
- Domain F: Dual Use and Bioethics

18.2.4 Codes of Conduct

In 2005 the InterAcademy Panel (IAP) – the Global Network of Science Academies issued the *IAP Statement on Biosecurity*[18] which laid out five fundamental principles intended to serve a basis for the development and promulgation of codes of conduct. Those principles are:

(1) Awareness.
(2) Safety and Security.
(3) Education and Information.
(4) Accountability.
(5) Oversight.

A number of states, including Indonesia, Malaysia, and the Netherlands, have also worked on the development and promulgation of codes of conduct for life scientists. In 2013 the Robert Koch Institute in Germany issued an internal policy on biosecurity titled, *Dual Use Potential of Life Sciences Research: Code of Conduct for Risk Assessment and Risk Mitigation.*[19] The policy features five parts, of which Part 5 is specifically dedicated to *Developing Awareness of the Dual Use Problem.* This policy makes explicit provisions for ensuring that members of staff are sensitised to the issue of dual use through a three-level educational programme. As such, it constitutes an integrated approach to engaging life scientists on dual use and biosecurity, highlighting the need for education and awareness – raising as an essential element for enhancing the effectiveness of codes of conduct, achieving compliance, and maintaining science integrity.

Further proposals for the development and promulgation of codes of conduct have been put forward by States Parties to the BTWC during the Intersessional Process in 2012–2015, as well as during the Preparatory Committee meetings in 2016. Relevant working papers include:

- Chile, Colombia, Costa Rica, Ecuador, El Salvador, Guatemala, Italy, Mexico, and Spain, *Code of Conduct for Scientists*, BWC/MSP/2014/WP.6;[20]
- China, *Proposal for the development of the template of biological scientist code of conduct under the Biological Weapons Convention*, BWC/MSP/2015/WP.9;[21]
- China and Pakistan, *Proposal for the Development of the Template of Biological Scientist Code of Conduct under the Framework of Biological Weapons Convention*, BWC/CONF.VIII/PC/WP.31.[22]

18.2.5 Review of Scientific and Technological Developments

Assessing the impact of scientific and technological advances on the BTWC is critical for ensuring both the relevance and adequate functioning of the Convention. This endeavour can hardly be carried out without the active engagement of the life science community. Hence, the Intersessional

Process was structured in such way as to enable contributions from life scientists. The BTWC Meeting of Experts typically held in summer provides a platform for considering the implications of the progress of science and technology, evaluating related risks and benefits, and suggesting measures for preventing the hostile misuse of the life sciences. During those meetings, upon an invitation by the Chair, scientists can serve as Guests of the Meeting and give an expert presentation on a topic of relevance to the proceedings of the meeting.

Additional options for participation of the life science community in the BTWC include the possibility of delivering an NGO statement during informal sessions, presenting a poster during the Meeting of Experts, and/or organising a side event on the margins of BTWC meetings and Review Conferences.

Prior to the Seventh Review Conference of the BTWC in 2011, the IAP—The Global Network of Science Academies—in collaboration with other national and international organisations held an international workshop in Beijing titled, *Trends in Science and Technology Relevant to the Biological Weapons Convention*. The aim of the workshop was to consider how to respond to the biosecurity implications of developments in the life sciences and assess trends in science and technology relevant to biological and chemical weapons nonprolifeation.[23] The final report of the workshop was presented to the States Parties of the BTWC. In 2015, a similar international workshop entitled *Assessing the Implications of Advances in Science and Technology for the Biological and Toxin Weapons Convention* was held in Warsaw in preparation for the Eight Review Conference of the Convention.[24] In the context of the CWC, similar initiatives have been undertaken by the International Union for Pure and Applied Chemistry (IUPAC) (see Chapter 17).

18.3 Implications for the Chemical Weapons Convention

Drawing upon the discussion in the preceding paragraphs, the present section aims to outline a set of key findings with a bearing on the efforts to prevent the development, proliferation, and use of chemical weapons.

18.3.1 Promoting Awareness-raising and Education

The importance of promoting awareness of the CWC among those in the chemical sciences is difficult to overstate. At the same time, it is worth noting that the implementation of education and outreach programmes is considered as part of a wide range of measures for engaging scientists and thus strengthens the Convention (see Chapter 17). Education is not a panacea, nor is it a 'silver bullet'. Rather, it is a critical but *per se* an insufficient step toward fostering and sustaining a norm of chemical security.

For the implementation of education programmes to be effective, it is crucial that attention is given both to the training content and the method of instruction. The experience of promoting biosecurity awareness is indicative in this regard. When developing training content, the following considerations are noteworthy. First, materials and resources need to be made widely available and accessible. Open-access online resources which are easy to adapt for different purposes in various educational settings and with diverse audiences can help correct the existing deficiency of training material. Second, in order to reach as many stakeholders as possible, it is recommended that the materials are made available in different languages. Third, it is recommended that training materials and resources are developed in a user-friendly manner, so as to enable as lecturers and educators to use them easily without the need for extensive prior preparation or training.

Efficient and effective awareness-raising requires that training content is coupled with adequate methods of instruction (see Chapter 17). Since the topics that chemical security education aims to address pertain not only to the technical dimensions, but also to the legal, social and ethical implications of the chemical sciences, their relevance to chemistry is not always immediately recognised. Hence, the development and implementation of effective chemical security education programmes depends as much on the adoption of appropriate approaches and methods of delivery as on the contents of such programmes. Given the substantial body of evidence suggesting that "humans are not adept at making connections between disparate fields or types of knowledge, unless they are specifically helped to do so through education",[25,26] it is of paramount importance that the delivery of chemical security education is tailored to the needs of those engaged in the chemical sciences. Within this context, the value of active learning needs to be highlighted and in particular when properly implemented, it enables students to acquire the skills necessary for the practical application of theories and concepts.

A critical component of active learning is that the learner, rather than the instructor, is at the centre and focus of the activities taking place in the classroom. As such, it is a learner-centred mode of instruction that stresses collaboration, enquiry and critical thinking. Active learning helps people take control of their own learning by enhancing people's abilities to recognise when they understand and when they need more information, thus allowing them to predict their performances on various tasks. Teaching practices congruent with active learning engage learners as active participants in their learning by focussing their attention on critical elements, fostering abstraction of common themes or procedures (principles), and evaluating their own progress toward understanding.[26] Sense-making, self-assessment and reflection on what worked and what needs improving are thus crucial elements of the active learning approach.

The US National Academy of Sciences' report[26] *How People Learn: Brain, Mind, Experience, and School* which appeared in 2000 provides an extensive overview of the value and practical uses of teaching approaches that encourage active learning. The report shows that there is a substantial body of evidence that active learning approaches enhance learning generally, enabling students to transfer and extend what they have learnt in one context to new contexts. In addition, active learning strategies to instruction *have been shown to increase the degree to which students will transfer to new situations without the need for explicit prompting*. Overall, the report[26] strongly endorsed the benefits of active learning strategies underscoring that the:

> "Integration of [active learning] instruction with discipline-based learning can enhance student achievement and develop in students the ability to learn independently. It should be consciously incorporated into curricula across disciplines and age levels."

In order to facilitate the implementation of awareness-raising and education programmes, the development of national, regional, and international networks need to be encouraged and supported. To this end, the use of the Team-based Learning format described above is recommended, not least because it is easy to replicate and easy to adapt in different settings and for different purposes. As such, it can be utilised for training trainers and building sustainable capability in chemical security.

A holistic approach to education and awareness-raising requires that those in chemical sciences are sensitised to their responsibilities under the CWC as early in their career as possible (see Chapter 17). That said, it is equally important that issues and topics related to the Convention are embedded in the education of chemical scientists at every level, starting from high school, if not even primary school, through undergraduate and postgraduate curricula, and continued professional development. The articulation of international professional competency standards can further promote the efforts to raise awareness of the CWC among those in the chemical sciences. Such standards can be of particular value for fostering awareness among practising scientists, for example, through the introduction of continued professional development courses and modules.

It needs to be recognised that building sustainable capacity in chemical security is a long-term endeavour which requires, *inter alia*, internationally coordinated action, a flexible combination of bottom-up and top-down approaches in accordance with individual States' circumstances, adequate financial support, and strategic planning underpinned with clear milestones and indicators for evaluating progress.

18.3.2 Promulgation of Codes of Conduct

Within the context of the BTWC, codes of conduct have been considered since 2005. By design, codes largely fall into three categories, namely aspirational, educational/advisory, and enforceable.[27] Aspirational codes, often designated as 'codes of ethics', set out ideals that practitioners should uphold, such as standards of research integrity, honesty, or objectivity. The Hague Ethical Guidelines described in Chapter 17 are indicative in this regard. Educational/advisory codes often designated as 'codes of conduct' provide guidelines suggesting how to act appropriately. Enforceable codes often designated as 'codes of practice' seek to codify acceptable behaviour and as such, are embedded within wider systems of professional and legal regulation.

Codes are important tools which can go a long way in promoting responsible conduct and fostering self-regulation. For codes to be effective though, attention needs to be given not only to their development, but also to the way in which they are promulgated. The process of developing codes needs to be all-inclusive, open, and transparent. Those who are expected to abide by the code should be directly involved in its development early on – they should see it as *their* initiative, *their* code. Having ownership of the code from its conception empowers scientists, as it allows them to see the code as an instrument of self-regulation aimed at preserving science integrity and academic freedom. Achieving this goal, however, requires that those in the chemical sciences are sensitised to their responsibilities under the CWC before codes are introduced. Education therefore is an essential prerequisite for the effective development and promulgation of codes.

It needs to be recognised that a code is a living document. Its effectiveness depends mainly on the vigour and enthusiasm of its architects to secure buy-in from everyone to whom the code should apply. Monitoring and maintaining compliance with the code thus becomes both an individual and collective responsibility. Codes need to be re-visited on a regular basis and, if necessary, revised to ensure that they are kept up-to-date and in line with scientific and technological advances.

18.3.3 Science Policy Advising

In contrast to the BTWC, the CWC has a far more sophisticated infrastructure for review of developments in science and technology. The Scientific Advisory Board (SAB) under the auspices of the OPCW, comprising 25 independent experts, provides specialised advice in science and technology to Member States (see Chapter 5).[28] The Board also has the authority to establish Temporary Working Groups (TWGs) on specific issues.

A case in point is the TWG on Convergence of Chemistry and Biology which was set up in 2011 following the growing recognition of the close integration of chemistry with biology and its implications for the BTWC and CWC:

> "With the increasing convergence of chemistry and biology, several conclusions were of joint concern for both the BWC and CWC. While chemical weapons and biological weapons are covered by their respective conventions, both conventions include toxins and other 'mid-spectrum' agents in their remit; although the CWC Schedules refer specifically only to saxitoxin and ricin for verification purposes. With convergence there is

increasing commonality of the science and technology that underpins the two treaties."[29]

The TWG drew upon a pool of wide-ranging expertise featuring industrial and academic scientists, defence laboratory scientists, toxicologists, analytical chemists and chemical engineers (including those with experience in bio-mediated processes). Among the guest speakers who attended the Group's meetings were practitioners with diverse backgrounds, including commercial biotechnology, informatics, and DIY biology.

In its Final Report issued in 2014, when considering the item 'On bringing together expertise on the BWC and CWC in order to discuss areas of common interest and share relevant knowledge', the TWG put forward the following list of recommendations:[29]

"Recommendation 13: A venue like the TWG on convergence of chemistry and biology should continue to exist, possibly as a temporary working group or a standing arrangement under the SAB.

Similar interactions could be valuable at the national level and the national authorities could play an important role in building the necessary bridges.

Recommendation 14: National Authorities could be encouraged to engage more actively on convergence issues, including interacting with relevant biological and chemical scientific communities and hosting relevant events. A standing item on science and technology at National Authority Days might provide an opportunity to promote and report back on such an activity. Adopting convergence as a major theme for a future National Authority Day would help draw attention to this issue.

There is continued value in bringing together expertise related to both treaties to consider developments in science and technology and their implications, education and outreach.

Recommendation 15: The SAB and TS should continue to work across areas of overlap between the CWC and the BWC. The Director-General might ask States to consider knowledge of the biological sciences when considering nominating experts to the SAB.

Side events held on the margins of BWC meetings and scientific conferences provided fora for the TS and the SAB to interact with experts from the BWC outside of the TWG meetings.

Recommendation 16: The TS, supported by the SAB, should continue to participate in such meetings and continue to address convergence.

Recommendation 17: The Director-General might consider meeting with the Chair of the BWC and heads of relevant international scientific bodies to explore issues around convergence.

The TWG identified a number of other options for discussing areas of common interest and sharing knowledge between CWC and BWC communities, including:

- Feeding the findings of this report back to the relevant scientific communities;
- Holding a conference involving both expert communities;
- Exploring opportunities for a special edition(s) of relevant scientific journals or 'crowd-sourced' engagement through written and visual media;
- Establishing a communications technology platform or online collaborative work space to allow interested experts to continue to work on these issues; and
- Increasing coordination and collaboration between the TS and the BWC ISU on their engagement with relevant scientific communities."

It is therefore evident that the life science community has an important role to play in strengthening the prohibitions of the BTWC and CWC, in order to prevent the hostile misuse of the knowledge and materials generated in the life sciences and chemical sciences. Closer engagement between the life science and chemical science communities is critical to ensure timely and adequate assessment of science and technology, as well as development of relevant measuring for minimising the potential risks arising from novel advances, thus upholding the norm against the development, proliferation, and use of chemical and biological weapons.

18.4 Conclusions

The purpose of this chapter has been to review and assess the role that the life science community can play in preventing the development, proliferation, and use of chemical weapons. Over the past decade, a considerable effort has been invested in attempts to engage the life science community with biosecurity in support of the BTWC. Relevant projects and initiatives designed to foster awareness of the Convention and facilitate the development of a biosecurity culture underpinned by continuous learning, codes of conduct, and the provision of expert advice have been carried out. Unfortunately, to date, the progress of this work has been severely hampered by the lack of international infrastructure for coordinating, facilitating, and supporting efforts to broader the involvement of life scientists in the BTWC.

By contrast, the CWC offers a well-established international framework for promoting effective action. The Advisory Board on Education and Outreach and the Scientific Advisory Board are key mechanisms which, if utilised systematically, can serve as important vehicles for enabling those in the chemical sciences to engage actively with the CWC and contribute their expertise to strengthening the Convention. The experience of promoting biosecurity in the life sciences constitutes an excellent model of the steps required for facilitating such an engagement in a consistent and all-inclusive manner. At the same time, the increasing convergence of chemistry and biology presents a unique opportunity for the life science and chemical science communities to collaborate closely and exchange knowledge, lessons learned and best practices, thus reinforcing the international norms against the development, spread, and use of chemical and biological weapons.

References

1. United Nations, *The Seventh Review Conference of the States Parties to the Convention on the Prohibition of the Development, Production and Stockpiling of Bacteriological (Biological) and Toxin Weapons and on Their Destruction, Final Declaration in Final Document*, BWC/CONF.VII/7, Palais des Nations, Geneva, 2012, available at: www.unog.ch, accessed 22 September 2016.
2. Scientific Advisory Board, *Organisation for the Prohibition of Chemical Weapons, Report of the Seventeenth Session of the Scientific Advisory Board*, SAB-17/1, OPCW, The Hague, 2011, available at: http://www.opcw.org, accessed 22 September 2016.
3. Conference of the States Parties, *Third Review Conference, Report of the Third Special Session of the Conference of the States Parties to Review the Operation of the Chemical Weapons Convention*, RC-3/3, OPCW, The Hague, 2013, available at: http://www.opcw.org, last accessed 22 September 2016.
4. Organisation for the Prohibition of Chemical Weapons, *Advisory Board on Education and Outreach*, https://www.opcw.org/about-opcw/subsidiary-bodies/advisory-board-on-education-and-outreach/, last accessed 22 September 2016.
5. S. Whitby, T. Novossiolova, G. Walther and M. Dando, *Preventing Biological Threats: What You Can Do*, University of Bradford, Bradford, 2015T. Novossiolova, *Biological Security Education Handbook: The Power of Team-based Learning*, University of Bradford, Bradford, 2016, available at https://www.unog.ch/80256EE600585943/(httpPages)/0A20E57D9F8424B8C-12581D8007EC32E?OpenDocument, accessed 27 March 2018.
6. T. Novossiolova, G. Mancini and M. Dando, *Effective and Sustainable Biosecurity Education for Those in the Life Sciences: The Benefits of Active Learning*, Bradford Briefing Papers, No.7, 3rd Series, University of Bradford, Bradford, June 2013.
7. The Handbook is available in English, French, Arabic and Russian at https://www.unog.ch/80256EE600585943/(httpPages)/0A20E57D9F-8424B8C12581D8007EC32E?OpenDocument, accessed 27 March 2018. The Handbook is also available in Ukrainian at: http://www.bsseducation.com.ua/uk/downloads, accessed 22 September 2016.

8. IFBA International Certification Programme in Biosecurity, http://www.internationalbiosafety.org/index.php/professional-certification/professional-certification/about-the-program, accessed 22 September 2016.
9. K. Almquist, J. Fernandez, S. Mantha and M. Kafenzakis, National implementation of biosecurity in Canada, in *Preventing Biological Threats: What You Can Do*, ed. S. Whitby, T. Novossiolova, G. Walther and M. Dando, University of Bradford, Bradford, 2015.
10. Centre for Biosecurity and Biopreparedness, *An Efficient and Practical Approach to Biosecurity*, CBB, Copenhagen, 2015, available at: https://www.biosikring.dk/biosecuritybook/, accessed 22 September 2016.
11. The Netherlands Biosecurity Office, http://www.bureaubiosecurity.nl/en, accessed 22 September 2016.
12. Draft Decree on Biosafety in Tunisia, https://bch.cbd.int/database/record.shtml?documentid=109161, accessed 22 September 2016.
13. W. So, The Federal Bureau of Investigation biosecurity Program: a case study of law enforcement and outreach, in *Preventing Biological Threats: What You Can Do*, ed. S. Whitby, T. Novossiolova, G. Walther and M. Dando, University of Bradford, Bradford, 2015.
14. US Federal Experts Security Advisor Panel, http://www.unog.ch/unog/website/disarmament.nsf/(httpPages)/3803c81a2e03e4c9c1257f6b00357a68?OpenDocument&ExpandSection=13#_Section13, accessed 22 September 2016.
15. Ukraine, Japan and the United Kingdom of Great Britain and Northern Ireland, *Awareness-Raising, Education and Outreach: Recent Developments,* BWC/MSP/2017/WP.22, Meeting of the States Parties to the Convention on the Prohibition of the Development, Production and Stockpiling of Bacteriological (Biological) Weapons and on Their Destruction, 6 December 2017, Geneva, Switzerland. Available at https://www.unog.ch/unog/website/disarmament.nsf/(httpPages)/9e2cf761a08a5e68c12581ee00318fa5?OpenDocument&ExpandSection=1#_Section.
16. IFBA Professional Certification Programme, http://www.internationalbiosafety.org/index.php/professional-certification/professional-certification/about-the-program, accessed 22 September 2016.
17. The Body of Knowledge (BOK), http://www.internationalbiosafety.org/images/IFBA_docs/Certification%20Tab/PC%20Biosecurity%20Exam%20Content.pdf, last accessed 22 September 2016.
18. IAP – A Global Network of Science Academies, *IAP Statement on Biosecurity*, 7 November 2005, available at: http://www.interacademies.net/File.aspx?id=5401, accessed 22 September 2016.
19. Robert Koch Institute Policy on Dual Use, http://www.rki.de/EN/Content/Institute/Dual_Use/code_of_conduct.html, accessed 22 September 2016.
20. Chile, *et al.*, Code of Conduct for Scientists, BWC/MSP/2014/WP.6, *Meeting of the States Parties to the Convention on the Prohibition of the Development, Production and Stockpiling of Bacteriological (Biological) and Toxin Weapons and on Their Destruction*, Geneva, Switzerland, 2 December 2014, available at: http://www.unog.ch/__80256ee600585943.nsf/(httpPages)/

e75aaa1d8234b830c1257d63002c7def?OpenDocument&ExpandSection=1#_Section1, accessed 22 September 2016.
21. China, Proposal for the development of the template of biological scientist code of conduct under the Biological Weapons Convention, BWC/MSP/2015/WP.9, *Meeting of the States Parties to the Convention on the Prohibition of the Development, Production and Stockpiling of Bacteriological (Biological) and Toxin Weapons and on Their Destruction*, Geneva, 15 December 2015, Switzerland, available at: http://www.unog.ch/80256EDD006B8954/(httpAssets)/77BBB042D87DD84AC1257F1D00457423/$file/Chinese+(Unofficial+English+translation+annexed).pdf, accessed 22 September 2016.
22. China and Pakistan, Proposal for the Development of the Template of Biological Scientist Code of Conduct under the Framework of Biological Weapons Convention, BWC/CONF.VIII/PC/WP.31, *Eighth Review Conference of the States Parties to the Convention on the Prohibition of the Development, Production and Stockpiling of Bacteriological (Biological) and Toxin Weapons and on Their Destruction*, Preparatory Committee, Geneva, Switzerland, 11 August 2016, available at: http://www.unog.ch/80256EDD006B8954/(httpAssets)/41F9E040F91BD007C125800D002FFFDF/$file/BWCCONF.VIIIPCWP.31.pdf, accessed 22 September 2016.
23. National Research Council, *Life Sciences and Related Fields: Trends Relevant to the Biological Weapons Convention*, The National Academies Press, Washington DC, 2011, available at: https://www.nap.edu/catalog/13130/life-sciences-and-related-fields-trends-relevant-to-the-biological, accessed 22 September 2016.
24. IAP – The Global Network of Science Academies, *The Biological and Toxin Weapons Convention (BTWC): Implications of Advances in Science and Technology*, The Royal Society, London, 2015, available at: https://royalsociety.org/topics-policy/projects/biological-toxin-weapons-convention/, accessed 22 September 2016.
25. J. Johnson, in Education and Ethics in the Life Sciences: Strengthening the Prohibition of Biological Weapons, ed. B. Rappert, ANU Press, Canberra, 2010, pp. 197–215.
26. Committee on Developments in the Science of Learning with additional material from the Committee on Learning Research and Educational Practice, National Research Council, *How People Learn: Brain, Mind, Experience, and School (Expanded Version)*, The National Academies Press, Washington, 2000.
27. B. Rappert, *Towards a Life Sciences Code: Countering the Threats from Biological Weapons*, Bradford Briefing Paper No.13, 2nd Series, University of Bradford, Bradford, 2004.
28. OPCW Scientific Advisory Board, https://www.opcw.org/about-opcw/subsidiary-bodies/scientific-advisory-board/, accessed 22 September 2016.
29. OPCW Scientific Advisory Board, *Convergence of Chemistry and Biology: Report of the Scientific Advisory Board's Temporary Working Group*, SAB/REP/1/14, OPCW, The Hague, 2014.

CHAPTER 19

The Role of the Non-governmental Medical Community in Combatting the Development, Proliferation and Use of Chemical Weapons

MICHAEL CROWLEY[a] AND VIVIENNE NATHANSON*[b]

[a]University of Bradford, Bradford, West Yorkshire, BD7 1DP, UK;
[b]Independent Medical Consultant
*E-mail: m.j.a.crowley1@bradford.ac.uk, viv.nathanson@gmail.com

19.1 Overarching Ethical Responsibilities of Health Professionals

Doctors share a collective ethical and moral code that has been steadily developed over centuries from a number of initial core documents. Over-arching ethical statements, including the original Hippocratic Oath (dating back over 2600 years)[1] and the 12th century Oath of Maimonides,[2] have established the relationship and duties of the physician firstly to their patients, and more broadly to humanity.

In the modern era, of particular importance are ethical codifications developed by the World Medical Association (WMA), the international

Preventing Chemical Weapons: Arms Control and Disarmament as the Sciences Converge
Edited by Michael Crowley, Malcolm Dando and Lijun Shang

Published by the Royal Society of Chemistry, www.rsc.org

organization representing physicians which was established after the Second World War to "ensure the independence of physicians, and to work for the highest possible standards of ethical behaviour and care by physicians, at all times."[3] The WMA has sought to establish and promulgate general standards in medical ethics that are applicable worldwide, and its first task was to update the Hippocratic Oath, which resulted in the 1948 *World Medical Association Declaration of Geneva* and the 1949 *WMA International Code of Medical Ethics*. These and other WMA Declarations and Codes gain their unique strength and credibility from being ethical standards written by doctors, for doctors. Under the Declaration of Geneva, physicians pledge to consecrate their lives to the service of humanity; ensure that the health of the patient will be their first consideration; maintain the utmost respect for human life; not use their medical knowledge to violate human rights and civil liberties, even under threat.[4] The *WMA International Code of Medical Ethics* enjoins physicians, *inter alia*, to be dedicated to providing competent medical service in full professional and moral independence, with compassion and respect for human dignity; strive to use health care resources in the best way to benefit patients and their community; act in the patient's best interest when providing medical care; and always bear in mind the obligation to respect human life.[5] The universal applicability of the WMA Code and other relevant WMA standards have been tested in courts around the world and there is general agreement that they bind all physicians regardless of whether they have formally sworn an ethical oath. In addition to these two foundational ethical statements, a major priority for the WMA since its inception has been the development of guidance regarding research on human subjects.

19.1.1 Medical Ethics and Scientific Research on Human Subjects

During 1946–7, the Nuremberg War Crimes Tribunal tried scientists and physicians accused of involvement in the horrific large-scale programme of experimentation conducted by the Nazi regime upon its prisoners. The evidence that came to light included "experiments" involving intentional infection with diseases and exposure to chemical warfare agents such as mustard gas, lewisite and phosgene; and the application of toxic chemicals, irradiation with X-rays, and surgery designed to forcibly sterilise patients.[6] In the light of such atrocity, the first substantial attempt to formulate universal ethical standards for human testing—the *Nuremberg Code*—was incorporated as part of the judgement of the Doctors Trial. A central tenet of the Code was that "the voluntary consent of the human subject is absolutely essential". In addition, the Code requires that "the experiment should be so conducted as to avoid all unnecessary physical and mental suffering and injury...[and] no experiment should be conducted, where there is an *a priori* reason to believe that death or disabling injury will occur."[7]

The international medical community has subsequently reaffirmed, revised and expanded the Code's injunctions in the light of ongoing advances in medicine and science creating a comprehensive ethical framework regulating medical involvement in all forms of scientific research on human subjects. Of particular importance has been the *WMA Declaration of Helsinki*,[8] originally adopted in 1964 after many years of discussion and debate and regularly amended, it states that: "medical research is subject to ethical standards that promote and ensure respect for all human subjects and protect their health and rights." It adds that whilst the "primary purpose of medical research is to generate new knowledge", this goal "can never take precedence over the rights and interests of individual research subjects". Furthermore, "it is the duty of physicians who are involved in medical research to protect the life, health, dignity, integrity, right to self-determination, privacy, and confidentiality of personal information of research subjects." In addition, "physicians must consider the ethical, legal and regulatory norms and standards for research involving human subjects in their own countries as well as applicable international norms and standards", but no such regulations should "reduce or eliminate any of the protections for research subjects set forth in this Declaration."

Although not specifically intended to address chemical weapons development, it is clear that such standards will severely constrain medical participation in research involving application of chemical agents upon human subjects. The importance of such considerations can be seen in the case study below.

19.1.1.1 UK Chemical Weapons Research, Human Experimentation and Medical Ethics in the 1960s

The UK's post World War II chemical weapons development programme conducted at the Chemical Defence Experimental Establishment at Porton Down, included research into potential incapacitating chemical agent (ICA) weapons and involved research upon volunteer human subjects. The main programme of human studies into such agents, which ran from 1959 to the early 1970s, included trials with physical ICAs such as oripavine derivates and psychological ICAs including LSD, LAE, BZ, tryptamines and MPIPG.[45,46] During the mid-1960s, members of Porton Down's Medical Division reportedly developed significant concerns about this chemical weapons research programme following serious ill-effects suffered by test subjects including a case of cardiovascular depression induced by T2636, a catatonic episode produced by T2833 and instances of toxic delirium in the BZ tests.[9] Two members of the Medical Division resigned in November 1964 and December 1964. Another, who played an important part in the programme, consequently fell sick in January 1965 and was admitted to hospital suffering from depression; although he subsequently recovered, he never rejoined Porton Down. His assistant was "very near breaking point" by mid-February 1965 and resigned a few days later.[9]

On February 2nd 1965, the Committee on the Safety of Human Experimentation (COHSE)—Porton Down's internal body for monitoring experiments—questioned the advisability of giving volunteers the ICA doses necessary to induce full incapacitation. The Head of the Medical Division at Porton Down thought that while ICAs were reputed to be safe, there appeared to be "strong reasons" for doubting this. If experiments were dangerous, even if the effects appeared to be innocuous, it was considered unethical and unjustified to conduct human tests with these agents in peacetime. The COSHE agreed that very serious consideration be given as to whether any more work with these agents should be conducted on humans.[9] On 12th February 1965, the Head of the Medical Division banned all further tests with hallucinogenic drugs, the ban being extended to oripavine derivatives in March 1965. The Director of Porton Down was concerned with this "question of medical conscience" and believed that the ban had caused "considerable embarrassment to senior officials in London".[9]

A meeting of senior external medical experts, with the Head of Porton Down's Medical Division, was held in August 1965 to address whether Porton Down medical officers were justified in "deliberately dosing healthy men with drugs specifically designed to induce some malfunction, either physiological or psychological?" The external medical experts were content that this could be done ethically if Porton Down were seeking to develop therapies against the chemical agents which enemies might use against the UK military. At this point, the Head of the Medical Division intervened: "[I] had therefore to disclose ... that we were in fact looking for an agent which we could use [as a weapon] under certain circumstances." His admission "changed the complexion [of the meeting] very considerably" and it was thought that scientists at Porton Down were being asked to do things that "went far beyond the Medical Research Council rules for human experiments."[10] One concrete result of these discussions was the establishment of the Applied Biology Committee (ABC)—an external group of respected scientists to monitor Porton Down's human experimentation—which should serve as "Father Confessor" for Porton Down staff and "damn the proposals [for experiments] of those who were trying to go too far and too fast."[10] It is notable that these ethical issues were being grappled with by UK medical researchers as the WMA was in the final stages of agreeing and then promoting the Declaration of Helsinki. Although the UK ICA weapons programme subsequently resumed, medical ethical considerations had clearly influenced both the actions of individual medical researchers and also delayed, increased regulation in, and constrained a State CW programme. This is even more notable given that these activities took place well before the coming into force of the BTWC or the CWC, when there was consequently no international legal prohibition on the research, development and stockpiling of chemical weapons.[†]

[†]At this time, the 1925 Geneva Protocol was in force, and there were concerns that certain uses of ICAs would breach its prohibitions. However, although it bans use of chemical and biological weapons in warfare, the Geneva Protocol does not explicitly address their research and development. For further discussion see Chapter 6 of this publication, and also ref. 46 (J. Walker) pp. 17–33.

19.1.2 Violations of Human Rights Law and Breaches of International Humanitarian Law

There is a range of declarations and regulations that guide health professionals in situations of conflict and unrest, and specifically prohibit their involvement in human rights abuse or breaches of international humanitarian law. Of particular importance are the *Regulations in Times of Armed Conflict and Other Situations of Violence*. Initially adopted by the WMA in 1956, they have subsequently been regularly reviewed and strengthened. These state that "Medical ethics in times of armed conflict is identical to medical ethics in times of peace" and that "[I]f, in performing their professional duty, physicians have conflicting loyalties, their primary obligation is to their patients... [and they]... should adhere to international conventions on human rights, international humanitarian law and WMA declarations on medical ethics." Consequently, the Regulations declare that it is "deemed unethical" for physicians to: "[G]ive advice or perform prophylactic, diagnostic or therapeutic procedures that are not justifiable for the patient's health care ... Weaken the physical or mental strength of a human being without therapeutic justification ... Employ scientific knowledge to imperil health or destroy life ... Condone, facilitate or participate in the practice of torture."[11] Furthermore, these Regulations state that "[D]uring times of armed conflict and other situations of violence, standard ethical norms apply, not only in regard to treatment but also to all other interventions, such as research. Research involving experimentation on human subjects is strictly forbidden on all persons deprived of their liberty, especially civilian and military prisoners and the population of occupied countries."[‡,11]

The WMA has subsequently conducted policy reviews and adopted more detailed guidance on specific human rights issues of grave and immediate concern to the medical community, enunciated in these Regulations, for example with regard to prevention of torture and ill-treatment. The *WMA Declaration of Tokyo* originally adopted in 1975 and most recently revised in 2016, states that "the physician shall not countenance, condone or participate ... [nor] ... provide any premises, instruments, substances or knowledge to facilitate the practice of torture or other forms of cruel, inhuman or degrading treatment or to diminish the ability of the victim to resist such treatment". Furthermore, "the physician shall not use nor allow to be used, as far as he or she can, medical knowledge or skills, or health information specific to individuals, to facilitate or otherwise aid any interrogation, legal or illegal, of those individuals."[§,47] Such provisions are of particular relevance to the considerations of this

‡Similarly, the Declaration of Helsinki requires physicians seeking informed consent for participation in a research study to be "particularly cautious" if the potential subject is "in a dependent relationship with the physician or may consent under duress". See ref. 8, clause 27.

§These guidelines have been widely promoted and operationalized by national medical associations. In certain countries, doctors accused of conducting or being complicit in torture have been investigated by their national medical association, and where found guilty have been disbarred.

chapter given historical and contemporary reports of medical involvement in the forced application of toxic chemical agents to prisoners, including the employment of psychotropic drugs, as part of abusive and unwarranted psychiatric treatment of political prisoners, and the employment of "truth drugs" in forced interrogations and attempted "brain-washing" procedures.[12]

Similarly, health professionals, their national associations and the WMA have been active in upholding, implementing and seeking to advance international humanitarian law, for example combatting development and employment of specific weapons which cause superfluous injury or unnecessary suffering, and of supporting attempts by the International Committee of the Red Cross (ICRC) to strengthen and codify regulations in this area.

19.1.3 The SIrUS Project and the Review of New Non-kinetic Weapons

The majority of weapons in common use by military forces are essentially kinetic – they have their effect by exerting force upon individuals or on the places in which they live and work. The impact on human health and survival is closely related to the level of force they are able to apply and is a factor of design and normal use. The medical community has a long experience in treating injuries arising from such weapons and planning provision of treatment for affected combatants and civilian populations. In contrast, advances in relevant areas of science and technology have been explored by weapons developers hoping to exploit alternative ways to interfere with the human body and thus disrupt the ability of individuals to function as combatants - including through application of chemical or biological agents, acoustic, electromagnetic or other forms of directed energy.

In the majority of States, the development, acquisition and introduction of new weapons does not include systematic reviews by independent physicians with the ability to assess the likely effects of the proposed weapons. Such weapon reviews would clearly be strengthened by the informed input of health professionals with experience in dealing with existing weapons employed, and acquainted with knowledge of the relevant proposed weapons technologies, expertise in assessing likely injuries caused by the proposed weapons, and the impact of a variety of different levels of access to post-injury trauma management. Without this type of assessment, there is a danger that the relevant elements of IHL *i.e.* the prohibition on the use of "means and methods of warfare of a nature to cause superfluous injury or unnecessary suffering"[13] will not be fully taken into account and that inherently cruel weapons will be introduced onto the battlefield. Consequently, health professionals and the broader medical and scientific communities have an important role to play assessing the consequences of damage to body functions of new (particularly non-kinetic) weapons.

In 1996, the ICRC established the SIrUS Project, comprising an inter-disciplinary group of experts in the fields of medicine, international law and weaponry, which attempted to develop objective criteria in this area. Analysing an ICRC database of over 26 000 military personnel wounded by existing permitted conventional weapons, the experts attempted to use both the casualty survival rates off the battlefield and the seriousness of the inflicted injury, as a yardstick to determine what is not 'superfluous injury or unnecessary suffering.' Consequently, four criteria were proposed by which the effects of any weapon can be compared to those of conventional weapons, and could be employed in determining which weapons cause "superfluous injury or unnecessary suffering".

The SIrUS Project proposed that what constituted superfluous injury and unnecessary suffering should "...be determined by design-dependent, foreseeable effects of weapons when they are used against human beings" and caused:

- "specific disease, specific abnormal physiological state, specific abnormal psychological state, specific and permanent disability or specific disfigurement; or
- field mortality of more than 25% or hospital mortality of more than 5%; or
- Grade 3 wounds as measured by the Red Cross wound classification; or
- effects for which there is no well-recognized and proven treatment."[14]

Criteria 1, 2 and 4, appear to be of potential applicability for any States considering the legality of the development or use of new weapons employing toxic chemicals.

The ICRC initiative gained strong and widespread backing from the medical community, with the 1998 WMA General Assembly adopting a Resolution on the issue.[15] This Resolution recognized previous WMA support of efforts to define "objective criteria which would measure the effects of current and future weapons, and which could be used to stop the development, manufacturing, sale and use of weapons", and called on "[N]ational Medical Associations formally to endorse the SirUS Project in order to contribute medical knowledge for the practical enforcement of international conventions concerning the limitation of the effects of weapons."

Unfortunately, a number of States were highly critical of the SIrUS Project. The US for example claimed that its methodology was well-intentioned but flawed and, in particular, did not take into consideration military necessity. Following opposition at the XXVIIth International Conference of the Red Cross and Red Crescent Movement, the Project was suspended in 2001.[¶]

19.2 Medical Involvement in Chemical and Biological Weapons Development

In addition to establishing generally applicable ethical standards prohibiting medical involvement in relevant breaches of international humanitarian law and human rights abuses, the WMA has adopted ethical guidelines

[¶]For a discussion from a perspective of those opposed to the project, see ref. 48.

specifically banning biomedical participation in the development of chemical and biological weapons. In its 1990 *Rancho Mirage Declaration on Chemical and Biological Weapons*, the WMA emphasised that "effects of exposure to chemical and biological weapons present a continuing threat to the health of human beings on a long term basis, possibly causing illness, injury, disease and defects in the population over a long period of time" and that "existing health care services, technology and manpower may be helpless to relieve the suffering caused by exposure to [such] weapons".[16] It consequently stated that: "the World Medical Association considers that it would be unethical for the physician, whose mission is to provide health care, to participate in the research and development of chemical and biological weapons, and to use his or her personal and scientific knowledge in the conception and manufacture of such weapons."[16] Furthermore, under this Declaration, the WMA: "condemn[ed] the development and use of chemical and biological weapons ... ask[ed] all governments to refrain from the development and use of chemical and biological weapons ... [and] ask[ed] all National Medical Associations to join [the] WMA in actively supporting this Declaration".[16]

In 2002, the WMA General Assembly adopted the *Washington Declaration on Biological Weapons*, which recommended that:"[T]he World Medical Association and National Medical Associations worldwide take an active role in promoting an international ethos condemning the development, production, or use of toxins and biological agents that have no justification for prophylactic, protective, or other peaceful purposes ... [and that] ... the World Medical Association urge all who participate in biomedical research to consider the implications and possible applications of their work and to weigh carefully in the balance the pursuit of scientific knowledge with their ethical responsibilities to society."[17]

In its 2013 *Resolution on the Prohibition of Chemical Weapons*, the WMA noted that "development, production, stockpiling and use of chemical weapons is banned under the CWC, and that use of such weapons is regarded by some to be a crime against humanity, regardless of whether the target populations are civilian or military". It consequently urged all relevant parties to "make active efforts to abide by the CWC ban on the development, production, stockpiling and use of chemical weapons; and supported "[U]N initiatives to identify anyone who is responsible for the use of chemical weapons and to bring them to justice".[18]

In addition, health professionals and their national associations have also played an important role in promoting a "culture of responsibility" amongst the broader bio-medical and life science research communities, particularly through the development of ethical codes regulating dual-use research that could be misused for malign intent; and the provision of attendant guidance for bodies funding such research and medical, and scientific journals publishing research findings. These issues are explored in Chapters 17 and 18 of this publication.

National medical bodies have also established mechanisms and structures to implement ethical standards, including ethics boards, which have the authority to suspend or disbar physicians from practising medicine in cases of extreme misconduct. In certain countries, such bodies have investigated the alleged participation of health professionals in chemical and biological weapons development programmes.

19.2.1 Investigating Medical Involvement in Apartheid South Africa's CBW Research Programme

In 2007, the Health Professions Council of South Africa (HPCSA) issued *Guidelines for Good Practice in the Health Professions: Research, Development and Use of Chemical and Biological Weapons*.[19] The preamble stated that:

"[T]he South African military authorities under the apartheid government sustained a covert programme for the development of chemical and biological weapons, and recruited health care practitioners and scientists to staff the programme. The evidence that emerged in the TRC's [Truth and Reconciliation Commission's] investigations into this clandestine project has pointed to the importance of developing clear guidelines for the health professions in regard to participation in such programmes. It was particularly evident that the secrecy surrounding the apartheid government's CBW programme enabled health professional scientists to conduct research outside of any ethical oversight."[20] The preamble further stated "It is completely contrary to the fundamental principles of the ethics of the health professions for a health care practitioner to participate in research activities directed at generating materials intended to cause harm to human health and well-being."[21]

The guidelines consequently established the following ethical obligations upon healthcare professions: "[A]ll research to develop CBWs designed to inflict harm on humans is unethical and health care practitioners should not participate in such activities. Should health care practitioners find themselves in dual loyalty situations where they are coerced or experience pressure or threats to comply from the military or other authorities, they should appeal to the HPCSA or any other appropriate professional body for support in resisting such pressures."[22]

While the guidelines recognized the potential for physicians to become involved in legitimate research to "protect military or civilian personnel against CBWs", such research "should be subject to open peer review and ethical oversight by a suitably appointed independent body", as elaborated in the guidelines.[23,49] The HPCSA considered that "the guidelines form an integral part of the standards of professional conduct against which a complaint of professional misconduct will be evaluated".[24] And, furthermore, that "Health care practitioners who decide not to follow the guidance... must be prepared to explain and justify their actions and decisions to patients and their families, their colleagues and, if necessary, to the courts and the HPCSA."[25]

The HPCSA have put these ethical principles into practice through its investigation of the activities of Dr Wouter Basson, former head of Project Coast, the Apartheid South African CBW R&D programme. The HPCSA investigation was initiated after it received a complaint submitted by over 40 doctors in 2000 about Basson's unethical conduct, but it was delayed pending Basson's criminal trial on related charges.[26] The HPCSA disciplinary hearing, which followed Basson's acquittal in 2002 on all criminal charges, was blocked following claims by Dr Basson that the HPCSA was biased against him. However, on 10th May 2010, the North Gauteng High Court of Pretoria found the claim to be unjustified, prompting the continuation of the HPCSA investigation.[27]

Dr Basson initially faced six charges of alleged professional misconduct as a result of his activities at Project Coast from 1981 to 1993.[28] Although two charges were dropped, the Professional Conduct Committee (PCC) of the HSPCA initiated proceedings against Dr Basson on the remaining four charges:[29]

- coordinating "the production of the following drugs and tear gases on a major scale ... methaqualone ... MDMA ... BZ ... CS ... CR ... "
- "weaponiz[ing] thousands of 120 mm mortars with tear gas", and of overseeing the filling of "some 120 mm mortars with CR ... which ... were supplied by the South African Defence Force to one Savimbi in Angola for use"
- providing "disorientation substances for over the border kidnapping ('grab') exercises, where the substances were used to tranquilise the person to be kidnapped"
- providing "cyanide capsules to operational officers commanding for distribution to members of its specialised units for suicidal usage."[30]

Although the substance of these charges was not in dispute, Dr Basson presented nine arguments as to why his conduct was not "unprofessional"; these were all rejected by the PCC.[31] On 18th December 2013, the PCC gave their ruling, concluding that the "conduct under discussion was in breach of established ethical rules" and that "the breaches of medical ethics amount to unprofessional conduct".[‖,31] The Chairperson of the PCC, Professor Jannie Hugo, stated that Dr Basson contravened international protocols and conventions: "[H]e confused ethics of a doctor with that of a soldier while discharging his duties. A doctor cannot rely on military orders to escape the consequences of his duties."[**,32]

‖Although Dr Basson's HPCSA sentencing hearing began in 2014, it was (and remains) interrupted by his successful application, to the High Court, for the recusal of two PCC members. In January 2018, the case was remitted to the High Court to decide upon the review application.

**Obligations of military medical personnel to act as physicians rather than soldiers have been examined and reaffirmed in other contexts and circumstances; see ref. 50 for further information.

19.3 Law Enforcement Weapons Employing Toxic Chemicals

Whilst important advances have been made by the international community to establish and enforce—through the CWC—an absolute prohibition on the development and use of weaponised toxic chemicals intended for armed conflict, States have been less successful in adequately constraining the use of weapons employing toxic chemicals ostensibly for law enforcement purposes. Consequently, physicians, nurses and the broader healthcare community have been at the forefront of civil society actions to prevent and respond to the misuse of such weapons. In addition to treating those effected, medical professionals have sought to document and alert the international community to the immediate and long-term health effects of such weapons, notably riot control agents and incapacitating chemical agent weapons.

19.3.1 Incapacitating Chemical Agent Weapons

In recent years, a number of States have explored the potential for developing weapons employing incapacitating chemical agents (ICAs) whose purported intended use is to cause prolonged but non-permanent disability; they include centrally acting agents producing loss of consciousness, sedation, hallucination, incoherence, paralysis, disorientation or other such effects. Potential candidate ICAs for weaponisation could include a broad range of pharmaceutical chemicals, many of which are legitimately used in medicine, as well as bio-regulators and toxins. Concern about State ICA weapons programmes grew markedly following the Russian Federation use of such weapons in a largescale anti-terrorist operation.

In October 2002, Russian security forces employed an ICA weapon to free 900 hostages held by heavily armed Chechen separatists in the Dubrovka theatre in Moscow. The Russian security forces pumped the aerosolised ICA into the theatre, putting hostages and hostage-takers into a 'deep sleep' and then stormed the theatre, killing all of the Chechen fighters. Although the hostages were freed, 125 hostages died due to the effects of the ICA and an unknown number of former hostages suffered long-term injury. Treatment of the hostages who had been poisoned was delayed and compromised by the refusal of the Russian authorities to immediately (or indeed ever) inform the relevant health care professionals of the nature and identity of the ICA employed.[††]

The British Medical Association (BMA) has taken the lead in the development of ethical guidance for the health community on the issue of ICA

[††]Whilst the Russian Government has to date refused to identify the specific agents employed, trace analysis undertaken by researchers from the UK Defence Science and Technology Laboratory (DSTL) at Porton Down of extracts of clothing and urine from survivors of the Moscow theatre siege indicated that the ICA weapon comprised a mixture of two anaesthetics, carfentanil and remifentanil, see ref. 51 for further information.

weapons, in 2006 holding a high-level round-table meeting on "the use of drugs as weapons" attended by experts from the fields of anaesthetics, pharmacology and toxicology, international law relating to weapons, and representatives from the ICRC and relevant BMA committees. The BMA subsequently conducted research and analysis of the implications for public health of continuing State ICA weapons research, development and use, publishing its findings in 2007. The "primary conclusion" of this publication was that the use of ICA weapons was "simply not feasible without generating a significant mortality among the target population". The report stated: "[T]he agent whereby people could be incapacitated without risk of death in a tactical situation does not exist and is unlikely to in the foreseeable future. In such a situation, it is and will continue to be almost impossible to deliver the right agent to the right people in the right dose without exposing the wrong people, or delivering the wrong dose. Countermeasures may be easy to apply if such an attack is expected. This brings into question whether drugs can be used for law enforcement or any other tactical situation in which deaths and injuries should be minimized."[33]

In addition, the BMA also addressed medical involvement in ICA weapons programmes, recognising that "[M]edical knowledge is required to select an agent and to calculate how the correct dose might be given. Medical knowledge is therefore required in planning an attack." However, the BMA categorically rejected the ethical legitimacy of any such involvement by health professionals. "[D]octors should not knowingly use their skills and knowledge for weapons development for the same reasons that these ethical considerations oppose doctors' involvement in torture and the development of more effective methods of execution. In other words, the duty to avoid doing harm rises above, for instance, a duty to contribute to national security."[34]

The report also highlighted the potentially serious implications for the ethos of medicine generally and the safety of physicians involved. "[D]octors risk losing their status as protected care-givers and will no longer be perceived as neutral but as active participants in combat."

Consequently the BMA called on national organizations that represent healthcare professionals to: "promote the norms prohibiting the use of poisons, and therefore the BTWC and the CWC ... [and] ... promote understanding that the use of drugs as weapons would violate such norms... Advocate against the use of drugs as weapons and not be involved in the training of military or law enforcement personnel in the administration of drugs as weapons."[35]

In 2013, as part of its *Resolution on the Prohibition of Chemical Weapons*, the WMA addressed the issue of "chemical agents used in policing actions", noting that such agents including when used by the military acting in a policing role, are allowed under the CWC. The WMA also noted that there was a "significant international dialogue underway on the definition of such agents and the situations in which they can be used." And, that "the CWC appears to assume such agents will not be lethal, but the use of any agent might have fatal consequences." Consequently, the WMA Resolution declared that "[T]hose using them, or authorising their use, must seek to ensure that they are not used in a manner which risks death or serious injury to targeted persons." The WMA consequently urged "[S]tates using chemical agents in riot control and related situations" to "carefully consider and minimise the risks and to, wherever possible, refrain from such use." Furthermore, "any use must follow the establishment of the necessary procedures to reduce the risk of death or serious injury...[and]... they should not be used in a manner, which deliberately increases the risk of injury, harm or death to their targets."[36] However, the 2013 WMA Resolution did not specifically identify the chemical agents of concern (*i.e.* whether it referred specifically to ICA weapons, RCAs, and/or other toxic chemicals) nor did it explicitly establish whether it was permissible or not for health professionals to become involved in the development, testing or employment of weapons employing such agents, and if so under what circumstances. Whilst greater clarity on these issues would be beneficial, the WMA position has subsequently significantly developed regarding use and misuse of RCAs in law enforcement operations.

19.3.2 Riot Control Agents

Riot control agents (RCAs) are potent sensory chemical irritants normally of low lethality that produce dose-and time-dependent acute site-specific toxicity. Targeted mainly towards the respiratory and mucosal surfaces, their short-term effects are well defined, although the potential for deleterious long-term effects is contested amongst the medical community. There is a small discrete range of agents, including CN, CS, OC and PAVA, widely stockpiled and regularly utilized by law enforcement officials in the majority of States for dispersal of crowds and/or incapacitating individuals.[37] Although RCAs are promoted as being safe when used according to manufacturers' instructions, serious health problems resulting from the inappropriate or abusive employment of RCAs by police and security forces have been documented. Of particular concern has been the employment of such agents in enclosed spaces or high concentrations, or against particularly vulnerable targets including infants, young children and the elderly, and those with underlying heath issues such as asthma, bronchitis, cardiac disease or hypertension. As well as treating those suffering from the effects of riot control agents, health professionals have played an increasingly active role in recording injuries resulting from the misuse of RCAs and advocating for an end to abusive practices (as illustrated in the case study below).

19.3.2.1 Turkey: Gezi Park Protests and the Misuse of Riot Control Agents

On 30th May 2013, police in Istanbul broke up a demonstration by several hundred environmentalists, using RCAs, beating protestors and burning their tents. The cause of the protestors, and the abusive response of the authorities, touched a nerve. Within days, tens of thousands of protestors had taken to the streets across the main cities of Turkey. By the middle of June, hundreds of thousands had taken part in "Gezi Park protests" that spanned almost every one of Turkey's 81 provinces. Turkish health and human rights organisations together with UN human rights bodies, and international non-governmental human rights organizations, highlighted the subsequent excessive use of force against peaceful demonstrators, including the unnecessary and disproportionate employment of RCAs.

In its September 2013 report, Physicians for Human Rights (PHR) documented how Turkish police "systematically used tear gas as a weapon on hundreds of thousands of demonstrators, firing tear gas canisters and capsules directly at protesters at close range, in confined spaces, and in other areas with no outlet for escape."[38] Amnesty International reported that approximately 130 000 tear gas canisters were used—equivalent to a year's supply—in the first 20 days alone of the protests. The police also reportedly employed water cannon laced with RCAs against peaceful protestors.[39]

In September 2013, the Turkish Medical Association (TMA) released a report summarizing medical information it had recorded detailing more than 8000 injuries as a result of RCAs, water cannon, rubber bullets, beatings, and live ammunition from 31st May till 25th June 2013. In all, 60 protesters were severely injured, including 11 people who lost one of their eyes, some as a result of being hit by tear gas canisters fired at short range. There were at least five civilian deaths caused by unnecessary and/or excessive use of force and other demonstration-related injuries.[40]

Doctors from the Human Rights Foundation of Turkey (HRFT), a torture rehabilitation organization, conducted medical evaluations of 200 injured protestors and concluded that the physical and psychological evidence in each case was consistent with torture and/or ill treatment intentionally and systematically inflicted on large populations of protesters. PHR reviewed the physical and psychological evidence collected by the HRFT that was available on 169 of the 200 clients they examined, and found that: nearly all (94%) of the 169 individuals were exposed to tear gas and experienced multiple symptoms, including respiratory problems; skin rashes; hyperemia (redness due to inflammation) of the pharynx, eyes, and tympanic membranes; hoarseness; hearing loss; high blood pressure regardless of age; asthma attacks; and allergic reactions. Furthermore, 87 of the 169 individuals were injured by the impact of RCA canisters, which resulted in soft tissue trauma. Specific injuries included head trauma and/or lacerations, skull fractures, eye trauma, fractured bones around the eyes, loss of vision, and bone fractures of the extremities or ribs.[41]

The TMA and human rights monitors also documented how law enforcement officials attacked clearly identifiable, independent medical personnel and medical facilities with RCAs, water cannon, and rubber bullets. Police beat and detained dozens of physicians and other medical personnel for providing emergency medical care to those injured during the demonstrations. In an interview with PHR, Dr Arpat described her experiences: "[W]e reassured ourselves that even in war doctors are allowed to provide medical treatment, so why would the police raid us? But the police started to swarm into the building [where the emergency medical center was located]. They shot tear gas at the infirmary entrance and then entered. ... We were choking with tear gas. We were just wearing surgical masks. Two of our colleagues went out in white coats and identification showing they were doctors. They were beaten harshly and insulted."[42]

On 4 June 2013 at the request of the TMA, the WMA together the German Medical Association, the Standing Committee of European Doctors and PHR sent a letter to the Turkish Prime Minister strongly condemning "crowd control or riot prevention technology—such as tear gas and water cannons—that is misused, including to perpetrate human rights abuses or that is used in a manner out of proportion with the need, or against populations with particular vulnerabilities." The organisations called upon the Turkish Government to "Immediately end *all* use of tear gas in Turkey, as it has been systematically used as a weapon on a massive scale with devastating health consequences" The organisations also called upon the international community to: "Immediately ban the sale of tear gas to Turkey".[43]

Of course the misuse of riot control agents by law enforcement officials, particularly in large-scale public order situations, has not been restricted to Turkey, but has been documented by health professionals and human rights organisations in a growing number of countries in recent years.[‡‡] The potentially deleterious health consequences for those individuals and populations targeted by such chemical agents has led to calls, particularly from local and national health professional organisations (such as the TMA) whose members have treated the victims of such misuse, for the international medical community to urgently address this issue. A highly significant advance in this regard was the adoption by the WMA General Assembly, in October 2015, of a *Statement on Riot Control Agents* which recognised that the "inappropriate use of riot control agents risks the lives of those targeted and exposes people around, amounting to a potential breach of human rights standards, in particular the right to life, the right to freedom of expression and of peaceful assembly". Consequently, the WMA urged States to: use RCAs in a manner designed to minimise the risk of serious harm to individuals, and to prohibit their use in the presence of vulnerable populations, such as children, older people or pregnant women; insisted that riot control agents should never be used in enclosed spaces where chemical concentrations may reach dangerous levels, and where people cannot move away from areas with high concentrations of the agent; insisted that governments train police and other security forces in the safe and legal use of riot control agents, in order to minimise the risk of harm when they are deployed; insisted that States penalise individuals who misuse riot control agents and who deliberately endanger human life and safety by using the agents; called for unimpeded and protected access of healthcare personnel to allow them to fulfil their

[‡‡]For further discussion and illustrative case studies see ref. 52.

duty of attending to the injured; and recommended that, because of the significant difficulties and risks to health and life associated with the use of such riot control agents, that States should refrain from using them in any circumstances.[44]

19.4 Conclusions

Led by the World Medical Association, the international medical community has developed a framework of ethical standards that guide health professionals' conduct in relation to their patients, colleagues and wider society, in times of peace and in times of armed conflict, many of which directly or indirectly constrain their involvement in research, development, testing or use of weapons employing toxic chemicals. Of particular relevance are the WMA Regulations and Declarations specifically opposing chemical and biological weapons programmes; injunctions upon health professionals to prevent human rights abuse and breaches of international humanitarian law; and guidance strictly regulating medical and scientific research upon human subjects.

Despite the ground-breaking work by certain national medical associations, such as the BMA, there are no widely accepted international guidelines specifically determining the permissibility or non-permissibility of medical involvement in the development of certain weapons employing toxic chemicals for law enforcement purposes, most notably ICA weapons and RCAs. Given the potential importance of medical participation to the development, testing and utilization of such weapons, the introduction of clear guidance by the WMA and professional associations representing other health professionals, such as anaesthetists and medical toxicologists, constraining the involvement of health professionals in such activities, is needed.

References

1. *The Hippocratic Oath: Text, Translation, and Interpretation, by Ludwig Edelstein*, Baltimore, The Johns Hopkins Press, 1943, text available from Bioethics. A modern version of this oath was developed in the 1960s, and it has since been adopted by a number of medical associations. See: L. Lasagna, Modern Hippocratic Oath, 1964, School of Medicine, Tufts University, Both texts available from Bioethics, McMaster University Health Sciences, http://hsl.mcmaster.libguides.com/c.php?g=306726&p=2044095, accessed 30 June 2017.
2. The Oath of Maimonides, *12th Century, Text Available from Bioethics*, McMaster University Health Sciences, http://hsl.mcmaster.libguides.com/c.php?g=306726&p=2044095, accessed 30 June 2017.
3. World Medical Association website, *What Is the WMA?* http://www.wma.net/en/60about/, accessed 30 June 2017.

4. World Medical Association, *Declaration of Geneva, Adopted by the 2nd General Assembly of the World Medical Association, Geneva, Switzerland, September 1948 and editorially revised by the 173rd WMA Council Session, Divonne-les-Bains, France, May 2006*, See also the accompanying Physician's Oath, which was added to the Declaration of Geneva and adopted by the General Assembly of the World Medical Association in September 1948.
5. *WMA International Code of Medical Ethics*, Adopted by the 3rd General Assembly of the World Medical Association, London, England, October 1949 and amended by the 57th WMA General Assembly, Pilanesberg, South Africa, October 2006.
6. *History and Theory of Human Experimentation the Declaration of Helsinki and Modern Medical Ethics*, ed. U. Schmidt and A. Frewer Stuttgart, Franz Steiner Verlag, 2007, and in particular Schmidt's chapter, The Nuremberg Doctors' Trial; and the Nuremberg Code, pp. 71–116; G. Edwards and M. Grodin, *The Nazi Doctors and the Nuremberg Code: Human Rights in Human Experimentation*, Oxford University Press, USA, 1992; V. Spitz, *Doctors from Hell: The Horrific Account of Nazi Experiments on Humans*, Sentient Publications, 2005.
7. *Trials of War Criminals before the Nuremberg Military Tribunals under Control Council Law No. 10*, U.S. Government Printing Office, Washington, D.C., 1949, vol. 2, pp. 181–182.
8. World Medical Association, *Declaration of Helsinki, Ethical Principles for Medical Research Involving Human Subjects*, Adopted by the 18th WMA General Assembly, Helsinki, Finland, June 1964, regularly amended, most recent amended version 64th WMA General Assembly, Fortaleza, Brazil, October 2013, text available: http://www.wma.net/en/30publications/10policies/b3/, accessed 30 June 2017.
9. A. Maclean, *Historical Survey of the Porton Down Volunteer Programme*, Ministry of Defence, June 2006, pp. 121–122, 134–135, 138.
10. A. Maclean, *Historical Survey of the Porton Down Volunteer Programme*, Ministry of Defence, June 2006, pp. 138–9.
11. World Medical Association, *Regulations in Times of Armed Conflict and Other Situations of Violence*, Adopted by the 10th World Medical Assembly, Havana, Cuba, October 1956, last revised by the 63rd WMA General Assembly, Bangkok, Thailand, October 2012, paragraph 2.
12. M. Crowley, *Chemical Control*, Palgrave, Basingstoke, UK, November 2015, pp. 178–181.
13. *Protocol Additional to the Geneva Conventions of 12th August 1949, and Relating to the Protection of Victims of International Armed Conflicts [Protocol I]*, 8th June 1977, Article 35.
14. *The SIRUS Project, Towards a Determination of Which Weapons Cause "superfluous Injury or Unnecessary Suffering"*, ed. R. Coupland, ICRC, 1997, p. 7.
15. WMA, *Resolution on the SirUS Project: Towards a Determination of Which Weapons Cause 'Superfluous Injury or Unnecessary Suffering'*, Adopted by

the 50th World Medical Assembly Ottawa, Canada, October 1998 and rescinded and archived by the WMA General Assembly, Seoul, Korea, October 2008.
16. World Medical Association, *Declaration on Chemical and Biological Weapons*, adopted by the 42nd World Medical Assembly Rancho Mirage, CA, USA, October 1990 and rescinded at the WMA General Assembly, Santiago, 2005.
17. World Medical Association, *Washington Declaration on Biological Weapons*, adopted by the 53rd WMA General Assembly, Washington, DC, USA, October 2002 and editorially revised by the 164th WMA Council Session, Divonne-les-Bains, France, May 2003, paragraphs 18 and 19.
18. World Medical Association, *Resolution on the Prohibition of Chemical Weapons*, adopted by the 64th General Assembly, Fortaleza, Brazil, October 2013.
19. Health Professions Council of South Africa, *Guidelines for Good Practice in the Health Professions: Research, Development and Use of Chemical and Biological Weapons*, Pretoria, 2nd edn, booklet 9, May 2007.
20. Health Professions Council of South Africa, *Guidelines for Good Practice in the Health Professions: Research, Development and Use of Chemical and Biological Weapons*, Pretoria, 2nd edn, booklet 9, May 2007, paragraph 1.2.
21. Health Professions Council of South Africa, *Guidelines for Good Practice in the Health Professions: Research, Development and Use of Chemical and Biological Weapons*, Pretoria, 2nd edn, booklet 9, May 2007, paragraph 1.3.
22. Health Professions Council of South Africa, *Guidelines for Good Practice in the Health Professions: Research, Development and Use of Chemical and Biological Weapons*, Pretoria, 2nd edn, booklet 9, May 2007, paragraph 3.1.
23. Health Professions Council of South Africa, *Guidelines for Good Practice in the Health Professions: Research, Development and Use of Chemical and Biological Weapons*, Pretoria, 2nd edn, booklet 9, May 2007, paragraph 3.2; see paragraphs 3.3 and 3.4 for further elaboration.
24. Health Professions Council of South Africa, *Guidelines for Good Practice in the Health Professions: Research, Development and Use of Chemical and Biological Weapons*, Pretoria, 2nd edn, booklet 9, May 2007, Introduction: The Spirit of Professional Guidelines.
25. Health Professions Council of South Africa, *Guidelines for Good Practice in the Health Professions: Research, Development and Use of Chemical and Biological Weapons*, Pretoria, 2nd edn, booklet 9, May 2007, paragraph 5.
26. SABC News, *Basson's Application to Bar HPCSA Inquiry Dismissed*, 10th May 2010.
27. HSPCA Press Release, *HSPCA Applauds High Court Ruling on Basson*, 10th May 2010.
28. HPCSA, *Before the Professional Conduct Committee of the Medical and Dental Professions Board*, In the matter of Dr W. J. Basson, accused, Charge sheet and annexures, undated.

29. J. Smith, *Wouter Basson to Face South Africa Misconduct Hearing*, The Guardian, 27th January 2012.
30. HPCSA, *Before the Professional Conduct Committee of the Medical and Dental Professions Board*, In the matter of Dr W. J. Basson, accused, Charge sheet and annexures, undated, charges 2.2, 4.1–2, 5 & 6.
31. HPCSA, *Professional Conduct Committee: Concerning DrW. Basson*, Reasons of the Committee, Issued by the HPCSA, 18th December 2013.
32. HPCSA, *Dr Wouter Basson Found Guilty of Unprofessional Conduct*, http://www.hpcsa-blogs.co.za/dr-wouter-basson-found-guilty-of-unprofessional-conduct/, accessed 30 June 2017.
33. British Medical Association, *The Use of Drugs as Weapons: The Concerns and Responsibilities of Healthcare Professionals*, BMA, London, 2007, p. x.
34. British Medical Association, *The Use of Drugs as Weapons: The Concerns and Responsibilities of Healthcare Professionals*, BMA, London, 2007, p. 20.
35. British Medical Association, *The Use of Drugs as Weapons: The Concerns and Responsibilities of Healthcare Professionals*, BMA, London, 2007, p. 24.
36. *WMA Resolution on the Prohibition of Chemical Weapons*, adopted by the 64th General Assembly, Fortaleza, Brazil, October 2013, http://www.wma.net/en/30publications/10policies/c12/, accessed 30 June 2017.
37. For further discussion see: M. Crowley, *Chemical Control*, Palgrave, Basingstoke, UK, November 2015, pp. 39–88; OPCW, Scientific Advisory Board, Response to the Director-General's Request to Provide Consideration on which RCAs are subject to declaration under the CWC, SAB-25/WP.1, March 2017.
38. Physicians for Human Rights, *Contempt for Freedom: State Use of Tear Gas as a Weapon and Attacks on Medical Personnel in Turkey*, September 2013.
39. Amnesty International, *Gezi Park Protests: Brutal Denial of the Right to Peaceful Assembly in Turkey*, Amnesty International, EUR 44/022/2013, October 2013.
40. Turkish Medical Association, *Report on the Assessment of Health Problems in Persons Exposed to Chemical Riot Control Agents*, Ankara, September 2013.
41. Physicians for Human Rights, *Contempt for Freedom: State use of tear gas as a weapon and attacks on medical personnel in Turkey*, September 2013, p. 8; see also U. Unuvar *et al.*, Usage of Riot Control Agents and Other Methods Resulting in Physical and Psychological Injuries Sustained During Civil Unrest in Turkey in 2013, *J. Forensic Leg. Med.*, 2016, **45**, 47–52.
42. Physicians for Human Rights, *Contempt for Freedom: State Use of Tear Gas as a Weapon and Attacks on Medical Personnel in Turkey*, September 2013, p. 18.
43. End Excessive Force in Turkey, *World Medical Association*, 12 July 2013, http://www.wma.net/en/45blogs/01wilson/2013_pblog65/, accessed 30 June 2017.

44. *WMA Statement Riot Control Agents*, adopted by the 69th WMA General Assembly, Moscow, Russia, October 2015.
45. A. Maclean, *Historical Survey of the Porton Down Volunteer Programme*, Ministry of Defence, June 2006, part 4: Human studies with incapacitating agents, pp. 109–142.
46. J. Walker, *"Inappropriately Hilarious": UK and Chemical Incapacitating Chemical Agents, in: Britain and Disarmament, the UK and Nuclear, Biological and Chemical Weapons Arms Control and Programmes 1956–1975*, Routledge, London, 2012; U. Schmidt, *Secret Science*, Oxford University Press, Oxford, 2015.
47. *WMA Declaration of Tokyo - Guidelines for Physicians Concerning Torture and Other Cruel, Inhuman or Degrading Treatment or Punishment in Relation to Detention and Imprisonment*, adopted by the 29th World Medical Assembly, Tokyo, Japan, October 1975, and most recently Revised by the 67th WMA General Assembly, Taipei, Taiwan, October 2016.
48. D. Verchio, Just Say No! The SIrUS Project: Well-Intentioned, But Unnecessary and Superfluous, *Air Force Law Review*, 2001, vol. 51, (183).
49. Health Professions Council of South Africa, *Guidelines for Good Practice in the Health Professions: Research, Development and Use of Chemical and Biological Weapons*, Pretoria, 2nd edn, booklet 9, May 2007, paragraphs 3.3 and 3.4 for further elaboration.
50. *The Baha Mousa Public Inquiry Report*, UK House of Commons, 8 September 2011.
51. J. Riches, R. Read, R. Black, N. Cooper and C. Timperley, Analysis of Clothing and Urine from Moscow Theatre Siege Casualties Reveals Carfentanil and Remifentanil Use, *J. Anal. Toxicol.*, 2012, **36**, 647–656.
52. M. Crowley, *Chemical Control*, Palgrave, Basingstoke, UK, November 2015, pp. 50–80, 90–94.

CHAPTER 20

The Role of Civil Society in Combating the Development, Proliferation and Use of Chemical Weapons

MICHAEL CROWLEY*[a], CAITRIONA MCLEISH[b] AND JAMES REVILL[b]

[a]University of Bradford, Bradford, West Yorkshire, BD7 1DP, UK; [b]Harvard Sussex Program, SPRU, University of Sussex, Pavilion Room, Hastings Building, Falmer, Brighton, BN1 9QE, UK
*E-mail: m.j.a.crowley1@bradford.ac.uk

20.1 Introduction

Previous chapters in this publication examined State-centric arms control and disarmament instruments, and legal and regulatory measures intended to prevent the hostile misuse of the chemical and life sciences. It is clear from such analysis that many of these instruments and accompanying regimes have ambiguities, weaknesses and limitations which, to date, the relevant Member States have been unable or unwilling to address. Furthermore, a number of States have failed to introduce and operate effective national implementation measures, and there remain limitations in the efforts of States Parties and relevant organizations to challenge reported treaty violations by certain Member States. In the light of such failings, it is important

Preventing Chemical Weapons: Arms Control and Disarmament as the Sciences Converge
Edited by Michael Crowley, Malcolm Dando and Lijun Shang

Published by the Royal Society of Chemistry, www.rsc.org

to explore the roles that global civil society has played and could play in facilitating effective implementation by States of their obligations, most notably those enshrined in the Chemical Weapons Convention, as they seek to respond to the ongoing developments in the chemical and life sciences, as well as changes in the global security environment.

Previous chapters have examined the nurturing of "cultures of responsibility" amongst the medical, chemical and life science communities, particularly through awareness raising, education, and the development of codes of conduct. In contrast, this chapter will focus upon other potential roles that civil society, broadly understood, can play to complement and strengthen existing official implementation and verification mechanisms, including those facilitated by the Organisation for the Prevention of Chemical Weapons (the OPCW). It will also outline constraints faced by civil society, and present recommendations to facilitate more effective and enhanced civil society participation in the OPCW's mission to prevent the re-emergence of chemical weapons.

20.1.1 Scope and Nature of Civil Society Engagement

Global civil society has been considered as encompassing "[T]he sphere of ideas, values, institutions, organizations, networks and individuals located between the family, the state, and the market, and operating beyond the confines of national societies, polities, and economies."[1] In this chapter, it will be treated as a broad church encompassing individuals, non-governmental organisations, informal associations and loose coalitions, "forming ... connections across national borders and inserting themselves into a wide range of decision-making processes."[2]

Previous analysis of the principal civil society actors engaged in chemical weapons disarmament has highlighted the relatively limited numbers and types of organisations involved, and the largely western-centric nature of such entities.[3] Feakes has argued that such civil society involvement in chemical weapons disarmament could be considered as an "epistemic community" namely "a network of professionals with recognized expertise and competence in a particular domain and an authoritative claim to policy-relevant knowledge within that domain or issue-area".[3] Furthermore, unlike comparable issues (such as nuclear disarmament) where the epistemic community formed as just one element of a broader civil society engagement which included public movement campaigns, non-violent direct action groups and other actors; for chemical weapons it was argued that civil society involvement has been largely limited to the members of the epistemic community.

Whilst such analysis certainly appeared to reflect the principal civil society actors engaged with the OPCW both during the negotiation of the CWC and in the first decade of its operation, this situation may be changing. As discussed in Section 20.5, partly as a result of the emphasis of OPCW Director General Ahmet Üzümcü upon "public diplomacy", the outreach activities of the Technical Secretariat and OPCW more broadly, and the development of the Chemical

Weapons Convention Coalition (CWCC) there has been a steady increase in the numbers of actors—as well as a broadening of the geographical spread and diversification in the types of NGOs—becoming involved in this issue.

In addition, the OPCW-focussed framing of previous analysis may underestimate both the range of engaged civil society actors and also the diverse nature of their activities that contribute to chemical weapons disarmament and the prevention of the re-emergence of such weapons. Accordingly, the following sections attempt to outline a wider range of actors and activities that have been undertaken—including the work of those documenting chemical attacks and giving voice to the victims—that have been under-represented in some accounts on this issue.

20.2 Societal Monitoring and Verification

The concept of "societal verification" has been discussed in the peace and disarmament literature for many years—particularly in relation to nuclear weapons—under terms such as "citizens reporting" or "social monitoring".[95] Although there is no agreed formal definition, Deiseroth[4] has described societal verification as:

> "connot[ting] the involvement of civil society in monitoring national compliance with, and overall implementation of, international treaties or agreements. One important element is citizens' reporting of violations or attempted violations of agreements by their own government or others in their own country."

Although the original vision of a global societal verification network—involving large numbers of civil society actors resident in all States party to relevant treaties—appears idealistic, if not utopian, in the near to medium term, the idea remains a powerful one. An adapted form of societal monitoring and verification could be envisaged, comprised of a smaller number of activist researchers, who have access to the relevant technical expertise and can, at the very least, undertake open source monitoring and analysis of the activities of relevant State and non-State actors. Due to resources, personnel, political and security constraints, such groups are likely to be limited in terms of the countries from which they can operate and, consequently, the quantity and quality of information they are able to receive, particularly in less accessible regions and from closed or semi-closed authoritarian countries. Despite such constraints, variant forms of such limited societal monitoring are presently carried out for certain environmental and human rights agreements, and have been taken up by a number of arms control and human security communities to monitor agreements prohibiting landmines and cluster munitions, and regulate small arms and light weapons.[†] Societal monitoring regarding weapons employing toxic chemicals appears,

[†]For further discussion, see ref. 96.

as of March 2018, to be at a rudimentary level, with *ad hoc*, limited and uncoordinated research activities being carried out by a small number of researchers. Some illustrative examples of research methodologies that have been or could be utilized for such work are explored in the following sections.

20.2.1 Open Source Monitoring and Analysis

20.2.1.1 Introduction

Historically civil society has played a valuable role in providing information and insight into chemical weapons investigations, in some cases alerting the world to chemical weapons development and use, in others challenging the veracity of government allegations of such activities. With advances in—and the democratisation of—communications technologies, such as social media, this role has the potential to increase.

20.2.1.2 The 'Yellow Rain' Case

In 1978, following press reports, US embassy staff in Bangkok, Thailand, began to examine alleged chemical weapons use by communist Pathet Lao and Vietnamese forces against Hmong villagers. Subsequently, from 1979 to 1982, an *ad hoc* investigation by US officials and a physician, supported by NGO volunteers, collected biomedical and environmental samples, including samples of what the Hmong claimed was "poison from the sky", and administered questionnaires to Hmong refugees in Thai camps. Following the detection of trichothecene mycotoxins in one of the environmental samples, in 1981 the US accused the Soviet-backed Laotian and Vietnamese forces of using 'lethal chemical weapons'.[5]

In 1982, a US Special National Intelligence Estimate (SNIE) based on "the results of an intensive review by the US intelligence community of all available evidence" including systematic interviews and "all the physical evidence available to the US", stated that "... there is no doubt that the deaths and casualties have resulted from chemical attacks ... [the] ... hypothesis that best fits all the evidence is that the trichothecene toxins were developed in the Soviet Union".[6]

However, whilst the allegations were "pressed vigorously by the US Government" over the course of the 1980s, the claims were challenged by academic scientists from civil society, such as Professor Matthew Meselson, whose ground-breaking investigation developed a counter-hypothesis to account for the so-called "yellow rain" phenomena. Based on open source monitoring and analysis of US government statements, laboratory analysis of collected samples, field study in Thailand and engagement with entomologists, Meselson made a compelling counter argument that the substance "the Hmong had frequently mistaken as poison from the sky and which [the US government] asserted was a chemical warfare agent, was, in fact, the faeces of Asian honey-bees".[7] Meselson and Robinson challenged the US claim on the grounds that it "failed to employ reliable methods of witness interrogation or of forensic laboratory investigation; it was further marred by the dismissal and withholding of contrary evidence and a lack of independent review".[7] Although the initial allegations dated from the late-1970s, this controversy remained unresolved decades later, with some, such as Katz, arguing as late as 2008 that the case was an act of chemical warfare supported by the Soviet Union,[8] although this claim became increasingly contested.

The role of challenging allegations of association with chemical and biological weapons on a scientific and technical basis, and where feasible, presenting reasoned counter-hypotheses, remains an important civil society function, not least because of the powerful place of chemical weapons in propaganda campaigns and (dis-) information warfare.[9]

This case also exemplifies how civil society can provide both "important information regarding allegations of CBW use" and "valuable expertise",[8] for instance, as Katz notes, "when medical information and background data on the health of the population must be collected and culturally competent professionals are needed to make interviews with alleged victims more comprehensive and reliable".[8]

The third role of civil society identifiable from this episode lies in shaping the practices of subsequent official investigations of alleged chemical weapons use. Sir David Omand, former UK Security and Intelligence Coordinator, stated in his article with Michael Goodman, *Teaching Intelligence Analysts in the UK*, published in a CIA journal, that the 'yellow rain' affair was "a well-documented case that illustrates the pitfalls" of verifying allegations of CBW use.[10] Such pitfalls, Goodman and Omand highlighted, included a lack of scientific rigour in the analysis of samples, and a failure to undertake wider consultations to explore different hypotheses. As a consequence of the 'yellow rain' case, Goodman and Omand claimed that changes to intelligence analysis of chemical incidents were subsequently developed.

Although the 'yellow rain' case illustrates a range of potential roles that have been played by civil society, it also highlights the difficulties faced by such actors. For example, the collection of samples for others to assess, an activity undertaken by many civil society actors, can become a politicised activity with potentially significant consequences; whereas the provision of culturally relevant expertise, is not something anyone, no matter how motivated, can undertake. Finally, the work of Meselson and others illustrates how mounting a credible and persistent challenge to allegations made by a State requires significant expertise, diligence and perseverance, not to mention more than a little bravery to confront official narratives.

20.2.1.3 Civil Society Reporting and Analysis of Alleged Use in Syria[‡]

Advances in technology have opened up new channels and new opportunities for civil society actors to engage in the reporting and analysis of alleged CW use. As a result, who, and what, might shape our basic knowledge of CW

[‡]For a more detailed discussion, see ref. 97–99.

allegations has evolved to now include citizens using mobile phones with cameras uploading images to online platforms such as Twitter, Facebook and YouTube.

The on-going conflict in Syria is one such place where this has occurred. Throughout the conflict, all sides have relied on internet-based technology platforms to organise, communicate and influence public opinion to the extent that it has been called the most "socially mediated civil conflict in history."[11] For those activists opposed to the regime, such as Rami Jarrah who is credited as being one of the first to pick up a camera, a reason for taking this action was because they had:

> "...expected that Al Jazeera would go down and film because that's what we saw them do in Egypt, and when we realized they wouldn't, a lot of people decided to film and upload these images themselves and send them to channels like France 24 and BBC Arabic."[12]

Increasingly supported,[12] including with technology donated from foreign governments, and becoming more organised, these citizen reporters became powerful figures in shaping what was 'understood' to be happening within Syria. This became especially so after foreign journalists were expelled from Syria in March 2011, leading, for example, to the BBC changing their copyright policy as they became increasingly reliant upon 'user generated content' from citizen reporters in Syria.[13]

20.2.1.3.1 Civil Society Reporting of Alleged CW Attacks. Allegations of CW use began appearing as early as April 2012[100] with the number of allegations increasing throughout 2013. These reports came in mixed formats – reports from non-governmental organisations,[101,102] press interviews with activists and defectors,[103,104] as well as the posting online of 'on the ground' videos. The very early allegations were set against a backdrop of increasing international concern about known Syrian government CW stockpiles, which had led US President Obama to issue his 'redline' speeches during 2012.[§] In her work, Storck[14] notes that after the first of these speeches, videos and images documenting alleged CW attacks poured out of the country including from four alleged 2013 attacks which were later investigated by the UN: Khan Al-Asal (Aleppo) and Uteibah (Damascus) on March 19th; Sheikh Maqsood neighbourhood (Aleppo) on April 13th; and at Saraqib (Idlib) on April 29th.[15]

The Sarin attack on August 21st 2013 on the Damascus suburb of Ghouta was different however, both in terms of the casualty rates being higher and also the volume of citizen reporting that ensued. News that 'something' involving toxic chemicals was happening was first broadcast on the Facebook and Twitter accounts of Local Coordinating Committees at 02:26 with the first visuals posted at 02:50.[16] Many hundreds more visuals were uploaded over the next twenty four hours.

[§]The two 'red line' speeches can be found in ref. 105 and 106.

Within hours of the first postings, the news that 'something' had happened in the Ghouta district had gone viral. US Secretary of State John Kerry later noted that: "just 90 minutes [after the first reports of an attack] all hell broke loose in the social media".[17]

20.2.1.3.2 Civil Society Analysis of Alleged CW Use. Using social media platforms to broadcast news from within Syria permitted commentary and analysis to be performed by interested observers from all over the world. In respect to analysing CW allegations, the chosen broadcasting platforms permitted new civil society experts to emerge and new methodologies to be applied.

One such new expert was British blogger, Eliot Higgins. Higgins quickly became considered to be an authoritative voice about the weapons being used in the Syrian conflict[18] with his almost daily posts on his blog *Brown Moses* seen as "an indispensable reference for a devoted international corps of Syria watchers…"[19] Scouring hundreds of YouTube channels[107] Higgins often worked collaboratively, "frankly acknowledge[ing] his amateur status",[18] and applied a crowdsourcing methodology to assist in his analysis.[18]

One striking feature of the analysis of uploaded material was the speed with which those not belonging to governments and international organisations were able to produce their reports. Higgins, for example, began his analysis of the Ghouta attacks on August 22nd and was able to post an analysis of videos on September 2nd. As was the norm of his working pattern, his analysis included input from specialist experts such as Nic Jenzen-Jones of *Rogueadventurer.com.*[20] In the September 2nd analysis, Higgins suggested that two types of munitions had been used during the August 21st attacks;[20] fourteen days later on September 16th official confirmation that these munitions had indeed been used during the attack came from the UN investigation team.[21]

In order to perform analysis based on online material—especially image-based material—verification methodologies were devised to check, amongst other things, provenance, source identity and location. Open platforms such as the *Syrian Civil War* subreddit and bloggers such as Higgins actively 'used the crowd' to verify authenticity. This, Matthew Ingram notes "isn't really all that new… It's just that social technologies…allow us to engage in this kind of shared decision-making process on a much larger and broader scale, and they allow us to do it faster as well."[22] As well as crowd sourcing, verification methodologies for images have been developed. The Dublin-based news agency *Storyful* for example, would verify imagery by "forensic frame-by-frame investigations and use other publicly available material to corroborate and confirm key details"[23] such as shadows to confirm time of day, weather, speakers' accents, and landmarks to compare with maps such as Wikimapia, a crowdsourced map of the world through satellite imagery, and GoogleEarth.[24]

Such methodologies enabled the debunking of claims from those supportive of the regime that YouTube videos about the Ghouta attacks were false, for example because some were time stamped August 20th.[25] As explained by Robert Mackay writing for *The Lede*, "[Y]ouTube has previously confirmed ... its computers automatically assign a date to each video based on the current time in California when the upload begins..."[26] and Storyful's[26] meticulous verification methodologies enabled them to say with authority that:

> "At Storyful, we extract the precise date and time of upload from YouTube videos as we add them to our system. The first video we have seen relating to the massacre was uploaded at 2:50:47 am on August 21 local time in Syria. This is approximately 20 minutes after the attacks are reported by eyewitnesses to have happened. The video was uploaded by a YouTube account active in the Erbin area, an uploader Storyful is familiar with and whose videos have been verified in the past. Very quickly, other videos and photos documenting similar scenes were uploaded to YouTube and other social media sites."

Although the false video thesis was debunked, it serves to remind us that the same online information can be used to support multiple hypotheses and that civil society is made up of diverse, sometime contradictory voices.

20.2.2 Evidence Derived from Field Missions

As illustrated in the following three case studies, independent scientists, health professionals and NGOs can sometimes collect their own information, first hand, from onsite investigations, or may be able to utilize information (*e.g.* witness testimony) and analyse materials (*e.g.*, munition shell fragments, clothing, blood, soil samples, *etc.*) obtained from other civil society actors operating in the field (*e.g.* journalists, national NGOs, *etc.*). There are several potential constraints upon such evidence gathering, including access, logistics and translation; safety considerations for researchers and witnesses; difficulties ensuring chain of custody of the evidence, preventing cross-contamination as well as establishing the representativeness of the information obtained. Despite such constraints, material and testimony collected during field missions can provide information that could not be obtained by any other means, potentially allowing identification of toxic chemical agents and associated delivery mechanisms utilized by State or non-State actors, or, conversely, casting doubt on allegations of such use.

In certain situations, civil society investigations may be able to gain access to locations that are 'out of bounds' to formal inter-governmental verification mechanisms, due to mandate constraints, political sensitivities, organisational prioritisation, or the refusal of the target country to cooperate with the international organisation responsible for carrying out the investigation.

20.2.2.1 Independent Investigations of the Use of Chemical Weapons Against the Kurdish Population

In September 1988, following reports from Kurdish NGOs and journalists alleging that Iraqi State planes had bombed a number of villages in the Kurdish part of Iraq with chemical weapons, ten UN Member States formally requested that the UN Secretary General initiate an investigation into the alleged attacks. However, the Iraqi government denied the UN Secretary General's request for access and stated that "[T]his is a question of sovereignty..."[27] A request, for access to Kurdish areas in Turkey, was also reportedly denied by the Turkish government.[28] Nevertheless, certain NGOs managed to conduct field missions in Kurdish areas in Turkey and Iraq, and in October 1988, a team of three medical doctors representing the US NGO Physicians for Human Rights (PHR) succeeded in gaining access to two Kurdish refugee camps. The PHR investigators gathered evidence using a standardised questionnaire survey; physical examinations of those suffering from the effects of the chemical weapons; videotaped in-depth interviews; and also undertook informal discussions with hundreds of camp residents, government officials, journalists, diplomats, and other observers to corroborate accounts of chemical weapons attacks.[28] The investigation team concluded that: "[E]yewitness accounts, the pattern of symptoms reported and physical evidence obtained by others point to the use of lethal poison gases including, but not restricted to sulphur mustard (Yperite)."[29]

In May–June 1992, Human Rights Watch working together with PHR sent a forensic investigation team into Iraqi Kurdistan. In the destroyed village of Birjinni the team took soil samples from craters where chemical weapons munitions were reported to have impacted. In 1993, the UK Chemical & Biological Defence Establishment (CBDE) at Porton Down analysed these samples, reportedly discovering traces of mustard gas and sarin, as well as their degradation products.[30] According to Human Rights Watch: "[T]his was the first instance of a chemical weapons attack being proved on the basis of chemical residues left behind at the impact site."[30]

As well as providing evidence of the nature and scale of the attack and identifying the chemical weapons employed, civil society has also been at the forefront of subsequent documenting and publicising of the immediate and long-term human cost of these attacks, and has lobbied for effective medical care to be provided to the survivors. It has been estimated that approximately 5000–7000 people out of the total population in Halabja of 80 000 died as immediate casualties of the attack and a further 30 000–40 000 were injured, many severely.[31] In 1998, an independent UK medical research scientist—Dr Christine Gosden—working together with Iraqi medical practitioners initiated the first detailed on-site investigation of the long-term health effects upon the citizens of Halabja of the Iraqi chemical weapons attack that had taken place 11 years previously.[32] In testimony before a US Senate Select Committee, Dr Gosden described being "shocked by the devastating effects of these weapons which have caused problems such as cancers, blindness and congenital malformations."[32] Dr Gosden also described how the "enormities of the health problems facing the population of Northern Iraq" were "magnified by an appalling lack of medical resources and infrastructure. Despite the fact that they were attacked eleven years ago, the survivors have received minimal, if any, humanitarian assistance."[32]

20.2.2.2 *Identification of a New Riot Control Agent Used in the West Bank*

In July 2005, the Israeli Army reportedly employed a new riot control agent (RCA) against Palestinian and Israeli civilians protesting about the erection of the 'Separation Wall' on the West Bank. The RCA, which was delivered *via* small plastic projectiles fired from launchers, caused severe skin injuries in the targets. The Israeli authorities subsequently refused to identify the RCA employed.[33]

Scientists based in the United Kingdom, led by Professor Alastair Hay, obtained one of the RCA projectiles utilized. They extracted the enclosed RCA, which was subsequently subjected to physical and chemical analysis at two independent UK laboratories,[¶,34] and was found to contain capsaicin with an inert carrier and a dispersal agent. Comparing the resulting data against the material safety data sheets of commercially available RCA ammunition, the results were found to correspond with "Pepperball Tactical Powder" marketed by the US company Pepperball Technologies, Inc.[33]

The UK researchers also obtained signed affidavits from Palestinians affected by exposure to the RCA. One individual (whose skin injuries were documented and photographed) was participating in a peaceful demonstration against the Separation Wall on 28 April 2005 when he was "shot at from a distance of not more than three metres". He described the effect of the RCA projectile as "being like an electric shock" and of becoming "extremely agitated". The authors noted that, although "[P]epperball tactical powder pellets are clearly designed to irritate the skin", the skin injuries that occurred in the two individuals detailed in their paper were "far more severe than the effects that material safety sheets for the product suggest may occur".[34] Such research clearly has utility in alerting clinicians (and future demonstrators) to the nature and effects of new weapons employing toxic chemicals they may have to face and/or treat in the future. As the research methodology described can be employed to identify specific RCAs and other toxic chemicals and associated means of delivery, it can also help to indicate the potential manufacturers and suppliers of such weapons.

[¶]The first laboratory was Analytichem, Solihull, West Midlands and the second was the Laboratory of the Government Chemist, Teddington, Middlesex.

20.2.2.3 *Field Mission Documents the Widespread Misuse of Riot Control Agents in Bahrain*

In April 2012, a two-person team comprising a medical doctor and lawyer from Physicians for Human Rights, conducted field research in Bahrain. The mission team investigated reported violations of medical neutrality and excessive use of force—particularly widespread and serious misuse of RCAs—by law enforcement officials alleged to have taken place in Bahrain during the previous year.[35]

The team conducted a total of 102 semi-structured interviews with Bahrainis who reported human rights violations, eyewitnesses to these alleged events, civil society leaders and also Government officials.[35] As well as locating and independently interviewing corroborating witnesses to certain alleged events, PHR conducted or utilized the following means to establish the veracity of witness testimony: (a) physical examinations by the PHR physician of victims of reported human rights violations; (b) evaluation of medical records of victims of reported human rights violations; (c) visual verification of locations of reported human rights violations; and (d) review of photographic, radiographic and video evidence. This on-sight evidence-gathering was subsequently analysed and supplemented by further extensive desk research employing medical and scientific literature databases.[36]

The PHR report highlighted the failure of Bahrani law enforcement officials to exercise restraint before resorting to force; their employment of disproportionate force when responding to protesters; and their failure to minimize damage and injury to demonstrators.[37] The PHR report specifically detailed numerous cases of misuse of tear gas to "methodically ... attack Shi'a civilians inside their homes and cars", which, it argued, constituted the "transformation of toxic chemical agents into weapons". PHR contended that "such unprovoked and flagrant assaults on families—who pose no threat to the safety of others—flout international human rights law and constitute[d] torture, cruel, and inhuman treatment".[38] The report also warned that the "persistent targeting" by law enforcement officials of civilians with RCAs in enclosed spaces "may lead to serious long-term health consequences", which could include "miscarriages, severe respiratory distress resulting in premature death, and a projected rise in asthma among the population".[39] The findings of the PHR field mission and subsequent desk-based study of RCA-related deaths were widely covered in the international media[40] and subsequently cited by the US State Department.[41]

The three case studies further illustrate the importance of field-based evidence-gathering, and of objectively recording and independently corroborating witness testimonies, supported by medical and other documentary evidence, and highlight the critical role that medical professionals, chemists and life scientists in particular can play in such documentation and in analysing evidence obtained through such research missions.

It is however, important to recognise the potential limitations of civil society investigations. The investigations are likely to be restricted in scope due to time, resource and access constraints—to both the chemical attack sites and to the victims—and may record only a snapshot or part of the picture. They may subsequently produce a limited or unrepresentative report of what happened due to research limitations or even organisational bias. Even where the investigations are conducted with rigour and the results are authoritative, they will likely be dismissed by the alleged perpetrators (and their allies) who may subsequently attempt to discredit the investigation and its finding amongst the international community as a whole, something visible in the current Syrian conflict.

Until very recently, the findings of such investigations were unlikely to be acknowledged or directly fed into relevant UN, OPCW or other inter-governmental investigation or verification processes, due to constraints upon the range of information relevant regimes could formally recognise. There are however indications that this situation may be changing with for example the OPCW–UN Joint Investigative Mechanism (JIM) operating practices explicitly including open source information amongst the data it actively sought and reviewed.[108]

20.2.3 Employment of National and International Reporting/Transparency Mechanisms

In the digital information age there are a number of open sources of information that can inform an assessment of a State's activities by civil society. As the previous sections have indicated, social media can play a significant role in the circulation of information on events of allegations, in addition to which satellite data can be used to monitor some facilities, trade databases provide a degree of transparency as to what chemicals or groups of chemicals are being imported, and publications databases can help in uncovering areas of research being undertaken.

However, there are also more traditional approaches to generating transparency that are available in some countries. One example is the freedom of information laws that have been passed by a number of States. In the case of certain Western States such laws have been used by civil society groups, such as the now defunct Sunshine Project, to acquire information on US (and other countries') governmental activities related to biological and chemical weapons.

20.2.3.1 The Sunshine Project

Established in 1999, the Sunshine Project was an NGO dedicated to upholding prohibitions against chemical and biological weapons and preventing the military abuse of related technologies. Based in Hamburg, Germany, and Texas, US, the project's focus was primarily on biological weapons; however, they played a significant role in efforts "to document and contain development of 'non-lethal' biochemical weapons",[42] exposing US efforts to rekindle research on such weapons and challenging such developments, in part through the use of hundreds of Freedom of Information Act (FOIA) requests to the US government.

The Sunshine project was a different—radical—organisation when compared with typical civil society organisations working on issues around chemical disarmament, and its publications were unflinchingly forthright in both content and tone. For example, one book chapter on "*The Pentagon's new Biochemical Warriors*" that was authored by the Sunshine Project[43] concluded by calling upon:

> "...the US Congress to investigate the [Joint Non-Lethal Weapons Directorate] JNLWD's arms control violations, to conduct public hearings, to hold the JNLWD and its superiors responsible for their actions, to freeze all ... funding... The Sunshine Project requests the [OPCW] sends a UN weapons inspection team to the US to investigate."

Such an approach may have reflected the founders' backgrounds in "public advocacy and grass-roots networking" and the Project's attempts to "re-frame CB disarmament so that it resonates more widely".[3] The project also exhibited a remarkable degree of persistence in efforts to acquire documents using the FOIA which led to the release of significant chunks of information on contracts, technical reports, presentations and reviews which formed part of an online clearing house of documents and which enabled the Sunshine Project to, *inter alia*, trace the evolution of then recent 'non-lethal' chemical weapons research in the US.[109]

Unfortunately, the Sunshine Project's operations were suspended in early 2008 due to a lack of funding. However, the legacy of their work lives on, firstly in the reports of subsequent scholars that have drawn upon the wealth of (previously unknown and hidden) materials uncovered by the Sunshine Project, and secondly through the work of NGOs and other researchers now utilising and adapting the Project's innovative methodology to acquire documents and uncompromisingly challenge governments over activities of concern.

20.2.3.2 The Bioweapons Prevention Project (BWPP)

A comparatively more conservative approach to national reporting is evident in the work of the Bioweapons Prevention Project (BWPP) which was supported and funded by the governments of Germany, Norway and Switzerland. Through the *BioWeapons Monitor*, an annual document produced by the BWPP, the project originally sought "to help monitor compliance with the international norm". It drew from the work of the International Campaign to Ban Landmines and the Cluster Munitions Coalition, specifically the development of the Landmine and Cluster Munitions Monitor, and represented an effort by civil society[44] "to hold governments accountable for their obligations to eliminate biological weapons permanently..."

Although it was primarily focused on biological weapons, many components of the *BioWeapons Monitor* had a bearing on chemical weapons related-issues with information on updates to national legislation and research activities reflecting the ongoing convergence of these technologies. Furthermore, the 2011 edition of the *Monitor* specifically included an analysis of eight countries activities related to "(bio) chemical non-lethal weapons"[45] as well as delving into political approaches to herbicides and defoliants in certain countries.

The BWPP *Bioweapons Monitor* incorporated national studies conducted by in-country researchers using open source data along with the active acquisition of information from respective government departments and facilities. By drawing on a wide range of sources, including government documentation, parliamentary debates, trade data, publications databases, and satellite data, the monitor sought to "complement BWC States Parties' reporting requirements under the BWC Confidence-Building-Measures (CBMs)" and provide "critical but constructive" analysis.[45]

The Sunshine Project and the BWPP, whilst different in their approach, suggest that one of the potential roles that civil society can play is through close examination and questioning of State policies and activities that might indicate the development, stockpiling or use of chemical weapons. In such work, monitoring of open source materials can be significantly strengthened by analysing national and international reporting and transparency mechanisms such as government reports to the BWC Implementation Support Unit (ISU) or the OPCW, briefings to parliament, responses to parliamentary questions, and by employing freedom of information requests.

20.2.4 Targeted Science and Technology Tracking

In addition to independent scientific involvement in the OPCW's five-yearly global reviews of S&T trends (examined in Chapter 17 of this publication), a range of academic institutes and University based entities such as the Biochemical Security 2030 Project, the Harvard Sussex Program (HSP), the Bradford Non-Lethal Weapons Research Project (BNLWRP) and the Hamburg Research Group have undertaken more narrowly focused research projects and promoted their findings directly to the CWC States Parties and the OPCW policy making organs on a more frequent and *ad hoc* basis.[‖] Such

[‖]Furthermore, certain scientific bodies have undertaken broader scope investigations. For example, in 2011 the Royal Society, the UK's national scientific academy, conducted the Brain Waves project which investigated developments in neuroscience and their implications for society and public policy. A major component of this work explored the implications for the CWC of the convergent trends in relevant science and technology, including those relating to incapacitating chemical agent weapons, synthetic biology, and nanotechnology. The Brain Wave reports were consequently disseminated amongst CWC States Parties and relevant OPCW structures, and the Royal Society also engaged in high-level advocacy to promote its findings and policy recommendations. For further details, see ref. 110.

targeted and in-depth research can potentially uncover contemporary scientific research activities of concern, identify the specific actors involved, and raise awareness of particular issues with the OPCW, the international arms control and scientific communities, and the wider public. In so doing, it can help inform and energise ongoing discussions amongst CWC States Parties as to the regulation of relevant science and technology, as illustrated by the case study below.

20.2.4.1 Open Source Investigation of Dual Use Research Potentially Applicable to Incapacitating Chemical Agent Weapons Development

In October 2002, Russian security forces employed a centrally acting incapacitating chemical agent (ICA) weapon in their attempt to save 900 hostages held in a Moscow theatre by armed Chechen fighters. Although the bulk of the hostages were freed, more than 120 were killed by the chemical agent and many more suffered long-term health problems. Russian authorities refused to publicly disclose the weapon they employed or provide details of the nature and levels of ICA weapons they may have developed or stockpiled.** Subsequently, concerns grew that other States may be seeking to exploit advances in neuroscience, medicinal pharmacology, synthetic biology and other relevant disciplines to develop such weapons.

Throughout 2014, the Bradford Non-Lethal Weapons Research Project (BNLWRP) conducted a study of contemporary dual-use chemical and life science research of potential relevance to the development of ICAs and associated means of delivery.†† The BNWLRP uncovered new evidence of continued relevant research by Russian scientists, including computer modelling of the distribution and movement of aerosolised incapacitating chemical agents in enclosed spaces,[113] and studies into the structure and nature of opiate receptors and their interaction with potential ICAs.[114]

The BNLWRP report also brought to light details of the development and marketing by Chinese companies[115] of ICA weapons employing an unknown anaesthetic agent specifically promoted for use by security forces against individuals (Figure 20.1), and the possession of such weapons by the Chinese Peoples' Liberation Army.[46] The BNLWRP highlighted previous research conducted by Israel into ICAs and the employment by Israeli Security Services in 1997 of an ICA weapon used on at least one occasion as an assassination tool.[47]

**Trace analysis subsequently undertaken by researchers from the UK Defence Science and Technology Laboratory (DSTL) at Porton Down of extracts of clothing and urine from survivors of the Moscow theatre siege indicated that the ICA weapon comprised a mixture of two anaesthetics, carfentanil and remifentanil, for further information see ref. 111.

††For further information about the survey methodology and the survey findings incorporating detailed country case studies, see ref. 112.

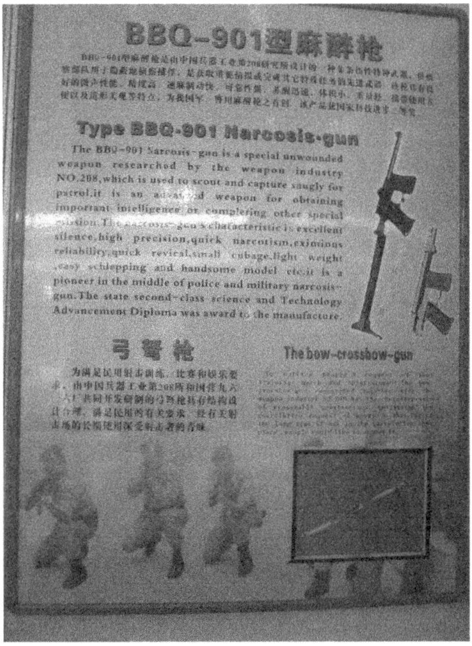

Figure 20.1 Poster for "BBQ-901 narcosis gun" on display on State 9616 Plant stand at. Asia Pacific China Police Expo 2006, Beijing, China, 24th–27th May 2006. © RobinBallantyne/Omega Research Foundation.

The BNLWRP also examined potentially relevant dual-use chemical and life-science research conducted since 1997 in the Czech Republic, India, Iran, the United Kingdom, and the United States.

Whilst China, Israel and Russia failed to respond to the BNLWRP information requests, the majority of States approached engaged in substantive correspondence providing clarification of the nature and purpose of research of potential concern. In addition, such engagement by the BNLWRP facilitated the annunciation of national positions constraining or prohibiting the development and use of ICA weapons and encouraged national promulgation activities in certain States. For example, the BNLWRP uncovered papers published by Czech scientists from 2005–2007, describing investigations relating to a range of pharmaceutical chemicals including opioids, ketamine, medetomidine and midazolam; which highlighted their potential utility as, in their own words, "pharmacological non-lethal weapons".[116] Research into such chemicals continued till at least 2012, but additional papers contained no explicit reference to their application as pharmacological non-lethal weapons.[117] The Czech Republic CWC National Authority subsequently investigated Czech research activities in this area and in 2014 correspondence with the BNLWRP stated that "[T]here was no connection [between] the research [and] creation of any sort of weapons or devices which could be used for military or police purposes."[48] With regard to the Czech pharmacological non-lethal weapons papers, the National Authority stated that "research programmes had justifiable medical goals, but their reporting in public media exceeded actual results of the research thus creating a false impression of possible development of some sort of chemical weapons."[49] Furthermore, the Czech National Authority confirmed that Czech CWC national implementation legislation "prohibits [ICA] use for law enforcement." The Czech National Authority also informed the BNLWRP of subsequent awareness-raising activities that they had undertaken including a series of presentations to the Ministries of Health, Defence and the Interior informing them of their obligations under the CWC and national implementation mechanisms.[49]

20.3 Promotion and Protection of Whistle-blowing

Any serious attempt by State or non-State actors to develop new or existing chemical or biological weapons on a significant scale would likely require the involvement of an array of scientists, engineers, technicians and other ancillary workers. Whilst some such staff are essential to the development and production of such weapons, they are also potentially capable of 'blowing the whistle' on these weapons programmes through public denunciations, leaking information to journalists, or by reporting concerns about potential or realized breaches of national regulations or violations of international treaties directly to the relevant national or international regulatory bodies. In his 1995 Nobel Peace Prize acceptance speech, Prof Joseph Rotblat[50] stated that:

> "The purpose of some government or industrial research is sometimes concealed, and misleading information is presented to the public. It should be the duty of scientists to expose such malfeasance. 'Whistleblowing' should become part of the scientist's ethos. This may bring reprisals; a price to be paid for one's convictions. The price may be very heavy."

Indeed, a number of scientists and technicians have suffered for alerting the world to clandestine State weapons programmes. This was certainly the case for Dr Vil Mirzayanov, the Russian chemist who was imprisoned in 1993 after publishing articles that revealed details of the Soviet Union's chemical weapons programme, including development of "novichock" binary chemical weapons, which continued despite the 1990 US–Soviet Accord to halt the production of chemical weapons.[51‡‡] If it is the duty of individual scientists to make known their concerns about the misuse of scientific research for activities that breach ethical standards or international law, then it is the responsibility of the scientific community as a whole to ensure that such whistle-blowers are fully protected. This was explicitly recognised by the

‡‡Concerns regarding such "novichok" weapons resurfaced following their apparent use in March 2018 in the attempted assassination of a former Russian spy then living in the UK.

ICRC in its 2004 statement, which declared that: "[T]hose working in life sciences who voice concern and take responsible action require and deserve political and professional support and protection",[52] and the corresponding action point, which called on all stakeholders in the life sciences to "ensure that adequate mechanisms exist for voicing such concerns without fear of retribution".[52]

Although certain States, such as South Africa,[53] the United Kingdom,[54] and the United States,[55] have legislation safe-guarding whistle-blowing activities on their statute books, the vast majority do not; and even where such legislation exists, its effectiveness and implementation has been variable. Martin, who has long experience of working with, and seeking to protect, whistle-blowers in many different spheres, has argued[56] that "the track record of whistle-blower protection measures—whistle-blower laws, hot-lines, ombudsmen and the like—is abysmal", and correspondingly recommended a more organic and grass-roots approach to the problem[56] in which whistle-blowing is better prepared for, including by "... gathering information, recruiting allies, developing skills and planning a course of action."

A fairly recent, albeit controversial, development has been the establishment of internet websites that host documents submitted by "whistle-blowers". The most notable of these is Wikileaks[57] which was established in 2007 as a non-profit media organization that publishes "material of ethical, political and historical significance while keeping the identity of our sources anonymous, thus providing a universal way for the revealing of suppressed and censored injustices."[58] It accepts "classified, censored or otherwise restricted material of political, diplomatic or ethical significance" and has made publicly available various documents related to chemical weapons.[59] Since it was established, Wikileaks has been severely criticised most notably by numerous States who have accused it of bias, and of publicising information that could endanger national security.

20.4 Track Two Scientist-to-scientist Relations

20.4.1 Introduction

Diplomatic scholarship typically divides negotiation into three distinct kinds of tracks: track one negotiations, in which official representatives of the parties participate; track two negotiations, in which unofficial but influential participants participate; and track three negotiations, which comprise grass-roots meetings involving people from both sides.[60] Within efforts to combat the development, proliferation and use of chemical weapons, the Pugwash Chemical Warfare Study Group (CWSG) is an important example of track two scientist-to-scientist negotiations.

CWSG was part of the larger Pugwash Conferences of Science and World Affairs which created a transnational group of scientists engaged politically in an enduring open framework for East–West dialogue on security-related issues. As such, Pugwash members—natural and social scientists as well as individuals, invited in their own capacity, from government and industry—functioned as "icebreakers"[61] during the Cold War "when exchanges of mutual suspicions and propaganda were clogging the more conventional channels of communication".[62]

One way of breaking the ice was by holding meetings away from public scrutiny. Although the existence of the meetings was not hidden, discussions were held under the Chatham House Rule of non-attribution so as to encourage openness and information sharing.[§§] By operating in such a way, Pugwash meetings became a space where "it [was] possible to explore, in a preliminary and noncommittal way, new approaches that could, at a later stage, become the basis for official negotiations."[63,64]

20.4.2 The Chemical Warfare Study Group

The CWSG which was established at the 23rd Pugwash Conference in August 1973, came to play an important role in nurturing the concepts behind a treaty to ban chemical weapons, and subsequently provided technical assistance to those diplomats engaged in negotiating what would become the Chemical Weapons Convention.

The CWSG's steering committee which included scientists from the USSR, Europe and the US initiated a series of workshops on different aspects of possible verification systems for the projected CWC, and also developed a series of roundtables on medical defence against chemical weapons.[65] Between 1974 and 1992, a total of 312 people from 38 countries attended these meetings.

The CWSG produced a succession of papers and reports which highlighted the necessity of international verification; explored particular techniques of verification that would minimise intrusion into areas of legitimate commercial secrecy; proposed strategies for responding to CBW-use allegations; and examined regional approaches to CW arms control. The papers were transmitted to the disarmament delegations in Geneva and to capitals stimulating State thinking in preparation for the commencement of CWC negotiations in 1984.[66]

The CWSG's work on the philosophy and design of on-site inspections in civilian chemical industry is of particular note. Robinson describes this as being enabled by building "on the cooperation it had received... from such companies as Shell, Bayer, the latter's US subsidiary Chemagro, and Albright & Wilson" as well as actively establishing working contacts with members of the European biotech industry and with individuals from the US, Swiss, and Japanese chemical manufacturers' associations.[67] As part of this work, a number of trial inspection visits were conducted at commercial chemical facilities in the US and European countries. As well as learning by doing, the

---

[§§]The Chatham House Rule is defined as follows: when a meeting, or part thereof, is held under the Chatham House Rule, participants are free to use the information received, but neither the identity nor the affiliation of the speaker(s), nor that of any other participant, may be revealed. For more see https://www.chathamhouse.org/

preparations for these visits and the visits themselves allowed prominent figures in the chemical industry to become aware of the demands that a treaty, if it were successfully concluded, would place on their corporations.

20.4.3 Pugwash and the CWC

The CWSG also provided expert advice once serious governmental attention was given to the idea of an international treaty banning chemical weapons. CWSG members developed working relationships with negotiating diplomats in Geneva and those in the capital covering the treaty negotiations. This was helped, somewhat, by some CWSG members being assigned as technical experts to delegations, a trend which changed the nature of the meetings.[68]

As well as performing this educational/advisory role, the CWSG continued to conduct research which fed into and influenced the nature of the resulting CWC. One particularly notable study was published by the pioneering Stockholm International Peace Research Institute (SIPRI) on the mustard gas precursor thiodiglycol, which Iraq had been purchasing for its chemical weapons program.[69] The chemical's primary use, however, is as a precursor for a range of commodity chemicals and as a specialised solvent in, for example, the textiles industry. Thiodiglycol had been tentatively placed on Schedule 2 Part A of the rolling text of the projected Convention, and the CWSG consequently examined the difficulties of developing effective non-intrusive verification and how these could be overcome. According to Robinson, the study, which took three years to complete had "concrete influence" on "the content of the treaty's control schedules".[68]

When negotiations of the CWC concluded in the late summer of 1992, the mandate of the study group as determined in 1973 effectively ended. However, from 1993, a new study group was formed which attended to issues of practical implementation of the CWC as well as those for the BWC.

20.5 Engagement with the OPCW and CWC States Parties

Scholars have discussed the responsibility that informed civil society actors—particularly from the medical and scientific communities—have to highlight the limitations and failings of those mechanisms, notably the CWC and the OPCW, in combating the development and use of chemical weapons, and to develop and promote appropriate strategies to strengthen these regimes. Robinson[70] has argued that:

> "When it comes to arms control, all of us. . . need reminding that treaties such as the CWC are engagements, not between governments, but between States Parties. Governments may represent States Parties in the [relevant regime fora]. . . but organs of civil society are also elements of those same states, no less responsible for proper implementation of the treaty."

Previously, however, CWC States Parties, and the OPCW as a whole, have been reluctant to receive information from, and interact with, civil society in a systematic manner. For example, whilst representatives of civil society organizations have been routinely invited to address plenary sessions at Review Conferences of the BWC and the Non-Proliferation Treaty, until recently no such opportunity existed at the CWC Review Conference.

Despite these restrictions, there have been some limited opportunities for civil society organizations to interact with State Party delegates, and a number of NGOs have used these opportunities to highlight concerns about limitations of the CWC and its implementation by States Parties, as well as highlighting certain controversial issues. For example, at both the First and Second Review Conferences, the Technical Secretariat hosted "Open Forums" on the Chemical Weapons Convention.[71] An editorial in the *CBW Conventions Bulletin* highlighted the benefit of such initiatives: "[T]he interest by delegations in the Open Forum ... seems to indicate that their [NGO] involvement is a valuable addition to the review process, not least for the ability to highlight sensitive topics that are politically untouchable by delegations."[72] Open Forums, subsequently held under the auspices of the CWC Coalition (CWCC) have become an established, though limited, mechanism for civil society engagement with States Parties at OPCW conferences. They have been held from 2008 at each annual Conference of States Parties (CSP14–CSP22) and during the Third Review Conference.

20.5.1 Advances in Civil Society Access and Engagement During the Third Review Conference

In recent years, the level and quality of civil society interaction with the OPCW has markedly improved, in large part due to the prioritization of public diplomacy by OPCW Director General Üzümcü. This was exemplified in the run-up to and during the Third CWC Review Conference. Firstly, through the promotional activities of the Technical Secretariat and the CWCC, far greater numbers of civil society organizations registered and attended this Conference than previously – whilst 22 and 28 NGOs attended the First and Second Review Conferences, respectively, this number had risen to 70 NGO attendees at the Third Review Conference.[118] This general upward trend has also been replicated for NGO attendance at the annual Conference of States Parties over the last ten years.[73]

Furthermore, with the support of Director General Üzümcü and a number of States Parties (notably Norway); the Chair of the Third Review Conference, Ambassador Paturej, brokered amendments to the Rules of Procedure and associated guidelines,[74] facilitating greater potential NGO participation. These measures included an opportunity for NGOs to address plenary sessions, greater access to Conference documents, provision of meeting rooms and technical equipment, a simplified registration process, and the creation of an "NGO Coordinator" to act as a focal point for communication with the Conference Chair and the Secretariat. Consequently, for the first time at a

CWC Review Conference, 14 NGOs made statements to a plenary session. Similarly, more side events were held during this Review Conference than for any of its predecessors. Accordingly, Guthrie suggested that non-governmental presence and presentations were "starting to approach the levels seen in the meetings associated with the Biological Weapons Convention".[75]

In a further significant advance, the Third Review Conference in its Final Document: "noted the importance of the contribution to the goals of the Convention that is made by the chemical industry, the scientific community, academia, and civil society organisations engaged with issues relevant to the Convention".[76] Consequently it "encouraged" the Technical Secretariat and States Parties to "improve interaction" with these civil society entities and to "develop a more open approach, in conformity with the Rules of Procedure of the policy-making organs..."[77]

20.5.1.1 Chemical Weapons Convention Coalition

One important development for civil society/OPCW interaction has been the establishment in December 2009 of the Chemical Weapons Convention Coalition (CWCC). The CWCC is an independent, international grouping of NGOs whose mission is to support the aims of the CWC with focused civil society action aimed at "achieving full membership of the CWC, the safe and timely elimination of all chemical weapons, preventing the misuse of chemicals for hostile purposes, and promoting their peaceful use".[78] The CWCC is unusual in that it was established with the active involvement of the OPCW Technical Secretariat. The then OPCW Director General, Ambassador Pfirter, was very supportive of this initiative, and this support was continued by Director General Üzümcü.[¶¶] The Technical Secretariat, particularly the Media and Public Affairs Department, has given valuable assistance and support to the CWCC and has tried to foster wider engagement of the OPCW with civil society. Michael Luhan, then the head of the OPCW Media and Public Affairs Department, has stated that both the OPCW and its Member States can "benefit from outside voices that can point out faults in the system ... There's a lot of things that go unsaid in more formal venues."[79]

After eight years, the CWCC is still in a formative period. As of March 2018, it had 86 members (predominately national organizations) in 42 countries. It has a functioning website and has helped to increase the numbers of civil society organizations attending and participating in OPCW meetings, most notably the Conferences of States Parties and the Third Review Conference. It does provide important (and sometimes the only) opportunities for civil society to interact with CWC States Parties and the Technical Secretariat in semi-official fora and has also played a role in promoting the CWC and the

[¶¶]For example, the Second Annual Meeting of the CWCC was hosted by the OPCW and was addressed by the Director General. The full details of the meeting, together with a video of the Director General's interventions, are available on the OPCW website. See ref. 119 for further information.

work of the OPCW to the wider world. However, despite these accomplishments, it is apparent that the CWCC has yet to become more than a loose grouping of civil society organizations with disparate objectives and priorities.

20.5.2 Raising Compliance Concerns

Since the CWC entry into force the OPCW has repeatedly shown itself unwilling or incapable of effectively addressing activities of Member States that potentially breach the Convention, when these are brought to their attention by the media or reputable civil society organisations. This serious systemic failure has been exacerbated by extremely limited (and in most countries non-existent) Parliamentary or other national oversight of State Party activities, something further compounded by patchy media reporting and the attendant low levels of public awareness and interest in these issues.

Opportunities for corrective action may present themselves in the area of implementation of Article VII of the Convention, which requires that all CWC States Parties: "adopt the necessary measures to implement [their] obligations under this Convention" and to "not permit in any place under [their] control any activity prohibited to a State Party under this Convention".[80] Consequently, there is space for concerned civil society actors to bring concerns regarding prohibited activities directly to the attention of the relevant National Authority of the State in question, or raise it with their own National Authority. However, if these State bodies are unwilling to take action, there are currently no mechanisms that enable civil society actors to formally present their concerns directly to the OPCW Technical Secretariat or an OPCW policy-making body for action.

As transition of the OPCW proceeds from destruction to preventing the (re)-emergence of chemical weapons, the need for increased accountability for compliance and implementation will require the creation of mechanisms which permit critical encouragement by civil society of proper implementation. It would be beneficial if relevant OPCW structures now began to consider the development of such mechanisms. The case study below illustrates how civil society can successfully bring compliance concerns to the attention of the OPCW.

20.5.2.1 Civil Society Investigations Led to the Destruction of Turkey's Stockpile of 120 mm Mortar Munitions Containing Riot Control Agents

In November 2003, *Jane's Defence Weekly* reported that the Turkish (State owned) arms manufacturer *Makina ve Kimya Endustrisi Kurumu* (MKEK) had developed the CS MKE MOD 251, a 120 mm mortar round filled with CS which weighed 17.34 kg and had a maximum range of 8132 metres.[81] It was promoted by MKEK on their website,[82] and at international security exhibitions, including the 7th International Defense Industry Fair (IDEF) held in Ankara, Turkey, in September 2005.

Given the nature of the RCA munition, it was clear that its use for riot control or other domestic law enforcement operations would have been inappropriate. Consequently, any manufacture, stockpiling and deployment of such munitions apparently breached Article I.1.(a), I.5 and II.1.(a) of the CWC. In addition, any promotion and transfer of such munitions would have potentially breached Article I.1.(a) and Article I.1.(d) of the Convention. Following correspondence from the Bradford Non-Lethal Weapons Research Project (BNLWRP) to the Turkish Government and MKEK highlighting concerns about this munition, all information concerning the CS MKE MOD 251 mortar round was subsequently removed from the MKEK website.[82]

However, a collaborative investigation by the BNLWRP, the Institute for Security Studies (ISS) and the Omega Research Foundation (ORF), uncovered and documented the subsequent promotion of the CS MKE MOD 251 mortar round by MKEK at the Africa Aerospace and Defence (AAD) exhibition held in Cape Town, South Africa, in September 2010 (Figure 20.2). The BNLWRP, ISS and ORF initiated correspondence with the Turkish Government, the South African Government and MKEK, raising concerns about the promotion of these munitions. On 29th November 2010, Dr Michael Crowley highlighted these issues during a presentation given at the OPCW Open Forum,[83] and during the subsequent question and answer session, a representative of the Turkish delegation stated that Turkey was investigating the issue and would publish its results in a transparent manner.

In February 2011, in correspondence with the BNLWRP, ISS and ORF, the Turkish OPCW Ambassador stated that 1000 CS MKE MOD 251 munitions had been produced in 1996, prior to Turkey's ratification of the Convention, and that "around 150 of the said ammunitions were used for testing purposes during the initial R&D phase in 1997".[84] In correspondence from July 2011, the Turkish Counsellor to the OPCW stated that: "[A]t the time of ratification, there remained 850 pieces of CS MKE MOD 251 type munitions in the inventory of the Turkish Armed Forces. The facility for their production was discontinued after 1997."[85] From the Counsellor's statement it appeared that Turkey had not formally declared the existence of the outstanding 850 munitions or provided details of the relevant production facilities to the OPCW Technical Secretariat. Such a declaration would appear to have been required under Article III, paragraphs 1(a) and (c) of the CWC.[86] It is unclear whether Turkey subsequently submitted a declaration. The Turkish OPCW Ambassador explained that: "[T]he remaining 850 [munitions], whose dates of expiry have passed, are stored at the Turkish Armed Forces ammunition destruction facility awaiting disposal."[84] Subsequently, in his July correspondence, the Turkish Counsellor reported that: "[T]he destruction of CS containing canisters of the remaining CS MKE MOD 251 munitions has now been completed at our state-of-the-art munitions disposal facility located near Ankara."[85] In addition to destroying all remaining 120 mm RCA mortar munitions, Turkey also put in place measures to halt the promotion and brokering of such munitions by Turkish companies or individuals.[85]

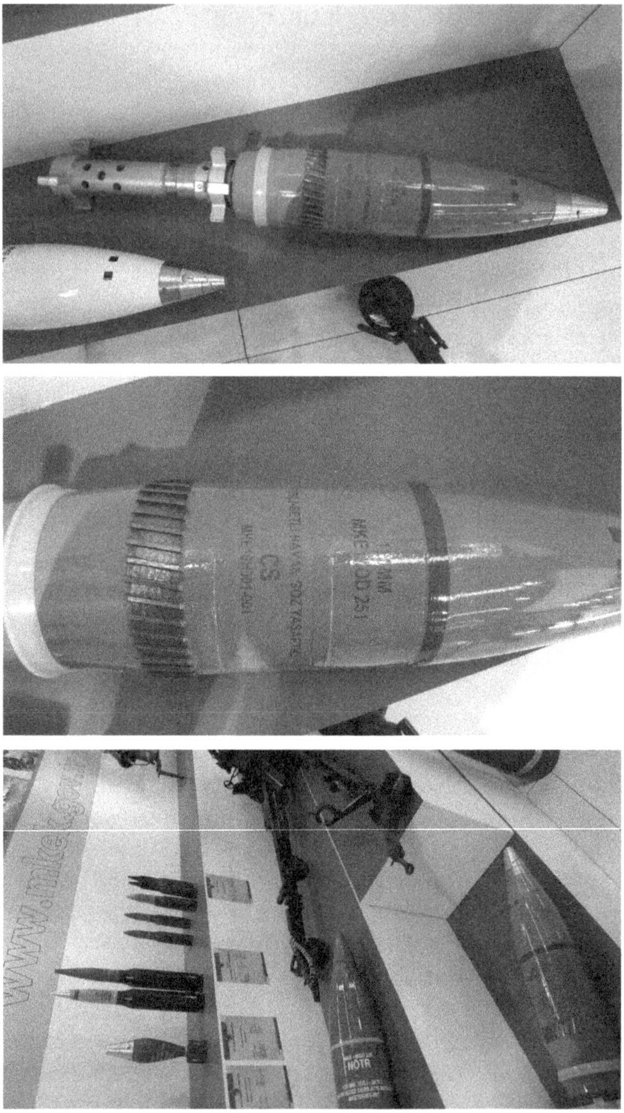

Figure 20.2 Images of MKEK Tactical CS 120 mm mortar round, photographed on display at AAD 2010, in Cape Town, South Africa, 21st-25th September 2010 (Photos: © Robin Ballantyne/Omega Research Foundation).

20.6 Facilitating More Effective Civil Society Engagement with the OPCW

20.6.1 Overcoming the State Security Paradigm

Scholars have explored how the inter-governmental discourse on chemical weapons has been framed largely in terms of a "State security discourse" and from within "a world view where States are perceived to be the primary agent for analysis and action"[87] with the consequent impact upon civil society engagement. Feakes identified this "[S]tate security discourse" with its framing of chemical (and biological) weapons as weapons of mass destruction (WMD) and as potential terrorist weapons, as the "main external factor influencing" effective civil society engagement in chemical and biological weapons disarmament.[88]

Whilst these factors are broadly relevant today, it is apparent that the OPCW's current period of transitioning from an organisation primarily focussed upon chemical weapons disarmament to one increasingly engaged upon a wider range of activities preventing the re-emergence of such weapons and ensuring the effective regulation of toxic chemicals, has begun to influence thinking by some of the key actors in this area.

In their 2011 report, the Advisory Panel on Future Priorities for the OPCW,[89] stated:

> "The engagement and contribution of the Convention's stakeholders is becoming ever more important. Their role (with the exception of chemical industry) was limited at the beginning of the Convention's operation—when the focus was on the elimination of State programmes—but is bound to increase and become more critical as the focus moves to prevention and cooperation."

Subsequently in April 2015, Director General Üzümcü circulated a Technical Secretariat paper intended to "set out an integrated vision and to develop a strategic direction for the Organisation", to all the States Parties for their consideration. A key, and potentially far reaching, message of the paper[90] was that:

> "Compliance with the requirements of the Convention cannot be achieved simply through a regulatory approach by national governments. It also requires the support and active collaboration of relevant stakeholders in the chemical industry, research, academia, NGOs, and civil society."

Consequently, the Technical Secretariat paper proposed committing the Organisation to work to "broaden its community of stakeholders and augment its engagement through enhanced education and outreach efforts" and that "[I]t will further develop effective networks and new partnerships with different stakeholder communities in order to leverage their expertise to help prepare responses to new implementation challenges as they emerge."[91] Furthermore, "by the middle of the next decade...the Organisation" will have established "a formal network of cooperation with the science and technology community of relevance to the Convention, especially scientific societies" which would be "supported by a broad-based education and outreach programme."[92]

Whilst such thinking reflects the position of Director-General Üzümcü and the Technical Secretariat, and has been supported by a number of States Parties, the vast majority of States have not made their positions publicly known.

It is consequently unclear as to whether certain States have reservations, simply have no view on the matter, or indeed oppose these proposals completely. Although it is currently unclear how consultations on these matters will be taken forward within the OPCW, the annual CSPs and the Fourth and Fifth Review Conferences in 2018 and (probably) 2023, respectively, are likely to be key forums where the future of the OPCW will be debated and any significant changes to the Organisation agreed. It is important that the voice (or rather voices) of key civil society actors be heard in these discussions and it is consequently imperative that the OPCW develop an appropriate process to facilitate such engagement. In addition, civil society, potentially through the CWCC, should initiate deliberations on these matters and explore whether, and if so how, an agreed civil society position can be developed and then enunciated.

20.6.2 Ensuring Greater Access and Establishing Forums for Substantive Dialogue

Despite the important advances made at the Third Review Conference and reinforced at subsequent CSPs to facilitate greater NGO access and participation during these meetings, NGOs still face substantial barriers to effective engagement with States Parties and the OPCW Technical Secretariat. For example, although they attend open sessions of the CSPs and Review Conferences, they cannot observe (never mind participate in) sessions of the subsidiary bodies, such as the Executive Council, where the real discussions occur; documentation available to NGOs does not include draft decisions or working papers; and documents submitted by civil society are not distributed as official documents of these conferences. Real access is important, for without it civil society is less aware of the substantive discussions occurring during these conferences, relying instead upon other partial sources of information, invariably derived from a sub-section of participating diplomats who do not always have an interest in providing a balanced account. As Feakes states: "[W]ithout access, civil society is little more than a disenfranchised observer able to listen to bland political statements but kept away from the real debate."[88]

Concerns relating to the continuing low levels of access extend beyond the CSPs, Review Conferences and other formal meetings, and are reflected in all dealings that civil society has with the OPCW. According to McLeish and Lak:[93]

> "...the limited access to meetings, documents and people... and the limited opportunities to interact lead to civil society (and industry) disenfranchisement, resulting in the creation of a knowledge gap between those on the inside (the TS and governments) and those on the outside (civil society and industry), which has the effect of 'keep out' signs posted around the OPCW to the CWC's non-governmental stakeholders."

It is important that the progress made at the Third Review Conference is built upon collectively by the OPCW, even if this is to be in a gradual manner (following negotiations and agreement of the States Parties) for example by placing more OPCW documents in the public domain and by allowing NGOs greater access to more aspects of meetings of the policy making organs. In addition, individual States Parties can do much to facilitate increased access and involvement by NGOs, for example, by sponsoring NGO meetings and encouraging State Party attendance during CSPs, holding regular briefings for NGOs to inform them of substantive OPCW discussions, and appointing NGO experts onto national delegations.

In addition, as part of its deliberations over the future of the Organisation in the post-destruction era, the OPCW should give due consideration to how it can improve access and engagement with relevant civil society bodies. There are indications of such progressive thinking within the Technical Secretariat, as evidenced in its April 2015 paper on the OPCW in 2025. One proposal in this regard was a commitment to hold "a major biannual engagement event to draw on the knowledge and expertise of NGOs, think tanks, academia, the science and technology community, and the chemical industry to further the object and purpose of the Convention."[92]

Worthy of note, in this context, are the proposals by McLeish and Lak for the development by the OPCW of a form of "consultative status" for NGOs in their relations to the Organisation, akin to that currently employed by the UN Economic and Social Council (ECOSOC).[‖‖] It would "allow NGOs to formally to participate in the system and to learn and contribute" and also very importantly "give NGOs and civil society members a level of legitimacy in the eyes of States Parties."[94] Such consultative status could, depending on its framing, ensure greater access to the OPCW building and facilitate increased interaction with the Technical Secretariat and States Parties; formalise enhanced participation in CSPs and Review Conferences, *etc.*; and provide a space for civil society contributions (*e.g.* research reports, policy papers, *etc.*) upon the OPCW website akin to the BWC's ThinkZone.[94]

20.6.3 Securing Funding and Resources

Writing in 2003, Feakes identified "[F]unding constraints [as] limit[ing] the activities which civil society can undertake and also restrict[ing] the entry of new talent into [chemical and biological weapons] disarmament, precisely at a time of generational change when new entrants are much needed."[88] At that time, the large bulk of funding for civil society chemical and biological weapons disarmament issues came from a small number of predominately US private foundations such as the MacArthur Foundation, as well as the Joseph Rowntree Charitable Trust in the UK.[88] Unfortunately, the funding environment has subsequently seriously deteriorated with a severe reduction

‖‖For information about ECOSOC status, see ref. 120.

in the numbers of organisations funding civil society work on chemical weapons disarmament and reduction in the amounts provided. Certain States and regional bodies notably the US, the UK and the EU have funded specific technical and service provision by NGOs, such as VERTIC's national implementation measures (NIM) programme. Furthermore, other States, notably Germany and Norway, have provided funding for the work of the Chemical Weapons Convention Coalition, particularly in facilitating attendance of NGOs from the Global South to OPCW events, such as the annual CSPs and the Third Review Conference. In contrast, there has been understandable wariness from both States and civil society for State funding of NGOs engaged in independent research, policy analysis, high-level lobbying or public advocacy work and consequently there has been significant contraction or suspension of such activities and the closure of a number of NGOs previously active in these areas. Regrettably, such activities are likely to be marginalised until a time when a significant re-engagement by major philanthropic foundations occurs or new funding streams (such as crowd-funding) can be utilised.

20.7 Conclusions

Civil society has played an important role in combating the development, proliferation and use of chemical weapons over the course of the last fifty years. This has been achieved by several different strands of activity undertaken by the diverse array of non-State actors that fall under the rubric of civil society.

As this chapter—and those of others in this volume have—demonstrated, such strands of activity include:

- responding to reported chemical weapons events variously through providing information pertinent to or supportive of investigations; challenging unverified claims; documenting the human consequences of chemical weapons attack; providing medical assistance to victims of such attacks; identifying agents used; and trying to hold manufacturers and governments to account;
- undertaking holistic technology-tracking of both generic trends in technologies of relevance to the CWC and targeted research into specific S&T developments or research trajectories of particular concern.
- providing expert scientific advice, for example in development and testing verification methodologies for governmental and IGO actors to consider;
- promoting and generating transparency into the activities of governmental and non-governmental research activities potentially relevant to the development of chemical weapons, exploiting both traditional and novel information gathering techniques;
- supporting, preparing and protecting whistle-blowers seeking to expose activities potentially relating to the development, proliferation and use of chemical weapons;

- fostering engagement between governmental actors across national boundaries through track two processes;
- seeking to reframe the discourse around the development, possession and use of chemical weapons, highlighting human security and human rights considerations.

However, it is also important to recognise that there are limits to what civil society actors can usefully do. Such limitations exist in part because of the lack of information on the dimmer dealings of chemical disarmament diplomacy and the information gap which renders constructive, useful contributions more difficult. In part, because of the time, access and, perhaps most significantly, resource constraints, imposed upon civil society by a seriously deteriorated funding environment, a trend which looks set to continue for the foreseeable future; but also, in large part, because chemical weapons issues have traditionally been predominantly framed in terms of a State security discourse with only narrow scope for the involvement of non-State actors.

As the OPCW shifts from a focus on the destruction of known stockpiles of chemical weapons to preventing the re-emergence of chemical warfare, in all its forms, it is possible that such a framing of chemical warfare may also shift. Indeed, dealing with the challenge of chemical attacks in the 21st century is unlikely to be something that can be achieved by States alone, rather it requires governments working with a much wider range of stakeholders, including the chemical industry, independent scientific and medical researchers, academics, human rights and humanitarian NGOs, and wider civil society, particularly those communities that have suffered from or are under threat from toxic chemicals employed as weapons. The need for such a paradigm shift has been recognised by some governments, however it remains unlikely that all governments will be amenable to opening up chemical disarmament diplomacy to a wider civil society audience. The Fourth CWC Review Conference will be an important forum in which these issues can be addressed. It will provide a litmus test as to whether the OPCW is willing to change longstanding institutional practices and processes in order to work more effectively with all relevant stakeholders to ensure "a world free of chemical weapons".

References

1. H. Anheier, M. Glasius and M. Kaldor, Introducing Global Civil Society, in *Global Civil Society 2001*, ed. H. Anheier, M. Glasius and M. Kaldor, Oxford University Press, Oxford, 2001, p. 17.
2. A. Florini, *The Third Force: The Rise of Transnational Civil Society*, Japan Center for International Exchange and Carnegie Endowment for International Peace, Washington D.C., 2000, pp. 7–9.
3. D. Feakes, Global Civil Society and Biological and Chemical Weapons, in *Global Civil Society 2003*, ed. M. Kaldor, H. Anheier and M. Glasius, Oxford University Press, Oxford, 2003, p. 97.

4. D. Deiseroth, Societal Verification, *Citizen Reporting and Whistleblowing as Integral Elements of the Disarmament Process and of Technological and Other Verification Systems*, Norderstedt: Books on Demand, 2nd edn, 2010, p. 265.
5. M. Haig Jr, *A Certain Idea of Man: The Democratic Revolution and Its Future, Current Policy, No. 311*, U.S. Department of State, Bureau of Public Affairs, Washington D.C., 13 September 1981.
6. Director of Central Intelligence, *Use of Toxins and Other Lethal Chemicals in Southeast Asia and Afghanistan*, Special National Intelligence Estimate (SNIE), 11/50/37-82, Feb 26 1982, vol. 2.
7. M. Meselson and J. P. Robinson, The Yellow Rain Affair Lessons from a Discredited Allegation, in *Terrorism, War, or Disease? Unravelling the Use of Biological Weapons*, ed. A. L. Clunan, P. R. Lavoy and S. B. Martin, Stanford University Press, Stanford, 2008, pp. 72–96.
8. R. Katz, Yellow Rain Biological Warfare Agent Use: Evidence and Remaining Questions, in *Terrorism, War, or Disease? Unravelling the Use of Biological Weapons*, ed. A. Clunan, P. Lavoy and S. Martin, Stanford University Press, Stanford, CA, 2008, pp. 97–119.
9. As Robinson has stated, Germ and chemical weapons may often be weak in their battlefield applications but they are always strong in their emotiveness. Accusations of association with them have for centuries, even millennia, been used by well-intentioned as well as unscrupulous people to vilify enemies and to calumniate rivals, J. P. Robinson, *Alleged Use of Chemical Weapons in Syria, HSPOP 4*, 2013, retrieved from: http://www.sussex.ac.uk/Units/spru/hsp/occasional-papers/HSPOP_4.pdf.
10. M. Goodman and D. Omand, *Teaching Intelligence Analysts in the UK What Analysts Need to Understand: The King's Intelligence Studies Program*, Studies in Intelligence (CIA), 2008, vol. 52, (4), 1–12.
11. M. Lynch, D. Lynch and S. Aday, *Syria's Socially Mediated War*, United States Institute of Peace, 2014, p. 5.
12. K. Leigh, *An Oral History of Syria's Citizen Journalism*, Syria Deeply, 4 Oct 2013, cited 18 Aug 2016, available from: http://www.syriadeeply.org/articles/2013/10/2517/oral-history-syrias-citizen-journalism/.
13. L. Johnston, UCG in the Newsroom: How BBC Journalists' Engagement with Internet Activists Has Altered Newsroom Practices, in *Media, Margins and Civic Agency*, ed. H. Savigny, E. Thorsen, D. Jackson and J. Alexander, Springer, 2015, pp. 182–194.
14. M. Storck, Streaming the Syrian War: A case study of the partnership between professional and citizen journalists in the Syrian Conflict, MSc Dissertation, London School of Economics, 2013, p. 23.
15. United Nations General Assembly, *Human Rights Council, Report of the Independent International Commission of Inquiry on the Syrian Arab Republic*, A/HRC/23/58, 13 Jul 2013, p. 2.
16. F. McMahon, *Alleged CW Attacks – Geography, Chronology as Posted on the Storyful Website*, 21 August 2013, p. 1517.

17. *US Secretary of State J. Kerry Statement on Syria*, 30 Aug 2013, cited 18 Aug 2016, available from: http://www.state.gov/secretary/remarks/2013/08/213668.htm.
18. P. Radden Keefe, *Rocket Man: How an Unemployed Blogger Confirmed that Syria Had Used Chemical Weapons*, The New Yorker, 25 November 2013, cited 18 Aug 2016, available from: http://www.newyorker.com/magazine/2013/11/25/rocket-man-2.
19. K. Leigh, *The Profile: The Man Behind Brown Moses*, Syria Deeply, 19 April 2013, cited 18 Aug 2016, at: https://www.newsdeeply.com/syria/articles/2013/04/19/the-profile-the-man-behind-brown-moses.
20. *A Detailed Summary of the Evidence on Munitions Linked to the August 21st Attacks*, Brown Moses, 2 Sep 2013, cited 18 Aug 2016, available from: http://brown-moses.blogspot.co.uk/2013/09/a-detailed-summary-of-evidence-on.html.
21. OPCW-UN Joint Investigative Mechanism, *First Report of the Organization for the Prohibition of Chemical Weapons-united Nations Joint Investigative Mechanism*, 2016.
22. M. Ingram, Putting the Crowd to Work, in *Verification Handbook: A Definitive Guide to Verifying Digital Content for Emergency Coverage*, ed. C. Silverman, January 2014, cited 18 Aug 2016, available from: http://verificationhandbook.com/book/.
23. *Forensic Verification of Social Content*, Storyful, cited 18 Aug 2016, available from: http://storyful.com/case-study/forensic-verification-of-social-content/.
24. M. Browne, Verifying Video, in *Verification Handbook: A Definitive Guide to Verifying Digital Content for Emergency Coverage*, ed. C. Silverman, January 2014, cited 18 Aug 2016, available from: http://verificationhandbook.com/book/.
25. Such claims were present, for example, in an investigative report compiled by the International Support Team for Mussalaha (Consilience) in Syria [ISTEAMS] later repeated by the Russian Government. See ISTEAMS Independent Investigation of Syria Chemical Attack Videos and Child Abductions, September 15th 2013http://tarpley.net/docs/20130915-ISTeams-Ghouta-Report.pdf and *Announcement by the Official Representative of the Ministry of Foreign Affairs of Russia Alexander Lukashevich Regarding the Messages about the Incident in the Damascus Suburb of Guta*, 23 August 2014, cited 18 Aug 2016, available from: http://www.mid.ru/en/kommentarii_predstavitelya/-/asset_publisher/MCZ7HQuMdqBY/content/id/99018.
26. R. Mackey, *Confused by How YouTube Assigns Dates, Russians Cite False Claim on Syria Videos*, The Lede, 23 August 2013, cited 18 Aug 2016, available from: http://thelede.blogs.nytimes.com/2013/08/23/confused-by-how-youtube-assigns-dates-russians-cite-false-claim-on-syria-videos/.
27. J. Tucker, *War of Nerves: Chemical Warfare from World War I to Al-Qaeda*, Pantheon Books, New York, 2006, p. 284.

28. Physicians for Human Rights, *Iraq: Winds of Death: Iraq's Use of Poison Gas Against its Kurdish Population, Report of a Mission to Turkish Kurdistan*, New York Times, February 1989.
29. Physicians for Human Rights, *Winds of Death: Iraq's Use of Poison Gas Against its Kurdish Population, Report of a Medical Mission to Turkish Kurdistan*, February 1989, pp. 1–2.
30. Human Rights Watch, *A Note on Methodology: The Methodological Approach to Documentary, Testimonial and Forensic Evidence Used in This Report*, 1993, cited 18 Aug 2016, at: https://www.hrw.org/reports/1993/iraqanfal/METHOD.htm; *A Middle East Watch Report, Genocide in Iraq: The Anfal Campaign Against the Kurds / Chapter Ten: Final Anfal: Badinan, August 25–September 6, 1988*, July 1993, cited 18 Aug 2016, available from: https://www.hrw.org/reports/1993/iraqanfal/; See also: J. Webb, Iraq caught out over nerve gas attack, *New Scientist*, 1993, 138, 1871.
31. C. Gosden, M. Amitay, D. Gardener and B. Amin, *Examining Long-Term Severe Health Consequences of CBW Use Against Civilian Populations*, Disarmament Forum, Open Forum, 1999, vol. 3, p. 67.
32. Testimony of Dr Christine Gosden before the Senate Judiciary Subcommittee on Technology, *Terrorism and Government and the Senate Select Committee on Intelligence on Chemical and Biological Weapons Threats to America: Are We Prepared? 1988 Congressional Hearings, Intelligence and Security*, 22 Apr 1998, cited 18 Aug 2016, http://www.globalsecurity.org/intell/library/congress/1998_hr/s980422-cg.htm.
33. A. Hay, R. Giacaman, R. Sansur and S. Rose Skin, Injuries Caused by New Riot Control Agent Used Against Civilians on the West Bank, *Med. Conflict Surviv.*, 2006, **22**(4), 283–291.
34. A. Hay, R. Giacaman, R. Sansur and S. Rose Skin, Injuries Caused by New Riot Control Agent Used Against Civilians on the West Bank, *Med. Conflict Surviv.*, 2006, **22**(4), 288.
35. Physicians for Human Rights, *Weaponizing Tear Gas: Bahrain's Unprecedented Use of Toxic Chemical Agents Against Civilians*, August 2012, cited 18 Aug 2016, available: https://s3.amazonaws.com/PHR_Reports/Bahrain-TearGas-Aug2012-small.pdf.
36. Physicians for Human Rights, *Weaponizing Tear Gas: Bahrain's Unprecedented Use of Toxic Chemical Agents Against Civilians*, August 2012, p. 5, cited 18 Aug 2016, available: https://s3.amazonaws.com/PHR_Reports/Bahrain-TearGas-Aug2012-small.pdf.
37. Physicians for Human Rights, *Weaponizing Tear Gas: Bahrain's Unprecedented Use of Toxic Chemical Agents Against Civilians*, August 2012, p. 20, cited 18 Aug 2016, available: https://s3.amazonaws.com/PHR_Reports/Bahrain-TearGas-Aug2012-small.pdf.
38. Physicians for Human Rights, *Weaponizing Tear Gas: Bahrain's Unprecedented Use of Toxic Chemical Agents Against Civilians*, August 2012, pp. 25–28, cited 18 Aug 2016, available: https://s3.amazonaws.com/PHR_Reports/Bahrain-TearGas-Aug2012-small.pdf.

39. Physicians for Human Rights, Physicians for Human Rights, *Weaponizing Tear Gas: Bahrain's Unprecedented Use of Toxic Chemical Agents Against Civilians*, August 2012, pp. 28–31, cited 18 Aug 2016, available: https://s3.amazonaws.com/PHR_Reports/Bahrain-TearGas-Aug2012-small.pdf.
40. BBC World News (UK), *Bahrain Authorities 'Weaponising' Tear Gas*, 1 August 2012, cited 18 Aug 2016, at: http://www.bbc.co.uk/news/world-middle-east-19078659.
41. US Department of State, *Country Reports on Human Rights Practices for 2012: Bahrain 2012 Human Rights Report*, Bureau of Democracy, Human Rights and Labor, cited 18 Aug 2016, available from: http://www.state.gov/documents/organization/204567.pdf.
42. The Sunshine Project, *Five Years Old, Sunshine Looks Ahead to the BWC Sixth Review Conference*, November/December 2004.
43. The Sunshine Project, The Pentagon's New Biochemical Warriors, in *Abuse Your Illusions: The Disinformation Guide to Media Mirages and Establishment Lies*, ed. R. Kick, Red Wheel Weiser, 2003.
44. BioWeapons Prevention Project (BWPP), *BioWeapons Monitor*, ed. I. Hunger, 2010, cited 18 Aug 2016, available at: http://www.bwpp.org/documents/BWM2010WEB.pdf.
45. BWPP, *BioWeapons Monitor*, ed. I. Hunger, 2011, cited 18 Aug 2016, available from: http://www.bwpp.org/documents/BWM2011WEB.pdf.
46. A. Arthur, *New Equipment in Hong Kong*, Defence Review Asia, 18 March 2012, p. 35.
47. Harvard Sussex Program, *News Chronology*, CBW Conventions Bulletin, 38, 29 December 1997; Physician Member of Hit Team, Paper Says, *Can. Med. Assoc. J.*, December 1997, **157**(11), 1504; L. Beyer, *Don't Try This at Home - Or in Aman*, Time Magazine, 1997 Oct 27, 150, (17) .
48. Czech Republic, National Authority of the CWC, Reply to the University of Bradford, Re: Request for Information Concerning Research Potentially Related to Incapacitating Chemical Agents, 2014 Jul 14, p. 4.
49. Czech Republic, *National Authority of the CWC*, Reply to the University of Bradford, Re: Request for Information Concerning Research Potentially Related to Incapacitating Chemical Agents, 2014 Jul 14, p. 5.
50. J. Rotblat, *Nobel Lecture: Remember Your Humanity*, Acceptance and Nobel Lecture, Oslo, 1995, cited 16 Aug 2016, available at: http://www.nobelprize.org/nobel_prizes/peace/laureates/1995/rotblat-lecture.html.
51. AAAS, *Vil Sultanovich Mirzayanov*, AAAS Archives and Records Centre, available at: http://archives.aaas.org/people.php?p_id=318; V. Mirzayanov, *State Secrets: An Insider's Chronicle of the Russian Chemical Weapons Program*, Outskirt's Press, Inc., Colorado, 2008.
52. International Committee of the Red Cross (ICRC), *Preventing Hostile Use of the Life Sciences: From Ethics and Law to Best Practices*, 2004 Nov 11, cited 4 Sep 2016, https://www.icrc.org/eng/resources/documents/misc/biotechnology-principles-of-practice-111104.htm.

53. *South African Protected Disclosures Act 26 of 2000*, cited 18 Aug 2016, available at: http://www.justice.gov.za/legislation/acts/2000-026.pdf.
54. UK Government, *Public Interest Disclosure Act*, PIDA, 1998.
55. *US Federal Whistleblower Protection Act (5 USC sec. 1201)*, 1989 Jul 9.
56. B. Martin, in *Whistle-blowing: Risks and Skills*, ed. B. Rappert and C. McLeish, A Web of Prevention: Biological Weapons, Life Sciences and the Governance of Research, Earthscan, London and Sterling VA, 2007, p. 47.
57. Other whistle-blower websites have subsequently been established including GlobaLeaks.com, RuLeaks.net and AJTransparency.com.
58. About WikiLeaks cited 30 Jul 2016, available on: http://wikileaks.ch/About.html.
59. J. Fildes, *What Is WikiLeaks?* BBC News, UK, 2010 Jul 25, cited 18 Aug 2016, available at: http://www.bbc.co.uk/news/technology-10757263.
60. Y. Benziman, Ingredients of a Successful Track Two Negotiation, *Negot. J.*, 2016, **32**, 51.
61. B. Kubbig, *Communicators in the Cold War: The Pugwash Conferences, the U.S.-Soviet Study Group and the ABM Treaty Natural Scientists as Political Actors. Historical Success and Lessons for the Future, PRIF-Report 44*, October 1996.
62. J. P. Robinson, The Impact of Pugwash on the Debates over Chemical and Biological Weapons, *Ann. N. Y. Acad. Sci.*, 1988, 866, p. 226.
63. A. Keynan, Cross Cutting Issues and Lessons Learned, *Ann. N. Y. Acad. Sci.*, 1988, 866, p. 273.
64. E. Garfield, Current Comments: The Pugwash Conferences on Science and World Affairs: Twenty-two Years in Search of Peace, *Essays of an Information Scientist*, 1979, vol. 4, pp. 200–207.
65. J. P. Robinson, The Impact of Pugwash on the Debates over Chemical and Biological Weapons, *Ann. N. Y. Acad. Sci.*, 1988, 241.
66. J. P. Robinson, *Contribution of the Pugwash Movement to the International Regime against Chemical and Biological Weapons, 10th Workshop of the Pugwash Study Group on the Implementation of the Chemical and Biological Weapons Conventions the BWC Protocol Negotiation: Unresolved Issues Geneva*, Switzerland, 1998 Nov 28–29, p. 20.
67. J. P. Robinson, The Impact of Pugwash on the Debates over Chemical and Biological Weapons, *Ann. N. Y. Acad. Sci.*, 1988, 242.
68. J. P. Robinson, The Impact of Pugwash on the Debates over Chemical and Biological Weapons, *Ann. N. Y. Acad. Sci.*, 1988, 243.
69. *Verification of Dual-use Chemicals under the Chemical Weapons Convention: The Case of Thiodiglycol*, ed. S. Lundin, SIPRI Chemical & Biological Warfare Studies, 1991, p. 13.
70. J. P. Robinson, in *Scientists and Chemical Weapons Policies*, ed. J. Finney and I. Slaus, IOS Press, 2010, Assessing the Threat of Weapons of Mass Destruction, p. 89.
71. Harvard Sussex Program, *Open Forum on the Chemical Weapons Convention: Challenges to the Chemical Weapons Ban*1 May 2003; OPCW, *Open Forum: Chemical Weapons Convention – Recent Experience and Future Prospects*, 9 April 2008.

72. Editorial, CBW Conventions Bulletin 60, *Harvard Sussex Program*, June 2003, pp. 4–5.
73. Analysis of the Final Reports for the Last 10 Conferences of States Parties Shows the Following Numbers of NGO Attendees: *CSP 13 (7); CSP 14 (30); CSP 14 (20); CSP 16 (29); CSP 17 (35); CSP 18 (39); CSP 19 (63); CSP 20 (52); CSP 21 (55); CSP 22 (52)*, Original documents available at: https://www.opcw.org/documents-reports/conference-states-parties/.
74. OPCW, Rule 33 of the Rules of Procedure was amended to read: "Representatives of non-governmental organisations may attend the plenary sessions of the Conference, and participate in the activities of the review conferences, in accordance with such rules or guidelines as the Conference has approved", *Report of the Third Review Conference, Part B*, 19 April 2013, paragraph 1.6.
75. R. Guthrie, *CWC Review Conference Report, No.1, the Third CWC Review Conference: Setting the Scene*, 8 Apr 2013.
76. OPCW, *Report of the Third Review Conference, Part B*, 19 April 2013, paragraph 9.144.
77. OPCW, *Report of the Third Review Conference, Part B*, 19 April 2013, paragraph 9.155.
78. Chemical Weapons Convention Coalition (CWCC), *Founding Document, Agreed by Consensus by the Participants at the Meeting to Establish the CWCC*, The Hague, Netherlands, 2–3 December 2009.
79. C. Schneidmiller, New Coalition Aims to Promote Chemical Weapons Disarmament, *Nonproliferation. Global Security Newswire*, 2 Jan 2010, cited 14 July 2015, available at: http://www.nti.org/gsn/article/new-coalition-aims-to-promote-chemicalweapons-disarmament-nonproliferation/.
80. OPCW, *Chemical Weapons Convention*, 1993, Article VII.
81. C. Foss, *Turkey Details 120mm Automatic Mortar. Jane's Defence Weekly*, 12 November 2003.
82. MKEK was promoting the CS mortar round till at least mid-November 2009 on its website, see: M. Crowley, Drawing the line: Regulation of "wide area" riot control agent delivery mechanisms under the Chemical Weapons Convention, BNLWRP/ORF, April 2013.
83. M. Crowley, *Regulation of Incapacitants, Riot Control Agents and Their Means of Delivery under the Chemical Weapons Convention*, OPCW, Open Forum Meeting, 29 Nov 2010.
84. Letter from Ambassador Dogan, *Permanent Representative of Turkey to the OPCW, to BNWLRP*, ORF and ISS, 25 February 2011.
85. Letter from Mr Utkan, Counsellor, *Permanent Representation of Turkey to the OPCW, to BNLWRP*, ORF and ISS, vol. 8, July 2011.
86. OPCW, *Chemical Weapons Convention*, 1993, Article III.1.
87. S. Carroll, *NGO Access to Multilateral Fora: Does Disarmament Lag Behind?*, Disarmament Forum, 2003, vol. 1, pp. 15–26.
88. D. Feakes, Global Civil Society and Biological and Chemical Weapons, in *Global Civil Society 2003*, ed. M. Kaldor, H. Anheier and M. Glasius, Oxford University Press, Oxford, 2003, p. 101.

89. OPCW, *Technical Secretariat. Note by the Director General Report of the Advisory Panel of the Future Priorities of the Organisation for the Prohibition of Chemical Weapons*, Office of the Director General, S/951/2011, 25 July 2011, paragraph 119.
90. OPCW, Note by the Technical Secretariat, *The OPCW in 2025: Ensuring a World Free of Chemical Weapons*, Office of Strategy and Policy, S/1252/2015, 6 March 2015, paragraph 33.
91. OPCW, Note by the Technical Secretariat, *The OPCW in 2025: Ensuring a World Free of Chemical Weapons*, Office of Strategy and Policy, S/1252/2015, 6 March 2015, paragraph 23.
92. OPCW, Note by the Technical Secretariat, *The OPCW in 2025: Ensuring a World Free of Chemical Weapons*, Office of Strategy and Policy, S/1252/2015, 6 March 2015, paragraph 120.
93. C. McLeish and M. Lak Civil, Society and Industrial Non-State Actors, in *The Chemical Weapons Convention: A Commentary*, ed. W. Krutzsch, E. Myjer and R. Trapp, Oxford University Press, Oxford, 2014, p. 47.
94. C. McLeish and M. Lak Civil, Society and Industrial Non-State Actors, in *The Chemical Weapons Convention: A Commentary*, ed. W. Krutzsch, E. Myjer and R. Trapp, Oxford University Press, Oxford, 2014, pp. 47–48.
95. J. Rotblat, Societal Verification, in *A Nuclear-weapon-free World: Desirable? Feasible?* ed. J. Rotblat, J. Steinberger and B. Udgaonkar, Westview Press, Boulder CO, 1993; D. Deiseroth, Societal Verification: Wave of the Future? in *2000 Verification Yearbook*, VERTIC, 2000.
96. M. Crowley and A. Persbo, The Role of Non-Governmental Organizations in the Monitoring and Verification of International Arms Control and Disarmament Agreements, in *Thinking outside the Box in Multilateral Disarmament and Arms Control Negotiations*, ed. J. Borrie and V. Martin-Randin, Institute for Disarmament Research (UNIDIR), United Nations, Geneva, 2006, pp. 225–252.
97. C. McLeish, *The Role of the Internet and Social Medial in the Immediate Aftermath of the August 21st Attacks on Ghouta*, 2016.
98. Human Rights Watch, *Analysis of Alleged Use of Chemical Weapons in Syria Attacks on Ghouta*, 2013.
99. B. Moses, *Who Was Responsible for the August 21st Attack?* Brown Moses Blog, 2013.
100. D. Iriarte, *The Syrian Opposition Accuses the Regime of Having Used Chemical Agents Against Civilians in Homs: A Doctor Shows a Video of a Patient Allegedly Affected by the Component*, ABC.es, 7 Apr 2012, cited 18 Aug 2016, available from: http://www.abc.es/20120406/internacional/abci-siria-armas-quimicas-201204052254.html.
101. Center for Documentation of Violations in Syria, *A Report of the Use of Chemical Weapons in Syria*, Jobar, Damascus, Apr 2013, cited 18 Aug 2016, available from: http://www.vdc-sy.info/index.php/en/reports/chemicaljobar#.V58woNArLVo.
102. Syrian Network for Human Rights, *Comprehensive Report of Using Chemical Weapons by Syrian Regime*, 14 May 2013, cited 18 Aug 2016,

available at: http://sn4hr.org/public_html/wp-content/pdf/english/2013/Usage-of-chemical-weapons-by-Syrian-Regime.pdf.

103. *Raji Rahmet Rabbou Talking to Al Jazeera as Quoted in Sarin-like Gas' Kills 7 in Homs*, Ynetnews, 24 Dec 2012, cited 18 Aug 2016, available from: http://www.ynetnews.com/articles/0,7340,L-4323666,00.html.
104. *Defecting Syrian Officer Brigadier-general Zaher Al-saket: I Was Ordered to Use Chemical Weapons*, The Middle East Media Research Institute, 27 Apr 2013, cited 18 Aug 2016, available from: http://www.memritv/clip_transcript/en/3822.htm#.
105. *Remarks by the President to the White House Press Corps*, 20 Aug 2012, cited 18 Aug 2016, at: http://www.whitehouse.gov/the-press-office/2012/08/20/remarks-president-white-house-press-corps.
106. *Remarks by the President at the Nunn-Lugar Cooperative Threat Reduction Symposium*, The National War College, Washington DC, 3 December 2012, cited 18 Aug 2016, at: https://www.whitehouse.gov/the-press-office/2012/12/03/remarks-president-nunn-lugar-cooperative-threat-reduction-symposium.
107. *Syrian Channel Listings*, cited 18 Aug 2016, http://brown-moses.blogspot.co.uk/2012/06/syrian-channel-listings.html.
108. *Third Report of the Organization for the Prohibition of Chemical Weapons-united Nations Joint Investigative Mechanism*, S/2016/738, 24 August 2016; For further discussion see Chapter 5 of this publication.
109. The Sunshine Project, *Non-lethal Weapons Research in the US: Calmatives and Malodorants*, 2001.
110. Royal Society website, *Brain Waves Project*, https://royalsociety.org/topics-policy/projects/brain-waves/.
111. J. Riches, R. Read, R. Black, N. Cooper, C. Timperley and Analysis, of Clothing and Urine from Moscow Theatre Siege Casualties Reveals Carfentanil and Remifentanil Use, *J. Anal. Toxicol.*, 2012, **36**, 647–656.
112. M. Crowley and M. Dando, Down the Slippery Slope? A Study of Contemporary Dual-Use Research Potentially Applicable to Incapacitating Agent Weapons, *Biochemical Security 2030 Policy Paper Series No.8, Biochemical Security 2030 Project: Bradford Non-lethal Weapons Research Project*, 2014 Oct, cited 14 Sep 2016, https://biochemsec2030dotorg.files.wordpress.com/2013/08/down-the-slippery-slope-final-web.pdf.
113. V. Klochikhin, A. Lushnikov, V. Zagaynov, A. Putilov, V. Selivanov and M. Zatekvakhin, Principles of Modelling of the Scenario of Calmative Application in a Building with Deterred Hostages, *Proceedings of the 3rd European Symposium on Non-lethal Weapons*, Fraunhofer ICT, Ettlingen, Germany, Pfinztal, 10–12 May 2005, vol. 17.
114. N. Kuzmina and V. Kuzmin, Development of Concepts on the Interaction of Drugs with Opioid Receptors, *Russ. Chem. Rev.*, 2011, **80**, 145–169.
115. State 9619 Plant Company Brochures, *Undated, Distributed during Asia Pacific China Police Expo 2006*, held at Beijing Exhibition Centre, Beijing, China, 24–27 May 2006.

116. L. Hess, J. Schreiberova and J. Fusek Pharmacological, *Non-lethal Weapons, Proceedings of the 3rd European Symposium on Non-lethal Weapons*, European Working Group on Non-Lethal Weapons, Fraunhofer ICT, Ettlingen, Germany, Pfinztal, 10–12 May 2005, vol. 23.
117. L. Hess, M. Votava, J. Schreiberová, J. Málek and M. Horáček, Experience with a Naphthylmedetomidine - Ketamine - Hyaluronidase Combination in Inducing Immobilization in Anthropoid Apes, *J. Med. Primatol.*, 2010, **39**(3), 151–159.
118. *Final Reports of the First, Second and Third CWC Review Conference*, original documents, cited 18 Aug 2016, https://www.opcw.org/documents-reports/conference-states-parties/.
119. *OPCW Hosts Second General Meeting of the Chemical Weapons Convention Coalition*, Apr 15 2011, cited 1 Sep 2012, http://www.opcw.org/news/article/opcw-hosts-2nd-general-meeting-ofchemical-weapons-convention-coalition/ and http://www.opcw.org/fileadmin/OPCW/video/DG-CWCC_01.mp4.
120. *Basic Facts about ECOSOC Status*, NGO Branch, Department of Economic and Social Affairs, United Nations, cited 16 Aug 2016, http://csonet.org/?menu=100.

Section VI

Conclusion

CHAPTER 21

Conclusions and Recommendations

MICHAEL CROWLEY*, RALF TRAPP, MALCOLM DANDO AND LIJUN SHANG

University of Bradford, UK
*E-mail: m.j.a.crowley1@bradford.ac.uk

21.1 Introduction

As outlined in Chapter 1, the approach we have taken in this publication has been informed by Holistic Arms Control (HAC) which consists of the following three stage process:

- *Stage one*: examination of the nature of the weapon and weapons-related technology under review; current and potential future scenarios of application, together with attendant national and human security concerns; and potential implications of relevant scientific and technological developments.
- *Stage two*: exploration of the full range of potentially applicable control mechanisms, analysing strengths, weaknesses and limitations.
- *Stage three*: development of a comprehensive strategy to improve existing mechanisms (and/or introduce additional mechanisms) for the effective regulation or prohibition of the weapon or weapons-related technology of concern.

Preventing Chemical Weapons: Arms Control and Disarmament as the Sciences Converge
Edited by Michael Crowley, Malcolm Dando and Lijun Shang

Published by the Royal Society of Chemistry, www.rsc.org

In this concluding chapter, we employ the HAC framework as a means of analysing the contributions of the preceding chapters and synthesising them into a proposed integrated approach which would prioritise the most appropriate and effective mechanisms to address the threats arising from the rapid advances in relevant science and technologies, notably the chemical and life sciences, computing and nanotechnology; and more generally the dangers of the development, proliferation and use of existing or new forms of weapons employing toxic chemicals.

Given the central importance of the Chemical Weapons Convention (CWC) and the accompanying implementation and verification regime, the chapter will highlight recommendations for both States and civil society to strengthen OPCW action in these areas, at the Fourth Review Conference and beyond.

21.2 Stage One: Review of Scientific and Technological Developments and Their Potential Applications

Repeated warnings were given by many authors throughout this publication, of the urgent need to address the increasingly diverse threats of the hostile use of toxic chemicals, by a broadening range of actors, some employing existing capabilities, others potentially being facilitated by advances in relevant science and technologies. For example, in Chapter 2, Paul Rogers explored how scientific and technological developments will take place in an increasingly fractured world in which 'revolts from the margins' by extremist non-State groups will intensify as resource limits and climate change weaken fragile economies. He argued that such groups could seek and eventually acquire the capability to produce and use weapons employing classic chemical agents, as well as industrial toxic chemicals. In Chapters 5, 6 and 12 Michael Crowley examined the contemporary use by State and non-State actors of weapons employing toxic chemicals against combatants and civilians in armed conflict, notably in Syria. Crowley also explored the potential desire from certain military, security and police forces for delivery systems employing toxic chemicals (*e.g.* riot control agents and incapacitating chemical agent weapons) for use in law enforcement or in certain military operations in populated areas. In such situations, there are concerns that the lines separating law enforcement, counterinsurgency, and low-level conflict may become blurred and the taboo against chemical weapons weakened.

Against such a background, Ralf Trapp in Chapter 7 considered the underlying convergence of chemical and life sciences at the turn of the 21st Century, *i.e.* the integration of theoretical concepts, experimental methods and scientific knowledge at the intersection of chemistry and biology. He noted that many new sub-disciplines have evolved, and transformational methods such as CRISPR Cas9 have spread rapidly in the scientific community.

Additionally, other technology developments, such as in nanotechnology, have acted as enablers in the practical application of scientific advances. Such developments clearly have overwhelimingly positive applications in many fields including agriculture, medicine, energy production, industrial and consumer goods manufacture. Furthermore, such advances have already resulted in, and are likely to further facilitate, important improvements in protection against and treatment of chemical agents, as detailed in Chapter 13's technical analysis by Xinbo Zhou, Wu Zhong, Pengbin Bu and Lijun Shang. However, some of these scientific advances have potentially grave malign applications and consequent implications for arms control, and the CWC in particular. Indeed, Trapp warned that “new discoveries could lead to novel chemical agents, new methods of chemical weapons production, or even new forms of chemical warfare.”

In Chapter 8, Malcolm Dando underlined and exemplified Trapp's concerns in his examination of previous and current State research of incapacitating chemical agents and highlighted potential future research trajectories as the revolution in neuroscience continues, and the dangers of their misuse for the development of weapons employing novel incapacitating agents intended for law enforcement and military operations. A similar point was made by Kathryn Nixdorff in her examination of the targeted delivery of agents such as toxins and bioregulators. As she noted in Chapter 8 “improved ways of delivering such agents are at hand and are being developed further at a rapid pace”, with the potential of their use for malign purposes. These concerns were exacerbated in Chapter 12 by Crowley's examination of the dangers of misuse of certain existing means of delivery and dispersal of toxic chemical agents, and the potential implications of the further technological advance of such delivery mechanisms. Crowley highlighted improvised chemical delivery mechanisms, misapplication of commercial general utility delivery mechanisms, wide area RCA delivery mechanisms and the adaption of a range of military delivery mechanisms.

In Chapters 10 and 11, Simon Whitby and Piers Millett discussed the much neglected issues of anti-plant and anti-animal chemical warfare. Following reviews of the historical hostile employment of chemical, toxin and biological agents in armed conflicts, both authors warned that the dangers that rapid changes in civilian chemical and life sciences and enabling technologies could open up capabilities for the development of further anti-animal or anti-plant weapons. Finally, in Chapter 14, Philippe Mauger and Ray Zilinskas examined the impact of technological change and its global diffusion on efforts to limit the proliferation of potential chemical warfare agents and delivery mechanisms through export controls and interdiction measures. Of particular note were the multifarious implications of the growth and application of information technology, with the authors illustrating the transformational role of the internet in the sourcing of dual-use items and information by would-be chemical weapons proliferators, and the consequent need for existing national and international export control measures to be rapidly enhanced to meet these challenges.

21.3 Stage Two: Review of Relevant Control Mechanisms

21.3.1 Chemical Weapons Convention

In Chapter 3, Trapp examined the significant achievements made following the entry into force of the CWC and establishment of the OPCW in 1997 towards freeing the world of the menace of chemical warfare. Yet, in the rapidly changing environment, 20 years after its entry into force he highlighted the need for organisational change stating that "the overall direction of the OPCW's work is gradually shifting towards a stronger emphasis on measures to prevent the re-emergence of chemical weapons in the future" and that one of the major challenges to the Convention is the radical developments in sciences and technology including the process of convergence.

Building upon this analysis, in Chapters 3 and 15, Trapp explored how, as the elimination of most of the past chemical weapons programmes nears completion, efforts to maintain and strengthen the comprehensive ban on chemical weapons must traverse and balance several parellel paths, including:

- Increased universalisation efforts to bring the remaining four non-Party States into the regime;
- Completing the elimination of past chemical weapons programmes and chemical weapons stockpiles;
- Ensuring that the Convention's implementation, compliance assurance and problem resolution mechanisms remain effective and adapt to change, in order to maintain confidence in the norm and to prevent the re-emergence of chemical weapons in whatever form;
- Improve international cooperation and assistance for peaceful purposes amongst States Parties, taking advantange of opportunities arising from advances in science and technology and the changes in the global landscape of chemicals manufacturing and trade, as well as the changing environment of science and industry collaboration;
- Strengthen the ability of States Parties to prevent, protect themselves against and respond to the hostile uses of toxic chemicals (as well as against natural and accidental releases of toxic chemicals), and the ability of the OPCW to provide assistance in such circumstances;
- Strengthen the intellectual, legal, institutional, scientific and technical capacity of the OPCW, and further enhance engagement with other stakeholders (non-governmental organisations, chemical industry, educational and scientific institutions and civil society more generally) to ensure continuing full compliance and effective implementation of the CWC provisions.

All this calls for an overarching vision of the future of the CWC and the OPCW – shared by its Member States, underpinned by a common strategy pursued by the policy-making organs and implemented by the Technical Secretariat and the States Parties themselves through their National Authorities. Significant progress towards such a shared vision has already been made, promoted by activities of Director-General Üzümcü and the Technical Secretariat, supported by the Executive Council and its Open-ended Working Group on Future OPCW Priorities; the Fourth Review Conference will provide an important opportunity to make progress in this regard; whether the States Parties seize that opportunity is of course another matter.

21.3.2 Biological and Toxin Weapons Convention

In Chapter 4, Jeremy Littlewood noted that in the five decades of its existence the Biological and Toxin Weapons Convention (BTWC) has embodied the normative constraint against the use of disease as a weapon and has survived a series of crises during that time. However, over this period "it has never thrived … and repeated efforts to strengthen the Convention have produced only incremental and piecemeal improvements." The chapter examined the evolution of the treaty and concluded that "[I]n essence, it is not the scope of the BTWC that is problematic: rather, implementation of the BTWC remains a work in progress and States Parties have been unable to agree upon the correct balance between the relationship of disarmament and development." The disappointing outcome of the Eighth Review Conference of the BTWC in 2016, where the States Parties were able to agree on very little except to meet again annually, reinforced his view that the difficulties that have prevented improving the strength of the Convention in the past remain. Although the December 2017 Meeting of States Parties did reach agreement on a new intersessional work programme up to the Ninth Review Conference in 2021, substantive collective advances through this process are unlikely.

In chapter 16, Littlewood argued that the most likely route of development will be the continuation of the very slow evolution of the BTWC along the lines of the past with stuttering change within the Convention being accompanied by action outside the auspices of the Convention by groups of like-minded States. He therefore argued that the most effective strategy for constructive and progressive development will be for BTWC States Parties to establish, or adapt, and adopt mechanisms that form a more dynamic and resilient network of legal and normative practices that are responsive to challenges, but place differential requirements on various actors.

21.3.3 United Nations Mechanisms

In Chapter 5, Crowley reviewed the United Nations (UN) mechanisms that have been and could be used to combat the development, acquisition and use of chemical weapons. The chapter examined the revivication of the UN Secretary General's Mechanism employed to investigate suspected chemical or biological weapons attacks, and its 2013 application in Syria. Crowley highlighted the significant advance in UN investigations, following establishment in 2015 of the OPCW–UN Joint Investigative Mechanism (JIM) which, unlike previous UN measures, was expressly tasked with identifying specific entities it deemed to be responsible for utilising chemical weapons in Syria; and which identified elements of the Syrian Armed Forces and Islamic State as the perpetrators of such attacks. Regrettably, due to Russian vetoes in the UNSC, the JIM's mandate was not renewed and it ceased investigations in November 2017. Crowley also explored sanctions that the UN can impose in response to confirmed threats to international peace and security, including the imposition of mandatory arms embargoes. The chapter then examined the rare occasions, such as Iraq in 1992 and Syria in 2013, when the threat to international security was deemed so grave and immediate that UN monitoring, verification and compliance enforcement mechanisms were tasked with entering the target States to oversee and facilitate the destruction of chemical weapons stockpiles and the elimination of chemical weapons production facilities. The chapter finally examined measures taken by the UN Security Council, particularly under UN Resolution 1540, to combat the proliferation and use of chemical weapons by non-State actors. Although the UNSC has, over time, created an impressive 'tool box' of measures, precedents and procedures to combat attempts by States and non-State actors to develop, acquire or employ such weapons, Crowley concluded that on many occasions when their use has been merited, their employment has been delayed, constrained or even prevented due to a lack of consensus between the veto-wielding members of the UNSC.

21.3.4 Relevant International Law

In Chapter 6, Crowley explored the full range of potentially applicable international law constraining the "weaponisation of toxic chemicals". In addition to the CWC and associated customary international law, the *lex specialis*, explicitly prohibiting States from developing, stockpiling and using chemical weapons, Crowley examined other relevant arms control and disarmament treaties, notably the Geneva Protocol and BTWC. Crowley then explored prohibitions on chemical weapons use in armed conflict derived from international humanitarian law (IHL), the further constraints that IHL imposes on the use of toxic chemicals by combatants, and the protection it affords, for example, to prisoners of war and civilian populations. However, Crowley highlighted how the potential utility of IHL to the regulation of toxic chemicals is curtailed in practice due to limitations in investigation and enforcement procedures, and extremely low levels of the State implementation of Article 36 legal reviews of new weapons. This was followed by an analysis of international human rights law (IHRL) and the wide-ranging obligations upon States to regulate the use of force (including weapons employing toxic chemicals) by law enforcement officials (potentially covering military and security personnel as well as police and prison officers). Whilst several human rights norms were potentially applicable to the regulation of such weapons, the rights to life; and to freedom from torture and other cruel, inhuman or degrading treatment or punishment, together with attendant obligations on the restraint of force, are the most relevant. An important positive aspect of human rights law is the existence of international and regional mechanisms monitoring adherence to relevant treaties, and the provision in certain treaties for the possibility of individual petition which can result in legally binding judgements for those States party to the relevant instrument. Whilst the preceding analysis concentrated upon international law constraining the activities of States, the chapter also explored an important alternative and complementary approach, *i.e.* directly addressing individual responsibility and culpability for crimes involving weaponised toxic chemicals through the employment of international criminal law. In practice however, there are considerable legal, logistical and political barriers to be overcome to make this potential recourse a reality. Notwithstanding such hurdles, in certain situations international criminal law could be applied by national courts or international courts or tribunals to those involved in the use of, and possibly the transfer of, certain weapons employing toxic chemicals.

21.3.5 Civil Society

In Chapter 17, Katherine Bowman, Alistair Hay and Jo Husbands examined the roles that the non-governmental chemical science community—and particularly the International Union of Pure and Applied Chemistry (IUPAC)—have played in support of the CWC, including through provision of independent expert advice to the OPCW concerning implications of developments in science and technology, both to undermine the Convention, as well as those that could assist in its implementation. IUPAC's role in raising awareness and knowledge about the Convention and its obligations amongst chemists, and in developing codes of conduct amongst the chemical science commuity were also examined. In contrast, in Chapter 18 Tatyana Novossiolova highlighted the comparative lack, historically, of the systematic input of science advice from the life science community to BTWC States Parties in support of that Convention. However, she noted that over the last ten years, a number of projects and initiatives designed to foster awareness of the BTWC and facilitate the development of a biosecurity culture underpinned by continuous learning, codes of conduct, and the provision of expert advice have been carried out with some success and perhaps with implications for taking such work forward in regard to the CWC.

The important role of the medical profession in combatting the development, proliferation and use of chemical weapons was examined in Chapter 19 by Crowley and Vivienne Nathanson. Through case studies in South Africa, Turkey and the UK, they examined how existing medical ethics have been applied in practice. They concluded that the international medical community has developed a framework of ethical standards that guide the conduct of health professionals in relation to their patients, colleagues and wider society, in times of peace and in times of armed conflict, many of which directly or indirectly constrain their involvement in the research, development, testing or use of weapons employing toxic chemicals. Of particular relevance were the World Medical Association's Regulations and Declarations specifically opposing chemical and biological weapons programmes; injunctions upon health professionals to prevent human rights abuse and breaches of international humanitarian law; and guidance strictly regulating medical and scientific research upon human subjects. However, they concluded that there were no widely accepted and applied international guidelines specifically determining the permissibility or non-permissibility of medical involvement in the development of certain weapons employing toxic chemicals for law enforcement purposes, most notably ICA weapons and RCAs, and argued that given the importance of medical participation to the development of such weapons, the introduction of clear guidance by professional associations representing health professionals constraining involvement in such activities is needed.

In Chapter 20, Crowley, Caitriona McLeish and James Revill considered the roles that activist civil society have played more broadly in combating the proliferation and use of chemical weapons, covering topics such as monitoring and analysing the development and use of toxic chemicals as weapons (including RCAs and ICA weapons), uncovering State activities of concern, and making independent assessments of relevant science and technology developments. The chapter additionally considered the role of "whistle blowers" and the forging of 'track two' relations between chemical and life scientists across national borders, and their role in nuturing the development of the CWC. The authors argued that civil society has had, and will continue to have, important roles in support of the CWC, but noted that "the extent to which organisations such as the OPCW will be willing and able to embrace greater civil society interaction as they shift from the destruction of known stockpiles, toward preventing the re-emergence of chemical weapons, remains to be seen." Consequently, the authors warned that civil society actors will continue to face a number of institutional, political and resource based limitations on the roles that they can play in this area.

21.4 Stage Three: A Proposed Strategy for Effective Regulation

The Holistic Arms Control (HAC) analytical framework seeks to explore the utility of employing a multiplicity of mechanisms, and to facilitate the active engagement of a variety of relevant actors, in the regulation of the weapons or weapons-related technology under review. Consequently, it is an axiomatic principle of HAC that a number of regulatory processes have elaborated upon as part of existing regimes (as well as those independent of such regimes) can be pursued in parallel. However, it is clearly apparent that relevant actors—be they international organizations, States or civil society bodies—have limited human, financial and temporal resources to devote to these issues, competing demands to address, as well as institutional mandates which may limit their freedom of action. When analysing the potential responsiveness of particular regimes, factors such as the existing political environment and previous negotiating history, relevant regulatory mechanisms and their application, and the informal and formal regime-forum schedules and "agenda space" will have to be considered by the relevant actors.

From the review of the regimes conducted under HAC stage two, it is clear that the CWC (and its attendant implementation and verification regime) is the most appropriate and receptive forum, for monitoring, evaluating and addressing potential malign application of the rapidly developing chemical and life sciences. However, the emergence of positive results of such discussions within the OPCW is by no means certain, and parallel processes should also be established to explore alternative complementary mechanisms with the BTWC, IHL, human rights and international criminal law instruments, and specific UN mechanisms, potentially yielding positive results in the next five to ten year period. Since these are predominately State-centric mechanisms, HAC stage three is drafted primarily in terms of policy proposals for States that are party to the relevant mechanism, with specific recommendations (below) for the Fourth CWC Review Conference. However, given its importance, the Section does conclude with recommendations for complementary civil society activity in these areas.

21.4.1 Chemical Weapons Convention: Recommendations for the Fourth CWC Review Conference and Beyond

The forthcoming quinquennial CWC Review Conference, with its explicit mandate to examine long-term issues of concern to the OPCW regarding implementation of the Convention in a strategic manner, and to "take into account any relevant scientific and technological developments", is clearly the most appropriate forum for the CWC States Parties to collectively review the nature and implications of the rapid developments in and convergence of the chemical, life and associated sciences, and establish appropriate measures to prevent their misuse. Consideration should be given by States Parties to the utility of exploring the policy options detailed below.

21.4.1.1 Ensuring CWC Universalisation and Complete Destruction of all Existing Chemical Weapons

Achieving full universality of the CWC will require recognising and understanding the specific political and security context that has prevented the four remaining non-Party States—Egypt, Israel, North Korea and South

Sudan—from joining the Convention. Consequently, the OPCW should design a targeted universalisation strategy for each State setting the Organisation's activities within the broader policy and security framework for that State, and coordinate its actions with other relevant actors, including key Member States, the UN, and regional organisations.

The primary focus of the Organisation in the short term will be to complete the ongoing tasks of eliminating past chemical weapons programmes, including the characterisation and elimination of any CW programmes of the remaining States not party once they join; the completion of the outstanding destruction and conversion activities of declared CW stockpiles, development and production facilities; and the resolution of concerns regarding possible undeclared stockpiles and production facilities.

With the completion on 27 September 2017 of the verified destruction of all declared Russian Federation chemical weapons stockpiles (totalling nearly 40 000 agent-tonnes), and that of Libya in January 2018, full attention must now be given to ensuring that the delayed and subsequently re-scheduled verified US destruction of its chemical weapons stockpiles (comprising 27 800 agent-tonnes) is completed, in a safe and environmentally responsible manner, by the planned US target date of 2023. Furthermore, the OPCW must investigate and address any concerns regarding potential undeclared chemical weapons activities, such as the alleged development, stockpiling and use of Novichok weapons.

Although the OPCW–UN Joint Mission in Syria successfully removed and destroyed all declared chemical weapons stockpiles, there are continuing concerns that the Syrian Government did not provide the UN and OPCW with full and accurate details of its chemical weapons programme and stockpiles. The Organisation through the work of its Declaration Assessment Team (DAT) has already successfully clarified a few inaccuracies resulting in a number of additional facilities and activities being declared by the Syrian Government, yet significant unanswered questions remain. This process must be continued until all uncertainties are fully resolved and this should be followed by the complete verified removal and destruction of any remaining chemical weapons stockpiles and the disarmament/conversion of any production or storage facilities discovered. In addition, the OPCW must continue investigations conducted by its Fact-finding Mission (FFM) into allegations of the use in Syria of sarin and improvised chemical weapons employing chlorine and other toxic chemicals. Where evidence of the development of such chemical weapons or other activities in breach of the Convention comes to light in Syria, the OPCW must act swiftly to halt such activities. All evidence of such activities must also be secured so that the perpetrators of such chemical weapons attacks can be identified through a UN-mandated investigation or other appropriate mechanism. Full cooperation in these processes must be given by the Syrian Government and all parties to the conflict in Syria. If the outstanding issues under its purview cannot be addressed and resolved within the OPCW, the Organisation should bring the unresolved matters to the UNSC or the UN General Assembly for their action.

However, these unresolved issues also pose more fundamental questions. The different *ad hoc* mechanisms established by the OPCW (DAT and FFM) are a sign of the need for pragmatic and flexible application of the provisions and principles of the CWC to a politically highly charged context. The standard tools of the Convention—challenge inspection and investigations of alleged use of chemical weapons under Article IX of the Convention—were seen as politically too sensitive under the circumstances to be invoked, and *ad hoc* mechansims were negotiated with the Syrian government instead, under the general authority of the Director-General and sanctioned by the Executive Council. But, there is a penalty for relying on *ad hoc* solutions: whilst the rules of the Convention have been painstakingly negotiated and form part of the Convention, the rules of engagement for *ad hoc* mechanisms are open to interpretation, and the standards used in the assessment of investigation results are not agreed. The conclusions of such investigations, which of course have political ramifications, are largely left to the Technical Secretariat to defend against any criticism, including by States Parties that have stakes in the matter. The price to be paid for flexibility and adaptability is that the standards of proof are not agreed, that procedures must be negotiated with the States Parties most concerned or involved, and that conclusions are open to challenge. There is therefore perhaps a limit as to how much the OPCW can rely on *ad hoc* arrangements *in lieu* of using the compliance assurance mechanisms of the Convention itself.

A related question is how the purpose of such an investigation is defined. The purpose of the Convention's compliance assurance mechanisms is not to conduct fact finding *per se*, but to create conditions that would compel States Parties to re-establish full compliance as quickly as possible. The Executive Council has a crucial role to play in this, and quiet diplomatic engagement of key actors is important to find common ground for how best to resolve a compliance matter. The decision on the elimination of the Syrian chemical weapons in 2013 was an example for how the necessary political support and unity can be achieved through effective diplomacy; the discussions in the Executive Council of the findings of the FFM and the JIM are examples for how the absence of such efforts can risk debates ending in divisiveness, political cross-fire and an increasing inability to act collectively.

As the Review Conference assesses the manner in which the Convention's compliance assurance mechanisms are working, it will be important to reflect critically on the (non-)employment of challenge inspection and investigations of alleged use to resolve serious compliance concerns, and on the Executive Council's decision making culture when addressing such matters.

21.4.1.2 Maintaining the Comprehensive Nature of the CW Prohibition

As the world moves away from 20th century chemical warfare concepts and weapons types, the risks associated with toxic chemicals and their potential misuse as weapons are becoming more diffuse, and less well defined in terms of chemical compounds or dissemination methods of concern. Such processes are exacerbated by the rapid advances and convergence of the chemical and life sciences and associated technologies. In order to respond effectively to such paradigmic shifts and maintain the comprehensive nature of the CW prohibition, the OPCW should:

Ensure effective implementation of the General Purpose Criterion: The Fourth Review Conference should reaffirm the importance of the General Purpose Criterion (GPC) as a vital safeguard ensuring the Convention's comprehensive and future-proofed prohibition of chemical weapons. Subsequently, the Organisation should ensure this reaffirmation is translated into effective and uniform implementation of the GPC by all States Parties at the national level. This is particularly important given long-standing concerns over divergent interpretation and inconsistent implementation of the GPC amongst State Parties with the consequent dangers of the weakening of its prohibition in practice. To facilitate effective and consistent implementation, the OPCW should establish a consultative process to develop guidelines on how the "types and quantities" principle should be applied in practice. Development of such guidelines should be informed by technical input from the Technical Secretariat and the Scientific Advisory Board (SAB), and be open to contributions from all State Parties. Such guidelines would be applicable to the employment of all toxic chemicals with weapons utility. The Technial Secretariat should integrate such guidance on GPC implementation into its continuing programme of national implementation assistance for CWC National Authorities, and should also be tasked with providing tailored assistance and clarification to individual States Parties on the application of the GPC in specific cases. States should consequently ensure that all relevant research activities potentially related to the development of weapons employing toxic chemicals are in conformity with the Convention. In addition, States should carry out necessary promulgation activities to ensure that all those engaged in such research, be they working in defence, law enforcement, industry, academia or other sectors, are aware of their obligations under the Convention.

There are specific challenges to the GPC arising from contested interpretation amongst States Parties as to the range of toxic chemicals and delivery mechanisms that could legitimately be employed for law enforcement purposes, and the nature of what constitutes legitimate use. Consequently, the Review Conference should establish an Open Ended Working Group (OEWG) or other mechanism to develop clear guidelines designed to prevent research, development and production and employment activities intended to support law enforcement that would undermine the prohibitions of the CWC. In its deliberations, the OEWG/mechanism should consider existing obligations under relevant international law, notably IHRL, and their bearing on the implementation of the CWC in this area. The OEWG/mechanism should specifically address:

- *Incapacitating chemical agent (ICA) weapons*: It is envisaged that a suitable mechanism would potentially come to a determination *either* that development, stockpiling, transfer and use of ICA weapons for law enforcement are prohibited under the CWC *or* that such actions are permitted, but should be severely restricted and stringently regulated under the Convention. If the latter position is taken, then the mechanism could also: clarify what agents and delivery mechanisms were permissible, under what limited circumstances and with what constraints such use would be allowed.

- *Riot control agents*: A suitable mechanism should clarify the nature and scope of activities consistent with "law enforcement including domestic riot control" and develop guidance to States Parties as to the "types and quantities" of riot control agents that can legitimately be employed in such circumstances, highlighting relevant obligations under IHRL, to ensure such RCA employment is proportionate, necessary, and does not endanger life or health.
- *Delivery mechanisms*: A suitable mechanism should develop criteria for determining which means of delivering and dispersing riot control agents and other toxic chemicals are inappropriate for law enforcement purposes and would consequently breach Article II.1 of the Convention. Subsequently, a guidance document should be developed detailing these criteria, and specifying those toxic chemical means of delivery that have been identified as inappropriate for law enforcement purposes and consequently prohibited. Such prohibited means of delivery should, as a minimum include: artillery shells, aerial bombs, mortar shells and cluster munitions. This guidance document and the list of delivery means identified as inappropriate for law enforcement purposes should be reviewed regularly in a suitable forum, such as a CSP or Review Conference, to take into account developments in relevant science and technology.

Improve OPCW monitoring and risk assessment of science and technology: In 2011, the report of a high-level expert panel convened by the OPCW Director General to explore the future priorities of the Organization recommended that the OPCW should "improve and widen the scope of monitoring and evaluating developments in chemical science and technology..." Subsequently, these issues were also included in the 2015 Director General's "Vision Paper". Consequently, the Organization should introduce measures to:

- Ensure more frequent and considered review by CWC States Parties of relevant advances in converging chemical and life sciences and technology, and the implications for the Convention. In addition to the broad-scope review currently undertaken every five years in preparation for and during the Review Conference, more limited reviews of specific issues or technologies of potential concern should be prepared by the Technical Secretariat/SAB for consideration by States Parties at the annual CSPs.
- Further enhance the Technical Secretariat's in-house capability to monitor advances in science and technology of potential concern to the Organization; suitable mechanisms should also be established to allow the Technical Secretariat to bring relevant concerns *proactively* to the attention of the Executive Council and the States Parties, in a timely manner.

21.4.1.3 Strengthing National Implementation and Verification

Convergence and the advances in science and technology may well have several effects on national implementation and verification, which the Fourth Review Conference should address:

Update industry verification measures: The existing Schedules of Chemicals will remain important triggers of verification measures for traditional chemical warfare agents identified from past State programmes. However, new production pathways to World War I, World War II, and 1950s era chemical warfare agents may become feasible as a result of technological advances. Alternatively. new potential chemical warfare agent types may become relevant involving intermediates and products not listed on any of the Schedules. Consequently, the industry verification regime as well as the analytical methods and databases available for challenge inspection and investigation of alleged use need to be adapted to these new technological and chemical realities. Consideration could be given to updating the Schedules themselves, at least to provide indicators of the new types of potential chemical agents (and their precursors) of concern.

Changing risk perceptions: Increasingly, in today's security environment, other types of toxic chemicals are also (and sometimes more) relevant than existing Scheduled chemicals. From a perspective of risks associated with State actors, convergence and advances in enabling technologies have increased the technical feasibility of developing chemical agents that are less lethal, more targeted in their physiological effect, and perceived to be more suitable for the emerging operational context with military operations in populated areas. In contrast, the choices of potential chemical agents by non-State actors are more likely to be driven by accessibility, as exemplified by terrorist attempts to develop improvised weapons employing toxic industrial chemicals or "home-made" chemical warfare agents. Consequently, the Fourth Review Conference should acknowledge that the emphasis in the conduct of industry inspections should gradually shift from verifying the accuracy of declared data and checking on the absence of undeclared Scheduled chemicals, and place more weight on the general verification aims set out in the Convention (*i.e.* that activities at an inspected industry facility are consistent with the obligations undertaken under the Convention).

Holistic verification and national implementation: The Organisation should seek to establish a closer association between verification evaluation results and assessments of the effectiveness of, as well as weaknesses in, national implementation systems; and provide augmented support for National Authorities for implementation measures that reach beyond the Scheduled chemicals and address risks related to other toxic chemicals, and relevant delivery mechanisms. To facilitate such activities, a more interactive system of monitoring and evaluating advances in science, technology and industrial applications should be developed involving the Technical Secretariat, the SAB, States Parties, industry and other non-governmental experts in order to create a more broadly-based and continuous mechanism for evaluating changes that the verification and national implementation systems ought to adapt to. One area where such a hoslitic approach would be particularly beneficial is transfer control; ensuring that continued assistance by the Technical Secretariat to National Authorities in effective implementation of CWC transfer control requirements, is informed by and responds to relevant developments in science, industrial processes and trade. At the international level, the OPCW should continue to develop its cooperation with the World Customs Organisation and other relevant organisations to enhance the effectiveness and efficiency of the oversight of trade in dual-use chemicals and relevant delivery mechanisms, and to combat illicit trade in this area.

In addition to these over-arching issues, the OPCW should resolve certain specific long-standing uncertainties limiting verification activities, including:

Other chemical production facilities (OCPFs): OCPFs are a category of chemical industry plants that do not currently, but are technically capable of, manufacturing chemical warfare agents or precursors listed in the Schedules as well as non-Scheduled chemical warfare agents. At present, only a small fraction of declared OCPFs are selected for verification; the Fourth Review Conference should consider authorising a significant increase in OCPF inspections per year. The OPCW should also be directed to refine the site-selection algorithm so as to target inspections on those multipurpose chemical plants that pose the greatest risk of being utilized for prohibited purposes.

Biological and biologically mediated processes for the production of discrete organic chemicals: The SAB has consistently recommended that the bio-manufacturing of chemical products should be covered under the scope of the Convention; States Parties however have yet to agree on how to treat these types of production processes and attendant facilitities, and show no signs of doing so. This is despite the fact that some of the products and processes used by the bio-manufacturing industry are as relevant to the CWC as those used by other OCPF facilities, including those the Technical Secretariat consider to pose significant risks, whilst other products and processes are not CWC-relevant at all. The Review Conference should follow SAB advice and establish measures to determine the differing relevance of the various types of bio-manufacturing processes and facilitites for CWC verification purposes.

21.4.1.4 Preventing and Responding to Hostile Uses of Toxic Chemicals

The Organisation should continue improvements in the operational and technical capacity of the TS to conduct challenge inspections and investigations of the alleged use of chemical weapons, with an increased focus upon chemical forensics. Ongoing work by the SAB into the opportunities and difficulties associated with chemial forensics will inform the Technical Secretariat of actions in this area and enable the Organisation to most effectively utilise potential new tools and methods arising from advances in science and technology. Such work must be complemented by efforts to compile, expand and properly curate the databases of reference spectra, and collections of reference materials, that will be needed for such forensic analysis. The Organisation should also build on the significant progress made towards developing a network of designated laboratories for the analysis of biomedical/biological samples.

The OPCW should also consider how best to contribute to strengthening the resilience of States Parties against hostile uses of toxic chemicals (as well as against natural and accidental releases of toxic chemicals). This could include: establishment of regional capacities for an effective response to the use of chemical weapons or the accidental release of toxic chemicals with OPCW-trained and certified experts, and supported by the regions or sub-regions with regard to equipment resources; an expansion of the number of States Parties that have a viable national protective programme, supported by OPCW training and capacity building where needed.

This should be complemented by continued efforts to maintain and strengthen the OPCW's capacity to engage in major international responses to the hostile use of toxic chemicals, in concert with other international organisations. Informed by its experiences in Iraq, Libya and Syria, a particular focus should be upon increasing the Technical Secretariat's capacity to quickly respond to contingencies such as missions to prevent the involvement of chemical weapons in conflict situations, or the delivery of assistance to victims of chemical attacks (whether involving the use of military chemical weapons or improvised devices that disseminate toxic industrial chemicals). An important advance in this area, upon which to build, has been the the establishment by the Technical Secretariat of a Rapid Response and Assistance Mission (RRAM) to respond to such situations.

21.4.2 Biological and Toxin Weapons Convention

BTWC States Parties should explore potential processes to collectively monitor and assess the risks of misuse of advances in relevant converging life and chemical science and technologies, and their implications for the Convention. Furthermore, the BTWC States Parties and the BTWC Implementation Support Unit (ISU) should improve coordination, and work more closely, with the OPCW to monitor, analyse and address the implications to both treaties of the convergence of the chemical and life sciences and related technologies.

Given the limited appetitite for and ability to undertake significant action within formal BTWC auspices, groups of like-minded BTWC States Parties should now seek to develop collaborative initiatives in support of the Convention at the national, regional and plurilateral levels, in conjunction with relevant life science organisations, and with the assistance—where appropriate—of the BTWC ISU. These could include activities to foster a "culture of responsibility" amongst the life science communtiy, notably by promulgating the BTWC, sensitising life scientists to the risks of malign application of their research, and developing and implementing codes of conduct to ameliorate such dangers.

21.4.3 UN Mechanisms

UN Investigatory Mechanisms: Investigations conducted under UN auspices into suspected chemical weapons attacks—specifically the UN Secretary General's Mechanism (UNSGM)—provide an important fallback when OPCW procedures cannot be initiated (notably when such weapons are used by and/or against CWC non-States Parties). With regard to investigating the alleged or suspected use of biological weapons, the UNSGM remains the only mechanism with a dedicated roster of experts and laboratories, albeit with limited operational capacity and capabilities. Consequently, the capability of the UNSGM to undertake investigations of the suspected use of chemical, biological or toxin weapons at short notice must be developed (taking into account the lesson learned from the exercise undertaken by UNODA into the 2013 UNSGM Syrian investigation). This should include measures to build up the capabilities of its international network of diagnostic and analytic laboratories, update and train the roster of chemical and biological experts that can be called upon to conduct investigations, and further develop and adopt standardised operating procedures for such investigations. Whilst the 2013 UNSGM Syrian investigation was not mandated to identify the perpetrators of the Ghouta chemical attacks, such an attribution is not normally precluded in UNSGM investigations, provided there is clear and verifiable evidence to support any such conclusions. Consequently, attention and resources should be given to developing the forensic capabilities, and UNSGM working practices to facilitate such identification by future UNSGM investigations.

The establishment by the UNSC in August 2015 of the Joint Investigative Mechanism (JIM) of the UN and OPCW specifically tasked to identify those responsible for the "use of chemicals as weapons" in Syria and its employment of innovative operating guidelines and procedures—including extensive use of open source information and civil society engagement, as well as specialised expert advice and forensic techniques—was a highly significant advance. However, although the OPCW–UN JIM subsequently determined that the Syrian Armed Forces were responsible for certain chemical weapons attacks employing chlorine in 2014 and 2015, and sarin in 2017, and that Islamic State were responsible for the use of sulfur mustard weapons in 2015 and 2016, the UNSC has proved itself incapable of acting on this evidence in view of vetoes wielded by Russia and China. Regrettably, on 24 October 2017 and again on 16 November 2017, the Russian Federation vetoed UN Security Council Resolutions which would have extended the lifespan of the JIM for another year. Consequently the Mechanism's mandate expired on 17 November 2017. Unfortunately, at the time of writing, all attempts to revive the JIM have failed due to lack of consensus among UNSC members.

Whatever the outcome of the debates in the UNSC about the future of the JIM, it is essential that investigations into Syrian chemical weapons attacks continue for the foreseeable future—collecting and preserving evidence and identifying perpetrators—thereby facilitating the possible initiation of legal measures by an appropriate judicial mechanism against those who conducted and facilitated these attacks.

Chemical weapons disarmament: There are continuing concerns that the Syrian Government did not provide the UN and OPCW with full details of its chemical weapons holdings. It is vitally important that the UNSC provides full support to the OPCW in its ongoing efforts to resolve all apparent discrepancies in the Syrian chemical weapons related declarations and in its efforts to oversee and verify the complete destruction of the final declared production and storage facilities. If sufficient cooperation is not received from Syria, consideration should be given to adopting Chapter VII measures to enforce complete chemical weapons disarmament and destruction. Such measures will no doubt be highly contentious and very difficult to agree, let alone implement – particularly given previous Russian and Chinese vetoes. However, the UNSC, and the international community more broadly, have a duty to act appropriately to address the Syrian Government's breach of its obligations under the CWC and previous UNSC Resolutions.

UN sanctions including arms embargoes: In the light of the OPCW–UN JIM determinations of chemical weapons attacks perpetrated by the Syrian Armed Forces, and also by Islamic State, renewed efforts must be made to impose binding UN sanctions against Syria and at the same time to ensure that all States comply with their obligations under UNSCR 1540 and Syria-related UNSC resolutions not to assist any party to the conflict in acquiring and using CW (or other WMD). Such measures should include binding UN embargoes against these perpetrators so as to prevent the direct or indirect supply, sale or transfer of all arms and related materials used to manufacture and deliver such weapons. If the UNSC remains unable to adopt such measures due to continued Russian and Chinese vetoes, concerned States should introduce national embargoes and explore the adoption of regional and plurilateral measures.

Combating non-State actor acquisition and use of chemical weapons: One of the key challenges preventing the universal and effective implementation of UNSCR 1540 is the inability of many States to introduce the 300 plus domestic legislation and enforcement measures required under the Resolution – because of limited technical and legal expertise, as well as financial and human resources. This is exacerbated by the failure of the Resolution to specify that any of its provisions are more important than others, and also by the lack of any dedicated resource provision under the Resolution to States to facilitate its implementation. Donor States and relevant international, multilateral and (sub) regional organisations—assisted by the 1540 Committee—should continue their concerted and coordinated efforts to build implementation capacity, particularly amongst developing States. Furthermore, the 1540 Committee and its panel of experts should be tasked with developing proposals for a hierarchy of priorities and associated implementation guidance, for consideration by the UNSC.

21.4.4 International Law Constraining the Weaponisation of Toxic Chemicals

21.4.4.1 International Humanitarian Law

States should explore the effective application of existing constraints imposed by international humanitarian law (IHL) instruments and customary IHL upon the development, acquisition or use of weapons employing toxic chemicals (including RCAs and ICA weapons) in armed conflict, and the consequent implications for State actions in this area. Of particular importance would be measures to promote and facilitate the effective implementation of the obligations of States under Article 36 of Additional Protocol I to conduct reviews of any new weapon developed or acquired, so as to determine its compatibility with the principles and rules of IHL. One possible forum for such deliberations could be the quadrennial International Conferences of the Red Cross and Red Crescent Movement, which bring together the States party to the Geneva Conventions, the ICRC, the International Federation of Red Cross and Red Crescent Societies, and all recognized National Red Cross and Red Crescent Societies, to examine and decide upon humanitarian matters of common interest; the next (33rd) International Conference will be held in 2019.

21.4.4.2 Human Rights Law

States could explore the constraints on the employment of weapons employing toxic chemicals (including ICA weapons and RCAs) arising from international and regional human rights instruments and customary international human rights law. States could also bring cases of reported weapons misuse to the attention of the appropriate human rights mechanisms, including: UN Special Procedures (notably the UN Special Rapporteur(s) on Torture, Extra-Judicial Executions, or Counter-Terrorism and Human Rights); the UN Human Rights Council; and relevant international and regional treaty bodies (*e.g.* the UN Human Rights Committee – under the International Covenant on Civil and Political Rights; and the UN Committee against Torture – under the UN Convention Against Torture). If appropriate, cases could be brought before the relevant regional judicial mechanisms capable of delivering binding legal judgements regarding violations of regional treaties (*e.g.* the European Court of Human Rights, the Inter-American Court, and the African Commission on Human and Peoples' Rights). Since a number of such regional judicial mechanisms are potentially open to individual petition, victims and their families can also directly seek redress in cases of toxic chemical agent misuse, and civil society organizations can attempt to employ such mechanisms to develop human rights case law on these issues.

21.4.4.3 International Criminal Law

Of the existing judicial bodies, the International Criminal Court—with its attendant review and amendment mechanisms—affords an important route for concerted international action and normative development of international criminal law in the long term, However, its potential utility is currently constrained, due to problems with establishing jurisdiction, unwillingness of certain States to prosecute suspects on their territory and the notable failure of the UNSC to refer cases to the Court *in extremis*. Consequently, no case involving the weaponisation of toxic chemicals has been referred to the Court so far, despite widespread international pressure to do so in response to the confirmed chemical attacks conducted by the Syrian Armed Forces in Syria, and by Islamic State in Syria and Iraq. It is important that pressure on the UNSC be maintained, for example through further UNGA Resolutions. In addition, individual ICC States Parties and the ICC Prosecutor should explore the possibility of initiating formal investigations into these chemical attacks, utilising, for example, evidence made available through OPCW, UN and UN-OPCW JIM mechanisms. Furthermore, States should explore the potential for initiating cases under their national legislation, where evidence involving direct or indirect complicity in the use of weaponised toxic chemicals, in Iraq, Syria or elsewhere, comes to light.

21.4.5 Civil Society

Concerned civil society actors—particularly those in the chemical, medical research and life science communities—can continue fostering a "culture of responsibility" in science and engineering by developing, promulgating and applying codes of conduct, pledges and conventions regulating "dual-use" research. While initiatives, to date, have focussed on combating the acquisition and misuse of research by non-State actors, increased attention should now be give to initiatives explicitly combating the misapplication of such research by State weapons programmes. Complementary activities include promoting the non-participation in and "whistle-blowing" on weapons programmes of concern; and education and awareness-raising amongst medical and scientific communities, State officials and the general public of the dangers of such weapons development and use.[1]

Individual chemical, engineering, medical research and life scientists and professional associations, as well as NGOs informed by such expertise, can play important constructive roles in highlighting existing limitations in the BTWC, CWC and attendant control regimes with regard to such weapons, and developing and promoting possible science-informed policy responses; undertaking societal monitoring and verification of existing implementation by States Parties of relevant instruments and highlighting R&D activities of concern; and, predicting research trajectories in relevant scientific disciplines and warning of potential future threats.

In certain CWC States Parties, there are limited mechanisms facilitating Government engagement with a narrow range of academics, NGOs and scientific bodies perceived to have relevant expertise in CBW arms control and disarmament. Where these exist, they provide important opportunities for at least some civil society actors to bring concrete proposals to address the implications of the advance and convergence of the chemical, life and related sciences and the dangers of their hostile use, to the attention of relevant State officials.

In order to foster more informed discussions and facilitate appropriate policy development in relevant national and OPCW forums, efforts can be made through coordinating bodies such as the Chemical Weapons Convention Coalition (CCWC) and by the OPCW itself, to interact more fully with a far wider range of civil society actors, including human rights, IHL and humanitarian aid organizations, and health professionals, as well as those local communities, pro-democracy and civil liberties movements that have suffered, are at risk from, or are concerned about weapons employing toxic chemicals. In addition, it would be beneficial if these civil society organizations and communities now sought to engage with the OPCW on these issues, for example by attending the Fourth Review Conference and subsequent Conferences of the States Parties, notwithstanding the OPCW's sporadic and rarely explained refusals to admit some of them.

Reference

1. Advisory Board on Education and Outreach, *Report on the Role of Education and Outreach in Preventing the Re-Emergence of Chemical Weapons*, ABEO-5/1, 12 February 2018, OPCW, The Hague.

Subject Index